Performance Analysis of Complex Networks and Systems

PIET VAN MIEGHEM

Delft University of Technology

CAMBRIDGE
UNIVERSITY PRESS

CAMBRIDGE
UNIVERSITY PRESS

University Printing House, Cambridge CB2 8BS, United Kingdom

One Liberty Plaza, 20th Floor, New York, NY 10006, USA

477 Williamstown Road, Port Melbourne, VIC 3207, Australia

314-321, 3rd Floor, Plot 3, Splendor Forum, Jasola District Centre, New Delhi-110025, India

79 Anson Road, #06-04/06, Singapore 079906

Cambridge University Press is part of the University of Cambridge.

It furthers the University's mission by disseminating knowledge in the pursuit of education, learning and research at the highest international levels of excellence.

www.cambridge.org
Information on this title: www.cambridge.org/9781107058606

© Cambridge University Press 2014

First published 2014

A catalogue record for this publication is available from the British Library

Library of Congress Cataloging in Publication data
Van Mieghem, Piet.
Performance analysis of complex networks and systems / Piet Van Mieghem,
Delft University of Technology.
pages   cm
ISBN 978-1-107-05860-6 (hardback)
1. System analysis–Mathematics.   2. Computational complexity.
3. Network analysis (Planning)
I. Title.
T57.V36 2014
003.01´5192–dc23   2014000241

ISBN 978-1-107-05860-6 Hardback

Waar een wil is, is een weg.
to my father

in memory of
my wife Saskia and my son Nathan

to my sons Vincent and Laurens

Contents

Preface

This book is a development of *Performance Analysis of Communications Networks and Systems* of 2006. Its current incarnation has a broader scope, extending to complex networks, the broad term covering all types of real-world networks, that encompass communications networks. Apart from the correction of numerous errors in the earlier book, nearly all chapters have been extended. Chapter 17 on Epidemics in networks has been added, while the appendix on algebraic graph theory has been deleted, because our book *Graph Spectra for Complex Networks* amply replaces this chapter. Similar to *Graph Spectra for Complex Networks*, **art. x** has been used to refer to article x in Appendix A. The number of problems, together with their solutions in Appendix B, has been doubled at least.

Performance analysis belongs to the domain of applied mathematics. In particular, the branches of mathematics as probability theory, stochastic processes and graph theory are exploited, besides analysis (calculus) and linear algebra that are omnipresent nearly everywhere. The major aim of this book is to offer several mathematical methods to address challenges in *network science*, the rapidly growing field of complex networking. The link with technology is kept shallow, on purpose, because most technical advances in micro-electronics, in communications protocols and services, standards, etc. have a more limited lifetime and a narrower scope compared to mathematical concepts.

This book aims to present methods rigorously, hence mathematically, with minimal resorting to intuition. It is my belief that intuition is often gained after the result is known and rarely before the problem is solved, unless the problem is simple. I have tried to interpret most of the important formulas in the sense of "What does this mathematical expression teach me?" This last step justifies the word "applied", since most mathematical treatises do not interpret as it contains the risk to be imprecise and incomplete.

As prerequisites, familiarity with elementary probability and the knowledge of the theory of functions of a complex variable are assumed. In particular, the beautiful book on the *Theory of Functions* by Titchmarsh (1964) is recommended for complex function theory. Appendix A briefly summarizes concepts of linear algebra used in this book. Parts in the text in small font refer to more advanced topics or to computations that can be skipped at first reading.

Chapters 2–3 in Part I briefly review probability theory; they are included to make the remainder self-contained. Chapter 4 discusses how to compute correlation between several random variables. The central role of the Gaussian distribution is emphasized. Since the Gaussian distribution is so widely known, I have included Gauss's own derivation of his distribution, whereas the Central Limit Theorem (studied in Chapter 6 together with other limit laws) is currently considered as the main law that produces a Gaussian distribution. Chapter 5 treats powerful inequalities at an introductory level, but skips the recent (and more difficult) inequalities, such as the FKG inequality due to Fortuin, Kasteleyn and Ginibre.

The book essentially starts with Chapter 7 (Part II) on Poisson processes. The Poisson process (independent increments and discontinuous sample paths) and Brownian motion (independent increments but continuous sample paths) are considered to be the most important basic stochastic processes. We briefly touch upon renewal theory to move to Markov processes. The theory of Markov processes is regarded as a fundament for many applications, particularly in queueing theory, epidemics on networks and some instances of shortest path routing. A large part of the book is consumed by Markov processes and its applications. The last chapters of Part II dive into queueing theory. Inspired by intriguing problems in telephony at the beginning of the twentieth century, Erlang has pushed queueing theory to the scene of sciences. Since his investigations, queueing theory has grown considerably. Especially during the last decade with the advent of the Asynchronous Transfer Mode (ATM) and the worldwide Internet, many early ideas have been refined (e.g. discrete-time queueing theory, large deviation theory, scheduling control of prioritized flows of packets) and new concepts (self-similar or fractal processes) have been proposed. Queuing theory is expected to emerge again in complex networks, when structural aspects are understood.

Part III covers the parts of network science that study the structure of and processes on complex networks. While the first decade of this century has predominantly focused on the topology and structure of complex networks, the next decade aims to understand the influence of the topology on the dynamic process(es) on the network. Chapter 15 overviews the topological properties of complex networks and invokes concepts of graph theory, as well as random graph theory. Both the famous Erdős-Rényi and the Barabási-Albert random graphs are introduced. The giant component of a network, which is the largest still connected subgraph, is studied, because the giant component can be considered as the operational heart of the complex network. Finally, interdependent networks are introduced and the surprisingly different nature of cascading failures between single and interdependent networks is discussed. Chapter 16 and 17 exemplify two dynamic processes on a network: the transport along the shortest path and the spread of epidemics in networks. Chapter 18 and 19 dive deeper into routing instances as multicasting and anycasting, originally proposed for internetworking, but extendable to the newer communications types as social networking and cloud computing.

Since network science is still developing at fast pace, Part III is undoubtedly the least mature and complete. Moreover, I have predominantly relied on my own research for the exposition of topics, without the ambition to thoroughly review (and cite) contributions in the field. The book of Remco van der Hofstad (2013) treats random graphs differently inspired by Erdős' probabilistic method. Remco seriously extends the material in Chapter 15 and I recommend his book for a deeper discussion about the giant component, phase transitions, inhomogeneous random graphs, the configuration model and preferential attachment models.

To Huijuan Wang, who has used the earlier book of 2006 in her classes, I am indebted for many suggestions, corrections and a shorter proof of the degree distribution in the URT in Section 16.7.2. Numerous people have pointed me to errors in the earlier 2006 book, while others gave suggestions for this book. I am very grateful to all of them: Chandrashekhar Pataguppe Suryanarayan Bhat, Ruud van de Bovenkamp, Eric Cator, Li Cong, Edwin van Dam, Michel Dekking, David Hemsley, Remco van der Hofstad, Gerard Hooghiemstra, David Hunter, Geurt Jongbloed, Merkouris Karaliopoulos, Rob Kooij, Javier Martin Hernandez, Jil Meier, Raphi Rom, Annalisa Socievole, Bart Steyaert, Siyu Tang, Stojan Trajanovski, Matthias Waehlisch, Huijuan Wang, and to the many students at Delft University of Technology that followed my course on Performance Analysis over the last 14 years.

Although this book is intended to be of practical use, in the course of writing it, I became more and more persuaded that mathematical rigor has ample virtues of its own.

Omnia sunt incerta, cum a
mathematicae discessum est

February 2014

PIET VAN MIEGHEM

Symbols

Only when explicitly mentioned will we deviate from the standard notation and symbols outlined here.

Random variables and matrices are written with capital letters, while complex, real, integer, etc., variables are in lower case. For example, X refers to a random variable, A to a matrix, whereas x is a real number and z is complex number. Usually, i, j, k, l, m, n are integers. Operations on random variables are denoted by $[.]$, whereas $(.)$ is used for real or complex variables. A set of elements is embraced by $\{.\}$.

Linear algebra

A $n \times m$ matrix
$$
\begin{bmatrix} a_{11} & \cdots & a_{1m} \\ \vdots & & \\ a_{n1} & \cdots & a_{nm} \end{bmatrix}
$$

$\det A$ determinant of a square matrix A; also denoted by $\begin{vmatrix} a_{11} & \cdots & a_{1n} \\ \vdots & & \\ a_{n1} & \cdots & a_{nn} \end{vmatrix}$

$\text{trace}(A)$ $= \sum_{j=1}^{n} a_{jj}$: sum of diagonal elements of A

$\text{diag}(a_k)$ $= \text{diag}(a_1, a_2, \ldots, a_n)$: a diagonal matrix with diagonal elements listed, while all off-diagonal elements are zero

A^T transpose of a matrix, the rows of A are the columns of A^T

A^* matrix in which each element is the complex conjugate of the corresponding element in A

A^H $= (A^*)^T$: Hermitian of matrix A

$c_A(x)$ $= \det(A - xI)$: characteristic polynomial of A

$\text{adj}A$ $= A^{-1} \det A$: adjugate of A

$Q(\lambda)$ $= \frac{c_A(\lambda)}{\lambda I - A}$: adjoint of A

e_j basic vector, all components are zero, except for component j that is 1

δ_{kj} Kronecker delta, $\delta_{kj} = 1$ if $k = j$, else $\delta_{kj} = 0$

Probability theory

$\Pr[X]$	probability of the event X
$E[X]$	$= \mu$: expectation of the random variable X
$\mathrm{Var}[X]$	$= \sigma_X^2$: variance of the random variable X
$f_X(x)$	$= \frac{dF_X(x)}{dx}$: probability density function of X
$F_X(x)$	probability distribution function of X
$\varphi_X(z)$	probability generating function of X
	$\varphi_X(z) = E\left[z^X\right]$ when X is a discrete r.v.
	$\varphi_X(z) = E\left[e^{-zX}\right]$ when X is a continuous r.v.
$\{X_k\}_{1\leq k\leq m}$	$= \{X_1, X_2, \ldots, X_m\}$
$X_{(k)}$	k-th order statistics, k-th smallest value in the set $\{X_k\}_{1\leq k\leq m}$
P	transition probability matrix (Markov process)
$1_{\{x\}}$	indicator function: $1_{\{x\}} = 1$ if the event or condition $\{x\}$ is true, else $1_{\{x\}} = 0$. For example, $\delta_{kj} = 1_{\{k=j\}}$
γ	$= 0.577\ 215\ldots$: Euler's constant
Ω	sample space
ω	sample point

Queuing theory

t_n	arrival time of the n-th packet
r_n	departure time of the n-th packet
$\tau_n = t_n - t_{n-1}$	n-th interarrival time
x_n	service time of n-th packet
w_n	waiting time of the n-th packet
$T_n = x_n + w_n$	system time of n-th packet
$v(t)$	virtual waiting time or unfinished work at time t
$\lambda = (E[\tau])^{-1}$	average arrival rate
$\mu = (E[x])^{-1}$	average service rate
$\rho = \frac{\lambda}{\mu}$	traffic intensity
$N_A(t)$	number of arrivals at time t
$N_S(t)$	number of packets in the system (queue plus server) at time t
$N_Q(t)$	number of packets in the queue at time t

1

Introduction

The aim of this first chapter is to motivate why stochastic processes, probability theory and graph theory are useful to solve problems in network science.

In any system or node in a network, there is always a non-zero probability of failure or of error penetration. A lot of problems in quantifying the failure rate, bit error rate or the computation of redundancy to recover from hazards are successfully treated by probability theory. Often we deal in communications with a large variety of signals, calls, source-destination pairs, messages, the number of customers per region, and so on. Often, precise information at any time is not available or, if it is available, deterministic studies or simulations are simply not feasible due to the large number of different parameters involved. For such problems, a stochastic approach is often a powerful vehicle, as has been demonstrated in the field of statistical physics or thermodynamics. Failure or attacks at the network level have reestablished the interest in network robustness analyses in relation to network security. In spite of the intuitively easy concept, a globally accepted definition as well as a framework to compute the robustness of a network is still lacking. Graph and probability theory are essential to address questions like: "Which are the vulnerable nodes?", "Is this network robust?", "Where do we need to add, remove or rewire links at minimum cost in order to maximize the network robustness?"

Perhaps the first impressing result of a stochastic approach was Boltzmann's and Maxwell's statistical theory. They studied the behavior of particles in an ideal gas and described how macroscopic quantities as pressure and temperature can be related to the microscopic motion of the huge amount of individual particles. Boltzmann also introduced the stochastic notion of the thermodynamic concept of entropy S,

$$S = k \log W$$

where W denotes the total number of ways in which the ensembles of particles can be distributed in thermal equilibrium and where $k = 1.380\ 65\ 10^{-23}$ J/K is a proportionality factor, afterwards attributed to Boltzmann as the Boltzmann constant. The pioneering work of these early physicists such as Boltzmann, Maxwell and others was the germ of a large number of breakthroughs in science. Shortly after their introduction of a stochastic theory in classical physics, the theory of quantum mechanics (see e.g. Cohen-Tannoudji *et al.*, 1977) was established. This theory proposes that the elementary building blocks of nature, the atom and electrons, can only be described in a probabilistic sense. The conceptually difficult notion of a wave function whose squared modulus expresses the probability that a set of particles is in a certain state and the Heisenberg's uncertainty relation exclude in a dramatic way our deterministic, macroscopic view of nature at the fine atomic scale. The quantum computer – the successor of our current digital computer – is looming around the corner as small quantum computing devices are currently built and tested. The next step is quantum networking, which will require different design rules in networking and will create new challenges for network science.

At about the same time as the theory of quantum mechanics was being created, Erlang applied probability theory to the field of telecommunications. Erlang succeeded to determine the number of telephone input lines m of a switch in order to serve N_S customers with a certain probability p. Perhaps his most used formula is the Erlang B formula (14.18), derived in Section 14.2.2,

$$\Pr\left[N_S = m\right] = \frac{\frac{\rho^m}{m!}}{\sum_{j=0}^{m} \frac{\rho^j}{j!}}$$

where the load or traffic intensity ρ is the ratio of the arrival rate of calls to the telephone local exchange or switch over the processing rate of the switch per line. By equating the desired blocking probability $p = \Pr\left[N_S = m\right]$, say $p = 10^{-4}$, the number of input lines m can be computed for each load ρ. Shannon, another pioneer in the field of communications, explored the concept of entropy S. He introduced (see e.g. Cover and Thomas, 1991; Walrand, 1998) the notion of the Shannon capacity of a channel, the maximum rate at which bits can be transmitted with arbitrary small (but non-zero) probability of errors, and the concept of the entropy rate of a source, which is the minimum average number of bits per symbol required to encode the output of a source. Many others have extended his basic ideas and so it is fair to say that Shannon founded the field of information theory.

An important driver in telecommunication is the concept of quality of service (QoS). Customers can use the network to transmit different types of information, such as pictures, files, voice, etc., by requiring a specific level of service depending on the type of transmitted information. For example, a telephone conversation requires that the voice packets arrive at the receiver D ms later, while a file transfer is mostly not time critical but requires an extremely low information loss probability. The value of the mouth-to-ear delay D is clearly related to the perceived quality of the voice conversation. As long as $D < 150$ ms, the voice conversation has toll quality, which is, roughly speaking, the quality that we are used to in classical telephony. When D exceeds 150 ms, rapid degradation is experienced and when $D > 300$ ms, most of the test persons have great difficulty in understanding the conversation. However, perceived quality may change from person to person and is difficult to determine, even for telephony. Therefore, QoS is both related to the nature of the information and to the individual's desire and perception. In the future, it is believed that customers may request a certain QoS for each type of information. Depending on the level of stringency, the network may either allow or refuse the customer. Since customers will also pay an amount related to this QoS stringency, the network function that determines to either accept or refuse a call for service will be of crucial interest to any network operator. Let us now state the connection admission control (CAC) problem for a voice conversation to illustrate the relation to stochastic analysis: "How many customers m are allowed in order to guarantee that the ensemble of all voice packets reaches the destination within D ms with probability p?" This problem is exceptionally difficult because it depends on the voice codecs used, the specifics of the network topology, the capacity of the individual network elements, the arrival process of calls from the customers, the duration of the conversation and other details. Therefore, we will simplify the question. Let us first assume that the delay is only caused by the waiting time of a voice packet in the queue of a router (or switch). As we will see in Chapter 13, this waiting time T of voice packets in a single queueing system depends on (a) the arrival process: the way voice packets arrive, and (b) the service process: how they are processed. Let us assume that the arrival process specified by the average arrival rate λ and the service process specified by the average service rate μ are known. Clearly, the arrival rate λ is connected to the number of customers m. A simplified statement of the CAC problem is, "What is the maximum λ allowed such that $\Pr[T > D] < \epsilon$?" In essence, the CAC problem consists in computing the tail probability of a quantity that depends on parameters of interest. We have elaborated on the CAC problem because it is a basic design problem that appears under several disguises. A related dimensioning problem is the determination of the buffer size in a router in order not to lose more than a certain number of packets with probability p, given the arrival and service process. The above-mentioned problem of Erlang is a third example. Another example treated in Chapter 19 is the server placement problem: "How many replicated servers m are needed to guarantee that any user can access the information within k hops with probability $\Pr[h_N(m) > k] \leq \epsilon$, where ϵ is certain level of stringency and $h_N(m)$ is the number of hops towards the most nearby of the m servers in a network with N routers."

Network science aims at understanding and at modeling complex networks such as the Internet, biological and brain networks, social networks and utility infrastructures for water, gas, electricity and transport (cars, trains, ships and airplanes). Since these networks consist of a huge number of nodes N and links L, classical and algebraic graph theory is often not suited to produce even approximate results. The beginning of probabilistic graph theory is commonly attributed to the appearance of papers by Erdős and Rényi in the late 1940s. They investigated a particularly simple growing model for a graph: start from N nodes and connect in each step an arbitrary random, not yet connected pair of nodes until all L links are used. After about $N/2$ steps, as shown in Section 16.9.1, they observed the birth of a giant component that, in subsequent steps, swallows the smaller ones at a high rate. The link density $p = L/\binom{N}{2}$ plays a crucial role the Erdős-Rényi random graph. Around a critical value $p_c \sim \frac{\log N}{N}$ (see Sections 15.7.4 and 15.7.5), the probability of connectivity jumps sharply. This phenomenon is called a phase transition and often occurs in nature. In physics it is studied in, for example, per-colation theory. The Internet is best regarded as a dynamic and growing network, whose graph is continuously changing. Yet, in order to deploy services over the Internet, an accurate graph model that captures the relevant structural properties is desirable. As shown in Part III, a probabilistic approach based on random graphs seems an efficient way to learn about the Internet's intriguing behavior. Although the Internet's topology is not a simple Erdős-Rényi random graph, results such as the hopcount of the shortest path and the size of a multicast tree deduced from the simple random graphs provide a first-order estimate for the Internet.

The popularity of the Internet gave birth to several developments in electronic banking, governing, publishing and other parts of society. On-line social networks, such as Twitter and Facebook, form open laboratories to study how humans use technology. Email forwarding and Twitter retweet times (see e.g. Doerr *et al.* (2013)) are not exponential, but rather lognormal, and thus refute a Markovian approach. Similarly as about 20 years earlier, Markovian-based design failed, when Internet traffic was shown to be "bursty" (long-range dependent, self-similar and even chaotic, non-Markovian (Veres and Boda, 2000)). As a consequence, new methods are needed to compute traffic or dimension servers and networks, based on non-Markovian human responses. Further interesting questions are: "How do we determine communities?", "Who are the key influential persons in a community?", "How does information spread, become popular and age?" or "What is the topo-logical structure of communities and how do communities grow and change over time?". The social embedding of the Internet produces many more such inspiring questions that ask for methods discussed in this book.

We hope that this brief overview motivates sufficiently to surmount the mathe-matical barriers. Skill with probability theory is deemed necessary to understand complex phenomena in network science. Once mastered, the power and beauty of mathematics will be appreciated.

Part I
Probability theory

2
Random variables

This chapter reviews basic concepts from probability theory. A random variable (rv) is a variable that takes certain values by chance. Throughout this book, this imprecise and intuitive definition suffices. The precise definition involves axiomatic probability theory (Billingsley, 1995).

Here, a distinction between discrete and continuous random variables is made, although a unified approach including also mixed cases via the Stieltjes integral (Hardy *et al.*, 1999, pp. 152–157), $\int g(x)df(x)$, is possible. In general, the distribution $F_X(x) = \Pr[X \leq x]$ holds in both cases, and

$$\int g(x)dF_X(x) = \sum_k g(k)\Pr[X = k] \qquad \text{where } X \text{ is a discrete rv}$$

$$= \int g(x)\frac{dF_X(x)}{dx}dx \qquad \text{where } X \text{ is a continuous rv}$$

In most practical situations, the Stieltjes integral reduces to the Riemann integral, otherwise, Lesbesgue's theory of integration and measure theory (Royden, 1988) is required.

2.1 Probability theory and set theory

Pascal (1623–1662) is commonly regarded as one of the founders of probability theory. In his days, there was much interest in games of chance[1] and the likelihood of winning a game. In most of these games, there was a finite number n of possible outcomes and each of them was equally likely. The probability of the event A of interest was defined as

$$\Pr[A] = \frac{n_A}{n} \tag{2.1}$$

---

[1] "La règle des partis", a chapter in Pascal's mathematical work (Pascal, 1954), consists of a series of letters to Fermat that discuss the following problem (together with a more complex question that is essentially a variant of the probability of gambler's ruin treated in Section 11.2.2): Consider the game in which two dice are thrown n times. How many times n do we have to throw the two dice to throw double six with probability $p = \frac{1}{2}$?

where n_A is the number of favorable outcomes (samples points of A). If the number of outcomes of an experiment is not finite, this classical definition of probability no longer suffices. In order to establish a coherent and precise theory, probability theory employs concepts of group or set theory.

The set of all possible outcomes of an experiment is called the sample space Ω. A possible outcome of an experiment is called a sample point ω that is an element of the sample space Ω. An event A consists of a set of sample points. An event A is thus a subset of the sample space Ω. The *complement* A^c of an event A consists of all sample points of the sample space Ω that are not in (the set) A, thus $A^c = \Omega \backslash A$. Clearly, $(A^c)^c = A$ and the complement of the sample space is the empty set, $\Omega^c = \emptyset$ or, vice versa, $\emptyset^c = \Omega$. A family $\mathcal{F}$ of events is a set of events and thus a subset of the sample space Ω that possesses particular events as elements. More precisely, a family $\mathcal{F}$ of events satisfies the three conditions that define a σ-field[2]: (a) $\emptyset \in \mathcal{F}$; (b) if $A_1, A_2, \ldots \in \mathcal{F}$, then $\cup_{j=1}^{\infty} A_j \in \mathcal{F}$; and (c) if $A \in \mathcal{F}$, then $A^c \in \mathcal{F}$. These conditions guarantee that $\mathcal{F}$ is closed under countable unions and intersections of events.

Events and the probability of these events are connected by a probability measure $\Pr[.]$ that assigns to each event of the family $\mathcal{F}$ of events of a sample space Ω a real number in the interval $[0, 1]$. As **Axiom 1**, we require that

$$\Pr[\Omega] = 1 \tag{2.2}$$

If $\Pr[A] = 0$, the occurrence of the event A is not possible, while $\Pr[A] = 1$ means that the event A is certain to occur. If $\Pr[A] = p$ with $0 < p < 1$, the event A occurs with probability p.

If the events A and B have no sample points in common, $A \cap B = \emptyset$, the events A and B are called *mutually exclusive events*. As an example, the event and its complement are mutually exclusive because $A \cap A^c = \emptyset$. **Axiom 2** of a probability measure is that, for mutually exclusive events A and B, it holds that $\Pr[A \cup B] = \Pr[A] + \Pr[B]$. The definition of a probability measure and the two axioms are sufficient to build a consistent framework on which probability theory is founded. Since $\Pr[\emptyset] = 0$ (which follows from Axiom 2 because $A \cap \emptyset = \emptyset$ and $A = A \cup \emptyset$), for mutually exclusive events A and B, it holds that $\Pr[A \cap B] = 0$.

As a classical example that explains the formal definitions, let us consider the experiment of throwing a fair die. The sample space consists of all possible outcomes: $\Omega = \{1, 2, 3, 4, 5, 6\}$. A particular outcome of the experiment, say $\omega = 3$, is

---

[2] A field $\mathcal{F}$ possesses the properties:

(i) $\emptyset \in \mathcal{F}$;
(ii) if $A, B \in \mathcal{F}$, then $A \cup B \in \mathcal{F}$ and $A \cap B \in \mathcal{F}$;
(iii) if $A \in \mathcal{F}$, then $A^c \in \mathcal{F}$.

This definition is redundant. For, we have by (ii) and (iii) that $(A \cup B)^c \in \mathcal{F}$. Further, by De Morgan's law $(A \cup B)^c = A^c \cap B^c$, which can be deduced from Figure 2.1 and again by (iii), the argument shows that the reduced statement (ii), if $A, B \in \mathcal{F}$, then $A \cup B \in \mathcal{F}$, is sufficient to also imply that $A \cap B \in \mathcal{F}$.

a sample point $\omega \in \Omega$. One may be interested in the event A where the outcome is even in which case $A = \{2, 4, 6\} \subset \Omega$ and $A^c = \{1, 3, 5\}$.

If A and B are events, the union of these events $A \cup B$ can be written using set theory as

$$A \cup B = (A \cap B) \cup (A^c \cap B) \cup (A \cap B^c)$$

because $A \cap B$, $A^c \cap B$ and $A \cap B^c$ are mutually exclusive events. The relation is immediately understood by drawing a Venn diagram as in Fig. 2.1. Taking the

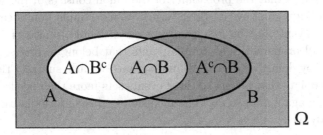

Fig. 2.1. A Venn diagram illustrating the union $A \cup B$.

probability measure of the union yields

$$\Pr[A \cup B] = \Pr[(A \cap B) \cup (A^c \cap B) \cup (A \cap B^c)]$$
$$= \Pr[A \cap B] + \Pr[A^c \cap B] + \Pr[A \cap B^c] \qquad (2.3)$$

where the last relation follows from Axiom 2. Figure 2.1 shows that $A = (A \cap B) \cup (A \cap B^c)$ and $B = (A \cap B) \cup (A^c \cap B)$. Since the events are mutually exclusive, Axiom 2 states that

$$\Pr[A] = \Pr[A \cap B] + \Pr[A \cap B^c]$$
$$\Pr[B] = \Pr[A \cap B] + \Pr[A^c \cap B]$$

Substitution into (2.3) yields the important relation

$$\Pr[A \cup B] = \Pr[A] + \Pr[B] - \Pr[A \cap B] \qquad (2.4)$$

Although derived for the measure $\Pr[.]$, relation (2.4) also holds for other measures, for example, the cardinality (the number of elements) of a set.

2.1.1 The inclusion-exclusion formula

A generalization of the relation (2.4) is the *inclusion-exclusion formula*,

$$\Pr\left[\cup_{k=1}^{n} A_k\right] = \sum_{k_1=1}^{n} \Pr\left[A_{k_1}\right] - \sum_{k_1=1}^{n} \sum_{k_2=k_1+1}^{n} \Pr\left[A_{k_1} \cap A_{k_2}\right]$$

$$+ \sum_{k_1=1}^{n} \sum_{k_2=k_1+1}^{n} \sum_{k_3=k_2+1}^{n} \Pr\left[A_{k_1} \cap A_{k_2} \cap A_{k_3}\right]$$

$$+ \cdots + (-1)^{n-1} \sum_{k_1=1}^{n} \sum_{k_2=k_1+1}^{n} \cdots \sum_{k_n=k_{n-1}+1}^{n} \Pr\left[\cap_{j=1}^{n} A_{k_j}\right] \qquad (2.5)$$

The formula shows that the probability of the union consists of the sum of probabilities of the individual events (first term). Since sample points can belong to
more than one event A_k, the first term possesses double countings. The second
term removes all probabilities of sample points that belong to precisely two event
sets. However, by doing so (draw a Venn diagram), we also subtract the probabilities of sample points that belong to three events sets more than needed. The third
term adds these again, and so on. The inclusion-exclusion formula can be written
more compactly as

$$\Pr\left[\cup_{k=1}^{n} A_k\right] = \sum_{j=1}^{n} (-1)^{j-1} \sum_{k_1=1}^{n} \sum_{k_2=k_1+1}^{n} \cdots \sum_{k_j=k_{j-1}+1}^{n} \Pr\left[\cap_{m=1}^{j} A_{k_m}\right] \qquad (2.6)$$

or with

$$S_j = \sum_{1 \le k_1 < k_2 < \cdots < k_j \le n} \Pr\left[\cap_{m=1}^{j} A_{k_m}\right]$$

as

$$\Pr\left[\cup_{k=1}^{n} A_k\right] = \sum_{j=1}^{n} (-1)^{j-1} S_j \qquad (2.7)$$

Although impressive, the inclusion-exclusion formula is useful when dealing with
dependent random variables because of its general nature. In particular, if

$$\Pr\left[\cap_{m=1}^{j} A_{k_m}\right] = a_j$$

and not a function of the specific indices k_m, the inclusion-exclusion formula (2.6)
becomes more attractive,

$$\Pr\left[\cup_{k=1}^{n} A_k\right] = \sum_{j=1}^{n} (-1)^{j-1} a_j \sum_{1 \le k_1 < k_2 < \cdots < k_j \le n} 1$$

$$= \sum_{j=1}^{n} (-1)^{j-1} \binom{n}{j} a_j$$

An application of the latter formula to multicast can be found in Chapter 18 and
many others are in Feller (1970, Chapter IV). Sometimes it is useful to reason with

the complement of the union $(\cup_{k=1}^n A_k)^c = \Omega \setminus \cup_{k=1}^n A_k = \cap_{k=1}^n A_k^c$. Applying Axiom 2 to $(\cup_{k=1}^n A_k)^c \cup (\cup_{k=1}^n A_k) = \Omega$,

$$\Pr\left[(\cup_{k=1}^n A_k)^c\right] = \Pr\left[\Omega\right] - \Pr\left[\cup_{k=1}^n A_k\right]$$

and using Axiom 1 and the inclusion-exclusion formula (2.7), we obtain

$$\Pr\left[(\cup_{k=1}^n A_k)^c\right] = 1 - \sum_{j=1}^n (-1)^{j-1} S_j = \sum_{j=0}^n (-1)^j S_j \qquad (2.8)$$

with the convention that $S_0 = 1$. The Boole's inequalities

$$\Pr\left[\cup_{k=1}^n A_k\right] \le \sum_{k=1}^n \Pr\left[A_k\right] \qquad (2.9)$$

$$\Pr\left[\cap_{k=1}^n A_k\right] \ge 1 - \sum_{k=1}^n \Pr\left[A_k^c\right]$$

are derived as consequences of the inclusion-exclusion formula (2.5). Only if all events are mutually exclusive, the equality sign in (2.9) holds whilst the inequality sign follows from the fact that possible overlaps in events are, in contrast to the inclusion-exclusion formula (2.5), not subtracted.

The inclusion-exclusion formula is of a more general nature and also applies to other measures on sets than $\Pr\left[.\right]$, for example to the cardinality as mentioned above. For the cardinality of a set A, which is usually denoted by $|A|$, the inclusion-exclusion variant of (2.8) is

$$|(\cup_{k=1}^n A_k)^c| = \sum_{j=0}^n (-1)^j |S_j| \qquad (2.10)$$

where the total number of elements in the sample space is $|S_0| = N$ and

$$|S_j| = \sum_{1 \le k_1 < k_2 < \ldots < k_j \le n} \left| \cap_{m=1}^j A_{k_m} \right|$$

A nice illustration of the above formula (2.10) applies to the *sieve of Eratosthenes* (Hardy and Wright, 1968, p. 4), a procedure to construct the table of prime numbers[3] up to N. Consider the increasing sequence of integers

$$\Omega = \{2, 3, 4, \ldots, N\}$$

and remove successively all multiples of 2 (even numbers starting from 4, 6, ...), all multiples of 3 (starting from 3^2 and not yet removed previously), all multiples of 5, all multiples of the next number larger than 5 and still in the list (which is the prime 7) and so on, up to all multiples of the largest possible prime divisor that is equal to or smaller than $\left[\sqrt{N}\right]$. Here, $[x]$ is the largest integer smaller than or equal to x. The remaining numbers in the list are prime numbers. Let us now compute the number of primes $\pi(N)$ smaller than or equal to N by using the inclusion-exclusion formula (2.10). The number of primes smaller than or equal to a real number x is $\pi(x)$

---

[3] An integer number p is prime if $p > 1$ and p has no other integer divisors than 1 and itself p. The sequence of the first primes are 2, 3, 5, 7, 11, 13, etc. If a and b are divisors of n, then $n = ab$ from which it follows that a and b cannot exceed both $\sqrt{n}$. Hence, any composite number n is divisible by a prime p that does not exceed $\sqrt{n}$.

and, evidently, if p_n denotes the n-th prime, then $\pi\left(p_n\right) = n$. Let A_k denote the set of the multiples of the k-th prime p_k that belong to Ω. The number of such sets A_k in the sieve of Eratosthenes is equal to the index of the largest prime number p_n smaller than or equal to $\left[\sqrt{N}\right]$, hence, $n = \pi\left(\sqrt{N}\right)$. If $q \in \left(\cup_{k=1}^n A_k\right)^c$, this means that q is not divisible by each prime number smaller than p_n and that q is a prime number lying between $\sqrt{N} < q \leq N$. The cardinality of the set $\left(\cup_{k=1}^n A_k\right)^c$, the number of primes between $\sqrt{N} < q \leq N$ is

$$|(\cup_{k=1}^n A_k)^c| = \pi(N) - \pi\left(\sqrt{N}\right)$$

On the other hand, if $r \in \cap_{m=1}^j A_{k_m}$ for $1 \leq k_1 < k_2 < \cdots < k_j \leq n$, then r is a multiple of $p_{k_1} p_{k_2} \cdots p_{k_j}$ and the number of multiples of the integer $p_{k_1} p_{k_2} \cdots p_{k_j}$ in Ω is

$$\left[\frac{N}{p_{k_1} p_{k_2} \cdots p_{k_j}}\right] = \left|\cap_{m=1}^j A_{k_m}\right|$$

Applying the inclusion-exclusion formula (2.10) with $|\Omega| = S_0 = N - 1$ and $n = \pi\left(\sqrt{N}\right)$ gives

$$\pi(N) - \pi\left(\sqrt{N}\right) = N - 1 - \sum_{j=1}^n (-1)^j \sum_{1 \leq k_1 < k_2 < \ldots < k_j \leq n} \left[\frac{N}{p_{k_1} p_{k_2} \cdots p_{k_j}}\right]$$

The knowledge of the prime numbers smaller than or equal to $\left[\sqrt{N}\right]$, i.e. the first $n = \pi\left(\sqrt{N}\right)$ primes, suffices to compute the number of primes $\pi(N)$ smaller than or equal to N without explicitly knowing the primes q lying between $\sqrt{N} < q \leq N$.

2.2 Discrete random variables

Discrete random variables are real functions X defined on a discrete probability space Ω as $X : \Omega \to \mathbb{R}$ with the property that the event

$$\{\omega \in \Omega : X\left(\omega\right) = x\} \in \mathcal{F}$$

for each $x \in \mathbb{R}$. The event $\{\omega \in \Omega : X\left(\omega\right) = x\}$ is further abbreviated as $\{X = x\}$. A discrete probability density function (pdf) $\Pr[X = x]$ has the following properties:

(i) $0 \leq \Pr[X = x] \leq 1$ for real x that are possible outcomes of an experiment. The set of values x can be finite or countably infinite and constitute the discrete probability space.

(ii) $\sum_x \Pr[X = x] = 1$.

In the classical example of throwing a die, the discrete probability space $\Omega = \{1, 2, 3, 4, 5, 6\}$ and, since each of the six edges of the (fair) die is equally possible as outcome, $\Pr[X = x] = \frac{1}{6}$ for each $x \in \Omega$.

2.2.1 The expectation

An important operator acting on a discrete random variable X is the expectation, defined as

$$E[X] = \sum_x x \Pr[X = x] \tag{2.11}$$

where the sum over x ranges over all possible discrete values x that the random variable X can attain. Most often this set consists of integers so that $x \in \mathbb{N}$ or $x \in \mathbb{Z}$. The expectation $E\left[X\right]$ is also called the mean[4] or first moment of X. More generally, if X is a discrete random variable and g is a function, then $Y = g(X)$ is also a discrete random variable with expectation $E\left[Y\right]$ equal to

$$E\left[g(X)\right] = \sum_x g(x)\Pr\left[X = x\right] \tag{2.12}$$

A special and often used function in probability theory is the indicator function 1_y, which is defined as 1 if the condition y is true and, otherwise zero. For example,

$$E\left[1_{X>a}\right] = \sum_x 1_{x>a}\Pr\left[X = x\right] = \sum_{x>a}\Pr\left[X = x\right] = \Pr[X > a]$$

$$E\left[1_{X=a}\right] = \Pr[X = a] \tag{2.13}$$

The higher moments of a random variable are defined as the case where $g(x) = x^n$,

$$E\left[X^n\right] = \sum_x x^n\Pr\left[X = x\right] \tag{2.14}$$

From the definition (2.11), it follows that the expectation is a linear operator,

$$E\left[\sum_{k=1}^n a_k X_k\right] = \sum_{k=1}^n a_k E\left[X_k\right]$$

The variance of X is defined as

$$\mathrm{Var}[X] = E\left[(X - E\left[X\right])^2\right] \tag{2.15}$$

The variance is always non-negative. Using the linearity of the expectation operator and $\mu = E\left[X\right]$, we rewrite (2.15) as

$$\mathrm{Var}[X] = E\left[X^2\right] - \mu^2 \tag{2.16}$$

Since $\mathrm{Var}[X] \geq 0$, relation (2.16) indicates that $E\left[X^2\right] \geq (E\left[X\right])^2$. Often the standard deviation, defined as $\sigma = \sqrt{\mathrm{Var}\left[X\right]}$, is used. An interesting variational principle of the variance follows, for the variable u, from

$$E\left[(X - u)^2\right] = E\left[(X - \mu)^2\right] + (u - \mu)^2$$

which is minimized at $u = \mu = E\left[X\right]$ with value $\mathrm{Var}[X]$. Hence, the best least-square approximation (see Section 4.6.3) of the random variable X is the number $E\left[X\right]$.

---

[4] The average of n real numbers $x_1, x_2, \ldots, x_n$ equals $\frac{1}{n}\sum_{k=1}^n x_k$. Hence, "average" is used for data or realizations of a stochastic random variable.

2.2.2 *The probability generating function*

The probability generating function (pgf) of a discrete random variable X is defined, for complex z, as

$$\varphi_X(z) = E\left[z^X\right] = \sum_x z^x \Pr\left[X = x\right] \qquad (2.17)$$

where the last equality follows from (2.12). If X is integer-valued and non-negative, thus $X \in \mathbb{N}$, then the pgf is the Taylor expansion of the complex function $\varphi_X(z)$. Commonly the latter restriction applies, otherwise the substitution $z = e^{it}$ can be used such that (2.17) expresses the Fourier series of $\varphi_X\left(e^{it}\right) = E\left[e^{itX}\right]$, which is also called the characteristic function of X when t is real. The characteristic function $\varphi_X\left(e^{it}\right)$ exists for any random variable X because, for real t, $\varphi_X\left(e^{it}\right)$ converges

$$\left|\varphi_X\left(e^{it}\right)\right| = \left|\sum_x e^{xit}\Pr\left[X = x\right]\right| \leq \sum_x \left|e^{xit}\right|\Pr\left[X = x\right] = \sum_x \Pr\left[X = x\right] = 1$$

Similarly, $\varphi_X(z) \leq 1$ converges for any random variable X, provided that $|z| \leq 1$. The importance of the pgf mainly lies in the fact that the theory of functions can be applied. Numerous examples of the power of analysis will be illustrated.

Concentrating on non-negative integer random variables X,

$$\varphi_X(z) = \sum_{k=0}^{\infty} \Pr\left[X = k\right] z^k \qquad (2.18)$$

and the Taylor coefficients obey

$$\Pr\left[X = k\right] = \frac{1}{k!}\left.\frac{d^k \varphi_X(z)}{dz^k}\right|_{z=0} \qquad (2.19)$$

$$= \frac{1}{2\pi i}\int_{C(0)}\frac{\varphi_X(z)}{z^{k+1}}dz \qquad (2.20)$$

where[5] $C(0)$ denotes a contour around $z = 0$. Both are inversion formulae[6]. Since the general form $E[g(X)]$ is completely defined when $\Pr[X = x]$ is known, the knowledge of the pgf results in a complete alternative description,

$$E\left[g(X)\right] = \sum_{k=0}^{\infty}\frac{g(k)}{k!}\left.\frac{d^k \varphi_X(z)}{dz^k}\right|_{z=0} \qquad (2.21)$$

---

[5] The integral (2.20) can be found by applying (2.19) to Cauchy's famous integral of a complex analytic function $f(z)$,

$$f(z) = \frac{1}{2\pi i}\int_{C(z)}\frac{f(\omega)}{\omega - z}d\omega$$

where the contour $C(z)$ enclosed the point z and $f(\omega)$ is analytic inside and on the contour $C(z)$. The beautiful theory of a complex function is treated by Titchmarsh (1964) and Whittaker and Watson (1996).

[6] A similar inversion formula for Fourier series exists (see e.g. Titchmarsh (1948)).

Sometimes it is more convenient to compute values of interest directly from (2.17) rather than from (2.21). For example, n-fold differentiation of $\varphi_X(z) = E\left[z^X\right]$ yields

$$\frac{d^n \varphi_X(z)}{dz^n} = E\left[X(X-1)\cdots(X-n+1)z^{X-n}\right] = n!E\left[\binom{X}{n}z^{X-n}\right]$$

such that

$$E\left[\binom{X}{n}\right] = \frac{1}{n!}\left.\frac{d^n \varphi_X(z)}{dz^n}\right|_{z=1} \tag{2.22}$$

Similarly, let $z = e^t$, then

$$\frac{d^n \varphi_X(e^t)}{dt^n} = E\left[X^n e^{tX}\right]$$

from which the moments follow as

$$E\left[X^n\right] = \left.\frac{d^n \varphi_X(e^t)}{dt^n}\right|_{t=0} \tag{2.23}$$

and, more generally,

$$E\left[(X-a)^n\right] = \left.\frac{d^n\left(e^{-ta}\varphi_X(e^t)\right)}{dt^n}\right|_{t=0} \tag{2.24}$$

2.2.3 *The logarithm of the probability generating function*

The logarithm of the probability generating function is defined as

$$L_X(z) = \log\left(\varphi_X(z)\right) = \log\left(E\left[z^X\right]\right) \tag{2.25}$$

from which $L_X(1) = 0$ because $\varphi_X(1) = 1$. The derivative $L'_X(z) = \frac{\varphi'_X(z)}{\varphi_X(z)}$ shows that $L'_X(1) = \varphi'_X(1)$, while from $L''_X(z) = \frac{\varphi''_X(z)}{\varphi_X(z)} - \left(\frac{\varphi'_X(z)}{\varphi_X(z)}\right)^2$, it follows that $L''_X(1) = \varphi''_X(1) - (\varphi'_X(1))^2$. These first few derivatives are interesting because they are related directly to probabilistic quantities. Indeed, from (2.23), we observe that

$$E[X] = \varphi'_X(1) = L'_X(1) \tag{2.26}$$

and from $E[X^2] = \varphi''_X(1) + \varphi'_X(1)$

$$\begin{aligned} \text{Var}[X] &= \varphi''_X(1) + \varphi'_X(1) - (\varphi'_X(1))^2 \\ &= L''_X(1) + L'_X(1) \end{aligned} \tag{2.27}$$

2.3 Continuous random variables

Although most of the concepts defined above for discrete random variables are readily transferred to continuous random variables, the calculus is in general more difficult. Indeed, instead of reasoning on the pdf, it is more convenient to work with the probability distribution function defined for both discrete and continuous random variables as

$$F_X(x) = \Pr[X \le x] \tag{2.28}$$

Clearly, we have $\lim_{x \to -\infty} F_X(x) = 0$, while $\lim_{x \to +\infty} F_X(x) = 1$. Further, $F_X(x)$ is non-decreasing in x and

$$\Pr[a < X \le b] = F_X(b) - F_X(a) \tag{2.29}$$

This relation follows from the observations $\{X \le a\} \cup \{a < X \le b\} = \{X \le b\}$ and $\{X \le a\} \cap \{a < X \le b\} = \emptyset$. For mutually exclusive events $A \cap B = \emptyset$, Axiom 2 in Section 2.1 states that $\Pr[A \cup B] = \Pr[A] + \Pr[B]$, which proves (2.29). As a corollary of (2.29), $F_X(x)$ is continuous at the right, which follows from (2.29) by denoting $a = b - \epsilon$ for any $\epsilon > 0$. Less precisely, it follows from the equality sign at the right, $X \le b$, and inequality at the left, $a < X$. Hence, $F_X(x)$ is not necessarily continuous at the left, which implies that $F_X(x)$ is not necessarily continuous and that $F_X(x)$ may possess jumps. But even if $F_X(x)$ is continuous, the pdf is not necessary continuous[7].

The pdf of a continuous random variable X is defined as

$$f_X(x) = \frac{dF_X(x)}{dx} \tag{2.30}$$

Assuming that $F_X(x)$ is differentiable at x, from (2.29), we have[8] for small, positive Δx

$$\Pr[x < X \le x + \Delta x] = F_X(x + \Delta x) - F_X(x)$$
$$= \frac{dF_X(x)}{dx} \Delta x + O\left((\Delta x)^2\right)$$

Using the definition (2.30) indicates that, if $F_X(x)$ is differentiable at x,

$$f_X(x) = \lim_{\Delta x \to 0} \frac{\Pr[x < X \le x + \Delta x]}{\Delta x} \tag{2.31}$$

If $f_X(x)$ is finite, then $\lim_{\Delta x \to 0} \Pr[x < X \le x + \Delta x] = \Pr[X = x] = 0$, which means that for well-behaved (i.e. $F_X(x)$ is differentiable for most x) continuous

---

[7] Weierstrass was the first to present a continuous non-differentiable function,

$$f(x) = \sum_{n=0}^{\infty} b^n \cos(a^n \pi x)$$

where $0 < b < 1$ and a is an odd positive integer. Since the series is uniformly convergent for any x, $f(x)$ is continuous everywhere. Titchmarsh (1964, Chapter IX) demonstrates for $ab > 1 + \frac{3\pi}{2}$ that $\frac{f(x+h)-f(x)}{h}$ takes arbitrarily large values such that $f'(x)$ does not exist. Another class of continuous non-differentiable functions are the sample paths of a Brownian motion. The Cantor function, which is discussed in Berger (1993, p. 21) and Billingsley (1995, p. 407), is an other classical, noteworthy function with peculiar properties.

[8] We will frequently use Landau's big O and small o notation. We write $f(x) = O(g(x))$ for $x \to \infty$ if there exist a positive real number c and a real number of x_0 for which $|f(x)| \le c|g(x)|$ for any $x > x_0$. The notation $f(x) = o(g(x))$ means that $\lim_{x \to \infty} \frac{f(x)}{g(x)} = 0$.

random variables X, the event that X precisely equals x is zero[9]. Hence, for well-behaved continuous random variables where $\Pr[X = x] = 0$ for all x, the inequality signs in the general formula (2.29) can be relaxed,

$$\Pr[a < X \le b] = \Pr[a \le X \le b] = \Pr[a \le X < b] = \Pr[a < X < b]$$

If $f_X(x)$ is not finite, then $F_X(x)$ is not differentiable at x such that

$$\lim_{\Delta x \to 0} F_X(x + \Delta x) - F_X(x) = \Delta F_X(x) \neq 0$$

This means that $F_X(x)$ jumps upwards at x over $\Delta F_X(x)$. In that case, there is a probability mass with magnitude $\Delta F_X(x)$ at the point x. Although the second definition (2.31) is strictly speaking not valid in that case, one sometimes denotes the pdf at $y = x$ by $f_X(y) = \Delta F_X(x)\delta(y - x)$ where $\delta(x)$ is the Dirac impulse or delta function with basic property that $\int_{-\infty}^{+\infty} \delta(y - x)dx = 1$. Even apart from the above-mentioned difficulties for certain classes of non-differentiable, but continuous functions, the fact that probabilities are always confined to the region $[0,1]$ may suggest that $0 \le f_X(x) \le 1$. However, the second definition (2.31) shows that $f_X(x)$ can be much larger than 1. For example, if X is a Gaussian random variable with mean μ and variance σ^2 (see Section 3.2.3) then $f_X(\mu) = \frac{1}{\sqrt{2\pi}\sigma}$ can be made arbitrarily large. In fact,

$$\lim_{\sigma \to 0} \frac{\exp\left(-\frac{(x-\mu)^2}{2\sigma^2}\right)}{\sqrt{2\pi}\sigma} = \delta(x - \mu)$$

2.3.1 Transformation of random variables

It frequently appears useful to know how to compute $F_Y(x)$ for $Y = g(X)$. Only if the inverse function g^{-1} exists, the event $\{g(X) \le x\}$ is equivalent to $\{X \le g^{-1}(x)\}$ if $\frac{dg}{dx} > 0$ and to $\{X > g^{-1}(x)\}$ if $\frac{dg}{dx} < 0$. Hence,

$$F_Y(x) = \Pr[g(X) \le x] = \begin{cases} F_X\left(g^{-1}(x)\right), & \frac{dg}{dx} > 0 \\ 1 - F_X\left(g^{-1}(x)\right), & \frac{dg}{dx} < 0 \end{cases} \qquad (2.32)$$

For well-behaved continuous random variables, we may rewrite (2.31) in terms of differentials,

$$f_X(x)\,dx = \Pr[x \le X \le x + dx]$$

and, similarly for $f_Y(y)$,

$$f_Y(y)\,dy = \Pr[y \le Y = g(X) \le y + dy]$$

If g is increasing, then the event $\{y \le g(X) \le y + dy\}$ is equivalent to

$$\{g^{-1}(y) \le X \le g^{-1}(y + dy)\} = \{x \le X \le x + dx\}$$

---

[9] In Lesbesgue measure theory (Titchmarsh, 1964; Billingsley, 1995), it is said that a countable, finite or enumerable (i.e. function evaluations at individual points) set is measurable, but its measure is zero.

such that

$$f_Y(y)\,dy = f_X(x)\,dx$$

If g is decreasing, we find that $f_Y(y)\,dy = -f_X(x)\,dx$. Thus, if g^{-1} and g' exist, then the relation between the pdf of a well-behaved continuous random variable X and that of the transformed random variable $Y = g(X)$ is

$$f_Y(y) = f_X(x)\left|\frac{dx}{dy}\right| = \frac{f_X(x)}{|g'(x)|} \tag{2.33}$$

This expression also follows by straightforward differentiation of (2.32). The chi-square distribution introduced in Section 3.3.5 is a nice example of the transformation of random variables.

2.3.2 The expectation

Analogously to the discrete case, we define the expectation of a continuous random variable as

$$E[X] = \int_{-\infty}^{\infty} x f_X(x)dx \tag{2.34}$$

In addition for the expectation to exist[10], we require $\int_{-\infty}^{\infty} |x|\,f_X(x)dx < \infty$. If X is a continuous random variable and g is a continuous function, then $Y = g(X)$ is also a continuous random variable with expectation $E[Y]$ equal to

$$E[g(X)] = \int_{-\infty}^{\infty} g(x) f_X(x)dx \tag{2.35}$$

It is often useful to express the expectation $E[X]$ of a non-negative random variable X in tail probabilities. Upon integration by parts,

$$E[X] = \int_0^{\infty} x f_X(x)dx = -x\int_x^{\infty} f_X(u)du\Big|_0^{\infty} + \int_0^{\infty} dx \int_x^{\infty} f_X(u)du$$

we find

$$E[X] = \int_0^{\infty} (1 - F_X(x))\,dx \tag{2.36}$$

---

[10] This requirement is borrowed from measure theory and Lebesgue integration (Titchmarsh, 1964, Chapter X; Royden, 1988, Chapter 4), where a measurable function is said to be integrable (in the Lebesgue sense) over A if $f^+ = \max(f(x),0)$ and $f^- = \max(-f(x),0)$ are both integrable over A. Although this restriction seems only of theoretical interest, in some applications (see the Cauchy distribution defined in (3.47)) the Riemann integral may exists where the Lesbesgue does not. For example, $\int_0^{\infty} \frac{\sin x}{x} dx$ equals, in the Riemann sense, $\frac{\pi}{2}$ (which is a standard excercise in contour integration), but this integral does not exist in the Lesbesgue sense. Only for improper integrals (integration interval is infinite), Riemann integration may exist where Lesbesgue does not. However, in most other cases (integration over a finite interval), Lesbesgue integration is more general. For instance, if $f(x) = 1_{\{x \text{ is rational}\}}$, then $\int_0^1 f(u)du$ does not exist in the Riemann sense (since upper and lower sums do not converge to each other). However, $\int_0^1 f(u)du = 0$ in the Lesbesgue sense (since there is only a set of measure zero different from 0, namely all rational numbers in $[0,1]$). In probability theory and measure theory, Lesbesgue integration is assumed.

The case for a non-positive random variable X is derived analogously,

$$E[X] = \int_{-\infty}^{0} x f_X(x) dx = x \int_{-\infty}^{x} f_X(u) du \Big|_{-\infty}^{0} - \int_{-\infty}^{0} dx \int_{-\infty}^{x} f_X(u) du$$

$$= - \int_{-\infty}^{0} F_X(x) dx$$

The general case follows by addition:

$$E[X] = \int_{0}^{\infty} (1 - F_X(x)) \, dx - \int_{-\infty}^{0} F_X(x) dx$$

A similar expression exists for discrete random variables. In general for any discrete random variable X, we can write

$$E[X] = \sum_{k=-\infty}^{\infty} k \Pr[X = k] = \sum_{k=-\infty}^{-1} k \Pr[X = k] + \sum_{k=0}^{\infty} k \Pr[X = k]$$

$$= \sum_{k=-\infty}^{-1} k \left(\Pr[X \le k] - \Pr[X \le k - 1] \right) + \sum_{k=0}^{\infty} k \left(\Pr[X \ge k] - \Pr[X \ge k + 1] \right)$$

$$= \sum_{k=-\infty}^{-1} k \Pr[X \le k] - \sum_{k=-\infty}^{-2} (k+1) \Pr[X \le k] + \sum_{k=1}^{\infty} k \Pr[X \ge k] - \sum_{k=1}^{\infty} (k-1) \Pr[X \ge k]$$

$$= - \Pr[X \le -1] - \sum_{k=-\infty}^{-2} \Pr[X \le k] + \sum_{k=1}^{\infty} \Pr[X \ge k]$$

or the mean of a discrete random variable X expressed in tail probabilities is[11]

$$E[X] = \sum_{k=1}^{\infty} \Pr[X \ge k] - \sum_{k=-\infty}^{-1} \Pr[X \le k] \qquad (2.37)$$

The special case of (2.37) for a non-negative discrete random variable X follows

---

[11] We remark that

$$E[X] = \sum_{k=-\infty}^{\infty} k \Pr[X = k] = \sum_{k=-\infty}^{\infty} k \left(\Pr[X \ge k] - \Pr[X \ge k + 1] \right)$$

$$\ne \sum_{k=-\infty}^{\infty} k \Pr[X \ge k] - \sum_{k=-\infty}^{\infty} k \Pr[X \ge k + 1] = \sum_{k=-\infty}^{\infty} \Pr[X \ge k]$$

because the series in the second line are diverging. In fact, there exists a finite integer k_ϵ such that, for any real arbitrarily small $\epsilon > 0$, it holds that $\Pr[X \ge k_\epsilon] = 1 - \epsilon$ and $\Pr[X \ge k_\epsilon] \le \Pr[X \ge k]$ for all $k < k_\epsilon$. Hence,

$$E[X] = \sum_{k=-\infty}^{k_\epsilon} \Pr[X \ge k] + \sum_{k=k_\epsilon}^{\infty} \Pr[X \ge k] \ge (1 - \epsilon) \sum_{k=-\infty}^{k_\epsilon} 1 + c \to \infty$$

where $\sum_{k=k_\epsilon}^{\infty} \Pr[X \ge k] = c$ is finite. Also, even for negative X, $\sum_{k=-\infty}^{\infty} \Pr[X \ge k]$ is always positive.

elegantly from the indicator identity,

$$X = \sum_{k=1}^{\infty} 1_{\{X \geq k\}}$$

after taking the expectation of both sides and using (2.13).

2.3.3 The probability generating function

The probability generating function (pgf) of a continuous random variable X is defined, for complex z, as the Laplace transform

$$\varphi_X(z) = E\left[e^{-zX}\right] = \int_{-\infty}^{\infty} e^{-zt} f_X(t)dt \tag{2.38}$$

Again, in some cases, it may be more convenient to use $z = iu$, in which case the double-sided Laplace transform reduces to a Fourier transform. The strength of these transforms is based on the numerous properties, especially the inverse transform,

$$f_X(t) = \frac{1}{2\pi i} \int_{c-i\infty}^{c+i\infty} \varphi_X(z) e^{zt} dz \tag{2.39}$$

where c is the smallest real value of $\text{Re}(z)$ for which the integral in (2.38) converges. Similarly as for discrete random variables, we have $e^{za}\varphi_X(z) = E\left[e^{-z(X-a)}\right]$

$$E[(X - a)^n] = (-1)^n \left. \frac{d^n\left(e^{za}\varphi_X(z)\right)}{dz^n} \right|_{z=0} \tag{2.40}$$

The main difference with the discrete case lies in the definition $E\left[e^{-zX}\right]$ (continuous) versus $E\left[z^X\right]$ (discrete). Since the exponential is an entire function[12] with power series around $z = 0$, $e^{-zX} = \sum_{k=0}^{\infty} \frac{(-1)^k X^k}{k!} z^k$, the expectation and summation can be reversed, leading to

$$E\left[e^{-zX}\right] = \sum_{k=0}^{\infty} \frac{(-1)^k E\left[X^k\right]}{k!} z^k \tag{2.41}$$

provided[13] $E\left[X^k\right] = O\left(k!\right)$, which is a necessary condition for the summation to converge for $z \neq 0$. Assuming convergence[14], the Taylor series of $E\left[e^{-zX}\right]$ around $z = 0$ is expressed as function of the moments of X, whereas in the discrete case, the Taylor series of $E\left[z^X\right]$ around $z = 0$ given by (2.18) is expressed in

---

[12] An entire (or integral) function is a complex function without singularities in the finite complex plane. Hence, a power series around any finite point has infinite radius of convergence. In other words, it exists for all finite complex values.

[13] The Landau big O-notation specifies the "order of a function" when the argument tends to some limit. Most often the limit is to infinity, but the O-notation can also be used to characterize the behavior of a function around some finite point. Formally, $f(x) = O(g(x))$ for $x \to \infty$ means that there exist positive numbers c and x_0 for which $|f(x)| \leq c|g(x)|$ for $x > x_0$.

[14] The lognormal distribution defined by (3.53) is an example where the summation (2.41) diverges for any $z \neq 0$.

terms of probabilities of X. This observation has led to $E\left[e^{-zX}\right]$ being called the moment generating function, while $E\left[z^X\right]$ is the probability generating function of the random variable X. If moments are desired, the substitution $z \to e^{-u}$ in $E\left[z^X\right]$ is appropriate.

On the other hand, series expansion of $E\left[z^X\right]$ around $z = 1$,

$$\varphi_X(z) = \sum_{k=0}^{\infty} \Pr\left[X = k\right](z+1-1)^k = \sum_{k=0}^{\infty} \Pr\left[X = k\right]\sum_{j=0}^{k}\binom{k}{j}(z-1)^j$$

$$= \sum_{j=0}^{\infty}\left[\sum_{k=j}^{\infty}\binom{k}{j}\Pr\left[X = k\right]\right](z-1)^j$$

shows with (2.22) that

$$E\left[\binom{X}{j}\right] = \sum_{k=j}^{\infty}\binom{k}{j}\Pr\left[X = k\right]$$

The converse follows in a similar way from

$$\varphi_X(z) = \sum_{k=0}^{\infty} E\left[\binom{X}{k}\right](z-1)^k = \sum_{k=0}^{\infty} E\left[\binom{X}{k}\right]\sum_{j=0}^{k}\binom{k}{j}(-1)^{k-j}z^j$$

$$= \sum_{j=0}^{\infty}\sum_{k=j}^{\infty}(-1)^{k-j}\binom{k}{j}E\left[\binom{X}{k}\right]z^j$$

After invoking (2.18) and equating corresponding powers of z, we obtain

$$\Pr\left[X = k\right] = \sum_{k=j}^{\infty}(-1)^{k-j}\binom{k}{j}E\left[\binom{X}{k}\right] = \sum_{k=j}^{\infty}(-1)^{k-j}\frac{E\left[X(X-1)\cdots(X-k+1)\right]}{j!\,(k-j)!}$$

Hence, we have demonstrated a one-to-one correspondence between the binomial moment $E\left[\binom{X}{k}\right]$ or the k-th factorial moment $E\left[X(X-1)\cdots(X-k+1)\right]$ and the probability density function $\Pr\left[X = k\right]$.

Besides defining the Laplace transform (2.38) as the probability generating function $\varphi_X(z) = E\left[e^{-zX}\right]$ of the continuous random variable X, we can, for *non-negative* continuous random variables, also consider the Mellin transform as the continuous variant of $E\left[z^{X-1}\right]$,

$$\psi_X(z) = \int_0^{\infty} t^{z-1} f_X(t)\,dt \tag{2.42}$$

The inverse Mellin transform (Titchmarsh, 1948, p. 7) is

$$f_X(t) = \frac{1}{2\pi i}\int_{c-i\infty}^{c+i\infty}\psi_X(z)\,z^{-z}dz \tag{2.43}$$

In some cases as illustrated in Section 2.5.4, the Mellin transform is useful.

2.3.4 The logarithm of the probability generating function

The logarithm of the probability generating function is defined as

$$L_X(z) = \log\left(\varphi_X(z)\right) = \log\left(E\left[e^{-zX}\right]\right) \tag{2.44}$$

from which $L_X(0) = 0$ because $\varphi_X(0) = 1$. Further, analogous to the discrete case, we see that $L_X'(0) = \varphi_X'(0)$, $L_X''(0) = \varphi_X''(0) - (\varphi_X'(0))^2$ and

$$E[X] = -\varphi_X'(0) = -L_X'(0)$$

However, the difference with the discrete case lies in the higher moments,

$$E[X^n] = (-1)^n \left.\frac{d^n\varphi_X(z)}{dz^n}\right|_{z=0} \tag{2.45}$$

because with $E[X^2] = \varphi_X''(0)$,

$$\begin{aligned}
\text{Var}[X] &= \varphi_X''(0) - (\varphi_X'(0))^2 \\
&= L_X''(0) \tag{2.46}
\end{aligned}$$

The latter expression makes $L_X(z)$ for a continuous random variable particularly useful. Since the variance is always positive, it demonstrates that $L_X(z)$ is convex (see Section 5.5) around $z = 0$. Finally, we mention that

$$E\left[(X - E[X])^3\right] = -L_X'''(0)$$

The third centered moment $E\left[(X - E[X])^3\right]$ measures the lack of symmetry of the distribution around the mean and $E\left[(X - E[X])^3\right]$ is negative (positive) if the left (right) tail of the distribution is longer than the right (left) tail. The normalized third moment

$$s_X = \frac{E\left[(X - E[X])^3\right]}{(\text{Var}[X])^{3/2}}$$

is called the skewness.

2.4 The conditional probability

The conditional probability of the event A given the event B (or on the hypothesis B) is defined as

$$\Pr[A|B] = \frac{\Pr[A \cap B]}{\Pr[B]} \tag{2.47}$$

The definition implicitly assumes that the event B has positive probability, otherwise the conditional probability remains undefined. We quote Feller (1970, p. 116):

Taking conditional probabilities of various events with respect to a particular hypothesis B amounts to choosing B as a new sample space with probabilities proportional to the original ones; the proportionality factor $\Pr[B]$ is necessary in order to reduce the total probability of the new sample space to unity. This formulation shows that all general theorems on probabilities are valid for conditional probabilities with respect to any particular hypothesis. For example, the law $\Pr[A \cup B] = \Pr[A] + \Pr[B] - \Pr[A \cap B]$ takes the form

$$\Pr[A \cup B|C] = \Pr[A|C] + \Pr[B|C] - \Pr[A \cap B|C]$$

The formula (2.47) is often rewritten in the form

$$\Pr[A \cap B] = \Pr[A|B] \Pr[B] \tag{2.48}$$

which easily generalizes to more events. For example, denote $A = A_1$ and $B = A_2 \cap A_3$, then

$$\Pr[A_1 \cap A_2 \cap A_3] = \Pr[A_1|A_2 \cap A_3] \Pr[A_2 \cap A_3]$$
$$= \Pr[A_1|A_2 \cap A_3] \Pr[A_2|A_3] \Pr[A_3]$$

Another application of the conditional probability occurs when a partitioning of the sample space Ω is known: $\Omega = \cup_k B_k$ and all B_k are mutually exclusive, which means that $B_k \cap B_j = \emptyset$ for any k and $j \neq k$. Then, with (2.48),

$$\sum_k \Pr[A \cap B_k] = \sum_k \Pr[A|B_k] \Pr[B_k]$$

The event $A_k = \{A \cap B_k\}$ is a decomposition (or projection) of the event A in the basis event B_k, analogous to the decomposition of a vector in terms of a set of orthogonal basis vectors that span the total state space. Indeed, using the associative property $A \cap \{B \cap C\} = A \cap B \cap C$ and $A \cap A = A$, the intersection $A_k \cap A_j = \{A \cap B_k\} \cap \{A \cap B_j\} = A \cap \{B_k \cap B_j\} = \emptyset$, which implies mutual exclusivity (or orthogonality). Using the distributive property $A \cap \{B_k \cup B_j\} = \{A \cap B_k\} \cup \{A \cap B_j\}$, we observe that

$$A = A \cap \Omega$$
$$= A \cap \{\cup_k B_k\} = \cup_k \{A \cap B_k\} = \cup_k A_k$$

Finally, since all events A_k are mutually exclusive,

$$\Pr[A] = \sum_k \Pr[A_k] = \sum_k \Pr[A \cap B_k]$$

Thus, if $\Omega = \cup_k B_k$ and in addition, for any pair j, k and $j \neq k$, it holds that $B_k \cap B_j = \emptyset$, we have proved the *law of total probability* or decomposability,

$$\Pr[A] = \sum_k \Pr[A|B_k] \Pr[B_k] \tag{2.49}$$

Conditioning on events is a powerful tool that will be used frequently. If the conditional probability $\Pr[A|B_k]$ is known as a function $g(B_k)$, the law of total probability can also be written in terms of the expectation operator defined in (2.12) as

$$\Pr[A] = E[g(B_k)] \tag{2.50}$$

The important memoryless property of the exponential distribution (see Section 3.2.2) is an example of the application of the conditional probability. Another classical example is Bayes' rule. Consider again the events B_k defined above. Using the definition (2.47) followed by (2.48),

$$\Pr[B_k|A] = \frac{\Pr[B_k \cap A]}{\Pr[A]} = \frac{\Pr[A \cap B_k]}{\Pr[A]} = \frac{\Pr[A|B_k]\Pr[B_k]}{\Pr[A]} \qquad (2.51)$$

Using (2.49), we arrive at Bayes' rule

$$\Pr[B_k|A] = \frac{\Pr[A|B_k]\Pr[B_k]}{\sum_j \Pr[A|B_j]\Pr[B_j]} \qquad (2.52)$$

where $\Pr[B_k]$ are called the *a-priori* probabilities and $\Pr[B_k|A]$ are the *a-posteriori* probabilities.

The conditional distribution function of the random variable Y given X is defined by

$$F_{Y|X}(y|x) = \Pr[Y \le y|X = x] \qquad (2.53)$$

for any x provided $\Pr[X = x] > 0$. This condition follows from the definition (2.47) of the conditional probability. The conditional probability density function of Y given X is defined by

$$f_{Y|X}(y|x) = \Pr[Y = y|X = x] = \frac{\Pr[X = x, Y = y]}{\Pr[X = x]}$$

$$= \frac{f_{XY}(x,y)}{f_X(x)} \qquad (2.54)$$

for any x such that $\Pr[X = x] > 0$ (and similarly for continuous random variables $f_X(x) > 0$) and where $f_{XY}(x,y)$ is the joint probability density function defined below in (2.62).

2.5 Several random variables and independence

2.5.1 Discrete random variables

Two events A and B are independent if

$$\Pr[A \cap B] = \Pr[A]\Pr[B] \qquad (2.55)$$

Alternatively, combining the definition of conditional probability (2.48) and of independence (2.55) shows that an event A is independent from the event B if

$$\Pr[A|B] = \Pr[A]$$

Similarly to (2.55), we define two discrete random variables to be independent if

$$\Pr[X = x, Y = y] = \Pr[X = x]\Pr[Y = y] \qquad (2.56)$$

If $Z = f(X, Y)$, then Z is a discrete random variable with

$$\Pr[Z = z] = \sum_{f(x,y)=z} \Pr[X = x, Y = y]$$

Applying the expectation operator (2.11) to both sides yields

$$E[f(X, Y)] = \sum_{x,y} f(x, y) \Pr[X = x, Y = y] \qquad (2.57)$$

If X and Y are independent and f is separable, $f(x, y) = f_1(x)f_2(y)$, then the expectation (2.57) reduces to

$$E[f(X, Y)] = \sum_x f_1(x) \Pr[X = x] \sum_y f_2(y) \Pr[Y = y] = E[f_1(X)] E[f_2(Y)]$$

$$(2.58)$$

The simplest example of the general function is $Z = X + Y$. In that case, the sum is over all x and y that satisfy $x + y = z$. Thus,

$$\Pr[X + Y = z] = \sum_x \Pr[X = x, Y = z - x] = \sum_y \Pr[X = z - y, Y = y]$$

If X and Y are independent, we obtain the convolution,

$$\Pr[X + Y = z] = \sum_x \Pr[X = x] \Pr[Y = z - x]$$

$$= \sum_y \Pr[X = z - y] \Pr[Y = y]$$

2.5.2 The covariance

The covariance of X and Y is defined as

$$\text{Cov}[X, Y] = E[(X - \mu_X)(Y - \mu_Y)] = E[XY] - \mu_X \mu_Y \qquad (2.59)$$

If $\text{Cov}[X, Y] = 0$, then the variables X and Y are *uncorrelated*. If X and Y are independent, then $\text{Cov}[X, Y] = 0$. Hence, independence implies uncorrelation, but the converse is not necessarily true. The classical example[15] is $Y = X^2$ where X has a normal distribution $N(0, 1)$ (Section 3.2.3) because $\mu_X = 0$ and $E[XY] = E[X^3] = 0$ as follows from (3.23). Although X and Y are perfect dependent, they are uncorrelated. Thus, independence is a stronger property than uncorrelation. The covariance $\text{Cov}[X, Y]$ measures the degree of dependence between two (or generally more) random variables. If X and Y are positively (negatively)

---

[15] Another example: let U be uniform on $[0, 1]$ and $X = \cos(2\pi U)$ and $Y = \sin(2\pi U)$. Using (2.35),

$$E[XY] = \int_0^1 \cos(2\pi u) \sin(2\pi u) \, du = 0$$

as well as $E[X] = E[Y] = 0$. Thus, $\text{Cov}[X, Y] = 0$, but X and Y are perfectly dependent because $X = \cos(\arcsin Y) = \pm\sqrt{1 - Y^2}$.

correlated, the large values of X tend to be associated with large (small) values of Y.

As an application of the covariance, consider the problem of computing the variance of a sum S_n of random variables $X_1, X_2, \ldots, X_n$. Let $\mu_k = E[X_k]$, then $E[S_n] = \sum_{k=1}^{n} \mu_k$ and

$$\text{Var}[S_n] = E\left[(S_n - E[S_n])^2\right] = E\left[\left(\sum_{k=1}^{n}(X_k - \mu_k)\right)^2\right]$$

$$= E\left[\sum_{k=1}^{n}\sum_{j=1}^{n}(X_k - \mu_k)(X_j - \mu_j)\right]$$

$$= E\left[\sum_{k=1}^{n}(X_k - \mu_k)^2 + 2\sum_{k=1}^{n}\sum_{j=k+1}^{n}(X_k - \mu_k)(X_j - \mu_j)\right]$$

Using the linearity of the expectation operator and the definition of the covariance (2.59) yields

$$\text{Var}[S_n] = \sum_{k=1}^{n}\text{Var}[X_k] + 2\sum_{k=1}^{n}\sum_{j=k+1}^{n}\text{Cov}[X_k, X_j] \qquad (2.60)$$

Observe that for a set of independent random variables $\{X_k\}$ the double sum with covariances vanishes.

The Cauchy-Schwarz inequality (5.22) derived in Chapter 5 indicates that

$$\left(E[(X - \mu_X)(Y - \mu_Y)]\right)^2 \leq E\left[(X - \mu_X)^2\right]E\left[(Y - \mu_Y)^2\right]$$

such that the covariance is always bounded by

$$|\text{Cov}[X, Y]| \leq \sigma_X \sigma_Y$$

2.5.3 The linear correlation coefficient

Since the covariance is not dimensionless, the *linear* correlation coefficient defined as

$$\rho(X, Y) = \frac{\text{Cov}[X, Y]}{\sigma_X \sigma_Y} \qquad (2.61)$$

and also called Pearson's correlation coefficient, is often convenient to relate two (or more) different physical quantities expressed in different units. The linear correlation coefficient remains invariant (possibly apart from the sign) under a linear transformation because

$$\rho(aX + b, cY + d) = \text{sign}(ac)\rho(X, Y)$$

This transform shows that the linear correlation coefficient $\rho(X, Y)$ is independent of the value of the mean μ_X and the variance σ_X^2 provided $\sigma_X^2 > 0$. Therefore, many computations simplify if we normalize the random variable properly. Let us introduce the concept of a *normalized random variable* $X^* = \frac{X - \mu_X}{\sigma_X}$. The normalized random variable has a zero mean and a variance equal to one. By the invariance under a linear transform, the correlation coefficient $\rho(X, Y) = \rho(X^*, Y^*)$ and also $\rho(X, Y) = \text{Cov}[X^*, Y^*]$. The variance of $X^* \pm Y^*$ follows from (2.60) as

$$\text{Var}[X^* \pm Y^*] = \text{Var}[X^*] + \text{Var}[Y^*] \pm 2\,\text{Cov}[X^*, Y^*]$$
$$= 2(1 \pm \rho(X, Y))$$

Since the variance is always positive, it follows that $-1 \le \rho(X, Y) \le 1$. The extremes $\rho(X, Y) = \pm 1$ imply a *linear* relation between X and Y. Indeed, $\rho(X, Y) = 1$ implies that $\text{Var}[X^* - Y^*] = 0$, which is only possible if $X^* = Y^* + c$, where c is a constant. Hence, $X = \frac{\sigma_X}{\sigma_Y} Y + c'$. A similar argument applies for the case $\rho(X, Y) = -1$. For example, in curve fitting, the goodness of the fit is often expressed in terms of the correlation coefficient. A perfect fit has correlation coefficient equal to 1. In particular, in linear regression where $Y = aX + b$, the regression coefficients a_R and b_R are the minimizers of the square distance $E\left[(Y - (aX + b))^2\right]$ and given by

$$a_R = \frac{\text{Cov}[X, Y]}{\sigma_X^2}$$
$$b_R = E[Y] - a_R E[X]$$

Since a correlation coefficient $\rho(X, Y) = 1$ implies $\text{Cov}[X, Y] = \sigma_X \sigma_Y$, we see that $a_R = \frac{\sigma_Y}{\sigma_X}$ as derived above with normalized random variables.

Although the linear correlation coefficient is a natural measure of the dependence between random variables, it has some disadvantages. First, the variances of X and Y must exist, which may cause problems with heavy-tailed distributions. Second, as illustrated above, dependence can lead to uncorrelation, which is awkward. Third, linear correlation is not invariant under non-linear strictly increasing transformations T such that $\rho(T(X), T(Y)) \ne \rho(X, Y)$. Common intuition expects that dependence measures should be invariant under these transforms T. This leads to the definition of *rank* correlation which satisfies that invariance property. Here, we merely mention Spearman's rank correlation coefficient, which is defined as

$$\rho_S(X, Y) = \rho(F_X(X), F_Y(Y))$$

where ρ is the linear correlation coefficient and where the non-linear strict increasing transform is the probability distribution. More details are found in Embrechts *et al.* (2001b) and in Chapter 4.

2.5.4 *Continuous random variables*

We define the joint distribution function by $F_{XY}(x, y) = \Pr[X \le x, Y \le y]$ and the joint probability density function by

$$f_{XY}(x, y) = \frac{\partial^2 F_{XY}(x, y)}{\partial x \partial y} \tag{2.62}$$

Hence,

$$F_{XY}(x, y) = \Pr[X \le x, Y \le y] = \int_{-\infty}^{x} \int_{-\infty}^{y} f_{XY}(u, v) du dv \tag{2.63}$$

The analogon of (2.57) is

$$E[g(X, Y)] = \int_{-\infty}^{\infty} \int_{-\infty}^{\infty} g(x, y) f_{XY}(x, y) dx dy \tag{2.64}$$

Most of the difficulties occur in the evaluation of the multiple integrals. The change of variables in multiple dimensions involves the Jacobian. Consider the transformed random variables $U = g_1(X, Y)$ and $V = g_2(X, Y)$ and denote the inverse transform by $x = h_1(u, v)$ and $y = h_2(u, v)$, then

$$f_{UV}(u, v) = f_{XY}(h_1(u, v), h_2(u, v)) |J(u, v)| \tag{2.65}$$

where the Jacobian $J(u, v)$ is

$$J(u, v) = \det \begin{bmatrix} \frac{\partial x}{\partial u} & \frac{\partial x}{\partial v} \\ \frac{\partial y}{\partial u} & \frac{\partial y}{\partial v} \end{bmatrix}$$

The absolute value of the Jacobian in (2.65) is required by similar arguments as in the one-dimensional case explained in Section 2.3.1.

For example, consider the random variable $Z = XY$, which is the product of two random variables X and Y. We choose the above random variables as $U = X$ and $V = XY$ such that $u = g_1(x, y) = x$ and $v = g_2(x, y) = xy$. The inverse transform is $x = h_1(u, v) = u$ and $y = h_2(u, v) = v/u$. The corresponding Jacobian is $J(u, v) = \frac{1}{u}$ and (2.65) becomes

$$f_{UV}(u, v) = f_{XY}\left(u, \frac{v}{u}\right) \left|\frac{1}{u}\right|$$

Because $Z = V$ and $\lim_{u \to \infty} \Pr[U \le u] = 1$, the definition (2.63) shows that

$$F_Z(w) = \lim_{u \to \infty} \Pr[U \le u, V \le w] = \int_{-\infty}^{\infty} \int_{-\infty}^{w} f_{UV}(u, v) du dv$$

and, by differentiation with respect to w, we find the probability density function (2.30) of $Z = XY$

$$f_Z(w) = \int_{-\infty}^{\infty} f_{UV}(u, w) du = \int_{-\infty}^{\infty} f_{XY}\left(u, \frac{w}{u}\right) \frac{du}{|u|} \tag{2.66}$$

If X and Y are independent, then $f_{XY}(x,y) = f_X(x)f_Y(y)$ and (2.66) reduces to

$$f_Z(w) = \int_{-\infty}^{\infty} f_X(u) f_Y\left(\frac{w}{u}\right) \frac{du}{|u|} \tag{2.67}$$

The quotient $Q = X/Y$ of two independent random variables X and Y, which is a product of the independent random variables X and Y^{-1}, elegantly follows from (2.67) and the transformation formula (2.33), which yields $f_{Y^{-1}}(u) = u^{-2}f_Y(u^{-1})$. Indeed, substituted in (2.67) gives

$$f_Q(w) = \frac{1}{w^2} \int_{-\infty}^{\infty} f_X(u) f_Y\left(\frac{u}{w}\right) |u| \, du$$

and, after changing the variable of integration into $t = \frac{u}{w}$, the probability density function of the quotient of two independent random variables becomes

$$f_Q(w) = \int_{-\infty}^{\infty} f_X(wt) f_Y(t) |t| \, dt \tag{2.68}$$

Finally, if X and Y are independent, non-negative random variables, then the lower bound in (2.67) reduces to zero and $|u| = u$. The Mellin transform (2.42) of Z then equals (Titchmarsh, 1948, p. 53)

$$\psi_Z(s) = \int_0^{\infty} t^{s-1} f_Z(t) \, dt = \int_0^{\infty} t^{s-1} \left\{ \int_0^{\infty} f_X(u) f_Y\left(\frac{t}{u}\right) \frac{du}{u} \right\} dt = \psi_X(s)\,\psi_Y(s) \tag{2.69}$$

Hence, the Mellin transform of the product of two independent, non-negative random variables X and Y equals the product of the Mellin transforms of X and Y.

If X and Y are independent and $Z = X + Y$, we obtain the convolution,

$$f_Z(z) = \int_{-\infty}^{\infty} f_X(x)f_Y(z-x)dx = \int_{-\infty}^{\infty} f_X(z-y)f_Y(y)dy \tag{2.70}$$

which is often denoted by $f_Z(z) = (f_X * f_Y)(z)$. If both $f_X(x) = 0$ and $f_Y(x) = 0$ for $x < 0$, then the definition (2.70) of the convolution reduces to

$$(f_X * f_Y)(z) = \int_0^z f_X(x)f_Y(z-x)dx$$

Since the event $\{Y < \infty\}$ is always true, we deduce from the definition (2.63) that

$$\Pr[X \le x] = \lim_{y \to \infty} F_{XY}(x, y) = \int_{-\infty}^{x} \int_{-\infty}^{\infty} f_{XY}(u, v)dudv$$

and, after derivation, that

$$f_X(x) = \int_{-\infty}^{\infty} f_{XY}(x, v)dv \tag{2.71}$$

Hence, the distribution function and pdf of a single random variable X can be found from a joint probability density function by integrating the remaining parameter y

over all possible values of the random variable Y. Both the student t and Fisher F distributions in Section 3.3.6 are nicely derived via the transformation (2.65) followed by (2.71).

2.5.5 The sum of independent random variables

Let $S_N = \sum_{k=1}^{N} X_k$, where the random variables X_k are all independent. We first concentrate on the case where $N = n$ is a (fixed) integer. Since $S_N = S_{N-1} + X_N$, direct application of (2.70) yields the recursion

$$f_{S_N}(z) = \int_{-\infty}^{\infty} f_{S_{N-1}}(z - y) f_{X_N}(y) dy \tag{2.72}$$

which, when written out explicitly, leads to the N-fold integral

$$f_{S_N}(z) = \int_{-\infty}^{\infty} f_{X_N}(y_N) dy_N \cdots \int_{-\infty}^{\infty} f_{X_1}(y_1) f_{X_0}(z - y_N - \cdots - y_1) dy_1 \tag{2.73}$$

In many cases, convolutions are more efficiently computed via generating functions. The generating function of S_n equals

$$\varphi_{S_n}(z) = E\left[z^{S_n}\right] = E\left[z^{\sum_{k=1}^{n} X_k}\right] = E\left[\prod_{k=1}^{n} z^{X_k}\right]$$

Since all X_k are independent, (2.58) can be applied,

$$\varphi_{S_n}(z) = \prod_{k=1}^{n} E\left[z^{X_k}\right]$$

or, in terms of generating functions,

$$\varphi_{S_n}(z) = \prod_{k=1}^{n} \varphi_{X_k}(z) \tag{2.74}$$

Hence, we arrive at the important result that the generating function of a sum of independent random variables equals the product of the generating functions of the individual random variables. We also note that the condition of independence is crucial in that it allows the product and expectation operator to be reversed, leading to the useful result (2.74). Often, the random variables X_k all possess the same distribution. In this case of independent identically distributed (i.i.d.) random variables with generating function $\varphi_X(z)$, the relation (2.74) further simplifies to

$$\varphi_{S_n}(z) = (\varphi_X(z))^n \tag{2.75}$$

In the case where the number of terms N in the sum S_N is a random variable with generating function $\varphi_N(z)$, independent of the X_k, we use the general definition of expectation (2.57) for two random variables,

$$\varphi_{S_N}(z) = E\left[z^{S_N}\right] = \sum_{k=0}^{\infty}\sum_x z^x \Pr\left[S_N = x, N = k\right]$$

$$= \sum_{k=0}^{\infty}\sum_x z^x \Pr\left[S_N = x | N = k\right] \Pr\left[N = k\right]$$

where the conditional probability (2.48) is used. Since the value of S_N depends on the number of terms N in the sum, we have $\Pr\left[S_N = x | N = k\right] = \Pr\left[S_k = x\right]$. Further, with

$$\sum_x z^x \Pr\left[S_N = x | N = k\right] = \varphi_{S_k}(z)$$

we have

$$\varphi_{S_N}(z) = \sum_{k=0}^{\infty} \varphi_{S_k}(z) \Pr\left[N = k\right] \tag{2.76}$$

The mean $E\left[S_N\right]$ follows from (2.26) as

$$E\left[S_N\right] = \sum_{k=0}^{\infty} \varphi'_{S_k}(1) \Pr\left[N = k\right] = \sum_{k=0}^{\infty} E\left[S_k\right] \Pr\left[N = k\right] \tag{2.77}$$

Since $E\left[S_k\right] = E\left[\sum_{j=1}^{k} X_j\right] = \sum_{j=1}^{k} E\left[X_j\right]$ and assuming that all random variables X_j have equal mean $E\left[X_j\right] = E\left[X\right]$, we have

$$E\left[S_N\right] = \sum_{k=0}^{\infty} k E\left[X\right] \Pr\left[N = k\right]$$

or

$$E\left[S_N\right] = E\left[X\right] E\left[N\right] \tag{2.78}$$

This relation (2.78) is commonly called *Wald's identity*. Wald's identity holds for any random sum of (possibly dependent) random variables X_j provided the number N of those random variables is independent of the X_j.

In the case, where $X_1, X_2, \ldots$ are i.i.d. random variables and N is independent of all $\{X_k\}_{k\geq 1}$, we apply (2.75) in (2.76) so that

$$\varphi_{S_N}(z) = \sum_{k=0}^{\infty} (\varphi_X(z))^k \Pr\left[N = k\right] = \varphi_N(\varphi_X(z)) \tag{2.79}$$

This expression is a generalization of (2.75).

2.6 Conditional expectation

The generating function (2.76) of a random sum of independent random variables can be derived using the conditional expectation $E[Y|X = x]$ of two random variables X and Y. We will first define the conditional expectation and derive an interesting property.

Suppose that we know that $X = x$, the conditional density function $f_{Y|X}(y|x)$ defined by (2.54) of the random variable $Y_c = Y|X$ can be regarded as only function of y. Using the definition of the expectation (2.34) for continuous random variables (the discrete case is analogous), we have

$$E[Y|X = x] = \int_{-\infty}^{\infty} y f_{Y|X}(y|x)\, dy \tag{2.80}$$

Since this expression holds for any value of x that the random variable X can take, we see that $E[Y|X = x] = g(x)$ is a function of x. In addition, $E[Y|X] = g(X)$ can be regarded as a random variable that is a function of the random variable X. Having identified the conditional expectation $E[Y|X]$ as a random variable, let us compute its expectation or the expectation of the slightly more general random variable $h(X) g(X)$ with $g(X) = E[Y|X]$. From the general definition (2.35) of the expectation, it follows that

$$E[h(X) g(X)] = \int_{-\infty}^{\infty} h(x) g(x) f_X(x)\, dx = \int_{-\infty}^{\infty} h(x) E[Y|X = x] f_X(x)\, dx$$

Substituting (2.80) yields

$$E[h(X) g(X)] = \int_{-\infty}^{\infty} \int_{-\infty}^{\infty} h(x) y f_{Y|X}(y|x) f_X(x)\, dy dx$$

$$= \int_{-\infty}^{\infty} \int_{-\infty}^{\infty} h(x) y f_{XY}(x, y)\, dy dx = E[h(X) Y]$$

where we have used (2.54) and (2.64). Thus, we find the interesting relation

$$E[h(X) E[Y|X]] = E[h(X) Y] \tag{2.81}$$

As a special case where $h(x) = 1$, the expectation of the conditional expectation follows as

$$E[Y] = E_X[E_Y[Y|X]] \tag{2.82}$$

where the index in E_Z clarifies that the expectation is over the random variable Z. Applying this relation to $Y = z^{S_N}$, where $S_N = \sum_{k=1}^{N} X_k$ and all X_k are independent, yields

$$\varphi_{S_N}(z) = E[z^{S_N}] = E_N[E_S[z^{S_N}|N]]$$

Since $E_S[z^{S_N}|N] = \tilde{\varphi}_{S_N}(z)$, where $\tilde{\varphi}$ denotes the random variable as a function of

N that is specified in (2.74), we end up with the pgf of S_N,

$$\varphi_{S_N}(z) = E_N\left[\tilde{\varphi}_{S_N}(z)\right] = \sum_{k=0}^{\infty} \varphi_{S_k}(z)\Pr\left[N=k\right]$$

which is (2.76). Other applications of the conditional expectation are found in Sections 12.1 and 12.2.

2.7 Problems

(i) *The Quizmaster problem.* Behind one of n opaque doors, the quizmaster tells you that there is a car placed and he asks you to choose one of the doors. If you have chosen correctly, the car behind the door is yours. Once you have made your choice, the quizmaster opens another door than the one chosen by you and behind that opened door, there is no car. Then, the quizmaster offers you the opportunity to alter your initial choice. The question now is: in order to enhance your chance of winning the car, would you choose another door or would you stick to your initial choice?

(ii) *The birthday paradox.* Compute the probability that in a group of n people, two of them have the same birthday. Explain why it is called a paradox.

(iii) *The inclusion-exclusion formula.* Prove (2.5).

(iv) If $X_1, X_2, \ldots$ is a set of i.i.d. random variables with the same distribution as the random variable X and $S_N = \sum_{k=1}^{N} X_k$, where the random variable N is independent from each X_k (for $k \geq 1$), show that

$$\text{Var}\left[S_N\right] = \text{Var}\left[X\right]E\left[N\right] + \text{Var}\left[N\right]\left(E\left[X\right]\right)^2 \tag{2.83}$$

3

Basic distributions

This chapter concentrates on the most basic probability distributions and their properties. From these basic distributions, other useful distributions are derived.

3.1 Discrete random variables

3.1.1 The Bernoulli distribution

A Bernoulli random variable X can only take two values: either 1 with probability p or 0 with probability $q = 1 - p$. The standard example of a Bernoulli random variable is the outcome of tossing a biased coin, and, more generally, the outcome of a trial with only two possibilities, either success or failure. The sample space is $\Omega = \{0, 1\}$ and $\Pr[X = 1] = p$, while $\Pr[X = 0] = q$. From this definition, the pgf follows from (2.17) as

$$\varphi_X(z) = E\left[z^X\right] = z^0 \Pr[X = 0] + z^1 \Pr[X = 1]$$

or

$$\varphi_X(z) = q + pz \tag{3.1}$$

Since $X \in \{0, 1\}$, it holds that $X^n = X$. From (2.23) or (2.14), the n-th moment is

$$E[X^n] = p$$

which shows that $E[X] = \Pr[X = 1]$. This property of Bernoulli random variables is useful in stochastic processes as illustated in Section 17.3. From (2.24), we find $E[(X - a)^n] = p(1 - a)^n + q(-a)^n$ such that the moments centered around the mean $\mu = E[X] = p$ are

$$E[(X - \mu)^n] = pq\left(q^{n-1} + (-1)^n p^{n-1}\right)$$

Explicitly, with $p + q = 1$, $\mathrm{Var}[X] = pq$ and $E\left[(X - \mu)^3\right] = pq(q - p)$.

3.1.2 The binomial distribution

A binomial random variable X is the sum of n independent Bernoulli random variables. The sample space is $\Omega = \{0, 1, \ldots, n\}$. For example, X may represent the number of successes in n independent Bernoulli trials such as the number of heads after n-times tossing a (biased) coin. Application of (2.75) with (3.1) gives

$$\varphi_X(z) = (q + pz)^n \tag{3.2}$$

Expanding the binomial pgf in powers of z, which justifies the name "binomial",

$$\varphi_X(z) = \sum_{k=0}^{n} \binom{n}{k} p^k q^{n-k} z^k$$

and comparing to (2.18) yields

$$\Pr[X = k] = \binom{n}{k} p^k q^{n-k} \tag{3.3}$$

The alternative, probabilistic approach starts with (3.3). Indeed, the probability that X has k successes out of n trials consists of precisely k successes (an event with probability p^k) and $n - k$ failures (with probability equal to q^{n-k}). The total number of ways in which k successes out of n trials can be obtained is precisely $\binom{n}{k}$.

The mean follows from (2.23) or from the definition $X = \sum_{j=1}^{n} X_j$, where X_j is a Bernoulli random variable, and the linearity of the expectation as $E[X] = np$. Higher-order moments around the mean can be derived from (2.24) as

$$E[(X - \mu)^m] = \left. \frac{d^m \left(e^{-tnp} (q + pe^t)^n \right)}{dt^m} \right|_{t=0} = \left. \frac{d^m}{dt^m} \sum_{k=0}^{n} \binom{n}{k} q^k p^{n-k} e^{t(qn-k)} \right|_{t=0}$$

$$= \sum_{k=0}^{n} \binom{n}{k} q^k p^{n-k} (qn - k)^m$$

In general, this form seems difficult to express more elegantly. It illustrates that, even for simple random variables, computations may rapidly become unattractive. For $m = 2$, the above differentiation leads to

$$\text{Var}[X] = npq \tag{3.4}$$

The variance (3.4) can also be obtained from (2.27), since $L_X(z) = n \log(q + pz)$, $L'_X(z) = \frac{np}{q+pz}$ and $L''_X(z) = -\frac{np^2}{(q+pz)^2}$.

3.1.3 The geometric distribution

The geometric random variable X returns the number of independent Bernoulli trials needed to achieve the first success. Here the sample space Ω is the infinite set of integers. The probability density function is

$$\Pr[X = k] = pq^{k-1} \tag{3.5}$$

because a first success (with probability p) obtained in the k-th trial is proceeded by $k-1$ failures (each having probability $q = 1 - p$). Clearly, $\Pr[X = 0] = 0$. The series expansion of the probability generating function,

$$\varphi_X(z) = pz \sum_{k=0}^{\infty} q^k z^k = \frac{pz}{1 - qz} \tag{3.6}$$

is the geometric series, justifying the name "geometric".

The mean $E[X] = \varphi'_X(1)$ equals $E[X] = \frac{1}{p}$. The higher-order moments can be deduced from (2.24) as

$$E[(X - \mu)^n] = p \frac{d^n}{dt^n} \left(\frac{e^{-tq/p}}{1 - qe^t} \right) \bigg|_{t=0} = p \sum_{k=0}^{\infty} q^k (k - q/p)^n$$

Similarly as for the binomial random variable, the variance

$$\operatorname{Var}[X] = \frac{q^2}{p^2} + \frac{q}{p} = \frac{q}{p^2} \tag{3.7}$$

most easily follows from (2.27) with $L_X(z) = \log p + \log(z) - \log(1 - qz)$, $L'_X(z) = \frac{1}{z} + \frac{q}{1-qz}$, $L''_X(z) = -\frac{1}{z^2} + \frac{q^2}{(1-qz)^2}$.

The distribution function $F_X(k) = \Pr[X \le k] = \sum_{j=1}^{k} \Pr[X = j]$ is obtained as

$$\Pr[X \le k] = p \sum_{j=0}^{k-1} q^j = p \frac{1 - q^k}{1 - q} = 1 - q^k$$

The tail probability is

$$\Pr[X > k] = q^k \tag{3.8}$$

Hence, the probability that the number of trials until the first success is larger than k decreases geometrically in k with rate q. Let us now consider an important application of the conditional probability. The probability that, given the success is not found in the first k trials, success does not occur within the next m trials, is with (2.47)

$$\Pr[X > k + m | X > k] = \frac{\Pr[\{X > k + m\} \cap \{X > k\}]}{\Pr[X > k]} = \frac{\Pr[X > k + m]}{\Pr[X > k]}$$

and with (3.8)

$$\Pr[X > k + m | X > k] = q^m \equiv \Pr[X > m]$$

This conditional probability turns out to be independent of the hypothesis, the event $\{X > k\}$, and reflects the famous *memoryless property*. Only because $\Pr[X > k]$ obeys the functional equation $f(x + y) = f(x)f(y)$, the hypothesis or initial knowledge does not matter. It is precisely as if past failures have never occurred or are forgotten and as if, after a failure, the number of trials is reset to 0. Furthermore, the only solution to the functional equation is an exponential function. Thus, the geometric distribution is the only discrete distribution that possesses the memory-less property.

3.1.4 The Poisson distribution

Often we are interested to count the number of occurrences of an event in a certain time interval, such as, for example, the number of IP packets during a time slot or the number of telephony calls that arrive at a telephone exchange per unit time. The Poisson random variable X with probability density function

$$\Pr[X = k] = \frac{\lambda^k e^{-\lambda}}{k!} \tag{3.9}$$

turns out to model many of these counting phenomena well as shown in Chapter 7. The corresponding generating function is

$$\varphi_X(z) = e^{-\lambda} \sum_{k=0}^{\infty} \frac{\lambda^k}{k!} z^k = e^{\lambda(z-1)} \tag{3.10}$$

and the mean number of occurrences in that time interval is

$$E[X] = \lambda \tag{3.11}$$

This mean determines the complete distribution. In applications it is convenient to replace the unit interval by an interval of arbitrary length t such that

$$\Pr[X = k] = \frac{(\lambda t)^k e^{-\lambda t}}{k!}$$

equals the probability that precisely k events occur in the interval with duration t. The probability that no events occur during t time units is $\Pr[X = 0] = e^{-\lambda t}$ and the probability that at least one event (i.e. one or more) occurs is $\Pr[X > 0] = 1 - e^{-\lambda t}$. The latter is equal to the exponential distribution. We will also see later in Theorem 7.3.2 that the Poisson counting process and the exponential distribution are intimately connected. The sum of n independent Poisson random variables each with mean λ_k is again a Poisson random variable with mean $\sum_{k=1}^{n} \lambda_k$ as follows from (2.74) and (3.10).

The higher-order moments can be deduced from (2.24) as

$$E[(X - \lambda)^n] = e^{-\lambda} \left. \frac{d^n \left(e^{-\lambda(t-e^t)} \right)}{dt^n} \right|_{t=0}$$

from which

$$E[X] = \text{Var}[X] = E[(X - \lambda)^3] = \lambda$$

The Poisson tail distribution equals

$$\Pr[X > m] = 1 - \sum_{k=0}^{m} \frac{\lambda^k e^{-\lambda}}{k!}$$

which precisely equals the sum of m exponentially distributed variables as demonstrated below in Section 3.3.1.

The Poisson density approximates the binomial density (3.3) if $n \to \infty$ but the mean $np = \lambda$. This phenomenon is often referred to as the *law of rare events*: in an arbitrarily large number n of independent trials each with arbitrarily small success $p = \frac{\lambda}{n}$, the total number of successes will approximately be Poisson distributed.

The classical argument is to consider the binomial density (3.3) with $p = \frac{\lambda}{n}$,

$$\Pr[X = k] = \frac{n!}{k!(n-k)!\, n^k} \lambda^k \left(1 - \frac{\lambda}{n}\right)^{n-k} = \frac{\lambda^k}{k!} \left(1 - \frac{\lambda}{n}\right)^{-k} \prod_{j=1}^{k-1} \left(1 - \frac{j}{n}\right) \left(1 - \frac{\lambda}{n}\right)^n$$

or

$$\log\left(\Pr[X = k]\right) = \log\left(\frac{\lambda^k}{k!}\right) - k\log\left(1 - \frac{\lambda}{n}\right) + \sum_{j=1}^{k-1} \log\left(1 - \frac{j}{n}\right) + n\log\left(1 - \frac{\lambda}{n}\right)$$

For large n, we use the Taylor expansion $\log\left(1 - \frac{x}{n}\right) = -\frac{x}{n} - \frac{x^2}{2n^2} + O\left(n^{-3}\right)$ to obtain, up to order $O\left(n^{-2}\right)$,

$$\log\left(\Pr[X = k]\right) = \log\left(\frac{\lambda^k}{k!}\right) + k\frac{\lambda}{n} + O\left(n^{-2}\right) - \frac{k(k-1)}{2n} + O\left(n^{-2}\right) - \lambda - \frac{\lambda^2}{2n} + O\left(n^{-2}\right)$$

$$= \log\left(\frac{\lambda^k}{k!}\right) - \lambda - \frac{1}{2n}\left((k - \lambda)^2 - k\right) + O\left(n^{-2}\right)$$

With $e^x = 1 + x + O(x^2)$, we finally obtain the approximation for large n,

$$\Pr[X = k] = \frac{\lambda^k e^{-\lambda}}{k!} \left(1 - \frac{1}{2n}\left[(k - \lambda)^2 - k\right] + O\left(n^{-2}\right)\right)$$

The coefficient of $\frac{1}{n}$ is negative if $k \in \left[\lambda + \frac{1}{2} - \sqrt{\lambda + \frac{1}{4}}, \lambda + \frac{1}{2} + \sqrt{\lambda + \frac{1}{4}}\right]$. In that k-interval, the Poisson density is a lower bound for the binomial density for large n and $np = \lambda$. The reverse holds for values of k outside that interval. Since for the Poisson density $\frac{\Pr[X=k]}{\Pr[X=k-1]} = \frac{\lambda}{k}$, we see that $\Pr[X = k]$ increases as $\lambda > k$ and decreases as $\lambda < k$. Thus, the maximum of the Poisson density lies around $k = \lambda = E[X]$. In conclusion, we can say that the Poisson density approximates the binomial density for large n and $np = \lambda$ from below in the region of about the standard deviation $\sqrt{\lambda}$ around the mean $E[X] = \lambda$ and from above outside this region (in the tails of the distribution).

A much shorter derivation anticipates results of Chapter 6 and starts from the probability generating function (3.2) of the binomial distribution after substitution of $p = \frac{\lambda}{n}$,

$$\lim_{n \to \infty} \varphi_X(z) = \lim_{n \to \infty} \left(1 + \frac{\lambda(z-1)}{n}\right)^n = e^{\lambda(z-1)}$$

Invoking the Continuity Theorem 6.1.3, comparison with (3.10) shows that the limit probability generating function corresponds to a Poisson distribution. The Stein–Chen (1975) Theorem[1] generalizes the law of rare events: this law even holds when the Bernoulli trials are weakly dependent.

---

[1] The proof (see e.g. Grimmett and Stirzacker (2001, pp. 130–132)) involves coupling theory of stochastic random variables. The degree of dependence is expressed in terms of the total

As a final remark, let S_n be the sum of i.i.d. Bernoulli trials each with mean p, then S_n is binomially distributed as shown in Section 3.1.2. If p is a constant and independent of the number of trials n, the Central Limit Theorem 6.3.1 states that $\frac{S_n-np}{\sqrt{np(1-p)}}$ tends to a Gaussian distribution. In summary, the limit distribution of a sum S_n of Bernoulli trials depends on how the mean p varies with the number of trials n when $n \to \infty$:

$$\text{if } p = \tfrac{\lambda}{n}, \text{ then} \qquad S_n \xrightarrow{d} \frac{\lambda^k e^{-\lambda}}{k!}$$

$$\text{if } p \text{ is constant, then} \qquad \frac{S_n-np}{\sqrt{np(1-p)}} \xrightarrow{d} \frac{e^{-\frac{x^2}{2}}}{\sqrt{2\pi}}$$

3.2 Continuous random variables

3.2.1 The uniform distribution

A uniform random variable X has equal probability to attain any value in the interval $[a, b]$ such that the probability density function is a constant. Since $\Pr[a \le X \le b] = \int_a^b f_X(x)dx = 1$, the constant value equals

$$f_X(x) = \frac{1}{b-a}1_{x\in[a,b]} \qquad (3.12)$$

where 1_y is the indicator function defined in Section 2.2.1. The distribution function then follows as

$$\Pr\left[a \le X \le x\right] = \frac{x-a}{b-a}1_{x\in[a,b]} + 1_{x>b}$$

The Laplace transform (2.38) is[2]

$$\varphi_X(z) = \int_{-\infty}^{\infty} e^{-zt}f_X(t)dt = \frac{e^{-za} - e^{-zb}}{z(b-a)} \qquad (3.13)$$

while the mean $\mu = E[X]$ follows from

$$E[X] = \int_{-\infty}^{\infty} \frac{xdx}{b-a}1_{x\in[a,b]} = \frac{a+b}{2}$$

---

variation distance. The total variation distance between two discrete random variables X and Y is defined as

$$d_{TV}(X,Y) = \sum_k |\Pr[X=k] - \Pr[Y=k]|$$

and satisfies

$$d_{TV}(X,Y) = 2 \sup_{A\subset\mathbb{Z}} |\Pr[X \in A] - \Pr[Y \in A]|$$

[2] Notice that $\frac{z\varphi_X(z)}{ab}$ equals the convolution $f * g$ of two exponential densities f and g with rates a and b, respectively.

The centered moments are obtained from (2.40) as

$$E\left[(X-\mu)^n\right] = \frac{(-1)^n}{b-a}\frac{d^n}{dz^n}\left.\frac{\left(e^{\frac{z}{2}(b-a)} - e^{-\frac{z}{2}(b-a)}\right)}{z}\right|_{z=0}$$

$$= \frac{2(-1)^n}{b-a}\frac{d^n}{dz^n}\left.\frac{\sinh(\frac{b-a}{2})z}{z}\right|_{z=0}$$

Using the power series

$$\frac{\sinh(\frac{b-a}{2})z}{z} = \sum_{k=0}^{\infty}\frac{(\frac{b-a}{2})^{2k+1}}{(2k+1)!}z^{2k}$$

leads to

$$E\left[(X-\mu)^{2n}\right] = \frac{(b-a)^{2n}}{(2n+1)2^{2n}} \tag{3.14}$$

$$E\left[(X-\mu)^{2n+1}\right] = 0$$

Let us define U as the uniform random variable in the interval $[0,1]$. If $W = 1 - U$ is a uniform random variable on $[0,1]$, then W and U have the same distribution denoted as $W \overset{d}{=} U$ because $\Pr[W \le x] = \Pr[1 - U \le x] = \Pr[U \ge 1 - x] = 1 - (1 - x) = x = \Pr[U \le x]$.

The probability distribution function $F_X(x) = \Pr[X \le x] = g(x)$ whose inverse exists can be written as a function of $F_U(x) = x1_{x\in[0,1]}$. Let $X = g^{-1}(U)$. Since the distribution function is non-decreasing, this also holds for the inverse $g^{-1}(.)$. Applying (2.32) yields with $X = g^{-1}(U)$

$$F_X(x) = \Pr\left[g^{-1}(U) \le x\right] = \Pr\left[U \le g(x)\right] = F_U\left(g(x)\right) = g(x)$$

For instance, $g^{-1}(U) = -\frac{\ln(1-U)}{\alpha} \overset{d}{=} -\frac{\ln U}{\alpha}$ are exponentially random variables (3.17) with parameter α; $g^{-1}(U) = U^{1/\alpha}$ are polynomially distributed random variables with distribution $\Pr\left[X \le x\right] = x^\alpha$; $g^{-1}(U) = \cot(\pi U)$ is a Cauchy random variable defined in (3.47) below. In addition, we observe that $U = g(X) = F_X(X)$, which means that any random variable X is transformed into a uniform random variable U on $[0,1]$ by its own distribution function.

The numbers a_k that satisfy congruent recursions of the form $a_{k+1} = (\alpha a_k + \beta)\,\mathrm{mod}\,M$, where M is a large prime number (e.g. $M = 2^{31} - 1$), α and β are integers (e.g. $\alpha = 397\,204\,094$ and $\beta = 0$) are to a good approximation uniformly distributed. The scaled numbers $y_k = \frac{a_k}{M-1}$ are nearly uniformly distributed on $[0,1]$. Since these recursions with initial value or seed $a_0 \in [0, M-1]$ are easy to generate with computers (Press *et al.*, 1992), the above property is very useful to generate arbitrary random variables $X = g^{-1}(U)$ from the uniform random variable U.

3.2.2 *The exponential distribution*

An exponential random variable X satisfies the probability density function

$$f_X(x) = \alpha e^{-\alpha x} \qquad\qquad \alpha, x \geq 0 \qquad\qquad (3.15)$$

where α is the rate at which events occur. The corresponding Laplace transform is

$$\varphi_X(z) = \alpha \int_0^\infty e^{-\alpha t} e^{-zt} dt = \frac{\alpha}{z + \alpha} \qquad\qquad (3.16)$$

and the probability distribution is, for $x \geq 0$,

$$F_X(x) = 1 - e^{-\alpha x} \qquad\qquad (3.17)$$

The mean follows from (2.34) or from $E[X] = -\varphi_X'(0)$ as $\mu = E[X] = \frac{1}{\alpha}$. Similarly, the n-th moment is derived as

$$E[X^n] = (-1)^n \left. \frac{d^n \varphi_X(z)}{dz^n} \right|_{z=0} = \frac{n!}{\alpha^n}$$

The centered moments are obtained from (2.40) as

$$E\left[\left(X - \frac{1}{\alpha}\right)^n\right] = (-1)^n \left. \frac{d^n \left(\frac{\alpha e^{z/\alpha}}{z + \alpha}\right)}{dz^n} \right|_{z=0}$$

Since the Taylor expansion of $\frac{\alpha e^{z/\alpha}}{z+\alpha}$ around $z = 0$ is

$$\frac{\alpha e^{z/\alpha}}{z + \alpha} = \sum_{k=0}^\infty \frac{1}{k!} \left(\frac{z}{\alpha}\right)^k \sum_{k=0}^\infty (-1)^k \left(\frac{z}{\alpha}\right)^k = \sum_{n=0}^\infty \left(\left(\frac{-1}{\alpha}\right)^n \sum_{k=0}^n \frac{(-1)^k}{k!}\right) z^n$$

we find that

$$E\left[\left(X - \frac{1}{\alpha}\right)^n\right] = \frac{n!}{\alpha^n} \sum_{k=0}^n \frac{(-1)^k}{k!} \qquad\qquad (3.18)$$

The variance follows from (3.18) for $n = 2$ as

$$\text{Var}[X] = \frac{1}{\alpha^2} = (E[X])^2$$

For large n, the centered moments are well approximated by

$$E\left[\left(X - \frac{1}{\alpha}\right)^n\right] \simeq \frac{n!}{e\alpha^n} = \frac{1}{e} E[X^n]$$

The exponential random variable possesses, just as its discrete counterpart, the geometric random variable, the *memoryless property*. Indeed, analogous to Section 3.1.3, consider

$$\Pr[X \geq t + T | X > t] = \frac{\Pr\left[\{X \geq t + T\} \cap \{X > t\}\right]}{\Pr[X > t]} = \frac{\Pr[X \geq t + T]}{\Pr[X > t]}$$

and since $\Pr[X > t] = e^{-\alpha t}$, the memoryless property

$$\Pr[X \geq t + T | X > t] = \Pr[X > T]$$

is established. Since the only non-zero solution (proved in Feller (1970, p. 459)) to the functional equation $f(x + y) = f(x)f(y)$, which implies the memoryless property, is of the form c^x, it shows that the exponential distribution is the only continuous distribution that has the memoryless property. As we will see later, this memoryless property is a fundamental property in Markov processes.

It is instructive to show the close relation between the geometric and exponential random variable (see Feller (1971, p. 1)). Consider the waiting time T (measured in integer units of Δt) for the first success in a sequence of Bernoulli trials where only one trial occurs in a timeslot Δt. Hence, $X = \frac{T}{\Delta t}$ is a (dimensionless) geometric random variable. From (3.8), $\Pr[T > k\Delta t] = (1 - p)^k$ and the mean waiting time is $E[T] = \Delta t E[X] = \frac{\Delta t}{p}$. The transition from the discrete to continuous space involves the limit process $\Delta t \to 0$ subject to a fixed mean waiting time $E[T]$. Let $t = k\Delta t$, then

$$\lim_{\Delta t \to 0} \Pr[T > t] = \lim_{\Delta t \to 0} \left(1 - \frac{\Delta t}{E[T]}\right)^{t/\Delta t} = e^{-t/E[T]}$$

For arbitrary small time units, the waiting time for the first success and with mean $E[T]$ turns out to be an exponential random variable.

3.2.3 *The Gaussian or normal distribution*

The Gaussian random variable X is defined for all x by the probability density function

$$f_X(x) = \frac{1}{\sigma\sqrt{2\pi}} \exp\left[-\frac{(x - \mu)^2}{2\sigma^2}\right] \tag{3.19}$$

which explicitly shows its dependence on the mean μ and variance σ^2. The importance of the Gaussian random variables stems from the Central Limit Theorem 6.3.1. Often a Gaussian – also called normal – random variable with mean μ and variance σ^2 is denoted by $N(\mu, \sigma^2)$. The distribution function is

$$F_X(x) = \frac{1}{\sigma\sqrt{2\pi}} \int_{-\infty}^{x} \exp\left[-\frac{(t - \mu)^2}{2\sigma^2}\right] dt \equiv \Phi\left(\frac{x - \mu}{\sigma}\right) \tag{3.20}$$

where[3] $\Phi(x) = \frac{1}{\sqrt{2\pi}} \int_{-\infty}^{x} e^{-\frac{t^2}{2}} dt$ is the normalized Gaussian distribution corresponding to $\mu = 0$ and $\sigma = 1$. The double-sided Laplace transform is

$$\varphi_X(z) = \frac{1}{\sigma\sqrt{2\pi}} \int_{-\infty}^{\infty} e^{-zt} \exp\left[-\frac{(t - \mu)^2}{2\sigma^2}\right] dt = e^{\frac{\sigma^2 z^2}{2} - \mu z} \tag{3.22}$$

---

[3] Abramowitz and Stegun (1968, Section 7.1.1) define the error function as

$$\text{erf}(z) = \frac{2}{\sqrt{\pi}} \int_0^z e^{-t^2} dt \tag{3.21}$$

and the centered moments (2.40) are

$$E\left[(X-\mu)^{2n}\right] = \left.\frac{d^{2n}\left(e^{\frac{\sigma^2 z^2}{2}}\right)}{dz^{2n}}\right|_{z=0} = \frac{(2n)!}{n!}\left(\frac{\sigma^2}{2}\right)^n$$

$$E\left[(X-\mu)^{2n+1}\right] = 0 \tag{3.23}$$

We note from (2.74) that a sum of independent Gaussian random variables $N(\mu_k, \sigma_k^2)$ is again a Gaussian random variable $N\left(\sum_{k=1}^n \mu_k, \sum_{k=1}^n \sigma_k^2\right)$. If $X = N(\mu, \sigma^2)$, then the scaled random variable $Y = aX$ is a $N(a\mu, (a\sigma)^2)$ random variable that is verified by computing $\Pr[Y \le y] = \Pr\left[X \le \frac{y}{a}\right]$. Similarly for translation, $Y = X + b$, then $Y = N(\mu + b, \sigma^2)$. Hence, a linear combination of Gaussian random variables, with all a_k and b real numbers, is again a Gaussian random variable,

$$\sum_{k=1}^n a_k N(\mu_k, \sigma_k^2) + b = N\left(\sum_{k=1}^n a_k \mu_k + b, \sum_{k=1}^n a_k^2 \sigma_k^2\right)$$

Let $y = F_X(x)$, then the inverse function $F_X^{-1}(y) = x$. In terms of the error function (3.21), satisfying $\mathrm{erf}(-x) = -\mathrm{erf}(x)$, we can write $y = \frac{1}{2}\left(1 + \mathrm{erf}\left(\frac{x-\mu}{\sqrt{2}\sigma}\right)\right)$ for real $0 \le y \le 1$, so that the normalized parameter $x^* = \frac{x-\mu}{\sigma}$ equals

$$x^* = \sqrt{2}\,\mathrm{erf}^{-1}(2y - 1)$$

There exists a beautiful Taylor series, due to Carlitz (1963) and Strecok (1968), for the inverse of the error function

$$\mathrm{erf}^{-1}(z) = \sum_{n=1}^\infty c_n z^n$$

that converges for $|z| < 1$ and where the Taylor coefficients $c_n = \frac{b_n}{n!}\left(\frac{\sqrt{\pi}}{2}\right)^n$ are non-negative, smaller than 1, all even $c_{2n} = 0$ and in which the integers b_n satisfy the recursion

$$b_k = 2\sum_{n=2}^{k-1}\binom{k-1}{n}b_{n-1}b_{k-n} \qquad \text{with } b_1 = 1$$

Hence, $\mathrm{erf}^{-1}(z) \ge \sum_{n=1}^K c_n z^n$ for finite K. Explicitly up to order 13, the Taylor series is, for $|z| < 1$,

$$\mathrm{erf}^{-1}\left(\frac{2}{\sqrt{\pi}}z\right) = z + \frac{z^3}{3} + \frac{7}{30}z^5 + \frac{127}{630}z^7 + \frac{4369}{22680}z^9 + \frac{34807}{178200}z^{11} + \frac{20036983}{97297200}z^{13} + O\left(z^{15}\right)$$

3.3 Derived distributions

From the basic distributions, a large number of other distributions can be derived as illustrated here.

---

such that (Abramowitz and Stegun, 1968, Section 7.1.22)

$$\frac{1}{\sigma\sqrt{2\pi}}\int_{-\infty}^x \exp\left[-\frac{(t-\mu)^2}{2\sigma^2}\right]dt = \frac{1}{2}\left(1 + \mathrm{erf}\left(\frac{x-\mu}{\sqrt{2}\sigma}\right)\right)$$

3.3.1 *The sum of independent exponential random variables*

By applying (2.74) and (2.39) a substantial amount of practical problems can be solved. For example, the sum of n independent exponential random variables, each with different rate $\alpha_k > 0$, has the generating function

$$\varphi_{S_n}(z) = \prod_{k=1}^{n} \frac{\alpha_k}{z + \alpha_k}$$

and probability density function

$$f_{S_n}(t) = \frac{\prod_{k=1}^{n} \alpha_k}{2\pi i} \int_{c-i\infty}^{c+i\infty} \frac{e^{zt}}{\prod_{k=1}^{n}(z + \alpha_k)} dz$$

The contour can be closed over the negative half plane for $t > 0$, where the integral has simple poles at $z = -\alpha_k$. From the Cauchy Integral Theorem, we obtain

$$f_{S_n}(t) = \left(\prod_{k=1}^{n} \alpha_k\right) \sum_{j=1}^{n} \frac{e^{-\alpha_j t}}{\prod_{k=1; k \neq j}^{n}(\alpha_k - \alpha_j)}$$

If all rates are equals $\alpha_k = \alpha$, the case reduces to $\varphi_{S_n}(z) = \left(\frac{\alpha}{z+\alpha}\right)^n$ with $E[S_n] = \frac{n}{\alpha}$ and with probability density function

$$f_{S_n}(t) = \frac{\alpha^n}{2\pi i} \int_{c-i\infty}^{c+i\infty} \frac{e^{zt}}{(z + \alpha)^n} dz$$

Again, the contour can be closed over the negative half plane and the n-th order poles are deduced from Cauchy's relation for the n-th derivative of a complex function

$$\frac{1}{k!} \frac{d^k f(z)}{dz^k}\bigg|_{z=z_0} = \frac{1}{2\pi i} \int_{C(z_0)} \frac{f(\omega)\, d\omega}{(\omega - z_0)^{k+1}}$$

as

$$f_{S_n}(t) = \frac{\alpha^n}{(n-1)!} \frac{d^{n-1} e^{zt}}{dz^{n-1}}\bigg|_{z=-\alpha} = \frac{\alpha(\alpha t)^{n-1}}{(n-1)!} e^{-\alpha t} \qquad (3.24)$$

For integer n, this density corresponds to the n-Erlang random variable. When extended to real values of $n = \beta$,

$$f_X(t; \alpha, \beta) = \frac{\alpha(\alpha t)^{\beta-1}}{\Gamma(\beta)} e^{-\alpha t} \qquad (3.25)$$

it is called the Gamma probability density function, with corresponding pgf

$$\varphi_X(z; \alpha, \beta) = \left(\frac{\alpha}{z+\alpha}\right)^\beta = \left(1 + \frac{z}{\alpha}\right)^{-\beta} \qquad (3.26)$$

and distribution

$$F_X(x; \alpha, \beta) = \frac{\alpha^\beta}{\Gamma(\beta)} \int_0^x t^{\beta-1} e^{-\alpha t}\, dt \qquad (3.27)$$

This integral, the incomplete Gamma-function, can only be expressed in closed analytic form if β is an integer. Hence, for the n-Erlang random variable X, the distribution follows after repeated partial integration as

$$F_X(x; \alpha, n) = \frac{\alpha^n}{(n-1)!} \int_0^x t^{n-1} e^{-\alpha t} dt = 1 - \sum_{k=0}^{n-1} \frac{(\alpha x)^k}{k!} e^{-\alpha x} \qquad (3.28)$$

We observe that $\Pr[X > x] = \sum_{k=0}^{n-1} \frac{(\alpha x)^k}{k!} e^{-\alpha x}$, which equals $\Pr[Y \leq n-1]$ where Y is a Poisson random variable with mean $\lambda = \alpha x$. Further, $\Pr[X > x] = \Pr[\frac{X^*}{\alpha} > x]$, where $E[X^*] = n$: the distribution of the sum of i.i.d. exponential random variables each with rate α follows by scaling $x \to \alpha x$ from the distribution of the sum of i.i.d. exponential random variables each with unit rate (or mean 1). Moreover, (2.74) and (3.26) show that a sum of n independent Gamma random variables specified by β_k (but with same α) is again a Gamma random variable with $\beta = \sum_{k=1}^{n} \beta_k$.

At last all centered moments follow from (2.40) by series expansion around $z = 0$ as

$$E\left[(X-a)^n\right] = (-1)^n \left. \frac{d^n \left(e^{za}\left(1+\frac{z}{\alpha}\right)^{-\beta}\right)}{dz^n} \right|_{z=0}$$

$$= (-1)^n n! a^n \sum_{m=0}^{n} \binom{-\beta}{m} \frac{(\alpha a)^{-m}}{(n-m)!}$$

In particular, since $E[X] = \mu = \frac{\beta}{\alpha}$, we find with $\binom{-z}{m} = (-1)^m \frac{\Gamma(z+m)}{m!\Gamma(z)}$

$$E\left[(X-\mu)^n\right] = (-1)^n n! \frac{\beta^n}{\alpha^n} \sum_{m=0}^{n} \binom{-\beta}{m} \frac{\beta^{-m}}{(n-m)!}$$

$$= (-1)^n \frac{\beta^n}{\alpha^n} \sum_{m=0}^{n} \binom{n}{m} (-1)^m \frac{\Gamma(\beta+m)}{\Gamma(\beta)\beta^m}$$

$$= (-1)^n \frac{\beta^n}{\alpha^n} (-\beta)^\beta U\left(\beta, \beta+1+n, -\beta\right)$$

where $U(a, b, z)$ is the confluent hypergeometric function (Abramowitz and Stegun, 1968, Chapter 13). For example, if $n = 2$, the variance equals $\sigma^2 = \frac{\beta}{\alpha^2}$ and further, $E\left[(X-\mu)^3\right] = \frac{2\beta}{\alpha^3}$, $E\left[(X-\mu)^4\right] = \frac{3\beta(\beta+2)}{\alpha^4}$ and $E\left[(X-\mu)^5\right] = \frac{4\beta(5\beta+6)}{\alpha^5}$.

3.3.2 The sum of independent uniform random variables

The sum $S_k = \sum_{j=1}^{k} U_j$ of k i.i.d. uniform random variables U_j has as distribution function the k-fold convolution of the uniform density function $f_U(x) = 1_{0 \leq x \leq 1}$ on

$[0, 1]$ denoted by $f_U^{(k*)}(x)$. The distribution function equals

$$\Pr\left[S_k \le x\right] = \sum_{j=0}^{[x]} \frac{(-1)^j}{j!(k-j)!}(x-j)^k \tag{3.29}$$

Indeed, from (2.75) and (3.13) the Laplace transform of S_k is

$$\varphi_{S_k}(z) = \left(\frac{1-e^{-z}}{z}\right)^k$$

The inverse Laplace transform determines, for $c > 0$,

$$f_U^{(k*)}(x) \triangleq \frac{d}{dx}\Pr\left[S_k \le x\right] = \frac{1}{2\pi i}\int_{c-i\infty}^{c+i\infty}\left(\frac{1-e^{-z}}{z}\right)^k e^{zx}dz$$

Using $(1-e^{-z})^k = \sum_{j=0}^{k}\binom{k}{j}(-1)^j e^{-jz}$ and the integral

$$\frac{1}{2\pi i}\int_{c-i\infty}^{c+i\infty}\frac{e^{sa}}{s^{n+1}}ds = \frac{a^n}{n!}1_{\operatorname{Re}(a)>0}$$

yields

$$f_U^{(k*)}(x) = \sum_{j=0}^{k}\binom{k}{j}(-1)^j\frac{(x-j)^{k-1}}{(k-1)!}1_{(x-j)\ge 0} \tag{3.30}$$

from which (3.29) follows by integration.

3.3.3 The product of two independent random variables

The probability density function of the product Z of two independent exponential random variables X_1 and X_2 with rate α_1 and α_2, respectively, follows directly from (2.67) as

$$f_Z(w) = \alpha_1\alpha_2\int_0^\infty e^{-\alpha_1 u}e^{-\alpha_2\frac{w}{u}}\frac{du}{u}$$

Before continuing, we introduce the integral, which is a Mellin transform (2.42), for $\operatorname{Re}(a) > 0$ and $\operatorname{Re}(b) > 0$ and for all s,

$$\int_0^\infty x^{s-1}e^{-ax-b/x}dx = 2\left(\frac{b}{a}\right)^{s/2}K_s\left(2\sqrt{ab}\right) \tag{3.31}$$

where $K_s(z)$ is the modified Bessel function of the second kind of order s (Abramowitz and Stegun, 1968; Watson, 1995). We also mention the formula (Watson, 1995, p. 206)

$$K_s(z) = \sqrt{\frac{\pi}{2z}}\frac{e^{-z}}{\Gamma\left(s+\frac{1}{2}\right)}\int_0^\infty e^{-u}u^{s-\frac{1}{2}}\left(1+\frac{u}{2z}\right)^{s-\frac{1}{2}}du \tag{3.32}$$

from which, by expanding $\left(1 + \frac{u}{2z}\right)^{s-\frac{1}{2}}$ in a power series, the asymptotic expansion of the modified Bessel function (Watson, 1995, p. 207) follows as

$$K_s(z) = \sqrt{\frac{\pi}{2z}} e^{-z} \left\{ 1 + \sum_{m=1}^{p-1} \frac{\Gamma(s+m+1/2)}{m!\Gamma(s-m+1/2)} \frac{1}{(2z)^m} + \frac{\Gamma(s+p+1/2)}{p!\Gamma(s-p+1/2)} \frac{\theta}{(2z)^p} \right\}$$
(3.33)

where s is real, $p \geq s - \frac{1}{2}$, $\mathrm{Re}(z) > 0$, and $0 \leq \theta \leq 1$. The expansion (3.33) shows that $K_s(z) \sim \sqrt{\frac{\pi}{2z}} e^{-z}$ for $z \to \infty$, independent of s. Moreover, for $s = n + \frac{1}{2}$ and $n \in \mathbb{N}$, the integral (3.32) can be evaluated exactly by using the binomial finite series

$$K_{n+\frac{1}{2}}(z) = \sqrt{\frac{\pi}{2z}} e^{-z} \sum_{k=0}^{n} \frac{(n+k)!}{k!(n-k)!} \frac{1}{(2z)^k}$$
(3.34)

Invoking (3.31), we find that

$$f_{Z;\exp}(w) = 2\alpha_1\alpha_2 K_0\left(2\sqrt{\alpha_1\alpha_2 w}\right)$$

Similarly, the probability density function of the product Z of two independent Gaussian random variables X_1 and X_2 with zero mean and variance σ_1 and σ_2, respectively, follows from (2.67) after substitution of $x = u^2$ as

$$f_Z(w) = \frac{1}{2\pi\sigma_1\sigma_2} \int_0^\infty x^{-1} \exp\left[-\frac{x}{2\sigma_1^2} - \frac{w^2}{2\sigma_2^2 x}\right] dx$$

With (3.31), we have

$$f_{Z;\text{Gaussian}}(w) = \frac{1}{\pi\sigma_1\sigma_2} K_0\left(\frac{w}{\sigma_1\sigma_2}\right)$$

More generally, since the exponential distribution is a special case of the Weibull distribution, defined in (3.48), with shape factor $b = 1$ and the Gaussian distribution is related to the Weibull distribution with $b = 2$, application of (2.67) yields, after substitution of $x = u^b$,

$$f_{Z;\text{Weibull}}(w) = \frac{1}{a_1 a_2 \Gamma^2\left(1 + \frac{1}{b}\right)} \int_0^\infty \exp\left(-\frac{x}{a_1^b}\right) \exp\left(-\frac{w^b}{a_2^b x}\right) \frac{1}{b} \frac{dx}{x}$$

By using (3.31), the probability density function of the product Z of two independent Weibull random variables X_1 and X_2 with parameters a_1 and a_2 and same shape factor b is

$$f_{Z;\text{Weibull}}(w) = \frac{2}{a_1 a_2 b \Gamma^2\left(1 + \frac{1}{b}\right)} K_0\left(\frac{2w^{\frac{b}{2}}}{(a_1 a_2)^{b/2}}\right)$$

The deep tail of $f_{Z;\text{Weibull}}(w)$ for large w decays according to (3.33) as

$$f_{Z;\text{Weibull}}(w) \sim \frac{\sqrt{\pi}\,(a_1 a_2)^{b/4-1}}{b\Gamma^2\left(1 + \frac{1}{b}\right) w^{\frac{b}{4}}} e^{-\frac{2w^{\frac{b}{2}}}{(a_1 a_2)^{b/2}}}$$

The generalization towards the Weibull distribution is a consequence of properties of the Mellin transform (2.42). For, if $\psi(z) = \int_0^\infty t^{z-1} f(t)\, dt$ is the Mellin transform of the function $f(x)$, then the Mellin transform of $f(x^a)$ for real a is $\frac{1}{|a|} \psi\left(\frac{z}{a}\right)$. The Mellin transform of e^{-at} is $\frac{\Gamma(z)}{a^z}$. The product $\psi_X(z)\, \psi_Y(z)$ of two Mellin transforms (2.69) of the probability density function of two independent non-negative random variables X and Y is always a Mellin transform $\psi_Z(z)$ of the probability density function of their product $Z = XY$. In Section 16.6.1, the product of two Gamma functions is shown to be the inverse Mellin transform of the modified Bessel function $K_s(z)$. In most cases, unfortunately, the probability density function $f_Z(x)$ in (2.67) cannot be evaluated analytically in closed form.

3.3.4 The quotient of two independent random variables

The probability density function of the quotient $Q = \frac{X_1}{X_2}$ of two independent exponential random variables X_1 and X_2 with rate α_1 and α_2, respectively, follows directly from (2.68) for $w \geq 0$ (because both X_1 and X_2 are non-negative) as

$$f_Q(w) = \alpha_1 \alpha_2 \int_0^\infty e^{-\alpha_1 w t} e^{-\alpha_2 t}\, dt = \frac{\alpha_1 \alpha_2}{(\alpha_1 w + \alpha_2)^2}$$

Comparing with (3.50), Q is recognized as a Pareto random variable.

Likewise, the probability density function of the quotient $Q = \frac{X_1}{X_2}$ of two independent Gaussian random variables $X_1 = N(\mu_1, \sigma_1^2)$ and $X_2 = N(\mu_2, \sigma_2^2)$ equals with (2.68)

$$f_Q(w) = \frac{1}{2\pi \sigma_1 \sigma_2} \int_{-\infty}^\infty \exp\left[-\frac{(wt - \mu_1)^2}{2\sigma_1^2} - \frac{(t - \mu_2)^2}{2\sigma_2^2} \right] |t|\, dt$$

The evaluation of the integral greatly simplifies when both means are $\mu_1 = \mu_2 = 0$, in which case

$$f_Q(w) = \frac{1}{\pi \sigma_1 \sigma_2} \int_0^\infty \exp\left[-\frac{1}{2}\left(\frac{w^2}{\sigma_1^2} + \frac{1}{\sigma_2^2} \right) t^2 \right] t\, dt$$

Using $\int t e^{-at^2}\, dt = -\frac{e^{-at^2}}{2a} + k$, we arrive at

$$f_Q(w) = \frac{1}{\pi \sigma_1 \sigma_2 \left(\frac{w^2}{\sigma_1^2} + \frac{1}{\sigma_2^2} \right)} = \frac{\frac{\sigma_1}{\sigma_2}}{\pi \left(w^2 + \left(\frac{\sigma_1}{\sigma_2} \right)^2 \right)}$$

which we recognize as a (scaled) Cauchy probability density function, defined in (3.46). An interesting observation is that $f_{Q^{-1}}(w)$ is also a Cauchy probability density function; a same observation applies also to the Pareto random variable with exponent $\alpha = 1$.

In contrast to the product (see Section 3.3.3), we find that the quotient of two random variables, with exponentially fast decreasing tails, possesses very "fat", power-law tails.

3.3.5 The chi-square distribution

Suppose that the total error of n independent measurements X_k, each perturbed by Gaussian noise, has to be determined. In order to prevent that errors may cancel out, the sum of the squared errors $S = \sum_{k=0}^{n} e_k^2$ is preferred rather than $\sum_{k=0}^{n} |e_k|$. For simplicity, we assume that all errors $e_k = X_k - x_k$, where x_k is the exact value of quantity k, have zero mean and unit variance. The corresponding distribution of S is known as the chi-square distribution. From the χ^2-distribution, the χ^2-test in statistics is deduced which determines the goodness of a model of a distribution to a set of measurements. We refer for a discussion of the χ^2-test to Leon-Garcia (1994, Section 3.8) or Allen (1978, Section 8.4).

We first deduce the distribution of the square $Y = X^2$ of a random variable X. The event $\{Y \leq y\}$ or $\{X^2 \leq y\}$ is equivalent to $\{-\sqrt{y} \leq X \leq \sqrt{y}\}$ and non-existent if $y < 0$. With (2.29) and $y \geq 0$,

$$\Pr[Y \leq y] = \Pr[-\sqrt{y} \leq X \leq \sqrt{y}] = F_X(\sqrt{y}) - F_X(-\sqrt{y})$$

and, after differentiation,

$$f_{X^2}(x) = \frac{f_X(\sqrt{x}) + f_X(-\sqrt{x})}{2\sqrt{x}} \tag{3.35}$$

If X is a Gaussian random variable $N(\mu, \sigma^2)$, then is, for $x \geq 0$,

$$f_{X^2}(x) = \frac{\exp\left[-\frac{(x+\mu^2)}{2\sigma^2}\right]}{\sigma\sqrt{2\pi x}} \cosh\left(\frac{\mu\sqrt{x}}{\sigma^2}\right)$$

In particular, for $N(0,1)$ random variables where $\mu = 0$ and $\sigma = 1$, $f_{X^2}(x) = \frac{e^{-\frac{x}{2}}}{\sqrt{2\pi x}}$ reduces to a Gamma distribution (3.25) with $\beta = \frac{1}{2}$ and $\alpha = \frac{1}{2}$. Since the sum of n independent Gamma random variables with (α, β) is again a Gamma random variable $(\alpha, n\beta)$, we arrive at the chi-square χ^2 probability density function,

$$f_{X^2}(x) = \frac{x^{\frac{n}{2}-1}}{2^{\frac{n}{2}}\Gamma\left(\frac{n}{2}\right)} e^{-\frac{x}{2}} \tag{3.36}$$

3.3.6 The Student t and the Fisher F distribution

Beside the chi-square distribution, also the Student t and the Fisher F distribution appear in statistical hypothesis testing (see e.g. Cramér (1999) and Lehmann (1999)).

As motivated in Section 4.6, the normalized sample mean $Y = \frac{\overline{X} - E[X]}{\sqrt{S_X^2/n}}$ is a useful estimator to approximately characterize some random variable X, measured from a finite sample of size n of a population. When X is a Gaussian random variable, Theorem 4.6.1 shows that the sample mean $\overline{X}$ and the sample variance S_X are independent. Let us compute here the distribution of an instance of a normalized sample mean $Y = \frac{X - E[X]}{\sqrt{R/n}}$, where X is a Gaussian random variable $N(\mu, \sigma^2)$ that is independent of the corresponding estimate for the variance R, that possesses a chi-square distribution with parameter n. In this case, Y is called a Student t random variable. The random variable $\tilde{X} = X - E[X]$ is a zero mean Gaussian random variable $N(0, \sigma^2)$ and we are interested in the distribution of $Y = \frac{\tilde{X}}{\sqrt{R/\sqrt{n}}}$. In order to apply the general transformation for two random variables introduced in Section 2.5.4, we add a second, auxiliary random variable W, for which we choose a very simple equation in terms of the known random variables $\tilde{X}$ and R. For example, $W = \tilde{X}$ or $W = R$. Here, we choose $W = R$ to ease the inverse transformation. The transformation of the random variables is

$$\begin{cases} W = R \\ Y = \frac{\tilde{X}}{\sqrt{R}/\sqrt{n}} \end{cases}$$

The joint probability density function $f_{WY}(w, y)$ can now be written using (2.65) in terms of the joint probability density function $f_{R\tilde{X}}(r, x)$, where the corresponding transformation for the parameters $r = h_1(w, y)$ and $x = h_2(w, y)$ is found by solving the above transformation after replacing the random variables by the corresponding parameters. The solution is

$$\begin{cases} r = w \\ x = \frac{1}{\sqrt{n}} y \sqrt{w} \end{cases}$$

and the Jacobian is

$$J(w, y) = \det \begin{bmatrix} \frac{\partial r}{\partial w} & \frac{\partial r}{\partial y} \\ \frac{\partial x}{\partial w} & \frac{\partial x}{\partial y} \end{bmatrix} = \det \begin{bmatrix} 1 & 0 \\ \frac{1}{\sqrt{n}} y \frac{1}{2\sqrt{w}} & \frac{\sqrt{w}}{\sqrt{n}} \end{bmatrix} = \frac{\sqrt{w}}{\sqrt{n}}$$

such that (2.65) becomes

$$f_{WY}(w, y) = f_{R\tilde{X}}\left(w, \frac{1}{\sqrt{n}} y \sqrt{w}\right) \frac{\sqrt{w}}{\sqrt{n}}$$

Since $\tilde{X}$ and $R = S_X^2$ are independent, the joint density is separable,

$$f_{WY}(w, y) = f_R(w) f_{\tilde{X}}\left(\frac{1}{\sqrt{n}} y \sqrt{w}\right) \frac{\sqrt{w}}{\sqrt{n}}$$

where, with (3.36) and (3.19), we have

$$f_R(x) = \frac{x^{\frac{n}{2}-1}}{2^{\frac{n}{2}} \Gamma\left(\frac{n}{2}\right)} e^{-\frac{x}{2}} \quad \text{and} \quad f_{\tilde{X}}(x) = \frac{1}{\sigma\sqrt{2\pi}} \exp\left[-\frac{x^2}{2\sigma^2}\right]$$

The joint probability density function is

$$f_{WY}(w, y) = \frac{w^{\frac{n}{2}-1}}{2^{\frac{n}{2}} \Gamma\left(\frac{n}{2}\right)} e^{-\frac{w}{2}} 1_{\{w \geq 0\}} \frac{1}{\sigma\sqrt{2\pi}} \exp\left[-\frac{1}{2n\sigma^2} w y^2\right] \frac{\sqrt{w}}{\sqrt{n}}$$

$$= \frac{w^{\frac{n-1}{2}} e^{-\frac{w}{2}\left(1 + \frac{y^2}{n\sigma^2}\right)}}{2^{\frac{n}{2}} \Gamma\left(\frac{n}{2}\right) \sigma\sqrt{2\pi n}} 1_{\{w \geq 0\}}$$

The pdf of the desired random variable Y follows from (2.71) as

$$f_Y(y) = \int_{-\infty}^{\infty} f_{WY}(w, y) dw$$

$$= \int_{-\infty}^{\infty} \frac{w^{\frac{n-1}{2}} e^{-\frac{w}{2}\left(1+\frac{y^2}{n\sigma^2}\right)}}{2^{\frac{n}{2}} \Gamma\left(\frac{n}{2}\right) \sigma \sqrt{2\pi n}} 1_{\{w \geq 0\}} dw$$

$$= \frac{1}{2^{\frac{n}{2}} \Gamma\left(\frac{n}{2}\right) \sigma \sqrt{2\pi n}} \int_0^{\infty} w^{\frac{n+1}{2}-1} e^{-\frac{w}{2}\left(1+\frac{y^2}{n\sigma^2}\right)} dw$$

Invoking the Euler integral of the Gamma function, $\Gamma(z) = \int_0^{\infty} t^{z-1} e^{-t} dt$ valid for $\text{Re}(z) > 0$, we arrive at the Student t probability density function

$$f_{T_n}(y) = \frac{\Gamma\left(\frac{n+1}{2}\right)}{\Gamma\left(\frac{n}{2}\right) \sigma \sqrt{\pi n}} \left(1 + \frac{y^2}{n\sigma^2}\right)^{-\frac{n+1}{2}} \tag{3.37}$$

where the random variable Y is replaced by the more standard notation T_n of the Student t random variable. The special case where $n = 1$ and $\sigma = 1$ reduces to the pdf of a Cauchy random variable (3.46). The other extreme for large n tends to a Gaussian distribution because $\lim_{n\to\infty} \left(1 + \frac{y^2}{n\sigma^2}\right)^{-\frac{n+1}{2}} = e^{-\frac{y^2}{2\sigma^2}}$. From (2.35), the k-th moment of T_n is

$$E\left[T_n^k\right] = \int_{-\infty}^{\infty} y^k f_{T_n}(y) dy = \frac{\Gamma\left(\frac{n+1}{2}\right)}{\Gamma\left(\frac{n}{2}\right) \sigma \sqrt{\pi n}} \int_{-\infty}^{\infty} \frac{y^k}{\left(1 + \frac{y^2}{n\sigma^2}\right)^{\frac{n+1}{2}}} dy$$

The fact that $f_{T_n}(y) = f_{T_n}(-y)$ is even, implies that all odd moments (including the mean) are zero, provided the integral converges. A necessary condition for convergence is that $k < n$, which shows that the mean (nor any moment) for the Cauchy distribution ($n = 1$) does not exist (see Section 3.5). The even moments are

$$E\left[T_n^{2k}\right] = \int_{-\infty}^{\infty} y^k f_{T_n}(y) dy = \frac{2\Gamma\left(\frac{n+1}{2}\right)}{\Gamma\left(\frac{n}{2}\right) \sigma \sqrt{\pi n}} \int_0^{\infty} \frac{y^{2k}}{\left(1 + \frac{y^2}{n\sigma^2}\right)^{\frac{n+1}{2}}} dy$$

$$= \frac{\Gamma\left(k + \frac{1}{2}\right) \Gamma\left(\frac{n}{2} - k\right) n^k \sigma^{2k}}{\Gamma\left(\frac{n}{2}\right)} \frac{1}{\sqrt{\pi}}$$

where we have used (Abramowitz and Stegun, 1968, Section 6.2.1) the Beta-function integral $\int_0^{\infty} \frac{t^{z-1} dt}{(1+t)^{z+w}} = \frac{\Gamma(z)\Gamma(w)}{\Gamma(z+w)}$, valid for $\text{Re}(z) > 0$, $\text{Re}(w) > 0$. In conclusion, the moments are

$$E\left[T_n^{2k-1}\right] = 0 \qquad\qquad \text{provided } 2k - 1 < n$$

$$E\left[T_n^{2k}\right] = \frac{\Gamma\left(k+\frac{1}{2}\right)\Gamma\left(\frac{n}{2}-k\right)}{\Gamma\left(\frac{n}{2}\right)} \frac{n^k \sigma^{2k}}{\sqrt{\pi}} \quad \text{provided } 2k < n$$

and, for $n > 2$,

$$\text{Var}\left[T_n\right] = E\left[T_n^2\right] = \frac{n\sigma^2}{n - 2}$$

The random variable $Z = \frac{R_m/m}{R_n/n}$, where R_k is a chi-square random variable with parameter (or degree of freedom) k and where both R_m and R_n are independent, is said to have a Fisher F distribution. Using the same transformation method, we choose the auxiliary r.v. $W = R_n$ such that $r_m = \frac{m}{n}zw$ and $r_n = w$ and the Jacobian $J(w, z) = -\frac{m}{n}w$. The transformation formula (2.65) becomes

$$f_{WZ}(w, z) = f_{R_m R_n}\left(\frac{m}{n}zw, w\right)\frac{m}{n}w = f_{R_m}\left(\frac{m}{n}zw\right)f_{R_n}(w)\frac{m}{n}w$$

$$= \frac{z^{\frac{m}{2}-1}\left(\frac{m}{n}\right)^{\frac{m}{2}} w^{\frac{m+n}{2}-1}}{2^{\frac{m}{2}}\Gamma\left(\frac{m}{2}\right)2^{\frac{n}{2}}\Gamma\left(\frac{n}{2}\right)}e^{-\frac{w}{2}\left(1+\frac{m}{n}z\right)}1_{\{w\geq 0;z\geq 0\}}$$

while the pdf of the desired random variable Z follows from (2.71) as

$$f_Z(z) = \int_0^\infty f_{WZ}(w, z)dw = \frac{z^{\frac{m}{2}-1}\left(\frac{m}{n}\right)^{\frac{m}{2}}}{2^{\frac{m}{2}}\Gamma\left(\frac{m}{2}\right)2^{\frac{n}{2}}\Gamma\left(\frac{n}{2}\right)}\int_0^\infty w^{\frac{m+n}{2}-1}e^{-\frac{w}{2}\left(1+\frac{m}{n}z\right)}dw$$

Finally, we obtain the pdf of the Fisher F distribution with (m, n) degrees of freedom, for $z \geq 0$,

$$f_{F_{m,n}}(z) = \frac{\Gamma\left(\frac{m+n}{2}\right)\left(\frac{m}{n}\right)^{\frac{m}{2}}}{\Gamma\left(\frac{m}{2}\right)\Gamma\left(\frac{n}{2}\right)}\frac{z^{\frac{m}{2}-1}}{\left(1+\frac{m}{n}z\right)^{\frac{m+n}{2}}} \tag{3.38}$$

Applying (3.35) to (3.37) with $\sigma = 1$ yields

$$f_{T_n^2}(z) = \frac{\Gamma\left(\frac{n+1}{2}\right)}{\Gamma\left(\frac{n}{2}\right)\sqrt{\pi n}}\frac{z^{-\frac{1}{2}}}{\left(1+\frac{z}{n}\right)^{\frac{n+1}{2}}}$$

Comparison with (3.38) remarkably shows that $T_n^2 = F_{1,n}$: the square of the Student t random variable equals the Fisher F random variable with $(1, n)$ degrees of freedom.

The moments of $F_{m,n}$ follow from the above integral of the Beta-function for $n > 2k$ as

$$E\left[F_{m,n}^k\right] = \frac{\Gamma\left(\frac{m+n}{2}\right)\left(\frac{m}{n}\right)^{\frac{m}{2}}}{\Gamma\left(\frac{m}{2}\right)\Gamma\left(\frac{n}{2}\right)}\int_0^\infty \frac{z^{\frac{m}{2}+k-1}dz}{\left(1+\frac{m}{n}z\right)^{\frac{m+n}{2}}}$$

$$= \frac{\Gamma\left(\frac{m}{2}+k\right)\Gamma\left(\frac{n}{2}-k\right)}{\Gamma\left(\frac{m}{2}\right)\Gamma\left(\frac{n}{2}\right)}\left(\frac{n}{m}\right)^k = \frac{n}{n-2}\left(\frac{n}{m}\right)^{k-1}\prod_{j=1}^{k-1}\frac{m+2j}{n-2(j+1)}$$

where in the last step the functional equation $\Gamma(z+1) = z\Gamma(z)$ has been repeatedly used. Thus, the mean is $E[F_{m,n}] = \frac{n}{n-2}$ and independent of m. The variance is

$$\text{Var}[F_{m,n}] = E\left[F_{m,n}^2\right] - (E[F_{m,n}])^2 = \frac{2n^2}{(n-2)^2}\left(\frac{n+m-2}{m(n-4)}\right)$$

The F-distribution is used to test the hypothesis that two normal distributions X and Y have the same variance (without the need to know the mean). The variance is a measure of uniformity and quality of production. In order to test that $\sigma_X = \sigma_Y$ for the two Gaussian random variables X and Y, we compute the sample variances S_X^2 and S_Y^2, defined in (4.23), of the samples $x_1, x_2, \ldots, x_m$ and $y_1, y_2, \ldots, y_n$. Theorem 4.6.1 shows that both S_X^2 and S_Y^2 have a chi-square distribution with $m-1$ and $n-1$ degrees of freedom, respectively. Next, we compute from (3.38) that value of q for which the tail probability $\Pr\left[F_{(m-1)(n-1)} > q\right] < \varepsilon$, where ε is a stringency, also called the significance level, that is about 1% to 5% in practice. If $\frac{S_X^2}{S_Y^2} < q$, we accept the hypothesis that $\sigma_X = \sigma_Y$, else, we reject it. The performance of statistical tests is treated in detail by Lehmann (1999, Chapter 3).

3.4 Functions of random variables

3.4.1 The maximum and minimum of a set of independent random variables

The minimum of m i.i.d. random variables $\{X_k\}_{1 \le k \le m}$ possesses the distribution[4]

$$\Pr\left[\min_{1 \le k \le m} X_k \le x\right] = \Pr\left[\text{at least one } X_k \le x\right] = \Pr\left[\text{not all } X_k > x\right]$$

or

$$\Pr\left[\min_{1 \le k \le m} X_k \le x\right] = 1 - \prod_{k=1}^{m} \Pr[X_k > x] \tag{3.39}$$

whereas for the maximum,

$$\Pr\left[\max_{1 \le k \le m} X_k > x\right] = \Pr\left[\text{not all } X_k \le x\right] = 1 - \prod_{k=1}^{m} \Pr[X_k \le x]$$

or

$$\Pr\left[\max_{1 \le k \le m} X_k \le x\right] = \prod_{k=1}^{m} \Pr[X_k \le x] \tag{3.40}$$

For example, the distribution function for the minimum of m independent exponential random variables follows from (3.17) as

$$\Pr\left[\min_{1 \le k \le m} X_k \le x\right] = 1 - \prod_{k=1}^{m} e^{-\alpha_k x} = 1 - \exp\left(-x \sum_{k=1}^{m} \alpha_k\right)$$

Thus, the minimum of m independent exponential random variables each with rate α_k is again an exponential random variable with rate $\sum_{k=0}^{m} \alpha_k$. In addition to the memoryless property, this property of the exponential distribution will determine the fundamentals of Markov processes.

---

[4] An alternative argument for independent random variables is that the event $\{\min_{1 \le k \le m} X_k > x\}$ is only possible if and only if $\{X_k > x\}$ for each $1 \le k \le m$. Similarly, the event $\{\max_{1 \le k \le m} X_k \le x\}$ is only possible if and only if all $\{X_k \le x\}$ for each $1 \le k \le m$.

3.4.2 Order statistics

The set $X_{(1)}, X_{(2)}, \ldots, X_{(m)}$ are called the order statistics of the set of random variables $\{X_k\}_{1 \leq k \leq m}$ if $X_{(k)}$ is the k-th smallest value of the set $\{X_k\}_{1 \leq k \leq m}$. Clearly, $X_{(1)} = \min_{1 \leq k \leq m} X_k$ while $X_{(m)} = \max_{1 \leq k \leq m} X_k$. If the set $\{X_k\}_{1 \leq k \leq m}$ consists of i.i.d. random variables with pdf f_X, the joint density function of the order statistics is, for only $x_1 < x_2 < \cdots < x_m$,

$$f_{\{X_{(j)}\}}(x_1, x_2, \ldots, x_m) = \frac{\partial^m}{\partial x_1 \ldots \partial x_m} \Pr\left[X_{(1)} \leq x_1, \ldots, X_{(m)} \leq x_m\right]$$

$$= m! \prod_{j=1}^{m} f_X(x_j) \tag{3.41}$$

Indeed, confining to discrete random variables for simplicity, if $x_1 < x_2 < \cdots < x_m$, then

$$\Pr\left[X_{(1)} = x_1, \ldots, X_{(m)} = x_m\right] = m! \Pr\left[X_1 = x_1, \ldots, X_m = x_m\right]$$

else

$$\Pr\left[X_{(1)} = x_1, X_{(2)} = x_2, \ldots, X_{(m)} = x_m\right] = 0$$

because there are precisely $m!$ permutations of the set $\{X_k\}_{1 \leq k \leq m}$ onto the given ordered sequence $\{x_1, x_2, \ldots, x_m\}$. If the sequence is not ordered such that $x_k > x_l$ for at least one couple of indices $k < l$, then the probability is zero because the event $\{X_{(k)} > X_{(l)}\}$ is, by definition, impossible. Finally, the product in (3.41) follows by independence.

If the set $\{X_k\}_{1 \leq k \leq m}$ is uniformly distributed over $[0, t]$, then

$$f_{\{X_{(j)}\}}(x_1, x_2, \ldots, x_m) = \frac{m!}{t^m} \qquad 0 \leq x_1 < x_2 < \cdots < x_m \leq t$$

$$= 0 \qquad \text{elsewhere}$$

while for exponential random variables with $f_X(x) = \alpha e^{-\alpha x}$

$$f_{\{X_{(j)}\}}(x_1, x_2, \ldots, x_m) = m! \alpha^m e^{-\alpha \sum_{j=1}^{m} x_j} \qquad 0 \leq x_1 < x_2 < \cdots < x_m$$

$$= 0 \qquad \text{elsewhere}$$

The order relation between the set $X_{(1)} \leq X_{(2)} \leq \cdots \leq X_{(m)}$ is preserved after a continuous, non-decreasing transform g, i.e. $g\left(X_{(1)}\right) \leq g\left(X_{(2)}\right) \leq \cdots \leq g\left(X_{(m)}\right)$. If the distribution function F_X is continuous (it is always non-decreasing), the argument shows that the order statistics of a general set of i.i.d. random variable $\{X_k\}_{1 \leq k \leq m}$ can be reduced to a study of the order statistics of the set of i.i.d. uniform random variables $\{U_k\}_{1 \leq k \leq m}$ on $[0,1]$ because $U = F_X(X)$.

The event $\{X_{(k)} \leq x\}$ means that at least k among the m random variables $\{X_j\}_{1 \leq j \leq m}$ are smaller than x. Since each of the m random variables is chosen independently from a same distribution F_X, the probability that precisely n of the m random variables is smaller than x is binomially distributed with parameter $p = \Pr[X \leq x]$. Hence,

$$\Pr[X_{(k)} \leq x] = \sum_{n=k}^{m} \binom{m}{n} (\Pr[X \leq x])^n (1 - \Pr[X \leq x])^{m-n} \qquad (3.42)$$

The probability density function can be obtained in the usual, though cumbersome, way by differentiation,

$$\begin{aligned}
f_{X_{(k)}}(x) &= \frac{d\Pr[X_{(k)} \leq x]}{dx} \\
&= \sum_{n=k}^{m} \binom{m}{n} \frac{d}{dx} (\Pr[X \leq x])^n (1 - \Pr[X \leq x])^{m-n} \\
&= f_X(x) \sum_{n=k}^{m} n \binom{m}{n} (\Pr[X \leq x])^{n-1} (1 - \Pr[X \leq x])^{m-n} \\
&\quad - f_X(x) \sum_{n=k}^{m} (m-n) \binom{m}{n} (\Pr[X \leq x])^n (1 - \Pr[X \leq x])^{m-n-1}
\end{aligned}$$

Using $n\binom{m}{n} = m\binom{m-1}{n-1}$, $(m-n)\binom{m}{n} = m\binom{m-1}{n}$ and lowering the upper index in the last summation, we have

$$\begin{aligned}
f_{X_{(k)}}(x) &= m f_X(x) \sum_{n=k}^{m} \binom{m-1}{n-1} (\Pr[X \leq x])^{n-1} (1 - \Pr[X \leq x])^{m-n} \\
&\quad - m f_X(x) \sum_{n=k}^{m-1} \binom{m-1}{n} (\Pr[X \leq x])^n (1 - \Pr[X \leq x])^{m-n-1} \\
&= m f_X(x) \sum_{n=k}^{m} \binom{m-1}{n-1} (\Pr[X \leq x])^{n-1} (1 - \Pr[X \leq x])^{m-n} \\
&\quad - m f_X(x) \sum_{n=k+1}^{m} \binom{m-1}{n-1} (\Pr[X \leq x])^{n-1} (1 - \Pr[X \leq x])^{m-n}
\end{aligned}$$

or, with $F_X(x) = \Pr[X \leq x]$,

$$f_{X_{(k)}}(x) = m f_X(x) \binom{m-1}{k-1} (F_X(x))^{k-1} (1 - F_X(x))^{m-k} \qquad (3.43)$$

The more elegant and faster argument is as follows: in order for $X_{(k)}$ to be equal to x, exactly $k-1$ of the m random variables $\{X_j\}_{1 \leq j \leq m}$ must be less than x, one equal to x and the other $m-k$ must all be greater than x. Abusing the notation $f_X(x) = \Pr[X_{(k)} = x]$ and observing that $m\binom{m-1}{k-1} = \frac{m!}{1!(k-1)!(m-k)!}$ is an instance of the multinomial coefficient $\frac{m!}{n_1! n_2! \cdots n_k!}$ which gives the number of ways of putting $m = n_1 + n_2 + \cdots + n_k$ different objects into k different boxes with n_j in the j-th box, leads alternatively to (3.43).

3.5 Examples of other distributions

3.5.1 Gumbel distribution

The **Gumbel distribution** appears in the theory of extremes (see Section 6.6) and is defined by the distribution function

$$F_{\text{Gumbel}}(x) = e^{-e^{-a(x-b)}} \tag{3.44}$$

The corresponding Laplace transform is

$$\varphi_{\text{Gumbel}}(z) = \int_{-\infty}^{\infty} e^{-zt} e^{-e^{-a(t-b)}} ae^{-a(t-b)} dt = e^{-bz}\Gamma\left(1 + \frac{z}{a}\right) \tag{3.45}$$

from which the mean follows as $E[X] = -\frac{d}{dz} e^{-bz}\Gamma\left(1 + \frac{z}{a}\right)\big|_{z=0} = b + \frac{\gamma}{a}$, where $\gamma = 0.577\,215...$ is the Euler constant. The variance is best computed with (2.46) resulting in $\text{Var}[X] = \frac{\pi^2}{6a^2}$.

3.5.2 Cauchy distribution

The **Cauchy distribution** has the probability density function

$$f_{\text{Cauchy}}(x) = \frac{1}{\pi(1+x^2)} \tag{3.46}$$

and corresponding distribution,

$$F_{\text{Cauchy}}(x) = \frac{1}{\pi}\left(\frac{\pi}{2} + \arctan x\right) \tag{3.47}$$

The Laplace transform

$$\varphi_{\text{Cauchy}}(z) = \frac{1}{\pi}\int_{-\infty}^{\infty} \frac{e^{-zx}dx}{1+x^2}$$

only converges for purely imaginary $z = i\omega$, in which case it reduces to a Fourier transform,

$$\varphi_{\text{Cauchy}}(i\omega) = \frac{1}{\pi}\int_{-\infty}^{\infty} \frac{e^{-i\omega x}dx}{1+x^2}$$

This integral is best evaluated by contour integration. If $\omega \leq 0$, we consider a contour C consisting of the real axis and the semi-circle that encloses the negative $\text{Im}(x)$-plane,

$$\int_C \frac{e^{-i\omega x}dx}{1+x^2} = \int_{-\infty}^{\infty} \frac{e^{-i\omega x}dx}{1+x^2} + \lim_{r\to\infty}\int_0^{\pi} \frac{e^{-i\omega re^{-i\theta}} d\left(re^{-i\theta}\right)}{1+r^2 e^{-2i\theta}}$$

Since $\left|e^{-i\omega re^{-i\theta}}\right| = e^{\omega r \sin\theta} = e^{-|\omega|r\sin\theta}$ and $\sin\theta \geq 0$ for $0 \leq \theta \leq \pi$, the limit of the last integral vanishes. The contour encloses the simple pole (zero of $x^2 + 1 = (x-i)(x+i)$) at $x = -i$. Applying Cauchy's residue theorem, we obtain

$$\int_{-\infty}^{\infty} \frac{e^{-i\omega x}dx}{1+x^2} = -2\pi i \lim_{x\to-i} \frac{e^{-i\omega x}(x+i)}{1+x^2} = \pi e^{-\omega}$$

If $\omega \geq 0$, we close the contour over the positive $\text{Im}(x)$-plane such that the contribution of the semi-circle to the contour C again vanishes. The resulting contour then encloses the simple pole at $x = i$ and

$$\int_{-\infty}^{\infty} \frac{e^{-i\omega x}dx}{1+x^2} = 2\pi i \lim_{x \to i} \frac{e^{-i\omega x}(x-i)}{1+x^2} = \pi e^{-\omega}$$

Combining both expressions results in

$$\varphi_{\text{Cauchy}}(i\omega) = E\left[e^{-i\omega X}\right] = e^{-|\omega|}$$

Since $|\omega|$ is not analytic around $\omega = 0$, none of the moments of the Cauchy distribution exists! Hence, the Cauchy distribution is an example of a distribution without mean (see the requirement for the existence of the expectation in Section 2.3.2), although the improper integral $\int_{-\infty}^{\infty} \frac{x\,dx}{1+x^2} = 0$ due to symmetry (in the Riemann sense), but both $\int_{-\infty}^{0} \frac{x\,dx}{1+x^2}$ and $\int_{0}^{\infty} \frac{x\,dx}{1+x^2}$ diverge. In addition, if $S_n = \sum_{k=1}^{n} X_k$ is the sum of i.i.d. Cauchy random variables X_k, the sample mean $\frac{S_n}{n}$ has the Fourier transform,

$$E\left[e^{-i\omega \frac{S_n}{n}}\right] = E\left[e^{-i\frac{\omega}{n}\sum_{k=1}^{n} X_k}\right] = \prod_{k=1}^{n} E\left[e^{-i\frac{\omega}{n}X_k}\right] = \left(E\left[e^{-i\frac{\omega}{n}X}\right]\right)^n = e^{-|\omega|}$$

Hence, the sample mean $\frac{S_n}{n}$ of i.i.d. Cauchy random variables is again a Cauchy random variable independent of n. This means that the law of large numbers (see Section 6.2) does not hold for the Cauchy random variable, as a consequence of the non-existence of the mean. Also, the sum S_n has Fourier transform $e^{-|n\omega|}$ and the pdf equals $f_{S_n}(x) = \frac{1}{n\pi\left(1+(x/n)^2\right)}$.

Finally, as shown[5] in Section 3.3.4, if X is a Cauchy random variable, then so is X^{-1}.

3.5.3 Weibull distribution

The **Weibull distribution** with pdf defined for $x \geq 0$ and $a, b > 0$

$$f_{\text{Weibull}}(x) = \frac{\exp\left(-\left(\frac{x}{a}\right)^b\right)}{a\Gamma\left(1+\frac{1}{b}\right)} \tag{3.48}$$

generalizes the exponential distribution (3.17) corresponding to $b = 1$ and $a = \frac{1}{\alpha}$. It is related to the Gaussian density function (3.19) if $b = 2$. Let X be a Weibull random variable. All higher moments can be computed from (2.35) as

$$E\left[X^k\right] = \frac{1}{a\Gamma\left(1+\frac{1}{b}\right)} \int_0^{\infty} x^k \exp\left(-\left(\frac{x}{a}\right)^b\right) dx = \frac{a^k \Gamma\left(\frac{k+1}{b}\right)}{\Gamma\left(\frac{1}{b}\right)}$$

---

[5] This observation also follows from the transformation formula (2.33), which yields $f_{X^{-1}}(u) = u^{-2} f_X(u^{-1})$.

Unfortunately, the distribution function $F_{\text{Weibull}}(x)$ of (3.48) is not available in closed form. The generating function possesses the expansion

$$\varphi_X(z) = E\left[e^{-zX}\right] = \sum_{k=0}^{\infty} \frac{(-z)^k}{k!} E\left[X^k\right] = \frac{1}{\Gamma\left(\frac{1}{b}\right)} \sum_{k=0}^{\infty} \Gamma\left(\frac{k+1}{b}\right) \frac{(-za)^k}{k!}$$

which cannot be summed in explicit form for general b.

Alternative definitions of the Weibull distribution appear in the literature. The original definition of Weibull (1951) himself is

$$F_{\text{Weibull}}(x) = 1 - e^{-a(x-x_0)^b} \qquad \text{for } x \geq x_0 \qquad (3.49)$$

with corresponding probability density function

$$f_{\text{Weibull}}(x) = ab(x-x_0)^{b-1} e^{-a(x-x_0)^b} \qquad \text{for } x \geq x_0$$

If X possesses the original Weibull distribution (3.49), the moments around x_0 are

$$E\left[(X-x_0)^k\right] = \int_{x_0}^{\infty} (x-x_0)^k f_{\text{Weibull}}(x)dx = a^{-\frac{k}{b}}\Gamma\left(1+\frac{k}{b}\right)$$

from which the mean follows as

$$E[X] = x_0 + a^{-\frac{1}{b}}\Gamma\left(1+\frac{1}{b}\right)$$

and the variance is

$$\text{Var}[X] = E\left[(X - x_0 + x_0 - E[X])^2\right] = E\left[(X-x_0)^2\right] - (E[X-x_0])^2$$
$$= a^{-\frac{2}{b}}\left\{\Gamma\left(1+\frac{2}{b}\right) - \Gamma^2\left(1+\frac{1}{b}\right)\right\}$$

Weibull (1951) proposed his Weibull distribution (3.49) as an empirical distribution that models a large variety of physical observations such the yield strength of steel and cotton, electrical insulation breakdown, and even the lifetime of man. The interest of the Weibull distribution in the Internet stems from the self-similar and long-range dependence of observables (i.e. quantities that can be measured such as the delay, the interarrival times of packets, etc.). Especially if the shape factor $b \in (0,1)$, the Weibull has a sub-exponential tail that decays more slowly than an exponential, but still faster than any power law.

3.5.4 *Pareto distribution*

Power-law behavior is often described via the **Pareto distribution** with pdf for $x \geq 0$ and $\tau > 0$,

$$f_{\text{Pareto}}(x) = \frac{\alpha}{\tau}\left(1+\frac{x}{\tau}\right)^{-\alpha-1} \qquad (3.50)$$

and "power-law" exponent α. The Pareto distribution function is

$$F_{\text{Pareto}}(x) = \frac{\alpha}{\tau} \int_0^x \left(1 + \frac{t}{\tau}\right)^{-\alpha-1} dt = 1 - \left(1 + \frac{x}{\tau}\right)^{-\alpha} \tag{3.51}$$

Since $\lim_{x\to\infty} F(x) = 1$, the exponent α must exceed 0. The higher moments are Beta-functions (Abramowitz and Stegun, 1968, Section 6.2.1)

$$E\left[X^k\right] = \frac{\alpha}{\tau} \int_0^\infty \frac{x^k \, dx}{(1 + \frac{x}{\tau})^{\alpha+1}} = k! \tau^k \frac{\Gamma(\alpha - k)}{\Gamma(\alpha)}$$

and show that $E\left[X^k\right]$ only exists if $\alpha > k$. Hence, the mean $E\left[X\right]$ only exists if $\alpha > 1$. The deep tail asymptotic for large x is $f_{\text{Pareto}}(x) = O\left(x^{-\alpha-1}\right)$ and $\Pr\left[X > x\right] = O\left(x^{-\alpha}\right)$. To exhibit the power-law behavior more explicitly, the Pareto distribution is sometimes defined, for $x > \tau$ and $\alpha > 0$, as

$$\Pr\left[X_{\text{Pareto}} \geq x\right] = \left(\frac{x}{\tau}\right)^{-\alpha} \tag{3.52}$$

For example, the distribution of the nodal degree in the Internet has an exponent around $\alpha = 2.4$ (see Section 15.3). When the discrete nature of the random variable is important, the discrete power law can be defined as in (15.5).

As overwhelmingly illustrated by Bak (1996), many observed phenomena exhibit power laws, such as Zipf's law,

$$\Pr\left[X_{\text{Zipf}} = k\right] = \frac{k^{-1}}{\sum_{j=1}^n j^{-1}} \qquad \text{for } 1 \leq k \leq n$$

for the ranking of words in the English language and equivalent to $\alpha \to 0$ case in (3.52), the Gutenberg-Richter law for the magnitude versus occurrence of earthquakes, the Mandelbrot (1977) fractal dimension of the length of a cost line, $1/f$ noise in signals, and so many more. In practice, power-law phenonema are recognized when the density or distribution function plotted on a log-log scale exhibits (approximately) a straight line.

3.5.5 Lognormal distribution

Another distribution with heavy tails is the **lognormal distribution** defined as the random variable $X = e^Y$, where $Y = N\left(\mu, \sigma^2\right)$ is a Gaussian or normal random variable. From (2.32), it follows that $\Pr\left[e^Y \leq x\right] = \Pr\left[Y \leq \log x\right]$ for $x \geq 0$, and with (3.20)

$$F_{\text{lognormal}}(x) = \frac{1}{\sigma\sqrt{2\pi}} \int_{-\infty}^{\log x} \exp\left[-\frac{(t-\mu)^2}{2\sigma^2}\right] dt \tag{3.53}$$

and, for $x > 0$,

$$f_{\text{lognormal}}(x) = \frac{\exp\left[-\frac{(\log x - \mu)^2}{2\sigma^2}\right]}{\sigma x \sqrt{2\pi}} \tag{3.54}$$

where $\mu \in \mathbb{R}$ and $\sigma \geq 0$ are called the parameters of the lognormal pdf. The pdf (3.54) can be rewritten as a function of only $u = \log x$,

$$f_{\text{lognormal}}(x) = f_{\text{lognormal}}(e^u) = e^{-\mu + \frac{\sigma^2}{2}} \frac{\exp\left[-\frac{(u - (\mu - \sigma^2))^2}{2\sigma^2}\right]}{\sigma\sqrt{2\pi}} \tag{3.55}$$

illustrating that the scaled lognormal pdf $e^{\mu - \frac{\sigma^2}{2}} f_{\text{lognormal}}(e^u)$ is a Gaussian pdf $N(\mu', \sigma^2)$ with mean $\mu' = \mu - \sigma^2$. The maximum of $f_{\text{lognormal}}(x)$ occurs at $x_{\max} = e^{\mu - \sigma^2}$ and equals $\max_{x \geq 0} f_{\text{lognormal}}(x) = \frac{e^{-\mu} e^{\frac{\sigma^2}{2}}}{\sigma\sqrt{2\pi}}$, which follows directly from (3.55). Moreover, we find easier from (3.55) than (3.54) that $f_{\text{lognormal}}(0) = \lim_{u \to -\infty} f_{\text{lognormal}}(e^u) = 0$ and that $f'_{\text{lognormal}}(0) = 0$. This means that any lognormal pdf starts at $x = 0$ from zero with zero slope, increases up to the maximum at $x_{\max} = e^{\mu - \sigma^2} > 0$ after which it decreases slowly towards zero at $x \to \infty$. Thus, the lognormal is bell-shaped, but, in contrast to the Gaussian, the lognormal pdf is not symmetric around its maximum at $x_{\max} = e^{\mu - \sigma^2}$ and can be seriously skewed. In particular, if the maximum at $x_{\max} = e^{\mu - \sigma^2}$ is small and close to zero, only the right-hand side heavy tail is observed in measurements. The skewness $s_X = \frac{E[(X - E[X])^3]}{(\text{Var}[X])^{3/2}}$ of the lognormal distribution equals $s_X = \left(e^{\sigma^2} + 2\right)\left(e^{\sigma^2} - 1\right)^{1/2}$, which illustrates that symmetry ($s_X = 0$) is only possible when $\sigma = 0$. In the limit $\sigma \to 0$, the pdf (3.54) reduces to a Dirac delta function at $x = e^{\mu}$, thus $\lim_{\sigma \to 0} f_{\text{lognormal}}(x) = \delta(x - e^{\mu})$.

The moments of the lognormal distribution are

$$E[X^k] = \frac{1}{\sigma\sqrt{2\pi}} \int_0^\infty x^{k-1} \exp\left[-\frac{(\log x - \mu)^2}{2\sigma^2}\right] dx$$

$$= \frac{1}{\sigma\sqrt{2\pi}} \int_{-\infty}^\infty e^{ku} \exp\left[-\frac{(u - \mu)^2}{2\sigma^2}\right] du$$

or, explicitly,

$$E[X^k] = \exp(\mu k) \exp\left(\frac{k^2 \sigma^2}{2}\right) \tag{3.56}$$

and

$$\text{Var}[X] = e^{2\mu}\left(e^{2\sigma^2} - e^{\sigma^2}\right) \tag{3.57}$$

Given the mean and variance, the parameters of the lognormal are found as

$$\sigma^2 = \log\left(1 + \frac{\text{Var}[X]}{(E[X])^2}\right) \tag{3.58}$$

and

$$\mu = \log E[X] - \frac{\sigma^2}{2} \tag{3.59}$$

Although $E[X] \geq 0$, we remark that the parameter μ can be negative and that, for $\sigma > 0$, the peak of the pdf occurs at

$$x_{\max} = e^{\mu - \sigma^2} = E[X] e^{-\frac{3\sigma^2}{2}} < E[X]$$

in contrast to the Gaussian, where $x_{\max} = E[X]$.

The product $Z_n = \prod_{j=1}^{n} X_j$ of lognormal random variables $X_j = e^{Y_j}$ is again a lognormal random variable because $\log Z_n = \sum_{j=1}^{n} Y_j$ is a sum of Gaussian random variables Y_j, which is a Gaussian random variable. Moreover, if X_j are i.i.d. random variables, so are $\log X_j$, then $\log Z_n$ converges to a Gaussian random variable for large n by the Central Limit Theorem 6.3.1, which implies that Z_n converges to a lognormal random variable. Theorem 6.4.1 demonstrates that the lognormal distribution is the fingerprint of the law of proportionate effect. Furthermore, a consequence of Marlow's Theorem 6.5.1 shows that a sum of independent lognormal random variables tends to a lognormal random variable.

The probability generating function is by definition (2.38)

$$\varphi_X(z; \mu, \sigma^2) = \frac{1}{\sigma\sqrt{2\pi}} \int_0^\infty e^{-zt} \frac{e^{-\frac{(\log t - \mu)^2}{2\sigma^2}}}{t} dt = \frac{1}{\sigma\sqrt{2\pi}} \int_{-\infty}^\infty e^{-ze^x} e^{-\frac{(x-\mu)^2}{2\sigma^2}} dx$$

$$(3.60)$$

Denoting $Z = ze^\mu$, we observe that $\varphi_X(z; \mu, \sigma^2)$ effectively depends upon two independent parameters $\varphi_X(z; \mu, \sigma^2) = \varphi_X(Z; 0, \sigma^2)$. The integral (3.60) indicates that $\varphi_X(z; \mu, \sigma^2)$ only exists for $\operatorname{Re}(z) \geq 0$. This means that $\varphi_X(z; \mu, \sigma^2)$ is not analytic at any point $z = it$ on the imaginary axis because the circle with arbitrary small but non-zero radius around $z = it$ necessarily encircles points with $\operatorname{Re}(z) < 0$ where $\varphi_X(z; \mu, \sigma^2)$ does not exist. Hence, the Taylor expansion (2.41) of the generating function around $z = 0$ does not exist, although all moments or derivatives at $z = 0$ exist. Indeed, the series

$$\sum_{k=0}^\infty \frac{(-1)^k E[X^k]}{k!} z^k = \sum_{k=0}^\infty \frac{(-ze^\mu)^k}{k!} \exp\left(\frac{k^2\sigma^2}{2}\right)$$

is a divergent series (except for $\sigma = 0$ or $z = 0$).

By substituting the inverse Mellin transform (16.41) of the exponential into (3.60), reversing the order of integration[6] and invoking (3.22), we arrive, with $\operatorname{Re}(Z) \geq 0$ and $c > 0$, at the important integral

$$\varphi_X(Z; 0, \sigma^2) = \frac{1}{2\pi i} \int_{c-i\infty}^{c+i\infty} Z^{-w} e^{\frac{\sigma^2 w^2}{2}} \Gamma(w)\, dw \qquad (3.61)$$

After expanding the entire function $e^{\frac{\sigma^2 w^2}{2}}$ in a Taylor series and using $\sum_{j=0}^\infty \frac{j^n}{j!} z^j = e^z \sum_{k=0}^n S_n^{(k)} z^k$ where $S_n^{(k)}$ are the Stirling Numbers of the Second Kind (Abramowitz and Stegun, 1968, 24.1.4), we obtain the exact series, valid for all positive Z,

$$\varphi_X(Z; 0, \sigma^2) = e^{-Z} + e^{-Z} \sum_{k=1}^\infty \frac{\left(\frac{\sigma^2}{2}\right)^k}{k!} \sum_{j=1}^{2k} S_{2k}^{(j)} (-Z)^j$$

---

[6] Allowed by Fubinni's Theorem (see e.g. Titchmarsh (1964)) because of absolute convergence.

but, due to the alternating j-sum, only numerically efficient for small Z. Also asymptotic expansions for large Z such as

$$\varphi_X(Z; 0, \sigma^2) = \frac{e^{-\frac{\log^2 Z}{2\sigma^2}} e^{-\frac{\log^2\left(\frac{\log Z}{\sigma^2}\right)}{2\sigma^2}}}{\sigma\sqrt{2\pi}} \Gamma\left(\frac{\log Z}{\sigma^2}\right)\left(1 + O\left(\frac{1}{\log Z}\right)\right)$$

can be deduced from (3.61).

The fact that the pgf (3.54) nor (3.61) is not available in closed form complicates the computation of the sum of i.i.d. lognormal random variables via (2.75). This sum appears in radio communications with several transmitters and receivers.

In radio communications, the received signal levels decrease with the distance between the transmitter and the receiver. This phenomenon is called *pathloss*. Attenuation of radio signals due to pathloss has been modeled by averaging the measured signal powers over long times and over various locations with the same distances to the transmitter. The mean value of the signal power found in this way is referred to as the *area mean power* $\mathbf{P}_a$ (in Watts) and is well-modeled as $\mathbf{P}_a(r) = c \cdot r^{-\eta}$, where c is a constant and η is the pathloss exponent[7]. In reality the received power levels may vary significantly around the mean power $\mathbf{P}_a(r)$ due to irregularities in the surroundings of the receiving and transmitting antennas. Measurements have revealed that the logarithm of the mean power $\mathbf{P}(r)$ at different locations on a circle with radius r around the transmitter is *approximately* normally distributed with mean equal to the logarithm of the area mean power $\mathbf{P}_a(r)$. The lognormal shadowing model assumes that the logarithm of $\mathbf{P}(r)$ is *precisely* normally distributed around the logarithmic value of the area mean power: $\log_{10}(\mathbf{P}(r)) = \log_{10}(\mathbf{P}_a(r)) + X$, where $X = N(0, \sigma)$ is a zero-mean normal distributed random variable (in dB) with standard deviation σ (also in dB and for severe fluctuations up to 12 dB). Hence, the random variable $\mathbf{P}(r) = \mathbf{P}_a(r) 10^X$ has a lognormal distribution (3.53) equal to

$$\Pr[\mathbf{P}(r) \le x] = \Pr\left[X \le \log_{10}\frac{x}{\mathbf{P}_a(r)}\right] = \frac{1}{\sigma\sqrt{2\pi}\log 10} \int_0^x \exp\left[-\frac{(\log_{10} u - \log_{10}(\mathbf{P}_a(r)))^2}{2\sigma^2}\right]\frac{du}{u}$$

The lognormal density also appears in a specific kind of online social news aggregators such as digg.com, delicious.com or reddit.com. In those online social networks, subscribers can vote on the submitted information (such as bookmarks, opinions, news, etc.), called "stories". The sum of all votes on a story, which is called the digg value of a story in digg.com, is advertised and reflects the impact or popularity of a story. The digg value of an arbitrary story follows a lognormal distribution as shown in Tang *et al.* (2011) and Van Mieghem *et al.* (2011a). A lognormal distribution is further observed in citation networks (Radicchi *et al.*, 2008), in scientific performance (Shockley, 1957) and Twitter retweet times (Doerr *et al.*, 2013).

The lognormal distribution is similar in shape to power-law distributions such as the Pareto distribution (3.52) with $\alpha \to 0$ or the Zipf distribution. Indeed, taking the logarithm of both sides of (3.54) yields

$$\log f_{\text{lognormal}}(x) = -\frac{(\log x)^2}{2\sigma^2} + \left(\frac{\mu}{\sigma^2} - 1\right)\log x - \log\left(\sigma\sqrt{2\pi}\right) - \frac{\mu^2}{2\sigma^2}$$

which is linear in $\log x$ for a large range of x values, provided σ is sufficiently large

---

[7] The constant c depends on the transmitted power, the receiver and the transmitter antenna gains and the wavelength. The pathloss exponent η depends on the environment and terrain structure and can vary between 2 in free space to 6 in urban areas.

and $\mu \neq \sigma^2$ such that the quadratic term is negligibly small. The expression for the lognormal pdf in (3.54) can be rewritten in a "power law"-like form as

$$f_{\text{lognormal}}(x) = \frac{e^{-\frac{\mu^2}{2\sigma^2}}}{\sigma\sqrt{2\pi}} x^{-\alpha(x)} \qquad (3.62)$$

where the exponent $\alpha(x)$ equals

$$\alpha(x) = 1 + \frac{\log x - 2\mu}{2\sigma^2} \qquad (3.63)$$

which illustrates that a lognormal random variable behaves as a power-law random variable, provided the last fraction in (3.63) is negligibly small, say ε. The latter happens when $\left|\frac{\log x - 2\mu}{2\sigma^2}\right| < \varepsilon$. Thus, when $x \in [e^{2\mu - 2\sigma^2\varepsilon}, e^{2\mu + 2\sigma^2\varepsilon}]$, the pdf of a lognormal random variable is almost indistinguishable from the pdf of a power-law random with power-law exponent $\alpha \approx 1 + \varepsilon$. The x-interval $[e^{2\mu - 2\sigma^2\varepsilon}, e^{2\mu + 2\sigma^2\varepsilon}]$ exceeds the maximum $x_{\max} = e^{\mu - \sigma^2}$ of the lognormal pdf and is clearly longer when σ is larger (as well as when the tolerated accuracy ε increases).

Another property of a lognormal random variable X is that scaling into the random variable $Y = bX$, where b is real and positive, changes the parameter $\mu_Y = \log b + \mu_X$, while (3.58) shows that $\sigma_Y = \sigma_X$, which is invariant to scaling! The maximum $x_{\max}$ of the pdf also shifts from $e^{\mu_X - \sigma^2}$ to $be^{\mu_X - \sigma^2}$. Hence, by measuring a desired quantity in a certain unit, e.g. seconds, resulting in a realization of X or in minutes, yielding a realization of Y, the maximum for Y is shifted towards zero by a factor of $b = \frac{1}{60}$. The measured histogram of Y (in minutes) thus exhibits more than that of X (in seconds) the right-hand side "power law"-like decreasing tail. Apart from the binning problem (on a log scale), these observations complicate to distinguish a real power law from lognormal behavior in data sets (see e.g. Clauset *et al.* (2009)), in which the deep tails are not (or cannot be) sufficiently sampled.

3.6 Summary tables of probability distributions

Although we have discussed in this chapter the most important probability distributions, many other distributions appear in physics and engineering. Among those omitted here, the great book of Papoulis and Unnikrishna Pillai (2002) covers the Beta distribution

$$f_{\text{Beta}}(x) = \frac{\Gamma(\alpha + \beta)}{\Gamma(\alpha)\Gamma(\beta)} x^{\alpha-1}(1-x)^{\beta-1} \qquad \alpha > 0, \beta > 0 \text{ and } x \in [0,1]$$

which generalizes the uniform distribution (3.12) on $[0,1]$, the Rayleigh distribution,

$$f_{\text{Rayleigh}}(x) = \frac{x}{\sigma^2} e^{-\frac{x^2}{2\sigma^2}} \qquad x \geq 0$$

and the hypergeometric distribution

$$\Pr\left[X_{\text{Hypergeometric}} = k\right] = \frac{\binom{M}{k}\binom{N-M}{n-k}}{\binom{N}{n}}$$

In addition, Papoulis and Unnikrishna Pillai (2002) provide many examples and exercises on random variables and distributions.

3.6.1 Discrete random variables

Name	$\Pr[X = k]$	$E[X]$	$\mathrm{Var}[X]$	$\varphi_X(z) = E[z^X]$
Bernoulli	$\Pr[X = 1] = p$	p	$p(1-p)$	$1 - p + pz$
Binomial	$\binom{n}{k}p^k(1-p)^{n-k}$	np	$np(1-p)$	$((1-p) + pz)^n$
Geometric	$p(1-p)^{k-1}$	$\frac{1}{p}$	$\frac{1-p}{p^2}$	$\frac{pz}{1-(1-p)z}$
Poisson	$\frac{\lambda^k}{k!}e^{-\lambda}$	λ	λ	$e^{\lambda(z-1)}$

3.6.2 Continuous random variables

Name	$f_X(x)$	$E[X]$	$\mathrm{Var}[X]$	$\varphi_X(z) = E[e^{-zX}]$
Uniform	$\frac{1_{a \le x \le b}}{b-a}$	$\frac{a+b}{2}$	$\frac{(b-a)^2}{12}$	$\frac{e^{-za}-e^{-zb}}{z(b-a)}$
Exponential	$\alpha e^{-\alpha x}$	$\frac{1}{\alpha}$	$\frac{1}{\alpha^2}$	$\frac{\alpha}{z+\alpha}$
Gaussian	$\frac{\exp\left[-\frac{(x-\mu)^2}{2\sigma^2}\right]}{\sigma\sqrt{2\pi}}$	μ	σ^2	$\exp\left[\frac{\sigma^2 z^2}{2} - \mu z\right]$
Gamma	$\frac{\alpha(\alpha x)^{\beta-1}}{\Gamma(\beta)}e^{-\alpha x}$	$\frac{\beta}{\alpha}$	$\frac{\beta}{\alpha^2}$	$\left(\frac{\alpha}{z+\alpha}\right)^\beta$
Gumbel	$e^{-x}e^{e^{-x}}$	$\gamma = 0.577...$	$\frac{\pi^2}{6}$	$\Gamma(z+1)$
Cauchy	$\frac{1}{\pi(1+x^2)}$	does not exist	does not exist	$e^{-|\mathrm{Im}(z)|}1_{\{\mathrm{Re}(z)=0\}}$
Weibull	$ab(x-x_0)^{b-1}e^{-a(x-x_0)^b}$	$x_0 + \frac{\Gamma(1+\frac{1}{b})}{a^{\frac{1}{b}}}$	$\frac{\Gamma(1+\frac{2}{b})-\Gamma^2(1+\frac{1}{b})}{a^{2/b}}$	
Pareto	$\frac{\alpha}{\tau}\left(1+\frac{x}{\tau}\right)^{-\alpha-1}$	$\frac{\tau 1_{\{\alpha>1\}}}{\alpha-1}$	$\frac{\tau^2\alpha 1_{\{\alpha>2\}}}{(\alpha-1)^2(\alpha-2)}$	
Lognormal	$\frac{\exp\left[-\frac{(\log x-\mu)^2}{2\sigma^2}\right]}{\sigma x\sqrt{2\pi}}$	$\exp(\mu)\exp\left(\frac{\sigma^2}{2}\right)$	$e^{2\mu}\left(e^{2\sigma^2}-e^{\sigma^2}\right)$	
Student t	$\frac{\Gamma\left(\frac{n+1}{2}\right)}{\Gamma\left(\frac{n}{2}\right)\sigma\sqrt{\pi n}\left(1+\frac{x^2}{n\sigma^2}\right)^{\frac{n+1}{2}}}$	0 for $n > 1$	$\frac{n\sigma^2}{n-2}$ for $n > 2$	
Fisher F	$\frac{\Gamma\left(\frac{m+n}{2}\right)\left(\frac{m}{n}\right)^{\frac{m}{2}}z^{\frac{m}{2}-1}}{\Gamma\left(\frac{m}{2}\right)\Gamma\left(\frac{n}{2}\right)(1+\frac{m}{n}z)^{\frac{m+n}{2}}}$	$\frac{n}{n-2}$ for $n > 2$	$\frac{2n^2}{(n-2)^2}\left(\frac{n+m-2}{m(n-4)}\right)$	
Poisson	$K^{-1}(\lambda)\frac{\lambda^x}{\Gamma(x+1)}$	$\lambda\left(1+\frac{r(\lambda)-r'(\lambda)}{K(\lambda)}\right)$		$\frac{K(\lambda e^{-z})}{K(\lambda)}$

3.7 Problems

(i) If $\varphi_X(z)$ is the probability generating function of a non-zero discrete random variable X, find an expression of $E[\log X]$ in terms of $\varphi_X(z)$.

(ii) Compute the mean value of the k-th order statistic in an ensemble of (a) m i.i.d. exponentially distributed random variables with mean $\frac{1}{\alpha}$ and (b) m i.i.d. polynomially distributed random variables on $[0,1]$.

(iii) *Histogram.* Discuss how a probability density function of a continuous random variable X can be approximated from a set $\{x_1, x_2, \ldots, x_n\}$ of n measurements or simulations.

(iv) In a circle with radius r around a sending mobile node, there are $N-1$ other mobile nodes uniformly distributed over that circle. The possible interference caused by these other mobile nodes depends on their distance to the sending node at the center. Derive for large N but constant density ν of mobile nodes the pdf of the distance of the m-th nearest node to the center.

(v) Let U and V be two independent random variables. What is the probability that the one is larger than the other?

(vi) Compute the probability density function of the product Z of two independent uniform random variables U_1 on $[a, b]$ and U_2 on $[c, d]$.

(vii) If X is a uniform, continuous random variable on the interval $[a, b]$ and Y is a uniform, discrete random variable on the interval $[k, l]$, where k and l are integers and $k < l$, then compute $\Pr[Z \le x]$, where $Z = X + Y$ given that X and Y are independent.

(viii) Suppose that the number N of pages in a fax transmission has a geometric probability distribution with mean $1/q = 4$. The number K of bits per page also has a geometric distribution with mean $1/p = 10^5$ bits, independent of any other page and of the number of pages. Show that the total number B of bits in a fax transmission is a geometric random variable with parameter pq.

(ix) *Random variables whose moments are also random variables.* Compute the probability density function of the random variable X that has

   (a) a lognormal distribution with zero mean where the variance is exponentially distributed with mean λ^{-1}.

   (b) a Poisson distribution whose mean λ is exponentially distributed with mean μ.

(x) Let X denote the number of packets in a flow F and let Y denote the number of packets that are sampled from that original flow F. The conditional distribution of Y, given that $X = l$, follows a binomial distribution. What is the probability that a sampled flow of length k is sampled from an original flow of length l? Original flow lengths are geometrically distributed.

(xi) *Uniform sampling.* Consider a set $\mathcal{N}$ of size N that contains items of type A and type B. The subset with m items of type A is denoted by $\mathcal{M}$. A uniform sampling strategy consists of testing the type of an arbitrary item of $\mathcal{N}$. Once tested, the item is not replaced in the set $\mathcal{N}$. Show that each drawing/test in the uniform sampling strategy has precisely probability $p = \frac{m}{N}$ to find an item of type A. An example is the test of infected hosts in the Internet by randomly chosing an IP address and testing whether the host with that IP address has immunization software for a certain type of malicious virus.

(xii) Consider the order statistics $\{w_{(k)}\}_{1 \le k \le m}$ of a set of real positive, i.i.d. random variables $\{w_k\}_{1 \le k \le m}$, each with distribution $F_w(x)$. For a fixed integer $h \in [1, m]$, we denote the sum $W_h = \sum_{k=1}^{h} w_{(k)}$. In any graph with i.i.d. link weights, the weight w_{sp} of any shortest path with h hops obeys $w_{\mathrm{sp}} \ge W_h$.

   (a) Show that

$$\Pr[W_h \le x] = \frac{m!}{(m-h)!} \int_0^{\frac{x}{h}} dt_1 \int_{t_1}^{\frac{x-t_1}{h-1}} dt_2 \cdots \int_{t_{h-1}}^{x - \sum_{k=1}^{h-1} t_k} dt_h \prod_{k=1}^{h} f_w(t_k)(1 - F_w(t_h))^{m-h} \quad (3.64)$$

   (b) If w_k is an exponential random variable, the integrals in (3.64) can be worked out. Compute the above integral for $h = 1, 2, 3$ and 4.

   (c) There is another, elegant way to compute the (3.64) for i.i.d exponential random variables by applying a theorem proved in Feller (1971, p. 19): if $w_1, w_2, \ldots, w_m$ are independent random variables, each exponentially distributed with mean $\frac{1}{a}$, then are the m random variables $w_{(1)}, w_{(2)} - w_{(1)}, \ldots, w_{(m)} - w_{(m-1)}$ independent and the density function of $y_{k+1} = w_{(k+1)} - w_{(k)}$ is $f_{y_{k+1}}(x) = (m-k)ae^{-(m-k)ax}$ for $y \ge 0$, with the convention that $w_{(0)} = 0$ or $y_1 = w_{(1)}$. Demonstrate that the probability density function $f_{W_h}(x)$ for exponential random variables is

$$f_{W_h}(x) = \frac{ae^{-ax}\binom{m}{h}}{(m-h)^{h-1}} \sum_{j=1}^{h} (-1)^{h-j} j^{h-1} \binom{h}{j} e^{-\frac{(m-h)ax}{j}} \quad (3.65)$$

(xiii) *Continuous Poisson distribution.* The *continuous* variant of the discrete Poisson random variable (Section 3.1.4) is the non-negative real, random variable X, that possesses the continuous Poisson probability density function

$$f_X(x) = K^{-1}(\lambda) \frac{\lambda^x}{\Gamma(x+1)} \qquad\qquad x \geq 0$$

where the normalization $K(\lambda)$ equals

$$K(\lambda) = \int_0^\infty \frac{\lambda^x}{\Gamma(x+1)} dx \qquad\qquad (3.66)$$

The corresponding probability generating function, defined in (2.38), is

$$\varphi_X(z) = K^{-1}(\lambda) \int_0^\infty \frac{(\lambda e^{-z})^x}{\Gamma(x+1)} dx \qquad\qquad (3.67)$$

The integral in (3.67) is thus the same as in (3.66) after letting $\lambda \to \lambda e^{-z}$, which underlines the importance of the function $K(\lambda)$.

(a) Show that the moments are

$$E[X^m] = K^{-1}(\lambda) \left.\frac{d^m}{dy^m} K(e^y)\right|_{y=\ln\lambda} \qquad\qquad (3.68)$$

from which the mean is

$$E[X] = \lambda \frac{K'(\lambda)}{K(\lambda)} = \lambda \frac{d\log K(\lambda)}{d\lambda} \qquad\qquad (3.69)$$

(b) As shown by Hardy (1978, p. 196), the characterizing function of the continuous Poisson distribution can be rewritten as

$$K(\lambda) = e^\lambda - r(\lambda) \qquad\qquad (3.70)$$

where

$$r(\lambda) = \int_0^\infty \frac{e^{-\lambda x}}{x\left\{\pi^2 + (\ln x)^2\right\}} dx = \int_{-\infty}^\infty \frac{e^{-\lambda e^t} dt}{\pi^2 + t^2} \qquad\qquad (3.71)$$

The function $r(\lambda)$ is strict decreasing in λ from 1 to 0 for all real, positive λ, as follows from

$$r(\lambda) = \int_{-\infty}^\infty \frac{e^{-\lambda e^t} dt}{\pi^2 + t^2} \leq \int_{-\infty}^\infty \frac{dt}{\pi^2 + t^2} = 1$$

Moreover, $r(\lambda)$ is convex (because $r''(\lambda) \geq 0$ for all $\lambda > 0$). Hence, $K(\lambda)$ is very close to e^λ for large λ and $e^\lambda - 1 \leq K(\lambda) < e^\lambda$. The mean (3.69), computed with (3.70), is always larger than the rate λ. The function $r(\lambda)$ is difficult to determine exactly, but good approximations are possible. By a change of the integration variable, we obtain

$$r(\lambda) = \int_{-\infty}^\infty \frac{e^{-e^{\log\lambda+t}} dt}{\pi^2 + t^2} = \int_{-\infty}^\infty \frac{e^{-e^u} du}{\pi^2 + (\log\lambda - u)^2}$$

which is recognized as a convolution integral in $x = \log\lambda$. Prove (3.70), by first computing the Laplace transform of $K(\lambda)$ and then the inverse Laplace transform.

(xiv) *One-sided stable distribution of index $\frac{1}{2}$.* Consider the random variable $X = 1/Y^2$, where Y is a Gaussian random variable with zero mean. Show that

(a) the pdf of X equals, for $x \geq 0$,

$$f_X(x) = \frac{1}{\sigma\sqrt{2\pi x^3}} e^{-\frac{1}{2\sigma^2 x}} \qquad\qquad (3.72)$$

(b) none of the moments of X exist and the probability generating function is

$$\varphi_X(z) = e^{-\frac{1}{\sigma}\sqrt{2z}} \qquad (3.73)$$

(c) If $X_1, X_2, \ldots, X_n$ are i.i.d. random variable with same distribution as X in (3.72), show that the maximum $X_{(n)}$ obeys the law for large n,

$$\Pr\left[\frac{X_{(n)}}{n^2} \leq x\right] \rightarrow e^{-\frac{1}{\sigma}\sqrt{\frac{2}{\pi x}}} \qquad (3.74)$$

(d) If $X_1, X_2, \ldots, X_n$ are i.i.d. random variable with same distribution as X in (3.72), show that the sum $S_n = \sum_{k=1}^{n} X_k$ obeys the law for large n,

$$\Pr\left[\frac{S_n}{n^2} \leq x\right] \rightarrow F_X(x) = \frac{1}{\sigma\sqrt{2\pi}} \int_0^x \frac{e^{-\frac{1}{2\sigma^2 t}}}{\sqrt{t^3}} dt = \frac{1}{\sigma}\sqrt{\frac{2}{\pi}} \int_{\frac{1}{\sqrt{x}}}^{\infty} e^{-\frac{u^2}{2\sigma^2}} du \qquad (3.75)$$

Hence, the random variable $\frac{S_n}{n^2}$ possesses, for large n, the same distribution as X with pdf (3.72). Feller (1971, p. 52) observes that the mean $\frac{S_n}{n}$ is likely of the order of magnitude of n. As demonstrated in Section 6.2, when $E[X] = \mu$ and finite, $\frac{S_n}{n} \rightarrow \mu$ for large n. The unbounded increase of $\frac{S_n}{n}$ agrees with the non-existence of the mean $E[X]$. Feller (1970, p. 246) remarks that the law of S_n is of the same character as the Central Limit Theorem 6.3.1 with the remarkable difference that here $\frac{S_n}{n^2}$ rather than $\frac{S_n}{n}$ possesses a limit distribution. The one-sided stable distribution with index $\frac{1}{2}$ occurs in waiting times of many physical and economical processes. Feller (1970, p. 90) also mentions that the probability that the first passage of a random walk through a point r occurs before epoch tr^2, for fixed t, tends to $F_X(t)$ with $\sigma = 1$ for $r \rightarrow \infty$.

4

Correlation

In this chapter methods to compute bi-variate correlated random variables are discussed. As a measure for the correlation, the linear correlation coefficient defined in (2.61) is used. The generation of n correlated Gaussian random variables is explained. The sequel is devoted to the construction of two correlated random variables with arbitrary distribution. Finally, we discuss the sampling of a population and the role of estimators. The statistical theory of sampling and estimation is treated in depth by Lehmann (1999) and, from an engineering perspective, by Papoulis and Unnikrishna Pillai (2002).

4.1 Generation of correlated Gaussian random variables

Due to the importance of Gaussian correlated random variables as an underlying system for generating arbitrary correlated random variables, as will be demonstrated in Section 4.3, we discuss how they can be generated in multiple dimensions. With the notation of Section 3.2.3, a Gaussian (normal) random variable with mean μ and variance σ^2 is denoted by $N(\mu, \sigma^2)$. By linearly combining Gaussian random variables, we can create a new Gaussian random variable with a desired mean μ and variance σ^2.

4.1.1 Generation of two independent Gaussian random variables

The fact that a linear combination of Gaussian random variables is again a Gaussian random variable allows us to concentrate on normalized Gaussian random variables $N(0, 1)$. Let X_1 and X_2 be two independent normalized Gaussian random variables. Independent random variables are not correlated and the linear correlation coefficient $\rho = 0$. The resulting joint probability distribution is $f_{X_1 X_2}(x, y; \rho) = f_{X_1}(x) f_{X_2}(y)$ and with (3.19),

$$f_{X_1 X_2}(x, y; 0) = \frac{e^{-\frac{x^2 + y^2}{2}}}{2\pi}$$

69

It is natural to consider a polar transformation and the transformed random variables $R = X_1^2 + X_2^2$ and $\Theta = \arctan\left(\frac{X_2}{X_1}\right)$. The inverse transform is $x = \sqrt{r}\cos\theta$ and $y = \sqrt{r}\sin\theta$, which differs slightly from the usual polar transformation in that we now define $r = x^2 + y^2$ instead of $r^2 = x^2 + y^2$. The reason is that the Jacobian is simpler for our purposes,

$$J(r,\theta) = \det\begin{bmatrix} \frac{\partial x}{\partial r} & \frac{\partial x}{\partial \theta} \\ \frac{\partial y}{\partial r} & \frac{\partial y}{\partial \theta} \end{bmatrix} = \det\begin{bmatrix} \frac{\cos\theta}{2\sqrt{r}} & -\sqrt{r}\sin\theta \\ \frac{\sin\theta}{2\sqrt{r}} & \sqrt{r}\cos\theta \end{bmatrix} = \frac{1}{2}$$

whereas the usual polar transformation has the Jacobian equal to the variable r. Using the transformation rules in Section 2.5.4,

$$f_{R\Theta}(r,\theta) = \frac{e^{-\frac{r}{2}}}{4\pi}$$

which shows that $f_{R\Theta}(r,\theta)$ does not depend on θ. Hence, we can write $f_{R\Theta}(r,\theta) = f_R(r)\,f_\Theta(\theta)$ with $f_\Theta(\theta) = c$, where c is a constant and $f_R(r) = \frac{e^{-\frac{r}{2}}}{4\pi c}$. This implies that Θ is a uniform random variable over an interval $1/c$. We also recognize from (3.15) that $f_R(r)$ is close to an exponential random variable with rate $\alpha = \frac{1}{2}$. Therefore, it is instructive to choose the constant c such that R is precisely an exponential random variable with rate $\alpha = \frac{1}{2}$. Thus, choosing $c = \frac{1}{2\pi}$, we end up with $f_R(r) = \frac{e^{-\frac{r}{2}}}{2}$ and $f_\Theta(\theta) = \frac{1}{2\pi}$. These two independent random variables R and Θ can each be generated separately from a uniform random variable U on [0,1], as discussed in Section 3.2.1, leading to

$$R = -2\ln(U_1)$$
$$\Theta = 2\pi U_2$$

and, finally, to the independent Gaussian random variables

$$X_1 = \sqrt{-2\ln(U_1)}\cos 2\pi U_2 \qquad X_2 = \sqrt{-2\ln(U_1)}\sin 2\pi U_2$$

The procedure can be used to generate a single Gaussian random variable, but also more independent Gaussians by repeating the generation procedure.

4.1.2 The n-joint Gaussian probability distribution function

A *random vector* $X = (X_1, X_2, \ldots, X_n)$ is an $n \times 1$ matrix with the random variables X_i as elements. The expectation of a random vector is a vector with components $E[X_i]$ for $1 \le i \le n$. The variance of a random vector

$$\mathrm{Var}[X] = E\left[(X - E[X])(X - E[X])^T\right] = E\left[XX^T\right] - E[X]\,(E[X])^T$$

is a matrix Σ_X with elements $(\Sigma_X)_{ij} = \mathrm{Cov}[X_i, X_j]$. Since the covariance is commutative, $\mathrm{Cov}[X_i, X_j] = \mathrm{Cov}[X_j, X_i]$, the covariance matrix Σ_X is real and symmetric,

$\Sigma_X = \Sigma_X^T$. The importance of real, symmetric matrices is that they have real eigenvectors (see Appendix A.3). Moreover, Σ_X is non-negative definite because, using vector norms defined in Appendix A.4,

$$x^T \Sigma_X x = E\left[x^T(X - E[X])(X - E[X])^T x\right]$$
$$= E\left[\left((X - E[X])^T x\right)^T (X - E[X])^T x\right]$$
$$= E\left[\|(X - E[X])^T x\|_2^2\right] \geq 0$$

which implies that all real eigenvalues λ_i are non-negative. Hence, there exists an orthogonal matrix U such that

$$\Sigma_X = U\mathrm{diag}(\lambda_i)U^T \tag{4.1}$$

If all random variables X_i are independent, which implies that $\mathrm{Cov}[X_i, X_j] = 0$ for $i \neq j$ and $\mathrm{Cov}[X_i, X_i] = \mathrm{Var}[X_i] \geq 0$, then $\Sigma_X = \mathrm{diag}(\mathrm{Var}[X_i])$.

Gaussian random variables are completely determined by the mean and the variance, i.e. by the first two moments. We will now show that the existence of an orthogonal transformation for any probability distribution such that $U^T \Sigma_X U = \mathrm{diag}(\lambda_i)$ implies that a vector of joint Gaussian random variables can be transformed into a vector of independent Gaussian random variables. Also the reverse holds, which will be used below to generate n jointly correlated Gaussian random variables. The multi-dimensional generating function of an n-joint Gaussian or n-joint normal random vector X is defined for the vector $z = (z_1, z_2, \ldots, z_n)^T$ as

$$\varphi_X(z) = E\left[e^{-zX}\right] = \exp\left(\frac{1}{2}z^T \Sigma_X z - E[X]^T z\right) \tag{4.2}$$

Using (4.1), and the fact that U is an orthogonal matrix such that $U^{-1} = U^T$ and $UU^T = I$,

$$\varphi_X(z) = \exp\left(\frac{1}{2}\left(U^T z\right)^T \mathrm{diag}(\lambda_i)U^T z - \left(U^T E[X]\right)^T U^T z\right)$$

Denote the vectors $w = U^T z$ and $m = U^T E[X]$. Then we have

$$\varphi_X(z) = \exp\left(\frac{1}{2}w^T \mathrm{diag}(\lambda_i)w - m^T w\right)$$
$$= \exp\left(\sum_{j=1}^{n} \frac{\lambda_j w_j^2}{2} - m_j w_j\right) = \prod_{j=1}^{n} e^{\frac{\lambda_j w_j^2}{2} - m_j w_j}$$

and $e^{\frac{\lambda_j w_j^2}{2} - m_j w_j} = \varphi_{X_j}(w_j)$ is the Laplace transform (3.22) of a Gaussian random variable X_j because all λ_j are real and positive[1]. With (2.74), this shows that a vector of joint Gaussian random variables can be transformed into a vector of

---

[1] We omit zero eigenvalues that lead to degenerate Gaussians which are a Dirac impulses (see Section 2.3).

independent Gaussian random variables. Reversing the order of the manipulations also justifies that (4.2) indeed defines a general n-joint Gaussian probability generating function. If $X_1, X_2, \ldots, X_n$ are joint normal and not correlated, then Σ_X is a diagonal matrix, which implies that $X_1, X_2, \ldots, X_n$ are independent. As discussed in Section 2.5.2, independence implies non-correlation, but the converse is generally not true. These properties make Gaussian random variables particularly suited to deal with correlations.

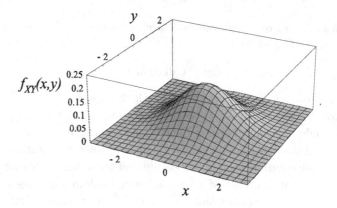

Fig. 4.1. The joint probability density function (4.4) with $\mu_X = \mu_Y = 0$ and $\sigma_X = \sigma_Y = 1$ and $\rho = 0$.

The corresponding n-joint Gaussian probability density function of the vector X can be derived after inverse Laplace transform for the vector $x = (x_1, x_2, \ldots, x_n)^T$ as

$$ f_X(x) = \frac{1}{(2\pi i)^n} \int_{c_1 - i\infty}^{c_1 + i\infty} dz_1 \int_{c_2 - i\infty}^{c_2 + i\infty} dz_2 \ldots \int_{c_n - i\infty}^{c_n + i\infty} dz_n \varphi_X(z) e^{x^T z} $$

Introducing (4.2) yields

$$ f_X(x) = \frac{1}{(2\pi i)^n} \int_{c_1 - i\infty}^{c_1 + i\infty} dz_1 \ldots \int_{c_n - i\infty}^{c_n + i\infty} dz_n \exp\left(\frac{1}{2} z^T \Sigma_X z + (x - E[X])^T z \right) $$

We now require that $z^T \Sigma_X z$ is positive definite such that $\exp\left(\frac{1}{2} z^T \Sigma_X z\right)$ is an entire function that tends to zero along each line parallel to the imaginary axis for each z_j. Furthermore, any finite shift of the line of integration parallel with the imaginary axis leads to the same integral such that we may choose all $c_j = 0$. Let $y = x - E[X]$, then evaluation of the contour yields

$$ f_X(x) = \frac{1}{(2\pi)^n} \int_{-\infty}^{\infty} dt_1 \int_{-\infty}^{\infty} dt_2 \ldots \int_{-\infty}^{\infty} dt_n \exp\left(-\frac{1}{2} t^T \Sigma_X t + i y^T t \right) $$

We introduce the diagonalization (4.1) of the covariance matrix Σ_X and obtain as above, using the same orthogonal transformation $w = U^T t$ with $v = U^T y$, while the Jacobian $J = |\det U| = |\pm 1| = 1$,

$$
f_X(x) = \frac{1}{(2\pi)^n} \int_{-\infty}^{\infty} dw_1 \int_{-\infty}^{\infty} dw_2 \cdots \int_{-\infty}^{\infty} dw_n \prod_{j=1}^{n} e^{-\frac{\lambda_j w_j^2}{2} + iv_j w_j}
$$

$$
= \frac{1}{(2\pi)^n} \prod_{j=1}^{n} \int_{-\infty}^{\infty} e^{-\frac{\lambda_j w_j^2}{2} + iv_j w_j} \, dw_j
$$

We rewrite $-\frac{\lambda_j w_j^2}{2} + iv_j w_j = -\frac{\lambda_j}{2}\left(w_j^2 - i\frac{2v_j}{\lambda_j}w_j\right) = -\frac{\lambda_j}{2}\left(w_j - i\frac{v_j}{\lambda_j}\right)^2 - \frac{1}{2}\frac{v_j^2}{\lambda_j}$ such that

$$
\int_{-\infty}^{\infty} e^{-\frac{\lambda_j w_j^2}{2} + iv_j w_j} \, dw_j = e^{-\frac{1}{2}\frac{v_j^2}{\lambda_j}} \int_{-\infty}^{\infty} \exp\left(-\frac{\lambda_j}{2}\left(w_j - i\frac{v_j}{\lambda_j}\right)^2\right) dw_j
$$

By substituting $s = w_j - i\frac{v_j}{\lambda_j}$ (i.e. shifting the line of integration parallel to the real axis as above), the integral becomes

$$
\int_{-\infty}^{\infty} \exp\left[-\frac{\lambda_j}{2}s^2\right] ds = 2 \int_{0}^{\infty} \exp\left[-\frac{\lambda_j}{2}s^2\right] ds = \frac{\sqrt{2}}{\sqrt{\lambda_j}} \int_{0}^{\infty} e^{-w} w^{-1/2} \, dw
$$

$$
= \frac{\sqrt{2}}{\sqrt{\lambda_j}} \Gamma\left(\frac{1}{2}\right) = \frac{\sqrt{2\pi}}{\sqrt{\lambda_j}}
$$

where we have used the Gamma function (Abramowitz and Stegun, 1968, Chapter 6) and the requirement that all eigenvalues $\lambda_j > 0$. Thus,

$$
f_X(x) = \frac{1}{(\sqrt{2\pi})^n} \prod_{j=1}^{n} \frac{e^{-\frac{1}{2}\frac{v_j^2}{\lambda_j}}}{\sqrt{\lambda_j}} = \frac{e^{-\frac{1}{2}\sum_{j=1}^{n}\frac{v_j^2}{\lambda_j}}}{(\sqrt{2\pi})^n \sqrt{\prod_{j=1}^{n} \lambda_j}}
$$

Finally, with $\sum_{j=1}^{n} \frac{v_j^2}{\lambda_j} = v^T \text{diag}\left(\frac{1}{\lambda_j}\right) v = y^T U \text{diag}\left(\frac{1}{\lambda_j}\right) U^T y$, (A.6) and (4.1), we arrive at the n-joint Gaussian probability density function of the vector X

$$
f_X(x) = \frac{1}{(\sqrt{2\pi})^n \sqrt{\det \Sigma_X}} \exp\left(-\frac{1}{2}(x - E[x])^T \Sigma_X^{-1} (x - E[x])\right) \tag{4.3}
$$

provided Σ_X is positive definite, i.e. all eigenvalues λ_j are strictly larger than zero.

After computing the inverse matrix and the determinant in (4.3) explicitly and using the definition of the correlation coefficient (2.61), the two-dimensional ($n = 2$) or bi-variate Gaussian probability density function is

$$
f_{XY}(x, y; \rho) = \frac{\exp\left[-\frac{\frac{(x-\mu_X)^2}{\sigma_X^2} - \frac{2\rho}{\sigma_X \sigma_Y}(x-\mu_X)(y-\mu_Y) + \frac{(y-\mu_Y)^2}{\sigma_Y^2}}{2(1-\rho^2)}\right]}{2\pi\sigma_X \sigma_Y \sqrt{1-\rho^2}} \tag{4.4}
$$

Figures 4.1–4.3 plot $f_{XY}(x, y; \rho)$ for various correlation coefficients ρ. If $\rho = 0$, we observe that $f_{XY}(x, y; 0) = f_X(x)f_Y(y)$, which indicates that uncorrelated Gaussian random variables are also independent.

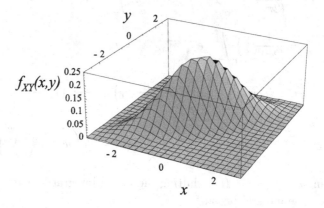

Fig. 4.2. The joint probability density function (4.4) with $\mu_X = \mu_Y = 0$ and $\sigma_X = \sigma_Y = 1$ and $\rho = 0.8$.

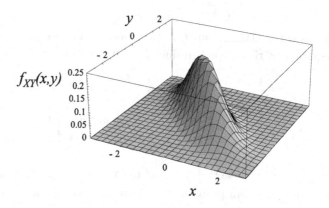

Fig. 4.3. The joint probability density function (4.4) with $\mu_X = \mu_Y = 0$ and $\sigma_X = \sigma_Y = 1$ and $\rho = -0.8$.

If we denote $x^* = \frac{(x-\mu_X)}{\sigma_X}$ and $y^* = \frac{(y-\mu_Y)}{\sigma_Y}$, the bi-variate normal density (4.4) reduces to

$$f_{XY}(x^*, y^*; \rho) = \frac{\exp\left[-\frac{(x^*)^2 - 2\rho x^* y^* + (y^*)^2}{2(1-\rho^2)}\right]}{2\pi\sigma_X\sigma_Y\sqrt{1-\rho^2}}$$

from which we can verify the partial differential equation

$$\frac{\partial f_{XY}(x^*, y^*; \rho)}{\partial \rho} = \frac{\partial f_{XY}(x^*, y^*; \rho)}{\partial x^* \partial y^*} \tag{4.5}$$

and the symmetry relations

$$f_{XY}(x^*, y^*; -\rho) = f_{XY}(-x^*, y^*; \rho) = f_{XY}(x^*, -y^*; \rho) \qquad (4.6)$$

4.1.3 Generation of n correlated Gaussian random variables

Let $\{X_i\}_{1 \le i \le n}$ be a set of n independent normal random variables, where each X_i is distributed as $N(0, 1)$. The vector X is rather easily simulated. The analysis above shows that $E[X] = 0$ (the null-vector) and $\Sigma_X = \text{diag}(\text{Var}[X_i]) = \text{diag}(1) = I$, the identity matrix. We want to generate the correlated normal vector Y with a given mean vector $E[Y]$ and a given covariance matrix Σ_Y. Since linear combinations of normal random variables are normal random variables, we consider the linear transformation $Y = AX + B$ where A and B are constant matrices. We will now determine A and B. First,

$$E[Y] = E[AX] + E[B] = AE[X] + E[B] = E[B]$$

Hence, the matrix B is a vector with components equal to the given components $E[Y_i]$ of the mean vector $E[Y]$. Second,

$$\begin{aligned}
\Sigma_Y &= E\left[(Y - E[Y])(Y - E[Y])^T\right] \\
&= E\left[AX(AX)^T\right] = E\left[AXX^T A^T\right] = AE\left[XX^T\right]A^T \\
&= A\Sigma_X A^T = AA^T
\end{aligned}$$

From the eigenvalue decomposition of $\Sigma_Y = U\text{diag}(\lambda_i)U^T$ with real eigenvalues $\lambda_i \ge 0$ and the fact that $\text{diag}(\lambda_i) = \text{diag}(\sqrt{\lambda_i})\left(\text{diag}(\sqrt{\lambda_i})\right)^T$, we obtain

$$AA^T = U\text{diag}(\sqrt{\lambda_i})\left(\text{diag}(\sqrt{\lambda_i})\right)^T U^T$$

Since any orthogonal matrix W satisfies $W^T W = I$ (see **art. 11**), we find a more general form

$$AA^T = U\text{diag}(\sqrt{\lambda_i})W^T W \left(\text{diag}(\sqrt{\lambda_i})\right)^T U^T$$

from which

$$A = U\text{diag}(\sqrt{\lambda_i})W^T$$

The matrix A is also called the *square root* matrix of Σ_Y, but it is clearly not unique because we can choose any orthogonal matrix W, such as, for example, $W = I$. If $W = U$, we construct a *symmetric* square root matrix $A = A^T = U\text{diag}(\sqrt{\lambda_i})U^T$, so that $\Sigma_Y = A^2$. The matrix A can be found from the singular value decomposition of Σ_Y or from Cholesky factorization (Press *et al.*, 1992).

Example Generate a normal vector Y with $E[Y] = (300, 300)^T$, with standard deviations $\sigma_1 = 106.066$, $\sigma_2 = 35.355$ and correlation $\rho_Y = 0.8$.

Solution: The covariance matrix of Y is obtained using the definition of the linear correlation coefficient (2.61),

$$\Sigma_Y = \begin{pmatrix} \sigma_1^2 & \rho_Y \sigma_1 \sigma_2 \\ \rho_Y \sigma_1 \sigma_2 & \sigma_2^2 \end{pmatrix} = \begin{pmatrix} 11250 & 3000 \\ 3000 & 1250 \end{pmatrix}$$

A square root matrix A of Σ_Y is

$$A = \begin{pmatrix} 63.640 & 84.853 \\ 0 & 35.355 \end{pmatrix}$$

which is readily checked by computing $AA^T = \Sigma_Y$. It remains to generate m independent draws for X_1 and X_2 from a normal distribution with zero mean and unit variance as explained in Section 4.1.1. Each pair (X_1, X_2) out of the m pairs is transformed as $Y = AX + E[Y]$. The result, component Y_2 versus Y_1, is shown in Fig. 4.4.

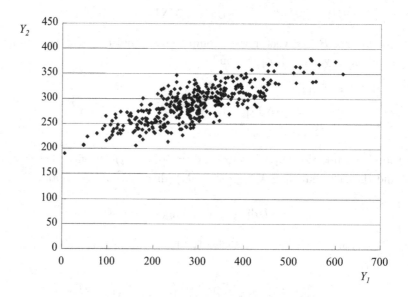

Fig. 4.4. The scatter diagram of the simulated vector Y.

4.2 Generation of correlated random variables

Let us consider the problem of generating two correlated random variables X and Y with given distribution functions F_X and F_Y. The correlation is expressed in terms of the *linear* correlation coefficient $\rho(X, Y) = \rho$ defined in (2.61). The need to generate correlated random variables often occurs in simulations. For example, as shown in Kuipers and Van Mieghem (2003), correlations in the link weight structure may significantly increase the computational complexity of multi-constrained routing, called in brief QoS routing. The importance of measures of dependence between quantities in risk and finance is discussed in Embrechts *et al.* (2001b).

In general, given the distribution functions F_X and F_Y, not all linear correlations from $-1 \leq \rho \leq 1$ are possible. Indeed, let X and Y be positive real random variables with infinite range which means that $F_X(x) = 1$ if $x \to \infty$ and that $F_X(x) = F_Y(x) = 0$ for $x < 0$. Consider $Y = aX + b$ with $a < 0$ and $b \geq 0$. For all finite $y < 0$,

$$F_Y(y) = \Pr\left[Y \leq y\right] = \Pr\left[aX + b \leq y\right] = \Pr\left[X \geq \frac{y-b}{a}\right] \geq \Pr\left[X > \frac{y-b}{a}\right]$$

$$= 1 - F_X\left(\frac{y-b}{a}\right) > 0$$

which contradicts the fact that $F_Y(y) = 0$ for $y < 0$. Hence, positive random variables with infinite range cannot be correlated with $\rho = -1$. The requirement that the range needs to be unbounded is necessary because two uniform random variables on $[0,1]$, U_1 and U_2, are negatively correlated with $\rho = -1$ if $U_1 = 1 - U_2$.

In summary, the set of all possible correlations is a closed interval $[\rho_{\min}, \rho_{\max}]$ for which $\rho_{\min} < 0 < \rho_{\max}$. The precise computation of $\rho_{\min}$ and $\rho_{\max}$ is, in general, difficult, as shown below.

4.3 The non-linear transformation method

The non-linear transformation approach starts from a given set of two random variables X_1 and X_2 that have a correlation coefficient $\rho_X \in [-1, 1]$. If the joint distribution function

$$F_{X_1 X_2}(x_1, x_2; \rho_X) = \Pr\left[X_1 \leq x_1, X_2 \leq x_2\right] = \int_{-\infty}^{x_1} \int_{-\infty}^{x_2} f_{X_1 X_2}(u, v; \rho_X) du dv$$

is known, the marginal distribution follows from (2.63) as

$$\Pr\left[X_1 \leq x_1\right] = \int_{-\infty}^{x_1} \int_{-\infty}^{\infty} f_{X_1 X_2}(u, v; \rho_X) du dv$$

Since for any random variable X, it holds that $F_X(X) = U$, where U is a uniform random variable on $[0, 1]$, it follows that $U_1 = F_{X_1}(X_1)$ and $U_2 = F_{X_2}(X_2)$ are uniformly correlated random variables with correlation coefficient ρ_U. As shown in Section 3.2.1, if U is a uniform random variable on $[0,1]$, any other random variable Y with distribution function $g(x)$ can be constructed as $g^{-1}(U)$. By combining the two transforms, we can generate $Y_1 = g_1^{-1}(F_{X_1}(X_1))$ and $Y_2 = g_2^{-1}(F_{X_2}(X_2))$ that are correlated because X_1 and X_2 are correlated. It may be possible to construct directly the correlated random variables $Y_1 = T_1(X_1)$ and $Y_2 = T_2(X_2)$ if the transforms T_1 and T_2 are known. More generally, if $X = (X_1, X_2, \ldots, X_n)$ is a random vector with continuous marginal distributions $F_{X_k}(x)$, then the random vector

$$U = (F_{X_1}(X_1), F_{X_2}(X_2), \ldots, F_{X_n}(X_n))$$

has uniform marginal distributions $U_k = F_{X_k}(X_k)$ for $1 \le k \le n$ and the *copula* of X is the joint distribution function

$$\Pr\left[U_1 \le u_1, \ldots, U_n \le u_n\right] = \Pr\left[X_1 \le F_{X_1}^{-1}(u_1), \ldots, X_n \le F_{X_n}^{-1}(u_n)\right]$$

In the sequel, we limit ourselves to $n = 2$ and refer to Nelsen (2006) for the theory and statistical applications of copulae.

The goal is to determine the linear correlation coefficient ρ_Y defined in (2.61),

$$\rho_Y = \frac{E[Y_1 Y_2] - E[Y_1] E[Y_2]}{\sqrt{\text{Var}[Y_1]}\sqrt{\text{Var}[Y_2]}}$$

as a function of ρ_X. Using (2.64),

$$E[Y_1 Y_2] = E\left[g_1^{-1}(F_{X_1}(X_1)) g_2^{-1}(F_{X_2}(X_2))\right]$$

$$= \int_{-\infty}^{\infty} \int_{-\infty}^{\infty} g_1^{-1}(F_{X_1}(u)) g_2^{-1}(F_{X_2}(v)) f_{X_1 X_2}(u, v; \rho_X) du dv \qquad (4.7)$$

This relation shows that ρ_Y is a continuous function in ρ_X and that the joint distribution function of X_1 and X_2 is needed. The main difficulty lies now in the computation of the integral appearing in $E[Y_1 Y_2]$. For X_1 and X_2, Gaussian correlated random variables are most often chosen because an exact analytic expression (4.4) exists for the joint distribution function $F_{X_2 X_2}(x_1, x_2; \rho_X)$.

4.3.1 Properties of ρ_Y as a function of ρ_X

From now on, we choose Gaussian correlated random variables for X_1 and X_2.

Theorem 4.3.1 *The correlation coefficient ρ_Y is a differentiable and increasing function of ρ_X.*

Proof: From the partial differential equation (4.5) of $f_{X_1 X_2}(u, v; \rho_X)$, it follows that

$$\frac{\partial E[Y_1 Y_2]}{\partial \rho_X} = \int_{-\infty}^{\infty} \int_{-\infty}^{\infty} g_1^{-1}(F_{X_1}(u)) g_2^{-1}(F_{X_2}(v)) \frac{\partial^2 f_{X_1 X_2}(u, v; \rho_X)}{\partial u \partial v} du dv$$

Partial integration with respect to u and v yields

$$\frac{\partial E[Y_1 Y_2]}{\partial \rho_X} = \int_{-\infty}^{\infty} \int_{-\infty}^{\infty} \frac{d g_1^{-1}(F_{X_1}(u))}{du} \frac{d g_2^{-1}(F_{X_2}(v))}{du} f_{X_1 X_2}(u, v; \rho_X) du dv$$

Applying the chain rule for differentiation and $\frac{d g^{-1}(x)}{dx} = \frac{1}{g'(g^{-1}(x))}$ gives

$$\frac{d g^{-1}(F_X(u))}{du} = \left.\frac{d g^{-1}(x)}{dx}\right|_{x = F_X(u)} \frac{dx}{du} = \frac{f_X(u)}{g'(g^{-1}(F_X(u)))}$$

Since $g'(x)$ and $f_X(u)$ are probability density functions and positive, $\frac{\partial E[Y_1 Y_2]}{\partial \rho_X} = \frac{\partial \rho_Y}{\partial \rho_X} > 0$. Hence, we have shown that ρ_Y is a differentiable, increasing function of ρ_X. □

Since $\rho_X \in [-1, 1]$, ρ_Y increases from $\rho_{Y\min}$ at $\rho_X = -1$ to $\rho_{Y\max}$ corresponding to $\rho_X = 1$. In the sequel, we will derive expressions to compute the boundary cases $\rho_X = -1$ and $\rho_X = 1$.

Theorem 4.3.2 (of Lancaster) *For any two strictly increasing real functions T_1 and T_2 that transform the correlated Gaussian random variables X_1 and X_2 to the correlated random variables $Y_1 = T_1(X_1)$ and $Y_2 = T_2(X_2)$, it holds that*

$$|\rho_Y| \le |\rho_X|$$

If two correlated random variables Y_1 and Y_2 can be obtained by separate transformations from a bi-variate normal distribution with correlation coefficient ρ_X, the correlation coefficient ρ_Y of the transformed random variables cannot in absolute value exceed ρ_X. The interest of the proof is that it uses powerful properties of orthogonal polynomials and that ρ_Y is expanded in a power series in ρ_X in (4.12).

Proof: The proof is based on the orthogonal Hermite polynomials $H_n(x)$ (see e.g. Rainville (1960) and Abramowitz and Stegun (1968, Chapter 22)) defined by the generating function

$$\exp\left(2xt - t^2\right) = \sum_{n=0}^{\infty} \frac{H_n(x)\, t^n}{n!} \tag{4.8}$$

After expanding $\exp\left(2xt - t^2\right)$ in a Taylor series and equating corresponding powers in t, we find that

$$H_n(x) = n! \sum_{k=0}^{\left[\frac{n}{2}\right]} \frac{(-1)^k\, (2x)^{n-2k}}{k!\,(n-2k)!} \tag{4.9}$$

with $H_0(x) = 1$. The Hermite polynomials satisfy the orthogonality relations

$$\int_{-\infty}^{\infty} e^{-x^2} H_n(x)\, H_m(x)\, dx = 0 \qquad m \ne n$$

$$\int_{-\infty}^{\infty} e^{-x^2} H_n^2(x)\, dx = 2^n n! \sqrt{\pi}$$

These orthogonality relations enable us to expand functions in terms of Hermite polynomials (similar to Fourier analysis). If the expansion of a function $f(x)$,

$$f(x) = \sum_{k=0}^{\infty} a_k H_k(x)$$

converges for all x, then it follows from the orthogonality relations that

$$a_k = \frac{1}{2^k k! \sqrt{\pi}} \int_{-\infty}^{\infty} e^{-x^2} f(x)\, H_k(x)\, dx$$

The joint normalized Gaussian density function can be expanded (Rainville, 1960, pp. 197–198) in terms of Hermite polynomials

$$\frac{\exp\left[-\frac{x^2 - 2\rho xy + y^2}{(1-\rho^2)}\right]}{\sqrt{1-\rho^2}} = e^{-x^2 - y^2} \sum_{n=0}^{\infty} H_n(x)\, H_n(y)\, \frac{\rho^n}{2^n n!} \tag{4.10}$$

In order for the covariance $\mathrm{Cov}[Y_1 Y_2]$ to exist, both $E\left[Y_1^2\right]$ and $E\left[Y_2^2\right]$ must be finite. Since $Y_j = T_j(X_j)$ for $j = 1, 2$, the mean is

$$E[Y_j] = \int_{-\infty}^{\infty} T_j(x) f_{X_j}(x)\, dx = \frac{1}{\sqrt{2\pi}\sigma_X} \int_{-\infty}^{\infty} T_j(x) \exp\left[-\frac{(x-\mu_X)^2}{2\sigma_X^2}\right] dx$$

$$= \frac{1}{\sqrt{\pi}} \int_{-\infty}^{\infty} T_j(\mu_X + \sqrt{2}\sigma_X u) e^{-u^2}\, du$$

Let

$$T_j \left(\mu_X + \sqrt{2}\sigma_X u \right) = \sum_{k=0}^{\infty} a_{k;j} H_k(u) \tag{4.11}$$

with

$$a_{k;j} = \frac{1}{2^k k! \sqrt{\pi}} \int_{-\infty}^{\infty} e^{-x^2} T_j \left(\mu_X + \sqrt{2}\sigma_X x \right) H_k(x) \, dx$$

then, since $H_0(x) = 1$,

$$E[Y_j] = a_{0;j}$$

The second moment $E\left[Y_j^2\right]$ follows from (2.35) as

$$E\left[Y_j^2\right] = \int_{-\infty}^{\infty} (T_j(x))^2 f_{X_j}(x) \, dx = \frac{1}{\sqrt{2\pi}\sigma_X} \int_{-\infty}^{\infty} T_j^2(x) \exp\left[-\frac{(x-\mu_X)^2}{2\sigma_X^2}\right] dx$$

$$= \frac{1}{\sqrt{\pi}} \int_{-\infty}^{\infty} T_j^2(\mu_X + \sqrt{2}\sigma_X u) e^{-u^2} \, du$$

Substituting (4.11) gives

$$E\left[Y_j^2\right] = \sum_{k=0}^{\infty} \sum_{m=0}^{\infty} a_{m;j} a_{k;j} \frac{1}{\sqrt{\pi}} \int_{-\infty}^{\infty} H_k(u) H_m(u) e^{-u^2} \, du = \sum_{k=0}^{\infty} a_{k;j}^2 2^k k!$$

which is convergent. Similarly, using (4.4),

$$E[Y_1 Y_2] = \int_{-\infty}^{\infty} \int_{-\infty}^{\infty} T_1(u) T_2(v) f_{X_1 X_2}(u,v;\rho_X) \, du \, dv$$

$$= \int_{-\infty}^{\infty} \int_{-\infty}^{\infty} T_1(u) T_2(v) \frac{\exp\left[-\dfrac{\dfrac{(u-\mu_X)^2}{\sigma_X^2} - \dfrac{2\rho_X}{\sigma_X \sigma_Y}(u-\mu_X)(v-\mu_Y) + \dfrac{(v-\mu_Y)^2}{\sigma_Y^2}}{2(1-\rho_X^2)}\right]}{2\pi \sigma_X \sigma_Y \sqrt{1-\rho^2}} \, du \, dv$$

$$= \int_{-\infty}^{\infty} \int_{-\infty}^{\infty} T_1 \left(\mu_X + \sqrt{2}\sigma_X x \right) T_2 \left(\mu_Y + \sqrt{2}\sigma_Y y \right) \frac{\exp\left[-\dfrac{x^2 - 2\rho_X xy + y^2}{(1-\rho_X^2)}\right]}{\pi\sqrt{1-\rho_X^2}} \, dx \, dy$$

$$= \frac{1}{\pi\sqrt{1-\rho_X^2}} \sum_{k=0}^{\infty} a_{k;1} \sum_{m=0}^{\infty} a_{m;2} \int_{-\infty}^{\infty} \int_{-\infty}^{\infty} H_k(x) H_m(y) \exp\left[-\frac{x^2 - 2\rho_X xy + y^2}{(1-\rho_X^2)}\right] dx \, dy$$

Using (4.10),

$$E[Y_1 Y_2] = \frac{1}{\pi} \sum_{k=0}^{\infty} a_{k;1} \sum_{m=0}^{\infty} a_{m;2} \sum_{n=0}^{\infty} \frac{\rho_X^n}{2^n n!} \int_{-\infty}^{\infty} e^{-x^2} H_k(x) H_n(x) \, dx \int_{-\infty}^{\infty} e^{-y^2} H_m(y) H_n(y) \, dy$$

Introducing the orthogonality relations for Hermite polynomials leads to

$$E[Y_1 Y_2] = \sum_{n=0}^{\infty} a_{n;1} a_{n;2} 2^n n! \rho_X^n$$

The correlation coefficient becomes

$$\rho_Y = \frac{\sum_{n=0}^{\infty} a_{n;1} a_{n;2} 2^n n! \rho_X^n - a_{0;1} a_{0;2}}{\sqrt{\sum_{k=0}^{\infty} a_{k;1}^2 2^k k! - a_{0;1}^2} \sqrt{\sum_{k=1}^{\infty} a_{k;2}^2 2^k k! - a_{0;2}^2}} = \frac{\sum_{n=1}^{\infty} a_{n;1} a_{n;2} 2^n n! \rho_X^n}{\sqrt{\sum_{k=1}^{\infty} a_{k;1}^2 2^k k! \sum_{k=1}^{\infty} a_{k;2}^2 2^k k!}}$$

Denote $\alpha_n = a_{n;1}\sqrt{2^n n!}$ and $\beta_n = a_{n;2}\sqrt{2^n n!}$, then $\text{Var}[Y_1] = \sum_{k=1}^{\infty} \alpha_k^2$ and $\text{Var}[Y_2] = \sum_{k=1}^{\infty} \beta_k^2$. Since the linear correlation coefficient $\rho(X, Y)$ equals the correlation coefficient of the corresponding normalized random variable with mean zero and variance 1, as shown in Section 2.5.3, we may choose $\sum_{k=1}^{\infty} \alpha_k^2 = \sum_{k=1}^{\infty} \beta_k^2 = 1$ such that

$$\rho_Y = \sum_{n=1}^{\infty} \alpha_n \beta_n \rho_X^n \tag{4.12}$$

If $\alpha_1^2 = 1$ and $\beta_1^2 = 1$, then $|\rho_Y| = |\rho_X|$ because all other α_k and β_k must then vanish. In all other cases, either $\alpha_1^2 < 1$ or $\beta_1^2 < 1$ or both, such that

$$\rho_Y = \alpha_1 \beta_1 \rho_X + \sum_{n=2}^{\infty} \alpha_n \beta_n \rho_X^n$$

and

$$\left| \sum_{n=2}^{\infty} \alpha_n \beta_n \rho_X^n \right| \leq \sum_{n=2}^{\infty} |\alpha_n \beta_n| |\rho_X|^n \leq \sqrt{\sum_{n=2}^{\infty} \alpha_n^2 |\rho_X|^n} \sqrt{\sum_{n=2}^{\infty} \beta_n^2 |\rho_X|^n}$$

where we have used the Cauchy-Schwarz inequality $\sum ab \leq \sqrt{\sum a^2} \sqrt{\sum b^2}$ (see Section 5.5). By partial summation,

$$\sum_{n=2}^{\infty} \alpha_n^2 |\rho_X|^n = (1 - |\rho_X|) \sum_{n=2}^{\infty} \sum_{k=2}^{n} \alpha_k^2 |\rho_X|^n \leq (1 - |\rho_X|) (1 - \alpha_1^2) \frac{|\rho_X|^2}{1 - |\rho_X|} = (1 - \alpha_1^2) |\rho_X|^2$$

because $\sum_{k=2}^{n} \alpha_k^2 < \sum_{k=2}^{\infty} \alpha_k^2 = 1 - \alpha_1^2$. Thus

$$|\rho_Y| \leq |\rho_X| \left(|\alpha_1 \beta_1| + \sqrt{1 - \alpha_1^2} \sqrt{1 - \beta_1^2} |\rho_X| \right)$$

Finally, for $\alpha_1^2 \leq 1$ and $\beta_1^2 \leq 1$, the inequality $|\alpha_1 \beta_1| + \sqrt{1 - \alpha_1^2} \sqrt{1 - \beta_1^2} \leq 1$ holds. This proves Lancaster's theorem because $|\rho_X| \leq 1$. $\qquad\square$

4.3.2 Boundary cases

Let us investigate some cases for special values of ρ_X.

1. $\rho_X = 0$. Since uncorrelated Gaussian random variables ($\rho_X = 0$) are independent, also $Y_1 = g_1^{-1}(F_{X_1}(X_1))$ and $Y_2 = g_2^{-1}(F_{X_2}(X_2))$ are independent such that $\rho_Y = 0$. Hence, uncorrelated Gaussian random variables with $\rho_X = 0$ lead to uncorrelated random variables Y_1 and Y_2 with $\rho_Y = 0$.

2. $\rho_X = 1$. Perfect positively correlated Gaussian random variables $X_1 = X_2 = X$ have joint distribution

$$f_{X_1 X_2}(u, v; 1) = \frac{1}{\sqrt{2\pi}\sigma_X} \exp\left[-\frac{(u - \mu_X)^2}{2\sigma_X^2} \right] \delta(u - v)$$

which follows from $\Pr[X_1 \leq x_1, X_2 \leq x_2] = \Pr[X \leq x_1, X \leq x_2] = \Pr[X \leq x]$ with

$x = \max(x_1, x_2)$. In that case,

$$E[Y_1 Y_2] = \int_{-\infty}^{\infty} g_1^{-1}\left(F_X(u)\right) g_2^{-1}\left(F_X(u)\right) dF_X(u)$$

$$= \int_0^1 g_1^{-1}(x) g_2^{-1}(x) dx \tag{4.13}$$

which may lead to $\rho_{Y\,\max} < 1$ depending on the specifics of g_1 and g_2.

By transforming $x = g_1(u)$, we obtain

$$E[Y_1 Y_2] = \int_{g_1^{-1}(0)}^{g_1^{-1}(1)} u g_2^{-1}\left(g_1(u)\right) g_1'(u) du$$

which shows that, if $g_1 = g_2 = g$,

$$E[Y_1 Y_2] = \int_{g^{-1}(0)}^{g^{-1}(1)} u^2 g'(u) du = E[Y^2]$$

Hence, if Y_1 and Y_2 have the same distribution function g as Y, the case $\rho_X = 1$ leads to

$$\rho_Y = \frac{E[Y^2] - (E[Y])^2}{\mathrm{Var}[Y]} = 1$$

3. $\rho_X = -1$. Perfect negatively correlated Gaussian random variables $X_1 = -X_2 = X$ have joint distribution

$$f_{X_1 X_2}(u, v; -1) = \frac{1}{\sqrt{2\pi}\sigma_X} \exp\left[-\frac{(u - \mu_X)^2}{2\sigma_X^2}\right] \delta(u + v)$$

which follows from the symmetry relations (4.6). In that case,

$$E[Y_1 Y_2] = \int_{-\infty}^{\infty} g_1^{-1}\left(F_X(u)\right) g_2^{-1}\left(F_X(-u)\right) dF_X(u)$$

$$= \int_{-\infty}^{\infty} g_1^{-1}\left(F_X(u)\right) g_2^{-1}\left(1 - F_X(u)\right) dF_X(u)$$

$$= \int_0^1 g_1^{-1}(x) g_2^{-1}(1 - x) dx \tag{4.14}$$

which may lead to $\rho_{Y\,\min} > -1$, depending on the specifics of g_1 and g_2.

4.4 Examples of the non-linear transformation method

4.4.1 Correlated uniform random variables

Let us first focus on the relation between ρ_X and ρ_U. Since $E[U] = \frac{1}{2}$ and $\sigma_U^2 = \frac{1}{12}$, the definition of the linear correlation coefficient (2.61) gives

$$\rho_U = \frac{E[U_1 U_2] - \frac{1}{4}}{\frac{1}{12}}$$

where, using (2.64),

$$E[U_1 U_2] = E[F_{X_1}(X_1) F_{X_2}(X_2)]$$
$$= \int_{-\infty}^{\infty} \int_{-\infty}^{\infty} F_{X_1}(u) F_{X_2}(v) f_{X_1 X_2}(u, v; \rho_X) du dv$$

In the case of Gaussian correlated random variables specified by (3.20) and (4.4), we must evaluate the integral

$$E[U_1 U_2] = \int_{-\infty}^{\infty} du \int_{-\infty}^{\infty} dv \int_{-\infty}^{u} dt\, e^{-\frac{(t-\mu_{X_1})^2}{2\sigma_{X_1}^2}} \int_{-\infty}^{v} d\tau\, e^{-\frac{(\tau-\mu_{X_2})^2}{2\sigma_{X_2}^2}}$$

$$\times \frac{\exp\left[-\dfrac{\frac{(u-\mu_{X_1})^2}{\sigma_{X_1}^2} - \frac{2\rho_X}{\sigma_{X_1}\sigma_{X_2}}(u-\mu_{X_1})(v-\mu_{X_2}) + \frac{(v-\mu_{X_2})^2}{\sigma_{X_2}^2}}{2(1-\rho_X^2)}\right]}{(2\pi)^2 \sigma_{X_1}^2 \sigma_{X_2}^2 \sqrt{1-\rho_X^2}}$$

Substituting successively $u' = \frac{u-\mu_{X_1}}{\sigma_{X_1}}$, $t' = \frac{t-\mu_{X_1}}{\sigma_{X_1}}$, $v' = \frac{v-\mu_{X_2}}{\sigma_{X_2}}$, $\tau' = \frac{\tau-\mu_{X_2}}{\sigma_{X_2}}$, we obtain

$$E[U_1 U_2] = \frac{1}{(2\pi)^2} \int_{-\infty}^{\infty} du' \int_{-\infty}^{\infty} dv' \int_{-\infty}^{u'} e^{-\frac{t'^2}{2}} dt' \int_{-\infty}^{v'} e^{-\frac{\tau'^2}{2}} d\tau' \frac{\exp\left[-\frac{u'^2 - 2\rho_X u'v' + v'^2}{2(1-\rho_X^2)}\right]}{\sqrt{1-\rho_X^2}}$$

We now use the partial differential equation (4.5),

$$\frac{\partial}{\partial \rho}\left(\frac{\exp\left[-\frac{u'^2 - 2\rho_X u'v' + v'^2}{2(1-\rho_X^2)}\right]}{\sqrt{1-\rho_X^2}}\right) = \frac{\partial^2}{\partial u' \partial v'}\left(\exp\left[-\frac{u'^2 - 2\rho_X u'v' + v'^2}{2(1-\rho_X^2)}\right]\right)\frac{1}{\sqrt{1-\rho_X^2}}$$

such that

$$\frac{\partial E[U_1 U_2]}{\partial \rho} = \int_{-\infty}^{\infty} du' \int_{-\infty}^{\infty} dv' \int_{-\infty}^{u'} e^{-\frac{t'^2}{2}} dt' \int_{-\infty}^{v'} e^{-\frac{\tau'^2}{2}} d\tau' \frac{\frac{\partial^2}{\partial u' \partial v'}\left(\exp\left[-\frac{u'^2 - 2\rho_X u'v' + v'^2}{2(1-\rho_X^2)}\right]\right)}{(2\pi)^2 \sqrt{1-\rho_X^2}}$$

Partial integration in the last integral v'

$$I_2 = \int_{-\infty}^{\infty} dv' \int_{-\infty}^{v'} e^{-\frac{\tau'^2}{2}} d\tau' \frac{\partial^2}{\partial u' \partial v'}\left(\exp\left[-\frac{u'^2 - 2\rho_X u'v' + v'^2}{2(1-\rho_X^2)}\right]\right)$$

$$= \int_{-\infty}^{\infty} dv\, e^{-\frac{v^2}{2}} \frac{\partial}{\partial u'}\left(\exp\left[-\frac{u'^2 - 2\rho_X u'v + v^2}{2(1-\rho_X^2)}\right]\right)$$

yields

$$\frac{\partial E[U_1 U_2]}{\partial \rho} = \int_{-\infty}^{\infty} dv\, e^{-\frac{v^2}{2}} \int_{-\infty}^{\infty} du' \int_{-\infty}^{u'} e^{-\frac{t'^2}{2}} dt' \frac{\frac{\partial}{\partial u'}\left(\exp\left[-\frac{u'^2 - 2\rho_X u'v' + v'^2}{2(1-\rho_X^2)}\right]\right)}{(2\pi)^2 \sqrt{1-\rho_X^2}}$$

and similarly in the u' integral,

$$
\frac{\partial E\left[U_1 U_2\right]}{\partial \rho}=\frac{1}{(2\pi)^2 \sqrt{1-\rho_X^2}} \int_{-\infty}^{\infty} dv \int_{-\infty}^{\infty} du \exp\left[-\frac{u^2\left(2-\rho_X^2\right)-2\rho_X uv + v^2\left(2-\rho_X^2\right)}{2\left(1-\rho_X^2\right)}\right]
$$

$$
=\int_{-\infty}^{\infty} dv \frac{\exp\left[-\left(\frac{\left(4-\rho_X^2\right)}{2\left(2-\rho_X^2\right)}\right) v^2\right]}{(2\pi)^2 \sqrt{1-\rho_X^2}} \int_{-\infty}^{\infty} du \exp\left[-\frac{\left(2-\rho_X^2\right)}{2\left(1-\rho_X^2\right)}\left(u-\frac{\rho_X v}{\left(2-\rho_X^2\right)}\right)^2\right]
$$

$$
=\frac{1}{(2\pi)^2 \sqrt{1-\rho_X^2}} \sqrt{\frac{2\pi\left(2-\rho_X^2\right)}{\left(4-\rho_X^2\right)}} \sqrt{\frac{2\pi\left(1-\rho_X^2\right)}{\left(2-\rho_X^2\right)}} = \frac{1}{2\pi}\frac{1}{\sqrt{4-\rho_X^2}}
$$

Thus, we find that

$$
\frac{\partial \rho_U}{\partial \rho_X}=\frac{6}{\pi}\frac{1}{\sqrt{4-\rho_X^2}}
$$

or that

$$
\rho_U=\frac{6}{\pi}\int \frac{1}{\sqrt{4-\rho_X^2}}+c=\frac{6}{\pi}\arcsin\left(\frac{\rho_X}{2}\right)+c
$$

It remains to determine the constant c. We have shown in Section 4.3.2 that random variables generated from uncorrelated Gaussian random variables are also uncorrelated implying that $\rho_U = 0$ if $\rho_X = 0$ and, hence, that the constant $c = 0$. This finally results in

$$
\rho_U=\frac{6}{\pi}\arcsin\left(\frac{\rho_X}{2}\right) \tag{4.15}
$$

In summary, two uniform correlated random variables U_1 and U_2 with correlation coefficient ρ_U are found by transforming two Gaussian correlated random variables X_1 and X_2 with correlation coefficient $\rho_X = 2\sin\left(\frac{\pi\rho_U}{6}\right)$. Equation (4.15) further shows that $\rho_U = \pm 1$ if $\rho_X = \pm 1$, which indicates that the whole range of the correlation coefficient ρ_U is possible.

4.4.2   *Correlated exponential random variables*

In Section 3.2.1, we have seen that, if U is a uniform random variable on $[0,1]$, $g^{-1}(U) = -\frac{1}{\alpha}\log U$ is an exponential random variable with mean $\frac{1}{\alpha}$. The correlation coefficient for two exponential random variables, Y_1 and Y_2, with mean $\frac{1}{\alpha_1}$ and $\frac{1}{\alpha_2}$ respectively, is

$$
\rho_Y=\frac{E\left[Y_1 Y_2\right]-\frac{1}{\alpha_1 \alpha_2}}{\frac{1}{\alpha_1 \alpha_2}}=E\left[\alpha_1 \alpha_2 Y_1 Y_2\right]-1
$$

As above, we generate $Y_1 = -\frac{1}{\alpha_1}\log F_{X_1}(X_1)$ and $Y_2 = -\frac{1}{\alpha_2}\log F_{X_2}(X_2)$, where X_1 and X_2 are correlated Gaussian random variables with correlation coefficient

ρ_X. Then,

$$E\left[\alpha_1\alpha_2 Y_1 Y_2\right] = \frac{\alpha_1\alpha_2}{\alpha_1\alpha_2} E\left[\log F_{X_1}(X_1)\log F_{X_2}(X_2)\right]$$

$$= \int_{-\infty}^{\infty}\int_{-\infty}^{\infty}\log F_{X_1}(u)\log F_{X_2}(v) f_{X_1 X_2}(u,v;\rho_X)\, du\, dv$$

In the general case for $\rho_X \neq 0$, the previous method can be followed, which yields after substitution towards normalized variables,

$$E\left[\alpha_1\alpha_2 Y_1 Y_2\right] = \int_{-\infty}^{\infty} du \int_{-\infty}^{\infty} dv \log\left(\int_{-\infty}^{u} e^{-\frac{t^2}{2}}\, dt\right)\log\left(\int_{-\infty}^{v} e^{-\frac{\tau^2}{2}}\, d\tau\right)\frac{\exp\left[-\frac{u^2-2\rho_X uv+v^2}{2(1-\rho_X^2)}\right]}{(2\pi)^2\sqrt{1-\rho_X^2}}$$

Unfortunately, we cannot evaluate this integral analytically.

Let us compute the upper bound $\rho_{Y\max}$ from (4.13) with $g_1^{-1}(x) = -\frac{1}{\alpha_1}\log x$ and $g_2^{-1}(x) = -\frac{1}{\alpha_2}\log x$,

$$E\left[\alpha_1\alpha_2 Y_1 Y_2; \rho_X = 1\right] = \int_0^1 \log^2 x\, dx = 2$$

and thus $\rho_{Y\max} = 1$. The lower boundary $\rho_{Y\min}$ follows from (4.14) as[2],

$$E\left[\alpha_1\alpha_2 Y_1 Y_2; \rho_X = -1\right] = \int_0^1 \log x \log(1-x)\, dx = 2 - \frac{\pi^2}{6}$$

Here, we find $\rho_{Y\min} = 1 - \frac{\pi^2}{6} = -0.644\,934....$

In summary, exponential correlated random variables can be generated from Gaussian correlated random variables, but the correlation coefficient ρ_Y is limited to the interval $\left[1 - \frac{\pi^2}{6}, 1\right]$. As explained in the introduction of Section 4.2, the exponential random variables are positive with infinite range for which not all negative correlations are possible. The analysis demonstrates that it is not possible to construct two exponential random variables with correlation coefficient smaller than $\rho_{Y\min} = 1 - \frac{\pi^2}{6} \simeq -0.645$.

---

[2] Substituting the Taylor expansion $\log(1-x) = -\sum_{k=1}^{\infty}\frac{x^k}{k}$ gives

$$\int_0^1 \log x \log(1-x)\, dx = -\sum_{k=1}^{\infty}\frac{1}{k}\int_0^1 x^k \log x\, dx$$

With $\int_0^1 x^k \log x\, dx = -\int_0^\infty e^{-(k+1)u} u\, du = -\frac{1}{(k+1)^2}$, we obtain

$$\int_0^1 \log x \log(1-x)\, dx = \sum_{k=1}^{\infty}\frac{1}{k(k+1)^2}$$

Since $\frac{1}{k(k+1)^2} = \frac{1}{k} - \frac{1}{k+1} - \frac{1}{(k+1)^2}$,

$$\int_0^1 \log x \log(1-x)\, dx = \sum_{k=1}^{\infty}\frac{1}{k} - \sum_{k=2}^{\infty}\frac{1}{k} - \sum_{k=2}^{\infty}\frac{1}{k^2} = 1 - \sum_{k=2}^{\infty}\frac{1}{k^2} = 2 - \sum_{k=1}^{\infty}\frac{1}{k^2} = 2 - \zeta(2) = 2 - \frac{\pi^2}{6}$$

4.4.3 Correlated lognormal random variables

Two correlated lognormal random variables Y_1 and Y_2 with distribution specified in (3.53) can be constructed directly from two correlated Gaussian random variables X_1 and X_2. In particular, let $Y_1 = e^{a_1 X_1}$ and $Y_2 = e^{a_2 X_2}$. The explicit scaling parameters can be used to determine the desired mean. From (4.7),

$$E\left[Y_1 Y_2\right] = \int_{-\infty}^{\infty} \int_{-\infty}^{\infty} e^{a_1 u} e^{a_2 v} f_{X_1 X_2}(u, v; \rho_X)\, du\, dv$$

$$= \exp\left[a_1 \mu_1 + a_2 \mu_2 + \frac{\sigma_1^2}{2} a_1^2 + a_1 a_2 \rho_X \sigma_1 \sigma_2 + \frac{\sigma_2^2}{2} a_2^2\right]$$

where the Laplace transform (4.2) for $n = 2$ has been used. Invoking (3.56) and (3.57) with $\mu_j \to a_j \mu_j$ and $\sigma_j^2 \to a_j^2 \sigma_j^2$, the correlation coefficient ρ_Y is

$$\rho_Y = \frac{e^{a_1 \sigma_1 a_2 \sigma_2 \rho_X} - 1}{\sqrt{\left(e^{a_1^2 \sigma_1^2} - 1\right)\left(e^{a_2^2 \sigma_2^2} - 1\right)}} \tag{4.16}$$

If at least one (but not all) of the quantities σ_1, σ_2, a_1 or a_2 grows large, ρ_Y tends to zero irrespective of ρ_X. Thus even if X_1 and X_2 and, hence also Y_1 and Y_2, have the strongest kind of dependence possible, i.e. $\rho_X = \pm 1$, the correlation coefficient ρ_Y can be made arbitrarily small. In case $a_1 \sigma_1 = a_2 \sigma_2 = \sigma$, (4.16) reduces to

$$\rho_Y = \frac{e^{\sigma^2 \rho_X} - 1}{e^{\sigma^2} - 1}$$

We observe that $\rho_{Y\,\max} = 1$, while $\rho_{Y\,\min} = -e^{-\sigma^2} > -1$ for $\sigma > 0$; again drawing that for positive random variables with infinite range not all negative correlations are possible.

4.5 Linear combination of independent auxiliary random variables

In spite of the generality of the non-linear transformation method, the involved computational difficulty suggests us to investigate simpler methods of construction. It is instructive to consider two independent random variables V and W with known probability generating functions $\varphi_V(z)$ and $\varphi_W(z)$ respectively. In the discussion of the uniform random variable in Section 3.2.1, it was shown how to generate by computer an arbitrary random variable from a uniform random variable. We thus assume that V and W can be constructed. Let us now write X and Y as a linear combination of V and W,

$$X = a_{11} V + a_{12} W + b_1$$
$$Y = a_{21} V + a_{22} W + b_2$$

which is specified by the matrix

$$A = \begin{bmatrix} a_{11} & a_{12} \\ a_{21} & a_{22} \end{bmatrix}$$

and compute the covariance defined in (2.59),

$$\text{Cov}\,[X,Y] = E\,[XY] - \mu_X \mu_Y$$
$$= a_{11}a_{21}\left[E\,[V^2] - (E\,[V])^2\right] + (a_{11}a_{22} + a_{12}a_{21})\,E\,[VW]$$
$$- (a_{11}a_{22} + a_{12}a_{21})\,E[V]\,E[W] + a_{12}a_{22}\left[E[W^2] - (E[W])^2\right]$$

Since V and W are independent, $E\,[VW] = E\,[V]\,E\,[W]$, and with the definition of the variance (2.16) and denoting $\sigma_V^2 = \text{Var}[V]$ and similarly for W, we obtain

$$\text{Cov}\,[X,Y] = a_{11}a_{21}\sigma_V^2 + a_{12}a_{22}\sigma_W^2$$

In the same way, we find

$$\sigma_X^2 = a_{11}^2\sigma_V^2 + a_{12}^2\sigma_W^2$$
$$\sigma_Y^2 = a_{21}^2\sigma_V^2 + a_{22}^2\sigma_W^2 \qquad (4.17)$$

such that the correlation coefficient, in general, becomes

$$\rho = \frac{a_{11}a_{21}\sigma_V^2 + a_{12}a_{22}\sigma_W^2}{\sqrt{a_{11}^2\sigma_V^2 + a_{12}^2\sigma_W^2}\,\sqrt{a_{21}^2\sigma_V^2 + a_{22}^2\sigma_W^2}}$$

which is independent of the constants b_1 and b_2 since for a centered moment $E\,(X - E\,[X])^2 = E\,(X + b - E\,[X + b])^2$.

In order to achieve our goal of constructing two correlated random variables X and Y, we can choose the coefficients of the matrix A to obtain an expression as simple as possible. If we choose $X = V$ or $a_{11} = 1$, $a_{12} = b_1 = 0$, the correlation coefficient reduces to

$$\rho = \frac{a_{21}\sigma_X^2}{\sqrt{\sigma_X^2}\,\sqrt{a_{21}^2\sigma_X^2 + a_{22}^2\sigma_W^2}} = \frac{1}{\sqrt{1 + \frac{a_{22}^2}{a_{21}^2}\frac{\sigma_W^2}{\sigma_X^2}}}$$

By rewriting this relation, we obtain

$$a_{21} = \pm\frac{\rho\sigma_W}{\sigma_X}\frac{a_{22}}{\sqrt{1 - \rho^2}}$$

If we choose $a_{22} = \sqrt{1 - \rho^2}$, the random variables X and Y are specified as

$$X = V$$
$$Y = \pm\frac{\rho\sigma_W}{\sigma_X}V + \sqrt{1 - \rho^2}W + b_2$$

and the corresponding variances (4.17) are $\sigma_X^2 = \sigma_V^2$ and $\sigma_Y^2 = \sigma_W^2$. Finally, we require that $E\,[W] = \mu_W = 0$, which specifies

$$b_2 = E\,[Y] \mp \frac{\rho\sigma_Y}{\sigma_X}E\,[X]$$

If W is a zero mean random variable with standard deviation $\sigma_W = \sigma_Y$, the random variables X and Y are correlated with correlation coefficient ρ

$$Y = \pm \frac{\rho \sigma_Y}{\sigma_X} X + \sqrt{1 - \rho^2} W + \mu_Y \mp \frac{\rho \sigma_Y}{\sigma_X} \mu_X \qquad (4.18)$$

In the sequel, we take the positive sign for ρ.

Let us now investigate what happens with the distribution functions of X and Y. Using the pgfs for continuous random variables $\varphi_X(z) = E\left[e^{-zX}\right]$ and $\varphi_Y(z) = E\left[e^{-zY}\right]$, the last relation (4.18) becomes

$$E\left[e^{-zY}\right] = e^{-z\left(\mu_Y - \frac{\rho\sigma_Y}{\sigma_X}\mu_X\right)} E\left[e^{-z\frac{\rho\sigma_Y}{\sigma_X}X}\right] E\left[e^{-z\sqrt{1-\rho^2}W}\right]$$

because $V = X$ and W are independent, or

$$\varphi_Y(z) = e^{-z\left(\mu_Y - \frac{\rho\sigma_Y}{\sigma_X}\mu_X\right)} \varphi_X\left(\frac{\rho\sigma_Y}{\sigma_X}z\right) \varphi_W\left(\sqrt{1-\rho^2}z\right) \qquad (4.19)$$

In order to produce two random variables X and Y that are correlated with correlation coefficient ρ, the pgf of the zero mean random variable W with variance σ_Y^2 must obey,

$$\varphi_W(z) = \frac{e^{\frac{\left(\mu_Y - \frac{\rho\sigma_Y}{\sigma_X}\mu_X\right)}{\sqrt{1-\rho^2}}z} \varphi_Y\left(\frac{z}{\sqrt{1-\rho^2}}\right)}{\varphi_X\left(\frac{\rho\sigma_Y}{\sigma_X\sqrt{1-\rho^2}}z\right)} \qquad (4.20)$$

which can be written in terms of the translated random variables $Y' = Y - \mu_Y$ and $X' = X - \rho\sigma_Y\mu_X$,

$$\varphi_W(z) = \frac{\varphi_{Y'}\left(\frac{z}{\sqrt{1-\rho^2}}\right)}{\varphi_{X'}\left(\frac{\rho\sigma_Y}{\sigma_X\sqrt{1-\rho^2}}z\right)}$$

This form shows that, if X' and Y' have a same distribution, W possesses, in general, a different distribution. Only the pgf of a Gaussian (with zero mean) obeys the functional equation

$$f(z) = \frac{f\left(\frac{z}{\sqrt{1-\rho^2}}\right)}{f\left(\frac{\rho}{\sqrt{1-\rho^2}}z\right)}$$

The joint probability generating function follows from (2.64) as

$$\varphi_{XY}(z_1, z_2) = E\left[e^{-z_1 X - z_2 Y}\right] = \int_{-\infty}^{\infty} \int_{-\infty}^{\infty} e^{-z_1 x - z_2 y} f_{XY}(x, y) dx dy$$

and the inverse is

$$f_{XY}(x, y) = \frac{1}{(2\pi i)^2} \int_{c_1-i\infty}^{c_1+i\infty} \int_{c_2-i\infty}^{c_2-i\infty} e^{z_1 x + z_2 y} \varphi_{XY}(z_1, z_2) dz_1 dz_2 \qquad (4.21)$$

Using (4.18), we have

$$\varphi_{XY}(z_1, z_2) = e^{-z_2\left(\mu_Y - \frac{\rho\sigma_Y}{\sigma_X}\mu_X\right)} E\left[e^{-\left(z_1 + z_2\frac{\rho\sigma_Y}{\sigma_X}\right)X}\right] E\left[e^{-z_2\sqrt{1-\rho^2}W}\right]$$

$$= e^{-z_2\left(\mu_Y - \frac{\rho\sigma_Y}{\sigma_X}\mu_X\right)} \varphi_X\left(z_1 + z_2\frac{\rho\sigma_Y}{\sigma_X}\right) \varphi_W\left(z_2\sqrt{1-\rho^2}\right) \qquad (4.22)$$

Introduced into the complex double integral (4.21), the joint probability density function of the two correlated random variables can be computed.

The main deficiency of the linear combination method is the implicit assumption that any joint distribution function $f_{XY}(x, y)$ can be constructed from two independent random variables X and W. The corresponding joint pgf (4.22) possesses a product form that cannot always be made compatible with the form of an arbitrary pgf $\varphi_{XY}(z_1, z_2)$. The examples below illustrate this deficiency.

4.5.1 Correlated Gaussian random variables

If X and Y are Gaussian random variables with Laplace transform $E\left[e^{-zY}\right]$ given in (3.22), the expression (4.20) for $\varphi_W(z)$ becomes

$$\varphi_W(z) = \exp\left[\left(\frac{\sigma_Y^2}{2}\right) z^2\right]$$

which shows that W is also a Gaussian random variable with mean $\mu_W = 0$ and standard deviation $\sigma_W = \sigma_Y$. Further, the joint pgf follows from (4.22) as

$$E\left[e^{-z_1 X - z_2 Y}\right] = \exp\left[-z_2\mu_Y - \mu_X z_1 + \frac{\sigma_X^2}{2}z_1^2 + z_1 z_2 \rho\sigma_Y\sigma_X + \frac{\sigma_Y^2}{2}z_2^2\right]$$

Since

$$\frac{\sigma_X^2}{2}z_1^2 + z_1 z_2 \rho\sigma_Y\sigma_X + \frac{\sigma_Y^2}{2}z_2^2 = \frac{1}{2}\begin{bmatrix} z_1 & z_2 \end{bmatrix}\begin{bmatrix} \sigma_X^2 & \rho\sigma_X\sigma_Y \\ \rho\sigma_X\sigma_Y & \sigma_Y^2 \end{bmatrix}\begin{bmatrix} z_1 \\ z_2 \end{bmatrix}$$

formula (4.2) indicates that $E\left[e^{-z_1 X - z_2 Y}\right]$ is the two-dimensional pgf of a joint Gaussian with pdf (4.4). The linear combination method thus provides the exact results for correlated Gaussian random variables.

4.5.2 Correlated exponential random variables

Let X and Y be two correlated, exponential random variables with rates α_x and α_y, respectively. Recall that $E[X] = \sigma_X = \frac{1}{\alpha_x}$. Using the Laplace transform (3.16) in (4.20), we obtain

$$\varphi_W(z) = e^{\frac{1-\rho}{\alpha_y\sqrt{1-\rho^2}}z}\,\frac{1+\frac{\rho z}{\alpha_y\sqrt{1-\rho^2}}}{1+\frac{z}{\alpha_y\sqrt{1-\rho^2}}}$$

The corresponding probability distribution function follows from (2.39) as

$$F_W(t) = \frac{1}{2\pi i}\int_{c-i\infty}^{c+i\infty}\frac{1+\frac{\rho z}{\alpha_y\sqrt{1-\rho^2}}}{1+\frac{z}{\alpha_y\sqrt{1-\rho^2}}}\,\frac{e^{z\left(t+\frac{1-\rho}{\alpha_y\sqrt{1-\rho^2}}\right)}}{z}\,dz \qquad c>0$$

Define the normalized time $T = \frac{1}{\alpha_y\sqrt{1-\rho^2}}$, then

$$F_W(t) = \frac{1}{2\pi i}\int_{c-i\infty}^{c+i\infty}\frac{1+T\rho z}{1+zT}\,\frac{e^{z(t+(1-\rho)T)}}{z}\,dz$$

Since $t+(1-\rho)T > 0$, the contour can be closed over the negative $\mathrm{Re}(z)$ plane encircling the poles at $z = -\frac{1}{T}$ and $z = 0$. By Cauchy's Residue Theorem,

$$F_W(t) = \lim_{z\to-\frac{1}{T}}\frac{(1+T\rho z)\left(z+\frac{1}{T}\right)}{(1+zT)\,z}e^{z(t+(1-\rho)T)} + \lim_{z\to 0}\frac{(1+T\rho z)\,z}{(1+zT)\,z}e^{z(t+(1-\rho)T)}$$

$$= 1 - (1-\rho)\,e^{-(1-\rho)}e^{-\frac{t}{T}}$$

Hence, for the generation of two exponential, correlated random variables, the auxiliary random variable W has an exponential distribution with an atom of size $(1-\rho)\,e^{-(1-\rho)}$ at $t=0$, which is fortunately easily to generate with a computer. It appears that only for $\rho \geq 0$, the linear combination method leads to correct results for exponential random variables. Moreover, the method does not give an indication of the validity in the range of ρ. We have shown above that $\rho \in \left[1-\frac{\pi^2}{6},1\right]$.

While the linear combination method applied to generate two exponential random variables still correctly treats a range of ρ, the application to correlated uniform random variables leads to bizarre results and definitely shows the deficiency of the method. The difficulties already encountered in this chapter in generating $n=2$ correlated random variables with arbitrary distribution suspects that the case for $n>2$ must be even more intractable.

4.6 Sampling and estimators

This section presents a little wandering into a subdomain of probability theory, called statistics, that basically tries to estimate the distribution of a random variable X, when only a sample, a finite set $x_1, x_2, \ldots, x_n$ of values from a population of all possible realizations of X, is known. The sample should have certain desirable properties in order to be a "good representative" of X. More generally, X can be a stochastic process, defined in Section 7.1, that we measure partially. Usually, there are many possible ways in which samples can be collected or obtained. Thus, rather than confining to one particular sample $x_1, x_2, \ldots, x_n$ of real values, it is instructive to concentrate on a *random* sample. A random sample $X_1, X_2, \ldots, X_n$ is a sequence of independent random variables, each with the same distribution as X. Clearly, a particular realization of our random sample is $X_1 = x_1, X_2 = x_2, \ldots, X_n = x_n$.

Given a sample of size n, one is often interested how closely the sample represents the underlying random variable, how large the deviations or bias is, and what can be concluded from this sample.

4.6.1 Estimators

An estimator $\hat{\theta}$ of a parameter θ of a random variable X is a random variable that depends upon a random sample $X_1, X_2, \ldots, X_n$. The most obvious estimator is the *sample mean* $\overline{X} = \frac{S_n}{n}$, where $S_n = \sum_{k=1}^{n} X_k$, to which a *sample variance* can be associated in the usual way as

$$S_X^2 = \frac{1}{n-1} \sum_{k=1}^{n} \left(X_k - \overline{X} \right)^2 \tag{4.23}$$

where $\beta = \frac{1}{n-1}$ is a normalization factor, to be determined. The sample mean $\overline{X}$ is an estimator for mean $E[X] = \mu$, while S_X^2 is an estimator for the variance $\mathrm{Var}[X] = \sigma^2$. We readily verify, using the linearity of the expectation, that $E[\overline{X}] = E[X]$. Similarly, it is desirable that $E[S_X^2] = \mathrm{Var}[X]$, and this requirement will determine the factor β. Taking the expection of both sides yields

$$E[S_X^2] = \beta \sum_{k=1}^{n} E\left[\left(X_k - \overline{X} \right)^2 \right]$$

and

$$E\left[\left(X_k - \overline{X} \right)^2 \right] = E[X_k^2] - 2E[X_k \overline{X}] + E[\overline{X}^2]$$

All random variables $\{X_k\}_{1 \leq k \leq n}$ are i.i.d. as X, such that $E[X_j X_k] = (E[X])^2$ if $j \neq k$ else $E[X_k^2] = E[X^2]$. Since

$$E[X_k \overline{X}] = \frac{1}{n} E\left[X_k \sum_{j=1}^{n} X_j \right] = \frac{1}{n} \sum_{j=1; j \neq k}^{n} E[X_j X_k] + \frac{1}{n} E[X_k^2]$$

$$= \frac{n-1}{n} (E[X])^2 + \frac{1}{n} E[X^2]$$

and

$$E[\overline{X}^2] = \frac{1}{n^2} E\left[\sum_{k=1}^{n} \sum_{j=1}^{n} X_k X_j \right] = \frac{1}{n^2} E\left[\sum_{k=1}^{n} X_k^2 + \sum_{k=1}^{n} \sum_{j=1; j \neq k}^{n} X_k X_j \right]$$

$$= \frac{1}{n^2} \left(\sum_{k=1}^{n} E[X_k^2] + \sum_{k=1}^{n} \sum_{j=1; j \neq k}^{n} E[X_k X_j] \right) = \frac{1}{n} E[X^2] + \frac{n-1}{n} (E[X])^2$$

we have that

$$E\left[(X_k - \overline{X})^2\right] = \frac{n-1}{n}\left(E\left[X^2\right] - (E\left[X\right])^2\right) = \frac{n-1}{n}\text{Var}\left[X\right]$$

Thus, we find that

$$E\left[S_X^2\right] = \beta\,(n-1)\,\text{Var}\left[X\right]$$

from which we need to choose $\beta = \frac{1}{n-1}$ (and not $\beta = \frac{1}{n}$ as a-priori expected) in order for $E\left[S_X^2\right]$ to be an *unbiased* estimator. In general, an estimator $\hat{\theta}$ is called unbiased if $E\left[\hat{\theta}\right] = \theta$.

The above computations show that

$$\text{Var}\left[\,\overline{X}\,\right] = \frac{1}{n}\text{Var}\left[X\right] = \frac{\sigma^2}{n}$$

meaning that the standard deviation of the sample mean, $\sigma_{\overline{X}} = \frac{\sigma}{\sqrt{n}}$, decreases proportional to $\frac{1}{\sqrt{n}}$ with the sample size n. The Central Limit Theorem 6.3.1 shows that the normalized random variable $\frac{\overline{X} - E[X]}{\sigma/\sqrt{n}} \xrightarrow{d} N(0,1)$ for large size n. Chebyshev's inequality (5.13) with $t = \alpha\sigma$ gives the bound

$$\Pr\left[\frac{|\overline{X} - E\left[X\right]|}{\sigma} \geq \alpha\right] \leq \frac{\text{Var}\left[\,\overline{X}\,\right]}{(\alpha\sigma)^2} = \frac{1}{\alpha^2 n}$$

that can be used to relate the size n of the sample to the probability that deviation of $\overline{X}$ from the real mean $E\left[X\right] = \mu$ in terms of α times the standard deviation σ is not larger than a prescribed stringency ε. For example, the requirement that the sample mean $\overline{X}$ lies within one standard deviation from the real mean with probability smaller than ε, $\Pr\left[|\overline{X} - E\left[X\right]| \geq \sigma\right] < \varepsilon$, is met when the number of independent samples is $n > \frac{1}{\varepsilon}$.

For large sample sizes n, the above analysis and the Central Limit Theorem 6.3.1 tell us that the specific details of the random variable X and of its pdf f_X are less important. However, the situation is significantly different for small sample sizes. Only for particular density functions f_X can the distribution of $f_{\overline{X}}$ (see Section 2.5.5) and of $f_{S_X^2}$ (see Section 3.3.5) be determined. Unfortunately, so far, exact results are known only for a Gaussian random variable.

Theorem 4.6.1 *For a Gaussian random variable X, the sample variance S_X^2 of a random sample $X_1, X_2, \ldots, X_n$ has a chi-square distribution with $n-1$ degrees of freedom. Moreover, the corresponding sample mean $\overline{X}$ is independent of the sample variance S_X^2.*

Proof: The n random variables $(X_k - \overline{X})^2$ in (4.23) are not independent because $\overline{X} = \frac{1}{n}\sum_{k=1}^{n} X_k$. In order to prove the theorem, we need to show that the sum of squares $s_n = \sum_{k=1}^{n}(X_k - \overline{X})^2$ can be represented as $s_n = \sum_{k=1}^{n-1} Y_k^2$, where all Y_k for $1 \leq k \leq n-1$ are independent Gaussian random variables $N\left(0, \sigma^2\right)$. Section 3.3.5 then demonstrates that s_n has a chi-square distribution with parameter (degree of freedom) equal to $n-1$.

We start from

$$s_n = \sum_{k=1}^{n} \left(X_k - \overline{X} \right)^2 = \sum_{k=1}^{n} X_k^2 - n\overline{X}^2$$

which we write, with $\overline{X} = \frac{1}{n} \sum_{k=1}^{n} X_k$, as a quadratic form

$$\sum_{k=1}^{n} X_k^2 = \sum_{k=1}^{n} \left(X_k - \overline{X} \right)^2 + \frac{1}{n} \left(\sum_{k=1}^{n} X_k \right)^2 \qquad (4.24)$$

Any orthogonal matrix C (**art. 11**), that transforms the vector x into the vector y as $y = Cx$, preserves the quadric form $y^T y = x^T C^T C x = x^T x$ because of the orthogonality property $C^T C = I$. Let the vector $x = \begin{bmatrix} X_1 & X_2 & \cdots & X_n \end{bmatrix}^T$, then (4.24) shows that $x^T x = Q_1 + Q_2$, where the non-negative quadratic forms $Q_1 = \sum_{k=1}^{n} \left(X_k - \overline{X} \right)^2$ and $Q_2 = \frac{1}{n} \left(\sum_{k=1}^{n} X_k \right)^2$. Suppose that there exists an orthogonal transformation $y = Cx$, with the vector $y = \begin{bmatrix} Y_1 & Y_2 & \cdots & Y_n \end{bmatrix}^T$, that changes Q_2 into Y_n^2, then that same orthogonal transformation will preserve the quadratic form

$$\sum_{k=1}^{n} X_k^2 = x^T x = y^T y = \sum_{k=1}^{n} Y_k^2$$

such that $Q_1 = \sum_{k=1}^{n} Y_k^2 - Y_n^2 = \sum_{k=1}^{n-1} Y_k^2$.

Consider the orthogonal matrix $C = \begin{bmatrix} c_1^T & c_2^T & \cdots & c_n^T \end{bmatrix}^T$, where c_j^T is a row vector (and c_j is a column vector) and where the orthogonality property (A.23) translates to $c_k^T c_j = \delta_{kj}$. Moreover, the quadratic form is, in vector notation,

$$y^T y = \sum_{j=1}^{n} x^T c_j c_j^T x = \sum_{j=1}^{n-1} \left(c_j^T x \right)^2 + \left(c_n^T x \right)^2$$

If $c_n = \frac{1}{\sqrt{n}} u$, where u is the all-one vector, then $Y_n = c_n^T x = \frac{1}{\sqrt{n}} u^T x = \frac{1}{\sqrt{n}} \sum_{k=1}^{n} X_k$ and $\left(c_n^T x \right)^2 = \frac{1}{n} \left(\sum_{k=1}^{n} X_k \right)^2 = Q_2$. Hence, that same orthogonal matrix[3] C will map $\sum_{k=1}^{n} \left(X_k - \overline{X} \right)^2$ into $\sum_{k=1}^{n-1} Y_k^2$. From (A.25), we deduce that Q_2 is of rank 1 and Q_1 is of rank $n-1$, which implies the $n-1$ degrees of freedom.

Next, we show that $E[Y_j] = 0$ for $1 \le j \le n-1$. Indeed,

$$E[Y_j] = E\left[c_j^T x \right] = c_j^T E[x] = \mu c_j^T u = 0$$

---

[3] An explicit form of the orthogonal matrix is

$$C = \begin{bmatrix} \frac{1}{\sqrt{2}} & -\frac{1}{\sqrt{2}} & 0 & \cdots & 0 \\ \frac{1}{\sqrt{2.3}} & \frac{1}{\sqrt{2.3}} & -\frac{2}{\sqrt{2.3}} & \cdots & 0 \\ \vdots & \vdots & \ddots & \ddots & \vdots \\ \frac{1}{\sqrt{(n-1)n}} & \frac{1}{\sqrt{(n-1)n}} & \frac{1}{\sqrt{(n-1)n}} & \cdots & -\frac{n-1}{\sqrt{(n-1)n}} \\ \frac{1}{\sqrt{n}} & \frac{1}{\sqrt{n}} & & \frac{1}{\sqrt{n}} & \frac{1}{\sqrt{n}} \end{bmatrix}$$

and the corrsponding transformation $y = Cx$ is

$$Y_j = \frac{\sum_{k=1}^{j} X_k - j X_{j+1}}{\sqrt{j(j+1)}} \quad \text{for } 1 \le j \le n-1$$

$$Y_n = \frac{1}{\sqrt{n}} \sum_{k=1}^{n} X_k = \sqrt{n}\overline{X}$$

because $c_n = \frac{1}{\sqrt{n}}u$ and all $c_j^T c_n = 0$ for $1 \leq j \leq n-1$. However, $E[Y_n] = \sqrt{n}\mu$. Similarly, we have that

$$E[Y_j Y_m] = E\left[c_j^T x.c_m^T x\right] = E\left[c_j^T x.x^T c_m\right] = c_j^T E\left[xx^T\right] c_m$$

because $c_m^T x = x^T c_m = Y_m$ is a scalar. The symmetric matrix xx^T is

$$xx^T = \begin{bmatrix} X_1^2 & X_1 X_2 & \cdots & X_1 X_n \\ X_2 X_1 & X_2^2 & \cdots & X_2 X_n \\ \vdots & \vdots & \ddots & \vdots \\ X_n X_1 & X_n X_2 & \cdots & X_n^2 \end{bmatrix}$$

and $E\left[xx^T\right]$ is a matrix, where all diagonal elements are equal to $E\left[X_j^2\right] = \sigma^2 + \mu^2$ and all others $E[X_m X_j] = \mu^2$, by independence of all X_j for $1 \leq j \leq n$. Thus,

$$E\left[xx^T\right] = (\sigma^2 + \mu^2)I + \mu^2(J - I)$$

$$= \sigma^2 I + \mu^2 J$$

where $J = u.u^T$ is the all-one matrix, such that

$$E[Y_j Y_m] = c_j^T (\sigma^2 I + \mu^2 J) c_m = \sigma^2 c_j^T c_m + \mu^2 c_j^T u.u^T c_m$$

If $j \neq m$, the orthogonality relation $c_j^T c_m = \delta_{jm}$ and $u = \sqrt{n}c_n$ show that $E[Y_j Y_m] = 0$. This means that all random variables Y_j are uncorrelated. Since they are Gaussian random variables, Section 4.1.2 shows that, only for Gaussian random variables, uncorrelations also implies independence. If $j = m$, then $E\left[Y_j^2\right] = \text{Var}[Y_j] = \sigma^2$ and $E[Y_n^2] = \sigma^2 + n\mu^2$, but $\text{Var}[Y_n] = \sigma^2$. This proves the theorem. $\qquad\square$

A major consequence of Theorem 4.6.1 is that, for a Gaussian random variable X, the distribution of the normalized sample mean $\overline{Y} = \frac{\overline{X} - E[X]}{S_X/\sqrt{n}}$, which can be regarded as another estimator, can be computed exactly. Indeed, only if X is a Gaussian random variable, Theorem 4.6.1 shows that $\overline{X}$ and S_X are independent such that the random variable Y has a Student t distribution as derived in Section 3.3.6. The proof of Theorem 4.6.1 relies on the fact that uncorrelation implies independence for Gaussian random variables. Since this property is unique for Gaussian random variables, independence between $\overline{X}$ and S_X only holds for a Gaussian random variable X.

4.6.2 Maximum likelihood

Assuming that the random variable X only depends on one parameter θ, then the probability

$$l(\theta) = \Pr[X_1 = x_1, X_2 = x_2, \ldots, X_n = x_n]$$

$$= \prod_{k=1}^{n} \Pr[X_k = x_k]$$

where the last step follows by independence of the random sample, is called the *likelihood function* l of θ. If X is continuous, then $\Pr[X_k = x_k]$ is replaced by the density function $f_X(x_k)$. The likelihood function depends upon a particularly selected sample $x_1, x_2, \ldots, x_n$ and the choice of θ. The maximum likelihood estimate of θ is that value of θ that maximizes $l(\theta)$, the joint probability of the selected sample. The maximum likelihood estimate was first introduced by (Gauss, 1809, p. 210-212) (see Section 4.6.3). If $l(\theta)$ is differentiable, a necessary condition is that $\frac{\partial l}{\partial \theta} = 0$ and $\frac{\partial^2 l}{\partial \theta^2} < 0$, whose solution $\theta_{\max}(x_1, x_2, \ldots, x_n)$ is the maximum likelihood estimate and the corresponding random variable $\hat{\theta}_{\max} = \theta_{\max}(X_1, X_2, \ldots, X_n)$ is called the maximum likelihood estimator of θ. The extension to several parameters $\theta_1, \theta_2, \ldots, \theta_m$ is conceptually straightforward, but computationally more complex.

For example, suppose that $X = N(\mu, \sigma^2)$, then there are two parameters, the mean μ and the variance σ^2, and

$$l(\mu, \sigma^2) = \left(\frac{1}{2\pi\sigma^2}\right)^{\frac{n}{2}} \exp\left(-\frac{1}{2\sigma^2}\sum_{k=1}^{n}(x_k - \mu)^2\right)$$

The maximum of $l(\mu, \sigma^2)$ is the same as that of

$$\log l(\mu, \sigma^2) = -\frac{n}{2}\log 2\pi\sigma^2 - \frac{1}{2\sigma^2}\sum_{k=1}^{n}(x_k - \mu)^2$$

When equating the partial derivatives

$$\frac{\partial \log l(\mu, \sigma^2)}{\partial \mu} = \frac{1}{\sigma^2}\sum_{k=1}^{n}(x_k - \mu)$$

and

$$\frac{\partial \log l(\mu, \sigma^2)}{\partial \sigma^2} = -\frac{n}{2\sigma^2} + \frac{1}{2\sigma^4}\sum_{k=1}^{n}(x_k - \mu)^2$$

equal to zero, the first equation leads to

$$\mu_{\max} = \frac{1}{n}\sum_{k=1}^{n}x_k \text{ and thus } \hat{\mu} = \frac{1}{n}\sum_{k=1}^{n}X_k = \overline{X}$$

while the second provides

$$\sigma^2_{\max} = \frac{1}{n}\sum_{k=1}^{n}(x_k - \mu)^2 \text{ and } \widehat{\sigma^2} = \frac{1}{n}\sum_{k=1}^{n}(X_k - \overline{X})^2 = \frac{n-1}{n}S^2$$

The last one is a biased estimator.

4.6.3 Gauss's derivation of the normal distribution

In his monumental treatise[4], Gauss (1809) has derived the Gaussian distribution (3.19) via a maximum likelihood argument, while discussing the *principle of the*

---

[4] In Sectio Tertia on "Determinatio orbitae observationibus quotcunque quam proxime satisfacientis".

least-squares applied to the computation of orbits of celestial bodies. Here, we will summarize Gauss's genial arguments, but we recommend to read the long[5], though well-written exposition on dealing with errors in measurements by the master himself.

Let $V_1, V_2, \ldots, V_n$ denote unknown functions of the parameters $u_1, u_2, \ldots, u_q$. By experiments, we can measure estimates $m_1, m_2, \ldots, m_n$ of those observables and we denote the errors by $\Delta_j = m_j - V_j$ for $1 \leq j \leq n$. Each error Δ_j (which is a random variable) is thus a function of the vector $u = (u_1, u_2, \ldots, u_q)$. Gauss assumes that each error Δ_j is independent from the others and has the same probability density function $\varphi(\Delta_j)$, in our notation $f_\Delta(x)$, where he writes $\Pr[v \leq \Delta_j \leq w] = \int_v^w \varphi(x)\,dx$. Neglecting systematic errors, Gauss only focuses on random errors for which it is reasonable to assume that positive and negative errors are equally likely to occur, hence, $f_\Delta(x) = f_\Delta(-x)$ with a maximum at $x = 0$. Next, Gauss considers the probability that all the specific values of $\Delta_1, \Delta_2, \ldots, \Delta_n$ simultaneously are produced by the measurements,

$$l(u) = \prod_{j=1}^{n} \varphi(\Delta_j(u)) \tag{4.25}$$

and he continues to find the maximum possible probability by computing $\frac{dl(u)}{du_k} = 0$ for $1 \leq k \leq q$, which is Gauss's maximum likelihood argument. Explicitly, denoting $\psi(x) = \frac{d}{dx} \log \varphi(x)$ and using $0 = \frac{dl(u)}{du_k} = \frac{d \log l(u)}{du_k}$, we obtain

$$0 = \frac{d \log l(u)}{du_k} = \sum_{j=1}^{n} \frac{d \log \varphi(\Delta_j(u))}{du_k} = \sum_{j=1}^{n} \frac{d \log \varphi(y)}{dy}\bigg|_{y=\Delta_j(u)} \frac{d\Delta_j(u)}{du_k}$$

and, for each $1 \leq k \leq q$,

$$\sum_{j=1}^{n} \psi(\Delta_j) \frac{d\Delta_j(u)}{du_k} = 0 \tag{4.26}$$

The corresponding matrix form,

$$\begin{bmatrix} \frac{d\Delta_1(u)}{du_1} & \frac{d\Delta_2(u)}{du_1} & \cdots & \frac{d\Delta_n(u)}{du_1} \\ \frac{d\Delta_1(u)}{du_2} & \frac{d\Delta_2(u)}{du_2} & \cdots & \frac{d\Delta_n(u)}{du_2} \\ \vdots & \vdots & \ddots & \vdots \\ \frac{d\Delta_1(u)}{du_q} & \frac{d\Delta_2(u)}{du_q} & \cdots & \frac{d\Delta_n(u)}{du_q} \end{bmatrix} \begin{bmatrix} \psi(\Delta_1) \\ \psi(\Delta_2) \\ \vdots \\ \psi(\Delta_n) \end{bmatrix} = 0$$

illustrates that this set cannot be solved without additional information. Gauss then continues by choosing $V_j = u_1$ for $1 \leq j \leq n$ so that $\Delta_j(u) = m_j - u_1$. After

---

[5] Gauss introduced and positioned his new ideas and basic assumptions in the known context of science at his time, when scientists still wrote in Latin, the world language ("lingua franca") since the Romans. The clarity and the line of reasoning of his explanations are remarkable and his treatment and analysis are surprisingly close to what currently is still taught about error estimation in measurements.

summing $\Delta_j = m_j - u_1$ over all $1 \le j \le n$ yields

$$u_1 = \frac{1}{n} \sum_{j=1}^{n} (m_j - \Delta_j)$$

Gauss argues that the best value for u_1 is the average $\frac{1}{n} \sum_{j=1}^{n} m_j$ over the measurements. Hence, in the above relation, we have $\sum_{j=1}^{n} \Delta_j = 0$, which likely leads to a minimum. The global minimum, that maximizes $l(u)$ in (4.25), is achieved when each $\Delta_j = 0$, because $\varphi(0)$ is a maximum. The heuristic argument to approximate each V_j by $u_1 = \frac{1}{n} \sum_{j=1}^{n} m_j$ is based on the hypothesis that the mean of a set of i.i.d random variables is the best estimator for each of them. The set of q equations in (4.26) reduces now to one equation, $\sum_{j=1}^{n} \psi(\Delta_j) = 0$. Explicitly, with $\Delta_j = m_j - u_1$, (4.26) simplifies to

$$\psi(m_1 - u_1) = - \sum_{j=2}^{n} \psi(m_j - u_1)$$

which becomes, after substituting $u_1 = \frac{1}{n} \sum_{j=1}^{n} m_j$ and some rearrangement,

$$\psi\left(\left(1 - \frac{1}{n}\right) m_1 - \frac{1}{n} \sum_{r=2}^{n} m_r\right) = - \sum_{j=2}^{n} \psi\left(-\frac{m_1}{n} + m_j - \frac{1}{n} \sum_{r=2}^{n} m_r\right)$$

In order to determine ψ, Gauss further confines to the special case, where $m_r = m$ for all $2 \le r \le n$, so that

$$\psi\left(\left(1 - \frac{1}{n}\right)(m_1 - m)\right) = -(n-1)\psi\left(-\frac{m_1 - m}{n}\right)$$

which, finally reduces with $x = \frac{m_1 - m}{n}$ to the functional equation, valid for any positive integer n and real x,

$$\psi((n-1)x) = -(n-1)\psi(-x)$$

Rewritten as $\frac{\psi((n-1)x)}{(n-1)x} = \frac{\psi(-x)}{-x}$, the functional equation shows that $\frac{\psi(x)}{x} = c$, where c is a real constant. With the definition $\psi(x) = \frac{d}{dx} \log \varphi(x)$, we find that $\log \varphi(x) = \frac{1}{2}cx^2 + cte$ or $\varphi(x) = \kappa e^{\frac{c}{2}x^2}$. Since the likelihood function $l(u)$ must attain a maximum, c must be negative. After letting $c = -\frac{1}{\sigma^2}$ and using the normalization of a probability density function, $\int_{-\infty}^{\infty} \varphi(x)\,dx = 1$, to determine κ, we arrive at (3.19) with $\mu = \frac{1}{n} \sum_{j=1}^{n} \Delta_j = 0$. Finally, after substituting "his" probability density function (3.19) in (4.25), which results in

$$l(u) = \frac{1}{\sigma^n (\sqrt{2\pi})^n} e^{-\frac{1}{2\sigma^2} \sum_{j=1}^{n} \Delta_j^2 (u)}$$

Gauss (1809, p. 213) concludes that the most probable system of the parameters $u_1, u_2, \ldots, u_q$ minimizes the sum of the squares of the differences between the observed and computed values of the functions $V_1, V_2, \ldots, V_n$, if the same degree of accuracy is to be presumed in all the observations. This principle of Gauss is known as the *method or principle of the least-squares.* Further, Gauss (1809, p. 214) discusses the special case when the set of $V_j = \sum_{k=1}^{q} \alpha_{jk} u_k$ for $1 \leq j \leq n$ is linear in the parameters $u_1, u_2, \ldots, u_q$. An application of this special case is the linear correlation coefficient, exactly computed in Section 2.5.3.

4.7 Problems

(i) Given the joint pdf $f_{XY}(x, y)$ of two random variables X and Y. Discuss how to generate a sequence $\{X_i, Y_i\}$ for $1 \leq i \leq n$ of random variables with the same pdf $f_{XY}(x, y)$.

(ii) Let $X_{\min}$ denote the minimum of m independent continuous, positive random variables $\{X_k\}_{1 \leq k \leq m}$, each with the same distribution $\Pr[X \leq x]$. Compute the linear correlation coefficient (2.61) between $X_{\min}$ and X, in terms of $\Pr[X \leq x]$.

5

Inequalities

Hardy *et al.* (1999) view the best-known inequalities from various angles, provide several different proofs and relate the nature of these inequalities. For example, starting from the most basic inequality between geometric and arithmetic mean[1],

$$\min(x, y) \le \sqrt{xy} \le \frac{x+y}{2} \le \max(x, y) \tag{5.1}$$

which directly follows from $\left(\sqrt{x} - \sqrt{y}\right)^2 \ge 0$, they masterly extend this relation to the theorem of the arithmetic and geometric mean in several real variables $x_1, x_2, \ldots, x_n$,

$$\prod_{k=1}^{n} x_k^{r_k} \le \sum_{k=1}^{n} r_k x_k \tag{5.2}$$

where $\sum_{k=1}^{n} r_k = 1$. In particular, for positive real numbers $a_1, a_2, \ldots, a_n$, the harmonic, geometric and arithmetic mean inequality is

$$\frac{n}{\sum_{k=1}^{n} \frac{1}{a_k}} \le \sqrt[n]{\prod_{k=1}^{n} a_k} \le \frac{1}{n} \sum_{k=1}^{n} a_k \tag{5.3}$$

with equality only if all a_j are equal. Another type of inequality is

$$\min_{1 \le k \le n} \frac{x_k}{a_k} \le \frac{x_1 + x_2 + \cdots + x_n}{a_1 + a_2 + \cdots + a_n} \le \max_{1 \le k \le n} \frac{x_k}{a_k} \tag{5.4}$$

which follows with $A = \sum_{k=1}^{n} a_k > 0$ and

$$\frac{x_1 + x_2 + \cdots + x_n}{a_1 + a_2 + \cdots + a_n} = \sum_{k=1}^{n} \frac{x_k}{A} = \sum_{k=1}^{n} \frac{x_k}{a_k} \frac{a_k}{A}$$

---

[1] The arithmetic-geometric mean $M(x, y)$ is the limit for $n \to \infty$ of the recursion $x_n = \frac{1}{2}(x_{n-1} + y_{n-1})$, which is an arithmetic mean, and $y_n = \sqrt{x_{n-1}y_{n-1}}$, which is a geometric mean, with initial values $x_0 = x$ and $y_0 = y$. Gauss's famous discovery on intriguing properties of $M(x, y)$ (which lead e.g. to very fast converging series for computing π) is narrated in a paper by Almkvist and Berndt (1988).

from

$$\min_{1 \le k \le n} \frac{x_k}{a_k} \sum_{k=1}^{n} \frac{a_k}{A} \le \sum_{k=1}^{n} \frac{x_k}{a_k} \frac{a_k}{A} \le \max_{1 \le k \le n} \frac{x_k}{a_k} \sum_{k=1}^{n} \frac{a_k}{A}$$

after noting that $\sum_{k=1}^{n} \frac{a_k}{A} = 1$.

Only a few inequalities are reviewed here and we recommend the classic treatise on inequalities by Hardy, Littlewood and Polya for those who search for more depth, elegance and insight.

5.1 The minimum (maximum) and infimum (supremum)

Since these concepts will be frequently used, we explain the difference by concentrating on the minimum and infimum (the maximum and the supremum follow analogously). Let Υ be a non-empty subset of $\mathbb{R}$. The subset Υ is said to be bounded from below by M if there exists a number M such that, for all $x \in \Upsilon$, it holds that $x \ge M$. The largest lower bound (largest number M) is called the infimum and is denoted by $\inf(\Upsilon)$. Further, if there exists an element $m \in \Upsilon$ such that $m \le x$ for all $x \in \Upsilon$, then this element m is called the minimum and is denoted by $\min(\Upsilon)$. If the minimum $\min(\Upsilon)$ exists, then $\min(\Upsilon) = \inf(\Upsilon)$. However, the minimum does not always exists. The classical example is the open interval (a, b), where $\inf((a, b)) = a$, but the minimum does not exist because $a \notin (a, b)$. On the other hand, for the closed interval $[a, b]$, we have that $\inf([a, b]) = \min([a, b]) = a$. This example also illustrates that every finite non-empty subset of $\mathbb{R}$ has a minimum.

5.2 Continuous convex functions

A continuous function f that satisfies, for each u and v in an interval I,

$$f\left(\frac{u + v}{2}\right) \le \frac{f(u) + f(v)}{2}$$

is called convex in that interval I. If $-f$ is convex, f is concave. Hardy *et al.* (1999, Section 3.6) demonstrate that this condition is fundamental from which the more general condition[2]

$$f\left(\sum_{k=1}^{n} q_k x_k\right) \le \sum_{k=1}^{n} q_k f(x_k) \tag{5.5}$$

---

[2] The convexity concept can be generalized (Hardy *et al.*, 1999, Section 98) to several variables in which case the condition (5.5) becomes

$$f\left(\sum_k q_k x_k, \sum_k q_k y_k\right) \le \sum_k q_k f(x_k, y_k)$$

where $\sum_{k=1}^{n} q_k = 1$, can be deduced. Moreover, they show that a convex function is either very regular or very irregular and that a convex function that is not "entirely irregular" is necessarily continuous. Current textbooks, in particular the book by Boyd and Vandenberghe (2004), usually start with the definition of convexity from (5.5) in case $n = 2$ where $q_1 = 1 - q_2 = q$ and $0 \le q \le 1$ as

$$f\left(qu + (1-q)v\right) \le qf(u) + (1-q)f(v) \tag{5.6}$$

where u and v can be vectors in an m-dimensional space.

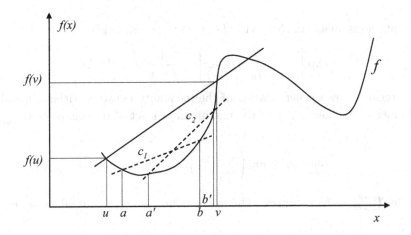

Fig. 5.1. The function f is convex between u and v.

Geometrically with $m = 1$ as illustrated in Fig. 5.1, relation (5.6) shows that each point on the chord between $(u, f(u))$ and $(v, f(v))$ lies above the curve f in the interval I. The more general form (5.5) asserts that the centre of gravity of any number of arbitrarily weighted points of the curve lies above or on the curve. Figure 5.1 illustrates that for any convex function f and points $a, a', b, b' \in [u, v]$ such that $a \le a' \le b'$ and $a < b \le b'$, the chord c_1 over (a, b) has a smaller slope than the chord c_2 over (a', b') or, equivalently, $\frac{f(b)-f(a)}{b-a} \le \frac{f(b')-f(a')}{b'-a'}$. Suppose that $f(x)$ is twice differentiable in the interval I, then a necessary and sufficient condition for convexity is $f''(x) \ge 0$ for each $x \in I$. This theorem is proved in Hardy et al. (1999, pp. 76–77). Moreover, they prove that the equality in (5.5) can only occur if $f(x)$ is linear.

Applied to probability, relation (5.5) with $q_k = \Pr[X = k]$ and $x_k = k$ is written with (2.12) as

$$f\left(E[X]\right) \le E[f(X)] \tag{5.7}$$

and is known as *Jensen's inequality*. The Jensen's inequality (5.7) also hold for continuous random variables. Indeed, if f is differentiable and convex, then $f(x) -$

$f(y) \geq f'(y)(x - y)$. Substitute x by the random variable X and $y = E[X]$, then

$$f(X) - f(E[X]) \geq f'(E[X])(x - E[X])$$

After applying the expectation operator to both sides, we obtain (5.7). An important application of Jensen's inequality is obtained for $f(x) = e^{-zx}$ with real z as

$$e^{-zE[X]} \leq E\left[e^{-zX}\right] = \varphi_X(z)$$

Any probability generating function $\varphi_X(z)$ is, for real z, bounded from below by $e^{-zE[X]}$.

A continuous analog of (5.5) with $f(x) = e^x$ (and similarly for $f(x) = -\log x$)

$$\exp\left[\frac{1}{v - u}\int_u^v f(x)dx\right] \leq \frac{1}{v - u}\int_u^v e^{f(x)}dx$$

can be regarded as a generalization of the inequality between arithmetic and geometric mean. An inequality for the maximum of a set of real numbers $\{x_k\}_{1 \leq k \leq n}$ is

$$\max_{1 \leq k \leq n} x_k \leq \log\left(\sum_{k=1}^n e^{x_k}\right) \leq \max_{1 \leq k \leq n} x_k + \log n$$

where $\log\left(\sum_{k=1}^n e^{x_k}\right)$ is convex. The upper bound is attained if all x_k are equal.

5.3 Inequalities deduced from the Mean Value Theorem

The Mean Value Theorem (Whittaker and Watson, 1996, p. 65) states that if $g(x)$ is continuous on $x \in [a, b]$, there exists a number $\xi \in [a, b]$ such that

$$\int_a^b g(u)du = (b - a)g(\xi)$$

or, alternatively, if $f(x)$ is differentiable on $[a, b]$, then

$$f(b) - f(a) = (b - a)f'(\xi) \tag{5.8}$$

The equivalence follows by putting $f(x) = \int_a^x g(u)du$. It is convenient to rewrite this relation for $0 \leq \theta \leq 1$ as

$$f(x + h) - f(x) = hf'(x + \theta h)$$

In this form, the Mean Value Theorem is nothing more than a special case for $n = 1$ of Taylor's theorem (Whittaker and Watson, 1996, p. 96),

$$f(x + h) - f(x) = \sum_{k=1}^{n-1} \frac{f^{(k)}(x)}{k!}h^k + \frac{h^n}{n!}f^{(n)}(x + \theta h) \tag{5.9}$$

where there exists a real number $\theta \in [0,1]$ such that equality in (5.9) is achieved. An important application of Taylor's theorem (5.9) to the exponential function,

$$e^x = \sum_{k=0}^{n-1} \frac{x^k}{k!} + \frac{x^n}{n!} e^{\theta x} \qquad (0 \leq \theta \leq 1) \tag{5.10}$$

gives a list of inequalities. Since $e^{\theta x} > 0$ for any finite x, we have for any $x \neq 0$ and $n = 1$,

$$e^x > 1 + x \tag{5.11}$$

and, in general for even $n = 2m$ and any $x \neq 0$,

$$e^x > \sum_{k=0}^{2m-1} \frac{x^k}{k!}$$

while for odd $n = 2m + 1$,

$$e^x > \sum_{k=0}^{2m} \frac{x^k}{k!} \qquad x > 0$$

$$e^x < \sum_{k=0}^{2m} \frac{x^k}{k!} \qquad x < 0$$

Estimates of the product $\prod_{k=0}^{n}(1 + a_k x)$, where $a_k x \neq 0$, are obtained from (5.11) as[3]

$$\prod_{k=0}^{n}(1 + a_k x) < \exp\left(\sum_{k=0}^{n} a_k x\right)$$

5.4 The Markov, Chebyshev and Chernoff inequalities

Consider first a non-negative random variable X. The expectation reads

$$E[X] = \int_0^\infty x f_X(x)dx = \int_0^a x f_X(x)dx + \int_a^\infty x f_X(x)dx$$

$$\geq \int_a^\infty x f_X(x)dx \geq a \int_a^\infty f_X(x)dx = a \Pr[X \geq a]$$

---

[3] A tighter bound is obtained if all $a_k > 0$ (e.g. a_k is a probability). The above relation indicates that $g(x) = \prod_{k=0}^{n}(1 + a_k x)$ is smaller than $f(x) = \exp\left(x \sum_{k=0}^{n} a_k\right)$ for any $x \neq 0$ and $g(0) = f(0) = 1$. Further, from $(1 + a_k x) < e^{a_k x}$ it can be verified that, for all Taylor coefficients $1 < k \leq n$, it holds that $0 < g_k \leq f_k$ and $g_2 < f_2$ such that $g(x) = \sum_{k=0}^{n} g_k x^k < \sum_{k=0}^{n} f_k x^k < \sum_{k=0}^{\infty} f_k x^k$ for $x > 0$. Thus, for $x = 1$, we have $g(1) < \sum_{k=0}^{n} f_k$ or

$$\prod_{k=0}^{n}(1 + a_k) < \sum_{k=0}^{n} \frac{1}{k!}\left(\sum_{k=0}^{n} a_k\right)^k$$

Hence, we obtain the Markov inequality

$$\Pr\left[X \geq a\right] \leq \frac{E\left[X\right]}{a} \tag{5.12}$$

Another proof of the Markov inequality follows after taking the expectation of the inequality $a1_{X \geq a} \leq X$ for $X \geq 0$. The restriction to non-negative random variables can be circumvented by considering the random variable $X = (Y - E\left[Y\right])^2$ and $a = t^2$ in (5.12),

$$\Pr\left[(Y - E\left[Y\right])^2 \geq t^2\right] \leq \frac{E\left[(Y - E\left[Y\right])^2\right]}{t^2} = \frac{\mathrm{Var}\left[Y\right]}{t^2}$$

From this, the Chebyshev inequality follows as

$$\Pr\left[|X - E\left[X\right]| \geq t\right] \leq \frac{\sigma^2}{t^2} \tag{5.13}$$

The Chebyshev inequality quantifies the spread of X around the mean $E\left[X\right]$. The smaller σ, the more concentrated X is around the mean.

Further extensions of the Markov inequality use the equivalence between the events $\{X \geq a\} \Leftrightarrow \{g(X) \geq g(a)\}$, where g is a monotonously increasing function. Hence, (5.12) becomes

$$\Pr\left[X \geq a\right] \leq \frac{E\left[g(X)\right]}{g(a)}$$

For example, if $g(x) = x^k$, then $\Pr\left[X \geq a\right] \leq \frac{E\left[X^k\right]}{a^k}$. An interesting application of this idea is based on the equivalence of the events $\{X \geq E\left[X\right] + t\} \Leftrightarrow \{e^{uX} \geq e^{u(E[X]+t)}\}$ provided $u \geq 0$. For $u \geq 0$,

$$\Pr\left[X \geq E\left[X\right] + t\right] = \Pr\left[e^{uX} \geq e^{u(E[X]+t)}\right] \leq e^{-u(E[X]+t)} E\left[e^{uX}\right] \tag{5.14}$$

where in the last step Markov's inequality (5.12) has been used. If the generating function or Laplace transform $E\left[e^{uX}\right]$ is known, the sharpest bound is obtained by the minimizer u^* in u of the right-hand side because (5.14) holds for any $u > 0$. In Section 5.7, we show that this minimizer u^* obeying $\mathrm{Re}\, u > 0$ indeed exists for probability generating functions. The resulting inequality

$$\Pr\left[X \geq E\left[X\right] + t\right] \leq e^{-u^*(E[X]+t)} E\left[e^{u^*X}\right] \tag{5.15}$$

is called the Chernoff bound.

The Chernoff bound of the binomial distribution Let X denote a binomial random variable with probability generating function given by (3.2) such that $E\left[e^{uX}\right] = E\left[(e^u)^X\right] = (q + pe^u)^n$. Then, with $E\left[X\right] = np$,

$$e^{-u(E[X]+t)} E\left[e^{uX}\right] = e^{-u(np+t)+n \log(q+pe^u)}$$

Provided $\frac{d^2}{du^2} e^{-u(E[X]+t)} E\left[e^{uX}\right]\Big|_{u=u^*} > 0$, the minimum u^* is solution of

$$\frac{d}{du} e^{-u(E[X]+t)} E\left[e^{uX}\right] = 0$$

Explicitly,

$$\frac{d}{du} e^{-u(E[X]+t)} E\left[e^{uX}\right] = e^{-u(np+t)+n\log(q+pe^u)} \left(-(np+t) + \frac{npe^u}{q+pe^u}\right)$$

from which u^* follows using $q = 1 - p$ as

$$u^* = \log\left(\frac{npq + qt}{npq - pt}\right)$$

Hence,

$$e^{-u^*(E[X]+t)} E\left[e^{u^*X}\right] = \frac{\left(1 - \frac{t}{nq}\right)^{t-nq}}{\left(1 + \frac{t}{np}\right)^{t+np}}$$

For large n, but p and t fixed, we observe[4] that

$$e^{-u^*(E[X]+t)} E\left[e^{u^*X}\right] = e^{-\frac{t^2}{npq}} \left(1 + O\left(\frac{1}{n}\right)\right)$$

Since $\mathrm{Var}[X] = npq$ and by denoting $y^2 = \frac{t^2}{\mathrm{Var}[X]}$, we find that the asymptotic regime for large n,

$$\Pr\left[\frac{|X - E[X]|}{\sqrt{\mathrm{Var}[X]}} \geq y\right] \leq e^{-y^2} \tag{5.16}$$

is in agreement with the Central Limit Theorem 6.3.1. The corresponding Chebyshev inequality,

$$\Pr\left[\frac{|X - E[X]|}{\sqrt{\mathrm{Var}[X]}} \geq y\right] \leq \frac{1}{y^2}$$

is considerably less tight for the binomial distribution than the Chernoff bound (5.16). When the Bernoulli random variables are not identically distributed, a related, though slightly different bound is

Lemma 5.4.1 *Let $X_1, X_2, \ldots, X_n$ be independent Bernoulli random variables with mean $E[X_i] = p_i$ for $1 \leq i \leq n$ and denote their sum by $S = \sum_{j=1}^{n} X_j$. Then, for $t > 0$,*

$$\Pr\left[S - E[S] \geq tE[S]\right] \leq e^{-E[S]h(t)} \tag{5.17}$$

---

[4] Write

$$e^{-u^*(E[X]+t)} E\left[e^{u^*X}\right] = \exp\left((t - nq)\log\left(1 - \frac{t}{nq}\right) - (t + np)\log\left(1 + \frac{t}{np}\right)\right)$$

and use the Taylor expansion of $\log(1 \pm x)$ around $x = 0$.

where $h(t) = (1+t)\ln(1+t) - t$.

Proof: Applying the Chernoff bound (5.15) to the sum of Bernoulli random variables yields

$$\Pr[S - E[S] \geq t] \leq e^{-u^* E[S](1+t)} E\left[e^{u^* S}\right]$$

Since all X_i are independent, we use (2.74),

$$E\left[e^{u^* S}\right] = \prod_{j=1}^{n} E\left[e^{u^* X_j}\right]$$

The pgf of a Bernoulli random variable (3.1) is $E\left[e^{u^* X_j}\right] = 1 + p_i\left(e^{u^*} - 1\right) \leq e^{p_i\left(e^{u^*} - 1\right)}$, where (5.11) has been used. Thus,

$$\Pr[S - E[S] \geq t] \leq e^{-u^* E[S](1+t)} \prod_{j=1}^{n} e^{p_i\left(e^{u^*} - 1\right)}$$

$$= e^{-u^* E[S](1+t)} e^{\left(e^{u^*} - 1\right) E[S]} = e^{-E[S]\left(-e^{u^*} + 1 + u^*(1+t)\right)}$$

The upper bound is minimized when $f(u) = -e^u + 1 + u(1+t)$ is maximized, which occurs for $u = u^* = \log(1+t)$ and $f(u^*) = (1+t)\log(1+t) - t$. This proves (5.17). □

More advanced and sharper inequalities than that of Chebyshev are surveyed by Janson (2002).

5.5 The Hölder, Minkowski and Young inequalities

The Hölder inequality is

$$E[XY] \leq (E[X^p])^{1/p} (E[Y^q])^{1/q} \qquad (5.18)$$

where $p > 1$ and $q > 1$ such that $\frac{1}{p} + \frac{1}{q} = 1$. The Hölder inequality can be deduced from the basic convexity inequality (5.6). Since $-\log x$ is a convex function for real $x > 0$, the basic convexity inequality (5.6) is with $0 \leq \theta \leq 1$,

$$\log(\theta u + (1-\theta)v) \geq \theta \log(u) + (1-\theta)\log(v)$$

After exponentiation, we obtain for $u, v > 0$ a more general inequality than (5.1), which corresponds to $\theta = \frac{1}{2}$,

$$u^\theta v^{1-\theta} \leq \theta u + (1-\theta)v$$

Substitute $u = \frac{|x_j|^p}{\sum_{j=1}^{n} |x_j|^p}$ and $v = \frac{|y_j|^q}{\sum_{j=1}^{n} |y_j|^q}$, then

$$\left(\frac{|x_j|^p}{\sum_{j=1}^{n} |x_j|^p}\right)^\theta \left(\frac{|y_j|^q}{\sum_{j=1}^{n} |y_j|^q}\right)^{1-\theta} \leq \theta \frac{|x_j|^p}{\sum_{j=1}^{n} |x_j|^p} + (1-\theta)\frac{|y_j|^q}{\sum_{j=1}^{n} |y_j|^q}$$

and summing over all j yields

$$\sum_{j=1}^{n} |x_j|^{p\theta} |y_j|^{q(1-\theta)} \leq \left(\sum_{j=1}^{n} |x_j|^p\right)^\theta \left(\sum_{j=1}^{n} |y_j|^q\right)^{1-\theta} \qquad (5.19)$$

By choosing $p = \frac{1}{\theta}$ and $q = \frac{1}{1-\theta}$, we arrive at the Hölder inequality with $p > 1$ and $\frac{1}{p} + \frac{1}{q} = 1$,

$$\sum_{j=1}^{n} x_j y_j \le \sum_{j=1}^{n} |x_j y_j| \le \left(\sum_{j=1}^{n} |x_j|^p \right)^{\frac{1}{p}} \left(\sum_{j=1}^{n} |y_j|^q \right)^{\frac{1}{q}} \tag{5.20}$$

Let $Y = 1$ in the Hölder inequality (5.18), then $E[X] \le (E[X^p])^{1/p}$ for $p > 1$. After substituting $X \to X^r$, with $r > 1$ and letting $q = rp > r > 1$, we arrive at the Lyapunov inequality

$$E[|X|^r]^{1/r} \le E[|X|^q]^{1/q} \qquad (q > r > 1) \tag{5.21}$$

For integer values of p, the Lyapunov inequality (5.21) leads to

$$E[|X|] \le E\left[|X|^2\right]^{1/2} \le E\left[|X|^3\right]^{1/3} \le \cdots \le E[|X|^p]^{1/p}$$

from which, for $1 \le s \le p$, it follows that

$$|E[X]|^s \le (E[|X|])^s \le E[|X|^s]$$

The first inequality is derived from the basic inequality $|a + b| \le |a| + |b|$.

A special important case of the Hölder inequality (5.18) for $p = q = 2$ is the Cauchy–Schwarz inequality,

$$(E[XY])^2 \le E[X^2] E[Y^2] \tag{5.22}$$

The Cauchy-Schwarz inequality follows immediately from the Cauchy identity (Gantmacher, 1959a, p. 10)

$$\sum_{j=1}^{n} x_j^2 \sum_{j=1}^{n} y_j^2 - \left(\sum_{j=1}^{n} x_j y_j \right)^2 = \frac{1}{2} \sum_{j=1}^{n} \sum_{k=1}^{n} (x_j y_k - x_k y_j)^2$$

It is of interest to mention that the Hölder inequality is of a general type in the following sense (Hardy et al., 1999, Theorem 101 (p. 82)). Suppose that $f(x)$ is convex (such that the inverse $g(x) = f^{-1}(x)$ is also convex) and that $f(0) = 0$. If $F(x) = \int_0^x f(u)du$ and $G(x) = \int_0^x g(u)du$, and if

$$\sum_{k=1}^{n} q_k a_k b_k \le F^{-1}\left(\sum_{k=1}^{n} q_k F(a_k) \right) G^{-1}\left(\sum_{k=1}^{n} q_k G(b_k) \right)$$

with $\sum_{k=1}^{n} q_k = 1$ holds for all positive a_k and b_k, then $f(x) = x^r$ and the above inequality is Hölder's inequality.

Another application of the Hölder inequality (5.20) to elements $a_{ij} \ge 0$ of an $m \times n$ non-negative matrix A follows by putting $x_j = a_{ij}$ and $y_j = (\sum_{k=1}^{m} a_{kj})^{p/q}$,

$$\sum_{j=1}^{n} a_{ij} \left(\sum_{k=1}^{m} a_{kj} \right)^{p/q} \le \left(\sum_{j=1}^{n} a_{ij}^p \right)^{\frac{1}{p}} \left(\sum_{j=1}^{n} \left(\sum_{k=1}^{m} a_{kj} \right)^p \right)^{\frac{1}{q}}$$

After summing i from 1 to m, we obtain

$$\sum_{i=1}^{m}\sum_{j=1}^{n} a_{ij} \left(\sum_{k=1}^{m} a_{kj}\right)^{p/q} \le \sum_{i=1}^{m}\left(\sum_{j=1}^{n} a_{ij}^p\right)^{\frac{1}{p}} \left(\sum_{j=1}^{n}\left(\sum_{k=1}^{m} a_{kj}\right)^p\right)^{\frac{1}{q}}$$

Using $\frac{p}{q} = p - 1$, we observe that

$$\sum_{i=1}^{m}\sum_{j=1}^{n} a_{ij}\left(\sum_{k=1}^{m} a_{kj}\right)^{p/q} = \sum_{j=1}^{n}\sum_{i=1}^{m} a_{ij}\left(\sum_{k=1}^{m} a_{kj}\right)^{p/q} = \sum_{j=1}^{n}\left(\sum_{k=1}^{m} a_{kj}\right)^{p}$$

which leads to the Minkowski inequality

$$\left(\sum_{j=1}^{n}\left(\sum_{k=1}^{m} a_{kj}\right)^{p}\right)^{\frac{1}{p}} \le \sum_{i=1}^{m}\left(\sum_{j=1}^{n} a_{ij}^p\right)^{\frac{1}{p}} \tag{5.23}$$

For $m = 2$, the Minkowski inequality (5.23) takes the form

$$\left(\sum_{j=1}^{n}|x_j + y_j|^p\right)^{\frac{1}{p}} \le \left(\sum_{j=1}^{n}|x_j|^p\right)^{\frac{1}{p}} + \left(\sum_{j=1}^{n}|y_j|^p\right)^{\frac{1}{p}}$$

or, in terms of the expectation of random variables,

$$(E\left[|X + Y|^p\right])^{1/p} \le (E\left[|X|^p\right])^{1/p} + (E\left[|Y|^p\right])^{1/p} \tag{5.24}$$

which is also known as the "triangle inequality", $\|x + y\|_p \le \|x\|_p + \|y\|_p$, for the vector norm (A.26).

Suppose that $f(x)$ is continuous and strictly increasing for $x \ge 0$ and $f(0) = 0$. Then the inverse function $g(x) = f^{-1}(x)$ satisfies the same conditions. The Young inequality states that, for $a \ge 0$ and $b \ge 0$, it holds that

$$ab \le \int_0^a f(u)du + \int_0^b g(u)du \tag{5.25}$$

with equality only if $b = f(a)$. The Young inequality follows by geometrical consideration. The first integral is the area under the curve $y = f(x)$ from $[0, a]$, while the second is the area under the curve $x = g(y) = f^{-1}(y)$ from $[0, b]$.

If two sequences are non-increasing, $x_1 \ge x_2 \ge \cdots \ge x_n$ and $y_1 \ge y_2 \ge \cdots \ge y_n$, then the Chebyshev's sum inequality is

$$\frac{1}{n}\sum_{j=1}^{n} x_j y_j \ge \left(\frac{1}{n}\sum_{j=1}^{n} x_j\right)\left(\frac{1}{n}\sum_{j=1}^{n} y_j\right) \tag{5.26}$$

which follows from

$$\sum_{k=1}^{n}\sum_{j=1}^{n}(x_k - x_j)(y_k - y_j) = 2n\sum_{k=1}^{n} x_k y_k - 2\sum_{k=1}^{n} x_k \sum_{j=1}^{n} y_j \ge 0$$

because both $(x_k - x_j)$ and $(y_k - y_j)$ have the same sign for any k and j due to the ordering of the sequences.

Applications of the Cauchy–Schwarz inequality

1. We will demonstrate that both the generating function $\varphi_X(z) = E\left[e^{-zX}\right]$ and its logarithm $L_X(z) = \log(\varphi_X(z))$ are convex functions of z. First, the second derivative is continuous and non-negative because $\varphi_X''(z) = E\left[X^2 e^{-zX}\right] \geq 0$. Further, since

$$L_X''(z) = \frac{\varphi_X(z)\varphi_X''(z) - (\varphi_X'(z))^2}{\varphi_X^2(z)}$$

it remains to show that $\varphi_X(z)\varphi_X''(z) - (\varphi_X'(z))^2 \geq 0$. The Cauchy–Schwarz inequality (5.22) with $X \to e^{-\frac{z}{2}X}$ and $Y = Xe^{-\frac{z}{2}X}$ shows that $(\varphi_X'(z))^2 = \left(E\left[Xe^{-zX}\right]\right)^2 \leq E\left[e^{-zX}\right]E\left[X^2 e^{-zX}\right] = \varphi_X(z)\varphi_X''(z)$. Hence, $L_X''(z) \geq 0$.

2. Let $Y = 1_{X>0}$ in (5.22) while X is a non-negative random variable, then with (2.13),

$$(E[X])^2 \leq E\left[X^2\right]E\left[1_{X>0}\right] = E\left[X^2\right](1 - \Pr[X=0])$$

such that an upper bound for $\Pr[X=0]$ is obtained,

$$\Pr[X=0] \leq 1 - \frac{(E[X])^2}{E[X^2]} \tag{5.27}$$

3. Consider the polynomial $p_n(x) = \sum_{k=0}^{n} a_k x^k = \prod_{j=1}^{n}(x - y_j)$ with $a_n = 1$. The *Newton identities* (Van Mieghem, 2011, p. 265)

$$a_l = -\frac{1}{n-l}\sum_{k=l+1}^{n} a_k Z_{k-l} \tag{5.28}$$

relate the coefficients a_k to sums $Z_m = \sum_{j=1}^{n} y_j^m$ of the positive powers of the zeros $\{y_j\}_{1 \leq j \leq n}$. Both the Newton identities and the Cauchy–Schwarz inequality are the basic ingredients in the proof of a remarkable theorem of Laguerre.

Theorem 5.5.1 (Laguerre) *If all the zeros* $y_1, \ldots, y_n$ *of a polynomial* $p_n(x) = \sum_{k=0}^{n} a_k x^k$ *with* $a_n = 1$ *are real, then they all lie in the interval* $[y_-, y_+]$ *where*

$$y_\pm = -\frac{a_{n-1}}{n} \pm \frac{n-1}{n}\sqrt{a_{n-1}^2 - \frac{2n}{n-1}a_{n-2}}$$

Proof: Let y_1 be an arbitrary zero of $p_n(x)$ because we can always relabel the zeros. In the Newton identities (5.28), we express the coefficients a_{n-1} and a_{n-2} in terms of y_1 as

$$-a_{n-1} = y_1 + \sum_{j=2}^{n} y_j$$

and

$$a_{n-1}^2 - y_1^2 - 2a_{n-2} = \sum_{j=2}^{n} y_j^2 \tag{5.29}$$

After applying the Cauchy–Schwarz inequality, (5.20) with $p = q = 2$, to $x_j = 1$ and the real zeros $y_j \neq y_1$, we obtain

$$\left(\sum_{j=2}^{n} |y_j| \right)^2 \leq (n-1) \sum_{j=2}^{n} y_j^2 = (n-1) \left(a_{n-1}^2 - y_1^2 - 2a_{n-2} \right)$$

Since $\sum_{j=2}^{n} y_j \leq \sum_{j=2}^{n} |y_j|$ and $\sum_{j=2}^{n} y_j = y_1 + a_{n-1}$, we arrive at the quadratic inequality in y_1

$$y_1^2 + \frac{2a_{n-1}}{n} y_1 + \frac{2(n-1) a_{n-2} - (n-2) a_{n-1}^2}{n} \leq 0$$

whose roots are $y_\pm$. The Cauchy–Schwarz inequality, now applied to all y_j and introduced in (5.29), shows that $a_{n-1}^2 \geq \frac{2n}{n-1} a_{n-2}$ implying that the roots $y_\pm$ are real. The quadratic inequality is obeyed if y_1, and hence all y_j, lie on or in between the roots $y_\pm$.   □

5.6 The Gauss inequality

In this section, we consider a continuous random variable X with even probability density function, i.e. $f_X(-x) = f_X(x)$, which is not increasing for $x > 0$. A typical example of such random variables are measurement errors due to statistical fluctuations.

In his epoch-making paper, Gauss (1821) established the method of the least squares (see e.g. Sections 2.2.1 and 2.5.3). In that same paper, Gauss (1821, pp. 10-11) also stated and proved Theorem 5.6.1, which is appealing because of its generality.

We define the probability m as

$$m = \Pr\left[-\lambda\sigma \leq X \leq \lambda\sigma\right] = \int_{-\lambda\sigma}^{\lambda\sigma} f_X(u)\, du \tag{5.30}$$

where $\sigma = \sqrt{\text{Var}[X]}$ is the standard deviation.

Theorem 5.6.1 (Gauss) *If X is a continuous random variable with even probability density function, i.e. $f_X(-x) = f_X(x)$, which is not increasing for $x > 0$, then*

$$\text{if } m < \tfrac{2}{3} \quad \text{then } \lambda \leq m\sqrt{3}$$
$$\text{if } m = \tfrac{2}{3} \quad \text{then } \lambda \leq \sqrt{\tfrac{4}{3}}$$
$$\text{if } m > \tfrac{2}{3} \quad \text{then } \lambda < \tfrac{2}{3\sqrt{1-m}}$$

and, conversely,

$$\text{if } \lambda < \sqrt{\tfrac{4}{3}} \quad \text{then } m \geq \tfrac{\lambda}{\sqrt{3}}$$
$$\text{if } \lambda > \sqrt{\tfrac{4}{3}} \quad \text{then } m \geq 1 - \tfrac{4}{\lambda^2}$$

Given a bound on the probability m, Gauss's Theorem 5.6.1 bounds the extent of the error X around its mean zero in units of the standard deviation σ or, equivalently, it provides bounds for the normalized random variable $X^* = \frac{X - E[X]}{\sigma}$. The proof of this theorem only uses real function theory and is characteristic for the genius of Gauss.

Proof: Consider the inverse function $x = g(y)$ of the integral $y = \int_{-x}^{x} f_X(u)\,du = F_X(x) - F_X(-x)$. An interesting general property of the inverse function is

$$\int_0^1 g^2(u)\,du = \int_{-\infty}^{\infty} x^2 f_X(x)\,dx$$

which is verified by the substitution $x = g(u)$. Since $E[X] = 0$ and $\mathrm{Var}[X] = E[X^2]$, we have

$$\int_0^1 g^2(u)\,du = \sigma^2 = \mathrm{Var}[X] \tag{5.31}$$

Beside $g(0) = 0$, the derivative

$$g'(y) = \frac{1}{F'_X(x) - F'_X(-x)} = \frac{1}{f_X(x) + f_X(-x)}$$

is increasing from $y = 0$ until $y = 1$ because $f_X(x)$ attains a maximum at $x = 0$ and is not increasing for $x > 0$. Hence, $g''(y) \geq 0$. From the differential

$$d\left(yg'(y)\right) = g'(y)\,dy + yg''(y)\,dy$$

we obtain by integration

$$yg'(y) - g(y) = \int_0^y ug''(u)\,du$$

Since $g''(y) \geq 0$, we have that $yg'(y) - g(y) \geq 0$ and since $yg'(y) > 0$ (for $y > 0$) that

$$h(y) = 1 - \frac{g(y)}{yg'(y)}$$

lies in the interval $[0, 1]$. From (5.30), it follows that $\lambda\sigma = g(m)$ and that $h(m) = 1 - \frac{\lambda\sigma}{mg'(m)}$ or

$$g'(m) = \frac{\lambda\sigma}{m(1 - h(m))}$$

With this preparation, consider now the following linear function

$$G(y) = \frac{\lambda\sigma}{m(1 - h(m))}(y - mh(m)) \tag{5.32}$$

Clearly, we have that $G(m) = \lambda\sigma$ and that $G'(y) = \frac{\lambda\sigma}{m(1-h(m))} = g'(m)$ is independent of y. Since $g'(y)$ is non-decreasing – which is the basic assumption of the theorem – the difference $g'(y) - G'(y)$ is negative if $y < m$, but positive if $y > m$. Since $g'(y) - G'(y) = \frac{d}{dy}(g(y) - G(y))$, the function $g(y) - G(y)$ is convex with minimum at $y = m$ for which $g(m) - G(m) = 0$. Hence, $g(y) - G(y) \geq 0$ for all $y \in [0, 1]$. Further, $G(y)$ is positive for $y \in (mh(m), 1]$. Especially in this interval, the inequality $g(y) \geq G(y)$ is sharp because $g(y)$ is positive in $(0, 1]$. Thus,

$$\int_{mh(m)}^1 G^2(y)\,dy \leq \int_{mh(m)}^1 g^2(y)\,dy < \int_0^1 g^2(y)\,dy$$

Using (5.31) and with (5.32), we have

$$\frac{\lambda^2\sigma^2}{m^2(1 - h(m))^2}\frac{(1 - mh(m))^3}{3} < \sigma^2$$

from which we arrive at the inequality

$$\lambda^2 < \frac{3m^2(1 - z)^2}{(1 - mz)^3} \tag{5.33}$$

where $z = h(m) \in [0, 1]$. The derivative of the right-hand side with respect to z,

$$\frac{d}{dz}\left(\frac{3(1-z)^2}{(1-mz)^3}m^2\right) = -\frac{3m^2(1-z)}{(1-mz)^4}(2 - 3m + mz)$$

shows that $\frac{3m^2(1-z)^2}{(1-mz)^3}$ is monotonously decreasing for all $z \in [0, 1]$ if $m < \frac{2}{3}$ with maximum at $z = 0$. Thus, if $m < \frac{2}{3}$, evaluating (5.33) at $z = 0$ yields $\lambda < \sqrt{3}m$. On the other hand, if $m > \frac{2}{3}$, then $\frac{3m^2(1-z)^2}{(1-mz)^3}$ is maximal provided $2 - 3m + mz = 0$ or for $z = 3 - \frac{2}{m}$. With that value of z, the inequality (5.33) yields $\lambda < \frac{2}{3\sqrt{1-m}}$. Both regimes $m > \frac{2}{3}$ and $m < \frac{2}{3}$ tend to a same bound $\lambda < \frac{2}{\sqrt{3}}$ if $m \to \frac{2}{3}$. The converse is similarly derived from (5.33).   □

If X has a symmetric uniform distribution with $f_X(x) = \frac{1}{2a}1_{x \in [-a,a]}$, then $m = \frac{\lambda\sigma}{a}$ and $\sigma = \frac{a}{\sqrt{3}}$ from which $m = \frac{\lambda}{\sqrt{3}}$. This example shows that Gauss's Theorem 5.6.1 is sharp for $m \le \frac{2}{3}$ in the sense that equality can occur in the first condition $\lambda \le \sqrt{3}m$.

5.7 The dominant pole approximation and large deviations

In this section, we relate asymptotic results of generating functions to the theory of large deviations. An asymptotic expansion in discrete-time is compared to established large deviations results.

The first approach using the generating function $\varphi_X(z)$ of the random variable X is an immediate consequence of Lemma 5.7.1.

Lemma 5.7.1 *If $\varphi_X(z)$ is meromorphic with residues r_k at the (simple) poles p_k ordered as $0 < |p_0| \le |p_1| \le |p_2| \le \cdots$ and if $\varphi_X(z) = o(z^{N+1})$ as $z \to \infty$, then holds*

$$\varphi_X(z) = \sum_{k=0}^{N}\Pr[X = k]z^k + \sum_{k=0}^{\infty}\frac{r_k}{p_k^{N+1}(z - p_k)}z^{N+1} \tag{5.34}$$

$$= \sum_{k=0}^{N}\Pr[X = k]z^k + \sum_{k=0}^{\infty}r_k\left(\frac{1}{z - p_k} + \sum_{m=0}^{N}\frac{z^m}{p_k^{m+1}}\right) \tag{5.35}$$

The normalization condition $\varphi_X(1) = 1$ implies that

$$\Pr[X > N] = 1 - \sum_{k=0}^{N}\Pr[X = k] = \sum_{k=0}^{\infty}\frac{r_k}{p_k^{N+1}(1 - p_k)} \tag{5.36}$$

The Lemma follows from Titchmarsh (1964, Section 3.21). Rewriting (5.35) gives,

$$\varphi_X(z) = \sum_{k=0}^{N}\Pr[X = k]z^k - \sum_{j=N+1}^{\infty}\left(\sum_{k=0}^{\infty}\frac{r_k}{p_k^{j+1}}\right)z^j \tag{5.37}$$

and hence,

$$\Pr[X = j] = -\sum_{k=0}^{\infty} \frac{r_k}{p_k^{j+1}} \qquad (j > N) \tag{5.38}$$

The cumulative density function for $K > N$ follows from (5.38) as

$$\Pr[X > K] = \sum_{j=K+1}^{\infty} \Pr[X = j] = \sum_{k=0}^{\infty} \frac{r_k}{p_k^{K+1}(1 - p_k)} \qquad (K > N) \tag{5.39}$$

Lemma 5.7.1 means that, if the plot $\Pr[X = j]$ versus j exhibits a kink at $j = N$, then $\varphi_X(z) = O\left(z^N\right)$ as $z \to \infty$. Alternatively[5], the asymptotic regime does not start earlier than $j \geq N$. For large K, only the pole with smallest modulus, p_0, will dominate. Hence,

$$\Pr[X > K] \approx \frac{r_0}{p_0^{K+1}(1 - p_0)} \tag{5.40}$$

This approximation is called the *dominant pole* approximation with the residue at the simple pole p_0 equal to $r_0 = \lim_{z \to p_0} \varphi_X(z)(z - p_0)$.

The second approach is a large deviations approximation in discrete-time:

$$-\log \Pr[X > K] = -\log \sum_{j=K+1}^{\infty} \Pr[X = j]$$

$$\geq -\log \sum_{j=K+1}^{\infty} x^{j-K-1} \Pr[X = j] \qquad (x \in \mathbb{R} \text{ and } x \geq 1)$$

$$\geq -\log \left(x^{-K-1} \sum_{j=0}^{\infty} x^j \Pr[X = j] \right)$$

$$= (K + 1) \log x - \log \varphi_X(x) \tag{5.41}$$

This inequality holds for all real $x \geq 1$. To get the tightest bound, we determine the maximizer $x_{\max}$ of (5.41), thus $I(K) = \sup_{x \geq 1}[(K + 1) \log x - \log \varphi_X(x)]$. There exists such a supremum on account of the convexity of $I(K)$ because $\varphi_X(x)$ and $\log \varphi_X(x)$ are convex for $x \geq 1$ as shown in Section 5.5. Assuming that the maximum, say $x_{\max}$ exists, then it is solution of $x_{\max} = (K + 1)\frac{\varphi_X(x_{\max})}{\varphi_X'(x_{\max})}$ and the large deviations estimate becomes

$$\Pr[X > K] \leq e^{-[(K+1) \log x_{\max} - \log \varphi_X(x_{\max})]} = \varphi_X(x_{\max}) \, x_{\max}^{-(K+1)} \tag{5.42}$$

Observe that (5.42) can be obtained directly from (5.14) with $K = t + E[X]$. Comparing (5.42) and (5.40) indicates, for large K, that $x_{\max} = p_0$ because

$$\lim_{K \to \infty} \frac{-\log \Pr[X > K]}{K} = \log p_0 = \log x_{\max}$$

---

[5] In terms of the queue occupancy in ATM, the initial $\Pr[X = j]$-regime for $j < N$ reflects the cell scale, while the asymptotic regime $j \geq N$ refers to the burst scale.

Example A frequently appearing "dominant pole" (see, for example, the extinction probability of a Poisson branching process in Section 12.3, the M/D/1 queue in Section 14.5 and the size of the giant component in the random graph in Section 15.7.6.1) is the real zero ζ different from 1 of $e^{\lambda(z-1)} - z$. The trivial zero is $z = 1$. The non-trivial solution $e^{\lambda(\zeta-1)} = \zeta$ can be expressed as a Lagrange series (Markushevich, 1985, p. 94) for[6] $\lambda > 1$,

$$\zeta = e^{-\lambda} \sum_{n=0}^{\infty} \frac{(n+1)^{n-1}}{n!} \left(\lambda e^{-\lambda}\right)^n \tag{5.43}$$

An exact and fast converging expansion for ζ around $\lambda = 1$,

$$\zeta = 1 + \frac{2}{\lambda}\left[(1-\lambda) + \frac{(1-\lambda)^2}{3} + \frac{2(1-\lambda)^3}{9} + \frac{22(1-\lambda)^4}{135} + \frac{52(1-\lambda)^5}{405} + \frac{20(1-\lambda)^6}{189}\right.$$
$$\left. + \frac{3824(1-\lambda)^7}{42525} + \frac{1424(1-\lambda)^8}{18225} + \frac{15856(1-\lambda)^9}{229635} + \frac{11714672(1-\lambda)^{10}}{189448875} + O\left((1-\lambda)^{11}\right)\right] \tag{5.44}$$

is derived in Van Mieghem (1996) as the zero of $\frac{e^{\lambda(z-1)} - z}{z-1}$. The numerical data show that the approximation $\zeta \simeq \frac{1}{\lambda^2}$, which can be deduced from the series, is within 1% accurate for $0.84 < \lambda \leq 1$.

5.8 Problems

(i) Show that, for any set of n random variables $X_1, X_2, \ldots, X_n$, it holds that

$$\Pr\left[a_1 X_1 + a_2 X_2 + \cdots + a_n X_n > t\right] \leq \sum_{k=1}^{n} \Pr\left[X_k > \frac{t}{n a_k}\right] \tag{5.45}$$

where each coefficient a_k is a positive real number.

(ii) By using the harmonic, geometric and arithmetic mean inequality (5.3), derive the inequality

$$\left(\prod_{k=1}^{n} (x_k + y_k)\right)^{1/n} \geq \left(\prod_{k=1}^{n} x_k\right)^{1/n} + \left(\prod_{k=1}^{n} y_k\right)^{1/n} \tag{5.46}$$

where the vector components $x_k \geq 0$, $y_k \geq 0$ and $n \in \mathbb{N}_0$. Equality is reached when the vector $y = \alpha x$ for some real α.

(iii) Let $f(x)$ be convex on an interval I and positive for all nonzero $x \in I$. If $f(\theta x) = \theta^p f(x)$ for some real $p > 1$ and for all $x \in I$ and $\theta \geq 0$, show that $h(x) = (f(x))^{1/p}$ also convex is on the interval I.

---

[6] From (14.44) we observe for $a = b = 1$ and $z = \lambda$ that in (5.43) $\zeta = 1$ for all $0 \leq \lambda < 1$.

6

Limit laws

Limit laws lie at the heart of analysis and probability theory. Solutions of problems often considerably simplify in limit cases. For example, in Section 16.6.1, the flooding time in the complete graph with N nodes and exponentially distributed link weights can be computed exactly. However, the expression is unattractive, but, fortunately, the limit result for $N \to \infty$ is appealing. Many more results and deep discussions are found in the books of Feller (1970, 1971). In this chapter, we will mainly be concerned with sums of independent random variables, $S_n = \sum_{k=1}^n X_k$.

6.1 General theorems from analysis

In this section, we define modes of convergence of sequences of random variables and state (without proof) some general theorems that will be used later on.

6.1.1 Summability

We will need results from the analysis on summability[1]. First the discrete case is presented and then the continuous case.

Lemma 6.1.1 *Let $\{a_n\}_{n \geq 1}$ be a sequence of numbers with $\lim_{n \to \infty} a_n = a$, then the average of the partial sums converges to a,*

$$\lim_{n \to \infty} \frac{1}{n} \sum_{m=1}^n a_m = a \tag{6.1}$$

Proof: The demonstration of (6.1) is short enough to include here. The fact that there is a limit a of the sequence $a_1, a_2, \ldots$ implies that, for an arbitrary $\varepsilon > 0$, there exist a finite number n_0 such that, for all $n > n_0$, it holds that $|a_n - a| < \varepsilon$. Consider the average partial sum $s_n = \frac{1}{n} \sum_{m=1}^n a_m$ or, rewritten,

$$s_n - a = \frac{1}{n} \sum_{m=1}^{n_0} (a_m - a) + \frac{1}{n} \sum_{m=n_0}^n (a_m - a)$$

---

[1] In his classical treatise on *Divergent Series*, Hardy (1948) discusses Césaro, Abel, Euler and Borel summability in depth.

115

Hence,

$$|s_n - a| \leq \frac{1}{n} \sum_{m=1}^{n_0} |a_m - a| + \frac{1}{n} \sum_{m=n_0}^{n} |a_m - a| < \frac{c}{n} + \left(\frac{n - n_0}{n}\right)\varepsilon < \frac{c}{n} + \varepsilon$$

Since c is a constant, $\frac{c}{n}$ can be made arbitrarily small for n large enough such that $|s_n - a| < \varepsilon$, which is equivalent to (6.1).   □

In fact, as illustrated by many examples in Hardy (1948, Chapters I and II), relation (6.1) converges in more cases than $\lim_{n \to \infty} a_n = a$ does. For example, if $a_{2n} = 1$ and $a_{2n+1} = 0$, the limit $\lim_{n \to \infty} a_n$ does not exist, but (6.1) tends to $\frac{1}{2}$. Probabilistically, Lemma 6.1.1 is closely related to the sample mean and the Law of Large Numbers (Section 6.2).

The continuous case distinguishes between $\lim_{t \to \infty} g(t)$, which is called the point-wise limit (for sufficiently large t, all points t will be arbitrarily close to that limit) and between the limit $\lim_{t \to \infty} \frac{1}{t} \int_0^t g(u)du$, which is called the time average[2] of g.

Lemma 6.1.2 *If the pointwise limit* $\lim_{t \to \infty} g(t) = g_\infty$ *exists, then the time average*

$$\lim_{t \to \infty} \frac{1}{t} \int_0^t g(u)du = g_\infty.$$

Proof: The proof is analogous to that of Lemma 6.1.1 in the discrete case since $\lim_{t \to \infty} g(t) = g_\infty$ means that for an arbitrary $\varepsilon > 0$, there exists a finite number T such that, for all $t > T$, it holds that $|g(t) - g_\infty| < \varepsilon$. For any $t > T$,

$$\frac{1}{t} \int_0^t g(u)du - g_\infty = \frac{1}{t} \int_0^T (g(u) - g_\infty)\, du + \frac{1}{t} \int_T^t (g(u) - g_\infty)\, du$$

and

$$\left| \frac{1}{t} \int_0^t g(u)du - g_\infty \right| \leq \frac{1}{t} \left| \int_0^T (g(u) - g_\infty)\, du \right| + \frac{1}{t} \int_T^t |g(u) - g_\infty|\, du < \frac{c}{t} + \varepsilon \frac{t - T}{t}$$

Since c is a constant, the lemma follows by letting $t \to \infty$.   □

Both in Markov theory (Section 9.3.2) and in Little's Law (Section 13.7) these lemmas will be used.

6.1.2 Convergence of a sequence of random variables

A sequence $\{X_k\}_{k \geq 0}$ of random variables may converge to a random variable X in several ways. If

$$\Pr\left[\lim_{k \to \infty} |X_k - X| = 0\right] = 1$$

then the sequence $\{X_k\}_{k \geq 0}$ converges to X with probability 1 (w.p. 1) or almost surely (a.s.). This mode of convergence is denoted by $X_k \to X$ w.p. 1 or a.s. as $k \to \infty$.

---

[2] In summability theory, it is also known as the Cesaro limit of g.

If, for any $\epsilon > 0$,

$$\lim_{k \to \infty} \Pr\left[|X_k - X| > \epsilon\right] = 0$$

then it is said that the sequence $\{X_k\}_{k \geq 0}$ converges in probability or in measure to X. This mode of convergence is denoted by $X_k \overset{p}{\to} X$ as $k \to \infty$. Convergence in probability is a weaker notion of convergence than almost sure convergence[3]. Almost sure convergence implies convergence in probability, whereas convergence in probability means that there exists a subsequence of $\{X_k\}_{k \geq 0}$ that converges almost surely. An equivalent criterion for almost sure convergence is

$$\Pr\left[|X_k - X| > \epsilon \text{ i.o.}\right] = 0$$

where "i.o." stands for "infinitely often", thus for an infinite number of k.

If, for all x with the possible exception of a set of measure zero where $F_{X_n}(x) = \Pr\left[X_n \leq x\right]$ is discontinuous, the distributions

$$\lim_{k \to \infty} F_{X_n}(x) = F_X(x)$$

then the sequence $\{X_k\}_{k \geq 0}$ converges in distribution to X, denoted as $X_k \overset{d}{\to} X$ as $k \to \infty$ or, sometimes, in mixed form as $X_k \overset{d}{\to} F_X$ as $k \to \infty$.

If, for $1 \leq q$,

$$\lim_{k \to \infty} E\left[|X_k - X|^q\right] = 0$$

then the sequence $\{X_k\}_{k \geq 0}$ converges to X in L^q, the space of all functions f for which $\int_{-\infty}^{\infty} |f(x)|^q dx < \infty$. The most common values of q are 1, 2 and $q = \infty$. This convergence is also called convergence in norm (see Appendix A.4).

The Markov inequality (5.12)

$$\Pr\left[|X_k - X| \geq \epsilon\right] \leq \frac{E\left[|X_k - X|\right]}{\epsilon}$$

shows that convergence in mean ($q = 1$) implies convergence in probability. Earlier, we mentioned that almost sure convergence implies convergence in probability: if $X_k \to X$ a.s, then $X_k \overset{p}{\to} X$ as $k \to \infty$. Moreover, convergence in probability implies convergence in distribution: if $X_k \overset{p}{\to} X$, then $X_k \overset{d}{\to} X$ as $k \to \infty$. The proof (see Billingsley (1995, p. 330)) relies on reasoning with the events $\{X < x - \epsilon\}$, $\{|X_k - X| > \epsilon\}$ and $\{X_k \leq x\}$ using the Boole's inequalities(2.9).

In general, it is fair to say that the convergence of sequences belong to the most complicated topics in both analysis and probability theory. In many limit theorems, for example, the Law of Large Numbers in Section 6.2 and Little's Law in Section 13.7, the art consists in proving the theorem with the least possible number of assumptions or in its most widely applicable form.

---

[3] See e.g. Billingsley (1995, p. 70) for a proof.

6.1.3 List of general theorems

Theorem 6.1.3 (Continuity Theorem) *Let $\{F_n\}_{n\geq 1}$ be a sequence of distribution functions with corresponding probability generating functions $\{\varphi_n\}_{n\geq 1}$. If $\lim_{n\to\infty} \varphi_n(z) = \varphi(z)$ exists for all z, and, in addition if φ is continuous at $z = 0$, then there exists a limiting distribution function F with generating function φ for which $F_n \xrightarrow{d} F$.*

Proof: See e.g. Berger (1993, p. 51).   □

Theorem 6.1.4 (Dominated Convergence Theorem) *Let $\{f_n\}_{n\geq 1}$ and f be real functions and suppose that for each x*

$$\lim_{n\to\infty} f_n(x) = f(x)$$

If there exists a real function $g(x)$ such that $|f_n(x)| < g(x)$ and for which the random variable $g(X)$ has finite expectation, then

$$\lim_{n\to\infty} E\left[f_n(X)\right] = E\left[f(X)\right]$$

Proof: See e.g. Royden (1988, Chapter 4).   □

6.2 Law of Large Numbers

Theorem 6.2.1 (Weak Law of Large Numbers) *Let $\{X_k\}$ be a sequence of independent random variables each with distribution identical to that of the random variable X with $\mu = E\left[X\right]$. If the expectation $\mu = E\left[X\right]$ exists, then, for any $\epsilon > 0$,*

$$\lim_{n\to\infty} \Pr\left[\left|\frac{S_n}{n} - \mu\right| \geq \epsilon\right] = 0 \tag{6.2}$$

Proof[4]: Replacing X_k by $X_k - \mu$ demonstrates that, without loss of generality, we may assume that $\mu = 0$. Denote $U_n = \frac{S_n}{n}$, then $\varphi_{U_n}(z) = E\left[e^{-zU_n}\right] = E\left[e^{-zS_n/n}\right]$. Since the set $\{X_k\}$ is independent with common distribution, applying relation (2.75) yields

$$\varphi_{U_n}(z) = \left(\varphi_X\left(\frac{z}{n}\right)\right)^n$$

Since the expectation exists ($\mu = 0$), the Taylor expansion (2.41) of φ_X around $z = 0$ is $\varphi_X(z) = 1 + o(z)$ and $\varphi_{U_n}(z) = \left(1 + o\left(\frac{z}{n}\right)\right)^n$. Taking the logarithm, $\log(\varphi_{U_n}(z)) = n\log\left(1 + o\left(\frac{z}{n}\right)\right) = n.o\left(\frac{z}{n}\right) = o(z)$ for large n, from which we deduce that $\lim_{n\to\infty} \varphi_{U_n}(z) = 1$. By the Continuity Theorem 6.1.3, $\varphi_U(z) = E\left[e^{-zU}\right] = 1$ which implies that $U_n \xrightarrow{d} 0$. Hence, the sequence $\frac{S_n}{n}$ converges in distribution to $\mu = 0$, which is equivalent to (6.2).   □

The Weak Law of Large Numbers is a general result of the behavior of the sample

---

[4] An alternative proof is given in Feller (1970, p. 247–248).

mean $\frac{S_n}{n}$ of independent random variables with same existing expectation μ. It is
weak in the sense that only convergence in probability is established. For large
n, the weak law of large numbers states that the sample mean $\frac{S_n}{n}$ will be close
(less than an arbitrary ϵ) to the expectation with high probability. It does not
imply that $\left|\frac{S_n}{n} - \mu\right|$ remains small for all large n. In fact, large fluctuations in
$\left|\frac{S_n}{n} - \mu\right|$ can happen; the Weak Law of Large Numbers only concludes that large
values of $\left|\frac{S_n}{n} - \mu\right|$ occur with (very) small probability. For example, in a coin-
tossing experiment with a fair coin such that $\Pr[X_k = 1] = \Pr[X_k = 0] = \mu = \frac{1}{2}$ in
n-trials, the sequence of always head $\{X_k = 1\}_{1 \le k \le n}$ is possible with probability
2^{-n} and $\frac{S_n}{n} = 1 > \mu$. But, only for $n \to \infty$, the probability of this "always head
sequence" is impossible ($\lim_{n \to \infty} 2^{-n} = 0$). For all finite n there is a non-zero
probability of having a large deviation from the mean.

If we assume in addition to the existence of the expectation that also the variance $\mathrm{Var}[X]$ exists, the Weak Law follows from the Chebyshev inequality (5.13).
This exemplifies the increasingly complexity if less restrictions in the theorems are
assumed. Indeed, using (2.60) for independent random variables, $\mathrm{Var}\left[\frac{S_n}{n}\right] = \frac{\mathrm{Var}[X]}{n}$
and the Chebyshev inequality (5.13) gives

$$\Pr\left[\left|\frac{S_n}{n} - \mu\right| \ge \epsilon\right] \le \frac{\mathrm{Var}[X]}{n\epsilon^2}$$

which tends to zero for any fixed ϵ and finite $\mathrm{Var}[X]$. In fact, with the additional
assumption of a finite variance $\mathrm{Var}[X]$, a much more precise result can be proved
known as the Central Limit Theorem (Section 6.3). We remark that the Weak Law
of Large Numbers also holds in the case $\mathrm{Var}[X]$ does not exist.

Theorem 6.2.2 (Strong Law of Large Numbers) *Let $\{X_k\}$ be a sequence of
independent random variables each with distribution identical to that of the random
variable X with $\mu = E[X]$. If the expectation $\mu = E[X]$ and variance $\mathrm{Var}[X]$
exists, then,*

$$\Pr\left[\lim_{n \to \infty} \frac{S_n}{n} = \mu\right] = 1 \tag{6.3}$$

Proof: See e.g. Feller (1970, p. 259–261), Berger (1993, pp. 46–48) or Wolff (1989, pp. 40–41).
Their proof is based on the Kolmogorov criterion: the convergence of $\sum_{k=1}^{\infty} \frac{\mathrm{Var}[X_k]}{k^2}$ is a sufficient
condition for the Strong Law of Large Numbers for independent random variables X_k with mean
$E[X_k]$ and variance $\mathrm{Var}[X_k]$. If the existence of $E[X^4]$ is assumed, Ross (1996, pp. 56–58) and
Billingsley (1995, p. 85) provide a different proof. Wolff (1989, pp. 41–42) remarks that both the
Weak and Strong Laws hold under much weaker conditions: it is only needed that the X_k are not
correlated. In other words, $\frac{S_n}{n} \to \mu$ w.p. 1 implies $E[X_k] = \mu$ even if $\mathrm{Var}[X_k] = \infty$.   □

The Strong Law of Large Numbers roughly states that $\left|\frac{S_n}{n} - \mu\right|$ remains small
for sufficiently large n with overwhelming probability. The importance of the Law
of Large Numbers is the mathematical foundation of the intuition that the sample
mean is the best estimator.

Theorem 6.2.3 (Law of the Iterated Logarithm) *Let $\{X_k\}$ be a sequence of independent random variables each with distribution identical to that of the random variable X with $\mu = E[X]$ and, if $Var[X]$ exists, then,*

$$\Pr\left[\limsup_{n\to\infty} \frac{S_n - n\mu}{\sigma\sqrt{2n\log\log n}} = 1\right] = 1 \tag{6.4}$$

Proof: See e.g. Billingsley (1995, p. 154–156) or Feller (1970, Section VIII.5)[5].□

In addition to the Weak and Strong Laws of Large Numbers, the Law of the Iterated Logarithm provides information about large values of $\left|\frac{S_n}{n} - \mu\right|$. Specifically, it states that the bound $\left|\frac{S_n}{n} - \mu\right| \le \sigma\sqrt{\frac{2\log\log n}{n}}$ holds almost surely. The latter means that it is satisfied infinitely often and only for a finite number of values of n, the converse $\left|\frac{S_n}{n} - \mu\right| > \sigma\sqrt{\frac{2\log\log n}{n}}$ may occur.

6.3 Central Limit Theorem

Theorem 6.3.1 (Central Limit Theorem) *Let $\{X_k\}$ be a sequence of independent random variables each with distribution identical to that of the random variable X with finite $\mu = E[X]$ and $\sigma^2 = Var[X]$. Then $\frac{S_n - n\mu}{\sigma\sqrt{n}} \xrightarrow{d} N(0,1)$ or, explicitly,*

$$\Pr\left[\frac{S_n - n\mu}{\sigma\sqrt{n}} \le x\right] \to \Phi(x) = \frac{1}{\sqrt{2\pi}} \int_{-\infty}^{x} e^{-\frac{t^2}{2}} dt$$

Proof: Without loss of generality, we may confine to normalized random variables – replace X_k by $\frac{X_k - \mu}{\sigma}$ – such that $\mu = 0$ and $\sigma = 1$. Consider the scaled random variable $U_n = \alpha_n S_n$, where α_n is a real number depending on n and to be determined later. Similarly as in the proof of the Weak Law of Large Numbers, we find that $\varphi_{U_n}(z) = (\varphi_X(\alpha_n z))^n$. Due to the existence of the variance, the Taylor expansion (2.41) of φ_X around $z = 0$ is known with higher precision as $\varphi_X(z) = 1 + \frac{z^2}{2} + o(z^2)$. For sufficiently small z, the logarithm

$$\log(\varphi_{U_n}(z)) = n\log\left(1 + \frac{\alpha_n^2 z^2}{2} + o(\alpha_n^2 z^2)\right) = n\frac{\alpha_n^2 z^2}{2} + o(n\alpha_n^2 z^2)$$

only converges to a finite (non-zero) number if $\alpha_n = O\left(\frac{1}{\sqrt{n}}\right)$. Choosing the simplest function that satisfies this condition, $\alpha_n = \frac{1}{\sqrt{n}}$, leads to $\lim_{n\to\infty}\log(\varphi_{U_n}(z)) = \frac{z^2}{2}$ or, since the logarithm is a continuous, increasing function, $\lim_{n\to\infty}\varphi_{U_n}(z) = \exp\left(\frac{z^2}{2}\right)$. The transform (3.22) shows that the corresponding limit random variable is a Gaussian $N(0,1)$. The theorem then follows by virtue of the Continuity Theorem 6.1.3. □

---

[5] Feller also mentions sharper bounds.

An alternative formulation of the Central Limit Theorem is that the k-fold convolution of any probability density function converges to a Gaussian probability distribution, $f_X^{(k*)}(x) \to \frac{1}{\sigma\sqrt{2\pi}} \exp\left[-\frac{(x-\mu)^2}{2\sigma^2}\right]$ with $\mu = kE[X]$ and $\sigma^2 = k\text{Var}[X]$. Both the Law of Large Numbers and the Central Limit Theorem can be shown to be valid for a surprisingly large class of sequences $\{X_k\}$ where each random variable may have a different distribution. The conditions for the extension of the Central Limit Theorem are summarized in the Lindeberg conditions (Feller, 1971, p. 263). An example where the sum of independent random variables tend to a different limit distribution than the Gaussian appears in Section 16.6.1.

If higher moments are known, the convergence to the Gaussian distribution can be bounded. Feller (1971, Chapter XVI) devotes a chapter on expansions related to the Central Limit Theorem culminating in the Berry–Esseen Theorem.

Theorem 6.3.2 (Berry–Esseen Theorem) *Let $\{X_k\}$ be a sequence of independent random variables each with distribution identical to that of the random variable X with finite $\mu = E[X]$, $\sigma^2 = Var[X]$ and $\rho = E\left[\frac{|X-\mu|^3}{\sigma^3}\right]$. Then, with $C = 3$,*

$$\sup_x \left| \Pr\left[\frac{S_n - n\mu}{\sigma\sqrt{n}} \leq x \right] - \Phi(x) \right| \leq \frac{C\rho}{\sqrt{n}} \tag{6.5}$$

Proof: See e.g. Feller (1971, Section XVI.5). The constant C can be slightly improved to $C \leq 2.05$. $\qquad\square$

As an example of the rate of convergence towards the Gaussian distribution, the k-fold convolutions of the uniform density given by (3.30) is plotted in Fig. 6.1 together with the Gaussian approximation (3.19).

6.4 The Law of Proportionate Effect

As mentioned in Crow and Shimizu (1988), Kapteyn considered in 1903 the equation

$$X_j - X_{j-1} = \alpha_j f(X_{j-1})$$

where the set $\{\alpha_j\}_{1 \leq j \leq n}$ of random variables is mutually independent and identically distributed, equal to the distribution of the random variable α with mean $E[\alpha]$ and variance $\text{Var}[\alpha]$. Moreover, the set $\{\alpha_j\}_{1 \leq j \leq n}$ of random variables is also independent of the random variables $X_1, X_2, \ldots, X_n$. The special case where $f(x) = x$ reduces to

$$X_j = (1 + \alpha_j) X_{j-1} \tag{6.6}$$

and the process that determines the sequence $X_1, X_2, \ldots, X_n$, given X_0, is said to obey the *law of proportionate effect*, which was first introduced by Gibrat (1930). After iterating the equation (6.6), we obtain

$$X_n = X_0 \prod_{j=1}^{n} (1 + \alpha_j) \tag{6.7}$$

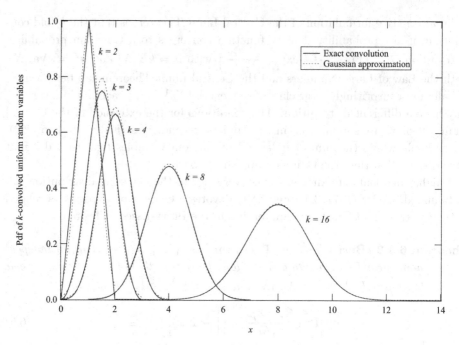

Fig. 6.1. Both the exact $f_U^{(k*)}(x)$ with $f_U(x) = 1_{0 \le x \le 1}$ and the Gaussian approximation for several values of k.

By the Central Limit Theorem 6.3.1 and assuming that any $\alpha_j > -1$, the sum $S_n = \sum_{j=1}^{n} \log(1 + \alpha_j)$ of the i.i.d. random variables $\{\log(1 + \alpha_j)\}_{j \ge 1}$, each with distribution identical to that of $\log(1 + \alpha)$ with (finite) mean $E[\log(1 + \alpha)] = \mu$ and variance $\sigma^2 = \text{Var}[\log(1 + \alpha)]$, converges to

$$\frac{S_n - n\mu}{\sigma \sqrt{n}} \xrightarrow{d} N(0, 1)$$

which implies that $S_n = \log\left(\prod_{j=1}^{n}(1 + \alpha_j)\right) \xrightarrow{d} N(n\mu, n\sigma^2)$. Equivalently, $e^{S_n} = \prod_{j=1}^{n}(1 + \alpha_j)$ tends, for large n, to a lognormal distribution (3.53) with parameters $n\mu$ and $n\sigma^2$. Hence, we have shown that, for large n, X_n is asymptotically lognormally distributed with parameters $n\mu$ and $n\sigma^2$, that are linear in n. In summary, we have proven:

Theorem 6.4.1 (Law of Proportionate Effect) *Let $\{\alpha_k\}_{k \ge 1}$ with $\alpha_k > -1$ be a sequence of independent random variables each with distribution identical to that of the random variable α with finite mean $E[\alpha]$ and variance $\text{Var}[\alpha]$. Let the set $\{\alpha_k\}_{k \ge 1}$ be independent of the set $\{X_k\}_{k \ge 1}$ of random variables, which obey the governing equation*

$$X_k - X_{k-1} = \alpha_k X_{k-1}$$

Then the random variable X_n tends, for large n, to a lognormal random variable

$$\Pr\left[X_n \leq x\right] \xrightarrow{d} \frac{1}{n\sigma\sqrt{2\pi}} \int_{-\infty}^{\log x} \exp\left[-\frac{(t-n\mu)^2}{2n^2\sigma^2}\right] dt$$

where $\mu = E\left[\log\left(1+\alpha\right)\right]$ and $\sigma^2 = Var[\log\left(1+\alpha\right)]$.

There is a continuous variant of the law of proportionate effect. In biology, the growth in the number $n\left(t\right)$ of items of a same species over time t can be modelled by the following first-order differential equation:

$$\frac{dn\left(t\right)}{dt} = r\left(t\right)n\left(t\right)$$

which relates the growth (change in the population) as proportional to the population $n\left(t\right)$ and the proportionality factor $r\left(t\right)$ is time dependent. The general solution is, for $t > a$,

$$\log n\left(t\right) = \log n\left(a\right) + \int_a^t r\left(u\right) du$$

When we additionally assume that $r\left(t\right)$ changes at some times $a = t_0 < t_1 < t_2 < \cdots < t_m = t$, where t_j are random time moments, then

$$\int_a^t r\left(u\right) du = \sum_{j=1}^m \int_{t_{j-1}}^{t_j} r\left(u\right) du = \sum_{j=1}^m R_j$$

where

$$R_j = \int_{t_{j-1}}^{t_j} r\left(u\right) du = r\left(\xi_j\right)\left(t_j - t_{j-1}\right) \text{ and } \xi_j \in [t_{j-1}, t_j]$$

is a random variable with mean $\mu_j = E\left[r\left(\xi_j\right)\left(t_j - t_{j-1}\right)\right]$. Assuming that the Central Limit Theorem can be applied, the set of random variables $\{R_j\}_{1 \leq j \leq m}$ tends to a Gaussian $N\left(m\mu, m\sigma^2\right)$, and the lognormal distribution of $n\left(t\right)$ for large t then follows in the usual way. Again, the mean is linear in the time t because $\frac{t-a}{m} = E\left[\Delta t\right]$, the average time-spacing.

6.5 Logarithm of a sum of random variables

Marlow (1967) provided the explanation of the numerical observation that a sum of independent lognormal random variables is approximately also a lognormal. In fact, Marlow proved a more general theorem from which the empirical observation follows. The intuition behind Marlow's Theorem 6.5.1 is that, if $S_n = \sum_{j=1}^n X_j$ with $E\left[X_j\right] = \mu$ and $Var[X_j] = \sigma^2$, then by the Law of Large Numbers (Section 6.2), we expect $\frac{S_n}{n\mu} \approx 1$. Around $x = 1$, we approximate $\log x \approx x - 1$ so that $\log \frac{S_n}{n\mu} \approx \frac{S_n - n\mu}{n\mu}$. After multiplying both sides by $\frac{n\mu}{\sqrt{n}\sigma}$, we have by the Central Limit Theorem 6.3.1 that

$$\frac{\mu\sqrt{n}}{\sigma} \log \frac{S_n}{n\mu} \approx \frac{S_n - n\mu}{\sqrt{n}\sigma} \xrightarrow{d} N\left(0, 1\right)$$

This suggestive argument, which results in (6.8), is proven precisely below.

Theorem 6.5.1 (Marlow) *Let $\{X_n\}_{n\geq 1}$ be a sequence of positive random variables. Suppose that there exist sequences of positive real numbers $\{a_n\}_{n\geq 1}$ and $\{b_n\}_{n\geq 1}$, and a distribution function F such that (i)*

$$\lim_{n\to\infty} \Pr\left[\frac{X_n - a_n}{b_n} \leq x\right] = F(x)$$

for each point x where $F(x)$ is continuous and (ii) $\lim_{n\to\infty} \frac{b_n}{a_n} = 0$. Then at each point x where $F(x)$ is continuous, it holds that

$$\lim_{n\to\infty} \Pr\left[\frac{a_n}{b_n} \log\left(\frac{X_n}{a_n}\right) \leq x\right] = F(x)$$

Proof: Let x be a continuity point of F and let $\varepsilon > 0$ be given. Since any distribution function F has at most a countable number of discontinuities, there exist a $\delta > 0$ such that F is continuous at $x + \delta$ and $F(x+\delta) - F(x) < \varepsilon$. Next, define

$$U_n = \frac{X_n - a_n}{b_n} \text{ and } V_n = \frac{a_n}{b_n} \log\left(\frac{X_n}{a_n}\right)$$

and consider

$$|\Pr[V_n \leq x] - F(x)| \leq \Delta_n(x) + |\Pr[U_n \leq x] - F(x)|$$

where

$$\Delta_n(x) = |\Pr[V_n \leq x] - \Pr[U_n \leq x]|$$

By assumption (i), we have that

$$\lim_{n\to\infty} |\Pr[V_n \leq x] - F(x)| \leq \lim_{n\to\infty} \Delta_n(x)$$

To complete the proof, it suffices to demonstrate that $\lim_{n\to\infty} \Delta_n(x) = 0$. The inequality[6] $\log x \leq x - 1$ for $x > 0$ shows that $V_n \leq U_n$ for all n. Hence, the event $\{U_n \leq x\}$ implies $\{V_n \leq x\}$ and

$$1 = \Pr[\{U_n \leq x\} \cup \{U_n > x\}] = \Pr[\{V_n \leq x\} \cup \{U_n > x\}]$$

Using (2.4) yields

$$1 = \Pr[V_n \leq x] + \Pr[U_n > x] - \Pr[\{V_n \leq x\} \cap \{U_n > x\}]$$

so that, with $\Pr[\{U_n > x\}] = 1 - \Pr[U_n \leq x]$,

$$\Delta_n(x) = \Pr[\{V_n \leq x\} \cap \{U_n > x\}]$$

The definition of V_n and U_n shows that $U_n = \frac{a_n}{b_n}\left(\exp\left(\frac{b_n V_n}{a_n}\right) - 1\right)$ so that

$$\Pr[\{V_n \leq x\} \cap \{U_n > x\}] = \Pr\left[x < U_n \leq \frac{a_n}{b_n}\left(\exp\left(\frac{b_n x}{a_n}\right) - 1\right)\right]$$

Applying the inequality[7] $e^y \leq 1 + y e^y$, valid for all y, yields

$$0 \leq \Delta_n(x) \leq \Pr\left[x < U_n \leq x\exp\left(\frac{b_n x}{a_n}\right)\right]$$

---

[6] Since $\log x$ is concave for $x > 0$, any tangent line lies above the curve, hence, also the tangent line at $x = 1$.

[7] Deduced from the Taylor expansion (5.10) for $n = 1$ combined with $\theta \leq 1$.

The definition $\frac{b_n}{a_n} > 0$ and the second assumption (ii) imply that there exist a natural number n_0 such that, for all $n \geq n_0$, $x < x \exp\left(\frac{b_n x}{a_n}\right) \leq x + \delta$. Hence, if $n \geq n_0$, then

$$0 \leq \Delta_n(x) \leq \Pr[x < U_n \leq x + \delta] = F(x + \delta) - F(x) < \varepsilon$$

After choosing ε arbitrary small, the proof is complete.   □

Theorem 6.5.1 provides sufficient conditions to transform limit theorems for sums of random variables to limit theorems for logarithms of sums. Indeed, when combining the Central Limit Theorem 6.3.1 with Marlow's Theorem 6.5.1, we find that each sum $S_n = \sum_{j=1}^n X_j$ of independent random variables, not necessarily identically distributed, but obeying the Lindeberg conditions of the Central Limit Theorem (Feller, 1971, p. 263), also satisfies

$$\lim_{n \to \infty} \Pr\left[\frac{\sum_{j=1}^n E[X_j]}{\sqrt{\sum_{j=1}^n \text{Var}[X_j]}} \log\left(\frac{S_n}{\sum_{j=1}^n E[X_j]}\right) \leq x\right] = \Phi(x) \qquad (6.8)$$

The particular example, where S_n is a sum of lognormals, was Marlow's driver, because the pdf of a sum of independent lognormal random variables is not known in closed form. Writing a lognormal random variable as $X = e^Y$, where $Y = N(\mu, \sigma^2)$ is a Gaussian or normal random variable, Marlow's Theorem 6.5.1 thus implies for large n, with $\mu_n = \log \sum_{j=1}^n E[X_j]$ and $s_n = \frac{\sqrt{\sum_{j=1}^n \text{Var}[X_j]}}{\sum_{j=1}^n E[X_j]}$, that

$$\sum_{j=1}^n e^{Y_j} \xrightarrow{d} e^{N(\mu_n, s_n^2)}$$

which demonstrates that a sum of lognormals tends to a lognormal. However, the Central Limit Theorem 6.3.1 also predicts convergence towards a Gaussian. In addressing this appearent anomaly, Marlow (1967) argues, based on simulations, that the convergence of a sum of lognormals towards a lognormal is better than towards a normal.

Marlow's Theorem 6.5.1 is a special case of the so-called *delta method*, which generalizes the Central Limit Theorem 6.3.1 to (well-behaved) functions of a random variable.

Theorem 6.5.2 (Delta method) *Let $\{X_n\}_{n \geq 1}$ be a sequence of random variables that satisfy $\sqrt{n}(X_n - \theta) \xrightarrow{d} N(0, \sigma^2)$ for $n \to \infty$, where θ and σ^2 are finite constants, then*

$$\sqrt{n}(g(X_n) - g(\theta)) \xrightarrow{d} N\left(0, (g'(\theta)\sigma)^2\right) \qquad (6.9)$$

for any function $g(x)$ with a non-zero, existing derivative $g'(\theta)$.

Proof: We sketch a proof of the theorem by assuming that $g'(x)$ is a continuous function and by omitting the precise convergence arguments as in the proof of Marlow's Theorem 6.5.1. By the mean-value theorem (5.8)

$$g(X_n) - g(\theta) = (X_n - \theta)g'(\xi)$$

where $\xi \in [X_n, \theta]$. Since $X_n \xrightarrow{p} \theta$, which is equivalent to $\lim_{n\to\infty} \Pr[|X_n - \theta| > \epsilon] = 0$ and since $|\xi - \theta| \leq |X_n - \theta|$, we conclude that $\xi \xrightarrow{p} \theta$ for $n \to \infty$. Since $g'(x)$ is continuous, the Continuity Theorem 6.1.3 implies that $g'(\xi) \xrightarrow{p} g'(\theta)$, from which

$$\sqrt{n}\left(g\left(X_n\right) - g\left(\theta\right)\right) \xrightarrow{p} \sqrt{n}(X_n - \theta)g'(\theta)$$

Finally, using $\sqrt{n}\left(X_n - \theta\right) \xrightarrow{d} N\left(0, \sigma^2\right)$, we arrive at (6.9). $\qquad \square$

6.6 Extremal distributions

6.6.1 Scaling laws

In this section, limit properties of the maximum and minimum of a set $\{X_k\}$ of independent random variables are discussed. For simplicity, we assume that all random variables X_k have identical distribution $F(x) = \Pr[X \leq x]$ such that (3.40) and (3.39) simplify to

$$\Pr\left[\max_{1\leq k\leq m} X_k \leq x\right] = F^m(x)$$

$$\Pr\left[\min_{1\leq k\leq m} X_k > x\right] = (1 - F(x))^m$$

Consider the limit process when $m \to \infty$. Let $\{x_m\}$ be a sequence of real numbers. Then, confining to the maximum first,

$$\log\left(\Pr\left[\max_{1\leq k\leq m} X_k \leq x_m\right]\right) = m \log F(x_m)$$

Since $0 \leq F(x_m) \leq 1$ and since the logarithm has a Taylor expansion $\log(1 - x) = -\sum_{k=1}^{\infty} \frac{x^k}{k}$ around $x = 0$ and convergent for $|x| < 1$, we rewrite the right-hand side as $\log F(x_m) = \log\left[1 - (1 - F(x_m))\right]$ and, after expansion,

$$\log\left(\Pr\left[\max_{1\leq k\leq m} X_k \leq x_m\right]\right) = -m\left(1 - F(x_m)\right) + o\left[m\left(1 - F(x_m)\right)\right]$$

If $\lim_{m\to\infty} m\left(1 - F(x_m)\right) = \xi$, we arrive at

$$\lim_{m\to\infty} \Pr\left[\max_{1\leq k\leq m} X_k \leq x_m\right] = e^{-\xi} \qquad (6.10)$$

Hence, by choosing an appropriate sequence $\{x_m\}$ such that ξ is finite (and preferably non-zero), a scaling law for the maximum of a sequence can be obtained and, similarly, for the minimum, if $\lim_{m\to\infty} m\left(F(x_m)\right) = \zeta$,

$$\lim_{m\to\infty} \Pr\left[\min_{1\leq k\leq m} X_k > x_m\right] = e^{-\zeta} \qquad (6.11)$$

The distribution of $\lim_{m\to\infty} \min_{1\leq k\leq m} X_k$ and $\lim_{m\to\infty} \max_{1\leq k\leq m} X_k$ are called extremal distributions.

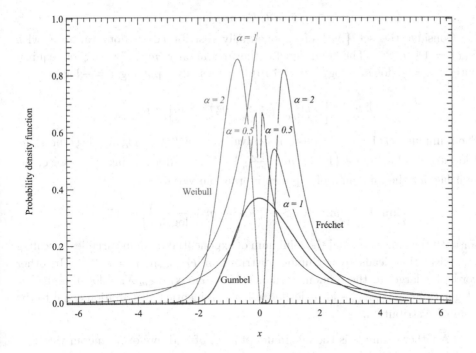

Fig. 6.2. The probability density function of the three types of extremal distributions.

6.6.2 The Law of Extremal Types

Two distribution functions F and G are said to be of the same type if there exist constants $a > 0$ and b for which $F(ax + b) = G(x)$ for all x.

Theorem 6.6.1 (Law of Extremal Types) *Any extremal distribution of a sequence of i.i.d random variables can only have one of three types:*

$$\textit{1. Gumbel} \quad F(x) = e^{-e^{-x}}$$
$$\textit{2. Fréchet} \quad F(x) = e^{-x^{-\alpha}} 1_{x \geq 0}$$
$$\textit{3. Weibull} \quad F(x) = e^{-(-x)^{\alpha}} 1_{x < 0} + 1_{x \geq 0}$$

where $\alpha > 0$.

Proof: See e.g. Berger (1993, pp. 65–69). $\qquad\qquad\qquad\qquad\qquad\qquad\qquad$ □

The generality of this theorem is appealing: any maximum or minimum of a set of i.i.d. random variables has (apart from the scaling constants a and b) one of the above three types. The corresponding probability density functions are plotted in Fig. 6.2.

6.6.3 Examples

1. Consider the set $\{X_k\}$ of exponentially distributed random variables with $F(x) = 1 - e^{-\alpha x}$. The condition for the maximum is $me^{-\alpha x_m} \to \xi$ or, equivalently, $x_m = \frac{1}{\alpha}(\log m - \log \xi)$ and (6.10) becomes, after putting $x = -\log \xi$,

$$\lim_{m \to \infty} \Pr \left[\max_{1 \le k \le m} X_k \le \frac{1}{\alpha}(\log m + x) \right] = e^{-e^{-x}}$$

The minimum $m(1 - e^{-\alpha x_m}) \to \zeta$ is equivalent to $e^{-\alpha x_m} \sim \log m - \log \zeta$ or $x_m \sim \frac{1}{\alpha}\log(\log m - \log \zeta) \sim \frac{1}{\alpha}\left(\log \log m - \frac{\log \zeta}{\log m}\right)$. Hence, after putting $x = -\log \zeta$, the limit law for the minimum of exponential random variables is

$$\lim_{m \to \infty} \Pr \left[\min_{1 \le k \le m} X_k > \frac{1}{\alpha}\left(\log \log m + \frac{x}{\log m}\right) \right] = e^{-e^{-x}}$$

For both the maximum and the minimum of exponential random variables, a scaling law exists that leads to a Gumbel distribution $F_{Gumbel}(x) = e^{-e^{-x}}$. In other words, for large m, the random variables $M = \alpha \max_{1 \le k \le m} X_k - \log m$ and $N = \alpha \log m \, (\min_{1 \le k \le m} X_k) - \log m \log \log m$ have an identical distribution equal to the Gumbel distribution.

2. Another example is the maximum of a set of i.i.d. uniform random variables $\{U_k\}$ in $[0, 1]$ with $F(x) = x$ for $0 \le x \le 1$. Since $m(1 - x_m) \to \xi$ or, equivalently, $x_m \to 1 - \frac{\xi}{m}$ with $0 \le x_m \le 1$ we have, after putting $x = \xi$ with $x \ge 0$,

$$\lim_{m \to \infty} \Pr \left[\max_{1 \le k \le m} U_k \le 1 - \frac{x}{m} \right] = e^{-x}$$

3. Consider a rectangular lattice with size z_1 and z_2 and with independent and identical, uniformly distributed link weights on $(0, 1]$ between each lattice point. The number of lattice points (nodes) equals $N = (z_1 + 1)(z_2 + 1)$ and the number of links is $L = 2z_1 z_2 + (z_1 + z_2)$. The shortest hop path between two diagonal corner points consists of $h = z_1 + z_2$ hops. The weight W_h of such a h hop path is the sum of h independent uniform random variables with distribution specified in (3.29),

$$F(x) = \Pr[W_h \le x] = \frac{1}{h!} \sum_{j=0}^{h} \binom{h}{j}(-1)^j (x - j)^h 1_{j \le x}$$

In particular, $\Pr[W_h \le h] = 1$ and for small $x < 1$ it holds that $F(x) = \frac{x^h}{h!}$. The precise computation of the minimum weight of a h hop path in a lattice is difficult due to dependence among those h hop paths and we content ourselves here with an approximate estimate. If we neglect the dependence of the h hops paths due to possible overlap, then the minimum weight among all h hop paths can be approximated by (6.11) because the number[8] $m = \binom{z_1 + z_2}{z_1} = \frac{h!}{z_1! z_2!}$ of those h hop

---

[8] Any path in a rectangular lattice can be represented by a sequence of r(ight), l(eft), u(p) and

paths is large. The limit sequence must obey $m(F(x_m)) \to \zeta$ for sufficiently large m, which implies that $F(x_m)$ must be small or, equivalently, x_m must be small. Hence, $m\frac{x_m^h}{h!} = \zeta$ or $x_m = \left(\frac{h!\zeta}{m}\right)^{\frac{1}{h}}$. The limit law (6.11) for the minimum weight $W = \min_{1 \le k \le m} W_{h,k}$ of the shortest hop path in a rectangular lattice is

$$\lim_{m \to \infty} \Pr\left[\min_{1 \le k \le m} W_{h,k} > \left(\frac{h!x}{m}\right)^{\frac{1}{h}}\right] = e^{-x}$$

In other words, the random variable $\frac{mW^h}{h!}$ tends to an exponential random variable with mean 1 for large $m = \frac{h!}{z_1!z_2!}$ or

$$\Pr[W \le y] \approx 1 - \exp\left(-m\frac{y^h}{h!}\right)$$

From (2.36), the mean shortest weight of a h hop path equals

$$E[W] = \int_0^\infty (1 - F_W(x))\, dx \approx \int_0^\infty \exp\left(-m\frac{x^h}{h!}\right) dx = \Gamma\left(1 + \frac{1}{h}\right)(z_1!z_2!)^{\frac{1}{h}}$$

For a square lattice where $z_1 = z_2 = \frac{h}{2}$, we have $E[W] = \Gamma\left(1 + \frac{1}{h}\right)\left(\left(\frac{h}{2}\right)!\right)^{\frac{2}{h}}$. Using Stirling's formula (Abramowitz and Stegun, 1968, Section 6.1.38) for the factorial $h! = \sqrt{2\pi}h^{h+\frac{1}{2}}e^{-h+\frac{\theta}{12h}}$, where $0 < \theta < 1$, for large h, the mean $E[W]$ increases approximately linearly in the number of hops h,

$$E[W] \simeq \left(\frac{h}{2e}\right)\left(\sqrt{\pi h}e^{\frac{\theta}{12h}}\right)^{\frac{2}{h}} \approx \frac{h}{2e}$$

The mean weight of a link of the shortest h hop path is roughly $\frac{1}{2e} \approx 0.184$.

In spite of the fact that path dependence (overlap) has been ignored in the computation of the minimum weight, $E[W] = O(h) = O\left(\sqrt{N}\right)$ is correct. However, the approximate analysis does *not* give the correct prefactor in $E[W]$ nor the correct limit pdf, which turns out to be Gaussian. Hence, if random variables are *not* independent, Theorem 6.6.1 does not apply. Finally, a shortest h hop path is not necessarily the overall shortest path because it is possible – though with small probability – that the overall shortest path has $h + 2j$ hops with $j > 0$.

4. The probability density function of the longest shortest path The most commonly used process that informs each node about changes in a network topology (e.g. an autonomous domain) is called flooding: every router forwards the packet on all interfaces except for the incoming one and duplicate packets are discarded. Flooding is particularly simple and robust since it progresses, in fact, along all possible paths from the emitting node to the receiving node.

---

d(own), which is called an encoded path word. The encoded path word of the shortest hop path between diagonal corner points consists of z_1 r's (or l's) and z_2 d's (or u's). The total number of these paths equals $\binom{z_1+z_2}{z_1}$. Two paths coincide in a same lattice point at $g \le h$ hops from the source node if their encoded path word has the same sum of r's and d's in the first g letters. The number of overlapping links between two paths equals the number of the same consecutive letters (r or d) in a block after the same sum of r's and d's in the encoded path words. Checking for overlap between h hop paths requires a comparison of $\binom{z_1+z_2}{z_1}!$ permutations in the encoded path words.

Hence, a flooded packet reaches a node in the network in the shortest possible time (if overhead in routers are ignored). Therefore, the interesting problem lies in the determination of the flooding time T_N, which is the minimum time needed to inform the last node in a network with N nodes. Only after T_N, all topology databases at each router in the network are again synchronized, i.e. all routers possess the same topology information. Rather than investigating the flooding time T_N (for which we refer to Section 16.6), the largest number of traversed routers (hops) or the longest shortest path from the emitting node to the furthermost node in its shortest path tree is computed.

The number of hops, in short the hopcount H_N, along the shortest path between two arbitrary nodes in a network containing N nodes is modeled subject to the following assumptions: (a) the hopcount H_N is a Poisson random variable with mean $E[H_N] = \lambda = \alpha \log N$ with $\alpha > 0$, which is motivated in Section 16.3.1; (b) the number of nodes N is very large[9]; (c) all shortest paths from the emitting node towards any other node in the network are independent. The problem reduces to compute the pdf of the random variable $\max_{1 \leq k \leq N-1} H_k$. The distribution function follows from (3.9) as

$$F_{H_N}(x) = \sum_{k=0}^{x} \frac{\lambda^k e^{-\lambda}}{k!} = 1 - \sum_{k=x+1}^{\infty} \frac{\lambda^k e^{-\lambda}}{k!}$$

The condition $\lim_{m \to \infty} m(1 - F(x_m)) = \xi$ becomes

$$\lim_{N \to \infty} N^{1-\alpha} \sum_{k=x_N+1}^{\infty} \frac{\lambda^k}{k!} = \xi$$

from which we must choose the appropriate x_N as function of N. Observe that the maximum term in the series has index $k = [\lambda]$, where the latter denotes the largest integer smaller or equal to λ. For, the ratio between two consecutive (positive) terms in the k-sum equals $\frac{a_k}{a_{k-1}} = \frac{\lambda}{k}$ such that, if $\lambda > k$, then $a_k > a_{k-1}$, implying that the terms increase, while if $\lambda < k$, the terms $a_k < a_{k-1}$ form a decreasing sequence. The series is rewritten as

$$\sum_{k=x_N+1}^{\infty} \frac{\lambda^k}{k!} = \frac{\lambda^{x_N+1}}{(x_N+1)!} \sum_{k=0}^{\infty} \frac{(x_N+1)!\lambda^k}{(x_N+1+k)!}$$

$$= \frac{\lambda^{x_N+1}}{(x_N+1)!} \left(1 + \frac{\lambda}{x_N+2} + \frac{\lambda^2}{(x_N+2)(x_N+3)} + \cdots\right)$$

We choose $x_N = [\lambda] + [\delta\lambda] \sim \lambda(1+\delta)$ for large N and, thus large λ, where δ must be related to ξ. The series then consists of decreasing terms. Moreover, for large λ,

$$\sum_{k=[\lambda]+[\delta\lambda]+1}^{\infty} \frac{\lambda^k}{k!} = \frac{\lambda^{\lambda(1+\delta)+1}}{(\lambda(1+\delta)+1)!} \left(1 + \frac{1}{(1+\delta)+2/\lambda} + \frac{1}{((1+\delta)+2/\lambda)((1+\delta)+3/\lambda)} + \cdots\right)$$

$$< \frac{\lambda^{\lambda(1+\delta)+1}}{(\lambda(1+\delta)+1)!} \left(1 + \frac{1}{(1+\delta)} + \frac{1}{(1+\delta)^2} + \cdots\right)$$

and thus,

$$\sum_{k=[\lambda]+[\delta\lambda]+1}^{\infty} \frac{\lambda^k}{k!} = \frac{\lambda^{\lambda(1+\delta)+1}}{(\lambda(1+\delta)+1)!} \frac{1+\delta}{\delta} \left(1 + O\left(\frac{1}{\lambda}\right)\right)$$

Using Stirling's formula (Abramowitz and Stegun, 1968, Section 6.1.38), $x! \sim \sqrt{2\pi} x^{x+\frac{1}{2}} e^{-x}$, for large x yields

$$\frac{\lambda^{\lambda(1+\delta)+1}}{(\lambda(1+\delta)+1)!} \frac{1+\delta}{\delta} \sim \frac{\lambda^{\lambda(1+\delta)+1} e^{\lambda(1+\delta)}}{(\lambda(1+\delta)+1)\sqrt{2\pi}\lambda^{\lambda(1+\delta)+\frac{1}{2}}(1+\delta)^{\lambda(1+\delta)+\frac{1}{2}}} \frac{1+\delta}{\delta}$$

$$\sim \frac{e^{\lambda(1+\delta)[1-\log(1+\delta)]}}{\sqrt{2\pi(1+\delta)\lambda}} \frac{1}{\delta}$$

---

[9] The size of the Internet is currently estimated at about $N \sim 10^5$.

For large N, the condition becomes

$$\xi \sim \frac{N^{\alpha(1+\delta)[1-\log(1+\delta)]+1-\alpha}}{\sqrt{2\pi(1+\delta)\alpha \log N}} \frac{1}{\delta} \left(1 + O\left(\frac{1}{\log N}\right)\right)$$

and, after taking the logarithm of both sides,

$$\log \xi \sim (\alpha(1+\delta)[1 - \log(1+\delta)] + 1 - \alpha) \log N - \frac{1}{2}\log\log N - \frac{1}{2}\log(2\pi(1+\delta)\alpha) - \log\delta + O\left(\frac{1}{\log N}\right)$$

or

$$\log \xi + (\alpha - 1)\log N + \frac{1}{2}\log\log N + O\left(\frac{1}{\log N}\right) \sim (\alpha(1+\delta)[1 - \log(1+\delta)])\log N$$

$$- \frac{1}{2}\log(2\pi(1+\delta)\alpha) - \log\delta \qquad (6.12)$$

At this point, we will assume that $\delta < 1$, which justifies the expansion $\log(1+\delta) = \delta + O(\delta^2)$. This assumption will be checked later. Thus,

$$\alpha(1+\delta)[1 - \log(1+\delta)] \sim \alpha(1+\delta)[1 - \delta] \sim \alpha(1 - \delta^2)$$

and δ must be solved from $R \sim \alpha(1 - \delta^2)\log N - \frac{\delta}{2} - \log\delta$ with $R = \log\left(\sqrt{2\pi\alpha}\xi\right) + (\alpha - 1)\log N + \frac{1}{2}\log\log N$. The Newton–Raphson iteration can be applied with starting value δ_0 to find the solution of the equation up to the leading order in $\log N$, i.e. $R \sim \alpha(1 - \delta^2)\log N$. Hence,

$$\delta_0 = \sqrt{1 - \frac{R}{\alpha \log N}} \sim 1 - \frac{R}{2\alpha \log N} \sim 1 - \frac{\alpha - 1}{2\alpha} - \frac{\log\left(\sqrt{2\pi\alpha}\xi\right) + \frac{1}{2}\log\log N}{2\alpha \log N}$$

which demonstrates that, for $\alpha \geq 1$, the assumption $\delta < 1$ is correct for large N. The case $\alpha < 1$ requires the application of Newton–Raphson's method on (6.12), which we omit here. The second iteration in Newton–Raphson's method leads to

$$\delta_1 = \delta_0 - \frac{\frac{\delta_0}{2} + \log\delta_0}{2\alpha\delta_0 \log N + \frac{1}{2} + \frac{1}{\delta_0}}$$

and shows that the k-th iteration improves the previous with a quantity of order $O\left(\log^{-k} N\right)$. Since (6.12) is only accurate up to $O\left(\log^{-1} N\right)$, a second iteration is superfluous and we obtain the choice $x_N = \lambda(1+\delta)$, or

$$x_N = \frac{3\alpha + 1}{2}\log N - \frac{1}{2}\log\left(\sqrt{2\pi\alpha}\xi\right) - \frac{1}{4}\log\log N$$

After substituting $x = -\log\xi$, we finally arrive for $\alpha \geq 1$ and large N at

$$\Pr\left[\max_{1 \leq k \leq N-1} H_k \leq \frac{x}{2} + \frac{3\alpha + 1}{2}\log N - \frac{1}{4}\log\log N - \frac{1}{4}\log(2\pi\alpha)\right] = e^{-e^{-x}}$$

from which the pdf of the hopcount of the longest shortest path (lsp) follows as

$$f_{lsp}(x) = 2e^{-e^{-2(x-c)}}e^{-2(x-c)} \qquad (6.13)$$

with

$$c = \frac{3\alpha + 1}{2}\log N - \frac{1}{4}\log\log N - \frac{1}{4}\log(2\pi\alpha) = \left(\frac{3}{2} + \frac{1}{2\alpha}\right)E[H_N] - \frac{1}{4}\log E[H_N] - \frac{1}{4}\log(2\pi)$$

and

$$E[lsp] = c + \frac{\gamma}{2} \approx \left(\frac{3}{2} + \frac{1}{2\alpha}\right)E[H_N] - \frac{1}{4}\log E[H_N] - 0.170$$

$$\text{Var}[lsp] = \frac{\pi^2}{24} \simeq 0.4112$$

Observe that the mean longest shortest path is about twice the mean hopcount if $\alpha = 1$ while the variance is small, constant and independent of the scaling parameter c or λ. Figure 6.3 compares the above approximate analysis with simulations.

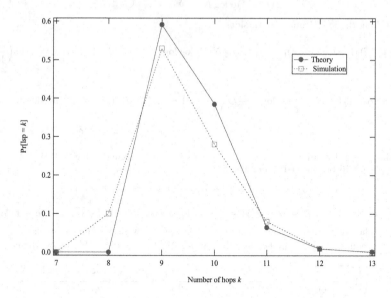

Fig. 6.3. The hopcount of the shortest path for $N = 4000$. Both simulations based on an Internet-like topology generator (with unit link weights) and theory $f_{lsp}(k)$ with $\alpha = 0.4786$ are shown.

6.7 Problem

(i) Consider a scale free graph with N nodes and with a power law degree distribution specified by $\Pr[D \geq x] = \left(\frac{x}{\tau}\right)^{-\alpha}$ for $x > \tau$. Show that, for large N, the scaling law for the highest degree node is $D_{\max} = O\left(N^{\frac{1}{\alpha}}\right)$.

Notes

(i) The classical theory of extremes, extremal properties of dependent sequences and extreme values in continuous-time are treated in detail in the book by Leadbetter *et al.* (1983).

(ii) A more recent book by Embrechts *et al.* (2001a) applies the theory of extremal events to problems in insurance and finance.

Part II
Stochastic processes

7

The Poisson process

The Poisson process is a prominent stochastic process, mainly because it frequently appears in a wealth of physical phenomena and because it is relatively simple to analyze. Therefore, we will first treat the Poisson process before considering the more general Markov processes.

7.1 A stochastic process

7.1.1 Introduction and definitions

A *stochastic*[1] *process*, formally denoted as $\{X(t), t \in T\}$, is a sequence of random variables $X(t)$, where the parameter t – most often the time – runs over an index set T. The *state space* of the stochastic process is the set of all possible values for the random variables $X(t)$ and each of these possible values is called a *state* of the process. If the index set T is a countable set, $X[k]$ is a discrete stochastic process. Often k is the discrete time or a time slot in computer systems. If T is a continuum, $X(t)$ is a continuous stochastic process. For example, the outcome of n tosses of a coin is a discrete stochastic process with state space {heads, tails} and the index set $T = \{0, 1, 2, \ldots, n\}$. The number of arrivals of packets in a router during a certain time interval $[a, b]$ is a continuous stochastic process because $t \in [a, b]$. Any realization of a stochastic process is called a *sample path*. For example, a sample path of the outcome of n tosses of a coin is {heads, tails, tails, ..., heads}, while a sample path of the number of arrivals in $[a, b]$ is $1_{a \leq t < a+h}, 3 \times 1_{a+h \leq t < a+4h}, 8 \times 1_{a+4h \leq t < a+5h}, \ldots, 13 \times 1_{a+(k-1)h \leq t < b}$, where $h = \frac{b-a}{k}$. Other examples are the measurement of the temperature each day, the notation of the value of a stock each minute or rolling a die and recording its value, which is illustrated in Fig. 7.1.

Especially in continuous stochastic processes, it is convenient to define increments as the difference $X(t) - X(u)$. The continuous time stochastic process $X(t)$ has *independent increments* if changes in the value of the process in different time intervals are independent, or, if for all $t_0 < t_1 < \cdots < t_n$, the random variables

---

[1] The word "stochastic" is derived from στοχαζεσθαι in Greek which means "to aim at, try to hit".

$X(t_1) - X(t_0), X(t_2) - X(t_1), \ldots, X(t_n) - X(t_{n-1})$ are independent. The continuous (time) stochastic process has *stationary increments* if $X(t+s) - X(s)$ possesses the same distribution for all s. Hence, changes in the value of the process only depend on the distance t between process events, not on the time point s.

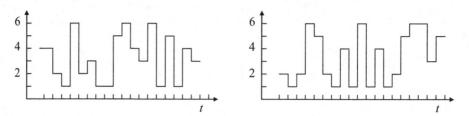

Fig. 7.1. Two different sample paths of the experiment: roll a die and record the outcome. The total number of different sample paths is 6^T where T is the number of times an outcome is recorded. The state space only contains six possible outcomes $\{1, 2, 3, 4, 5, 6\}$.

Stochastic processes are distinguished by (a) their state space, (b) the index set T and (c) by the dependence relations between random variables $X(t)$. For example, a standard Brownian motion (or Wiener process)[2] is defined as a stochastic process $X(t)$ having continuous sample paths, stationary independent increments[3] and $X(t)$ has a normal distribution $N(0, t)$. A Poisson process, defined in more detail in Section 7.2, is a stochastic process $X(t)$ having discontinuous sample paths, stationary independent increments and $X(t)$ has a Poisson distribution. A generalization of the Poisson process is a counting process. A counting process is defined as a stochastic process $N(t) \geq 0$ with discontinuous sample paths, stationary independent increments, but with arbitrary distribution. A counting process $N(t)$ represents the total number of events that have occurred in a time interval $[0, t]$. Examples of a counting process are the number of telephone calls at a local exchange during an interval, the number of failures in a telecommunication network, the number of corrupted bits after transmission due to channel errors, etc.

7.1.2 Modeling a stochastic process from measurements

In practice, understanding observed phenomena often asks for a stochastic model that captures the main characteristics of the studied phenomena and that enables computations of diverse quantities of interest. Examples in the field of data communications networks are the determination of the arrival process at a switch or router in order to dimension the number of buffer (memory) places, the modeling of the graph of the Internet, the distribution of the duration of a telephone call or

---

[2] Harrison (1990) shows that the converse is also true: if Y is a continuous process with stationary independent increments, then Y is a Brownian motion.

[3] Strictly speaking, the increment $X(t+s) - X(s)$ has a normal distribution $N(0, t)$. Since stationarity implies that this holds for any time $s \geq 0$ and since we assume, in addition, that $X(0) = 0$, we arrive at $X(t) \stackrel{d}{=} N(0, t)$, where $\stackrel{d}{=}$ means equality in distribution.

web browsing session, the number of visits to certain websites, the number of links that refer to a web page, the amount of downloaded information, the number of traversed routers by an email, etc. Accurate modeling is in general difficult and often trades off complexity against accuracy of the model.

Let us illustrate some aspects of *modeling* by considering Internet delay measurements. A motivation for obtaining an end-to-end delay model for (a part of) the Internet is the question whether massive service deployment of voice over IP (VoIP) can substitute classical telephony with a comparable quality. Specifically, classical telephony requires that the end-to-end delay of an arbitrary telephone conversation hardly exceeds 100 ms.

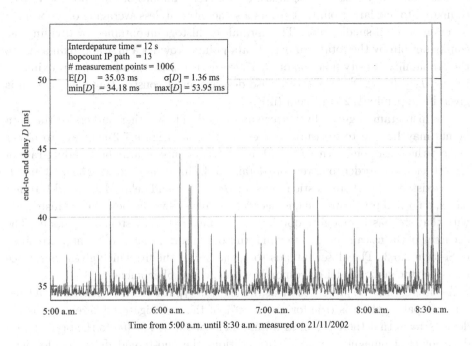

Fig. 7.2. The raw data of the end-to-end delay of IP test packets along a same path of 13 hops in the Internet measured during 3.5 hours.

The end-to-end delay along a fixed path between source and destination measured during some interval is an example of a continuous time stochastic process. We have received data of the delay measured at RIPE-NCC as illustrated in Fig. 7.2. Figure 7.2 shows a *sample path* of this continuous stochastic process. The precise details of the measurement configuration are for the present purpose not relevant. It suffices to add that Fig. 7.2 shows the time difference between the departure of an IP test packet of 100 bytes at the sending box and its arrival at the destination box, to an accuracy of less than 10 μs. The average sending rate of IP test packets is $\frac{1}{12}$ packets per second. Each IP test packet is assumed to follow the same path from sending to receiving box. The steadiness of the path is checked by trace-route measurements every 6 minutes.

Usually, in the next step, the histogram of the raw data is made. A histogram counts the number of data points that lie in an interval of ΔD ms, which is often called the bin size. Most graphical packages allow a choice of the bin size. Fig. 7.3 shows two different histograms with bin size $\Delta D = 0.5$ ms and $\Delta D = 0.1$ ms. In general, there is no universal rule regarding the choice of the bin size ΔD. Clearly, the bin size is bounded below by the measurement accuracy, in our case $\Delta D > 10$ μs. A finer bin size provides more detail, but the resulting histogram exhibits also more stochastic variations because there are fewer data points in a small bin and adjacent bins may possess a significantly different amount of data points. Hence, compared to one larger bin that covers a same interval, less averaging or smoothing occurs in a set of smaller bins. The normalized histogram obtained by dividing the counts per bin by the total number of data points provides a first approximation to the probability density function of D. However, it is still discrete and approximates $\Pr[k < D \le k + \Delta D]$. A more precise description of constructing a histogram is given in Appendix B.2 (Problem (iii)).

The histogram is generally better suited to decide whether outliers in the data points may be due to measurement errors or not. Figure 7.3 suggests to either neglect the data points with $D > 40$ ms or to measure at a higher sending rate of IP test packets in order to have more details in the intervals exceeding 38 or 40 ms. If there existed a good[4] stochastic model for the end-to-end delay along fixed Internet paths, a normal procedure[5] in engineering and physics would be to fit the histogram with that stochastic model to obtain the parameters of that stochastic model. The accuracy of the fit can be expressed in terms of the correlation coefficient ρ explained in Section 2.5.3. The closer ρ tends to 1, the better the fit, which gives confidence that the stochastic model corresponds with the real phenomenon.

Assuming that the presented measurement is a typical measurement along a fixed Internet path (which is true for about 80% of the investigated different paths), it demonstrates that there is a clear minimum at about 34 ms due to the propagation delay of electromagnetic waves. In addition, the end-to-end delay lies for 99% between 34 and 38 ms. However, there is insufficient data to pronounce claims in the tail behavior ($\Pr[D > x]$ for $x > 40$ ms). Just this region is of interest to compute the quality of service expressed as the probability that the end-to-end delay exceeds x ms is smaller than 10^{-a}, where a specifies the stringency on the quality requirement. Toll quality in classical telephony sets x at 100 ms and a in the range of 4 to 5. The existence of a good stochastic model covering the whole possible range of the end-to-end delay D would enable us to compute tail probabilities based on the parameters that can be fitted from the measurements.

---

[4] Which is still lacking at the time of writing.

[5] Other more difficult methods in the realm of statistics must be invoked in case the measurement data are so precious and rare that any additional measurement point has a far larger cost than the cost of extensive additional computations.

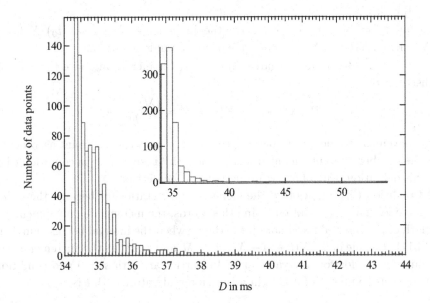

Fig. 7.3. The histogram of the end-to-end delay with a bin size of 0.1 ms (the insert has bin size of 0.5 ms).

The histogram is in fact a projection of the raw measurement data onto the ordinate (end-to-end delay axis). All time information (the abscissa in Fig. 7.2) is lost. Usually, the time evolution and the dependencies or correlations over time of a stochastic phenomenon are difficult and most analyses are only tractable under certain simplifying conditions. For example, often only a steady-state analysis is possible and the increments $X(t_k) - X(t_{k-1})$ of the process for all $t_0 < \cdots < t_{k-1} < t_k < \cdots < t_n$ are assumed to be independent or weakly dependent. The study of Markov processes (Chapters 9–11) basically tries to compute and analyze the process in steady-state. Figure 7.2 is measured over a relatively long period of time and indicates that after 8.00 a.m. the background traffic increases. The background traffic interferes with the IP test packets and causes them to queue longer in routers such that larger variations are observed. However, it is in general difficult to ascertain that (a part of) the measurement is performed while the system operates in a certain stable regime (or steady-state).

We have touched upon some aspects in the art of modeling to motivate the importance of studying stochastic processes. In the sequel of this chapter, one of the most basic and simplest stochastic processes is investigated.

7.2 The Poisson process

A Poisson process with parameter or rate $\lambda > 0$ is an integer-valued, continuous time stochastic process $\{X(t), t \geq 0\}$ satisfying:

(i) $X(0) = 0$;

(ii) for all $t_0 = 0 < t_1 < \cdots < t_n$, the increments $X(t_1) - X(t_0), X(t_2) - X(t_1), \ldots, X(t_n) - X(t_{n-1})$ are independent random variables;

(iii) for $t > 0$, $s \geq 0$ and non-negative integers k, the increments have the Poisson distribution

$$\Pr\left[X(t+s) - X(s) = k\right] = \frac{(\lambda t)^k e^{-\lambda t}}{k!} \tag{7.1}$$

It is convenient to view the Poisson process $X(t)$ as a special counting process, where the number of events in any interval of length t is specified via condition (iii). From this definition, the following properties can be derived:

(a) Condition (iii) implies that the increments are stationary because the right-hand side does not dependent on s. In other words, the increments only depend on the length of the interval t and not on the time s when the interval begins. Further, with (3.11), the mean $E\left[X(t+s) - X(s)\right] = \lambda t$ and because the increments are stationary, this holds for any value of s. In particular, with $s = 0$ and condition (i), the expected number of events in a time interval with length t is

$$E\left[X(t)\right] = \lambda t \tag{7.2}$$

Relation (7.2) explains why λ is called the rate of the Poisson process, namely, the derivative over time t or the number of events per time unit.

(b) The probability that exactly one event occurs in an arbitrarily small time interval of length h follows from condition (iii) as

$$\Pr\left[X(h+s) - X(s) = 1\right] = \lambda h e^{-\lambda h} = \lambda h + o(h)$$

while the probability that no event occurs in an arbitrarily small time interval of length h is

$$\Pr\left[X(h+s) - X(s) = 0\right] = e^{-\lambda h} = 1 - \lambda h + o(h)$$

Similarly, the probability that more than one event occurs in an arbitrarily small time interval of length h is

$$\Pr\left[X(h+s) - X(s) > 1\right] = o(h)$$

This relation means that the simultaneous occurrence of multiple Poisson events is almost surely not possible. Hence, in a Poisson process, events occur sequentially in time, the one after the other, and almost never together at a same time.

Example 1 A conversation in a wireless ad-hoc network is severely disturbed by interference signals according to a Poisson process of rate $\lambda = 0.1$ per minute. (a) What is the probability that no interference signals occur within the first two minutes of the conversation? (b) Given that the first two minutes are free of disturbing effects, what is the probability that in the next minute precisely one interfering signal disturbs the conversation?

(a) Let $X(t)$ denote the Poisson interference process, then $\Pr[X(2) = 0]$ needs to be computed. Since $X(0) = 0$ and with (7.1), we can write $\Pr[X(2) = 0] = \Pr[X(2) - X(0) = 0] = e^{-2\lambda}$, which equals $\Pr[X(2) = 0] = e^{-0.2} = 0.8187$.

(b) The events during two non-overlapping intervals of a Poisson process are independent. Since the event $\{X(2) - X(0) = 0\}$ is independent from the event $\{X(3) - X(2) = 1\}$, the conditional probability that has been asked for then equals $\Pr[X(3) - X(2) = 1 | X(2) - X(0) = 0] = \Pr[X(3) - X(2) = 1]$. From (7.1), we obtain $\Pr[X(3) - X(2) = 1] = 0.1e^{-0.1} = 0.0905$.

Example 2 During a certain time interval $[t_1, t_1 + 10 \text{ s}]$, the number of IP packets that arrive at a router is on average 40 s^{-1}. A service provider asks us to compute the probability that there arrive 20 packets in the period $[t_1, t_1 + 1 \text{ s}]$ and 30 IP packets in $[t_1, t_1 + 3 \text{ s}]$. We may regard the arrival process as a Poisson process.

We are asked to compute $\Pr[X(1) = 20, X(3) = 30]$ knowing that $\lambda = 40 \text{ s}^{-1}$. Using the independence of increments and (7.1), we rewrite

$$
\begin{aligned}
\Pr[X(1) = 20, X(3) = 30] &= \Pr[X(1) - X(0) = 20, X(3) - X(1) = 10] \\
&= \Pr[X(1) - X(0) = 20]\Pr[X(3) - X(1) = 10] \\
&= \frac{(\lambda)^{20}e^{-\lambda}}{20!}\frac{(2\lambda)^{10}e^{-2\lambda}}{10!} = 10^{-26} \approx 0
\end{aligned}
$$

which means that the request of the service provider does not occur in practice.

7.3 Properties of the Poisson process

The first theorem is the converse of the above property (b) that immediately followed from the definition. The theorems presented here reveal the methodology of how stochastic processes are studied.

Theorem 7.3.1 *A counting process $N(t)$ that satisfies the conditions (i) $N(0) = 0$, (ii) the process $N(t)$ has stationary and independent increments, (iii) $\Pr[N(h) = 1] = \lambda h + o(h)$ and (iv) $\Pr[N(h) > 1] = o(h)$ is a Poisson process with rate $\lambda > 0$.*

Proof: We must show that conditions (iii) and (iv) are equivalent to condition (iii) in the definition of the Poisson process. Denote

$$
P_n(t) = \Pr[N(t) = n] \tag{7.3}
$$

and consider first the case $n = 0$, then

$$
P_0(t + h) = \Pr[N(t + h) = 0] = \Pr[N(t + h) - N(t) = 0, N(t) = 0]
$$

Invoking independence via (ii)

$$
P_0(t + h) = \Pr[N(t + h) - N(t) = 0]\Pr[N(t) = 0]
$$

and using the definition (7.3), we obtain

$$P_0(t + h) = \Pr[N(t + h) - N(t) = 0] P_0(t)$$

The stationarity in (ii) implies that

$$\Pr[N(t + h) - N(t) = 0] = \Pr[N(h) - N(0) = 0] = \Pr[N(h) = 0]$$

where the last step follows from (i). From (iii), (iv) and the fact that

$$\sum_{k=0}^{\infty} \Pr[N(h) = k] = 1$$

it follows that

$$\Pr[N(h) = 0] = 1 - \lambda h + o(h) \tag{v}$$

Combining all yields

$$P_0(t + h) = P_0(t)(1 - \lambda h + o(h))$$

or

$$\frac{P_0(t + h) - P_0(t)}{h} = -\lambda P_0(t) + \frac{o(h)}{h}$$

from which, in the limit $h \to 0$, the differential equation

$$P_0'(t) = -\lambda P_0(t)$$

is immediate. The solution is $P_0(t) = Ce^{-\lambda t}$ and the integration constant C follows from (i) and $P_0(0) = \Pr[N(0) = 0] = 1$ as $C = 1$. This establishes condition (iii) in the definition of the Poisson process for $k = 0$ in (7.1).

The verification for $n > 0$ is more involved. Applying the law of total probability (2.49),

$$P_n(t + h) = \Pr[N(t + h) = n]$$

$$= \sum_{j=0}^{n} \Pr[N(t + h) - N(t) = j | N(t) = n - j] \Pr[N(t) = n - j]$$

By independence (ii),

$$\Pr[N(t + h) - N(t) = j | N(t) = n - j] = \Pr[N(t + h) - N(t) = j]$$

and by the definition (7.3), $\Pr[N(t) = n - j] = P_{n-j}(t)$, we have

$$P_n(t + h) = \sum_{j=0}^{n} \Pr[N(t + h) - N(t) = j] P_{n-j}(t)$$

By the stationarity (ii)

$$\Pr[N(t + h) - N(t) = j] = \Pr[N(h) - N(0) = j]$$

we obtain using (i)

$$P_n(t+h) = \sum_{j=0}^{n} \Pr[N(h) = j] P_{n-j}(t)$$

while (v) and (iii) suggest to write the sum as

$$P_n(t+h) = P_n(t)\Pr[N(h) = 0] + P_{n-1}(t)\Pr[N(h) = 1] + \sum_{j=2}^{n} P_{n-j}(t)\Pr[N(h) = j]$$

Since $P_n(t) \leq 1$ and using (iv),

$$\sum_{j=2}^{n} P_{n-j}(t)\Pr[N(h) = j] \leq \sum_{j=2}^{n} \Pr[N(h) = j] = \Pr[N(h) > 1] = o(h)$$

we arrive with (v), (iii) at

$$P_n(t+h) = P_n(t)(1 - \lambda h + o(h)) + P_{n-1}(t)(\lambda h + o(h)) + o(h)$$

or

$$\frac{P_n(t+h) - P_n(t)}{h} = -\lambda P_n(t) + \lambda P_{n-1}(t) + \frac{o(h)}{h}$$

which leads, after taking the limit $h \to 0$, to the differential equation

$$P_n'(t) = -\lambda P_n(t) + \lambda P_{n-1}(t) \tag{7.4}$$

with initial condition $P_n(0) = \Pr[N(0) = n] = 1_{\{n=0\}}$.

This differential equation is rewritten as

$$\frac{d}{dt}\left(e^{\lambda t} P_n(t)\right) = \lambda e^{\lambda t} P_{n-1}(t) \tag{7.5}$$

In case $n = 1$, the differential equation reduces with $P_0(t) = e^{-\lambda t}$ to $\frac{d}{dt}\left(e^{\lambda t} P_1(t)\right) = \lambda$. The general solution is $e^{\lambda t} P_1(t) = \lambda t + C$ and, from the initial condition $P_1(0) = 0$, we have $C = 0$ and $P_1(t) = \lambda t e^{-\lambda t}$. The general solution to (7.5) is proved by induction. Assume that $P_n(t) = \frac{(\lambda t)^n e^{-\lambda t}}{n!}$ holds for n, then the case $n + 1$ follows from (7.5) as

$$\frac{d}{dt}\left(e^{\lambda t} P_{n+1}(t)\right) = \lambda \frac{(\lambda t)^n}{n!}$$

and integrating from 0 to t using $P_{n+1}(0) = 0$, yields $P_{n+1}(t) = \frac{(\lambda t)^{n+1} e^{-\lambda t}}{(n+1)!}$ which establishes the induction and finalizes the proof of the theorem.

We present a generating function approach that elegantly solves the differential equation (7.4). Let

$$\varphi_{N(t)}(z) = E\left[z^{N(t)}\right] = \sum_{n=0}^{\infty} \Pr[N(t) = n] z^n = \sum_{n=0}^{\infty} P_n(t) z^n$$

denote the probability generating function of the counting process $N(t)$, with boundary condition $\varphi_{N(0)}(z) = 1$. After multiplying (7.4) by z^n and summing over all n, we obtain

$$\frac{\partial \varphi_{N(t)}(z)}{\partial t} = \lambda (z - 1) \varphi_{N(t)}(z)$$

whose solution, using $\varphi_{N(0)}(z) = 1$, is $\varphi_{N(t)}(z) = e^{\lambda t(z-1)}$, from which $P_n(t) = \frac{(\lambda t)^n e^{-\lambda t}}{n!}$ follows. □

The second theorem has very important applications since it relates the number of events in non-overlapping intervals to the interarrival time between these events.

Theorem 7.3.2 *Let $\{X(t); t \geq 0\}$ be a Poisson process with rate $\lambda > 0$ and denote by $t_0 = 0 < t_1 < t_2 < \cdots$ the successive occurrence times of events. Then, the interarrival times $\tau_n = t_n - t_{n-1}$ for $n > 0$ are independent identically distributed exponential random variables with mean $\frac{1}{\lambda}$.*

Proof: For any $s \geq 0$ and any $n \geq 1$, the event $\{\tau_n > s\}$ is equivalent to the event $\{X(t_{n-1} + s) - X(t_{n-1}) = 0\}$. Indeed, the n-th interarrival time τ_n can only be longer than s time units if and only if the n-th event has not yet occurred s time units after the occurrence of the $(n-1)$-th event at t_{n-1}. Since the Poisson process has independent increments (condition (ii) in the definition of the Poisson process), changes in the value of the process in non-overlapping time intervals are independent. By the equivalence in events, this implies that the set of interarrival times $\tau_1, \ldots, \tau_n, \ldots$ are independent random variables. Further, by the stationarity of the Poisson process (deduced from condition (iii) in the definition of the Poisson process),

$$\Pr[\tau_n > s] = \Pr[X(t_{n-1} + s) - X(t_{n-1}) = 0] = e^{-\lambda s}$$

which implies that any interarrival time has an identical, exponential distribution,

$$F_{\tau_n}(x) = \Pr[\tau_n \leq x] = 1 - e^{-\lambda x}$$

This proves the theorem. □

The converse of Theorem 7.3.2 also holds: if the interarrival times $\{\tau_n\}$ of a counting process $\{N(t), t \geq 0\}$ are i.i.d. exponential random variables with mean $\frac{1}{\lambda}$, then $\{N(t), t \geq 0\}$ is a Poisson process with rate λ.

An association to the exponential distribution is the memoryless property,

$$\Pr[\tau_n > s + t | \tau_n > s] = \Pr[\tau_n > t]$$

By the equivalence of the events, for any $t, s \geq 0$,

$$\Pr[\tau_n > s + t | \tau_n > s] = \Pr[X(t_{n-1} + s + t) - X(t_{n-1}) = 0 | X(t_{n-1} + s) - X(t_{n-1}) = 0]$$
$$= \Pr[X(t_{n-1} + s + t) - X(t_{n-1} + s) = 0 | X(t_{n-1} + s) - X(t_{n-1}) = 0]$$

By the independence of increments (in non-overlapping intervals),

$$\Pr[\tau_n > s + t | \tau_n > s] = \Pr[X(t_{n-1} + s + t) - X(t_{n-1} + s) = 0]$$

and by the stationarity of the increments, the memoryless property is established,

$$\Pr[\tau_n > s + t | \tau_n > s] = \Pr[X(t_{n-1} + t) - X(t_{n-1}) = 0] = \Pr[\tau_n > t]$$

Hence, the assumption of stationary and independent increments is equivalent to asserting that, at any time s, the process probabilistically restarts again with the same distribution and is independent of occurrences in the past (before s). Thus, the process has no memory and, since the only continuous distribution that satisfies the memoryless property is the exponential distribution, exponential interarrival times τ_n are a natural consequence.

The arrival time of the n-th event or the waiting time until the n-th event is $W_n = \sum_{k=1}^{n} \tau_k$. In Section 3.3.1, it is shown that the probability distribution of the sum of independent exponential random variables has a Gamma distribution or Erlang distribution (3.24). Alternatively, the equivalence of the events, $\{W_n \leq t\} \Longleftrightarrow \{N(t) \geq n\}$, directly leads to the Erlang distribution,

$$F_{W_n}(t) = \Pr\left[W_n \leq t\right] = \Pr\left[N(t) \geq n\right] = \sum_{k=n}^{\infty} \frac{(\lambda t)^k e^{-\lambda t}}{k!} \qquad (7.6)$$

The equivalence of the events, $\{W_n \leq t\} \Longleftrightarrow \{N(t) \geq n\}$, is a general relation and a fundamental part of the theory of renewal processes, which we will study in the next Chapter 8.

Theorem 7.3.3 *If $X(t)$ and $Y(t)$ are two independent Poisson processes with rates λ_x and λ_y, then $Z(t) = X(t) + Y(t)$ is also a Poisson process with rate $\lambda_x + \lambda_y$.*

Proof: It suffices to demonstrate that the counting process $N_Z(t) = N_X(t) + N_Y(t)$ has exponentially distributed interarrival times τ_Z. Suppose that $N_Z(t_n) = n$, it remains to compute the next arrival at time $t_{n+1} = t_n + s$ for which $N_Z(t_n + s) = n + 1$. Due to the memoryless property of the Poisson process, the interarrival time of an event from t_n on for each random variable X and Y is again exponentially distributed with parameter λ_x and λ_y, respectively. In other words, it is irrelevant which process X or Y has previously caused the arrival at time t_n. Further, the event that the interarrival time of the sum processes $\{\tau_Z > s\}$ is equivalent to $\{\tau_X > s\} \cap \{\tau_Y > s\}$ or

$$\Pr\left[\tau_Z > s\right] = \Pr\left[\tau_X > s, \tau_Y > s\right] = \Pr\left[\tau_X > s\right] \Pr\left[\tau_Y > s\right] = e^{-(\lambda_x + \lambda_y)s}$$

where the independence of $X(t)$ and $Y(t)$ has been used. This proves the theorem. $\square$

A direct consequence is that any sum of independent Poisson processes is also a Poisson process with aggregate rate equal to the sum of the individual rates. This theorem is in correspondence with the sum property of the Poisson distribution.

7.4 The Poisson process and the uniform distribution

We will show that, given a precise number of Poisson events have occurred in the interval $[0, t]$, their occurrence time is uniformly distributed over that interval. The association of a Poisson process with the uniform distribution illustrates that the Poisson process behaves as an entire random process. If we encounter a process, such as the number of rain drops in a certain area, whose occurrences are independent and uniformly distributed over a space, that process is very likely a Poisson process.

Theorem 7.4.1 *Given that exactly one event of a Poisson process $\{X(t); t \geq 0\}$ has occurred during the interval $[0, t]$, the time of occurrence of this event is uniformly distributed over $[0, t]$.*

Proof: Immediate application of the conditional probability (2.47) yields for $0 \leq s \leq t$,

$$\Pr\left[\tau_1 \leq s | X(t) = 1\right] = \frac{\Pr\left[\{\tau_1 \leq s\} \cap \{X(t) = 1\}\right]}{\Pr\left[X(t) = 1\right]}$$

Using the equivalence $\{\tau_1 \leq s\} \iff \{X(t_0 + s) - X(t_0) = 1\}$ and the fact that $\{X(t_0 + s) - X(t_0) = 1\} = \{X(s) = 1\}$ by the stationarity of the Poisson process gives

$$\{\tau_1 \leq s\} \cap \{X(t) = 1\} = \{X(s) = 1\} \cap \{X(t) = 1\}$$
$$= \{X(s) = 1\} \cap \{X(t) - X(s) = 0\}$$

Applying the independence of increments over non-overlapping intervals and (7.1) yields

$$\Pr\left[\tau_1 \leq s | X(t) = 1\right] = \frac{\Pr\left[X(s) = 1\right] \Pr\left[X(t) - X(s) = 0\right]}{\Pr\left[X(t) = 1\right]} = \frac{(\lambda s)e^{-\lambda s}e^{-\lambda(t-s)}}{(\lambda t)e^{-\lambda t}} = \frac{s}{t}$$

which completes the proof. $\qquad\qquad\qquad\qquad\qquad\qquad\qquad\qquad\qquad\qquad\qquad\square$

Application The arrival process of most real-time applications (such as telephony calls, interactive video, ...) in a network is well approximated by a Poisson process. Suppose a measurement configuration is built to collect statistics of the arrival process of telephony calls in some region. During a period $[0, T]$, precisely one telephony call has been measured. What can be said of the time $x \in [0, T]$ at which the telephony call has arrived at the measurement device? Theorem 7.4.1 tells us that any time in that interval is equally probable.

A related example is the conditional probability where $0 < s < t$ and $0 \leq k \leq n$,

$$\Pr\left[X(s) = k | X(t) = n\right] = \frac{\Pr\left[\{X(s) = k\} \cap \{X(t) = n\}\right]}{\Pr\left[X(t) = n\right]}$$
$$= \frac{\Pr\left[\{X(s) = k\} \cap \{X(t) - X(s) = n - k\}\right]}{\Pr\left[X(t) = n\right]}$$

Invoking the independence of increments yields

$$\Pr\left[X(s) = k | X(t) = n\right] = \frac{\Pr\left[X(s) = k\right]\Pr[X(t) - X(s) = n - k]}{\Pr\left[X(t) = n\right]}$$

$$= \frac{n!(\lambda s)^k e^{-\lambda s}}{k!(\lambda t)^n e^{-\lambda t}} \frac{(\lambda(t-s))^{n-k} e^{-\lambda(t-s)}}{(n-k)!} = \binom{n}{k} \frac{s^k}{t^n} (t-s)^{n-k}$$

Hence, if $p = \frac{s}{t}$, the conditional probability becomes

$$\Pr\left[X(s) = k | X(t) = n\right] = \binom{n}{k} p^k (1-p)^{n-k}$$

Given that a total number of n Poisson events have occurred in time interval $[0,t]$, the chance that precisely k events have taken place in the sub-interval $[0,s]$ is binomially distributed with parameter n and $p = \frac{s}{t}$. Observe that also this conditional probability is independent of the rate λ. In addition, since $\lim_{t\to\infty} X(t) = \infty$ such that $n \to \infty$, applying the law of rare events (see Section 3.1.4) results in

$$\lim_{t\to\infty} \Pr\left[X(s) = k | X(t) = n\right] = \frac{s^k}{k!} e^{-s}$$

Given an everlasting Poisson process, the chance that precisely k events occur in the interval $[0,s]$ is Poisson distributed with mean equal to the length of the interval.

Theorem 7.4.1 can be generalized to n events. For any set of real variables s_j satisfying $0 = s_0 < s_1 < s_2 < \cdots < s_n < t$ and given that n events of a Poisson process $\{X(t); t \geq 0\}$ have occurred during the interval $[0,t]$, the probability of the successive occurrence times $0 < t_1 < t_2 < \cdots < t_n < t$ of these n Poisson events is

$$\Pr\left[t_1 \leq s_1, \ldots, t_n < s_n | X(t) = n\right] = \frac{\Pr\left[\{t_1 \leq s_1, \ldots, t_n < s_n\} \cap \{X(t) = n\}\right]}{\Pr\left[X(t) = n\right]}$$

Using a similar argument as in the proof of Theorem 7.4.1,

$$p = \Pr\left[\{t_1 \leq s_1, t_2 \leq s_2, \ldots, t_n < s_n\} \cap \{X(t) = n\}\right]$$

$$= \Pr\left[X(s_1) - X(s_0) = 1, \ldots, X(s_n) - X(s_{n-1}) = 1, X(t) - X(s_n) = 0\right]$$

$$= \left(\prod_{j=1}^{n} \Pr\left[X(s_j) - X(s_{j-1}) = 1\right]\right) \Pr[X(t) - X(s_n) = 0]$$

$$= \left(\prod_{j=1}^{n} e^{-\lambda(s_j - s_{j-1})} \lambda (s_j - s_{j-1})\right) e^{-\lambda(t - s_n)}$$

$$= \lambda^n \prod_{j=1}^{n} (s_j - s_{j-1}) e^{-\lambda \sum_{k=1}^{n}(s_k - s_{k-1}) - \lambda(t - s_n)} = \lambda^n e^{-\lambda t} \prod_{j=1}^{n} (s_j - s_{j-1})$$

Thus,

$$\Pr\left[t_1 \le s_1, \ldots, t_n < s_n | X(t) = n\right] = \frac{\lambda^n e^{-\lambda t} \prod_{j=1}^{n} (s_j - s_{j-1})}{\frac{(\lambda t)^n e^{-\lambda t}}{n!}} = \frac{n!}{t^n} \prod_{j=1}^{n} (s_j - s_{j-1})$$

from which the density function

$$f_{\{t_j\}}(s_1, \ldots, s_n | X(t) = n) = \frac{\partial^n}{\partial s_1 \ldots \partial s_n} \Pr\left[t_1 \le s_1, \ldots, t_n < s_n | X(t) = n\right]$$

follows as

$$f_{\{t_j\}}(s_1, s_2, \ldots, s_n | X(t) = n) = \frac{n!}{t^n}$$

which is independent of the rate λ. If $0 < t_1 < t_2 < \cdots < t_n < t$ are the successive occurrence times of n Poisson events in the interval $[0, t]$, then the random variables $t_1, t_2, \ldots, t_n$ are distributed as a set of order statistics, defined in Section 3.4.2, of n uniform random variables in $[0, t]$. In other words, if n i.i.d. uniform random variables on $[0, t]$ are assorted in increasing order, they may represent n successive occurrence times of a Poisson process. The mean spacing between these n ordered i.i.d. uniform random variables is $\frac{t}{n+1}$ as computed in Problem (ii) of Section 3.7.

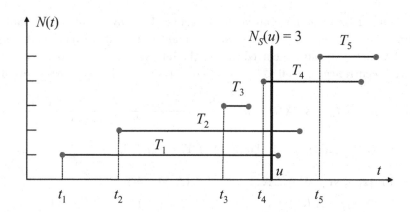

Fig. 7.4. A Poisson counting process $\{N(t), t \ge 0\}$ is sketched. The k-th request arrives at a webserver at time t_k according to a Poisson process and requires a processing time T_k. The number $N_S(u)$ of active requests in the server at time u is the desired quantity.

Example Requests arrive at a webserver (or multiprocessor) according to a Poisson process with intensity λ. Each request k needs T_k time units to be processed and each processing time is independent and distributed with distribution function $F_T(x) = \Pr\left[T \le x\right]$. What is the distribution of the number $N_S(t)$ of active request in the server at time t?

Fig. 7.4 sketches the Poissonean arrivals of requests and their processing time. The number of active requests at time t equals

$$N_S(t) = \sum_{k=1}^{N(t)} 1_{\{t_k + T_k \geq t\}}$$

where $N(t)$ is the number of Poisson events in the interval $[0, t]$ as defined in Theorem 7.3.1. We invoke the law of total probability (2.49) by conditioning on $N(t)$,

$$\Pr[N_S(t) = m] = \sum_{n=0}^{\infty} \Pr[N_S(t) = m | N(t) = n] \Pr[N(t) = n]$$

The conditional probability equals

$$\Pr[N_S(t) = m | N(t) = n] = \Pr\left[\sum_{k=1}^{n} 1_{\{t_k + T_k \geq t\}} = m | N(t) = n\right]$$

Since the successive occurrence times $t_1, t_2, \ldots, t_n$ of n Poisson events in the interval $[0, t]$ are distributed as a set of order statistics of n uniform random variables on $[0, t]$ and since the order relation is irrelevant in the sum, we can replace t_k by the uniform random variable U_k on $[0, t]$ such that

$$\Pr\left[\sum_{k=1}^{n} 1_{\{t_k + T_k \geq t\}} = m | N(t) = n\right] = \Pr\left[\sum_{k=1}^{n} 1_{\{U_k + T_k \geq t\}} = m\right]$$

All $U_1, U_2, \ldots, U_n$ as well as all $T_1, T_2, \ldots, T_n$ are i.i.d. random variables and, for each $1 \leq k \leq n$, $U_k \overset{d}{=} U$, a uniform random variable on $[0, t]$, while $T_k \overset{d}{=} T$ with given distribution function $F_T(x)$. Each indicator represents a Bernoulli random variable with probability

$$p = \Pr[U_k + T_k \geq t] = \Pr[U + T \geq t]$$

and the sum of i.i.d. Bernoulli random variables is a binomial random variable as shown in Section 3.1.2,

$$\Pr\left[\sum_{k=1}^{n} 1_{\{U_k + T_k \geq t\}} = m\right] = \binom{n}{m} p^m (1-p)^{n-m}$$

We compute the probability p by using the continuous variant of the law of total probability (2.49)

$$p = \Pr[U + T \geq t] = \int_0^t \Pr[U + T \geq t | U = u] \frac{dF_U(u)}{du} du = \frac{1}{t} \int_0^t \Pr[T \geq t - u] du$$

because U and T are independent. Expressed in terms of the given distribution function $F_T(x) = \Pr[T \leq x]$ yields

$$p = \frac{1}{t} \int_0^t (1 - F_T(t - u)) du = \frac{1}{t} \int_0^t (1 - F_T(x)) dx$$

We return to $\Pr[N_S(t) = m]$ by collecting all parts,

$$\Pr[N_S(t) = m] = \sum_{n=0}^{\infty} \Pr[N_S(t) = m|N(t) = n]\Pr[N(t) = n]$$

$$= \sum_{n=0}^{\infty} \frac{n!}{m!(n-m)!} p^m (1-p)^{n-m} \frac{(\lambda t)^n}{n!} e^{-\lambda t}$$

Simplified,

$$\Pr[N_S(t) = m] = \frac{p^m e^{-\lambda t}}{m!} \sum_{n=m}^{\infty} \frac{(\lambda t)^n}{(n-m)!} (1-p)^{n-m}$$

$$= \frac{p^m e^{-\lambda t}}{m!} \sum_{n=0}^{\infty} \frac{(\lambda t)^{n+m}}{n!} (1-p)^n = \frac{(p\lambda t)^m e^{-\lambda t}}{m!} e^{(1-p)\lambda t}$$

Finally, we find that

$$\Pr[N_S(t) = m] = \frac{(p\lambda t)^m}{m!} e^{-p\lambda t}$$

which illustrates that $N_S(t)$, the number of active requests in a webserver at t, has a Poisson distribution with mean

$$p\lambda t = \lambda \int_0^t (1 - F_T(x)) \, dx$$

and that $E[N_S(t)]$ increases with t. We observe from (2.36) that $\lim_{t \to \infty} E[N_S(t)] = \lambda E[T]$, which agrees[6] with Little's Law in Section 13.7.

Assuming that $E[T]$ is finite, in order to dimension the webserver safely, we determine the number m of available requests that the server can accept and process so that $\Pr[N_S(\infty) > m] \le \varepsilon$. The stringency ε is typically of the order of 10^{-4} to 10^{-6} and the tail probability $\Pr[N_S(\infty) > m]$ is given in (3.28). Unfortunately, we cannot solve $\Pr[N_S(\infty) > m] \le \varepsilon$ for m analytically as a function of ε, given $\lambda E[T]$.

7.5 The non-homogeneous Poisson process

As will be shown later in Section 11.3.2, the Poisson process is a special case of a birth-and-death process, which is in turn a special case of a Markov process. Hence, it seems more instructive to discuss these special processes as applications of the Markov process. Therefore, only associations to the Poisson process are treated here. In many cases, the rate is a time variant function $\lambda(t)$ and such a process is termed *a non-homogeneous or non-stationary Poisson process*. For example, the arrival rate of a large number m of individual IP-flows at a router is well approximated by a non-homogeneous Poisson process, where the rate $\lambda(t)$

---

[6] Anticipating queuing theory, the current process can be classified as an M/G/∞ queue with infinitely long queue (see Section 13.1.4), in which each request has its own server, there is no waiting time and each request or job is directly processed upon arrival.

varies over the day depending on the number m and the individual rate of each flow of packets. Since the sum of independent Poisson random variables is again a Poisson random variable, we have $\lambda(t) = \sum_{j=1}^{m(t)} \lambda_j(t)$.

If $X(t)$ is a non-homogeneous Poisson process with rate $\lambda(t)$, the increment $X(t) - X(s)$ reflects the number of events in an interval $(s, t]$ and increments of non-overlapping intervals are still independent.

Theorem 7.5.1 *If $\Lambda(t) = \int_0^t \lambda(u)du$ and $s < t$, then $X(t) - X(s)$ is Poisson distributed with mean $\Lambda(t) - \Lambda(s)$.*

The demonstration is analogous to the proof of Theorem 7.3.1.

Proof (partly): Denote by $P_n(t) = \Pr\left[N(t) - N(s) = n\right]$, then

$$P_0(t + h) = \Pr\left[N(t + h) - N(s) = 0\right]$$
$$= \Pr\left[N(t + h) - N(t) = 0, N(t) - N(s) = 0\right]$$

Invoking the independence of the increments,

$$P_0(t + h) = \Pr\left[N(t + h) - N(t) = 0\right]\Pr[N(t) - N(s) = 0]$$
$$= P_0(t)(1 - \lambda(t)h + o(h))$$

or

$$\frac{P_0(t + h) - P_0(t)}{h} = -\lambda(t)P_0(t) + \frac{o(h)}{h}$$

from which, in the limit $h \to 0$, the differential equation

$$P_0'(t) = -\lambda(t)P_0(t)$$

is immediate. Rewritten as $\frac{d}{dt}\log P_0(t) = -\lambda(t)$, after integration over $(s, t]$, we find $\log P_0(t) = -(\Lambda(t) - \Lambda(s))$ since $P_0(s) = \Pr\left[N(s) - N(s) = 0\right] = 1$. Thus, for the case $n = 0$, we find $P_0(t) = \exp\left[-(\Lambda(t) - \Lambda(s))\right]$, which proves the theorem for $n = 0$.

The remainder of the proof $(n > 0)$ uses the same ingredients as the proof of Theorem 7.3.1 and is omitted. □

Theorem 7.5.1 thus shows that the increments of a non-homogeneous Poisson process satisfy

$$\Pr\left[X(t + s) - X(s) = k\right] = \frac{(\Lambda(t + s) - \Lambda(s))^k}{k!}e^{-(\Lambda(t+s)-\Lambda(s))}$$

Since the right-hand side is generally a function of s in contrast to the homogeneous variant (7.1), the non-homogeneous Poisson process is generally *not* stationary. Consequently, the set of interarrival times $\{\tau_n = t_n - t_{n-1}\}_{n \geq 1}$ are still independent as in the homogeneous case (Theorem 7.3.2), but not identically distributed. In fact, it follows from the equivalence $\{\tau_n > s\} \Leftrightarrow \{X(t_{n-1} + s) - X(t_{n-1}) = 0\}$ derived in Theorem 7.3.2 that

$$\Pr\left[\tau_n > s\right] = e^{-(\Lambda(t_{n-1}+s)-\Lambda(t_{n-1}))}$$

which clearly depends on the occurrence time t_{n-1} of the $(n-1)$-th Poisson event. This unsatisfactory relation illustrates that inhomogeneous Poisson processes are better reformulated within the realm of continuous-time Markov processes, where Theorem 10.2.3 nicely specifies the interarrival relation between events. Roughly speaking, the relation with continuous-time Markov processes suggests to approximate the integral $\Lambda(t)$ for small Δt by a Riemann sum $\Lambda(t) \approx \sum_{k=0}^{n} \lambda(\xi_k)\Delta t$, where ξ_k lies in the interval $[k\Delta t, (k+1)\Delta t]$ and each (constant) rate $\lambda(\xi_k)$ corresponds to a rate q_k of state k in the continuous-time Markov chain with n states.

Finally, a non-homogeneous Poisson process $X(t)$ with rate $\lambda(t)$ can be transformed to a homogeneous Poisson process $Y(u)$ with rate 1 by the time transform $u = \Lambda(t)$. For, $Y(u) = Y(\Lambda(t)) = X(t)$, and $Y(u + \Delta u) = Y(\Lambda(t) + \Delta\Lambda(t)) = X(t + \Delta t)$ because $\Delta\Lambda(t) = \Lambda(\Delta t)$ for small Δt such that

$$\Pr\left[Y(u + \Delta u) - Y(u) = 1\right] = \Pr\left[X(t + \Delta t) - X(t) = 1\right]$$
$$= \lambda(t)\Delta t + o(\Delta t) = \Delta u + o(\Delta u)$$

because $\Delta u = \lambda(t)\Delta t + o(\Delta t)$. Hence, all problems concerning non-homogeneous Poisson processes can be reduced to the homogeneous case treated above.

7.6 The failure rate function

Previous sections have shown that the Poisson process is specified by a rate function $\lambda(t)$. In this section, we consider the failure rate function of some object or system. Often it is interesting to know the probability that an object will fail in the interval $[t, t + \Delta t]$ given that the object was still functioning well up to time t. Let X denote the lifetime of an object[7], then this probability can be written with (2.47) as

$$\Pr\left[t \le X \le t + \Delta t | X > t\right] = \frac{\Pr\left[\{t \le X \le t + \Delta t\} \cap \{X > t\}\right]}{\Pr\left[X > t\right]}$$
$$= \frac{\Pr\left[t < X \le t + \Delta t\right]}{\Pr\left[X > t\right]}$$

If $f_X(t)$ is the probability density function of X and $F_X(t) = \Pr\left[X \le t\right]$, then for small Δt and assuming that $f_X(t)$ is well behaved[8] such that $\Pr\left[t < X \le t + \Delta t\right] = f_X(t)\Delta t + o(\Delta t)$, and

$$\Pr\left[t \le X \le t + \Delta t | X > t\right] = \frac{f_X(t)}{1 - F_X(t)}\Delta t + o(\Delta t)$$

This expression shows that

$$r(t) = \frac{f_X(t)}{1 - F_X(t)} \tag{7.7}$$

---

[7] In medical sciences, X can represent in general the time for a certain event to occur. For example, the time it takes for an organism to die, the time to recover from illness, the time for a patient to respond to a therapy and so on.

[8] Recall the discussion in Section 2.3 to led to (2.31).

can be interpreted as the intensity or rate that a t-year-old object will fail. It is called *the failure rate $r(t)$* and

$$R(t) = 1 - F_X(t) = \Pr[X > t] \tag{7.8}$$

is usually termed[9] *the reliability function*. Since $r(t) = \frac{\Pr[t \leq X \leq t + \Delta t | X > t]}{\Delta t}$ for small Δt, the failure rate $r(t) > 0$ because $r(t) = 0$ would imply an infinite lifetime X. Using the definition (2.30) of a probability density function, we observe that

$$r(t) = -\frac{\frac{dR(t)}{dt}}{R(t)} = -\frac{d \ln R(t)}{dt} \tag{7.9}$$

Since $R(0) = 1$, the corresponding integrated relation is

$$R(t) = \exp\left[-\int_0^t r(u)du \right] \tag{7.10}$$

The expressions (7.9) and (7.10) are inverse relations that specify $r(t)$ as function of $R(t)$ and vice versa. The reliability function $R(t)$ is non-increasing with maximum at $t = 0$ since it is a probability distribution function. On the other hand, the failure rate $r(t)$ being a probability density function can take any positive real value. From (7.7) we obtain the density function of the lifetime X in terms of failure rate $r(t)$ as

$$f_X(t) = r(t)R(t) = r(t)\exp\left[-\int_0^t r(u)du \right]$$

with $f_X(0) = r(0)$. Using the tail relation (2.36) for the expectation of the lifetime X immediately gives the *mean time to failure*,

$$E[X] = \int_0^\infty R(t)dt \tag{7.11}$$

In case $F_X(T) = 1$ and $f_X(T) \neq 0$ for a finite time T, which is the maximum lifetime, the definition (7.7) demonstrates that $r(t)$ has a pole at $t = T$. In practice, the failure rate $r(t)$ is relatively high for small t due to initial imperfections that cause a number of objects to fail early and $r(t)$ is increasing towards the maximum lifetime T due to aging or wear and tear. This shape of $r(t)$ as illustrated in Fig. 7.5 is called a "bath-tub" curve, which is convex.

An often used model for the failure rate is $r(t) = a\lambda t^{a-1}$ with corresponding reliability function $R(t) = \exp[-\lambda t^a]$ and where the lifetime X has a Weibull distribution function $F_X(t) = 1 - R(t)$ as in (3.49). In case $a = 1$, the failure rate $r(t) = \lambda$ is constant over time, while $a > 1$ ($a < 1$) reflects an increasing (decreasing) failure rate over time. Hence, a "bath-tub" shaped (realistic) failure function as in Fig. 7.5 can be modeled by $r(t) = a\lambda t^{a-1}$ with $a < 1$ in the beginning, $a = 1$ in the middle and $a > 1$ at the end of the lifetime.

For an exponential lifetime where $f_X(t) = \lambda e^{-\lambda t}$, the failure rate (7.7) equals

---

[9] In biology, medical sciences and physics, $R(t)$ is called the survival function and $r(t)$ is the corresponding mortality rate or hazard rate.

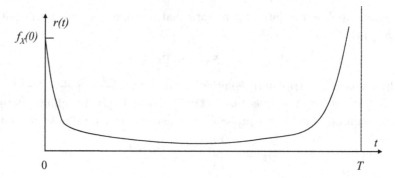

Fig. 7.5. Example of a "bath-tub" shaped failure rate function $r(t)$.

$r(t) = \lambda$ and is independent of time. This means that the failure rate for a t-year-old object is the same as for a new object, which is a manifestation of the memoryless property of the exponential distribution. It also explains why λ in both the exponential distribution and the Poisson process is often called a "rate".

7.7 Problems

(i) A series of test strings each with a variable number N of bits all equal to 1 are transmitted over a channel. Due to transmission errors, each 1-bit can be effected independently from the others and only arrives non-corrupted with probability p. The length N of the test strings (words) is a Poisson random variable with mean length λ bits. In this test, the sum Y of the bits in the arriving words is investigated to determine the channel quality via p. Compute the pdf of Y.

(ii) At a router, four QoS classes are supported and for each class packets arrive according to a Poisson process with rate λ_j for $j = 1, 2, 3, 4$. Suppose that the router had a failure at time t_1 that lasted T time units. What is the probability density function of the total number of packets of the four classes that have arrived during that period?

(iii) Let $N(t) = N_1(t) + N_2(t)$ be the sum of two independent Poisson processes with rates λ_1 and λ_2. Given that the process $N(t)$ had an arrival, what is the probability that that arrival came from the process $N_1(t)$?

(iv) Peter has been monitoring the highway for nearly his entire life and found that the cars pass his house according to a Poisson process. Moreover, he discovered that the Poisson process in one lane is independent from that in the other lanes. The rate of these independent processes differs per lane and is denoted by $\lambda_1, \lambda_2, \lambda_3$, where λ_j is expressed in the number of cars on lane j per hour.

  (a) Given that one car passed Peter, what is the probability that it passed in lane 1?
  (b) What is the probability that n cars pass Peter in 1 hour ?
  (c) What is the probability that in 1 hour n cars have passed and that they all have used lane 1?

(v) In a game, audio signals arrive in the interval $(0, T)$ according to a Poisson process with rate λ, where $T > 1/\lambda$. The player wins only if at least one audio signal arrives in that interval, and if he or she pushes a button upon the last of the signals. There is only one push allowed and we assume that a user can react without delay. The player uses the following strategy: after a fixed time $s \leq T$, he or she pushes the button upon the arrival of the first signal (if any).

  (a) What is the probability that the player wins?
  (b) Which value of s maximizes the probability of winning, and what is the probability in that case?

(vi) The arrivals of voice over IP (VoIP) packets to a router is close to a Poisson process with rate $\lambda = 0.1$ packets per minute. Due to an upgrade to install weighted fair queueing as priority scheduling rule, the router is switched off for 10 minutes.

    (a) What is the probability of receiving no VoIP packets when switched off?

    (b) What is the probability that more than ten VoIP packets will arrive during this upgrade?

    (c) If there was one VoIP in the meantime, what is the most probable minute of the arrival?

(vii) A link of a packet network carries on average ten packets per second. The packets arrive according to a Poisson process. A packet has a probability of 30 % to be an acknowledgment (ACK) packet independent of the others. The link is monitored during an interval of 1 second.

    (a) What is the probability that at least one ACK packet has been observed?

    (b) What is the expected number of all packets given that five ACK packets have been spotted on the link?

    (c) Given that eight packets have been observed in total, what is the probability that two of them are ACK packets?

(viii) An ADSL helpdesk treats exclusively customer requests of one of three types: (i) login-problems, (ii) ADSL hardware and (iii) ADSL software problems. The opening hours of the helpdesk are from 8:00 until 16:00. All requests are arriving at the helpdesk according to a Poisson process with different rates: $\lambda_1 = 8$ requests with login problems/hour, $\lambda_2 = 6$ requests with hardware problems/hour, and $\lambda_3 = 6$ requests with software problems/hour. The Poisson arrival processes for different types of requests are independent.

    (a) What is the expected number of requests in one day?

    (b) What is the probability that in 20 minutes exactly three requests arrive, and that all of them have hardware problems?

    (c) What is the probability that no requests will arrive in the last 15 minutes of the opening hours?

    (d) What is the probability that one request arrives between 10:00 and 10:12 and two requests arrive between 10:06 and 10:30?

    (e) If at the moment $t + s$ there are $k + m$ requests, what is the probability that there were k requests at the moment t?

(ix) Arrival of virus attacks to a PC can be modeled by a Poisson process with rate $\lambda = 6$ attacks per hour.

    (a) What is the probability that exactly one attack will arrive between 1 p.m. and 2 p.m.?

    (b) Suppose that at the moment the PC is turned on there were no attacks on PC, but at the shut-down time precisely 60 attacks have been observed. What is the expected amount of time that the PC has been on?

    (c) Given that six attacks arrive between 1 p.m. and 2 p.m., what is the probability that the fifth attack will arrive between 1:30 p.m. and 2 p.m.?

    (d) What is the expected arrival time of that fifth attack?

(x) *Competing independent Poisson processes.* Consider two, independent Poisson processes X and Y with rate λ and μ, respectively.

    (a) Compute the probability that the k-th event in Poisson process X occurs before the m-th event in Poisson Y process.

    (b) What is the probability that the number of X-events in the time interval $[0, t]$ is larger than the number of Y-events in the time interval $[0, u]$?

(xi) Consider a system S consisting of n subsystems in series as shown in Fig. 7.6. The system

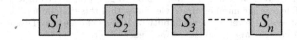

Fig. 7.6. A system consisting of n subsystems in series.

S operates correctly only if all subsystems operate correctly. Assume that the probability that a failure in a subsystem S_i occurs is independent of that in subsystem S_j. Given the reliability functions $R_j(t)$ of each subsystem S_j, show that the reliability function $R(t)$ of the system S is

$$R_S(t) = \prod_{j=1}^{n} R_j(t)$$

In many cases is $R_j(t)$ the same for each subsystem. The fact that $R_j(t) = \exp\left[-\int_0^t r(u)du\right]$ and $R_S(t) = \exp\left[-n\int_0^t r(u)du\right]$ are presented by essentially the same kind of distribution led Weibull (1951) to propose his famous Weibull distribution in (3.49) where $\int_0^t r(u)du = a(t-t_0)^b$ and t_0 is not necessarily zero. Weibull (1951) considered a chain consisting of n identical links. Suppose that the probability that a single link fails at load x is found by testing to be p. Then, the probability of non-failure of the chain, $1 - P_n$, is equal to the probability that none of the single links fails, which is equal to $(1-p)^n$. Weibull (1951) showed that the empirical data of various failure phenomena on a $\log\log\frac{1}{1-p}$ versus $\log(t-t_0)$ plot was surprisingly linear, which was the observational justification of (3.49).

(xii) Same question as in previous problem but applied to a system S consisting of n subsystems in parallel as shown in Fig. 7.7.

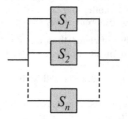

Fig. 7.7. A system consisting of n subsystems in parallel.

(xiii) *Interference in ad-hoc networks.* The interference Y at a certain point, which we further consider as the orgin $r = 0$, consists of the sum of the power signals g from X emitting terminals in an ad-hoc network,

$$Y = \sum_{j=1}^{X} g(R_j, W_j)$$

where R_j is the distance of the j-th terminal from the origin. The random variables W_j reflect stochastic variations in the radio signal and all W_j are independent and identically distributed with probability density function $f_W(r)$. The terminals are distributed over the plane as a Poisson point process. The number X of the terminals in the plane at a uniform position R_j is a Poisson random variable with mean λA, where A is the area of the region of interest.

(a) Derive a general expression for the pgf of the interference Y.
(b) Compute from that general expression, the specific case where all W_j are negligibly small, $g(r,w) = g(r) + h(w)$ and $g(r) = r^{-\eta}$ with $\eta > 2$. In addition, assume that the area A is a circle with large radius.

(xiv) The number of viewers watching TV from 19:00 to 20:00 on a certain day follows a Poisson distribution. The average number of TV viewers within this time interval is 1000. Among the TV channel list, each viewer has a probablity of 10% to choose the "Discovery" TV channel. Viewers of "Discovery" are independent from each other.

(a) Prove that the number of viewers of "Discovery" follows a Poisson distribution.
(b) What is the probablity that there are fewer than four viewers of "Discovery" from 19:00 to 20:00 on that certain day?

8

Renewal theory

A renewal process is a counting process for which the interarrival times τ_n are i.i.d. random variables with distribution $F_\tau(t)$. Hence, a renewal process generalizes the exponential interarrival times in the Poisson process (see Theorem 7.3.2) to an arbitrary distribution. Since the interarrival times are i.i.d. random variables, at each event (or renewal) the process probabilistically restarts. The classical example of a renewal process is the successive replacement of light bulbs: the first bulb is installed at time W_0, fails at time $W_1 = \tau_1$, and is immediately exchanged for a new bulb, which in turn fails at $W_2 = \tau_1 + \tau_2$, and thereafter replaced by a third bulb, and so on. How many light bulbs are replaced in a period of t time units given the lifetime distribution $F_\tau(t)$?

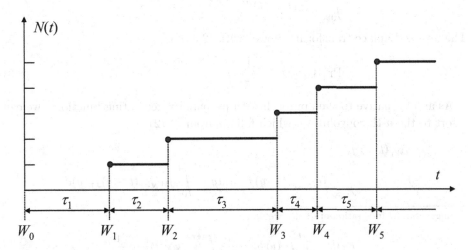

Fig. 8.1. The relation between the renewal counting process $N(t)$, the interarrival time τ_n and the waiting time W_n.

8.1 Basic notions

As illustrated in Fig. 8.1, the waiting time $W_n = \sum_{k=1}^{n} \tau_k$ (for $n \geq 1$, with $W_0 = 0$ by convention) is related to the counting process $\{N(t), t \geq 0\}$ by the equivalence $\{N(t) \geq n\} \Longleftrightarrow \{W_n \leq t\}$: the number of events (renewals) up to time t is at least n if and only if the n-th renewal occurred on or before time t. Alternatively, the number of events by time t equals the largest value of n for which the n-th event occurs before or at time t, $N(t) = \max [n : W_n \leq t]$. The convention that $W_0 = 0$ implies that $N(0) = 0$: the counting process starts counting from zero at time 0. The main objective of renewal theory is to deduce properties of the process $\{N(t), t \geq 0\}$ as a function of the interarrival distribution $F_\tau(t) = \Pr[\tau \leq t]$.

8.1.1 The distribution of the waiting time W_n

If we assume that the interarrival times are i.i.d. having a Laplace transform

$$\varphi_\tau(z) = \int_0^\infty e^{-zt} dF_\tau(t) = \int_0^\infty e^{-zt} f_\tau(t) dt$$

the waiting time W_n is the sum of n i.i.d. random variables specified by (2.75) as

$$\varphi_{W_n}(z) = \int_0^\infty e^{-zt} f_{W_n}(t) dt = \varphi_\tau^n(z). \tag{8.1}$$

By partial integration, we find the Laplace transform of the distribution $F_{W_n}(t) = \Pr[W_n \leq t] = \int_0^t f_{W_N}(u) du$

$$\int_0^\infty e^{-zt} F_{W_n}(t) dt = \frac{\varphi_{W_n}(z)}{z} = \frac{\varphi_\tau^n(z)}{z} \tag{8.2}$$

The inverse Laplace transform follows[1] with (2.39) as

$$\Pr[W_n \leq t] = \frac{1}{2\pi i} \int_{c-i\infty}^{c+i\infty} \frac{\varphi_\tau^n(z)}{z} e^{zt} dz \tag{8.3}$$

As an alternative to the approach with probability generating functions, we can resort to the n-th convolution, which follows from (2.72) as

$$f_{W_1}(t) = f_\tau(t)$$

$$f_{W_N}(t) = \int_{-\infty}^\infty f_{W_{N-1}}(t-y) f_\tau(y) dy = \int_0^t f_{W_{N-1}}(t-y) f_\tau(y) dy$$

---

[1] In general, by integration of (2.39), we find

$$F_X(t) = \int_0^t f_X(u) du = \frac{1}{2\pi i} \int_{c-i\infty}^{c+i\infty} \varphi_X(z) \frac{e^{zt} - 1}{z} dz$$

whose form seems different from (8.3). However, $\frac{1}{2\pi i} \int_{c-i\infty}^{c+i\infty} \frac{\varphi_X(z)}{z} dz = 0$ because the contour can be closed over the positive $\mathrm{Re}(z) > c$ plane where $\varphi_X(z)$ is analytic and because $\lim_{R \to \infty} \varphi_X(Re^{i\theta}) = 0$ for $-\frac{\pi}{2} < \theta < \frac{\pi}{2}$, which follows from the existence of the Laplace integral $\int_0^\infty e^{-zt} f_X(t) dt$.

Integrated,

$$\Pr\left[W_n \le t\right] = \int_{-\infty}^{t} du \int_{-\infty}^{\infty} f_{W_{N-1}}(u - y) f_\tau(y) dy$$

$$= \int_{-\infty}^{\infty} \left(\int_{-\infty}^{t-y} f_{W_{N-1}}(u) du \right) f_\tau(y) dy$$

$$= \int_{0}^{t} \Pr\left[W_{n-1} \le t - y\right] f_\tau(y) dy$$

By denoting $\Pr\left[W_n \le t\right] = F_\tau^{(n*)}(t)$, we have

$$F_\tau^{(1*)}(t) = F_\tau(t)$$

$$F_\tau^{(n*)}(t) = \int_{0}^{t} F_\tau^{((n-1)*)}(t - y) f_\tau(y) dy$$

These equations also show that we can define $F_\tau^{(0*)}(t) = 1$.

Let us define $U_n(t) = \sum_{k=1}^{n} F_\tau^{(k*)}(t)$, which is the finite sum that will appear in the definition (8.7) of the renewal function below. By summing both sides in the last equation, we obtain

$$U_n(t) = \int_{0}^{t} \sum_{k=1}^{n} F_\tau^{((k-1)*)}(t - y) f_\tau(y) dy = \int_{0}^{t} \sum_{k=0}^{n-1} F_\tau^{((k)*)}(t - y) f_\tau(y) dy$$

With the definition $F_\tau^{(0*)}(t) = 1$, we arrive at

$$U_n(t) = \int_{0}^{t} U_{n-1}(t - y) dF_\tau(y) + F_\tau(t) \tag{8.4}$$

or, written in terms of convolutions,

$$U_n(t) = (U_{n-1} * F_\tau)(t) + F_\tau(t)$$

Finally, we mention the interesting bound on the convolution $F_\tau^{(n*)}(t)$ for a non-negative random variable τ,

$$F_\tau^{(n*)}(t) = \int_{0}^{t} F_\tau^{((n-1)*)}(t - y) dF_\tau(y)$$

$$\le F_\tau^{((n-1)*)}(t) \int_{0}^{t} dF_\tau(y) = F_\tau^{((n-1)*)}(t) F_\tau(t)$$

which follows from the monotone increasing nature of any distribution function. By iteration on n starting from $F_\tau^{(0*)}(t) = 1$, it is immediate that

$$F_\tau^{(n*)}(t) \le (F_\tau(t))^n \tag{8.5}$$

Since $(F_\tau(t))^n$ is the distribution of the maximum (3.40) of a set of n i.i.d. random variables $\{\tau_k\}_{1 \le k \le n}$, the bound (8.5) means that, for $\tau_k \ge 0$,

$$\Pr\left[\sum_{k=1}^{n} \tau_k \le x\right] \le \Pr\left[\max_{1 \le k \le n} \tau_k \le x\right]$$

which is rather obvious because $\sum_{k=1}^{n} \tau_k \geq \max_{1 \leq k \leq n} \tau_k$. The equality sign is only possible if $n-1$ of the τ_k are zero.

8.1.2 The renewal function $m\,(t) = E\,[N\,(t)]$

From the equivalence $\{N(t) \geq n\} \Longleftrightarrow \{W_n \leq t\}$, we directly have

$$\Pr\,[N(t) \geq n] = \Pr\,[W_n \leq t] = F_\tau^{(n*)}(t) \tag{8.6}$$
$$\Pr\,[N(t) = n] = \Pr\,[N(t) \geq n] - \Pr\,[N(t) \geq n+1]$$
$$= F_\tau^{(n*)}(t) - F_\tau^{((n+1)*)}(t)$$

The expected number of events in $(0, t]$ expressed via the tail probabilities (2.37) follows with (8.6) as

$$m(t) = E\,[N(t)] = \sum_{k=1}^{\infty} F_\tau^{(k*)}(t) \tag{8.7}$$

and $m(t)$ is called *the renewal function*. According to a property of the counting process, $N\,(0) = 0$, the number of events in $(0, t]$ when $t \to 0$ is assumed to be zero such that $m(0) = 0$. From (8.5), it follows at each point t for which $F_\tau(t) < 1$ that

$$m(t) \leq \sum_{k=1}^{\infty} (F_\tau(t))^k = \frac{1}{1 - F_\tau(t)} - 1$$

Hence, for finite t where $F_\tau(t) < 1$, the renewal function $m(t)$ converges at least as fast as a geometric series and is bounded. In the limit $t \to \infty$, where $\lim_{t \to \infty} F_\tau(t) = 1$, we see that $m(t)$ is no longer bounded. Intuitively, the number of repeated events (renewals) in an infinite time interval is clearly infinite.

The renewal function $m(t)$ completely characterizes the renewal process. Indeed, if $\varphi_m(z)$ is the Laplace transform of $m(t)$, then after taking the Laplace transform of both sides in (8.7) and using the definition $\Pr\,[W_n \leq t] = F_\tau^{(n*)}(t)$ together with (8.2), we obtain

$$\varphi_m(z) = \frac{1}{z} \sum_{k=1}^{\infty} \varphi_\tau^k(z) = \frac{1}{z} \frac{\varphi_\tau(z)}{1 - \varphi_\tau(z)} \tag{8.8}$$

provided $|\varphi_\tau(z)| < 1$. From this expression, the interarrival time can be found from

$$\varphi_\tau(z) = \frac{z\varphi_m(z)}{1 + z\varphi_m(z)}$$

after inverse Laplace transform. By taking the inverse Laplace transform (2.39), $m(t)$ is written as a complex integral

$$m(t) = \frac{1}{2\pi i} \int_{c-i\infty}^{c+i\infty} \frac{\varphi_\tau(z)}{1 - \varphi_\tau(z)} \frac{e^{zt}}{z} dz$$

8.1.3 *The renewal equation*

After taking the inverse Laplace transform of $\varphi_m(z) = \varphi_m(z)\varphi_\tau(z) + \frac{\varphi_\tau(z)}{z}$, which is deduced from (8.8), a third relation for $m(t)$ that often occurs is

$$m(t) = \int_0^t m(t-u)dF_\tau(u) + F_\tau(t)$$

$$= \int_0^t F_\tau(t-u)dm(u) + F_\tau(t) \qquad (8.9)$$

and is called *the renewal equation*. Taking the limit $n \to \infty$ in (8.4) also leads to the renewal equation. Since $m(0) = 0$, the renewal equation implies that $F_\tau(0) = \Pr[\tau \leq 0] = 0$ or that processes where a zero interarrival time is possible (e.g. in simultaneous events) are ruled out. For a Poisson process, Theorem 7.3.1 states that the occurrence of simultaneous events ($h \to 0$) is zero. The requirement $m(0) = 0$ generalizes the exclusion of simultaneous events in any renewal process.

The probabilistic argument that leads to the renewal equation is as follows. By conditioning on the first renewal for $k > 0$,

$$\Pr[N(t) = k | W_1 = s] = 0 \qquad\qquad t < s$$

$$= \Pr[N(t-s) = k-1] \qquad t \geq s$$

where in the last case for $t \geq s$ the event $\{N(t) = k\}$ is only possible if $k-1$ renewals occur in time interval $(s, t]$, which is, due to the stationarity of the renewal process, equal to $k-1$ renewals in $(0, t-s]$. By the law of total probability (2.49), we uncondition to find for $k \geq 1$,

$$\Pr[N(t) = k] = \int_0^\infty \Pr[N(t) = k | W_1 = s] \frac{d\Pr[W_1 \leq s]}{ds} ds$$

$$= \int_0^t \Pr[N(t-s) = k-1] f_\tau(s) ds \qquad (8.10)$$

Multiplying both sides by k and summing over all $k \geq 1$ gives the mean at the left-hand side,

$$E[N(t)] = \sum_{k=1}^\infty k \Pr[N(t) = k]$$

The sum at the right-hand side is

$$\sum_{k=1}^\infty k \Pr[N(t-s) = k-1] = \sum_{k=0}^\infty (k+1) \Pr[N(t-s) = k]$$

$$= E[N(t-u)] + 1$$

Combining both sides yields

$$E[N(t)] = F_\tau(t) + \int_0^t E[N(t-u)] dF_\tau(s)$$

which is again the renewal equation (8.9) since $m(t) = E[N(t)]$.

8.1.4 A generalization of the renewal equation

The renewal equation (8.9) is a special case of the more general class of integral equations

$$Y(t) = h(t) + \int_0^t Y(t-u)dF(u), \qquad t \geq 0 \tag{8.11}$$

in the unknown function $Y(t)$, where $h(t)$ is a known function and $F(t)$ is a distribution function. This equation can be written using the convolution notation as

$$Y(t) = h(t) + Y * F(t)$$

By conditioning on the first renewal as shown above, many renewal problems can be recast into the form of the general renewal equation (8.11). An example is the derivation of the residual life or waiting time given in Section 8.3. Therefore, it is convenient to present the solution to the general renewal equation (8.11).

Lemma 8.1.1 *If $h(t)$ is bounded for all t, then the unique solution of the general renewal equation (8.11) is*

$$Y(t) = h(t) + \int_0^t h(t-u)dm(u) \tag{8.12}$$

where $m(t) = \sum_{k=1}^{\infty} F^{(k*)}(t)$ is the renewal function.

Proof: Let us first concentrate on the formal solution. In general, convolutions are best treated in the transformed domain. After taking the Laplace transform of the general renewal equation (8.11), we obtain

$$\varphi_Y(z) = \varphi_h(z) + \varphi_Y(z)\varphi_F(z)$$

such that

$$\varphi_Y(z) = \frac{\varphi_h(z)}{1 - \varphi_F(z)}$$

There always exists a region in the z-domain where $|\varphi_F(z)| < 1$ such that the geometric series applies,

$$\varphi_Y(z) = \varphi_h(z) \sum_{k=0}^{\infty} (\varphi_F(z))^k = \varphi_h(z) + \varphi_h(z) \sum_{k=1}^{\infty} (\varphi_F(z))^k$$

Back transforming and taking into account that $(\varphi_F(z))^k$ is the transform of a k-fold convolution yields

$$Y(t) = h(t) + h * \sum_{k=1}^{\infty} F^{(k*)}(t) = h(t) + h * m(t)$$

This formal manipulation demonstrates[2] that (8.12) is a solution of the general renewal equation (8.11).

Suppose now that there are two solutions $Y_1(t)$ and $Y_2(t)$. Their difference $V(t) = Y_1(t) - Y_2(t)$ obeys

$$V(t) = \int_0^t V(t-u)dF(u) = V * F(t)$$

By convolving both sides with F and using the original equation, we deduce that $V(t) = V * F * F(t)$. Continuing this process, for each k, we have that $V(t) = V * F^{(k*)}(t)$. Since $F^{(k*)}(t) \to 0$ for all finite t and $k \to \infty$ (because $m(t)$ exists for all finite t), and if $V(t)$ is bounded, this implies that $V(t) = 0$ for all finite t. This demonstrates the uniqueness and motivates the requirement that $h(t)$ should be bounded. □

A further generalization of (8.11) with applications in time-dependent branching processes (Section 12.7.1) is

$$Y(t) = h(t) + \xi \int_0^t Y(t-u)dF(u), \qquad t \geq 0 \tag{8.13}$$

where ξ is a positive real number. Clearly, if $\xi = 1$, we find (8.11) again. Similar arguments as in the proof of Lemma 8.1.1 lead to the solution

$$Y(t) = h(t) + \int_0^t h(t-u)dm_\xi(u) \tag{8.14}$$

where the generalized renewal function is $m_\xi(t) = \sum_{k=1}^\infty \xi^k F^{(k*)}(t)$.

8.1.5 The renewal function and a Poisson process

Before showing below that the renewal function $m(t)$ can be specified in detail as $t \to \infty$, we consider first the Poisson process where the interarrival times $\{\tau_n\}_{n \geq 1}$ are i.i.d. exponentially distributed with rate λ. Since $\varphi_\tau(z) = \frac{\lambda}{z+\lambda}$,

$$m(t) = \frac{1}{2\pi i} \int_{c-i\infty}^{c+i\infty} \frac{\lambda e^{zt}}{z^2} dz \qquad c > 0$$

---

[2] Alternatively, by substituting the solution into the equation, a check is

$$Y(t) = h(t) + Y * F(t) = h(t) + h * F(t) + h * m * F(t)$$

$$= h(t) + h * \left[F(t) + \sum_{k=1}^\infty F^{(k*)}(t) * F(t) \right]$$

$$= h(t) + h * \left[F(t) + \sum_{k=2}^\infty F^{(k*)}(t) \right] = h(t) + h * \left[\sum_{k=1}^\infty F^{(k*)}(t) \right]$$

$$= h(t) + h * m(t)$$

The contour can be closed over the negative $\text{Re}(z)$-plane (because $t \geq 0$). The only singularity of the integrand is a double pole at $z = 0$ with residue $m(t) = \lambda \frac{de^{zt}}{dz}\Big|_{z=0} = \lambda t$. This result, of course, follows directly from the definition of the Poisson process given in (7.2). We see that the renewal function $m(t)$ for the Poisson process is linear for all t. Moreover, the Poisson process is the only continuous-time renewal process with a linear renewal function $m(t)$. Indeed, if[3] $m(t) = \alpha t$, the renewal equation is

$$\alpha t = \int_0^t \left(\alpha(t-u)\right) dF_\tau(u) + F_\tau(t) = \alpha \int_0^t F_\tau(u) du - t F_\tau(0) + F_\tau(t)$$

By differentiation with respect to t and assuming non-zero interarrival times such that $F_\tau(0) = \Pr[\tau \leq 0] = 0$, we obtain a differential equation

$$\alpha = \alpha F_\tau(t) + \frac{dF_\tau(t)}{dt}$$

whose solution is $F_\tau(t) = 1 - e^{-\alpha t}$. By Theorem 7.3.2, exponential interarrival times characterize a Poisson process with rate $\lambda = \alpha$.

We end this section by presenting a general bound for the maximum deviation of a Poisson process from its mean, which is proved in Draief and Massoulié (2010, p. 55).

Theorem 8.1.2 *Let $N(t)$ be a Poisson process with rate λ, then it holds for any $\varepsilon > 0$ and $T > 0$ that*

$$\Pr\left[\sup_{0 \leq t \leq T} (N(t) - \lambda t) \geq \varepsilon\right] \leq 2 \exp\left(-\lambda T h\left(\frac{\varepsilon}{\lambda T}\right)\right)$$

where $h(x) = (1+x)\log(1+x) - x$.

Since $xh\left(\frac{\varepsilon}{x}\right)$ is positive and decreases with positive x towards zero, Theorem 8.1.2 can be useful for small T. The maximum difference between the Poisson counting process $N(t)$ and its mean $E[N(t)] = \lambda t$ increases probabilistically with the interval length T as well as with its rate λ, which is rather intuitive.

8.2 Limit theorems

In the limit $t \to \infty$, the equivalence relation (8.6) indicates that, for any fixed value of n, $\Pr[N(t) \geq n] = 1$, which means that the number of events $N(t) \to \infty$ as $t \to \infty$. Let us consider $\frac{W_{N(t)}}{N(t)}$, which is the sample mean of the first $N(t)$ interarrival times in the interval $(0, t]$. The Strong Law of Large Numbers (6.3) indicates that $\Pr\left[\lim_{n \to \infty} \frac{W_n}{n} = \mu\right] = 1$ and, because $N(t) \to \infty$ as $t \to \infty$, we have that $\frac{W_{N(t)}}{N(t)} \to \mu = E[\tau]$ as $t \to \infty$. Since $W_{N(t)} \leq t < W_{N(t)+1}$, we obtain the

---

[3] A linear form $m(t) = \alpha t + \beta$ with $\beta \neq 0$ is impossible because $m(0) = 0$.

inequality

$$\frac{W_{N(t)}}{N(t)} \le \frac{t}{N(t)} < \frac{W_{N(t)+1}}{N(t)}$$

Since both lower and upper bound tend to μ, we arrive at the important result that $\lim_{t \to \infty} \frac{N(t)}{t} = \frac{1}{\mu}$. The random variable counting the number of events in $(0, t]$ per interval length t, converges to the mean interarrival time $\mu = E[\tau]$. Unfortunately[4], we cannot simply deduce the intuitive result that also the expectation, $E\left[\frac{N(t)}{t}\right]$ tends to $\frac{1}{\mu}$. On the other hand, the expectation of $W_{N(t)}$ is obtained from Wald's identity (2.78) as $E\left[W_{N(t)}\right] = E[N(t)]E[\tau]$. Taking the expectation in the inequality $W_{N(t)} \le t < W_{N(t)+1}$, leads to $\frac{E[N(t)]}{t} \le \frac{1}{E[\tau]} < \frac{E[N(t)]}{t} + \frac{1}{t}$ from which, after the limit $t \to \infty$, the intuitive result follows. Thus, we have proved[5] the following theorem:

Theorem 8.2.1 (Elementary Renewal Theorem) *If $\mu = E[\tau]$ is the mean interarrival time of events in the renewal process, then*

$$\lim_{t \to \infty} \frac{E[N(t)]}{t} = \lim_{t \to \infty} \frac{m(t)}{t} = \frac{1}{\mu} \qquad (8.15)$$

$$\lim_{t \to \infty} \frac{N(t)}{t} = \frac{1}{\mu}$$

The left-hand side in (8.15) describes the long-run average number of events (renewals) per unit time. The right-hand side is the reciprocal of the mean interarrival rate (or lifetime). For example, in the light bulb replacement process, a bulb lasts on average μ time units, then, in the long-run or steady-state, the light bulbs must be replaced at rate $\frac{1}{\mu}$ per time unit.

The extension[6] of the Elementary Renewal Theorem is the Key Renewal Theorem. The Key Renewal Theorem gives the limit $t \to \infty$ of the solution (8.12) of the general renewal equation (8.11).

---

[4] As remarked by Ross (1996, p. 108), if U is uniformly distributed on $(0, 1)$, consider the random variables Y_n defined as $Y_n = n1_{U \le \frac{1}{n}}$. For large n, $U > 0$ with probability 1, whence $Y_n \to 0$ if $n \to \infty$. However, $E[Y_n] = nE\left[1_{U \le \frac{1}{n}}\right] = n\frac{1}{n} = 1$, for all n. The sequence of random variables Y_n converges to 0, although the expected values of Y_n are all precisely 1.

[5] The elementary renewal theorem can be proved only by resorting to complex function theory and using Laplace–Stieltjes transforms (Cohen, 1969, p. 100). The limit argument provided by the Strong Law of Large Numbers follows then from a Tauberian theorem.

[6] In the sequel we assume that the distribution of the interarrival times $F_\tau(t)$ is not periodic in the sense that there exists no integer d such that $\sum_{n=0}^{\infty} \Pr[\tau = nd] = 1$. Thus, the random variable τ does not only take integer units of some integer d.

Theorem 8.2.2 (Key Renewal Theorem) *If $h(t)$ is directly[7] Riemann integrable over $[0, \infty)$, then*

$$\lim_{t \to \infty} \int_0^t h(t-u) dm(u) = \frac{1}{\mu} \int_0^\infty h(u) du \qquad (8.16)$$

The proof[8], which is more complicated and based on analysis, is found in Feller (1971, Section XI.1). The essential difficulty is demonstrating that the limit at the right-hand side indeed exists. Due to the importance of the Key Renewal Theorem, we give supporting, analytic arguments for the expression (8.16).

Outline of arguments: In the proof of Lemma 8.1.1, it was shown that the Laplace transform $\varphi_Y(z)$ of $Y(t)$ equals

$$\varphi_Y(z) = \frac{\varphi_h(z)}{1 - \varphi_F(z)}$$

which also suggests the following formal argument. By inverse Laplace transformation, we obtain

$$Y(t) = \frac{1}{2\pi i} \int_{c-i\infty}^{c+i\infty} \frac{\varphi_h(z)}{1 - \varphi_F(z)} e^{zt} dz$$

where c is smallest real value of z for which the integral $\varphi_Y(z) = \int_0^\infty Y(t) e^{-zt} dt$ exists. Since $\varphi_F(z) = E\left[e^{-zF}\right]$ is a probability generating function, we know that $|\varphi_F(z)| < 1$ for $\mathrm{Re}(z) > 0$, implying that the integrand has no singularities for $\mathrm{Re}(z) > 0$, because $h(t)$ is bounded so that $\varphi_h(z)$ is analytic for $\mathrm{Re}(z) > 0$. Hence, $c > 0$. Furthermore, when $z \to 0$, then $1 - \varphi_F(z) = 0$ and the singularity at $z = 0$ in the integrand is a simple pole, provided $\varphi_F'(0) = -E[F] \neq 0$. Henceforth, we assume that $E[F] \neq 0$. When we move the line of integration from $\mathrm{Re}(z) = c > 0$ to $\mathrm{Re}(z) = \delta < 0$, where δ is sufficiently small (but independent of t) so that the resulting integral exists, then, by Cauchy's residue theorem,

$$Y(t) = \frac{1}{2\pi i} \int_{C(0)} \frac{\varphi_h(z)}{1 - \varphi_F(z)} e^{zt} dz + \frac{1}{2\pi i} \int_{\delta-i\infty}^{\delta+i\infty} \frac{\varphi_h(z)}{1 - \varphi_F(z)} e^{zt} dz$$

Since $z = 0$ is a simple pole, Cauchy's integral theorem yields

$$\frac{1}{2\pi i} \int_{C(0)} \frac{\varphi_h(z)}{1 - \varphi_F(z)} e^{zt} dz = \frac{1}{2\pi i} \int_{C(0)} \frac{z}{1 - \varphi_F(z)} \frac{\varphi_h(z) e^{zt}}{z} dz$$

$$= \frac{\varphi_h(0)}{E[F]}$$

because $\lim_{z \to 0} \frac{1 - \varphi_F(z)}{z} = -\varphi_F'(0) = E[F]$. Evaluating the second integral along the line $z = \delta + iu$ yields

$$\int_{\delta-i\infty}^{\delta+i\infty} \frac{\varphi_h(z)}{1 - \varphi_F(z)} e^{zt} dz = i \int_{-\infty}^\infty \frac{\varphi_h(\delta + iu)}{1 - \varphi_F(\delta + iu)} e^{t(\delta+iu)} du$$

and

$$\left| \int_{\delta-i\infty}^{\delta+i\infty} \frac{\varphi_h(z)}{1 - \varphi_F(z)} e^{zt} dz \right| \leq e^{t\delta} \int_{-\infty}^\infty \left| \frac{\varphi_h(\delta + iu)}{1 - \varphi_F(\delta + iu)} \right| du = C(\delta) e^{t\delta}$$

---

[7] The concept is introduced to avoid widly oscillating functions that are still integrable over $[0, \infty)$, such as $h(t) = t1_{\{|t-n| < \frac{1}{n^2}\}}$. The precise definition is given in Feller (1971). A sufficient condition for direct Riemann integrability is (a) $h(t) \geq 0$ for all $t \geq 0$, (b) $h(t)$ is non-increasing and (c) $\int_0^\infty h(u) du < \infty$.

[8] Based on the relatively new probabilistic concept of "coupling", alternative proofs of the Key Renewal Theorem exist (see e.g. Grimmett and Stirzacker (2001, pp. 429–430)).

where the value $C\left(\delta\right)$ is finite. In summary, if the integrand $\frac{\varphi_h(z)}{1-\varphi_F(z)}$ exists for $\mathrm{Re}\,z = \delta < 0$ so that $C\left(\delta\right)$ is finite, then we find that

$$\lim_{t\to\infty} Y\left(t\right) = \frac{\varphi_h\left(0\right)}{E\left[F\right]} = \frac{\int_0^\infty h\left(t\right)dt}{\int_0^\infty \left(1 - F\left(t\right)\right)dt}$$

These formal arguments support (8.16). □

Applications of the Key Renewal Theorem are presented in Section 8.3 and in Section 12.7 on the time-dependent branching process, that can be regarded as a generalization of a renewal process. Here we consider Blackwell's Theorem.

Blackwell's Theorem follows from the Key Renewal Theorem when choosing $h\left(t\right) = 1_{t\in[0,T)}$ in the general renewal equation (8.11). The corresponding solution (8.12) for $t > T$ is

$$Y\left(t\right) = \int_0^t 1_{t-u\in[0,T)}dm(u) = \int_{t-T}^t dm(u) = m(t) - m(t - T)$$

while the Key Renewal Theorem states that $\lim_{t\to\infty} Y\left(t\right) = \frac{1}{\mu}\int_0^\infty h(u)du = \frac{T}{\mu}$. Hence, we arrive at Blackwell's Theorem, for any fixed $T > 0$,

$$\lim_{t\to\infty} \frac{m(t) - m(t - T)}{T} = \frac{1}{\mu}$$

The interpretation of Blackwell's Theorem is that the number of expected renewals in an interval with length T sufficiently far from the origin (or in steady-state regime) is approximately equal to $\frac{T}{\mu}$. It can be shown that the reverse, i.e. the Key Renewal Theorem can be deduced from Blackwell's Theorem, also holds. Hence, the Key Renewal Theorem is equivalent to Blackwell's Theorem.

Similarly to the Key Renewal Theorem the difficulty in Blackwell's Theorem is the proof that the limit exists. If the existence of the limit is proved, which means that $\lim_{t\to\infty} m(t) - m(t - T) = a(T)$ exists, the Elementary Renewal Theorem suffices to prove that the limit has value $\frac{1}{\mu}$. Following the argument of Ross (1996, p. 110), we can write, for finite x and y,

$$a(x + y) = \lim_{t\to\infty} [m(t) - m(t - x - y)]$$
$$= \lim_{t\to\infty} [m(t) - m(t - x)] + \lim_{t\to\infty} [m(t - x) - m(t - x - y)]$$
$$= a(x) + a(y)$$

Apart from the trivial solution $a(x) = 0$, the only other[9] solution of $a(x + y) = a(x) + a(y)$ is $a(x) = cx$, where c is a constant. Hence, given that $\lim_{t\to\infty} m(t) -$

---

[9] The proof is as follows: (i) if $y = 0$, we see that $a(x + 0) = a(x) + a(0)$ or $a(0) = 0$. (ii) $a(nx) = na(x)$ for integer n. (iii) Using (ii), we have that $a(nx + my) = na(x) + ma(y)$. By choosing $nx + my = 0$ it follows from (i) that $a\left(-\frac{m}{n}y\right) = -\frac{m}{n}a(y)$ such that (ii) holds for rational numbers. Thus, $a(q_1 x + q_2 y) = q_1 a(x) + q_2 a(y)$ for rational numbers q_1 and q_2. (iv) Recalling the definition $f\left(\frac{u+v}{2}\right) \leq \frac{f(u)+f(v)}{2}$ of a convex function in Section 5.2 and the fact that a function that is both concave and convex is a linear function, it follows that $a(x)$ is linear and with (i) that $a(x) = cx$.

$m(t - T) = a(T)$ exists, this is equivalent to the fact that the sequence $\{b_n\}_{n \geq 0}$ where $b_n = \frac{m(t_n) - m(t_n - T)}{T}$ and $t_n > t_{n-1}$ converges to a constant c. The simplest sequence with this property is $\{b_n^*\}_{n \geq 0}$ where $b_n^* = m(n) - m(n - 1)$ and $T = 1$. Lemma 6.1.1 states that

$$c = \lim_{n \to \infty} \frac{1}{n} \sum_{k=1}^{n} b_k^* = \lim_{n \to \infty} \frac{1}{n} \sum_{k=1}^{n} m(k) - m(k - 1) = \lim_{n \to \infty} \frac{m(n)}{n} = \frac{1}{\mu}$$

where the last equality follows from the Elementary Renewal Theorem (8.15).

Theorem 8.2.3 (Asymptotic Renewal Distribution) *If the mean $\mu = E[\tau]$ and variance $\sigma^2 = Var[\tau]$ of the interarrival time of the events in a renewal process exist, then*

$$\lim_{t \to \infty} \Pr\left[\frac{N(t) - \frac{t}{\mu}}{\sigma\sqrt{\frac{t}{\mu^3}}} < x\right] = \frac{1}{\sqrt{2\pi}} \int_{-\infty}^{x} e^{-u^2/2} du \qquad (8.17)$$

Proof: The Elementary Renewal Theorem states that $N(t) \sim \frac{t}{\mu}$ for large t, which suggests to consider the random variable $U(t) = N(t) - \frac{t}{\mu}$ with $E[U(t)] \to 0$. From the equivalence $\{N(t) < n\} \iff \{W_n > t\}$, we have $\{U(t) < x_t\} \iff \{W_{x_t + \frac{t}{\mu}} > t\}$ where x_t is such that $x_t + \frac{t}{\mu}$ is a positive integer. Then,

$$\Pr[U(t) < x_t] = \Pr\left[W_{x_t + \frac{t}{\mu}} > t\right]$$

$$= \Pr\left[\frac{W_{x_t + \frac{t}{\mu}} - \left(x_t + \frac{t}{\mu}\right)\mu}{\sigma\sqrt{x_t + \frac{t}{\mu}}} > \frac{t - \left(x_t + \frac{t}{\mu}\right)\mu}{\sigma\sqrt{x_t + \frac{t}{\mu}}}\right]$$

The waiting time W_n consists of a sum of i.i.d. random variables with mean μ and variance σ^2. By the Central Limit Theorem 6.3.1,

$$\lim_{n \to \infty} \Pr\left[\frac{W_n - n\mu}{\sigma\sqrt{n}} > x\right] = \frac{1}{\sqrt{2\pi}} \int_x^{\infty} e^{-u^2/2} du$$

which implies that

$$\lim_{t \to \infty} \Pr\left[\frac{W_{x_t + \frac{t}{\mu}} - \left(x_t + \frac{t}{\mu}\right)\mu}{\sigma\sqrt{x_t + \frac{t}{\mu}}} > \frac{t - \left(x_t + \frac{t}{\mu}\right)\mu}{\sigma\sqrt{x_t + \frac{t}{\mu}}}\right] = \frac{1}{\sqrt{2\pi}} \int_y^{\infty} e^{-u^2/2} du$$

provided $\lim_{t \to \infty} \frac{t - \left(x_t + \frac{t}{\mu}\right)\mu}{\sigma\sqrt{x_t + \frac{t}{\mu}}} = y$. Hence, we must determine x_t such that, for large t,

$$\frac{\mu x_t}{\sigma\sqrt{x_t + \frac{t}{\mu}}} = -y$$

which is satisfied if $x_t = \frac{y^2\sigma^2}{2\mu^2}\left(1 \pm \sqrt{1 + 4\left(\frac{\mu}{y\sigma}\right)^2 \frac{t}{\mu}}\right)$ and provided the negative

sign is chosen. For large t, we see that $x_t \sim -\frac{y\sigma}{\mu}\sqrt{\frac{t}{\mu}} + O(1)$. Thus,

$$\lim_{t\to\infty} \Pr\left[U(t) < -\frac{y\sigma}{\mu}\sqrt{\frac{t}{\mu}}\right] = \frac{1}{\sqrt{2\pi}} \int_y^\infty e^{-u^2/2} du$$

which is equivalent to

$$\lim_{t\to\infty} \Pr\left[\frac{N(t) - \frac{t}{\mu}}{\sigma\sqrt{\frac{t}{\mu^3}}} < x\right] = \frac{1}{\sqrt{2\pi}} \int_{-x}^\infty e^{-u^2/2} du$$

Noting that $\int_{-x}^\infty e^{-u^2/2} du = \int_{-\infty}^x e^{-u^2/2} du$ finally proves (8.17).   □

Comparing Theorem 8.2.3 to the Central Limit Theorem 6.3.1 shows that the asymptotic variance of $N(t)$ behaves as

$$\lim_{t\to\infty} \frac{\text{Var}\,[N(t)]}{t} = \frac{\sigma^2}{\mu^3} \tag{8.18}$$

Moreover, Theorem 8.2.3 is a central limit theorem for the *dependent* random variables $N(W_n)$ where dependence is obvious from $N(W_n) = N(W_{n-1}) + 1$.

8.3 The residual waiting time

Suppose we inspect a renewal process at time t and ask the question "How long do we have to wait on average to see the next renewal?" This question frequently arises in renewal problems. For instance, the arrivals of taxis at a station is a renewal process and, often, we are interested in knowing how long we might have to wait until the next taxi. Also, packets arriving at a router may find an earlier packet that is partially served. In order to compute the total time spent in the system, it is desirable to know the residual service time of that packet. In addition, this problem is a classical example of how misleading intuition in probability problems can be. There are two different arguments to the question above leading to two different answers:

(i) Since my inspection of the process does not alter or influence the process, the distribution of my waiting time should not depend on the time t; hence, my mean waiting time equals the mean interarrival time of the renewal process.

(ii) The time t of the inspection is chosen at random in (i.e. uniformly distributed over) the interval between two consecutive renewals; hence my expected waiting time should be half of the mean interarrival time.

Both arguments seem reasonable although it is plain that one of them must be wrong. Let us try to sort out the correct answer to this apparent paradox, which,

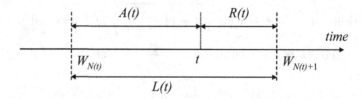

Fig. 8.2. Definition of the random variables the age $A(t)$, the lifetime $L(t)$ and the residual life (or waiting time) $R(t)$.

according to Feller (1971, pp. 12–13), puzzled many before its solution was properly understood.

Figure 8.2 defines the setting of the renewal problem and the quantities of interest: $A(t)$ is the age at time t, which is the total time elapsed since the last renewal before t at time $W_{N(t)}$, the residual waiting time (or residual life or excess life) $R(t)$ is the remaining time at t until the next renewal at time $W_{N(t)+1}$ and $L(t)$ is the total waiting time (or lifetime). From Fig. 8.2, we verify that

$$A(t) = t - W_{N(t)}$$
$$R(t) = W_{N(t)+1} - t$$
$$L(t) = W_{N(t)+1} - W_{N(t)} = A(t) + R(t)$$

The distribution of the residual waiting time, $F_{R(t)}(x) = \Pr[R(t) \leq x]$ will be derived. Similar to the probabilistic argument before, we condition on the first renewal. If $W_1 = s \leq t$, then the first renewal occurs before time t and the event $\{R(t) > x | W_1 = s\}$ has the same probability as the event $\{R(t - s) > x\}$ because the renewal process restarts from scratch at time s. If $s > t$, the residual waiting time $R(t)$ lies in the first renewal interval $[0, s]$. In this case, we have either that the residual waiting time $R(t)$ is certainly shorter than x if s is contained in the interval $[t, t + x]$, else the residual waiting time $R(t)$ is surely larger than x. In summary,

$$\Pr[R(t) > x | W_1 = s] = \begin{cases} \Pr[R(t - s) > x] & \text{if } 0 \leq s \leq t \\ 0 & \text{if } t < s \leq t + x \\ 1 & \text{if } s > x + t \end{cases}$$

Using the law of total probability (2.49),

$$\Pr[R(t) > x] = \int_0^\infty \Pr[R(t) > x | W_1 = s] \frac{d\Pr[W_1 \leq s]}{ds} ds$$
$$= \int_0^t \Pr[R(t - s) > x] f_\tau(s) ds + \int_{x+t}^\infty \frac{d\Pr[\tau \leq s]}{ds} ds$$
$$= \int_0^t \Pr[R(t - s) > x] dF_\tau(s) + 1 - F_\tau(x + t)$$

This relation is an instance of the general renewal equation (8.11). Since $1-F_\tau(x+t)$ is monotonously decreasing, for all x, it holds with (2.36) that

$$\int_0^\infty (1 - F_\tau(x+t))\, dt \leq \int_0^\infty (1 - F_\tau(t))\, dt = E[\tau] < \infty$$

which also implies that $\lim_{t\to\infty} 1 - F_\tau(x+t) = 0$. Hence, $h(t) = 1 - F_\tau(x+t)$ is bounded for all $t \geq 0$ and Lemma 8.1.1 is applicable, yielding

$$\Pr[R(t) > x] = 1 - F_\tau(x+t) + \int_0^t [1 - F_\tau(x+t-s)]\, dm(s)$$

Also, the conditions for direct Riemann integrability in the Key Renewal Theorem 8.2.2 for $h(t) = 1 - F_\tau(x+t)$ are satisfied such that

$$\lim_{t\to\infty} \Pr[R(t) > x] = \lim_{t\to\infty} \int_0^t [1 - F_\tau(x+t-s)]\, dm(s)$$

$$= \frac{1}{E[\tau]} \int_0^\infty (1 - F_\tau(x+t))\, dt \qquad \text{with (8.16)}$$

$$= \frac{1}{E[\tau]} \int_x^\infty (1 - F_\tau(t))\, dt$$

In other words, the steady-state or equilibrium distribution function for the residual waiting time equals

$$\lim_{t\to\infty} \Pr[R(t) \leq x] = \Pr[R \leq x] = F_R(x) = \frac{1}{E[\tau]} \int_0^x (1 - F_\tau(t))\, dt \qquad (8.19)$$

Similarly, for $t > y$, the event $\{A(t) > y\}$ is equivalent to the event $\{$no renewals in $[t-y,t]\}$, which is equivalent to $\{R(t-y) > y\}$. Hence,

$$\lim_{t\to\infty} \Pr[A(t) > y] = \lim_{t\to\infty} \Pr[R(t-y) > y] = \lim_{t\to\infty} \Pr[R(t) > y]$$

$$= \frac{1}{E[\tau]} \int_y^\infty (1 - F_\tau(t))\, dt$$

or, both the residual waiting time R and the age A have the same distribution in steady-state $(t \to \infty)$. Intuitively, when reversing the time axis in steady-state or looking backward in time, an identically distributed renewal process is observed in which the role of the age A and the residual life R are interchanged. Thus, by a time symmetry argument, both distributions must be the same in steady-state.

It is instructive to compute the mean residual waiting time $E[R] = E[A]$ in steady-state. Using the expression of the mean in terms of tail probabilities (2.36), we have

$$E[R] = \int_0^\infty (1 - F_R(x))\, dx$$

$$= \frac{1}{E[\tau]} \int_0^\infty dx \int_x^\infty (1 - F_\tau(t))\, dt$$

Reversing the order of the x- and t-integration yields

$$E[R] = \frac{1}{E[\tau]} \int_0^\infty dt \, (1 - F_\tau(t)) \int_0^t dx$$

$$= \frac{1}{E[\tau]} \int_0^\infty t \, (1 - F_\tau(t)) \, dt$$

After partial integration, we end up with

$$E[R] = \frac{1}{2E[\tau]} \int_0^\infty t^2 f_\tau(t) dt = \frac{E[\tau^2]}{2E[\tau]} = \frac{\mathrm{Var}[\tau] + (E[\tau])^2}{2E[\tau]}$$

or

$$E[R] = \frac{E[\tau]}{2} + \frac{\mathrm{Var}[\tau]}{2E[\tau]} \tag{8.20}$$

This expression shows that the mean remaining waiting time equals half of the mean interarrival time plus the ratio of the variance over the mean of the interarrival time. The last term is always positive. Since $E[A] = E[R]$ and $E[L] = E[A] + E[R]$, we observe the curious result that

$$E[L] = E[\tau] + \frac{\mathrm{Var}[\tau]}{E[\tau]} \geq E[\tau]$$

or that the mean total waiting $E[L]$ is longer than the mean interarrival time $E[\tau]$, contrary to intuition. This fact is referred to as the *inspection paradox*: the steady-state interrenewal time, $L(t) = W_{N(t)+1} - W_{N(t)}$, containing the inspection point at time t, exceeds on average the generic interarrival time, say W_1. The explanation is that the inspection point at time t is uniformly chosen over the time axis and every inspection point is thus equally likely. The chance that the inspection point t lies in a renewal interval is proportional to the length of that interval. Hence, it has higher probability to fall in a long interval, which explains[10] why $E[L] \geq E[\tau]$. Only for deterministic interarrival times where $\mathrm{Var}[\tau] = 0$ holds the equality sign, $E[L] = E[\tau]$. For exponential interarrival times, application of (3.18) gives $\mathrm{Var}[\tau] = (E[\tau])^2$ and $E[R] = E[\tau]$ while $E[L] = 2E[\tau]$: the fact of being inspected at time t changes the lifetime distribution and even doubles the expected total lifetime for exponentially distributed failure or interoccurrence times.

Returning to the initial question, we observe that the intuitive result that my waiting time $E[R] = \frac{E[\tau]}{2}$ is only correct for deterministic processes. Thus, the variability in the interarrival process causes the paradox. We will see later, in queueing theory in Section 14.3.1, that also in queueing systems the variability in service discipline causes the mean waiting time to increase. At last, Feller (1971, p. 187) remarks that an apparently unbiased inspection plan may lead to false conclusions because the actual observations are not typical of the population as a whole. When

---

[10] A similar type of reasoning is used in the computation of the waiting of the GI/D/m queueing system in Section 14.4.2.

people complain that buses or trains start running irregularly, the inspection para-dox shows that above-average interarrival times are experienced more often. The inspection paradox thus implies that complaints may be erroneously based on an overestimation of the real deviations from the regular time schedule of buses or trains.

By separating each renewal interval into two non-overlapping subintervals $A(t)$ and $R(t)$, we have described an alternating renewal process. An alternating renewal process models a system that can be in on- or off-period with a repeating pattern $X_1, Y_1, X_2, Y_2, \ldots$, where each on-period X_n has a same distribution F_{on} and is followed by an off-period Y_n. Each off-period has also a same distribution F_{off}. The off-period Y_n may dependent on the on-period X_n, but the n-th renewal cycle with duration $X_n + Y_n$ is independent of any other cycle. An alternating renewal process can be used to model a data stream of packets, where the on-period reflects the time to store or process an arriving packet and the off-period a (random) delay between two packets. Another example is the modeling of the end-to-end delay from a source s to a destination d in the Internet, where the off-period describes a queueing in a router due to other interfering traffic along that path. During the on-period, a packet is not blocked by other packets. The on-period equals the propagation delay to travel from the output port of one router to the output port of the next-hop router. The end-to-end delay along a path with h hops equals the sum of h consecutive off-periods augmented by the propagation time from s to d.

8.4 The renewal reward process

The renewal reward process associates at each renewal at time W_n a certain cost or reward R_n, which may vary over time and can be negative. For example, each time a light bulb fails, it must be replaced at a certain cost (negative reward) or each customer in a restaurant pays for his meal (positive reward). The reward R_n may depend on the interarrival time τ_n or length of the n-th renewal interval, but it is independent of other renewal epochs (different from the n-th). Thus, the pairs (R_n, τ_n) are assumed to be independent and identically distributed. Most often one is interested in the total reward $R(t)$ over a period t (not to be confused with the residual lifetime) defined as

$$R(t) = \sum_{n=1}^{N(t)} R_n \tag{8.21}$$

In this setting, the renewal reward process is a generalization of the counting process where $R_n = 1$.

By slightly rewriting the total reward $R(t)$ earned over an interval t as

$$\frac{R(t)}{t} = \frac{\sum_{n=1}^{N(t)} R_n}{N(t)} \frac{N(t)}{t}$$

and taking the limit $t \to \infty$, the first fraction tends with probability one to the mean reward $E[R]$ per renewal period by the Strong Law of Large Numbers (Theorem 6.2.2), while the second fraction tends to $\frac{1}{\mu} = \frac{1}{E[\tau]}$ by the Elementary Renewal Theorem 8.2.1. Hence, with probability one,

$$\lim_{t\to\infty} \frac{R(t)}{t} = \frac{E[R]}{E[\tau]} \tag{8.22}$$

which means that the time average reward rate equals the average award per renewal period multiplied by the interarrival rate of renewals (or divided by the average length of a renewal interval).

Similarly as in the proof of the Elementary Renewal Theorem 8.2.1, the inequality for any t,

$$\sum_{n=1}^{N(t)} R_n \le R(t) \le \sum_{n=1}^{N(t)} R_n + R_{N(t)+1}$$

leads, after taking the expectations and using Wald's identity (2.78), to an inequality for the means,

$$E[N(t)]\, E[R] \le E[R(t)] \le E[N(t)]\, E[R] + E\left[R_{N(t)+1}\right]$$

Dividing by t, the limit $t \to \infty$ becomes

$$E[R]\lim_{t\to\infty} \frac{E[N(t)]}{t} \le \lim_{t\to\infty} \frac{E[R(t)]}{t} \le E[R]\lim_{t\to\infty} \frac{E[N(t)]}{t} + \lim_{t\to\infty} \frac{E\left[R_{N(t)+1}\right]}{t}$$

Since the mean reward per renewal period is finite and $E\left[R_{N(t)+1}\right] = E[R]$, we obtain by the Elementary Renewal Theorem 8.2.1 that

$$\lim_{t\to\infty} \frac{E[R(t)]}{t} = \frac{E[R]}{E[\tau]} \tag{8.23}$$

Hence, by comparing (8.23) and (8.22), the time average of the average reward rate equals the time average of the reward rate.

Example 1 The hard disc in a network server is replaced at cost C_1 at time T. The lifetime or age of this mass storage has pdf f_A. If the hard disc fails earlier, the cost of the repair and the penalties for service disruption is C_2. What is the long-run cost of the hard disc in the server per unit time?

Consider the replacement of hard discs as a renewal process with i.i.d. interarrival times τ and with distribution

$$\Pr[\tau < t] = \begin{cases} \int_0^t f_A(u)\,du & \text{if } t < T \\ 1 & \text{if } t \ge T \end{cases}$$

The mean replacement time follows from the tail expression (2.36),

$$E[\tau] = \int_0^\infty (1 - \Pr[\tau < t])\,dt = \int_0^T (1 - \Pr[\tau < t])\,dt$$

The replacement cost $C(\tau)$ equals $C(\tau) = C_1 1_{\tau < T} + C_2 1_{\tau \geq T}$ and the mean cost is with (2.13),

$$E[C] = C_1 \Pr[\tau < T] + C_2 (1 - \Pr[\tau < T])$$

The Elementary Renewal Reward Theorem (8.23) and (8.22) states that the long-run cost of replacements equals

$$\frac{E[C]}{E[\tau]} = \frac{C_1 \Pr[\tau < T] + C_2 (1 - \Pr[\tau < T])}{\int_0^T (1 - \Pr[\tau < t]) \, dt}$$

Usually the replacement time T is chosen to minimize this long-run cost.

Example 2 *Sequential quality control.* We consider a transmission process where each received packet has a probability $q = 1 - p$ to be corrupted during transmission, independently of the other packets. Since checking the quality of every packet is too expensive or delays the transmission of the entire message unnecessarily, the following checking rule with two regimes is adopted. Each packet is checked until k subsequent error-free packets are found (100 % inspection regime). Thereafter (in regime 2), we only check a packet with probability α. If a corrupted packet is found, we return again to the 100 % inspection regime. The period between two consecutive times, at which the 100 % inspection regime starts, is called a cycle. Our aim is to assess the quality of our checking rule.

Let X denote the number of transmitted packets during a cycle and R the number of checked packets during a cycle. The fraction of checked packets during a cycle or renewal period follows from (8.23) as

$$P(k, \alpha, p) = \frac{E[R]}{E[X]}$$

Let N_k denote the number of packets that needs to be checked until k subsequent error-free packets appear. Thus N_k includes those last k error-free packets. The number of checked packets equals N_k augmented with a geometric number of packets until the first one fails. Each checked packet has probability q to be corrupted. Hence, the mean number of checked packets during a cycle equals $E[R] = E[N_k] + \frac{1}{q}$ (see Section 3.1.3). Likewise, the mean number of transmitted packets during a cycle equals $E[X] = E[N_k] + \frac{1}{q\alpha}$, because only with probability α a packet is checked, leading to a geometric number of tranmitted packets with parameter αq until the first packet is corrupted. It remains to deduce N_k for which the recursion holds

$$N_k = (N_{k-1} + 1) 1_{\{\text{no error packet } k\}} + \left(N_{k-1} + 1 + \widetilde{N}_k\right) 1_{\{\text{error packet } k\}}$$

The number of packets N_k until k subsequent error-free ones are transmitted equals N_{k-1} increased with 1, provided that k-th in sequence packet is error-free, else, N_k equals N_{k-1} plus an corrupted packet after which we need again to restart with counting. Thus, we need to add a new random variable $\widetilde{N}_k$, which is distributed as N_k. After taking the expectation of both sides and using independence between the number N_k and the fact that a packet is corrupted, we have

$$E\left[N_k\right] = \left(E\left[N_{k-1}\right] + 1\right) p + \left(E\left[N_{k-1}\right] + 1 + E\left[N_k\right]\right) q$$

or

$$E\left[N_k\right] = \frac{1}{p}\left(E\left[N_{k-1}\right] + 1\right)$$

Iterating from $E\left[N_0\right] = 0$, we find that $E\left[N_k\right] = \sum_{j=1}^{k} \frac{1}{p^j} = \frac{1}{q}\left(p^{-k} - 1\right)$ so that

$$P\left(k, \alpha, p\right) = \frac{E\left[N_k\right] + \frac{1}{q}}{E\left[N_k\right] + \frac{1}{q\alpha}} = \frac{1}{1 + \left(\frac{1}{\alpha} - 1\right) p^k}$$

Let $k = 5$, $\alpha = \frac{1}{20}$ and $p = 0.9$, then $P\left(5, \frac{1}{20}, 0.9\right) \simeq 0.082$. Suppose that the total nuber of transmitted packets equals 1100, then, on average, $1100q = 110$ are corrupted. If the checked corrupted are removed, then

$$1100q\left(1 - P\left(5, \frac{1}{20}, 0.9\right)\right) \simeq 110\left(1 - 0.082\right) \simeq 101$$

corrupted packets on average are still transmitted. If there are now two separate, but sequential transmission processes, where the first process produces 1000 packets and has an error-free probability $p_1 = 0.95$, while the second bad process only produces 100 packets with $p_2 = 0.4$. Packets are not distinguishable so that the overall failure probability of a packet equals $\frac{1000*0.95+100*0.4}{1100} = 0.9$. Nevertheless, the average amount of corrupted transmitted packets is now

$$1000q_1\left(1 - P\left(5, \frac{1}{20}, p_1\right)\right) + 100q_2\left(1 - P\left(5, \frac{1}{20}, p_2\right)\right) \simeq 57$$

illustrating that the latter batch transmission process outperforms a single one with $p = 0.9$.

8.5 Problems

More worked examples can be found in Karlin and Taylor (1975, Chapter 5).

(i)   Calculate $\Pr\left[W_{N(t)} \le x\right]$.

(ii) Derive a recursion equation for the generating function $\varphi_{N(t)}\left(z\right) = E\left[z^{N(t)}\right]$ of the number of renewals in the interval $[0, t]$ and deduce from that equation the renewal equation (8.9) and a relation for $\mathrm{Var}[N(t)]$.

(iii) In a TCP session from A to B, IP data packets and IP acknowledgement packets travel a distance of 2000 km over precisely the same bi-directional path. In case of congestion, the average speed is 40 000 km/s and without congestion the speed is three times higher. Congestion only occurs in 20% of the travels. What is the average speed of IP packets in the TCP session?

(iv) The production of digitalized speech samples depends primarily on the codec, with an effective average rate r (bits/s). Since this rate is low compared to the ATM capacity C (bits/s), UMTS will use AAL2 mini-cells in which 1 ATM cell is occupied by N users. The financial cost of an UMTS operator increases at nc euro per unit time whenever there are $n < N$ speech samples are waiting for transmission and an additional cost of K euro each time an ATM cell is transmitted. What is the mean cost per unit time for the UMTS operator?

(v) The cost of replacing a router that has failed is A euro. However, one can decide to replace a router that has been in service for a period of time T. The advantage of this approach is that the cost of replacing a working router is only B euro, where $B < A$. The policy CHANGEROUTER consists of replacing a router either upon failure or upon reaching the age T, whichever occurs first. Replacement of the current router by a new one occurs instantaneously and at each time there can only be one router in the network. Let $\{X_j\}$ be a sequence of i.i.d. random variables, where X_j is the lifetime of a router j.

   (a) Find the time average cost rate C of the policy CHANGEROUTER.
   (b) Compute C if $T = 5$ years, the cost of replacing the failed router is $A = 10\,000$ euro and the cost of replacing a working router is $B = 7000$ euro. The independent random variables X_j are exponentially distributed and the mean lifetime of a router is 10 years.

(vi) The mean duration D of a telephone call in the Netherlands is about $\mu = 4$ minutes. To a good approximation, the duration D has an exponential distribution. Suppose you telephone somebody in the Netherlands and hear a busy-tone.

   (a) What is the pdf of your waiting time?
   (b) You waited some time T after the first dialling and then dialling again the same number, what is the distribution of the waiting time of this second trial?
   (c) Suppose that the waiting time is uniformly distributed over $[0, 2\mu]$, what is the distribution of the second waiting time given that your first, unsuccessful trial was at time T?

(vii) A highly loaded FTP server is providing access to a 1 Mbyte video file through a 1 Mbps link. Due to the poor design of the server, only one client can download the file at the same time, and one out of twenty requests is served at 200 kbps. When a video download starts, a user at home quickly calls his brother to come and watch the incoming video. What is the mean time the brother is going to wait for the download to finish? (*Hint:* Assume that the brother arrives at a random time after the call, but always before the download finishes.)

9

Discrete-time Markov chains

A large number of stochastic processes belong to the important class of Markov processes. The theory of Markov chains and Markov processes is well established and furnishes powerful tools to solve practical problems. This chapter will be mainly devoted to the theory of discrete-time Markov chains, while the next chapter concentrates on continuous-time Markov chains. The theory of Markov processes will be applied in later chapters to compute or formulate queueing, routing and epidemics on networks.

9.1 Definition

A stochastic process $\{X(t), t \in T\}$ is a Markov process if the future state of the process only depends on the current state of the process and not on its past history. Formally, a stochastic process $\{X(t), t \in T\}$ is a continuous-time Markov process if, for all $t_0 < t_1 < \cdots < t_{n+1}$ of the index set T and for any set $\{x_0, x_1, \ldots, x_{n+1}\}$ of the state space, it holds that

$$\Pr[X(t_{n+1}) = x_{n+1} | X(t_0) = x_0, \ldots, X(t_n) = x_n] = \Pr[X(t_{n+1}) = x_{n+1} | X(t_n) = x_n]$$
(9.1)

Similarly, a discrete-time Markov chain $\{X_k, k \in T\}$ is a stochastic process whose state space is a finite or countably infinite set with index set $T = \{0, 1, 2, \ldots\}$ obeying

$$\Pr[X_{k+1} = x_{k+1} | X_0 = x_0, \ldots, X_k = x_k] = \Pr[X_{k+1} = x_{k+1} | X_k = x_k] \quad (9.2)$$

A Markov process is called a Markov chain if its state space is discrete. The conditional probabilities $\Pr[X_{k+1} = x_{k+1} | X_k = x_k]$ are called the transition probabilities of the Markov chain. In general, these transition probabilities can depend on the (discrete) time k. A Markov chain is entirely defined by the transition probabilities (9.2) and the initial distribution $\Pr[X_0 = x_0]$ of the Markov chain. Indeed, by the definition of conditional probability (2.48), we obtain

$$\Pr\left[X_0 = x_0, \ldots, X_k = x_k\right] = \Pr\left[X_k = x_k | X_0 = x_0, \ldots, X_{k-1} = x_{k-1}\right]$$
$$\times \Pr\left[X_0 = x_0, \ldots, X_{k-1} = x_{k-1}\right]$$

and, by the definition of the Markov chain (9.2),

$$\Pr\left[X_0 = x_0, \ldots, X_k = x_k\right] = \Pr\left[X_k = x_k | X_{k-1} = x_{k-1}\right]$$
$$\times \Pr\left[X_0 = x_0, \ldots, X_{k-1} = x_{k-1}\right]$$

This recursion relation can be iterated resulting in

$$\Pr\left[X_0 = x_0, \ldots, X_k = x_k\right] = \Pr\left[X_0 = x_0\right] \prod_{j=1}^{k} \Pr\left[X_j = x_j | X_{j-1} = x_{j-1}\right] \quad (9.3)$$

which demonstrates that the complete information of the Markov chain is obtained if, apart from the initial distribution, all transition probabilities are known.

9.2 Discrete-time Markov chains

We assume that the set of possible states $\{x_0, x_1, \ldots, x_{n+1}\}$ is countable and we restrict ourselves to the case where each state $x_j = j$ for $1 \leq j \leq N$, mainly to ease matrix computations. Hence, the set of possible states is $\{1, \ldots, N\}$ so that the state space S has N states. If the transition probabilities are independent of time k,

$$\Pr\left[X_{k+1} = j | X_k = i\right] = P_{ij} \quad (9.4)$$

the Markov chain is called *stationary*. In the sequel, we will confine ourselves to stationary Markov chains. Since the discrete-time Markov chain is conceptually simpler than the continuous counterpart, we start the discussion with the discrete case.

It is convenient to introduce a vector notation[1]. Since X_k can only take N possible values, we denote the corresponding state vector at discrete-time k by $s[k] = [s_1[k] \ s_2[k] \ \cdots \ s_N[k]]$ with $s_i[k] = \Pr\left[X_k = i\right]$. Hence, $s[k]$ is a $1 \times N$ vector. Since the state X_k at discrete-time k must be in one of the N possible states (Axiom 1, p. 8), we have that $\sum_{i=1}^{N} \Pr\left[X_k = i\right] = 1$ or, in vector notation, $s[k]u = \sum_{i=1}^{N} s_i[k] = 1$, where $u^T = [1 \ 1 \ \cdots \ 1]$. This fact is also written as $\|s[k]\|_1 = 1$, where $\|a\|_1$ is the $q = 1$ norm of vector a defined in Appendix A.4. In a stationary Markov chain, the states X_{k+1} and X_k are connected via the law of

---

[1] Unfortunately, a vector in Markov theory is represented as a single row matrix which deviates from the general theory in linear algebra, followed in Appendix A, where a vector is represented as a single column matrix. In order to be consistent with the literature on Markov processes, we have chosen to follow the notation of Markov theory here, but elsewhere we adhere to the general convention of linear algebra.

total probability (2.49),

$$\Pr[X_{k+1} = j] = \sum_{i=1}^{N} \Pr[X_{k+1} = j | X_k = i] \Pr[X_k = i]$$

$$= \sum_{i=1}^{N} P_{ij} \Pr[X_k = i] \tag{9.5}$$

which holds for all j, or, in vector notation,

$$s[k+1] = s[k]P \tag{9.6}$$

where the transition probability matrix P is

$$P = \begin{bmatrix} P_{11} & P_{12} & P_{13} & \cdots & P_{1;N-1} & P_{1N} \\ P_{21} & P_{22} & P_{23} & \cdots & P_{2;N-1} & P_{2N} \\ P_{31} & P_{32} & P_{33} & \cdots & P_{3;N-1} & P_{3N} \\ \vdots & \vdots & \vdots & \cdots & \vdots & \vdots \\ P_{N-1;1} & P_{N-1;2} & P_{N-1;3} & \cdots & P_{N-1;N-1} & P_{N-1;N} \\ P_{N1} & P_{N;2} & P_{N3} & \cdots & P_{N;N-1} & P_{NN} \end{bmatrix} \tag{9.7}$$

Since (9.6) must hold for any initial state vector $s[0]$, by choosing $s[0]$ equal to a base vector $[0 \cdots 0\, 1\, 0 \cdots\, 0]$ (all entries zero except for entry i) which expresses that, if the Markov chain starts from one of the possible states, say state i, then $s[1] = [P_{i1}\ P_{i2}\ \cdots\ P_{iN}]$. Furthermore, since $\|s[k]\|_1 = 1$ for any k, it must hold, for any state i, that

$$\sum_{j=1}^{N} P_{ij} = 1 \tag{9.8}$$

Alternatively and more generally based on Axiom 1 of probability theory on p. 8, right-multiplication of both sides in (9.6) by the all-one vector u and using $s[k]\,u = 1$, valid for each discrete time k, yields

$$1 = s[k]Pu$$

which is only possible for all k provided

$$Pu = u \tag{9.9}$$

Hence, the probabilistic nature that requires $s[k]\,u = 1$ dictates that the all-one vector u is the right-eigenvector of P belonging to the eigenvalue $\lambda = 1$. The relation $(Pu)_i = 1$, equivalent to (9.8), means that, at each discrete-time k, there certainly occurs a transition in the Markov chain, possibly to the same state as at time $k-1$. The $N \times N$ transition probability matrix P thus consists of at most $N^2 - N$ independent transition probabilities P_{ij} and at each row, one transition probability can be expressed in terms of the others, e.g. $P_{ik} = 1 - \sum_{j=1;j\neq k}^{N} P_{ij}$. A matrix with elements $0 \leq P_{ij} \leq 1$ obeying (9.8) is called a *stochastic matrix* whose properties are investigated in Appendix A. Apart from the matrix representation, Markov chains are often described by a directed graph (as illustrated in the figure below), where P_{ij} is represented by an edge from state i to j provided $P_{ij} > 0$. The Markov graph enables to deduce structural properties of the Markov chain (such as, e.g., communicating states) elegantly.

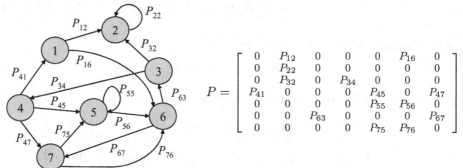

Given the initial state vector $s[0]$, the general solution of (9.6) is

$$s[k] = s[0]P^k \tag{9.10}$$

Similarly, when knowledge of the Markov chain at discrete-time k is available, we obtain from (9.6) that

$$s[k+n] = s[k]P^n$$

The elements of the matrix P^n are called the n-step transition probabilities,

$$(P^n)_{ij} = \Pr[X_{k+n} = j | X_k = i] \tag{9.11}$$

for $k \geq 0$ and $n \geq 0$. Since the discrete Markov chain must be surely in one of the N states n time units later, given that it started at time k in state i, we obtain an extension of (9.8), for all $n \geq 1$,

$$\sum_{j=1}^{N} (P^n)_{ij} = 1 \tag{9.12}$$

which follows after iteratively left-multiplying both sides in (9.9) by P leading to $P^n u = u$.

9.2.1 *Inequalities for matrix elements of P*

We add some consequences of the non-negative matrix P. From the matrix multiplication $P^{n+m} = P^n P^m$, we obtain the matrix element

$$\left(P^{n+m}\right)_{ik} = \sum_{l=1}^{N} (P^n)_{il} (P^m)_{lk} = (P^n)_{ij} (P^m)_{jk} + \sum_{l=1;l\neq j}^{N} (P^n)_{il} (P^m)_{lk}$$

The non-negativity of P (i.e. $P_{ij} \geq 0$), and thus of any power P^n, leads to the bound

$$\left(P^{n+m}\right)_{ik} \geq \left(P^n\right)_{ij}\left(P^m\right)_{jk} \tag{9.13}$$

In particular, for $i = j = k$, it holds that

$$\left(P^{n+m}\right)_{jj} \geq \left(P^n\right)_{jj}\left(P^m\right)_{jj} \tag{9.14}$$

Moreover, by iterating an instance $\left(P^n\right)_{jj} \geq P_{jj}\left(P^{n-1}\right)_{jj}$ of the inequality (9.14), we find that

$$\left(P^n\right)_{jj} \geq P_{jj}^n \tag{9.15}$$

In particular, if $P_{jj} > 0$, then $\left(P^n\right)_{jj} > 0$ and, if $P_{jj} = 1$, then $\left(P^n\right)_{jj} = 1$. Incidently, the notation also shows that care is needed by using powers of a matrix and of its corresponding elements.

Similarly, from $P^{n+l+m} = P^n P^l P^m$, we find that

$$\left(P^{n+l+m}\right)_{ij} = \sum_{r=1}^{N}\sum_{k=1}^{N}\left(P^n\right)_{ir}\left(P^l\right)_{rk}\left(P^m\right)_{kj}$$

from which the triple element inequality is deduced

$$\left(P^{n+l+m}\right)_{ij} \geq \left(P^n\right)_{ir}\left(P^l\right)_{rk}\left(P^m\right)_{kj} \tag{9.16}$$

9.2.2 Definitions and classification

9.2.2.1 Irreducible Markov chains

A state j in a Markov chain is said to be *reachable* from state i if it is possible to proceed from state i to state j in a finite number of transitions which is equivalent to $\left(P^n\right)_{ij} > 0$ for finite $n \geq 0$. If every state is reachable from every other state, the Markov chain is said to be *irreducible*. The example of the Markov graph above is not irreducible because state 2 is absorbing, i.e., $P_{22} = 1$. Markov theory is considerably more simplified if we know that the chain is irreducible, which justifies to investigate methods to determine irreducibility.

An equivalent requirement for the Markov chain to be irreducible is that the associated directed graph is strongly connected, i.e. if there is a path from node i to node j for any pair of distinct nodes (i, j). Let us review some basic notions from graph theory (see Van Mieghem (2011)). Denote by A the adjacency matrix of P where all non-zero elements in P are replaced by 1. A walk of length k from state i to state j is a succession of k arcs of the form $(n_0 \rightarrow n_1)(n_1 \rightarrow n_2)\cdots(n_{k-1} \rightarrow n_k)$, where $n_0 = i$ and $n_k = j$. A path is a walk in which all nodes are different, i.e. $n_l \neq n_m$ for all $0 \leq l \neq m \leq k$. The number of walks of length k from state i to state j is equal to the element $\left(A^k\right)_{ij}$.

A directed graph is strongly connected if and only if each non-diagonal element of the matrix $\sum_{k=1}^{N-1} P^k$ or, equivalently, of $B = \sum_{k=1}^{N-1} A^k$ is positive. Since P has

N states, the longest possible path between two states consists of $N - 1$ hops. By summing over all powers of $1 \leq k \leq N - 1$, the element b_{ij} of the matrix B equals the number of all possible walks (of any possible length) between i and j. Hence, if $b_{ij} > 0$ for all $i \neq j$, there exist walks from any state i to any other state j. The converse is readily verified.

Another way to determine irreducibility follows from the definition of reducibility in the Appendix A.5. However, the methods for strongly connectivity or irreducibility are still algebraic in that they require matrix operations. A computationally more efficient method consists of applying all-pair shortest path algorithms on the Markov graph. Examples of all-pair shortest path algorithms are that of Floyd-Warshall (with computational complexity $C_{\text{Floyd-Warschall}} = O(N^3)$) or the algorithm of Johnson (complexity $C_{\text{Johnson}} = O(N^2 \log N + NL)$, where L is the total number of links in the Markov graph). These algorithms are nicely discussed in Cormen *et al.* (1991).

9.2.2.2 Communicating states

If two states i and j are reachable from one to each other, they are said to *communicate*, which is often denoted by $i \longleftrightarrow j$.

The concept of communication is an equivalence relation: (a) reflexivity: $i \longleftrightarrow i$ since $P^0 = I$ or $P_{ij}^0 = \delta_{ij}$; (b) symmetry: if $i \longleftrightarrow j$, then $j \longleftrightarrow i$, which follows from the definition of communication; and (c) transitivity: if $i \longleftrightarrow j$ and $j \longleftrightarrow k$, then $i \longleftrightarrow k$. By definition of $i \longleftrightarrow j$ and $j \longleftrightarrow k$, we have $(P^n)_{ij} > 0$ and $(P^m)_{jk} > 0$ for some finite n and m. Hence, (9.13) shows that $(P^{n+m})_{ik} > 0$, which implies $i \longleftrightarrow k$.

As an application, the total state space can be partitioned into equivalence classes. States in one equivalence class communicate with each other. If there is a possibility to start in one class and then enter another class, with no return possible to the first (otherwise the two classes would form one class), the Markov chain is reducible. In other words, a Markov chain is irreducible if the equivalence relations result in one class.

9.2.2.3 Periodic and aperiodic Markov chains

Consider a state j in a Markov chain with $(P^n)_{jj} > 0$ for some $n \geq 1$. The *period* d_j is defined as the greatest common divisor of those n for which $(P^n)_{jj} > 0$. The figure below illustrates a Markov chain with period $d = N$.

Since the greatest common divisor of a set is the largest integer d that divides any integer in a set, it is smaller than the minimum element in the set. Thus,

$$1 \leq d_j \leq \min_n \{(P^n)_{jj} > 0\}$$

Hence, it follows from (9.15) that, if $P_{jj} > 0$, then all $(P^k)_{jj} > 0$ for $k > 1$ and thus $d_j = 1$.

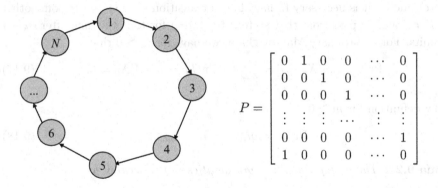

$$P = \begin{bmatrix} 0 & 1 & 0 & 0 & \cdots & 0 \\ 0 & 0 & 1 & 0 & \cdots & 0 \\ 0 & 0 & 0 & 1 & \cdots & 0 \\ \vdots & \vdots & \vdots & & \ddots & \vdots & \vdots \\ 0 & 0 & 0 & 0 & \cdots & 1 \\ 1 & 0 & 0 & 0 & \cdots & 0 \end{bmatrix}$$

Lemma 9.2.1 *If two states i and j communicate $(i \longleftrightarrow j)$, then $d_i = d_j$.*

Proof: Let n and m be integers such that $(P^n)_{ij} > 0$ and $(P^m)_{ji} > 0$. Then, it follows from (9.13) that $P_{ii}^{n+m} > 0$. By definition of a period, $d_i|(n+m)$. An instance of (9.16) is $(P^{n+l+m})_{ii} \geq (P^n)_{ij} (P^l)_{jj} (P^m)_{ji}$. Now, if $(P^l)_{jj} > 0$, which implies by definition that $d_j|l$, then we also have that $(P^{n+l+m})_{ii} > 0$ from which $d_i|(n+m+l)$. The combination of the conditions $d_i|(n+m)$ and $d_i|(n+m+l)$ implies that $d_i|l$. But since d_j is the largest such divisor $d_j \geq d_i$. By symmetry of the communication relation (replace $i \to j$ and $j \to i$), $d_i \geq d_j$ which proves the lemma. □

The consequence of Lemma 9.2.1 is that all the states in an irreducible Markov chain have common period d. The irreducible Markov chain is periodic with period d if $d > 1$ else it is aperiodic ($d = 1$). A simple sufficient condition for an irreducible chain to be aperiodic is that $P_{ii} > 0$ for some state i. Most Markov chains of practical interest are aperiodic.

9.2.3 The hitting time

Let A be a subset of states, $A \subset S$. The hitting time T_A is the first time the Markov chain is in a state of the set A, thus for any discrete time $k \geq 0$, $T_A = \min_{k \geq 0}(X_k \in A)$. Linear relations involving the hitting time T_A to a set A are given in (9.53) and (9.54) on p. 202. Here, we confine to the hitting time[2] of a state j, which follows from the definition if $A = \{j\}$. For irreducible Markov chains, the hitting time $T_j \stackrel{\Delta}{=} T_{\{j\}}$ is finite, for any state j.

From the definition of the hitting time, the recursion

$$\Pr\left[T_j = m | X_0 = i\right] = \sum_{k \neq j} \Pr\left[X_1 = k | X_0 = i\right] \Pr\left[T_j = m - 1 | X_0 = k\right]$$

is immediate. Indeed, in order to have the transition from state i to state j at

---

[2] The hitting time T_j is also called the first passage time into a state j.

discrete-time m, it is necessary to have first a transition from state i to some other state $k \neq j$ and to pass from that state k to state j for the first time after $m - 1$ time units. For a stationary Markov chain, we have for $m > 0$ that

$$\Pr\left[T_j = m | X_0 = i\right] = \sum_{k \neq j} P_{ik} \Pr\left[T_j = m - 1 | X_0 = k\right] \tag{9.17}$$

and, by definition for $m \geq 0$,

$$\Pr\left[T_j = m | X_0 = j\right] = \delta_{0m} \tag{9.18}$$

Lemma 9.2.2 *The n-step transition probabilities can be written as*

$$\left(P^n\right)_{ij} = \sum_{m=0}^{n} \Pr\left[T_j = m | X_0 = i\right] \left(P^{n-m}\right)_{jj} \tag{9.19}$$

Proof: The event $\{X_n = j\}$ can be decomposed in terms of the hitting time T_j. Indeed, since the events $\{T_j = m, X_n = j\}$ are disjoint for $1 \leq m \leq n$,

$$\{X_n = j\} = \cup_{m=1}^{n}\{T_j = m, X_n = j\}$$

Applied to the n-step transition probabilities,

$$\Pr\left[X_n = j | X_0 = i\right] = \Pr\left[\cup_{m=1}^{n}\{T_j = m, X_n = j\} | X_0 = i\right]$$

$$= \sum_{m=1}^{n} \Pr\left[T_j = m, X_n = j | X_0 = i\right]$$

$$= \sum_{m=1}^{n} \Pr\left[T_j = m | X_0 = i\right] \Pr\left[X_n = j | X_0 = i, T_j = m\right]$$

By definition of the hitting time, $\{T_j = m\} = \cup_{k=1}^{m-1}\{X_k \neq j\}\{X_m = j\}$ such that

$$\Pr\left[X_n = j | X_0 = i, T_j = m\right] = \Pr\left[X_n = j | X_0 = i, \cup_{k=1}^{m-1}\{X_k \neq j\}\{X_m = j\}\right]$$

$$= \Pr\left[X_n = j | X_m = j\right]$$

where the last step follows from the Markov property (9.2). Thus we obtain

$$\Pr\left[X_n = j | X_0 = i\right] = \sum_{m=1}^{n} \Pr\left[T_j = m | X_0 = i\right] \Pr\left[X_n = j | X_m = j\right]$$

Written in terms of n-step transition probabilities with (9.11), we arrive at (9.19), in which the case $\Pr\left[T_j = 0 | X_0 = i\right] = \delta_{ij}$ is incorporated. $\square$

For an absorbing state j where $P_{jj} = 1$, relation (9.19) simplifies to

$$\left(P^n\right)_{ij} = \Pr\left[T_j \leq n | X_0 = i\right] \tag{9.20}$$

Theorem 9.2.3 *An explicit expression for the conditional hitting time probability, in terms of the n-step transition probabilities P_{ij}^n, is*

$$\Pr\left[T_j = m | X_0 = i\right] = \sum_{q=0}^{m} \left(P^{m-q}\right)_{ij} \frac{1}{q!} \frac{d^q}{dz^q} \frac{1}{\sum_{k=0}^{\infty} \left(P^k\right)_{jj} z^k} \Bigg|_{z=0} \tag{9.21}$$

Proof[3]: By multiplying both sides in (9.19) by z^n, summing over all $n \geq 0$ and taking into account that $P^0 = I$ or $P_{ij}^0 = \delta_{ij}$, we have

$$\delta_{ij} + \sum_{n=1}^{\infty} (P^n)_{ij} z^n = \sum_{n=0}^{\infty} \sum_{m=0}^{n} \Pr[T_j = m|X_0 = i] (P^{n-m})_{jj} z^n$$

$$= \sum_{m=0}^{\infty} \Pr[T_j = m|X_0 = i] z^m \sum_{n=0}^{\infty} (P^n)_{jj} z^n$$

The generating function of the hitting time $T_j|X_0 = i$ is thus found as

$$\varphi_{T_j|X_0=i}(z) = \sum_{m=0}^{\infty} \Pr[T_j = m|X_0 = i] z^m = \frac{\delta_{ij} + \sum_{n=1}^{\infty} (P^n)_{ij} z^n}{\sum_{n=0}^{\infty} (P^n)_{jj} z^n}$$

Taking the m-th derivative with respect to z evaluated at $z = 0$ yields

$$\Pr[T_j = m|X_0 = i] = \frac{1}{m!} \sum_{n=0}^{\infty} (P^n)_{ij} \left. \frac{d^m}{dz^m} \frac{z^n}{\sum_{k=0}^{\infty} (P^k)_{jj} z^k} \right|_{z=0}$$

Applying Leibniz' rule

$$\frac{d^m}{dz^m} \frac{z^n}{\sum_{k=1}^{\infty} (P^k)_{jj} z^k} = \sum_{q=0}^{m} \binom{m}{q} \frac{n!}{(n-q)!} z^{n-q} \frac{d^{m-q}}{dz^{m-q}} \frac{1}{\sum_{k=0}^{\infty} (P^k)_{jj} z^k}$$

and evaluation at $z = 0$ leads to

$$\Pr[T_j = m|X_0 = i] = \sum_{n=0}^{\infty} (P^n)_{ij} \frac{1}{(m-n)!} \left. \frac{d^{m-n}}{dz^{m-n}} \frac{1}{\sum_{k=0}^{\infty} (P^k)_{jj} z^k} \right|_{z=0}$$

Since $\frac{1}{(m-n)!} = 0$ for $n > m$ because the Gamma function has poles at non-positive integers, the general form (9.21) is established after letting $q = m - n$. □

9.2.4 Transient and recurrent states

The probability that a Markov chain, initiated at state i, will ever come into state j is denoted as

$$r_{ij} = \Pr[T_j < \infty|X_0 = i] \qquad (9.22)$$

If the starting state i equals the target state j, then r_{ii} is the probability of ever returning to state i. If $r_{ii} = 1$, the state i is a *recurrent* state, while, if $r_{ii} < 1$, state i is a *transient* state. If i is a recurrent state, the Markov chain started at i will definitely (i.e. with probability 1) return to state i after some time. On the other hand, if i is a transient state, the Markov chain started at i has probability $1 - r_{ii}$ of never returning to state i. For an absorbing state i defined by $P_{ii} = 1$, we have by (9.20) that $r_{ij} = 1_{\{j=i\}}$, implying that an absorbing state is a recurrent state. Further, the mean return time to state j when the chain started in j is denoted by

$$m_j = E[T_j < \infty|X_0 = j] \qquad (9.23)$$

Concepts of renewal theory will now be applied to a Markov process. Let $N_k(j)$

---

[3] An alternative proof starting from the recursion (9.17) is presented on p. 608.

denote the number of times that the Markov chain is in state j during the time interval $[1, k]$ given that the chain started in state i. In terms of indicator functions, we have

$$N_k(j) = \sum_{n=1}^{k} 1_{\{X_n=j|X_0=i\}}$$

Using (2.13), the mean number of visits to state j in the time interval $[1, k]$ is

$$E\left[N_k(j)|X_0 = i\right] = E\left[\sum_{n=1}^{k} 1_{\{X_n=j\}}|X_0 = i\right] = \sum_{n=1}^{k} E\left[1_{\{X_n=j\}}|X_0 = i\right]$$

$$= \sum_{n=1}^{k} \Pr\left[X_n = j|X_0 = i\right]$$

or, in terms of n-step transition probabilities (9.11),

$$E\left[N_k(j)|X_0 = i\right] = \sum_{n=1}^{k} (P^n)_{ij} \qquad (9.24)$$

The mean number of times that the Markov chain is ever in state j given that it started from state i, is with $N(j) = \lim_{k\to\infty} N_k(j)$,

$$E\left[N(j)|X_0 = i\right] = \sum_{n=1}^{\infty} \Pr\left[X_n = j|X_0 = i\right] = \sum_{n=1}^{\infty} (P^n)_{ij}$$

Hence, if state j is reachable from state i, by definition, there is some k for which $(P^k)_{ij} > 0$, which implies that $E\left[N(j)|X_0 = i\right] > 0$. Further, consider the probability $\Pr\left[N(j) \geq n|X_0 = i\right]$ that the number of returns to state j exceeds n, given the Markov chain started from state i. The event $\{N(j) \geq n\}$ is equivalent to the occurrence of the event $\{N(j) \geq n - 1\}$ and the event that the Markov chain will return to j again given that it started from j. The probability of the latter event is precisely r_{jj}. Thus, we obtain the recursion

$$\Pr\left[N(j) \geq n|X_0 = i\right] = r_{jj} \Pr\left[N(j) \geq n - 1|X_0 = i\right]$$

with solution for $n \geq 1$,

$$\Pr\left[N(j) \geq n|X_0 = i\right] = (r_{jj})^{n-1} \Pr\left[N(j) \geq 1|X_0 = i\right]$$

Now, $\Pr\left[N(j) \geq 1|X_0 = i\right] = \Pr\left[T_j < \infty|X_0 = i\right] = r_{ij}$, such that

$$\Pr\left[N(j) \geq n|X_0 = i\right] = (r_{jj})^{n-1} r_{ij} \qquad (9.25)$$

The mean computed with (2.37) yields,

$$E\left[N(j)|X_0 = i\right] = \frac{r_{ij}}{1 - r_{jj}} = \sum_{n=1}^{\infty} (P^n)_{ij} \qquad (9.26)$$

provided $r_{ij} > 0$. If $r_{ij} = 0$, then (9.25) vanishes for every n and $E\left[N(j)|X_0 = i\right] = 0$, which means that state j is not reachable from state i. In summary:

- For a recurrent state j for which $r_{jj} = 1$, we obtain from (9.26) that

$$E\left[N(j)|X_0 = i\right] \to \infty$$

(if $r_{ij} \neq 0$ else $E\left[N(j)|X_0 = i\right] = 0$) and from (9.25),

$$\Pr\left[N(j) = \infty|X_0 = i\right] = \lim_{n \to \infty} \Pr\left[N(j) \geq n|X_0 = i\right] = r_{ij}$$

A state j is recurrent if and only if $\sum_{n=1}^{\infty} (P^n)_{jj}$ is infinite.
- For a transient state j for which $r_{jj} < 1$, it holds that $E\left[N(j)|X_0 = i\right]$ will be finite and $\Pr\left[N(j) = \infty|X_0 = i\right] = 0$ or, equivalently,

$$\Pr\left[N(j) < \infty|X_0 = i\right] = 1$$

A state j is transient if and only if $\sum_{n=1}^{\infty} (P^n)_{jj}$ is finite.

These relations illustrate the difference between a recurrent and a transient state. When the Markov chain starts at a recurrent state, it returns infinitely often to that state because $r_{jj} = \Pr\left[N(j) = \infty|X_0 = j\right] = 1$. If the chain starts at some other state i that is reachable from state j ($r_{ij} > 0$), then the chain will visit state j infinitely often. From this analysis, some consequences arise.

Using the bound (9.15), we have that

$$\sum_{n=1}^{\infty} (P^n)_{jj} \geq \sum_{n=1}^{\infty} P_{jj}^n = \frac{P_{jj}}{1 - P_{jj}} \geq 0$$

Hence, the fact that $\sum_{n=1}^{\infty} (P^n)_{jj}$ is finite, implies that $P_{jj} < 1$. A transient state can never be an absorbing state ($P_{jj} = 1$). No additional insight on the matrix elements P_{jj} can be gained for a recurrent state, for which $\sum_{n=1}^{\infty} (P^n)_{jj}$ is infinite.

Corollary 9.2.4 *A finite-state Markov chain must have at least one recurrent state.*

Proof: Suppose the contrary that, for a finite state space S, all states are transient states. For a transient state j it follows from (9.26) that $\sum_{k=1}^{\infty} (P^k)_{ij}$ is finite, which implies that $\lim_{k \to \infty} (P^k)_{ij} = 0$ for any other state i. If the state space is finite and all states are transient states, then

$$\sum_{j \in S} \lim_{k \to \infty} (P^k)_{ij} = 0$$

Since the summation has a finite number of terms, the limit and summation operator can be reversed,

$$\lim_{k \to \infty} \sum_{j \in S} (P^k)_{ij} = 0$$

But the law of total probability (9.12) requires that $\sum_{j \in S} (P^k)_{ij} = 1$ (for any time k), which leads to a contradiction. $\qquad \square$

Theorem 9.2.5 *If i is a recurrent state that is reachable from state j, then the state j is also a recurrent state and $r_{ij} = r_{ji} = 1$.*

Proof: Clearly, the theorem is true if $i = j$. Suppose that, for $i \neq j$, it holds that $\Pr[T_i < \infty | X_0 = j]$ $r_{ji} < 1$. This implies that the Markov chain starting from state j has probability $1 - r_{ji} > 0$ of never hitting state i, which is impossible because i is a recurrent state that will be visited infinitely often. Hence $r_{ji} = 1$.

Since state $j \neq i$ is reachable from state i, by definition $r_{ij} > 0$ and there is a minimum discrete-time n such that $(P^n)_{ij} > 0$ and $\Pr[X_k = j | X_0 = i] = 0$ for $k < n$. Similarly, since $r_{ji} = 1$, there exists a minimum discrete-time m to have a transition from $j \to i$ given the chain started in state j, thus, $\Pr[X_m = i | X_0 = j] = (P^m)_{ji} > 0$. From (9.16),

$$\left(P^{n+l+m}\right)_{jj} \geq (P^m)_{ji} \left(P^l\right)_{ii} (P^n)_{ij}$$

and summing over all l, we have

$$\sum_{l=1}^{\infty} \left(P^{n+l+m}\right)_{jj} \geq (P^n)_{ij} (P^m)_{ji} \sum_{l=1}^{\infty} \left(P^l\right)_{ii}$$

Since $\sum_{l=1}^{\infty} \left(P^{n+l+m}\right)_{jj} = \sum_{l=1+n+m}^{\infty} \left(P^l\right)_{jj} \leq \sum_{l=1}^{\infty} \left(P^l\right)_{jj}$, it holds that

$$\sum_{l=1}^{\infty} \left(P^l\right)_{jj} \geq (P^n)_{ij} (P^m)_{ji} \sum_{l=1}^{\infty} \left(P^l\right)_{ii}$$

Invoking the fact that $(P^n)_{ij} > 0$ and $(P^m)_{ji} > 0$ and the fact that state i is a recurrent state for which $\sum_{l=1}^{\infty} \left(P^l\right)_{ii}$ diverges, we conclude that $\sum_{l=1}^{\infty} \left(P^l\right)_{jj}$ must diverge. Relation (9.26) then indicates that $r_{jj} = 1$ or, that j must be a recurrent state.   □

A non-empty set $C \subset S$ of states is said to be *closed* if no state $i \in C$ leads to (or is reachable from) a state $j \notin C$. Thus, $r_{ij} = 0$ for any $i \in C$ and $j \notin C$, which implies that $P_{ij} = 0$. If the set C is closed, the Markov chain starting in C will remain, with probability 1, in C all the time. For example, if i is an absorbing state, $C = \{i\}$ is closed. A closed set C is irreducible if state i is reachable from state j for all $i, j \in C$. Theorem 9.2.5 together with Corollary 9.2.4 implies that, if C is a finite, irreducible closed set, all states are recurrent.

After a proper relabeling of the states in the Markov chain and assuming that there is one set of closed states, the $N \times N$ transition probability matrix can be transformed[4] to the block matrix

$$P = \begin{bmatrix} (P_c)_{n \times n} & O_{n \times m} \\ R_{m \times n} & (P_t)_{m \times m} \end{bmatrix} \tag{9.27}$$

where the $n \times n$ matrix P_c is the transition probability (stochastic) matrix among the closed states, $m \times m$ matrix P_t is the substochastic matrix of the transient states and the rectangular matrix R describes the transitions from the closed states to the transient states, while there are no transitions from the transient to the closed states as follows from the zero block matrix $O_{n \times m}$. Since (9.9) holds for

---

[4] Appendix A.5 represents a reducible matrix in the convention of linear algebra, by essentially describing the transpose of the matrix P.

each stochastic matrix P, we find that $P_c u_n = u_n$ (where u_n is the all-one vector with n components), illustrating that P_c is also stochastic, and that

$$P_t u_m = u_m - R u_n$$

Hence, since the non-negative $m \times n$ matrix R does contain non-zero elements, some row sum of P_t is smaller than (or equal) to one. Such a matrix P_t is called *substochastic*, satisfying $P_t u < u$ and, consequently, the largest eigenvalue of P_t is smaller than one (**art.** 18). Thus, $\lim_{k \to \infty} (P_t)^k = O$, which is a necessary condition for $\sum_{n=1}^{\infty} (P_t)^n$ to be finite, in agreement with the nature of transient states. The steady-state of the general stochastic matrix is analyzed in Appendix A.5. Finally, if there are m sets of closed states, then (9.27) is recast (omitting the possible different dimensions of each closed set) in the most general form

$$P = \begin{bmatrix} P_{c_1} & O & \cdots & O & O \\ O & P_{c_2} & \cdots & O & O \\ \vdots & \vdots & \ddots & \vdots & \vdots \\ O & O & \cdots & P_{c_m} & O \\ R_1 & R_2 & \cdots & R_m & P_t \end{bmatrix}$$

9.3 The steady-state of a Markov chain

9.3.1 The irreducible Markov chain

For an aperiodic, irreducible Markov chain, the steady-state vector $\pi = \lim_{k \to \infty} s[k]$ follows, after taking the limit $k \to \infty$ in (9.6), as

$$\pi = \pi P \tag{9.28}$$

or, for each component π_j,

$$\pi_j = \sum_{k=1}^{N} P_{kj} \pi_k \tag{9.29}$$

with $\pi u = 1$ or $\|\pi\|_1 = 1$. Equation (9.28) shows that the steady-state vector π does not depend on the initial state $s[0]$.

Theorem 9.3.1 *For an aperiodic, irreducible Markov chain, all rows in the matrix $A = \lim_{k \to \infty} P^k$ are equal to the steady-state vector π.*

Proof: The adjoint matrix of P is defined in **art.** 8 as $Q(\lambda) = (\lambda I - P)^{-1} c(\lambda)$, where $c(\lambda)$ is the characteristic polynomial. The adjoint matrix thus satisfies $(\lambda I - P) Q(\lambda) = Q(\lambda)(\lambda I - P) = c(\lambda)$, and $c(\lambda) = 0$ if and only if λ is an eigenvalue of P. In that case, every non-zero column of the adjoint matrix $Q(\lambda)$ is an eigenvector belonging to the eigenvalue λ. In view of (9.10), we trivially write $P^k - P^{k-1} P = 0$ or $P P^{k-1} - P^k = 0$ and if $A = \lim_{k \to \infty} P^k$ exists, then $A - AP = 0$ or $PA - A = 0$. This implies that $A(I - P) = (I - P)A = 0$. Hence, $A = Q(1)$, the adjoint matrix belonging to the eigenvalue 1. Consequently, the non-zero columns (or rows) of the adjoint matrix $Q(\lambda)$ consist of the (unscaled) eigenvector(s) belonging to eigenvalue λ. By (9.12) the rows of P^k for any k are normalized and so must $A = Q(1)$. Since there is only one eigenvector π belonging to $\lambda = 1$ for a stochastic matrix P as follows from Frobenius' Theorem A.5.2, the rows of $A = \lim_{k \to \infty} P^k$ must all be the same and equal to $\pi = [a_{11}\ a_{12}\ \cdots\ a_{1N}]$. $\square$

Furthermore, only if all rows of A are equal to the steady-state vector π, then the dependence on the initial state $s[0]$ vanishes since relation (9.10) becomes $\pi_j = \sum_{i=1}^{N} s_i[0] a_{ij} = a_{1j} \sum_{i=1}^{N} s_i[0] = a_{1j}$. Hence, $A = \lim_{k \to \infty} P^k = u\pi$ or, componentwise, for all $1 \le j \le N$,

$$\lim_{k \to \infty} \left(P^k\right)_{ij} = \pi_j \qquad (9.30)$$

The sequence of matrices $P, P^2, P^3, \ldots, P^k$ thus converges to $A = u\pi$ for $k \to \infty$. Instead of multiplying the last matrix P^k in the sequence by P to obtain the next one P^{k+1}, with the same computational effort, the sequence $P, P^2, P^4, \ldots, P^{2^k}$, obtained by successively squaring, converges considerably faster to $A = u.\pi$ and may be useful for sparse P.

On the other hand, relation (9.28) is an eigenvalue equation with eigenvalue $\lambda = 1$ and left-eigenvector π. Recall from (9.9) that u is the right-eigenvector of P belonging to eigenvalue $\lambda = 1$ (see **art.** 2 on p. 540). The Frobenius Theorem A.5.2 states that the transition probability matrix P has one eigenvalue $\lambda = 1$ with corresponding eigenvector π. Since in (9.28) the set $(P - I)^T \pi^T = 0$ has rank $N - 1$, the normalization condition $\|\pi\|_1 = 1$ furnishes the (last) remaining equation. Except for the trivial case where P is the identity matrix I, the solution of π is obtained from

$$\begin{bmatrix} P_{11} - 1 & P_{21} & P_{31} & \cdots & P_{N-1;1} & P_{N1} \\ P_{12} & P_{22} - 1 & P_{32} & \cdots & P_{N-1;2} & P_{N2} \\ P_{13} & P_{23} & P_{33} - 1 & \cdots & P_{N-1;3} & P_{N3} \\ \vdots & \vdots & \vdots & \ddots & \vdots & \vdots \\ P_{1;N-1} & P_{2;N-1} & P_{3;N-1} & \cdots & P_{N-1;N-1} - 1 & P_{N;N-1} \\ 1 & 1 & 1 & \cdots & 1 & 1 \end{bmatrix} \cdot \begin{bmatrix} \pi_1 \\ \pi_2 \\ \pi_3 \\ \vdots \\ \pi_{N-1} \\ \pi_N \end{bmatrix} = \begin{bmatrix} 0 \\ 0 \\ 0 \\ \vdots \\ 0 \\ 1 \end{bmatrix}$$

$$(9.31)$$

In practice, this method is used, especially if the number of states N is small and the transition probability matrix P does not exhibit a special matrix structure.

In summary, for irreducible Markov chains, there are in general two ways[5] of computing the state distribution π: via the limiting process (9.30) or via solving the set of linear equations (9.31). Recall that we have invoked the Frobenius Theorem A.5.2, which is only applicable for irreducible Markov chains. There exist cases of practical interest where (9.30) fails to hold. For example, in the two-state Markov chain studied in Section 9.4, there is a chain where the limit process bounces back and forth between state 0 and state 1. It is of importance to know whether the

---

[5] A third method consists of a directed graph solution of linear, algebraic equations discussed by Chen (1971, Chapter 3) and is applied to the steady-state equation (9.28) by Hooghiemstra and Koole (2000).

steady-state distribution π exists in the sense that $\pi_j \neq 0$ for at least one j. If $\pi_j = 0$ for all j, then there is no stationary (or equilibrium or steady-state) probability distribution.

9.3.2 The mean number of visits to a recurrent state

A direct application of Lemma 6.1.1 to the steady-state of a Markov chain is that, if (9.30) holds, then

$$\lim_{n \to \infty} \frac{1}{n} \sum_{m=1}^{n} (P^m)_{ij} = \pi_j$$

Invoking (9.24) where $\frac{N_n(j)}{n}$ is the fraction of the time the chain is in state j during the interval $[1, n]$, the relation is equivalent to

$$\lim_{n \to \infty} \frac{E\left[N_n(j) | X_0 = i\right]}{n} = \pi_j \tag{9.32}$$

The time average of the mean number of visits to state j, given the Markov chain started in state i, converges to the steady-state distribution. In other words, the long-run mean fraction of time that the chain spends in state j equals π_j and is independent of the initial state i. From (9.26), it immediately follows that, if j is a transient state, $\pi_j = 0$. Only recurrent states j have a non-zero steady-state probability π_j. Lemma 6.1.1 and its consequence (9.32) suggests to investigate $\frac{N_n(j)}{n}$ for recurrent states j.

If the Markov chain starts in a recurrent state j, we know from (9.26) that the chain returns to state j infinitely often. Let $W_k(j)$ denote the time of the k-th visit of the Markov chain to state j. Then,

$$W_k(j) = \min_{m \geq 1}(N_m(j) = k)$$

The interarrival time between the k-th and $(k-1)$-th visit is $\tau_k(j) = W_k(j) - W_{k-1}(j)$. The interarrival times $\{\tau_k(j)\}_{k \geq 1}$ are independent and identically distributed random variables as follows from the Markov property. Indeed, every time t the Markov chain returns to state j, it behaves from that time onwards as if the Markov process would have started from state j, ignoring the past before time t. Moreover, they have a common mean $E\left[\tau(j)\right] = E\left[\tau_1(j)\right]$ equal to the mean return time to j given by $E\left[T_j | X_0 = j\right] = m_j$ because the hitting time is $T_j = \tau_1(j)$. In other words, just as in renewal theory in Chapter 8, we have a counting process $\{N_m(j), m \geq 1\}$ with associated waiting times $W_k(j)$ and i.i.d. interarrival times $\tau_k(j)$, specified by the equivalence

$$\{N_m(j) < k\} \Longleftrightarrow \{W_k(j) > m\}$$

Invoking the Elementary Renewal Theorem 8.15, we obtain with $m_j = E\left[T_j | X_0 = j\right]$

$$\lim_{n \to \infty} \frac{N_n(j)}{n} = \frac{1}{m_j} \tag{9.33}$$

Thus, the chain returns to state j on average every m_j time units and, hence, the fraction of time the chain is in state j is roughly $\frac{1}{m_j}$. These results are summarized as follows:

Theorem 9.3.2 (Limit Law of Markov Chains) *If j is a recurrent state and the Markov chain starts in state i, then, with probability 1,*

$$\lim_{n \to \infty} \frac{N_n(j)}{n} = \frac{1_{\{T_j < \infty | X_0 = i\}}}{m_j} \qquad (9.34)$$

and

$$\lim_{n \to \infty} \frac{E[N_n(j)|X_0 = i]}{n} = \frac{r_{ij}}{m_j} \qquad (9.35)$$

Proof: Above we have proved the case (9.33) where the initial state $X_0 = j$. In that case $1_{\{T_j < \infty\}} = 1$. For an arbitrary initial distribution, it is possible that the chain will never reach the recurrent state j. In that case $1_{\{T_j < \infty\}} = 0$ given $X_0 = i$. It remains to prove (9.35). By definition, $0 \le N_n(j) \le n$ or, $0 \le \frac{N_n(j)}{n} \le 1$, which demonstrates that, for any n, $\frac{N_n(j)}{n}$ is bounded. From the Dominated Convergence Theorem 6.1.4, we have

$$\lim_{n \to \infty} E\left[\frac{N_n(j)}{n}\middle| X_0 = i\right] = E\left[\lim_{n \to \infty} \frac{N_n(j)}{n}\middle| X_0 = i\right]$$

$$= E\left[\frac{1_{\{T_j < \infty\}}}{m_j}\middle| X_0 = i\right]$$

$$= \frac{\Pr[T_j < \infty | X_0 = i]}{m_j} = \frac{r_{ij}}{m_j}$$

which completes the proof.   □

Theorem 9.3.2 introduces the need for an additional definition. A recurrent state j is called *null recurrent* if $m_j = \infty$, in which case (9.35) reduces to

$$\lim_{n \to \infty} \frac{E[N_n(j)|X_0 = i]}{n} = 0 \qquad (9.36)$$

By Tauberian theorems (which investigate conditions for the converse of Lemma 6.1.1 but which are far more difficult, as illustrated in the book by Hardy (1948)), it can be shown that, for null recurrent states, the stronger result $\lim_{n \to \infty} (P^n)_{ij} = 0$ also holds. A recurrent state j called *positive recurrent* if $m_j < \infty$. The difference between a transient and a null recurrent state that both obey (9.36) lies in the fact that, for a transient state, the limit $\lim_{n \to \infty} E[N_n(j)|X_0 = j]$ is finite while, for a null recurrent state, $\lim_{n \to \infty} E[N_n(j)|X_0 = j] = \infty$. Relation (9.36) indicates that for a null recurrent state

$$E[N_n(j)|X_0 = j] = O(n^a)$$

where $0 < a < 1$, while for a positive recurrent state

$$E[N_n(j)|X_0 = j] = \pi_j n + o(n)$$

The growth of $E[N_n(j)|X_0 = j]$ with n leads to coin "positive recurrent states" also as *strongly ergodic* states, while null recurrent states are called *weakly ergodic*. Figure 9.1 sketches the classification of states in a Markov process.

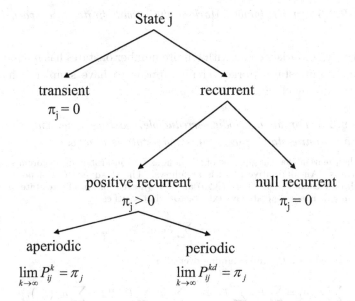

Fig. 9.1. Classification of the states in a Markov process with the corresponding steady-state vector π_j.

With these additional definitions, Corollary 9.2.4 can be sharpened as follows:

Corollary 9.3.3 *A finite-state Markov chain must have at least one positive recurrent state.*

Proof: By summing (9.12) over n and dividing by n, we find

$$\frac{1}{n} \sum_{j=1}^{N} \sum_{m=1}^{n} (P^m)_{ij} = 1$$

Using (9.24) yields

$$\sum_{j=1}^{N} \frac{E\left[N_n(j)|X_0 = i\right]}{n} = 1$$

When taking the limit $n \to \infty$ of both sides, the summation and limit operator can be reversed because the summation involves a finite number of terms. Hence,

$$\sum_{j=1}^{N} \lim_{n \to \infty} \frac{E\left[N_n(j)|X_0 = i\right]}{n} = 1$$

which is only possible if at least one state j is positive recurrent because transient and null recurrent states obey (9.36). □

Similarly, Theorem 9.2.5 and the combined consequence can be sharpened:

Theorem 9.3.4 *If i is a positive recurrent state that leads to a state j, then the state j is also a positive recurrent state.*

Theorem 9.3.5 *An irreducible Markov chain with finite-state space is positive recurrent.*

Alternatively, a Markov chain with a finite number of states has no null recurrent states. Thus, finite-state Markov chains appear to have simpler behavior than infinite-state Markov chains.

Theorem 9.3.6 *For an aperiodic, irreducible, positive recurrent Markov chain (even with an infinite-state space), the steady-state is unique.*

Proof: The theorem is a consequence of the famous Perron-Frobenius Theorem A.5.2 for non-negative matrices. An alternative proof is as follows. The steady-state of a positive recurrent irreducible Markov chain satisfies both (9.29) and (9.30), even for an infinite-state Markov chain. Suppose that $a \neq \pi$ is a second steady-state vector, that satisfies

$$a_j = \sum_{k=1}^{N} P_{kj} a_k \tag{9.37}$$

Multiplying both sides by P_{ji} and summing over all j

$$\sum_{j=1}^{N} P_{ji} a_j = \sum_{j=1}^{N} P_{ji} \sum_{k=1}^{N} P_{kj} a_k = \sum_{k=1}^{N} a_k \sum_{j=1}^{N} P_{kj} P_{ji} = \sum_{k=1}^{N} a_k \left(P^2 \right)_{ki}$$

The reversal in j- and k-summation is always allowed (even for $N \to \infty$) by absolute convergence. Using (9.37),

$$a_i = \sum_{k=1}^{N} a_k \left(P^2 \right)_{ki}$$

Repeating this process leads, for any $n \geq 1$ and $i \geq 1$ to

$$a_i = \sum_{k=1}^{N} a_k \left(P^n \right)_{ki}$$

In the limit for $n \to \infty$, application of (9.30) yields

$$a_i = \pi_i \sum_{k=1}^{N} a_k$$

such that $\pi_i = c^{-1} a_i$ with $c^{-1} = \sum_{k=1}^{N} a_k$. After normalization $\tilde{a}_i = c a_i$ to satisfy $\|\tilde{a}\|_1 = 1$ as required for any state vector $s[k]$ of a Markov process, the uniqueness $\pi = \tilde{a}$ is demonstrated. $\square$

Theorem 9.3.7 *For an irreducible, positive recurrent Markov chain holds*

$$\lim_{n \to \infty} \frac{E\left[N_n(j)|X_0 = i\right]}{n} = \lim_{n \to \infty} \frac{N_n(j)}{n} = \frac{1}{m_j} = \pi_j \tag{9.38}$$

and

$$\frac{N_n(j) - n\pi_j}{\sigma_j \pi_j^{3/2} \sqrt{n}} \xrightarrow{d} N(0,1) \tag{9.39}$$

where $\sigma_j^2 = Var[T_j|X_0 = j]$

Proof: For an irreducible, finite-state Markov chain where $r_{ij} = 1$, Theorem 9.3.2 and Theorem 9.3.5 together with (9.32) lead to the fundamental relation (9.38). Relation (9.39) is an application of the Asymptotic Renewal Distribution Theorem 8.2.3. We have shown that the interarrival times $\{\tau_k(j)\}_{k\geq 1}$ are i.i.d. with mean $E\left[\tau(j)\right] = E\left[T_j|X_0 = j\right] = m_j$ and (assumed finite) variance $Var[\tau(j)] = Var[T_j|X_0 = j] = \sigma_j^2$. $\square$

As a corollary, from (8.18), we have

$$\lim_{n\to\infty} \frac{Var\left[N_n(j)|X_0 = i\right]}{n} = \sigma_j^2 \pi_j^3 \tag{9.40}$$

A Markov chain that is irreducible and for which all states are positive recurrent is said to be *ergodic*. Ergodicity implies that the steady-state distribution π and the long-run probability distribution $\lim_{k\to\infty} s[k]$ are the same. Ergodic Markov chains are basic stochastic processes in the study of queueing theory.

9.4 Example: the two-state Markov chain

The two-state Markov chain is defined by

$$P = \begin{bmatrix} 1-p & p \\ q & 1-q \end{bmatrix}$$

and illustrated in Fig. 9.2. A matrix computation of the two-state Markov chain is

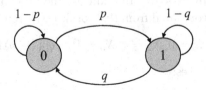

Fig. 9.2. A two-state Markov chain.

presented in Appendix A.5.2. Here, we follow a probabilistic approach. Since there are only two states, at any discrete-time k, it holds that $\Pr\left[X_k = 0\right] = 1 - \Pr\left[X_k = 1\right]$. Hence, it suffices to compute $\Pr\left[X_k = 0\right]$. By the law of total probability and the Markov property (9.2), we have

$$\Pr\left[X_{k+1} = 0\right] = \Pr\left[X_{k+1} = 0|X_k = 1\right]\Pr\left[X_k = 1\right]$$
$$+ \Pr\left[X_{k+1} = 0|X_k = 0\right]\Pr\left[X_k = 0\right]$$

or, from Fig. 9.2, the Markov chain can only be in state 0 at time $k+1$, if it is in state 0 at time k and the next event at time $k+1$ brings it back to that same state

0, or if it is in state 1 at time k and the next event at time $k+1$ induces a transfer to state 0. Introducing the transition probabilities,

$$\Pr\left[X_{k+1} = 0\right] = q\Pr\left[X_k = 1\right] + (1 - p)\Pr\left[X_k = 0\right]$$
$$= q\left(1 - \Pr\left[X_k = 0\right]\right) + (1 - p)\Pr\left[X_k = 0\right]$$
$$= (1 - p - q)\Pr\left[X_k = 0\right] + q$$

This recursion can be iterated back to $k = 0$,

$$\Pr\left[X_k = 0\right] = (1 - p - q)^k \Pr\left[X_0 = 0\right] + q\sum_{j=0}^{k-1}(1 - p - q)^j$$

Using the finite geometric series $\sum_{j=0}^{k-1} x^j = \frac{1-x^k}{1-x}$ for any $x \neq 1$ else $\sum_{j=0}^{k-1} x^j = k$,

$$\Pr\left[X_k = 0\right] = \frac{q}{p+q} + (1 - p - q)^k \left(\Pr\left[X_0 = 0\right] - \frac{q}{p+q}\right) \qquad (9.41)$$

With $\Pr\left[X_k = 1\right] = 1 - \Pr\left[X_k = 0\right]$,

$$\Pr\left[X_k = 1\right] = \frac{p}{p+q} + (1 - p - q)^k \left(\Pr\left[X_0 = 1\right] - \frac{p}{p+q}\right) \qquad (9.42)$$

If $|1 - p - q| < 1$, the steady-state $\pi = \begin{bmatrix} \Pr\left[X_\infty = 0\right] & \Pr\left[X_\infty = 1\right] \end{bmatrix}$ follows as

$$\pi = \begin{bmatrix} \frac{q}{p+q} & \frac{p}{p+q} \end{bmatrix} \qquad (9.43)$$

Observe from (9.41) and (9.42) that, if $\Pr\left[X_0 = 0\right] = \frac{q}{p+q} = \Pr\left[X_\infty = 0\right]$ and $\Pr\left[X_0 = 1\right] = \frac{p}{p+q} = \Pr\left[X_\infty = 1\right]$, the Markov chain starts and remains the whole time (for all k) in the steady-state. In addition, the probability of a particular sequence of states can be computed from (9.3) or directly from Fig. 9.2. For example,

$$\Pr\left[X_0 = 1, X_1 = 0, X_2 = 1, X_3 = 1\right] = qp(1 - q)\Pr\left[X_0 = 1\right]$$

We distinguish three cases:

(i) $p = q = 0$: The Markov chain consists of two separate states that do not communicate. Each state can be considered as a single state, irreducible, Markov chain. Any real number belonging to $[0, 1]$ is a steady-state solution of each separate set. Also, $P = I$ and, hence, $P^\infty = \lim_{k \to \infty} P^k = I$.

(ii) $0 < p + q < 2$: The Markov chain is aperiodic irreducible positive recurrent with steady-state π given in (9.43). This is the regular case.

(iii) $p = q = 1$: The Markov chain is periodic with period 2, but still irreducible positive recurrent with steady-state $\pi = \begin{bmatrix} \frac{1}{2} & \frac{1}{2} \end{bmatrix}$ given above. However, $P^{2n} = I$ and $P^{2n+1} = P$ such that $\lim_{k \to \infty} P^k$ does not exist, but

$$\lim_{k \to \infty} \frac{1}{k}\sum_{j=1}^{k} P^j = \begin{bmatrix} \frac{1}{2} & \frac{1}{2} \\ \frac{1}{2} & \frac{1}{2} \end{bmatrix}$$

9.5 A generating function approach

In addition to the algebraic formulation presented so far, Markov chains can also be studied via function theory. The starting point is the definition of the generating function of the Markov matrix

$$\Phi_P(z) = \sum_{n=0}^{\infty} P^n z^n \tag{9.44}$$

as well as its (i,j)-element

$$\varphi_P(i,j;z) = \sum_{n=0}^{\infty} (P^n)_{ij} z^n \tag{9.45}$$

where the n-step transition probabilities $0 \le (P^n)_{ij} \le 1$ are defined in (9.11). Thus, $\varphi_P(i,j;0) = \delta_{ij} = 1_{\{i=j\}}$. The generating function (9.45) already appeared in (9.21). Relations between different generating functions are treated by Woess (2009).

9.5.1 Functional equations

When we rewrite (9.44) as

$$\Phi_P(z) = \sum_{n=0}^{\infty} P^n z^n = I + \sum_{n=1}^{\infty} P^n z^n = I + Pz \sum_{n=0}^{\infty} P^n z^n$$

from which

$$\Phi_P(z) = I + Pz\Phi_P(z) \tag{9.46}$$

we find (for finite N) that $\Phi_P(z) = (I - Pz)^{-1}$ is the adjoint matrix of P (see **art**.8) and, consequently, that $\Phi_P(z)$ is a rational function in z. The corresponding relation for $\varphi_P(i,j;z)$, which is the (i,j)-element in (9.46), is

$$\varphi_P(i,j;z) = \delta_{ij} + z \sum_{k=1}^{N} P_{ik}\varphi_P(k,j;z) \tag{9.47}$$

from which

$$\varphi_P(i,j;z) = \frac{\delta_{ij} + z \sum_{k=1;k\ne i}^{N} P_{ik}\varphi_P(k,j;z)}{1 - zP_{ii}}$$

illustrating the existence of a possible pole at $z = \frac{1}{P_{ii}}$.

Multiplying both sides of the matrix product identity $(P^{n+m})_{ij} = \sum_{k=1}^{N}(P^n)_{ik}(P^m)_{kj}$ by $x^n y^m$ and summing over all $n \ge 0$ and $m \ge 0$ yields

$$\sum_{n=0}^{\infty}\sum_{m=0}^{\infty}(P^{n+m})_{ij} x^n y^m = \sum_{k=1}^{N}\sum_{n=0}^{\infty}(P^n)_{ik} x^n \sum_{m=0}^{\infty}(P^m)_{kj} y^m = \sum_{k=1}^{N}\varphi_P(i,k;x)\,\varphi_P(k,j;y)$$

Let $s = n + m$, then $s \ge 0$. Combining $m = s - n$ with $m \ge 0$ yields that $s \ge n \ge 0$ so that

$$\sum_{n=0}^{\infty}\sum_{m=0}^{\infty}(P^{n+m})_{ij} x^n y^m = \sum_{s=0}^{\infty}\sum_{n=0}^{s}(P^s)_{ij} x^n y^{s-n} = \sum_{s=0}^{\infty}(P^s)_{ij} y^s \sum_{n=0}^{s}\left(\frac{x}{y}\right)^n$$

$$= \sum_{s=0}^{\infty}(P^s)_{ij}\frac{y^{s+1} - x^{s+1}}{y - x} = \frac{y\varphi_P(i,j;y) - x\varphi_P(i,j;x)}{y - x}$$

Finally, we arrive at the functional relation

$$\frac{y\varphi_P(i,j;y) - x\varphi_P(i,j;x)}{y - x} = \sum_{k=1}^{N}\varphi_P(i,k;x)\,\varphi_P(k,j;y) \tag{9.48}$$

that holds for *any* matrix P. When $y \to x$, then, after using de l'Hospital's rule, we find

$$\varphi_P(i,j;x) + x \left.\frac{d\varphi_P(i,j;y)}{dy}\right|_{y=x} = \sum_{k=1}^{N} \varphi_P(i,k;x)\,\varphi_P(k,j;x)$$

When the matrix P is symmetric, $P = P^T$, then $\varphi_P(j,k;x) = \varphi_P(k,j;x)$ and the above equation becomes, for $i = j$ and real x,

$$\varphi_P(j,j;x) + x \left.\frac{d\varphi_P(j,j;y)}{dy}\right|_{y=x} = \sum_{k=1}^{N} \varphi_P^2(j,k;x) \geq 0$$

which resembles the Christoffel-Darboux formula (Van Mieghem, 2011, p. 319) for orthogonal polynomials from which the interlacing property of zeros of orthogonal polynomials is derived.

Incorporating the stochastic requirement (9.12) for the matrix P results, for $|z| < 1$, in

$$\sum_{j=1}^{N} \varphi_P(i,j;z) = \frac{1}{1-z} \tag{9.49}$$

9.5.2 Radius of convergence

The radius of convergence R_{ij} of the power series in (9.45) is given by[6]

$$R_{ij}^{-1} = \lim_{n \to \infty} \left((P^n)_{ij}\right)^{\frac{1}{n}} \tag{9.50}$$

or by

$$R_{ij}^{-1} = \lim_{n \to \infty} \frac{(P^{n+1})_{ij}}{(P^n)_{ij}} \tag{9.51}$$

From (9.13), it follows that $(P^{n+1})_{ij} \geq (P^n)_{ij}\,P_{jj}$ and we directly find from (9.51) that $R_{ij} \leq \max_{1 \leq j \leq N} \left(\frac{1}{P_{jj}}\right)$. On the other hand[7], since $(P^n)_{ij} \leq 1$, (9.50) shows that $R_{ij} \geq 1$ so that

$$1 \leq R_{ij} \leq \max_{1 \leq j \leq N} \left(\frac{1}{P_{jj}}\right)$$

For a transient Markov chain, $R_{ij} > 1$ and R_{ij} can be different from R_{kl} (see Woess (2009)).

After multiplying both sides of the general bound (9.16) by z^l and summing over all $l \geq 0$, we obtain for irreducible Markov chains

$$\sum_{l=0}^{\infty} \left(P^{n+l+m}\right)_{ij} z^l \geq (P^n)_{ir}\,(P^m)_{kj} \sum_{l=0}^{\infty} \left(P^l\right)_{rk} z^l$$

---

[6] The power series $\sum_{k=0}^{\infty} f_k z^k$ converges for $|z| < R$, where the radius of convergence R satisfies $\frac{1}{R} = \limsup_{k \to \infty} |f_k|^{1/k}$ or $\frac{1}{R} = \lim_{k \to \infty} \left|\frac{f_{k+1}}{f_k}\right|$ when the latter exists (see e.g. Titchmarsh (1964)).

[7] Since $\varphi_{P^k}(i,j;z) = \sum_{n=0}^{\infty} \left(P^{kn}\right)_{ij} z^n$, we have, excluding periodic Markov chains, that

$$R^{-1}\left(P_{ij}^k\right) = \lim_{n \to \infty} \left(\left(P^{kn}\right)_{ij}\right)^{\frac{1}{n}} = \lim_{n \to \infty} \left(\left(\left(P^{kn}\right)_{ij}\right)^{\frac{1}{nk}}\right)^k = \left(\lim_{kn \to \infty} \left(\left(P^{kn}\right)_{ij}\right)^{\frac{1}{nk}}\right)^k$$

and that $R^{-1}\left(P_{ij}^k\right) = \left(R^{-1}(P_{ij})\right)^k$ or that $R\left(P_{ij}^k\right) = R^k(P_{ij})$.

Invoking the definition (9.45) and with

$$\sum_{l=0}^{\infty} \left(P^{n+l+m}\right)_{ij} z^l = z^{-n-m} \sum_{l=n+m}^{\infty} \left(P^l\right)_{ij} z^l = z^{-n-m} \left(\varphi_P\left(i,j;z\right) - \sum_{l=0}^{n+m-1} \left(P^l\right)_{ij} z^l \right)$$

we find, for irreducible Markov chains, the inequality

$$\varphi_P\left(i,j;z\right) \geq z^{n+m} \left(P^n\right)_{ir} \left(P^m\right)_{kj} \varphi_P\left(r,k;z\right)$$

Hence, for irreducible Markov chains, the generating function $\varphi_P\left(i,j;z\right)$ for any pair (i,j) can always be lower bounded by any other generating function $\varphi_P\left(r,k;z\right)$ at a pair (r,k) and we conclude that the series in (9.45) for a specific z either converges or diverges for any pair (i,j) simultaneously. Consequently, all series $\varphi_P\left(i,j;z\right)$ for $1 \leq i \leq N$ and $1 \leq j \leq N$ possess the same radius of convergence $R\left(P\right)$. Relation (9.26) is equivalent to $\varphi_P\left(i,j;1\right) = E\left[N(j)|X_0 = i\right] + 1$, which implies that $\varphi_P\left(i,j;1\right)$ diverges for an irreducible (finite) Markov chain. Hence, for an irreducible (finite) Markov chains, the radius of convergence is $R\left(P\right) = 1$. Relation (9.49), combined with the above argument that all $\varphi_P\left(i,j;z\right)$ converge or diverge for a same z, shows that $R\left(P\right) = 1$ for any, finite stochastic matrix P.

Assuming that the eigenvalues of P are distinct, then **art.** 4 indicates that $P = X\mathrm{diag}(\lambda_k)Y^T$, from which

$$\sum_{n=0}^{\infty} P^n z^n = X\mathrm{diag}\left(\sum_{n=0}^{\infty} \lambda_k^n z^n\right) Y^T = X\mathrm{diag}\left(\frac{1}{1-\lambda_k z}\right) Y^T$$

and

$$\varphi_P\left(i,j;z\right) = \left(X\mathrm{diag}\left(\frac{1}{1-\lambda_k z}\right) Y^T \right)_{ij} = \sum_{k=1}^{N} \frac{\left(x_k y_k^T\right)_{jj}}{1-z\lambda_k} \tag{9.52}$$

This expression illustrates that $\varphi_P\left(i,j;z\right)$ converges for all pairs (i,j), provided for each k, it holds that $|\lambda_k z| \leq 1$. Thus, $|z| \leq \min_{1 \leq k \leq N} \frac{1}{|\lambda_k|}$. By the Perron-Frobenius Theorem for non-negative matrices (Van Mieghem, 2011, p. 235), the largest eigenvalue (also called the spectral radius of P) is real and positive so that the radius of convergence satisfies $R_{ij} = \frac{1}{\lambda_1}$. Alternatively, this shows that the radius of convergence for an irreducible Markov chain equals $R\left(P\right) = 1$.

The generating function approach proves to be useful when $R_{ij} > 1$ as illustrated by Woess (2000) in his study on infinite random walks on graphs.

9.6 Problems

(i) Given the transition probability matrix P,

$$P = \begin{bmatrix} 0.8 & 0.2 & 0.0 \\ 0.8 & 0.0 & 0.2 \\ 0.0 & 0.8 & 0.2 \end{bmatrix}$$

(a) draw the Markov chain, (b) compute the steady-state vector in three different ways.

(ii) Consider the discrete-time Markov chain with N states and with transition probabilities at each state j,

$$P_{j,j+1} = 1 - \frac{1}{j}$$

$$P_{j1} = \frac{1}{j}$$

(a) draw the Markov chain, (b) show that the drift is positive, but that the Markov chain is nevertheless recurrent.

(iii) Assume that trees in a forest fall into four age groups. Let $b[k]$, $y[k]$, $m[k]$ and $u[k]$ denote the number of baby trees, young trees, middle-aged trees and old trees, respectively, in the forest at a given time period k. A time period lasts 15 years. During a time period, the total number of trees remains constant, but a certain percentage of trees in each age group dies and is replaced with baby trees. All surviving trees in the baby, young and middle-aged group enter into the next age group. Surviving old trees remain old. Let $0 < p_b$, p_y, p_m, $p_o < 1$ denote the loss rates in each age group in percent.

(a) Make a discrete Markov chain presentation of the process of aging and replacement in the forest.

(b) The distribution of tree population amongst different age categories in time period k is represented by

$$x[k] = \begin{bmatrix} b[k] & y[k] & m[k] & u[k] \end{bmatrix}^T$$

If $x[k+1] = Px[k]$, what is the transition probability matrix P?

(c) Let $p_b = 0.1$, $p_y = 0.2$, $p_m = 0.3$, $p_o = 0.4$ and suppose that $x[0] = \begin{bmatrix} 5000 & 0 & 0 & 0 \end{bmatrix}^T$. What is the number of trees in each category after 15 and after 30 years?

(d) What is the steady-state situation?

(iv) A faulty digital video conferencing system shows a clustered error pattern. If a bit is received correctly, then the chance to receive the next bit correctly is 0.999. If a bit is received incorrectly, then the next bit is incorrect with probability 0.95.

(a) Model the error pattern of this system using the discrete-time Markov chain.

(b) How many communicating classes does the Markov chain have? Is it irreducible?

(c) In the long-run, what is the fraction of correctly received bits and the fraction of incorrectly received bits?

(d) After the system is repaired, it works properly for 99.9% of the time. A test sequence after repair shows that, when always starting with a correctly received bit, the next 10 bits are correctly received with probability 0.9999. What is the probability now that a correctly (and analogously incorrectly) received bit is followed by another correct (incorrect) bit?

(v) The Frobenius Theorem A.5.2 shows that the largest eigenvalue of a stochastic matrix P is equal to 1. Can you give another simple argument that the eigenvalue equation $xP = \lambda x$ cannot hold for any $\lambda > 1$, where $x > 0$ is a probability vector?
(*Hint:* Show that $xP \geq \lambda x$ cannot hold with $x > 0$ for any $\lambda > 1$ by using (9.8)).

(vi) Prove the Theorem: The multiplicity of the eigenvalue 1 of a $N \times N$ stochastic matrix P is equal to the number of recurrent classes in the associated Markov chain. (*Hint:* see Van Mieghem (2011, p. 73-74)).

(vii) Find a general expression for the hitting time in terms of the transition probability matrix elements by solving (9.17).

(viii) Based on the previous solution, derive the probability generated function of the hitting time and estimate its deep tail behavior.

(ix) Let S denote the state-space, i.e. the total set of states in which the discrete-time Markov process $\{X_k, k \geq 0\}$ and $X_k \in S$ can be. Show that the hitting time T_A into the set A, given that the stationary Markov process started in state i, obeys the equations

$$\begin{cases} \Pr[T_A < \infty | X_0 = i] = 1 & \text{for } i \in A \\ \Pr[T_A < \infty | X_0 = i] = \sum_{j \in S} P_{ij} \Pr[T_A < \infty | X_0 = j] & \text{for } i \notin A \end{cases} \quad (9.53)$$

The mean hitting time to reach a state of the set A, given that the stationary Markov process started in state j, satisfies the linear set of equations

$$\begin{cases} E[T_A | X_0 = i] = 0 & \text{for } i \in A \\ E[T_A | X_0 = i] = 1 + \sum_{j \notin A} P_{ij} E[T_A | X_0 = j] & \text{for } i \notin A \end{cases} \quad (9.54)$$

(x) During the downloading of a file from a server to a client, a single path is followed. The available capacity $C(t)$ along that path varies over time due to interfering traffic of other users. Compute the download time T of file of constant size B for the following cases:

(a) $C(t)$ is a stationary stochastic process with independent increments;

(b) $C(t)$ equal to a uniform random variable on $[c_1, c_2]$;

(c) $C(t) = C[k\Delta]$ is a stationary, discrete stochastic process where Δ is a unit time during which $C(t)$ is constant.

(xi) Consider a recurrent Markov chain where state k is the only absorbing state.

    (a) Show that the steady-state vector has the components $\pi_j = \delta_{k,j}$. In other words, the steady-state is the absorbing state k.

    (b) Of what type are the other states?

(xii) A distributed peer-to-peer network uses the following simplified scheme to find content. Firstly, a peer constructs a list of peers with whom it can communicate. Afterwards, the peer periodically contacts these peers to update its knowledge about the network, i.e. about the active peers in the network and about the content discovered so far. Assume peer A obtains a list of 20 peers. Each time, A randomly selects a peer in its list to which it sends a message. If the selected peer does not answer the message after a certain time, A considers that peer as inactive or off-line and deletes that peer from the list. If the selected peer replies to A, it remains in A's list, and it can be selected again. Suppose that six peers in A's list are off-line; thus they cannot reply when selected. Peer A repeats the selection process for peers k times. Let X_k be the number of inactive peers in the list.

    (a) Verify that X_k is a Markov process. Write down the transition probability matrix P for X_k.

    (b) What will the state vector be after all the inactive peers have been deleted from the list? Determine $\lim_{k \longrightarrow \infty} P^k$.

10
Continuous-time Markov chains

Just as it was convenient in Chapter 2 to treat discrete and continuous random variables distinctly, the same recipe is advised for discrete-time and continuous-time Markov chains. Here also, it appears that the continuous case is, in general, more intricate than the discrete counterpart.

10.1 Definition

For the continuous-time Markov chain $\{X(t), t \geq 0\}$ with N states, the Markov property (9.1) can be written as

$$\Pr[X(t+\tau)=j|X(\tau)=i, X(u)=x(u), 0 \leq u < \tau] = \Pr[X(t+\tau)=j|X(\tau)=i]$$

which reflects the fact that the future state at time $t+\tau$ only depends on the current state at time τ. Similarly as for the discrete-time Markov chain, we assume that the transition probabilities for the continuous-time Markov chain $\{X(t), t \geq 0\}$ are stationary, i.e. independent of a point τ in time,

$$P_{ij}(t) = \Pr[X(t+\tau) = j|X(\tau) = i] = \Pr[X(t) = j|X(0) = i] \qquad (10.1)$$

Analogous to (9.5) and (9.6), the state vector $s(t)$ in continuous-time with components $s_k(t) = \Pr[X(t) = k]$ obeys

$$s(t + \tau) = s(\tau)P(t) \qquad (10.2)$$

Immediately, it follows from (10.2) that

$$s(t + u + \tau) = s(\tau)P(t + u)$$
$$s(t + u + \tau) = s(\tau + u)P(t) = s(\tau)P(u)P(t)$$
$$= s(\tau + t)P(u) = s(\tau)P(t)P(u)$$

such that, for all $t, u \geq 0$, the $N \times N$ transition probability matrix $P(t)$ satisfies

$$P(t + u) = P(u)P(t) = P(t)P(u) \qquad (10.3)$$

This fundamental relation[1] (10.3) is called the Chapman-Kolmogorov equation. Furthermore, since the Markov chain must be at any time in one of the N states, the analogon of (9.8) is, for any state i,

$$\sum_{j=1}^{N} P_{ij}(t) = 1 \tag{10.4}$$

For continuous-time Markov chains, it is convenient (and consistent with $t = 0$ in (10.2)) to postulate the initial condition of the transition probability matrix

$$P(0) = I \tag{10.5}$$

where $P(0) = \lim_{t \downarrow 0} P(t)$. The relations (10.1), (10.3), (10.4) and (10.5) are sufficient to describe the continuous-time Markov process completely.

10.2 Properties of continuous-time Markov processes

We will now concentrate on typical properties of a continuous-time Markov process.

10.2.1 The infinitesimal generator Q

Lemma 10.2.1 *The transition probability matrix $P(t)$ is continuous for all $t \geq 0$.*

Proof: Continuity is proved if $\lim_{h \downarrow 0} P(t + h) = \lim_{h \downarrow 0} P(t - h) = P(t)$. From (10.3) and (10.5), we have for $h > 0$,

$$\lim_{h \downarrow 0} P(t + h) = P(t) \lim_{h \downarrow 0} P(h) = P(t)I = P(t)$$

The other limit follows for $t > 0$ and $0 < h < t$ from $P(t) = P(t - h)P(h)$. $\qquad \square$

If a function is differentiable, it is continuous. However, the converse is not generally true. Therefore, we include the additional assumption that the matrix

$$\lim_{h \downarrow 0} \frac{P(h) - I}{h} = P'(0) = Q \tag{10.6}$$

exists. This matrix Q is called the *infinitesimal generator* of the continuous-time Markov process and Q plays an important role as shown below. The infinitesimal generator Q corresponds to $P - I$ in discrete-time. From (10.4),

$$\sum_{j=1, j \neq i}^{N} P_{ij}(h) = 1 - P_{ii}(h)$$

---

[1] On a higher level of abstraction, $P(t)$ can be viewed as a linear operator acting upon the vector space defined by all possible state vectors $s(t)$. Relation $P(t + u) = P(u)P(t)$ is known as the semigroup property. The family of these commuting operators possesses an interesting algebraic structure (see e.g. Schoutens (2000)).

and, dividing both sides by h and letting h approach zero, we find for each i with the definition of Q that[2]

$$q_i = \sum_{j=1, j \neq i}^{N} q_{ij} = -q_{ii} \geq 0 \qquad (10.7)$$

Hence, the sum of the rows in Q is zero (implying that $Qu = 0$ and the determinant $\det Q = 0$), $q_{ij} = \lim_{h \downarrow 0} \frac{P_{ij}(h)}{h} \geq 0$ and $q_{ii} \leq 0$. The elements q_{ij} of Q are derivatives of probabilities and reflect a change in transition probability from state i towards state j, which suggests us to call them "rates". The definition (10.7) of q_i shows that $\sum_{j=1}^{N} |q_{ij}| = 2q_i$, which demonstrates that Q is bounded if and only if the rates q_i are bounded. Karlin and Taylor (1981, p. 140) show that q_{ij} is always finite. For finite-state Markov processes, the rates q_j are finite (since q_{ij} is finite), but, in general, q_j can be infinite. If $q_j = \infty$, the state is called *instantaneous* because when the process enters this state, it immediately leaves the state. In the sequel, we confine the discussion to non-instantaneous states, thus $0 \leq q_j < \infty$. Continuous-time Markov chains with all states non-instantaneous are coined *conservative*.

Applying Taylor's theorem to (10.1) indicates that, for small h,

$$P_{ij}(h) = P_{ij}(0) + P'_{ij}(0) h + o(h)$$

Invoking the initial condition (10.5) and the definition (10.6) of the infinitesimal generator, then shows that, probabilistically,

$$\Pr[X(t+h) = j | X(t) = i] = q_{ij}h + o(h) \qquad (i \neq j)$$
$$\Pr[X(t+h) = i | X(t) = i] = 1 - q_i h + o(h) \qquad (10.8)$$

The relations (10.8) clearly generalize the Poisson process (see Theorem 7.3.1) and motivate us to call q_i the rate corresponding to state i. In fact, given that the process is in state i, the continuous-time Markov chain can be interpreted as consisting of $N-1$ competing and independent Poisson processes that attempt to jump to another state j, each with own rate q_{ij} and the next transition towards state k implies that the Poisson process with rate q_{ik} was the fastest among all $N-1$ to generate the next event (see also Section 10.4.1 below). Moreover, as in the Poisson process, there can occur a.s. only one event in an arbitrary small time interval h.

Lemma 10.2.2 *Given the infinitesimal generator Q, the transition probability matrix $P(t)$ is differentiable for all $t \geq 0$,*

$$P'(t) = P(t)Q \qquad (10.9)$$
$$= QP(t) \qquad (10.10)$$

---

[2] That definition is inspired by the Laplacian matrix (see Van Mieghem (2011)), which is defined for a graph G with N nodes as

$$Q = \Delta - A$$

where $\Delta = \text{diag}(d_1, d_2, \ldots, d_N)$, d_j the degree of node j and where A is the adjacency matrix. The infinitesimal generator can thus be interpreted as *minus* a weighted Laplacian.

These equations are called the forward (10.9) and backward (10.10) equation.

Proof: For $t = 0$, the lemma follows from the existence of $Q = P'(0)$. The derivative $P'(t)$ is defined, for $t > 0$, as

$$P'(t) = \lim_{h \to 0} \frac{P(t+h) - P(t)}{h}$$

where the derivative of the matrix has elements $P'_{ij}(t) = \frac{dP_{ij}(t)}{dt}$. Using (10.3),

$$P(t+h) - P(t) = P(t)P(h) - P(t) = P(t)\left(P(h) - I\right)$$
$$= P(h)P(t) - P(t) = \left(P(h) - I\right)P(t)$$

we obtain

$$P'(t) = P(t) \lim_{h \to 0} \frac{P(h) - I}{h} = P(t)Q$$
$$= \lim_{h \to 0} \frac{P(h) - I}{h} P(t) = QP(t)$$

which proves the lemma.   □

Suppose we are interested in the probabilities $s_k(t) = \Pr\left[X(t) = k\right]$ of finding the system in state k at time t. Each component of the state vector $s(t)$ is determined by (10.2) as

$$s_k(t+h) = \sum_{j=1}^{N} s_j(t) P_{jk}(h)$$

from which

$$\frac{s_k(t+h) - s_k(t)}{h} = s_k(t) \frac{P_{kk}(h) - 1}{h} + \sum_{j=1, j \neq k}^{N} s_j(t) \frac{P_{jk}(h)}{h}$$

In the limit $h \downarrow 0$, we find with $q_{jk} = \lim_{h \downarrow 0} \frac{P_{jk}(h)}{h}$ and $q_k = \lim_{h \downarrow 0} \frac{1 - P_{kk}(h)}{h}$ the differential equation for $s_k(t)$,

$$s'_k(t) = -q_k s_k(t) + \sum_{j=1, j \neq k}^{N} q_{jk} s_j(t) \tag{10.11}$$

which, together with the initial condition $s_k(0)$, completely determines the probability $s_k(t)$ that the Markov process is in state k at time t. Alternatively, differentiating (10.2) with respect to t and using (10.9) leads to $s'(t+\tau) = s(t+\tau)Q$ and, for $\tau = 0$,

$$s'(t) = s(t)Q \tag{10.12}$$

Selecting the k-th vector component reduces to (10.11).

10.2.2 Algebraic properties of the infinitesimal generator Q

Equation (10.10) is a matrix differential equation in t that can be similarly solved as the scalar differential equation $f'(t) = qf(t)$. With the initial condition (10.5), the solution is

$$P(t) = e^{Qt} \tag{10.13}$$

which demonstrates the importance of the infinitesimal generator Q, explicitly given by

$$Q = \begin{bmatrix} -q_1 & q_{12} & q_{13} & \cdots & q_{1;N-1} & q_{1N} \\ q_{21} & -q_2 & q_{23} & \cdots & q_{2;N-1} & q_{2N} \\ q_{31} & q_{32} & -q_3 & \cdots & q_{3;N-1} & q_{3N} \\ \vdots & \vdots & \vdots & \ddots & \vdots & \vdots \\ q_{N-1;1} & q_{N-1;2} & q_{N-1;3} & \cdots & -q_{N-1} & q_{N-1;N} \\ q_{N1} & q_{N;2} & q_{N3} & \cdots & q_{N;N-1} & -q_N \end{bmatrix} \tag{10.14}$$

Moreover, if all eigenvalues λ_k of Q are distinct, **art.** 4 and **art.** 9 indicate that

$$P(t) = e^{Qt} = X \mathrm{diag}(e^{\lambda_k t}) Y^T \tag{10.15}$$

where X and Y contain as columns the right- and left-eigenvectors of Q respectively. Written explicitly in terms of the right-eigenvectors x_k and left-eigenvectors y_k (which both are $N \times 1$ matrices or column vectors as common in vector algebra), (10.15) reads

$$P(t) = \sum_{k=1}^{N} e^{\lambda_k t} x_k y_k^T$$

where the inner or scalar vector product $y_k^T x_k = 1$ while the outer product $x_k y_k^T$ is a $N \times N$ matrix,

$$x_k y_k^T = \begin{bmatrix} x_{k1} y_{k1} & x_{k1} y_{k2} & x_{k1} y_{k3} & \cdots & x_{k1} y_{kN} \\ x_{k2} y_{k1} & x_{k2} y_{k2} & x_{k2} y_{k3} & \cdots & x_{k2} y_{kN} \\ x_{k3} y_{k1} & x_{k3} y_{k2} & x_{k3} y_{k3} & \cdots & x_{k3} y_{kN} \\ \vdots & \vdots & \vdots & \vdots & \vdots \\ x_{kN} y_{k1} & x_{kN} y_{k2} & x_{kN} y_{k3} & \cdots & x_{kN} y_{kN} \end{bmatrix}$$

If we further assume (thus omitting pathological cases) that $P(t)$ is a stochastic, irreducible matrix for any time t, Frobenius' Theorem A.5.2 indicates that all eigenvalues $\left| e^{\lambda_k t} \right| < 1$ and that only the largest eigenvalue is precisely equal to 1, say $e^{\lambda_1 t} = 1$, which corresponds to the steady-state eigenvectors $y_1^T = \pi$ and $x_1 = u$, where $u^T = [1 \ 1 \ \cdots \ 1]$. Frobenius' Theorem A.5.2 implies that all eigenvalues of Q have a negative real part, except for the steady-state eigenvalue $\lambda_1 = 0$. Hence, we may write

$$P(t) = u\pi + \sum_{k=2}^{N} e^{-|\mathrm{Re}\, \lambda_k t| + \mathrm{Im}\, \lambda_k t} x_k y_k^T \tag{10.16}$$

where $P_\infty = u\pi$ is the $N \times N$ matrix with each row containing the steady-state vector π. The expression (10.16) is called the spectral or eigen decomposition of the transition probability matrix $P(t)$.

Apart from the eigen decomposition method and the Taylor expansion

$$e^{Qt} = \sum_{k=0}^{\infty} \frac{(Qt)^k}{k!} \tag{10.17}$$

the matrix equivalent of $e^x = \lim_{n\to\infty}(1 + x/n)^n$ can be used,

$$P(t) = e^{Qt} = \lim_{n\to\infty} \left(I + \frac{Qt}{n}\right)^n \tag{10.18}$$

Since Q has negative diagonal elements and positive off-diagonal elements, computing the powers Q^k as required in (10.17) suffers from numerical rounding-off error propagation. Relation (10.18) circumvents this problem by choosing n sufficiently high, $\max_i q_i t < n$, such that $I + \frac{Qt}{n}$ has non-negative elements smaller than 1 everywhere. For stochastic matrices P, the sequence $P, P^2, P^4, \ldots, P^{2^k}$ rapidly converges. Yet another useful representation (10.26) of $P(t)$ is discussed in Section 10.4.1.

Combining (10.13) and the definition (10.2) for $\tau = 0$, the state vector $s(t)$ can be written as

$$s(t) = s(0)e^{Qt} \tag{10.19}$$

which can be regarded as the integrated form of (10.12).

10.2.3 Exponential sojourn times

We end this section on properties by proving a remarkable and important characteristic of continuous-time Markov processes.

Theorem 10.2.3 *The sojourn times τ_j of a continuous-time Markov process in a state j are independent, exponential random variables with mean $\frac{1}{q_j}$.*

Proof: The independence of the sojourn times follows from the Markov property (see the renewal argument in Section 9.3.2). The exponential sojourn time is proved in two different ways.

1. The proof consists in demonstrating that the sojourn times τ_j satisfy the memoryless property. In Section 3.2.2, it has been shown that the only continuous distribution that satisfies the memoryless property is the exponential distribution.

The event $\{\tau_j > t + T | \tau_j > T\}$ for any $T \geq 0$ and $t \geq 0$ is equivalent to the event $\{X(t + T + u) = j | X(T + u) = j, X(u) = j\}$ for $u \geq 0$. According to the Markov property (9.1) and with (10.1),

$$
\begin{aligned}
\Pr\left[\tau_j > t + T \,|\, \tau_j > T\right] &= \Pr\left[X(t + T + u) = j \,|\, X(T + u) = j, X(u) = j\right] \\
&= \Pr\left[X(t + T + u) = j \,|\, X(T + u) = j\right] \\
&= P_{jj}(t)
\end{aligned}
$$

which is independent of T illustrating the memoryless property. Using the definition of conditional probability (2.47),

$$
\Pr\left[\tau_j > t + T \,|\, \tau_j > T\right] = \frac{\Pr\left[\tau_j > t + T\right]}{\Pr\left[\tau_j > T\right]} = P_{jj}(t)
$$

which holds for any T and thus also for $T = 0$, where $\Pr\left[\tau_j > 0\right] = 1$. The distribution of the sojourn time at state j satisfies

$$
\Pr\left[\tau_j > t\right] = e^{-\alpha_j t} = P_{jj}(t)
$$

After differentiation evaluated at $t = 0$, we find $\alpha_j = q_j$.

    2. An alternative demonstration of the exponential sojourn times starts by considering for an initial state j, the probability H_n that the process remains in state j during an interval $[0, t]$. The idea is to first sample the continuous-time interval with step $\frac{t}{n}$ and afterwards proceed to the limit $n \to \infty$, which corresponds to a sampling with infinitesimally small step,

$$
\begin{aligned}
H_n &= \Pr\left[X(0) = j, X\left(\frac{t}{n}\right) = j, X\left(\frac{2t}{n}\right) = j, \ldots, X(t) = j\right] \\
&= \prod_{m=0}^{n-1} \Pr\left[X\left(\frac{(m+1)t}{n}\right) = j \,\middle|\, X\left(\frac{mt}{n}\right) = j\right] \Pr\left[X(0) = j\right] \\
&= \left(\Pr\left[X\left(\frac{t}{n}\right) = j \,\middle|\, X(0) = j\right]\right)^n \Pr\left[X(0) = j\right] \\
&= \left[P_{jj}\left(\frac{t}{n}\right)\right]^n \Pr\left[X(0) = j\right]
\end{aligned}
$$

where (9.3) and (10.1) are used. For large n, $P_{jj}\left(\frac{t}{n}\right)$ can be expanded in a Taylor series around the origin,

$$
P_{jj}\left(\frac{t}{n}\right) = P_{jj}(0) + P_{jj}'(0)\frac{t}{n} + O\left(\frac{1}{n^2}\right) = 1 - q_j \frac{t}{n} + O\left(\frac{1}{n^2}\right)
$$

such that

$$
\left[P_{jj}\left(\frac{t}{n}\right)\right]^n = \exp\left[n \log\left(1 - q_j \frac{t}{n} + O\left(\frac{1}{n^2}\right)\right)\right]
$$

For large n, the logarithm can be expanded to first order as

$$
\log\left(1 - q_j \frac{t}{n} + O\left(\frac{1}{n^2}\right)\right) = -q_j \frac{t}{n} + O\left(\frac{1}{n^2}\right)
$$

which shows that

$$\lim_{n \to \infty} \left[P_{jj} \left(\frac{t}{n} \right) \right]^n = e^{-q_j t}$$

On the other hand,

$$\lim_{n \to \infty} H_n = \Pr\left[X(u) = j, 0 \le u \le t \right]$$

Hence, the probability that the process remains in state j at least for a duration t equals

$$\Pr\left[X(u) = j, 0 \le u \le t \right] = e^{-q_j t} \Pr\left[X(0) = j \right]$$

Conditioned to the initial state with (2.47),

$$\Pr\left[X(u) = j, 0 \le u \le t \mid X(0) = j \right] = \Pr\left[\tau_j > t \right] = e^{-q_j t} \qquad (10.20)$$

Without resorting to the memoryless property, Theorem 10.2.3 has been proved. $\square$

In summary, the continuous-time Markov process $\{X(t), t \in T\}$ can be described in two equivalent ways, either by the transition probability matrix $P(t)$ or by the infinitesimal generator Q. In the first description, the process starts at time $t = t_0 = 0$ in state x_0, where it stays until a transition occurs at $t = t_1$, which makes the process jump to state x_1. In state x_1, the process stays until $t = t_2$ at which time it jumps to state x_2, and so on. The sequence of states $x_0, x_1, x_2, \dots$ is a discrete Markov process and is called the *embedded* Markov chain. The embedded Markov chain is further discussed in Section 10.4. The infinitesimal description based on Q formulates the evolution of the process in terms of rates. The process waits in a state j until a jump or trigger occurs with rate q_j and the mean waiting time in state j is $\frac{1}{q_j}$. If $q_j = 0$, the Markov process stays infinitely long in state j, implying that state j is an absorbing state.

10.3 Steady-state

Theorems 9.3.5 and 9.3.7 demonstrate that, when a finite-state Markov chain is aperiodic and irreducible (all states communicate and $P_{ij}(t) > 0$ for $t > 0$), the steady-state π exists. Similarly to the discrete-time Markov chain, we describe the steady-state in two different ways as the limit case for $t \to \infty$ of (a) the time-dependent system (either forward (10.9) or backward (10.10)) equation and of (b) the state vector $s(t)$.

By definition, the steady-state does not change over time, which means that

$$\lim_{t \to \infty} P'(t) = 0$$

It then follows from (10.9) and (10.10) that

$$Q P_\infty = P_\infty Q = 0$$

where $\lim_{t \to \infty} P(t) = P_\infty$. This relation implies that P_∞ is the adjoint matrix of Q

belonging to eigenvalue $\lambda = 0$, which plays a role analogous to $\lambda = 1$ in the discrete case. By the same arguments as in the discrete case and as shown in Section 10.2.2, all rows of $P_\infty = u\pi$ are proportional to the eigenvector of Q belonging to $\lambda = 0$. Thus, the steady-state (row) vector π is solution of

$$\pi Q = 0 \tag{10.21}$$

which means that π is orthogonal to any column vector of Q such that necessarily $\det Q = 0$ in order for a non-zero solution to exist. A single component of π in (10.21) obeys, using (10.7),

$$\pi_i q_i = \sum_{j=1, j \neq i}^{N} \pi_j q_{ji} \tag{10.22}$$

This equation has a continuity or conservation law interpretation. The left-hand side reflects the long-run rate at which the process leaves state i. The right-hand side is the sum of the long-run rates of transitions towards the state i from other states $j \neq i$ or the aggregate long-run rate towards state i. Both the inward and outward fluxes at any state i are in steady-state precisely in balance. Therefore, relations (10.22) for each $1 \leq i \leq N$ are called the *balance equations*. The balance equation (10.22) directly follows from the differential equation (10.11) of the state probabilities $s_k(t)$, because $\lim_{t \to \infty} s_k(t) = \pi_k$ and $\lim_{t \to \infty} s'_k(t) = 0$.

Alternatively, the steady-state vector π obeys (10.2),

$$\pi = \lim_{t \to \infty} s(t) = s(0) P_\infty = s(0) \lim_{t \to \infty} e^{Qt}$$

which, together with (10.15), implies that all eigenvalues of Q must have non-positive real parts and that only $\lambda = 0$ can correspond to the steady-state. This stability condition on the eigenvalues corresponds to that in a linear, time-variant system. Since all rows in P_∞ are equal (see also (10.16)), the dependence of the steady-state vector π on the initial state drops out. For, analogous to the discrete-time case and recalling the normalization $\|s(0)\|_1 = 1$, a single component becomes

$$\pi_j = \sum_{k=1}^{N} s_k(0) (P_\infty)_{kj} = (P_\infty)_{1j} \sum_{k=1}^{N} s_k(0) = (P_\infty)_{1j}$$

We end this section by making some additional observations. The notion of probability flux conservation in the steady-state, reflected by the balance equations (10.22), can be translated to the discrete case as well. For, we can rewrite the steady-state equation (9.29) for state i as

$$\pi_i P_i = \sum_{j=1; j \neq i}^{N} \pi_j P_{ji}$$

where $1 - P_{ii} = P_i \geq 0$, and which is formally equivalent to (10.22). Moreover, instead of considering balance between the inward and outward fluxes at a particular state i as in (10.22), we can further combine a subset S out of the total number N of states and express equality between the probability flux into S and leaving S, by summing (10.22) over all states $i \in S$. This means that, in the directed Markov graph that represents the transitions between states (or nodes), a cut along transitions (or links) can be made that isolate the set S from the rest of the graph. As shown in Section 11.2.1, the balance equations applied to that set S may lead to a more elegant solution of the steady-state vector π or may ease the interpretation of probability flows in a particular Markov graph.

10.4 The embedded Markov chain

The main difference between discrete and continuous-time Markov chains lies, apart from the concept of time, in the determination of the number of transitions. The sojourn time in a discrete chain is deterministic and all times are equal to 1. In other words, if $F_{ij}(t)$ denotes the distribution function of the time until a transition from state i to state j occurs, then it is plain that, for a discrete-time process,

$$F_{ij}(t) = 1_{t \geq 1}$$

Even though the process remains in state j with probability P_{jj}, there has been a transition precisely after $t = 1$ units. Theorem 10.2.3, on the other hand, demonstrates that the sojourn times in state j are exponentially distributed with mean $\frac{1}{q_j}$. After, on average $\frac{1}{q_j}$ time units, a transition from state j to another state occurs. In contrast to discrete-time Markov chains, after an exponentially distributed sojourn time in state j the process makes a transition to other states $i \neq j$. The *embedded* Markov chain of a continuous-time Markov process is the corresponding discrete Markov chain that follows the same state transitions, but that abstracts the (exponentially distributed) sojourn time relation.

Let us investigate this fact in more detail. Let us denote

$$V_{ij}(h) = \Pr[X(h) = j | X(h) \neq i, X(0) = i]$$

which describes the probability that, if a transition occurs, the process moves from state i to a different state $j \neq i$. Using the definition of conditional probability (2.47),

$$V_{ij}(h) = \frac{\Pr[\{X(h) = j\} \cap \{X(h) \neq i\} | X(0) = i]}{\Pr[X(h) \neq i | X(0) = i]} = \frac{P_{ij}(h)}{1 - P_{ii}(h)}$$

In the limit $h \downarrow 0$, we have

$$V_{ij} = \lim_{h \downarrow 0} V_{ij}(h) = \lim_{h \downarrow 0} \frac{\frac{P_{ij}(h)}{h}}{\frac{1 - P_{ii}(h)}{h}} = \frac{q_{ij}}{q_i}$$

By (10.7), we see that $\sum_{j=1, j \neq i}^{N} V_{ij} = 1$, demonstrating that, given a transition, it is a transition out of state i to another state j. The quantities V_{ij} correspond to the transition probabilities of the embedded Markov chain. Alternatively, we can write the rate q_{ij} in terms of the transition probabilities V_{ij} of the embedded Markov chain as

$$q_{ij} = q_i V_{ij} \qquad (10.23)$$

Since q_i is the rate (i.e. the number of transitions per unit time) of the process in state i, relation (10.23) shows that the transition rate q_{ij} from state i to state j equals the rate of transitions in state i multiplied by the probability that a transition from state i to state j occurs. By definition, $V_{ii} = 0$. For, if we assume that $V_{ij} > 0$, relation (10.23) would result in $q_{ii} = V_{ii}q_i > 0$ which contradicts the definition $q_{ii} = -q_i$. Hence, in the embedded Markov chain specified by the transition probability matrix V, there are no self-transitions ($V_{ii} = 0$), which is equivalent to the fact that the sum of the eigenvalues of V is zero (A.7), since trace$(V) = 0$.

From the steady-state equation or balance equation (10.22), (10.23) and $V_{ii} = 0$, we observe that

$$\pi_i q_i = \sum_{j=1}^{N} \pi_j q_j V_{ji}$$

On the other hand, the embedded Markov chain has a steady-state vector v that obeys (9.28) or (9.29)

$$v_i = \sum_{j=1}^{N} v_j V_{ji}$$

the normalization $\|v\|_1 = vu = 1$ and $v_i = \lim_{n \to \infty} (V^n)_{ji}$. The relations between the steady-state vectors of the continuous-time Markov chain, π, and of its corresponding embedded discrete-time Markov chain, v, are deduced analogously to the proof of Theorem 9.3.6 with $a_j = \pi_j q_j$ in (9.37) as

$$v_i = \frac{\pi_i q_i}{\sum_{j=1}^{N} \pi_j q_j} \qquad (10.24)$$

The inverse of (10.24), with $c = \sum_{j=1}^{N} \pi_j q_j$, is $c\frac{v_i}{q_i} = \pi_i$. Summing over all states i and using $\|\pi\|_1 = 1$ yields $c^{-1} = \sum_{j=1}^{N} v_j/q_j$ and

$$\pi_i = \frac{v_i/q_i}{\sum_{j=1}^{N} v_j/q_j} \qquad (10.25)$$

The classification of the discrete-time case into transient and recurrent states can be transferred via the embedded Markov chain to continuous-time Markov processes.

10.4.1 Uniformization

The restriction $V_{ii} = 0$ or $q_{ii} = 0$, which means that there are no self-transitions from a state into itself, can be removed. Indeed, we can rewrite the basic relation (10.13) between the transition probability matrix $P(t)$ and the infinitesimal generator Q for all β as

$$P(t) = \exp\left[-\beta I t + \beta t\left(I + \frac{Q}{\beta}\right)\right] = e^{-\beta t} \exp\left[\beta t\left(I + \frac{Q}{\beta}\right)\right]$$

Defining $T(\beta) = I + \frac{Q}{\beta}$ and $\beta \geq \max_i q_i$, an alternative description to (10.16), (10.17) and (10.18) appears

$$P_{ij}(t) = e^{-\beta t} \sum_{k=0}^{\infty} \frac{(\beta t)^k}{k!} T_{ij}^k(\beta) \tag{10.26}$$

where $T(\beta)$ is a stationary transition probability matrix and, hence, a stochastic matrix.

We also observe that $\beta T(\beta) = Q + \beta I$ can be regarded as a rate matrix, with the property that, for each state i,

$$\sum_{j=1}^{N} \beta T_{ij}(\beta) = \sum_{j=1}^{N} Q_{ij} + \beta \sum_{j=1}^{N} \delta_{ij} = \beta$$

the transition rate in any state i is precisely the same, equal to β. Whereas the embedded Markov chain defined by (10.23) has no self-transitions ($V_{ii} = 0$), we see for any i and j, that $T_{ii}(\beta) = 1 - \frac{1}{\beta}\sum_{j=1;j\neq i}^{N} q_{ij} = 1 - \frac{q_i}{\beta} \geq 0$ while $T_{ij}(\beta) = \frac{q_{ij}}{\beta}$. Hence, $T(\beta)$ can be interpreted as an embedded Markov chain that allows self-transitions. In view of (10.23), the embedded structure of $T(\beta)$ is summarized as

$$q_{ij} = \beta T_{ij}(\beta) \qquad \text{for } i \neq j$$
$$q_{ii} = 1 - \beta T_{ii}(\beta)$$

where the constant rate $q_i = \beta$ for any state i is, besides self-transitions $q_{ii} \neq 0$, the characterizing property. These properties also reveal that, starting from the embedded chain V where $V_{ii} = 0$, we can add self-transitions $q_{ii} > 0$ with the effect that, on (10.7), the transition rate $q_i \to q_i + q_{ii}$. The opposite figure illustrates an embedded Markov chain with self-transitions and the corresponding transition probability matrix, where the transition rates q_i follow from $\sum_{j=1}^{6} V_{ij} = 1$ with $V_{ij} = \frac{q_{ij}}{q_i}$. This change in transition rate will change the steady-state vector since the balance equations (10.22) change. However, the Markov process $\{X(t), t \geq 0\}$ is not modified because a self-transition does not change $X(t)$ nor the distribution of the time until the next transition to a different state. But self-transitions clearly change the number of transitions during some period of time. When the transition rate q_j at each state j is the same, the embedded Markov chain $T(\beta)$ is called a *uniformized* chain.

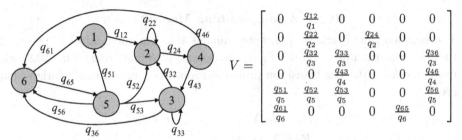

$$V = \begin{bmatrix} 0 & \frac{q_{12}}{q_1} & 0 & 0 & 0 & 0 \\ 0 & \frac{q_{22}}{q_2} & 0 & \frac{q_{24}}{q_2} & 0 & 0 \\ 0 & \frac{q_{32}}{q_3} & \frac{q_{33}}{q_3} & 0 & 0 & \frac{q_{36}}{q_3} \\ 0 & 0 & \frac{q_{43}}{q_4} & 0 & 0 & \frac{q_{46}}{q_4} \\ \frac{q_{51}}{q_5} & \frac{q_{52}}{q_5} & \frac{q_{53}}{q_5} & 0 & 0 & \frac{q_{56}}{q_5} \\ \frac{q_{61}}{q_6} & 0 & 0 & 0 & \frac{q_{65}}{q_6} & 0 \end{bmatrix}$$

In addition, in a uniformized chain, the steady-state vector $t(\beta)$ of $T(\beta)$ is the same as the steady-state vector π. Indeed, from (9.29),

$$t_j(\beta) = \sum_{k=1}^{N} T_{kj}(\beta) t_k(\beta)$$

we have, with $T(\beta) = I + \frac{Q}{\beta}$,

$$t_j(\beta) = \sum_{k=1}^{N} \left(\delta_{kj} + \frac{q_{kj}}{\beta} \right) t_k(\beta)$$

$$= t_j(\beta) + \frac{1}{\beta} \sum_{k=1}^{N} t_k(\beta) q_{kj}$$

or,

$$t_k(\beta) q_k = \sum_{k=1; k \neq j}^{N} t_k(\beta) q_{kj}$$

where $t_k(\beta) = \pi_k$ (independent of β), since it satisfies the balance equation (10.22), and Theorem 9.3.6 assures that the steady-state of a positive recurrent chain is unique.

We will now interpret (10.26) probabilistically. Let $N(t)$ denote the total number of transitions in $[0, t]$ in the uniformized (discrete) process $\{X_k(\beta)\}$. Since the transition rates $q_i = \beta$ are all the same, $N(t)$ is a Poisson process with rate β because, for any continuous-time Markov chain, the inter-transition or sojourn times are i.i.d. exponential random variables. Thus, $\Pr[N(t) = k] = e^{-\beta t} \frac{(\beta t)^k}{k!}$ is recognized as the probability that the number of transitions that occur in $[0, t]$ in the uniformized Markov chain with rate β equals k. With (9.11), $T_{ij}^k(\beta) = \Pr[X_k(\beta) = j | X_0(\beta) = i]$ is the k-step transition probability of that discrete $\{X_k(\beta)\}$ uniformized Markov process. Relation (10.26) can be interpreted as

$$P_{ij}(t) = \sum_{k=0}^{\infty} \Pr[X_k(\beta) = j | X_0(\beta) = i, N(t) = k] \Pr[N(t) = k]$$

or, the probability that the continuous Markov process moves from state i to state j in a time interval of length t, can be decomposed in an infinite sum of probabilities. Each probability corresponds to a transition from state i to state j in k-steps, where the number of intermediate transitions k is a Poisson counting process with rate β.

10.4.2 A sampled-time Markov chain

The sampled-time Markov chain approximates the continuous Markov process in that the transition probabilities $P_{ij}(t)$ are expanded to first order as in (10.8) with fixed step $h = \Delta t$. The transition probabilities of the sampled-time Markov chain are

$$P_{ij} = q_{ij}\Delta t \qquad (i \neq j)$$
$$P_{ii} = 1 - q_i\Delta t$$

Clearly, the sampled-time Markov chain also allows self-transitions, as illustrated in Fig. 10.1.

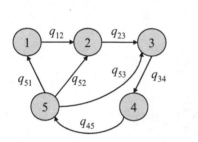

Continuous-time Markov process

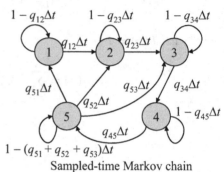

Sampled-time Markov chain

Fig. 10.1. A continuous-time Markov process and its corresponding sampled-time Markov chain.

From (10.8), we observe that the approximation lies in two facts: (a) Δt is fixed such that $q_{ij}\Delta t \approx \Pr\left[X(t + \Delta t) = j | X(t) = i\right]$ is increasingly accurate as $\Delta t \to 0$ and (b) transitions occur at discrete times every Δt time units. The sampling step Δt should be chosen such that the transition probabilities obey $0 \leq P_{ij} \leq 1$, from which we find that $\Delta t \leq \frac{1}{\max_i q_i}$. Hence, the sampling rate $(\Delta t)^{-1}$ must be higher than the fastest possible transition rate $\max_i q_i$ in the Markov process.

Let $\varkappa$ denote the steady-state vector of the sampled-time Markov chain with $\|\varkappa\|_1 = 1$. Being a discrete Markov chain, the steady-state vector components $\varkappa_j$ satisfy (9.29) for each component j,

$$\varkappa_j = \sum_{k=1}^{N} P_{kj}\varkappa_k = \Delta t \sum_{k=1;k\neq j}^{N} q_{kj}\varkappa_k + (1 - q_j\Delta t)\varkappa_j$$

or

$$q_j\varkappa_j = \sum_{k=1;k\neq j}^{N} q_{kj}\varkappa_k$$

By comparing with the balance equation (10.22) and on the uniqueness of the steady-state (Theorem 9.3.6), we observe that $\varkappa = \pi$ or, the steady-state of the sampled-time Markov chain is *exactly* (not approximately) equal to the steady-state of the continuous Markov chain for any sampling step $\Delta t \leq \frac{1}{\max_i q_i}$. Although we can possibly miss by sampling every Δt time units the smaller-scale dynamics of the continuous Markov chain, the long-run behavior or steady-state is exactly captured!

10.5 The transitions in a continuous-time Markov chain

Based on the embedded Markov chain, there exists a framework that deduces all properties of the continuous-time Markov chain, by postulating, as a defining characteristic, the exponential sojourn times (Theorem 10.2.3) of a continuous-time Markov chain.

Theorem 10.5.1 *Let V_{ij} denote the transition probabilities of the embedded Markov chain and q_{ij} the rates of the infinitesimal generator. The transition probabilities of the corresponding continuous-time Markov chain are found as*

$$P_{ij}(t) = \delta_{ij}e^{-q_i t} + q_i \sum_{k \neq i} V_{ik} \int_0^t e^{-q_i u} P_{kj}(t-u)du \qquad (10.27)$$

Proof: If i is an absorbing state ($q_i = 0$), then, by definition, $P_{ij}(t) = \delta_{ij}$ for all $t \geq 0$. For a non-absorbing state i and a process starting from state i, the event $\{\tau \leq t, X(\tau) = k\} \cap \{X(t) = j\}$ is possible if and only if the first transition from i to k occurs at some time $u \in [0, t]$ and the next transition from k to j takes place in the remaining time $t - u$. The probability density function of the sojourn time is $f_{\tau_i}(t) = \frac{d}{dt} \Pr[\tau_i \leq t] = q_i e^{-q_i t}$ for $t \geq 0$ and for infinitesimally small ϵ, we have

$p = \Pr[\tau \leq t, X(\tau) = k, X(t) = j | X(0) = i]$

$\quad = \int_0^t du \Pr[\tau = u, X(u - \epsilon) = i | X(0) = i] \Pr[X(u) = k | X(u - \epsilon) = i] \Pr[X(t) = j | X(u) = k]$

$\quad = \int_0^t du f_{\tau_i}(u) V_{ik} P_{kj}(t-u) = V_{ik} \int_0^t q_i e^{-q_i u} P_{kj}(t-u)du$

Furthermore,

$$\Pr[\tau \leq t \text{ and } X(t) = j | X(0) = i] = \sum_{k \neq i} \Pr[t \geq \tau, X(\tau) = k, X(t) = j | X(0) = i]$$

and

$$\Pr[\tau > t \text{ and } X(t) = j | X(0) = i] = \delta_{ij} \Pr[\tau_i > t] = \delta_{ij} e^{-q_i t}$$

Finally,

$$P_{ij}(t) = \Pr[X(t) = j | X(0) = i]$$
$$= \Pr[\tau \leq t \text{ and } X(t) = j | X(0) = i] + \Pr[\tau > t \text{ and } X(t) = j | X(0) = i]$$

Combining all above relations into the last one proves the theorem. $\qquad\square$

By a change of variable $s = t - u$ in (10.27), we have

$$P_{ij}(t) = \delta_{ij}e^{-q_i t} + q_i \sum_{k \neq i} V_{ik} e^{-q_i t} \int_0^t e^{q_i s} P_{kj}(s)ds$$

and, after differentiation with respect to t, we find for $t \geq 0$,

$$P'_{ij}(t) = -q_i \delta_{ij} e^{-q_i t} - q_i \sum_{k \neq i} q_k V_{ik} e^{-q_i t} \int_0^t e^{q_i s} P_{kj}(s) ds + q_i \sum_{k \neq i} V_{ik} P_{kj}(t)$$

$$= -q_i \delta_{ij} e^{-q_i t} - q_i \left(P_{ij}(t) - \delta_{ij} e^{-q_i t} \right) + q_i \sum_{k \neq i} V_{ik} P_{kj}(t)$$

$$= -q_i P_{ij}(t) + q_i \sum_{k \neq i} V_{ik} P_{kj}(t)$$

Evaluated at $t = 0$, recalling that $P'(0) = Q$ and $P(0) = I$,

$$P'_{ij}(0) = -q_i P_{ij}(0) + q_i \sum_{k \neq i} V_{ik} P_{kj}(0)$$

$$q_{ij} = -q_i \delta_{ij} + q_i \sum_{k \neq i} V_{ik} \delta_{kj} = -q_i \delta_{ij} + q_i V_{ij}$$

which is precisely relation (10.23). With $q_i = -q_{ii}$ and (10.23), we arrive at

$$P'_{ij}(t) = \sum_{k=1}^N q_{ik} P_{kj}(t)$$

which is precisely the backward equation (10.10). Hence, (10.27) can be interpreted as an integrated form of the backward equation and thus of the entire continuous-time Markov process.

10.6 Example: the two-state Markov chain in continuous-time

The continuous-time two-state Markov chain is defined by the infinitesimal generator

$$Q = \begin{bmatrix} -\lambda & \lambda \\ \mu & -\mu \end{bmatrix}$$

where $\lambda, \mu \geq 0$. We will solve $P(t)$ from the forward equation (10.9),

$$\begin{bmatrix} P'_{11}(t) & P'_{12}(t) \\ P'_{21}(t) & P'_{22}(t) \end{bmatrix} = \begin{bmatrix} P_{11}(t) & P_{12}(t) \\ P_{21}(t) & P_{22}(t) \end{bmatrix} \begin{bmatrix} -\lambda & \lambda \\ \mu & -\mu \end{bmatrix}$$

which actually contains two independent transition probabilities because $P_{12}(t) = 1 - P_{11}(t)$ and $P_{21}(t) = 1 - P_{22}(t)$. The forward equation simplifies to

$$P'_{11}(t) = -(\lambda + \mu) P_{11}(t) + \mu$$
$$P'_{22}(t) = -(\lambda + \mu) P_{22}(t) + \lambda$$

Only the first equation needs to be solved since, by symmetry, the solution of $P_{11}(t)$ equals that of $P_{22}(t)$ after changing the role of $\lambda \rightarrow \mu$ and $\mu \rightarrow \lambda$. The linear, first-order, non-homogeneous differential equation consists of the solution to the corresponding homogeneous differential equation and a particular solution.

The solution of the homogeneous differential equation, $P'_{11}(t) = -(\lambda + \mu)P_{11}(t)$, is $P_{11}(t) = Ce^{-(\lambda+\mu)t}$. The particular solution is generally found by variation of the constant C, which proposes $P_{11}(t) = C(t)e^{-(\lambda+\mu)t}$ as general solution, where $C(t)$ needs to satisfy the original differential equation. Hence,

$$C'(t) = \mu e^{(\lambda+\mu)t}$$

or, after integration, $C(t) = \frac{\mu}{\lambda+\mu}e^{(\lambda+\mu)t} + c$. The integration constant c follows from the initial condition (10.5), $P_{11}(0) = 1$. Finally, we arrive at

$$P_{11}(t) = \frac{\mu}{\lambda + \mu} + \frac{\lambda}{\lambda + \mu}e^{-(\lambda+\mu)t}$$

$$P_{22}(t) = \frac{\lambda}{\lambda + \mu} + \frac{\mu}{\lambda + \mu}e^{-(\lambda+\mu)t}$$

from which the steady-state vector is immediate,

$$\pi = \begin{bmatrix} \frac{\mu}{\lambda+\mu} & \frac{\lambda}{\lambda+\mu} \end{bmatrix}$$

10.7 Time reversibility

In this section, we consider only ergodic Markov chains for which the steady-state probability for each state j is strictly positive, $\pi_j > 0$. Suppose the Markov process operates already in the steady-state, or, in other words, the Markov process is stationary. We are interested in the time-reversed process defined by the sequence $X_n, X_{n-1}, \ldots$ We will show that this time-reversed sequence again constitutes a Markov process.

Theorem 10.7.1 *The time-reversed Markov process is a Markov chain.*

Proof: It suffices to demonstrate that the time-reversed process satisfies the Markov property

$$\Pr\left[X_n = x_n | X_{n+1} = x_{n+1}, \ldots, X_{n+k} = x_{n+k}\right] = \Pr\left[X_n = x_n | X_{n+1} = x_{n+1}\right]$$

By definition of the conditional probability (2.47),

$$R = \Pr\left[X_n = x_n | X_{n+1} = x_{n+1}, X_{n+2} = x_{n+2}, \ldots, X_{n+k} = x_{n+k}\right]$$

$$= \Pr\left[X_n = x_n | \cap_{m=1}^{k}\{X_{n+m} = x_{n+m}\}\right]$$

$$= \frac{\Pr\left[\cap_{m=0}^{k}\{X_{n+m} = x_{n+m}\}\right]}{\Pr\left[\cap_{m=1}^{k}\{X_{n+m} = x_{n+m}\}\right]}$$

Since the intersection is commutative $A \cap B = B \cap A$, the indices can be reversed,

$$R = \frac{\Pr\left[\cap_{m=k}^{0}\{X_{n+m} = x_{n+m}\}\right]}{\Pr\left[\cap_{m=k}^{1}\{X_{n+m} = x_{n+m}\}\right]}$$

$$= \frac{\Pr\left[X_{n+k} = x_{n+k} | \cap_{m=k-1}^{0}\{X_{n+m} = x_{n+m}\}\right]\Pr\left[\cap_{m=k-1}^{0}\{X_{n+m} = x_{n+m}\}\right]}{\Pr\left[\cap_{m=k}^{1}\{X_{n+m} = x_{n+m}\}\right]}$$

The original stationary process is a Markov process that satisfies (9.2). Using (9.2) and (9.3) we have

$$\Pr\left[X_{n+k} = x_{n+k} | \cap_{m=k-1}^{0}\{X_{n+m} = x_{n+m}\}\right] = \Pr\left[X_{n+k} = x_{n+k} | X_{n+k-1} = x_{n+k-1}\right]$$

and

$$\Pr\left[\cap_{m=k-1}^{0}\{X_{n+m}=x_{n+m}\}\right]=\Pr[X_n=x_n]\prod_{m=1}^{k-1}\Pr[X_{n+m}=x_{n+m}|X_{n+m-1}=x_{n+m-1}]$$

and, similarly,

$$\Pr\left[\cap_{m=k}^{1}\{X_{n+m}=x_{n+m}\}\right]=\Pr[X_{n+1}=x_{n+1}]\prod_{m=2}^{k}\Pr[X_{n+m}=x_{n+m}|X_{n+m-1}=x_{n+m-1}]$$

Hence,

$$R=\frac{\Pr\left[X_{n+k}=x_{n+k}|X_{n+k-1}=x_{n+k-1}\right]\Pr\left[X_{n+1}=x_{n+1}|X_n=x_n\right]\Pr\left[X_n=x_n\right]}{\Pr\left[X_{n+k}=x_{n+k}|X_{n+k-1}=x_{n+k-1}\right]\Pr\left[X_{n+1}=x_{n+1}\right]}$$

$$=\frac{\Pr\left[X_{n+1}=x_{n+1}|X_n=x_n\right]\Pr\left[X_n=x_n\right]}{\Pr\left[X_{n+1}=x_{n+1}\right]}$$

Applying Bayes' rule (2.51) to the last relation finally proves the theorem. □

Consider the transition probability of the time-reversed Markov process

$$R_{ij}=\Pr\left[X_n=j|X_{n+1}=i\right]$$

With Bayes' rule (2.51),

$$\Pr\left[X_n=j|X_{n+1}=i\right]=\frac{\Pr\left[X_{n+1}=i|X_n=j\right]\Pr\left[X_n=j\right]}{\Pr\left[X_{n+1}=i\right]}$$

and, since the original, forward-time Markov process is stationary,

$$\Pr\left[X_n=j\right]=\pi_j,\quad\Pr\left[X_{n+1}=i\right]=\pi_i$$

the transition probability of the time-reversed process is

$$R_{ij}=\frac{\pi_j P_{ji}}{\pi_i}\tag{10.28}$$

A Markov chain is said to be *time reversible* if, for all i and j, $P_{ij}=R_{ij}$. From (10.28), the condition for time reversibility is

$$\pi_i P_{ij}=\pi_j P_{ji}\tag{10.29}$$

This condition means that, for all states i and j, the rate $\pi_i P_{ij}$ from state $i\to j$ equals the rate $\pi_j P_{ji}$ from state $j\to i$. An interesting property of time-reversible Markov chains is that any vector x satisfying $\|x\|_1=1$ and $x_i P_{ij}=x_j P_{ji}$ is a steady-state vector of a time-reversible Markov chain. Indeed, summing over all i,

$$\sum_i x_i P_{ij}=x_j\sum_i P_{ji}=x_j$$

Theorem 9.3.6 indicates that the steady-state is unique and, thus, $x=\pi$. As a side remark, we note that a transition matrix is only equal to its transpose $P=P^T$ if the Markov process is time reversible and doubly stochastic (i.e. $\pi_i=\frac{1}{N}$ for all i, as shown in Appendix A.6.1).

The continuous-time analogon can be immediately deduced from the discrete-time embedded Markov chain defined by the transition probabilities V_{ij}. Let U_{ij} denote the transition probabilities of the time-reversed embedded Markov chain and r_{ij} the rates of the corresponding continuous Markov chain, then by (10.23)

$$r_{ij} = r_i U_{ij} \tag{10.30}$$

We will now show that the rates r_i of the time-reversed continuous Markov process are indeed exponential random variables. Assume that the time-reversed process is in state i at time t. The probability that the process is still in state i at reversed time $t - u$ is, using Theorem 10.2.3,

$$\Pr\left[X(\tau) = i, t - u \le \tau \le t | X(t) = i\right] = \frac{\Pr\left[X(\tau) = i, t - u \le \tau \le t\right]}{\Pr\left[X(t) = i\right]}$$
$$= \frac{\Pr\left[X(t - u) = i\right] e^{-q_i u}}{\Pr\left[X(t) = i\right]} = e^{-q_i u}$$

because, in the steady state $t \to \infty$, $\Pr\left[X(t - u) = i\right] = \Pr\left[X(t) = i\right] = \pi_i$ for any finite u. Thus, the sojourn time in state i of the time-reversed process is exponentially distributed with precisely the same rate $r_i = q_i$ as the forward time process. The steady-state vector v of the embedded Markov chain can be written in terms of the steady-state vector of the continuous Markov chain via (10.24). By (10.28), we obtain

$$U_{ij} = \frac{v_j V_{ji}}{v_i} = \frac{\pi_j q_j V_{ji}}{\pi_i q_i}$$

With (10.23) and (10.30)

$$\frac{\pi_i r_{ij}}{r_i} = \frac{\pi_j q_{ji}}{q_i}$$

but, since $r_i = q_i$, we finally arrive at

$$\pi_i r_{ij} = \pi_j q_{ji} \tag{10.31}$$

Comparing (10.31) with the discrete case (10.28), we see that the transition probabilities P_{ij} and R_{ij} are changed for the rates q_{ij} and r_{ij}. We know that π_j is the portion of time the process (both forward and reversed) spends in state j and that q_{ij} is the rate at which the process makes transitions from state i to state j. Equation (10.31) has again a balance interpretation: $\pi_j q_{ji}$ is the rate at which the forward process moves from state j to i, while $\pi_i r_{ij}$ is the rate of the time-reversed process from state i to j and both rates are equal. Intuitively, when a process jumps from state $i \to j$ in forward time, it is plain that the process makes, in reversed time, just the opposite transition from $j \to i$. Similarly as above, a continuous-time Markov chain is time reversible if, for all i and j, it holds that $r_{ij} = q_{ij}$. For these processes (which occur often in practice, as demonstrated in the chapters on queueing), the rate from $i \to j$ is equal to the rate from $j \to i$ since $\pi_i q_{ij} = \pi_j q_{ji}$.

10.8 Problems

(i) Consider a computer that has two identical and independent processors. The time between failures has an exponential distribution. The mean value of this distribution is 1000 hours. The repair time for a damaged processor is exponentially distributed as well, with a mean value of 100 hours. We assume damaged processors can be repaired in parallel. There are clearly three states for this computer: (1) both processors work, (2) one processor is damaged and (3) both processors are damaged.

   (a) Make a continuous Markov chain presentation of these states.
   (b) What is the infinitesimal generator matrix Q for this Markov chain? Give the relation between the state probability at time t and its derivative.
   (c) Calculate the steady-state of this process.
   (d) What is the availability of the computer if (i) both processors are required to work, or (ii) at least one processor should work?

(ii) Consider two identical servers that are working in parallel. When one server fails, the other has to do the whole job alone under a higher load. The failure times of servers are exponentially distributed: $\mu_H = 3 \times 10^{-4}\ h^{-1}$, when the servers are equally loaded and $\mu_F = 7 \times 10^{-4}\ h^{-1}$, when one of the servers works under the "full load". In addition, due to external influences, both servers may fail at the same time with a failure rate of $\mu_E = 6 \times 10^{-5}\ h^{-1}$. This type of external stress affects the system irrespective of how many of its units are operating. As soon as one of the servers fails, the repair is initiated. The average downtime of a server is $\lambda^{-1} = 15$ hours. However, if both servers are damaged, the whole system must be shut down. The average time needed to repair both damaged servers is $\lambda_B^{-1} = 20$ hours.

   (a) Draw the Markov chain for this system.
   (b) Determine the infinitesimal generator matrix Q.
   (c) Determine the steady-state probabilities.
   (d) Determine the mean lifetime of different states.
   (e) What is the average number of server repairs needed during a period of one year?

(iii) Consider a private network that has two access points towards two different ISPs. The access points are non-identical routers that cannot fail at the same time. If one router fails, the other processes the whole job alone. In this situation, the router is prone to failures. The time between failures of a router has an exponential distribution. When both routers are working, the mean failure time for the first router is 100 hours and 200 hours for the second. When only one router is working, the mean failure time for the first router is 50 hours and 100 hours for the second. The routers are not repaired until both are down and they are repaired at the same time. The repair time of both damaged routers is exponentially distributed as well, with a mean reparation time of 50 hours.

   (a) Draw a continuous Markov chain of these states and determine the infinitesimal generator matrix Q.
   (b) Calculate the steady-state probabilities of this failure and reparation process.
   (c) What is the probability in steady-state that the first router is working?
   (d) How many hours on average is the second router down?

(iv) In a protected data communications network, a disjoint backup route is associated to each primary connection route. Under normal conditions, all data flows are transmitted along the primary route. However, if there is failure along the primary route, the data flows are rerouted using the backup route. As soon as the primary route is repaired, the data is again sent along the primary route, thereby freeing the backup route for another connection that may be sharing the backup route. Two connections c_1 and c_2 share the same backup route b. Any primary route has only two states: "up" (U) and "down" (D), whereas the backup route is always "up". For each primary route, the failure and repair times are independent and exponentially distributed with rates λ and μ, respectively.

   (a) Make a continuous-time Markov chain presentation of this problem. The states should show which connection is currently using the backup route.
   (b) What is the infinitesimal generator matrix Q for this Markov chain?
   (c) Compute the steady-state of this process.
   (d) Given that both c_1 and c_2 have failed, what is the probability that c_1 is using the backup route?

(v) Consider a ring network between a source node s and a destination node t. There are two link-disjoint paths from the source node s to the destination node t. Initially, a connection between s and t is setup using one of the paths. As soon as a failure is detected along that path, the connection is rerouted to the other path. The failure rate for each path is λ. The failure along a path is either a failure at an intermediate node, that occurs with probability p_2, or a link failure having with probability p_2. Along the path on which the connection is setup, the mean delay (switching delay) before a node failure is sensed equals $1/\delta_1$, whereas the mean switching delay for a link failure is $1/\delta_2$. We further assume that all component failure events are mutually independent and the occurrence of each component failure event is exponentially distributed in time. Different types of failures have the same repair rate μ. The switching delay is small compared to the mean time of path failures, so that during the switching delay no additional failure event occurs. Finally, the source and destination nodes are assumed not to fail.

(a) Make a continuous-time Markov chain presentation of this problem.
(b) What is the infinitesimal generator matrix Q for this Markov chain?
(c) Compute the steady-state of this process.
(d) What is the steady-state connection availability?

11

Applications of Markov chains

This chapter illustrates the theory of Markov chains with several examples. Applications of Markov theory to queueing problems and epidemics on networks are deferred to later chapters. Generally, Markov processes can be solved explicitly provided the transition probability matrix P or the infinitesimal generator Q has a special structure. Only in a very small number of problems is the entire time dependence of the process available in analytic form.

11.1 Discrete Markov chains and independent random variables

This section gives examples of some simple Markov chains. Consider a set $\{Y_n\}_{n\geq 1}$ of positive integer, independent random variables that are identically distributed with $\Pr[Y = k] = a_k$.

The discrete-time Markov process, defined by $X_n = Y_n$ for $n \geq 1$, possesses the (infinite) transition probability matrix,

$$
P = \begin{bmatrix}
a_1 & a_2 & a_3 & a_4 & \cdots \\
a_1 & a_2 & a_3 & a_4 & \cdots \\
a_1 & a_2 & a_3 & a_4 & \cdots \\
a_1 & a_2 & a_3 & a_4 & \cdots \\
\cdots & \cdots & \cdots & \cdots & \cdots
\end{bmatrix}
$$

All rows are identical and $\Pr[X_{n+1} = j | X_n = i] = a_j$ shows that the states X_{n+1} and X_n are independent from each other.

Another, more interesting, discrete-time Markov process is defined by

$$
X_n = \max[Y_1, Y_2, Y_3, \ldots, Y_n].
$$

Hence, the process X_n mirrors the maxima of the first n random variables. Clearly, $X_{n+1} = \max[X_n, Y_{n+1}]$ reflects the Markov property: the next state is only dependent on the previous state of the process. From $X_{n+1} = \max[X_n, Y_{n+1}]$, we observe that $\Pr[X_{n+1} = j | X_n = i] = 0$ if $j < i$ because the maximum does not decrease by adding a new random variable in the list. If $j > i$, the state j is determined by

$Y_{n+1} = j$ with probability a_j. If $j = i$, then $Y_{n+1} \leq j$, which has probability

$$\Pr\left[Y_{n+1} \leq j\right] = \sum_{k=1}^{j} \Pr\left[Y_{n+1} = k\right] = \sum_{k=1}^{j} a_k = A_j$$

The corresponding probability matrix is

$$P = \begin{bmatrix} A_1 & a_2 & a_3 & a_4 & \cdots \\ 0 & A_2 & a_3 & a_4 & \cdots \\ 0 & 0 & A_3 & a_4 & \cdots \\ 0 & 0 & 0 & A_4 & \cdots \\ \cdots & \cdots & \cdots & \cdots & \cdots \end{bmatrix}$$

A related discrete-time Markov chain is $X_n = \sum_{k=1}^{n} Y_k$, which obeys $X_{n+1} = X_n + Y_{n+1}$. Furthermore, if $j \leq i$, $\Pr\left[X_{n+1} = j | X_n = i\right] = 0$ because the random variables Y_n are non-negative such that the sum cannot decrease by adding a new member. If $j > i$, then

$$\Pr\left[X_{n+1} = j | X_n = i\right] = \Pr\left[X_n + Y_{n+1} = j | X_n = i\right]$$
$$= \Pr\left[Y_{n+1} = j - i\right] = a_{j-i}$$

The corresponding probability matrix has a Toeplitz structure possessing the same elements on diagonal lines

$$P = \begin{bmatrix} 0 & a_1 & a_2 & a_3 & \cdots \\ 0 & 0 & a_1 & a_2 & \cdots \\ 0 & 0 & 0 & a_1 & \cdots \\ 0 & 0 & 0 & 0 & \cdots \\ \cdots & \cdots & \cdots & \cdots & \cdots \end{bmatrix}$$

This list can be extended by considering other integer functions of the set $\{Y_n\}_{n \geq 1}$ (such as $X_{n+1} = \min\left[X_n, Y_{n+1}\right]$ or $X_n = \prod_{k=1}^{n} Y_k$ etc.).

11.2 The general random walk

The general random walk is an important model that describes the motion of an item that is constrained to move either one step forwards, stay at the position where it currently is or move one step backwards. In general, this "three-possibility motion" has transition probabilities that depend on the position j. Fig. 11.1 illustrates that, if the process is in state j, it has three possible choices: remain in state j with probability $r_j = \Pr\left[X_{k+1} = j | X_k = j\right]$, move to the next state $j + 1$ with probability $p_j = \Pr\left[X_{k+1} = j + 1 | X_k = j\right]$ or jump back to state $j - 1$ with probability $q_j = \Pr\left[X_{k+1} = j - 1 | X_k = j\right]$. A general random walk is defined by the $(N+1) \times (N+1)$ band matrix

$$P = \begin{bmatrix} r_0 & p_0 & 0 & 0 & \cdots & 0 & 0 & 0 \\ q_1 & r_1 & p_1 & 0 & \cdots & 0 & 0 & 0 \\ 0 & q_2 & r_2 & p_2 & \cdots & 0 & 0 & 0 \\ \vdots & \vdots & \vdots & \vdots & \vdots & \vdots & \vdots & \vdots \\ 0 & 0 & 0 & 0 & \cdots & q_{N-1} & r_{N-1} & p_{N-1} \\ 0 & 0 & 0 & 0 & \cdots & 0 & q_N & r_N \end{bmatrix} \qquad (11.1)$$

where $p_j > 0$, $q_j > 0$, $r_j \geq 0$ and $q_j + r_j + p_j = 1$ for all $0 < j < N$. The bordering states zero and N are special: $p_0 \geq 0$, $r_0 \geq 0$ and $p_0 + r_0 = 1$ and $q_N \geq 0$, $r_N \geq 0$ and $q_N + r_N = 1$.

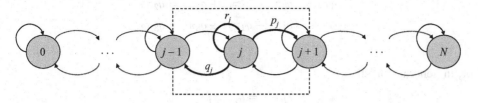

Fig. 11.1. A transition graph of the general random walk.

The general random walk serves as model for a number of phenomena:

- The one-dimensional motion of physical particles, electrons that hop from one atom to another. In this case, the number of states N can be very large.
- The gambler's ruin problem[1]: a state j reflects the capital of a gambler, where p_j is the chance that the gambler wins and q_j is the probability that he loses. The gambler achieves his target when he reaches state N, but he is ruined at state 0. In that case, both states are absorbing states with $r_0 = r_N = 1$. In games, most often the probabilities are independent of the state and simplify to $p_j = p$, $q_j = q$ and $r_j = 1 - p - q$.
- The continuous-time counterpart, the birth and death process (Section 11.3), has applications to queueing processes. For a wealth of examples and applications of the random walk, we refer to the classical treatise of Feller (1970, Chapters III and XIV).

11.2.1 The steady-state

The steady-state equation (9.29) for the vector component π_j becomes, for $1 \leq j < N$,

$$\pi_j = p_{j-1}\pi_{j-1} + r_j\pi_j + q_{j+1}\pi_{j+1} \qquad (11.2)$$

---

[1] Another variant is a game with two players a and b in which player a starts with capital j and has winning chance of p, while player b starts with capital $N - j$ and wins with probability $q = 1 - p$.

and, for $j = 0$ and $j = N$,

$$\pi_0 = r_0 \pi_0 + q_1 \pi_1$$
$$\pi_N = p_{N-1} \pi_{N-1} + r_N \pi_N$$

We rewrite these equations using $r_j = 1 - q_j - p_j$, $r_0 = 1 - p_0$ and $r_N = 1 - q_N$ as

$$p_0 \pi_0 = q_1 \pi_1 \tag{11.3}$$
$$p_j \pi_j = (p_{j-1} \pi_{j-1} - q_j \pi_j) + q_{j+1} \pi_{j+1} \tag{11.4}$$
$$p_{N-1} \pi_{N-1} = q_N \pi_N$$

Explicitly, for a few values of j in (11.4), we observe that

$$p_1 \pi_1 = (p_0 \pi_0 - q_1 \pi_1) + q_2 \pi_2 = q_2 \pi_2$$
$$p_2 \pi_2 = (p_1 \pi_1 - q_2 \pi_2) + q_3 \pi_3 = q_3 \pi_3$$
$$\ldots$$

or, in general, for all j,

$$p_j \pi_j = q_{j+1} \pi_{j+1}$$

Before continuing, we note that the above equation can be found at once by applying the balance equations to a particular set of nodes as explained in Section 10.3. Indeed, consider a cut in the Markov graph of Fig. 11.1 that separates the states/nodes 0 up to i, forming set S_1, from the states $i + 1$ to N, that constitute set S_2. The probability flux from S_1 towards S_2 is $p_j \pi_j$, while that from S_2 towards S_1 is $q_{j+1} \pi_{j+1}$. The probability flux traversing the cut needs to be in balance in the steady-state so that $p_j \pi_j = q_{j+1} \pi_{j+1}$. The nature of the Markov graph in Fig. 11.1, which is a directed path with self-loops, further shows that such a cut can be made between any consecutive pair of nodes, which establishes the truth of $p_j \pi_j = q_{j+1} \pi_{j+1}$ for all nodes j in the Markov graph.

By iteration of $\pi_{j+1} = \frac{p_j}{q_{j+1}} \pi_j$ starting at $j = 0$, we find

$$\pi_{j+1} = \frac{p_j}{q_{j+1}} \frac{p_{j-1}}{q_j} \cdots \frac{p_0}{q_1} \pi_0 = \pi_0 \prod_{m=0}^{j} \frac{p_m}{q_{m+1}}$$

The normalization $\|\pi\|_1 = 1$ yields a condition for π_0,

$$\pi_0 = \left(1 + \sum_{j=1}^{N} \prod_{m=0}^{j-1} \frac{p_m}{q_{m+1}} \right)^{-1} \tag{11.5}$$

which determines the complete steady-state vector π for the general random walk as

$$\pi_j = \frac{\prod_{m=0}^{j-1} \frac{p_m}{q_{m+1}}}{1 + \sum_{k=1}^{N} \prod_{m=0}^{k-1} \frac{p_m}{q_{m+1}}} \tag{11.6}$$

These relations remain valid even when the number of states N tends to infinity provided the infinite sum converges.

In the simple case, where $p_m = p$ and $q_{m+1} = q$ for all $m \geq 0$, we obtain from (11.6) with $\rho = \frac{p}{q}$,

$$\pi_j = \frac{(1-\rho)\rho^j}{1-\rho^{N+1}} \tag{11.7}$$

The spectral decomposition of the matrix P in (11.1) is computed in Appendix A.6.3, where it is shown that all eigenvalues of P are real and lie within the interval $(-1, 1]$.

11.2.2 The probability of gambler's ruin

The probability of gambler's ruin is defined as $u_j = \Pr[X_T = 0 | X_0 = j]$ where $T = \min_k \{X_k = 0\}$ is the hitting time to state 0, which is equivalent to $u_j = \Pr[T_0 < \infty | X_0 = j]$. By definition, $u_0 = 1$ and since the gambler achieves his goal at state N, he stops and never gets ruined, $u_N = 0$. The law of total probability (2.49) gives the situation after the first transition,

$$\Pr[X_T = 0 | X_0 = j] = \sum_{k=0}^{N} \Pr[X_T = 0 | X_0 = j, X_1 = k]\Pr[X_1 = k | X_0 = j]$$

Invoking the Markov property (9.1),

$$\Pr[X_T = 0 | X_0 = j, X_1 = k] = \Pr[X_T = 0 | X_1 = k]$$

yields

$$\Pr[X_T = 0 | X_0 = j] = \sum_{k=0}^{N} \Pr[X_T = 0 | X_1 = k]P_{jk}$$
$$= q_j \Pr[X_T = 0 | X_1 = j-1] + r_j \Pr[X_T = 0 | X_1 = j]$$
$$+ p_j \Pr[X_T = 0 | X_1 = j+1]$$

After the first transition, the probability $\Pr[X_T = 0 | X_1 = j] = u_j$ remains the same as the initial $\Pr[X_T = 0 | X_0 = j]$ because T is a random variable depending on the state and not on the discrete-time. Hence, we obtain the equations

$$u_0 = 1$$
$$u_j = q_j u_{j-1} + r_j u_j + p_j u_{j+1} \qquad (1 \leq j < N)$$

which are different from the corresponding steady-state equations (11.2). The difference lies in left-multiplication of P, yielding $u = Pu$ instead of right-multiplication in $\pi = \pi P$. Substituting $r_j = 1 - p_j - q_j$ gives after some modification

$$u_{j+1} = -\frac{q_j}{p_j}u_{j-1} + \left(1 + \frac{q_j}{p_j}\right)u_j \tag{11.8}$$

Iteration on j for the first few values using $u_0 = 1$ yields

$$u_2 = -\frac{q_1}{p_1} + \left(1 + \frac{q_1}{p_1}\right) u_1$$

$$u_3 = -\frac{q_2}{p_2} u_1 + \left(1 + \frac{q_2}{p_2}\right) u_2 = -\left(\frac{q_1}{p_1} + \frac{q_1 q_2}{p_1 p_2}\right) + \left(1 + \frac{q_1}{p_1} + \frac{q_1 q_2}{p_1 p_2}\right) u_1$$

which suggests

$$u_j = -\sum_{k=1}^{j-1} \prod_{m=1}^{k} \frac{q_m}{p_m} + \left(1 + \sum_{k=1}^{j-1} \prod_{m=1}^{k} \frac{q_m}{p_m}\right) u_1$$

as readily verified by substitution in (11.8). The unknown u_1 is determined by the last relation, $u_N = 0$. Finally, the probability of gambler's ruin is

$$\Pr\left[X_T = 0 | X_0 = j\right] = \frac{\sum_{k=j}^{N-1} \prod_{m=1}^{k} \frac{q_m}{p_m}}{1 + \sum_{k=1}^{N-1} \prod_{m=1}^{k} \frac{q_m}{p_m}} \tag{11.9}$$

Similarly, the mean hitting time $\eta_j = E\left[T | X_0 = j\right]$ follows by reasoning on the possible transitions. From state j, there is a transition to state $j-1$ with probability q_j. In case of a transition from state j to state $j - 1$, the hitting time T consists of the first transition plus the remaining time from state $j - 1$ on which is η_{j-1}. Using the law of total probability or reading all possible transitions from the transition graph (see Fig. 11.1), we find

$$\eta_j = 1 + q_j \eta_{j-1} + r_j \eta_j + p_j \eta_{j+1}$$

The boundary equations for the state 0 and state N (which are absorbing states in the gambler's ruin problem) are $\eta_0 = 0$ and $\eta_N = 0$ because, if $X_0 = N$, the gambler has reached his goal and does not play. In that case, there is no game nor hitting time and the event $\{T | X_0 = N\}$ does not exist. With $r_j = 1 - p_j - q_j$,

$$\eta_{j+1} = -\frac{1}{p_j} - \frac{q_j}{p_j} \eta_{j-1} + \left(1 + \frac{q_j}{p_j}\right) \eta_j$$

By iteration,

$$\eta_2 = -\frac{1}{p_1} + \left(1 + \frac{q_1}{p_1}\right) \eta_1$$

$$\eta_3 = -\frac{1}{p_2} - \frac{q_2}{p_2} \eta_1 + \left(1 + \frac{q_2}{p_2}\right) \eta_2 = -\frac{1}{p_1} - \frac{1}{p_2}\left(1 + \frac{q_2}{p_1}\right) + \left(1 + \frac{q_1}{p_1} + \frac{q_1 q_2}{p_1 p_2}\right) \eta_1$$

$$\eta_4 = -\frac{1}{p_3} - \frac{q_3}{p_3} \eta_2 + \left(1 + \frac{q_3}{p_3}\right) \eta_3$$

$$= -\frac{1}{p_1} - \frac{1}{p_2}\left(1 + \frac{q_2}{p_1}\right) - \frac{1}{p_3}\left(1 + \frac{q_3}{p_2} + \frac{q_2 q_3}{p_1 p_2}\right) + \left(1 + \frac{q_1}{p_1} + \frac{q_1 q_2}{p_1 p_2} + \frac{q_1 q_2 q_3}{p_1 p_2 p_3}\right) \eta_1$$

or

$$\eta_j = -\sum_{k=1}^{j-1} \frac{1}{p_k}\left(1 + \sum_{n=1}^{k-1} \prod_{m=1}^{n} \frac{q_{k-m+1}}{p_{k-m}}\right) + \left(1 + \sum_{k=1}^{j-1} \prod_{m=1}^{k} \frac{q_m}{p_m}\right) \eta_1$$

Eliminating η_1 from $\eta_N = 0$, finally leads to the mean hitting time to ruin or the mean duration of the game

$$
\eta_j = -\sum_{k=1}^{j-1} \frac{1}{p_k} \left(1 + \sum_{n=1}^{k-1} \prod_{m=1}^{n} \frac{q_{k-m+1}}{p_{k-m}} \right)
$$

$$
+ \left(1 + \sum_{k=1}^{j-1} \prod_{m=1}^{k} \frac{q_m}{p_m} \right) \frac{\sum_{k=1}^{N-1} \frac{1}{p_k} \left(1 + \sum_{n=1}^{k-1} \prod_{m=1}^{n} \frac{q_{k-m+1}}{p_{k-m}} \right)}{1 + \sum_{k=1}^{N-1} \prod_{m=1}^{k} \frac{q_m}{p_m}} \tag{11.10}
$$

In spite of the relatively simple difference equations, the solution rapidly grows unattractively. The particular case of the Markov chain where $q_k = q$ and $p_k = p$ simplifies considerably. The probability of gambler's ruin (11.9) becomes

$$
\Pr\left[X_T = 0 | X_0 = j\right] = \frac{\sum_{k=j}^{N-1} \left(\frac{q}{p} \right)^k}{\sum_{k=0}^{N-1} \left(\frac{q}{p} \right)^k} = \frac{\left(\frac{q}{p} \right)^j - \left(\frac{q}{p} \right)^N}{1 - \left(\frac{q}{p} \right)^N}
$$

and, if $p = q = \frac{1}{2}$, via de l'Hospital's rule, $\Pr\left[X_T = 0 | X_0 = j\right] = \frac{N-j}{N}$. If the target fortune N (at which the game ends) is infinitely large and for a fixed j (finite starting capital),

$$
\Pr\left[X_T = 0 | X_0 = j\right] = 1 \qquad \text{if } q \geq p
$$

$$
= \left(\frac{q}{p} \right)^j \qquad \text{if } q < p
$$

which demonstrates that the gambler surely will loose all his money if his chances p on winning are smaller than those q on losing. Even in a fair game where $p = q$, he will be defeated surely. Making profit with probability 1 in a fair game is a strong business advantage of a casino. In a favorable game ($p > q$) and with start capital j, ruin is possible with probability $\left(\frac{q}{p} \right)^j$.

Similarly, the mean duration of the game (11.10) simplifies to

$$
\eta_j = -\sum_{k=1}^{j-1} \left(\frac{1 - \left(\frac{q}{p} \right)^k}{p - q} \right) + \left(\frac{1 - \left(\frac{q}{p} \right)^j}{1 - \left(\frac{q}{p} \right)^N} \right) \sum_{k=1}^{N-1} \frac{1 - \left(\frac{q}{p} \right)^k}{p - q}
$$

or

$$
E\left[T | X_0 = j\right] = \frac{1}{p - q} \left[N \left(\frac{1 - \left(\frac{q}{p} \right)^j}{1 - \left(\frac{q}{p} \right)^N} \right) - j \right]
$$

11.3 Birth and death process

A birth and death process[2] is defined by the infinitesimal generator matrix

$$
Q = \begin{bmatrix}
-\lambda_0 & \lambda_0 & 0 & 0 & 0 & 0 & 0 & \cdots \\
\mu_1 & -(\lambda_1+\mu_1) & \lambda_1 & 0 & 0 & 0 & 0 & \cdots \\
0 & \mu_2 & -(\lambda_2+\mu_2) & \lambda_2 & 0 & 0 & 0 & \cdots \\
0 & 0 & \mu_3 & -(\lambda_3+\mu_3) & \lambda_3 & 0 & 0 & \cdots \\
0 & 0 & 0 & \mu_4 & -(\lambda_4+\mu_4) & \lambda_4 & 0 & \cdots \\
\vdots & \vdots & \vdots & \vdots & \vdots & \vdots & \vdots & \vdots
\end{bmatrix}
$$

The transition graph is shown in Fig. 11.2. Although the theory in the previous chapter was derived for finite-state Markov chains, the birth and death process is a generalization to an infinite number of states. The general random walk (Section 11.2) forms the embedded Markov chain of the birth and death process with transition probabilities specified by (10.23) resulting in $V_{i,i-1} = \frac{\mu_i}{\lambda_i+\mu_i}$, $V_{i,i+1} = \frac{\lambda_i}{\lambda_i+\mu_i}$ and $V_{ik} = 0$ when $k \neq i-1$ and $k \neq i+1$. The transition probability matrix is a tri-band diagonal matrix which is irreducible if all $\lambda_j > 0$ and $\mu_j > 0$.

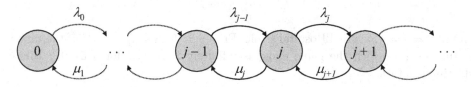

Fig. 11.2. The transition graph of a birth and death process.

The basic system of differential equations that completely describes the birth and death process follows from the general state probability equations (10.11) for $s_k(t) = \Pr[X(t) = k]$ as

$$s_0'(t) = -\lambda_0 s_0(t) + \mu_1 s_1(t) \tag{11.11}$$

$$s_k'(t) = -(\lambda_k + \mu_k) s_k(t) + \lambda_{k-1} s_{k-1}(t) + \mu_{k+1} s_{k+1}(t) \tag{11.12}$$

with initial condition $s_k(0) = \Pr[X(0) = k]$. Exact analytic solutions for any λ_k and μ_k are not possible. Indeed, let us denote the Laplace transform of $s_k(t)$ by

$$S_k(z) = \int_0^\infty e^{-zt} s_k(t)dt. \tag{11.13}$$

Since $s_k(t)$ is a continuous and bounded function ($|s_k(t)| \leq 1$ for all $t > 0$), the Laplace transform exists for $\mathrm{Re}(z) > 0$. The Laplace transform of (11.11) and (11.12) becomes

$$(\lambda_0 + z) S_0(z) = s_0(0) + \mu_1 S_1(z) \tag{11.14}$$

$$(\lambda_k + \mu_k + z) S_k(z) = s_k(0) + \lambda_{k-1} S_{k-1}(z) + \mu_{k+1} S_{k+1}(z) \tag{11.15}$$

---

[2] Kleinrock (1975, p. ix) mentions that William Feller was the father of the birth and death process.

which is a set of difference equations more complex due to the initial condition $s_k(0)$ than the set studied in Appendix A.6.3.

The infinite set (11.11) and (11.12) of differential equations has been thoroughly studied over years under several simplifying conditions for λ_k and μ_k, for example, $\lambda_k = \lambda$ and $\mu_k = \mu$ for all k. As shown in Chapter 13, they form the basis for the simplest set of queueing models of the family M/M/m/K.

11.3.1 The steady-state

The steady-state follows from (10.21) as solution of the set

$$-\lambda_0 \pi_0 + \mu_1 \pi_1 = 0$$

$$\lambda_{j-1} \pi_{j-1} - (\lambda_j + \mu_j) \pi_j + \mu_{j+1} \pi_{j+1} = 0$$

This set is identical to (11.3) and (11.4) provided p_j and q_j are changed for λ_j and μ_j. After this modification in (11.6), the steady-state of the birth and death process is

$$\pi_0 = \frac{1}{1 + \sum_{k=1}^{\infty} \prod_{m=0}^{k-1} \frac{\lambda_m}{\mu_{m+1}}} \tag{11.16}$$

$$\pi_j = \frac{\prod_{m=0}^{j-1} \frac{\lambda_m}{\mu_{m+1}}}{1 + \sum_{k=1}^{\infty} \prod_{m=0}^{k-1} \frac{\lambda_m}{\mu_{m+1}}} \qquad j \geq 1 \tag{11.17}$$

Theorem 9.3.5 states that an irreducible Markov chain with a finite number of states is necessarily recurrent. However, it is in general difficult to decide whether an irreducible Markov chain with an infinite number of states is recurrent or transient. In case of the birth and death process, it is possible to determine when the process is transient or recurrent. The process is transient if and only if the embedded Markov chain (determined above) is transient. Section 9.2.4 discusses that, for a recurrent chain, $r_{ij} = \Pr[T_j < \infty | X_0 = i]$ equals 1: a finite hitting time means every state j is certainly visited, when starting from initial state i. Applied to the embedded Markov chain, it follows from the gambler's ruin (11.9) that

$$\Pr[T_0 < \infty | X_0 = j] = 1 - \frac{\sum_{k=0}^{j-1} \prod_{m=1}^{k} \frac{q_m}{p_m}}{\sum_{k=0}^{N-1} \prod_{m=1}^{k} \frac{q_m}{p_m}}$$

Thus, for any fixed initial state j, the condition for a recurrent chain

$$\Pr[T_j < \infty | X_0 = i] = 1$$

is, apart from $i = j$, only possible in the limit $N \to \infty$ if $\lim_{N\to\infty} \sum_{k=0}^{N-1} \prod_{m=1}^{k} \frac{q_m}{p_m} = \infty$. Transformed to the birth and death rates, the condition for recurrence becomes $\Sigma_2 = \sum_{k=1}^{\infty} \prod_{m=0}^{k-1} \frac{\mu_m}{\lambda_m} = \infty$. Furthermore, we observe from (11.17) that the infinite series $\Sigma_1 = \sum_{k=1}^{\infty} \prod_{m=0}^{k-1} \frac{\lambda_m}{\mu_{m+1}}$ must converge to have a stationary or steady-state distribution.

In summary, if $\Sigma_1 < \infty$ and $\Sigma_2 = \infty$, the birth and death process is positive recurrent. If $\Sigma_1 = \infty$ and $\Sigma_2 = \infty$, it is null recurrent. If $\Sigma_2 < \infty$, the birth and death process is transient.

11.3.2 A pure birth process

A pure birth process is defined as a process $\{X(t), t \geq 0\}$ for which in any state i it holds that $\mu_i = 0$. It follows from Fig. 11.2 that a birth process can only jump to higher states such that $P_{ij}(t) = 0$ for $j < i$. Similarly, in a pure death process $\{X(t), t \geq 0\}$ all birth rates $\lambda_i = 0$.

11.3.2.1 The Poisson process

Let us first consider the simplest case where all birth rates are equal $\lambda_i = \lambda$ and where $P_{jj}(t) = \Pr[\tau_j > t] = e^{-\lambda t}$ (see proof of Theorem 10.2.3). Using either the back or forward equation or (10.27) with $V_{i,i+1} = \delta_{i,i+1}$, yields

$$P_{ij}(t) = \delta_{ij} e^{-\lambda t} + \lambda \int_0^t e^{-\lambda u} P_{i+1,j}(t - u) du$$

or, for $j = i + k$ with $k > 0$

$$P_{i,i+k}(t) = \lambda e^{-\lambda t} \int_0^t e^{\lambda u} P_{i+1,i+k}(u) du$$

Explicitly, for $k = 1$ and using $P_{jj}(t) = e^{-\lambda t}$,

$$P_{i,i+1}(t) = \lambda e^{-\lambda t} \int_0^t e^{\lambda u} e^{-\lambda u} du = \lambda t e^{-\lambda t}$$

which is independent of i and, thus, it holds for any $i \geq 1$. For $k = 2$,

$$P_{i,i+2}(t) = \lambda e^{-\lambda t} \int_0^t e^{\lambda u} P_{i+1,i+2}(u) du = \lambda e^{-\lambda t} \int_0^t e^{\lambda u} \lambda u e^{-\lambda u} du = \frac{(\lambda t)^2}{2} e^{-\lambda t}$$

This suggests us to propose for any $i \geq 0$,

$$P_{i,i+k}(t) = \frac{(\lambda t)^k}{k!} e^{-\lambda t} \tag{11.18}$$

which is verified inductively as

$$P_{i,i+k}(t) = \lambda e^{-\lambda t} \int_0^t e^{\lambda u} P_{i+1,i+k}(u) du = \lambda e^{-\lambda t} \int_0^t e^{\lambda u} P_{i,i+k-1}(u) du$$

$$= \lambda e^{-\lambda t} \int_0^t e^{\lambda u} \frac{(\lambda u)^{k-1}}{(k-1)!} e^{-\lambda u} du = \frac{(\lambda t)^k}{k!} e^{-\lambda t}$$

Hence, the transition probabilities of a pure birth process have a Poisson distribution (11.18) and are only function of the difference in states $k = j - i \geq 0$ for any

$t \geq 0$. Moreover, for $0 \leq u \leq t$, consider the increment $X(t) - X(u)$,

$$
\Pr\left[X(t) - X(u) = k\right] = \sum_{i \geq 0} \Pr\left[X(u) = i, X(t) = i + k\right]
$$

$$
= \sum_{i \geq 0} \Pr\left[X(u) = i\right] \Pr\left[X(t) = i + k | X(u) = i\right]
$$

$$
= \sum_{i \geq 0} \Pr\left[X(u) = i\right] P_{i,i+k}(t - u)
$$

$$
= \frac{(\lambda(t-u))^k}{k!} e^{-\lambda(t-u)} \sum_{i \geq 0} \Pr\left[X(u) = i\right]
$$

Thus, the increment $X(t) - X(u)$ has a Poisson distribution,

$$
\Pr\left[X(t) - X(u) = k\right] = \frac{(\lambda(t-u))^k}{k!} e^{-\lambda(t-u)} \tag{11.19}
$$

and, since $X(0) = 0$ and the increments are independent (Markov property), we conclude that the pure birth process is a Poisson process (Section 7.2).

11.3.2.2 The general birth process

In case the birth rates λ_k depend on the actual state k, the pure birth process can be regarded as the simplest generalization of the Poisson. The Laplace transform difference equations (11.14) and (11.15) reduce to the set

$$
S_0(z) = \frac{s_0(0)}{\lambda_0 + z}
$$

$$
S_k(z) = \frac{s_k(0)}{\lambda_k + z} + \frac{\lambda_{k-1}}{\lambda_k + z} S_{k-1}(z)
$$

which, by the usual iteration, has the solution,

$$
S_k(z) = \sum_{j=0}^{k} \frac{s_j(0) \prod_{m=j}^{k-1} \lambda_m}{\prod_{m=j}^{k} (\lambda_m + z)} \tag{11.20}
$$

with the convention that $\prod_{m=a}^{b} f(m) = 1$ if $a > b$. The validity of this general solution is verified by substitution into the difference equation for $S_k(z)$. The form of $S_k(z)$ is a ratio that can always be transformed back to the time-domain provided that λ_k is known. If all $\lambda_k > 0$ are distinct, using (2.39) with $c > 0$, we find

$$
s_k(t) = \sum_{j=0}^{k} s_j(0) \prod_{m=j}^{k-1} \lambda_m \frac{1}{2\pi i} \int_{c-i\infty}^{c+i\infty} \frac{e^{zt}}{\prod_{m=j}^{k} (\lambda_m + z)} dz
$$

By closing the contour over the negative real plane $(\mathrm{Re}(z) < 0)$, only simple poles at $z = -\lambda_n$ are encountered,

$$
\frac{1}{2\pi i} \int_{c-i\infty}^{c+i\infty} \frac{e^{zt}}{\prod_{m=j}^{k} (\lambda_m + z)} dz = \sum_{n=j}^{k} \frac{e^{-\lambda_n t}}{\prod_{m=j; m \neq n}^{k} (\lambda_m - \lambda_n)}
$$

resulting in

$$s_k(t) = \sum_{j=0}^{k} s_j(0) \sum_{n=j}^{k} \frac{e^{-\lambda_n t} \prod_{m=j}^{k-1} \lambda_m}{\prod_{m=j;m\neq n}^{k} (\lambda_m - \lambda_n)} \tag{11.21}$$

If some $\lambda_k = \lambda_j$, multiple poles occur and a slightly more complex result appears that still can be computed in exact analytic form.

11.3.2.3 The Yule process

A classical example of a process with distinct birth rates is the Yule process, where $\lambda_k = k\lambda$. In that case, (11.21) can be simplified. With

$$\frac{\prod_{m=j}^{k-1} \lambda_m}{\prod_{m=j;m\neq n}^{k} (\lambda_m - \lambda_n)} = \frac{(k-1)!}{(j-1)! \prod_{m=j}^{n-1} (m-n) \prod_{m=n+1}^{k} (m-n)}$$

and with $\prod_{m=j}^{n-1} (m-n) = (-1)^{n-j} \prod_{l=1}^{n-j} l = (-1)^{n-j}(n-j)!$ and $\prod_{m=n+1}^{k} (m-n) = \prod_{l=1}^{k-n} l = (k-n)!$ we find

$$\frac{\prod_{m=j}^{k-1} \lambda_m}{\prod_{m=j;m\neq n}^{k} (\lambda_m - \lambda_n)} = \frac{(-1)^{n-j}(k-1)!}{(j-1)!(n-j)!(k-n)!}$$

such that

$$\sum_{n=j}^{k} \frac{e^{-\lambda_n t} \prod_{m=j}^{k-1} \lambda_m}{\prod_{m=j;m\neq n}^{k} (\lambda_m - \lambda_n)} = \frac{(k-1)!}{(j-1)!} \sum_{n=0}^{k-j} \frac{e^{-(n+j)\lambda t}(-1)^n}{n!(k-j-n)!}$$

$$= \frac{(k-1)!e^{-j\lambda t}}{(k-j)!(j-1)!} \sum_{n=0}^{k-j} \binom{k-j}{n} \left(-e^{-\lambda t}\right)^n$$

$$= \binom{k-1}{j-1} e^{-j\lambda t} \left(1 - e^{-\lambda t}\right)^{k-j}$$

Finally, for the Yule process, we obtain from (11.21) the evolution of the state probabilities over time

$$s_k(t) = \sum_{j=0}^{k} s_j(0) \binom{k-1}{j-1} e^{-j\lambda t} \left(1 - e^{-\lambda t}\right)^{k-j} \tag{11.22}$$

In practice, $s_j(0) = \delta_{jn}$ if the process starts from state n (implying $s_k(t) = 0$ for $k < n$ because the process moves to the right for $t \geq 0$) and the general form simplifies to

$$s_k(t) = \binom{k-1}{n-1} e^{-n\lambda t} \left(1 - e^{-\lambda t}\right)^{k-n} \tag{11.23}$$

The Yule process has been used as a simple model for the evolution of a population in which each individual gives birth at exponential rate λ and $X(t)$ denotes the number of individuals in the population (which never decreases as there are no deaths) as a function of time t. At each state k the population has precisely k individuals that each generate births at rate $\lambda_k = k\lambda$, the birth rate of the population. If the population starts at $t = 0$ with one individual $n = 1$, the evolution over time has the distribution $s_k(t) = e^{-\lambda t} \left(1 - e^{-\lambda t}\right)^{k-1}$, which is recognized from (3.5) as a geometric distribution with mean $e^{\lambda t}$. Since the sojourn times of a Markov process are i.i.d. exponential random variables, the mean time T_k to reach k individuals from one ancestor equals $E[T_k] = \sum_{j=1}^{k} \frac{1}{\lambda_j} = \frac{1}{\lambda} \sum_{j=1}^{k} \frac{1}{j}$ which is well approximated (Abramowitz and Stegun, 1968, Section 6.3.18) as $E[T_k] \approx \frac{\log(k+1)+\gamma}{\lambda}$, where $\gamma = 0.577\,215\ldots$ is Euler's constant. If the population starts with n individuals, the distribution (11.23) at time t consists of a sum of n i.i.d. geometric random variables, which is a negative binomial distribution. The Yule process has been employed as a crude model to estimate the spread of a disease and the split of molecules in new species by cosmic rays.

11.3.3 Constant rate birth and death process

In a constant rate birth and death process, both the birth rate $\lambda_k = \lambda$ and death rate $\mu_k = \mu$ are constant for any state k. From (11.17), the steady-state for all states j with $\rho = \frac{\lambda}{\mu} < 1$,

$$\pi_j = (1 - \rho)\,\rho^j \qquad\qquad j \geq 0 \qquad\qquad (11.24)$$

only depends on the ratio of birth over death rate. The time-dependent constant rate birth and death process can still be computed in analytic form. In this case, the matrix form of the infinitesimal generator Q has the tri-band Toeplitz structure, which can be diagonalized in analytic form as shown in Appendix A.6.2.1. In this section, we present an alternative approach. Instead of dealing with an infinite set of difference equations, a generating function approach seems more convenient. Let us denote the generating function of the Laplace transforms $S_k(z)$ by

$$\varphi(x, z) = \sum_{k=0}^{\infty} S_k(z) x^k \qquad\qquad (11.25)$$

Using (11.13) into (11.25) gives

$$\varphi(x, z) = \sum_{k=0}^{\infty} \int_0^{\infty} e^{-zt} s_k(t) x^k \, dt = \int_0^{\infty} e^{-zt} \sum_{k=0}^{\infty} s_k(t) x^k \, dt$$

where the reversal of summation and integration is allowed because all terms are positive. Since $0 \leq s_k(t) \leq 1$, the sum is at least convergent for $|x| < 1$, which shows that $\varphi(x, z)$ is analytic inside the unit circle $|x| < 1$ for any $\mathrm{Re}(z) > 0$.

After multiplying (11.15) by x^k and summing over all k, we obtain

$$(\lambda + z) S_0(z) + (\lambda + \mu + z) \sum_{k=1}^{\infty} S_k(z) x^k = \sum_{k=0}^{\infty} s_k(0) x^k + \lambda \sum_{k=1}^{\infty} S_{k-1}(z) x^k$$

$$+ \mu S_1(z) + \mu \sum_{k=1}^{\infty} S_{k+1}(z) x^k$$

and, written in terms of $\varphi(x, z)$,

$$\varphi(x, z) = \frac{-\sum_{k=0}^{\infty} s_k(0) x^{k+1} + \mu (1 - x) S_0(z)}{\lambda x^2 - x (\lambda + \mu + z) + \mu}$$

Note that for general λ_k and μ_k, an expression in terms of $\varphi(x, z)$ is not possible. Before continuing with the computations, we make the additional simplification that $s_k(0) = \delta_{kj}$: we assume that the constant rate birth and death process starts in state j. With this initial condition, the generating function

$$\varphi(x, z) = \frac{\mu (1 - x) S_0(z) - x^{j+1}}{\lambda x^2 - x (\lambda + \mu + z) + \mu} \tag{11.26}$$

still depends on the unknown function $S_0(z)$. The following derivation involving the theory of complex functions demonstrates a standard procedure that will also be useful in other problems.

The denominator in (11.26) has two roots,

$$x_1 = \frac{\lambda + \mu + z}{2\lambda} + \frac{1}{2\lambda} \sqrt{(\lambda + \mu + z)^2 - 4\lambda\mu}$$

$$x_2 = \frac{\lambda + \mu + z}{2\lambda} - \frac{1}{2\lambda} \sqrt{(\lambda + \mu + z)^2 - 4\lambda\mu}$$

We need the powerful theorem of Rouché (Titchmarsh, 1964, p. 116) to deduce more on the location of x_1 and x_2.

Theorem 11.3.1 (Rouché) *If $f(z)$ and $g(z)$ are analytic inside and on a closed contour C, and $|g(z)| < |f(z)|$ on C, then $f(z)$ and $f(z) + g(z)$ have the same number of zeros inside C.*

Choose $f(x) = \mu - x (\lambda + \mu + z)$ and $g(x) = \lambda x^2$ such that $f(x) + g(x) = \lambda x^2 - x (\lambda + \mu + z) + \mu$, the denominator in (11.26). Since both $f(x)$ and $g(x)$ are polynomials, they are analytic everywhere in the complex x-plane. We know that $\varphi(x, z)$ is analytic inside the unit disk. If the roots x_1 or x_2 lie inside the unit disk, the numerator in (11.26) must have zeros at precisely the same place in order for $\varphi(x, z)$ to be analytic inside the unit disk. Hence, we consider as contour C in Rouché's Theorem, the unit circle $|x| = 1$. Clearly, $f(x)$ has one single zero $\frac{\mu}{\lambda + \mu + z}$ inside the unit circle (because $\lambda > 0$, $\mu > 0$ and $\mathrm{Re}(z) > 0$). Furthermore, on the unit circle $|x| = 1$,

$$|\mu - x (\lambda + \mu + z)| \geq |\mu - |x| |(\lambda + \mu + z)|| = |\mu - |(\lambda + \mu + z)|| > \lambda = \lambda |x^2|$$

which shows that $|g(z)| < |f(z)|$ on the unit circle. Rouché's Theorem then tells us that $f(x) + g(x)$ has precisely one zero inside the unit circle. This implies that $|x_1| > 1$ and $|x_2| < 1$ and that the numerator in (11.26) has a zero x_2,

$$\mu \left(1 - x_2\right) S_0(z) - x_2^{j+1} = 0$$

This relation determines the unknown function $S_0(z)$ as

$$S_0(z) = \frac{x_2^{j+1}}{\mu \left(1 - x_2\right)} \tag{11.27}$$

such that (11.26) becomes

$$\varphi(x, z) = \frac{x_2^{j+1} \frac{1-x}{1-x_2} - x^{j+1}}{\lambda \left(x - x_1\right)\left(x - x_2\right)} = \frac{x_2^{j+1} \left(1 - x\right) - x^{j+1} \left(1 - x_2\right)}{\lambda \left(1 - x_2\right)\left(x - x_1\right)\left(x - x_2\right)}$$

We know that the numerator can be divided by $(x - x_2)$, or explicitly,

$$x_2^{j+1} \left(1 - x\right) - x^{j+1} \left(1 - x_2\right) = x_2^{j+1} - x^{j+1} + x_2 x(x^j - x_2^j)$$

$$= (x - x_2) \left[-\sum_{k=0}^{j} x_2^{j-k} x^k + x_2 x \sum_{k=0}^{j-1} x_2^{j-1-k} x^k \right]$$

$$= (x - x_2) \left[-x_2^j + (x_2 - 1) \sum_{k=1}^{j} x_2^{j-k} x^k \right]$$

Finally,

$$\varphi(x, z) = \frac{-x_2^j + (x_2 - 1) \sum_{k=1}^{j} x_2^{j-k} x^k}{\lambda \left(1 - x_2\right)\left(x - x_1\right)} \tag{11.28}$$

By expanding the denominator in a Taylor series around $x = 0$ and denoting $a_0 = -x_2^j$, $a_k = (x_2 - 1)x_2^{j-k}$ for $1 \le k \le j$ and $a_k = 0$ for $k > j$, we have

$$\varphi(x, z) = \frac{-1}{\lambda x_1 \left(1 - x_2\right)} \sum_{k=0}^{\infty} a_k x^k \sum_{m=0}^{\infty} x_1^{-m} x^m = \frac{-1}{\lambda x_1 \left(1 - x_2\right)} \sum_{k=0}^{\infty} \left(\sum_{m=0}^{k} \frac{a_{k-m}}{x_1^m} \right) x^k$$

Comparing with (11.25) and equating the corresponding powers in x, we find an explicit form of the Laplace transforms of the probabilities that the birth and death process is in state $k \ge 0$,

$$S_k(z) = \frac{-1}{\lambda x_1 \left(1 - x_2\right)} \sum_{m=0}^{k} \frac{a_{k-m}}{x_1^m} = \frac{x_2^{j-k}}{\lambda x_1} \left(\frac{1}{\left(1 - x_2\right) x_1^k} + \sum_{m=0}^{k-1} \frac{x_2^k}{x_1^m} \right) 1_{k<j}$$

$$= \frac{1}{\lambda} \left(\frac{1}{\left(1 - x_2\right) x_1^{k+1}} + \frac{x_2^{j-k} - x_2^j x_1^{-k}}{x_1 - x_2} \right) 1_{k<j}$$

This expression can be put in different forms by using relations among the zeros x_1 and x_2, such as $x_1 + x_2 = \frac{\lambda + \mu + z}{\lambda}$ and $x_1 x_2 = \frac{\mu}{\lambda}$, from which, for $k = 0$, (11.27) is recovered. Some ingenuity is required to recognize in $S_k(z)$ a known Laplace transform. Otherwise, the inverse Laplace transform needs to be computed by

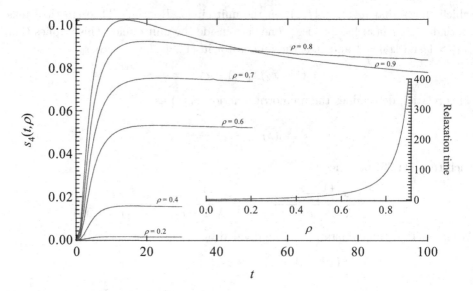

Fig. 11.3. The probability $s_4(t)$ that the process is in state 4 given that it started from state 0 as function of time (in units of average death time, $\mu = 1$) for various $\rho = \lambda$. The insert shows the relaxation time $\tau = (1 - \sqrt{\rho})^{-2}$ (in units of average death time, $\mu = 1$). The corresponding steady state probability π_4 are 0.0012, 0.015, 0.051, 0.072, 0.082, 0.065 for $\rho = 0.2, 0.4, 0.6, 0.7, 0.8, 0.9$ respectively. Observe that for $\rho = 0.9$, the plotted 100 time units are smaller than the relaxation time, which is 379 time units.

contour integration via (2.39). In any case, the computation needs advanced skills in complex function theory and we content ourselves here with presenting the result without derivation (see e.g. Cohen (1969, pp. 80–82)),

$$s_k(t) = e^{-(\lambda+\mu)t}\left[\rho^{(k-j)/2}I_{k-j}\left(at\right) + \rho^{(k-j-1)/2}I_{k+j+1}\left(at\right)\right] \qquad (11.29)$$

$$+ e^{-(\lambda+\mu)t}(1-\rho)\rho^k \sum_{m=k+j+2}^{\infty} \rho^{-m/2}I_m(at)$$

where $\rho = \frac{\lambda}{\mu}$, $a = 2\mu\sqrt{\rho}$ and where $I_s(z)$ denotes the modified Bessel function (Abramowitz and Stegun, 1968, Section 9.6.1). Using the asymptotic formulas for the modified Bessel function, the behavior of $s_k(t)$ for large t and $\rho < 1$ can be derived (see e.g. Cohen (1969, p. 84)),

$$s_k(t) = (1-\rho)\rho^k + \frac{\rho^{(k-j)/2}e^{-\left(1-\sqrt{\rho}\right)^2\mu t}}{2\sqrt{\pi}\left(\sqrt{\rho}\mu t\right)^{3/2}}\left[\left(k - \frac{\sqrt{\rho}}{1-\sqrt{\rho}}\right)\left(j - \frac{\sqrt{\rho}}{1-\sqrt{\rho}}\right) + O(t^{-1})\right]$$

and only if $\rho = 1$,

$$s_k(t) = \frac{1}{\sqrt{\pi\mu t}}\left[1 + O(t^{-1})\right]$$

This expression demonstrates that the constant rate birth-death process converges to the steady-state $(1-\rho)\rho^k$ with a relaxation rate $\left(1-\sqrt{\rho}\right)^2\mu$. Clearly, the higher ρ, the lower the relaxation rate and the slower the process tends to equilibrium as illustrated in Fig. 11.3. Intuitively, two effects play a role. Since the probability that states with large k are visited increases with increasing ρ, the built-up time for this occupation will be larger. In addition, the variability of the number of visited states (further derived for the M/M/1 queue in Section 14.1) increases with increasing ρ, which suggests that larger oscillations of the sample paths around the steady-state are likely to occur, enlarging the convergence time.

11.3.4 Linear rate birth and death process

We now consider a birth and death process where the birth rate $\lambda_k = \lambda(t)k$ and death rate $\mu_k = \mu(t)k$ are linear in k for any state k and depending on time t. This "linear" birth and death process generalizes the Yule process of Section 11.3.2.3, where $\mu(t) = 0$ and $\lambda(t) = \lambda$. The time-dependent evolution of the linear birth and death process is described from (11.11) and (11.12) by the following set of differential equations

$$s_0'(t) = \mu(t)s_1(t) \tag{11.30}$$

$$s_k'(t) = -\left(\lambda(t)+\mu(t)\right)ks_k(t) + \lambda(t)(k-1)s_{k-1}(t) + \mu(t)(k+1)s_{k+1}(t) \tag{11.31}$$

with the initial condition $s_k(0) = \delta_{km}$. The "linear" birth and death process can model the changes in the size $X(t)$ of a population, where births and deaths occur proportional to its size. Initially, at time $t = 0$, the size of the population is $X(0) = m$. We will first present the beautiful solution of the linear birth and death process due to Kendall (1948) and then discuss a few applications.

Similarly to the previous Section 11.3.3, we start by defining the probability generating function

$$\varphi(x,t) = \sum_{k=0}^{\infty} s_k(t)x^k \tag{11.32}$$

After multiplying (11.31) by x^k and summing over all k, we obtain[3] the first-order partial differential equation

$$\frac{\partial\varphi}{\partial t} = (x-1)\left(\lambda(t)x - \mu(t)\right)\frac{\partial\varphi}{\partial x} \tag{11.33}$$

with boundary condition $\varphi(x,0) = \sum_{k=0}^{\infty}\delta_{km}x^k = x^m$. When the transition rates are proportional to k^2, i.e. $\lambda_k = \lambda(t)k^2$ and $\mu_k = \mu(t)k^2$, a second-order partial differential equation is found, whose solution is more difficult than for first order. The equations for higher-order rates, where $\lambda_k = \lambda(t)k^r$ and $\mu_k = \mu(t)k^s$ (for integers s and $r > 2$), are expected to be analytically intractable.

The next step is to solve this partial differential equation (11.33) of first order.

---

[3] See Appendix B.10.

We start by briefly reviewing the theory of first-order partial differential equations and refer to the famous treatise by Courant and Hilbert (1953a) for an entire analysis. The most general first-order partial differential in two variables x and t has the form

$$a(x,t)\frac{\partial \varphi}{\partial x} + b(x,t)\frac{\partial \varphi}{\partial t} = c(x,t)\varphi + d(x,t) \qquad (11.34)$$

where a, b, c and d are given functions of x and t. Suppose that x and t are a function of the parameter s, then the chain rule of differentiable functions results in

$$\frac{d\varphi}{ds} = \frac{\partial \varphi}{\partial x}\frac{dx}{ds} + \frac{\partial \varphi}{\partial t}\frac{dt}{ds}$$

Comparison with the left-hand side of (11.34) shows that, when equating

$$\frac{dx}{ds} = a(x,t) \quad \text{and} \quad \frac{dt}{ds} = b(x,t)$$

these last equations specify a family of curves $(x(s), t(s))$ whose tangent vector $(x'(s), t'(s))$ precisely equals the vector (a, b) for each value of the parameter s (provided (a, b) is defined and non-zero). Moreover, we also have from (11.34) that

$$\frac{d\varphi}{ds} = c(x,t)\varphi + d(x,t)$$

The family of curves $(x(s), t(s), \varphi(s))$, determined by the ordinary differential equations, are called the characteristic curves of the partial differential equation (11.34). Subject to certain conditions for a, b, c and d, the existence and uniqueness theory of ordinary differential equations guarantees that exactly one characteristic curve $(x(s), t(s), \varphi(s))$ passes through a given point (x_0, t_0, φ_0). In most cases, we are not interested in the general solution of (11.34), but rather in a specific solution $\varphi(x, t)$ that contains a curve C given by the boundary condition. In summary, every surface $\varphi(x, t)$ generated by a one-parameter family of characteristic curves is a solution of the partial differential equation (11.34) and vice versa.

Applying the theory to (11.33), we first parametrize the initial curve C, determined by initial condition $\varphi(x, 0) = x^m$, by using the parameter τ as

$$x = \tau \quad t = 0 \quad \varphi = \tau^m$$

The set of ordinary differential equation corresponding to (11.33) is

$$\frac{dx}{ds} = -(x-1)(\lambda(t)x - \mu(t)) \quad \frac{dt}{ds} = 1 \quad \frac{d\varphi}{ds} = 0$$

whose solution is $t(s, \tau) = s$, $\varphi(s, \tau)$ is independent of s and thus equal to $\varphi(s, \tau) = \varphi(0, \tau) = \tau^m$. We need to solve

$$\frac{dx}{ds} = -\mu(t) + (\lambda(t) + \mu(t))x - \lambda(t)x^2$$

Using $t = s$, this differential equation becomes

$$\frac{dx}{dt} = -\mu(t) + (\lambda(t) + \mu(t))x - \lambda(t)x^2$$

which is a generalized Riccati equation, analyzed in Watson (1995, p. 92-94), who showed that the general solution is

$$x = x(s, \tau) = \frac{f_1(t) + cf_2(t)}{f_3(t) + cf_4(t)}$$

where f_1, f_2, f_3 and f_4 are functions of t to be determined from the Riccati differential equation and $c = x(0, \tau)$ is an integration constant. Since $c = x(0, \tau) = \tau$ (by our parametrization), we solve c from the general Riccati solution, substitute the result into $\varphi(s, \tau) = \tau^m$ and arrive at the general solution of (11.33)

$$\varphi(x, t) = \left(\frac{xf_3(t) - f_1(t)}{f_2(t) - xf_4(t)}\right)^m$$

We can divide both numerator and denominator by $f_2(t)$ and denote $h_j = \frac{f_j}{f_2}$ for $1 \leq j \leq 4$. Since

$\varphi(x,t)$ is a probability generating function, $\varphi(0,t) = 1$ so that $\frac{h_3(t)-h_1(t)}{1-h_4(t)} = 1$ or $h_3 = 1+h_1-h_4$. Hence, we have reduced the four unknown functions f_j $(1 \leq j \leq 4)$ to only two functions,

$$\frac{xf_3(t) - f_1(t)}{f_2(t) - xf_4(t)} = \frac{x(1+h_1-h_4) - h_1(t)}{1 - xh_4(t)}$$

To simplify the notation, we proceed with $f = -h_1$ and $g = h_4$, so that

$$\varphi(x,t) = \left(\frac{f(t) + x(1 - f(t) - g(t))}{1 - xg(t)}\right)^m \tag{11.35}$$

and we will determine f and g from the partial differential equation (11.33). The initial condition $\varphi(x,0) = x^m$ implies that $f(0) = 0$ and $g(0) = 0$. Substitution of the solution (11.35) into (11.33) leads, after a tedious computation, to

$$f' + x(g'f - gf' - f' - g') + x^2(g' - g'f + gf') = \mu(t)(fg + 1 - f - g)$$
$$- x(\lambda(t) + \mu(t))(fg + 1 - f - g)$$
$$+ x^2\lambda(t)(fg + 1 - f - g)$$

and equating corresponding powers in x yields

$$\begin{cases} f' = \mu(fg + 1 - f - g) \\ g'f - gf' - f' - g' = -(\lambda + \mu)(fg + 1 - f - g) \\ g' - g'f + gf' = \lambda(fg + 1 - f - g) \end{cases}$$

Since the second equation is minus the sum of the first and third one, we end up with two differential equations

$$\begin{cases} f' = \mu(1 - f)(1 - g) \\ gf' - g'f + g' = \lambda(1 - f)(1 - g) \end{cases}$$

Let $r(t) = 1 - f(t)$ and $q(t) = 1 - g(t)$, then

$$\begin{cases} -r' = \mu qr \\ -r' + qr' - q'r = \lambda rq \end{cases}$$

Substituting r' from the first equation into the second yields

$$\begin{cases} \frac{r'}{r} = -\mu q \\ q' = (\mu - \lambda)q - \mu q^2 \end{cases} \tag{11.36}$$

The last differential equation (of the Bernoulli type) can be transformed elegantly after letting $w = \frac{1}{q}$ into a linear first-order differential equation

$$w' + w(\mu - \lambda) = \mu$$

The initial conditions $f(0) = g(0) = 0$ lead to $r(0) = q(0) = w(0) = 1$ and the solution of the linear w-differential equation is

$$w(t) = e^{-\rho(t)}\left\{1 + \int_0^t e^{\rho(u)}\mu(u)\,du\right\} \tag{11.37}$$

where

$$\rho(t) = \int_0^t (\mu(u) - \lambda(u))\,du \tag{11.38}$$

Using the definition of $w = \frac{1}{q}$ and (11.37) into (11.36) yields

$$\frac{r'}{r} = -\frac{\mu}{w} = -\frac{w' + w(\mu - \lambda)}{w}$$

and, since $\rho'(t) = \mu(t) - \lambda(t)$, the logarithmic derivative $\frac{d}{dt}\ln r(t) = \frac{r'(t)}{r(t)}$ satisfies

$$\frac{r'}{r} = -\frac{w'}{w} - \rho'$$

After integration, using $r(0) = w(0) = 1$, we obtain

$$r = \frac{e^{-\rho}}{w}$$

Finally, we arrive with the definition of $\rho(t)$ in (11.38) at

$$f(t) = \frac{\int_0^t e^{\rho(u)}\mu(u)du}{1 + \int_0^t e^{\rho(u)}\mu(u)du} \quad \text{and} \quad g(t) = 1 - \frac{e^{\rho(t)}}{1 + \int_0^t e^{\rho(u)}\mu(u)du} \tag{11.39}$$

which completely determined the probability generating function $\varphi(x,t)$ in (11.35).

Next, we expand (11.35) in a Taylor series around $x = 0$, using $(1-x)^{-m} = \sum_{k=0}^{\infty}\binom{k+m-1}{m-1}x^k$ for $|x| < 1$,

$$\varphi(x,t) = (f(t) + x(1 - f(t) - g(t)))^m (1 - xg(t))^{-m}$$

$$= \sum_{l=0}^{\infty}\binom{m}{l}(1 - f(t) - g(t))^l f^{m-l}(t)x^l \sum_{k=0}^{\infty}\binom{k+m-1}{m-1}g^k(t)x^k$$

$$= \sum_{k=0}^{\infty}\left\{\sum_{l=0}^{k}\binom{m}{l}(1 - f(t) - g(t))^l f^{m-l}(t)\binom{k-l+m-1}{m-1}g^{k-l}(t)\right\}x^k$$

Using the definition (11.32) yields, after equating corresponding powers in x, the exact expression of the time-dependent probability $s_k(t) = \Pr[X(t) = k]$,

$$s_k(t) = f^m(t)g^k(t)\sum_{l=0}^{k}\binom{m}{l}\binom{k-l+m-1}{m-1}\left(\frac{1 - f(t) - g(t)}{f(t)g(t)}\right)^l \tag{11.40}$$

while the probability of extinction at time t equals

$$s_0(t) = \varphi(0,t) = f^m(t)$$

It follows from (11.39) that $f(t) > 0$, except if the death rate $\mu(t) = 0$. Thus, for any non-zero death rate $\mu(t)$, there is always a non-zero chance $s_0(t) = \Pr[X(t) = 0]$ that the population dies out. Furthermore, almost certain extinction of the population is only possible if the integral $\int_0^{\infty} e^{\rho(u)}\mu(u)\,du$ diverges. The mean $E[X(t)]$ of the population at time t follows from $\left.\frac{\partial\varphi}{\partial x}\right|_{x=1} = \sum_{k=0}^{\infty}ks_k(t)$ and (11.35) as

$$E[X(t)] = m\frac{1 - f(t)}{1 - g(t)} = me^{-\rho(t)} \tag{11.41}$$

while the variance is obtained from (2.27) and (11.35) as

$$\text{Var}[X(t)] = m\frac{(1 - f(t))(f(t) + g(t))}{(1 - g(t))^2} = me^{-2\rho(t)}\int_0^t e^{\rho(u)}\{\lambda(u) + \mu(u)\}\,du \tag{11.42}$$

The special case with time-independent birth and death rates, $\lambda(t) = \lambda$ and $\mu(t) = \mu$, for which $\rho(t) = (\mu - \lambda)t$, has a mean population size $me^{-(\mu-\lambda)t}$ and an extinction probability $s_0(t) = \left(\frac{\mu e^{(\mu-\lambda)t} - \mu}{\mu e^{(\mu-\lambda)t} - \lambda}\right)^m$. A particularly interesting observation is that, for $\lambda \le \mu$, $\lim_{t\to\infty} s_0(t) = 1$, implying that the population is almost certain to die out, even though in the critical case ($\lambda = \mu$), the expected population size equals m.

As an application of the linear birth and death process, Kendall (1948) mentions the cascade showers associated with cosmic radiation, in which the birth rate $\lambda(t) = \lambda$ and the death rate $\mu(t) = \mu t$, so that, from (11.38), $\rho(t) = \frac{1}{2}\mu t^2 - \lambda t$. The mean growth of this process is, with (11.41), Gaussian. The linear birth and death process can also model an epidemic in a population with infected and healthy, but susceptible individuals. Let $X(t)$ denote the size of the subpopulation of infected individuals. Each infected individual can infect a susceptible member of the population with infection rate $\lambda(t)$. Hence, given that $X(t) = k$, the infected subpopulation can grow at rate $k\lambda(t)$. At the same time, an infected individual can be cured at rate $\mu(t)$ and leave the infected subpopulation. At state k, the size of the infected subpopulation decreases at rate $k\mu(t)$. The overall dynamics obeys the linear birth and death equations (11.30) and (11.31). The major potential of the analytic solution of the linear birth and death process lies in the full generality of the time-dependence of both birth $\lambda(t)$ and death $\mu(t)$ rate, which allows us to fit epidemic data (see, e.g., van den Broek and Heesterbeek (2007)).

11.4 Slotted Aloha

The Aloha protocol is a basic example of a multiple access communication scheme of which Ethernet[4] is considered as the direct descendant. Aloha – which means "hello" in the Hawaiian language – was invented by Norman Abramson at the university of Hawaii in the beginning of 1970s to provide packet-switched radio communication between a central computer and various data terminals at the campus. Slotted Aloha is a discrete-time version of the pure Aloha protocol, where all transmitted packets have equal length and where each packet requires one timeslot for transmission.

Consider a network consisting of N nodes that can communicate with each other via a shared communication channel (e.g. a radio channel) using the slotted Aloha protocol. The simplest arrival process A of packets at each node is a Poisson process. We assume that these Poisson arrivals at a node are independent from the Poisson arrivals at another node and that all Poisson arrivals at a node have the same rate $\frac{\lambda}{N}$, where λ is the overall arrival rate at the network of N nodes. The idea of the Aloha protocol is that, upon receipt of a packet, the node transmits that newly arrived packet in the next timeslot. If two nodes happen to transmit a packet at the same timeslot, a collision occurs, which results in a retransmission of the packets. A node with a packet that must be retransmitted is said to be *backlogged*. Even if new packets arrive at a backlogged node, the retransmitted packet is the first one to be transmitted and, for simplicity (to ignore queueing of packets at a node), we assume that those new packets are discarded. If backlogged nodes retransmit the packet in

---

[4] The essential difference with the Ethernet's CSMA/CD (carrier sense multiple access with collision detection) is that Aloha does not use carrier sensing and does not stop transmitting when collisions are detected. Carrier sensing is only adequate if the nodes are near to each other (as in a local area network) such that collisions can be detected before the completion of transmission. Only then is a timely reaction possible.

the next timeslot, surely a new collision would occur. Therefore, backlogged nodes wait for some random number of timeslots before retransmitting. We assume, for simplicity, that p_r is the probability (which is the same for all backlogged nodes) that a transmission occurs in the next time slot. Moreover, the probability p_r of retransmission is the same for each timeslot. The number of time slots between the occurrence of a collision and a retransmission is a geometric random variable T_r (see Section 3.1.3) with parameter p_r such that $\Pr\left[T_r = k\right] = p_r \left(1 - p_r\right)^{k-1}$.

11.4.1 The Markov chain

The slotted Aloha protocol can be described by a discrete-time Markov chain X_k with state space $\{0, 1, \ldots, N\}$, where a state j counts the number of backlogged nodes out of the N nodes in total and the subscript k refers to the k-th timeslot. Each of the j backlogged nodes retransmits a packet in the next time slot with probability p_r, while each of the $N - j$ unbacklogged nodes will transmit surely a packet in the next time slot provided a packet arrives in the current timeslot. The latter event (at least one arrival A) occurs with probability $p_a = \Pr\left[A > 0\right] = 1 - \Pr\left[A = 0\right]$. If we assume that the arrival process is Poissonean, then $p_a = 1 - \exp\left(-\frac{\lambda}{N}\right)$, but the computations in this section are more generally valid.

The probability that n backlogged nodes in state j retransmit in the next time slot is binomially distributed

$$b_n(j) = \binom{j}{n} p_r^n \left(1 - p_r\right)^{j-n}$$

and, similarly, the probability that n unbacklogged nodes in state j transmit in the next time slot is

$$u_n(j) = \binom{N-j}{n} p_a^n \left(1 - p_a\right)^{N-j-n}$$

A packet is transmitted successfully if and only if (a) one new arrival and no backlogged packet or (b) no new arrival and one backlogged packet is transmitted. The probability of successful transmission in state j and per time slot equals

$$p_s(j) = u_1(j) b_0(j) + u_0(j) b_1(j)$$

The transition probability $P_{j,j+m}$ equals

$$P_{j,j+m} = \begin{cases} u_m(j) & 2 \leq m \leq N - j \\ u_1(j)\left(1 - b_0(j)\right) & m = 1 \\ u_1(j) b_0(j) + u_0(j)\left(1 - b_1(j)\right) & m = 0 \\ u_0(j) b_1(j) & m = -1 \end{cases}$$

The state with j backlogged nodes jumps to the state $j-1$ with one backlogged node less if no new packets are sent and there is precisely one successful retransmission. The state j remains in the state j if there is one new arrival and no retransmissions or if there are no new arrivals and none or more than one retransmissions. The state

j jumps to the state $j+1$ if there is one new arrival from a non-backlogged node and at least one retransmission because then there are surely collisions and the number of backlogged nodes increases by one. Finally, the state j jumps to the state $j+m$ if $m \geq 2$ new packets arrive from m different non-backlogged nodes, which always causes collisions irrespective of how many backlogged nodes also retransmit in the next time slot.

The Markov chain is illustrated in Fig. 11.4, which shows that the state can only decrease by at most 1.

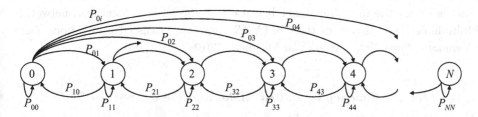

Fig. 11.4. Graph of the Markov chain for slotted Aloha. Each state j counts the number of backlogged nodes.

The transition probability matrix P has the structure

$$P = \begin{bmatrix} P_{00} & P_{01} & P_{02} & \cdots & & \cdots & & P_{0N} \\ P_{10} & P_{11} & P_{12} & \cdots & & \cdots & & P_{1N} \\ 0 & P_{21} & P_{22} & \cdots & & \cdots & & P_{2N} \\ \vdots & \vdots & \vdots & & \ddots & & \vdots & \vdots \\ 0 & 0 & \cdots & P_{N-1,N-2} & & P_{N-1;N-1} & & P_{N-1;N} \\ 0 & 0 & \cdots & & 0 & & P_{N;N-1} & P_{NN} \end{bmatrix}$$

whose eigenstructure is computed in Appendix A.6.4.

In the asymptotic regime when $N \to \infty$, slotted Aloha has the peculiar property that the steady-state vector π does not exist. Although for a small number of nodes N, the steady-state equations can be solved, when the number N grows, slotted Aloha turns out to be instable. It seems difficult to prove that $\lim_{N \to \infty} \pi = 0$, but there is another argument that suggests the truth of this awkward Aloha property. The expected change in backlog per time slot is equivalent to

$$E[X_{k+1} - X_k | X_k = j] = (N - j) p_a - p_s(j) \tag{11.43}$$

and equals the expected number of new arrivals minus the expected number of successful transmissions. This quantity $E[X_{k+1} - X_k | X_k = j]$ is often called the *drift*. If the drift is positive for all timeslots k, the Markov chain moves (on average) to higher states or to the right in Fig. 11.4. Since $p_s(j) \leq 1$ and $p_a = 1 - \exp\left(-\frac{\lambda}{N}\right)$, it follows that

$$\lim_{N \to \infty} E[X_{k+1} - X_k | X_k = j] = \infty$$

Thus, the drift tends to infinity, which means that, on average, the number of back-logged nodes increases unboundedly and suggests (but does not prove, a counter example is given in problem (ii) of Section 9.6) that the Markov chain is transient for $N \to \infty$.

A more detailed discussion and engineering approaches to cure this instability are found in Bertsekas and Gallager (1992, Chapter 4). The interest of the analysis of slotted Aloha lies in the fact that other types of multiple access protocols, such as the important class of carrier sense multiple access (CSMA) protocols, can be deduced in a similar manner. Of the CSMA class with collision detection, Ethernet is by far the most important because it is the basis of local area networks. Multiple access protocols of the CSMA/CD type are discussed in our book *Data Communications Networking* (Van Mieghem, 2010a).

11.4.2 Efficiency of slotted Aloha and the offered traffic G

We now investigate the probability of a successful transmission in state j in more detail,

$$
\begin{aligned}
p_s\left(j\right) &= u_1(j)b_0\left(j\right) + u_0\left(j\right)b_1\left(j\right) \\
&= \left(N-j\right)p_a\left(1-p_a\right)^{N-j-1}\left(1-p_r\right)^j + jp_r\left(1-p_r\right)^{j-1}\left(1-p_a\right)^{N-j} \\
&= \left[\frac{\left(N-j\right)p_a}{1-p_a} + \frac{jp_r}{1-p_r}\right]\left(1-p_a\right)^{N-j}\left(1-p_r\right)^j
\end{aligned}
$$

For small arrival probability p_a and small retransmission probability p_r, the probability of successful transmission in state j can be approximated by using the Taylor expansions of $(1-x)^\alpha = e^{\alpha \ln(1-x)} = e^{-\alpha x}\left(1+o\left(1\right)\right)$ and $\frac{x}{1-x} = x + o\left(x^2\right)$ as

$$
\begin{aligned}
p_s\left(j\right) &= \left[\left(N-j\right)p_a + jp_r + o\left(p_a^2 + p_r^2\right)\right]e^{-\left[(N-j)p_a+jp_r\right]}\left(1+o\left(1\right)\right) \\
&= \left[\left(N-j\right)p_a + jp_r\right]\exp\left(-\left[\left(N-j\right)p_a + jp_r\right]\right)\left(1+o\left(1\right)\right)
\end{aligned}
$$

Similarly, the probability that no packet is transmitted in state j equals

$$
\begin{aligned}
p_{no}\left(j\right) &= u_0\left(j\right)b_0\left(j\right) = \left(1-p_r\right)^j\left(1-p_a\right)^{N-j} \\
&= \exp\left(-\left[\left(N-j\right)p_a + jp_r\right]\right)\left(1+o\left(1\right)\right)
\end{aligned}
$$

Hence, for small p_a and small p_r, the probability of successful transmission and of no transmission in state j is well approximated by

$$
p_s\left(j\right) \simeq t\left(j\right)e^{-t(j)}
$$

$$
p_{no}\left(j\right) \simeq e^{-t(j)}
$$

Now, $t\left(j\right) = \left(N-j\right)p_a + jp_r$ is the expected number of arrivals and retransmissions in state j or, equivalently, the total rate of transmission attempts in state j. That total rate of transmissions in state j, $t(j)$, is also called the offered traffic G. The analysis shows that, for small p_a and small p_r, $p_s\left(j\right)$ and $p_{no}\left(j\right)$ are closely approximated in terms of a Poisson random variable with rate $t(j)$. Moreover, the probability of successful transmission $p_s(j)$ can be interpreted as the departure rate from state j or the throughput $S_{\text{SAloha}} = Ge^{-G}$, which is maximized if $G = t(j) = 1$. By controlling p_r to achieve $t(j) = (N-j)p_a + jp_r = 1$, slotted Aloha performs with highest throughput. The *efficiency* η_{SAloha} of slotted Aloha with many nodes $N > 1$ is defined as the maximum fraction of time during which packets are transmitted successfully which is $\max p_s(j) = e^{-1}$. Hence, $\eta_{\text{SAloha}} = 36\%$.

Pure Aloha (Bertsekas and Gallager, 1992), where the nodes can start transmitting at arbitrary times instead of only at the beginning of time slots, only performs half as efficiently as slotted Aloha with $\eta_{\text{PAloha}} = 18\%$. Recall that each packet is assumed to have an equal length that corresponds with the length of one timeslot. In pure Aloha, a transmitted packet at time t is successful if no other packet is sent during $(t-1, t+1)$. This time interval is precisely equal to two timeslots in slotted Aloha, which explains why $\eta_{\text{PAloha}} = \frac{1}{2}\eta_{\text{SAloha}}$. The same observation tells us that, in pure Aloha, $p_{no}(j) \simeq e^{-2t(j)}$ because in the successful interval the expected number of arrivals and retransmissions is twice that in slotted Aloha. The throughput S roughly equals the total rate of transmission attempts G (which is the same as in slotted Aloha) multiplied by $p_{no}(j) \simeq e^{-2t(j)}$, hence, $S_{\text{PAloha}} = Ge^{-2G}$.

11.5 Ranking of webpages

To retrieve webpages related to a user's query, current websearch engines first perform a search similar to that in text processors to find all webpages containing the query terms. Due to the massive size of the World Wide Web, this first action can result in a huge number of retrieved webpages. Several thousand webpages related to a query are not uncommon. To reduce the list of webpages, many search engines apply a ranking criterion to sort this list. In this section, we discuss PageRank, the hyperlink-based ranking system used by the Google search engine. PageRank elegantly exploits the power of discrete Markov theory.

11.5.1 A Markov model of the web

The hyperlink structure of the World Wide Web can be viewed as a directed graph with N nodes. Each node in the webgraph represents a certain webpage and the directed edges represent hyperlinks. Let us consider a small collection of webpages as in Fig. 11.5 to illustrate the underlying idea of PageRank, invented by Brin and Page, the founders of Google.

The topology of any graph is determined by its adjacency matrix. A reasonable criterion to assess the importance of a webpage is the number of times that this webpage is visited. This criterion suggests us to consider a discrete Markov chain whose transition probability matrix P corresponds to the adjacency matrix of the

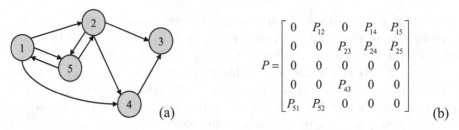

(a)   (b)

Fig. 11.5. A subgraph of the World Wide Web (a) and the corresponding transition probability matrix P (b).

webgraph (shown in Fig. 11.5 (b)). The element P_{ij} of the Markov transition probability matrix is the probability of moving from webpage i (state i) to webpage j (state j) in one time step. The component $s_i[k]$ of the corresponding state vector $s[k]$ denotes the probability that at time k the webpage i is visited. The long-run mean fraction of time that webpage i is visited equals the steady-state probability π_i of the Markov chain. This probability π_i is the ranking measure of the importance of webpage i used in Google. The basic idea is indeed simple, but we have not shown yet how to determine the elements P_{ij} nor whether the steady-state probability vector π exists. In particular, we will demonstrate that guaranteeing that the steady-state vector π exists and that π can be computed for a Markov chain containing some billion of states – the order of magnitude of the number N of webpages – requires a deeper knowledge of discrete Markov chains.

To start determining the elements P_{ij}, we assume that, given we are on webpage i, any hyperlink on that webpage has equal probability to be clicked on. This assumption implies that $P_{ij} = \frac{1}{d_i}$, where the degree d_i of a node i (see also Section 15.3) equals the number of adjacent neighbors of node i in the webgraph. This number d_i is thus equal to the number of hyperlinks on webpage i. The transition probability matrix in Fig. 11.5 then becomes

$$P = \begin{bmatrix} 0 & \frac{1}{3} & 0 & \frac{1}{3} & \frac{1}{3} \\ 0 & 0 & \frac{1}{3} & \frac{1}{3} & \frac{1}{3} \\ 0 & 0 & 0 & 0 & 0 \\ 0 & 0 & 1 & 0 & 0 \\ \frac{1}{2} & \frac{1}{2} & 0 & 0 & 0 \end{bmatrix}$$

The uniformity assumption is in most cases the best we can make if no additional information is available. If, for example, web usage information is available showing that a random surfer accessing page 2 is twice as likely to jump to page 4 than to any other neighboring webpage of 2, then the second row can be replaced by $\left(\begin{array}{ccccc} 0 & 0 & \frac{1}{4} & \frac{1}{2} & \frac{1}{4} \end{array} \right)$.

When solely adopting the adjacency matrix of the webgraph as underlying structure of the Markov transition probability matrix P, we cannot assure that P is a stochastic matrix. For example, it often occurs that a node such as node 3 in our example in Fig. 11.5 does not contain outlinks. Such nodes are called *dangling nodes*. For example, many webpages may point to an important document on the web, which itself does not refer to any other webpage. The corresponding row in P possesses only zero elements, which violates the basic law (9.8) of a stochastic matrix. To rectify the deviation from a stochastic matrix, each zero row must be replaced by a particular non-zero row vector[5] v^T that obeys (9.8), i.e. $\|v\|_1 = v^T u = 1$, where $u^T = [1 \ 1 \ \cdots \ 1]$. Again, the simplest recipe is to invoke uniformity and to replace any zero row by $v^T = \frac{u^T}{N}$. In our example, we replace the third row by ($\frac{1}{5}$ $\frac{1}{5}$ $\frac{1}{5}$ $\frac{1}{5}$ $\frac{1}{5}$) and obtain

$$\bar{P} = \begin{bmatrix} 0 & \frac{1}{3} & 0 & \frac{1}{3} & \frac{1}{3} \\ 0 & 0 & \frac{1}{3} & \frac{1}{3} & \frac{1}{3} \\ \frac{1}{5} & \frac{1}{5} & \frac{1}{5} & \frac{1}{5} & \frac{1}{5} \\ 0 & 0 & 1 & 0 & 0 \\ \frac{1}{2} & \frac{1}{2} & 0 & 0 & 0 \end{bmatrix}$$

However, this adjustment is not sufficient to insure the existence of a steady-state vector π. In Section 9.3.1, we have shown that, if the Markov chain is irreducible, the steady-state vector exists. In an irreducible Markov chain any state is reachable from any other (Section 9.2.2.1). By its very nature, the World Wide Web leads almost surely to a reducible Markov chain. In order to create an irreducible matrix, Brin and Page have considered

$$\bar{\bar{P}} = \alpha \bar{P} + (1 - \alpha) \frac{uu^T}{N}$$

where $0 < \alpha < 1$, $uu^T = J$ is a $N \times N$ matrix with each element equal to 1 and $\bar{P}$ is the previously adjusted matrix without zero-rows. The linear combination of the stochastic matrix $\bar{P}$ and a stochastic perturbation matrix J ensures that $\bar{\bar{P}}$ is an irreducible stochastic matrix. Every node is now directly connected (reachable in one step) to any other (because of J), which makes the Markov chain irreducible with aperiodic, positive recurrent states (see Fig. 9.1). Slightly more generally, we can replace the matrix $\frac{J}{N}$ by uv^T, where v is a probability vector as above but where we must additionally require that each component of v is non-zero in order to guarantee reachability. Brin and Page have called v^T the *personalization* vector, which enables us to deviate from non-uniformity. Hence, we arrive at the Brin and Page Markov transition probability matrix

$$\bar{\bar{P}} = \alpha \bar{P} + (1 - \alpha)uv^T \tag{11.44}$$

For $v^T = [\ \frac{1}{16} \ \ \frac{4}{16} \ \ \frac{6}{16} \ \ \frac{4}{16} \ \ \frac{1}{16} \ ]$ and $\alpha = \frac{4}{5}$, the probability transition matrix in our example becomes

---

[5] We use the normal vector algebra convention, but remark that the stochastic vectors π and $s[k]$ are *also* row vectors (without the transpose sign)!

$$\bar{\bar{P}} = \begin{bmatrix} \frac{1}{80} & \frac{19}{60} & \frac{3}{40} & \frac{19}{60} & \frac{67}{240} \\ \frac{1}{80} & \frac{1}{20} & \frac{41}{120} & \frac{19}{60} & \frac{67}{240} \\ \frac{1}{16} & \frac{1}{4} & \frac{3}{8} & \frac{1}{4} & \frac{1}{16} \\ \frac{1}{80} & \frac{1}{20} & \frac{7}{8} & \frac{1}{20} & \frac{1}{80} \\ \frac{33}{80} & \frac{9}{20} & \frac{3}{40} & \frac{1}{20} & \frac{1}{80} \end{bmatrix}$$

If the presented method were implemented, the initially very sparse matrix P would be replaced by the dense matrix $\bar{\bar{P}}$, which for the size N of the web would increase storage dramatically. Therefore, a more effective way is to define a special vector r whose component $r_j = 1$ if row j in P is a zero-row or node j is dangling node. Then, $\bar{P} = P + rv^T$ is a rank-one update of P and so is $\bar{\bar{P}}$ because

$$\bar{\bar{P}} = \alpha\left(P + rv^T\right) + (1-\alpha)u.v^T = \alpha P + (\alpha r + (1-\alpha)u)\,v^T$$

11.5.2 Computation of the PageRank steady-state vector

The steady-state vector π obeys[6] the eigenvalue equation (9.28), thus $\pi = \pi\,\bar{\bar{P}}$. Rather than solving this equation, Brin and Page propose to compute the steady-state vector from $\pi = \lim_{k\to\infty} s[k]$. Specifically, for any starting vector $s[0]$ (usually $s[0] = \frac{u^T}{N}$), we iterate the equation (9.6) m-times and choose m sufficiently large such that $\|s[m] - \pi\| \le \epsilon$, where ϵ is a prescribed tolerance. Before turning to the convergence of the iteration process that actually computes powers of $\bar{\bar{P}}$ as observed from (9.10), we first concentrate on the basic iteration (9.6),

$$s[k+1] = s[k]\bar{\bar{P}} = s[k]\left(\alpha P + (\alpha r + (1-\alpha)u)\,v^T\right)$$

Since $s[k]u = 1$, we find

$$s[k+1] = \alpha s[k]P + (\alpha s[k]r + (1-\alpha))\,v^T \tag{11.45}$$

This formula indicates that only the product of $s[k]$ with the (extremely) sparse matrix P needs to be computed and that $\bar{P}$ and $\bar{\bar{P}}$ are never formed nor stored. As shown in Appendix A.5.3, the rate of convergence of a Markov chain towards the steady-state is determined by the second largest eigenvalue. Furthermore, Lemma A.5.4 demonstrates that, for any personalization vector v^T, the second largest eigenvalue of $\bar{\bar{P}}$ is $\alpha\lambda_2$, where λ_2 is the second largest eigenvalue of $\bar{P}$. Lemma A.5.4 thus shows that by choosing α in (11.44) appropriately, the convergence of the iteration (11.45) tends at least as α^k (since $\lambda_2 < 1$ for irreducible and $\lambda_2 = 1$ for reducible Markov chains) towards the steady-state vector π. Brin and Page report that only

---

[6] For a stochastic matrix $P = \Delta^{-1}A$, the j-th component of the steady state vector is $\pi_j = \frac{d_j}{2L}$ as shown in Van Mieghem (2011). This means that the PageRank of webpage j is proportional to its degree d_j. Section 15.3 argues that the degree distribution in the webgraph follows a power law such that $\Pr[d_j = k] = ck^{-\tau}$, where τ for the webgraph is around 2.1.

50 to 100 iterations of (11.45) for $\alpha = 0.85$ are sufficient. Clearly, a fast convergence is found for small α, but then (11.44) shows that the true characteristics of the webgraph are suppressed.

This brings us to a final remark concerning the irreducibility approach. The original method of Brin and Page that resulted in (11.44) by enforcing that each node is connected to each other alters the true nature of the webgraph even though the "connectivity strength" to create irreducibility is extremely small, $\frac{1}{N}$ in the case $v^T = \frac{u^T}{N}$. Instead of maximally connecting all nodes, another irreducibility approach of minimally connecting nodes investigated by Langville and Meyer (2005) consists of creating one dummy node that is connected to all other nodes and to which all other nodes are connected to ensure overall reachability. Such an approach changes the webgraph less. The large size N of the web introduces several challenges such as storage, stability and updating of the PageRank vector, choosing the personalization vector v and other implementational considerations for which we refer to Langville and Meyer (2005).

11.6 Problems

(i) Determine the steady-state probability distribution for the birth-death processes with the following transition intensities:

(a) $\lambda_i = \lambda$ and $\mu_i = i\mu$;

(b) $\lambda_i = \frac{\lambda}{i+1}$ and $\mu_i = \mu$

where λ and μ are constants.

(ii) Derive the partial differential equation (11.33).

(iii) Consider a slotted Aloha in Section 11.4. There are eight stations that compete for slots by transmitting with probability 0.12 each in one slot. Assume that the stations always have packets to transmit. Compute the mean time for one station to transmit seven packets.

12

Branching processes

A branching process is an evolutionary process that starts with an initial set of items that produce several other items with a certain probability distribution. These generated items in turn again produce new items and so on. If we denote by X_k the number of items in the k-th generation and by $Y_{k,j}$ the number of items produced by the j-th item in generation k, then the basic law between the number of items in the k-th and $k + 1$-th generation is, for $k \geq 0$,

$$X_{k+1} = \sum_{j=1}^{X_k} Y_{k,j} \tag{12.1}$$

Figure 12.1 illustrates the basic law (12.1) of a branching process.

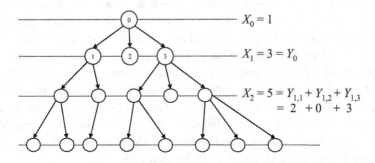

Fig. 12.1. A branching process with one root ($X_0 = 1$) drawn as a tree in which all nodes of generation k lie at a same distance k from the root (label 0).

In general, the production process in each generation k can be different, but most often and in the sequel it is assumed that all generations produce items with the same probability distribution such that all random variables $Y_{k,j}$ are independent and have the same distribution as Y. The branching process is entirely defined by the basic law (12.1) and the distribution of the initial set X_0. The basic law (12.1) indicates that the number of items X_{k+1} in generation $k + 1$ is only dependent on

257

the number of items X_k in the previous generation k. The Markov property (9.2)

$$\Pr\left[X_{k+1} = x_{k+1} | X_0 = x_0, \ldots, X_k = x_k\right] = \Pr\left[X_{k+1} = x_{k+1} | X_k = x_k\right]$$

$$= \Pr\left[\sum_{j=1}^{x_k} Y_j = x_{k+1}\right]$$

is obeyed, which shows that the branching process $\{X_k\}_{k \geq 0}$ is a Markov chain with transition probabilities $P_{ij} = \Pr\left[\sum_{k=1}^{i} Y_k = j\right] = \left(\Pr\left[Y = j\right]\right)^{i*}$, the i-fold convolution of $\Pr\left[Y = j\right]$ for $i \geq 0$, and $P_{0j} = \delta_{0j}$ for all $j \geq 0$.

The discrete branching process can be extended to a continuous-time branching process in which items are produced continuously in time, rather than by generations. The continuous-time branching process is treated in Section 12.7, but we refer, for more details, to the book of Harris (1963) and of Athreya and Ney (1972) and to a simple example, the Yule process, in Section 11.3.2.3.

There are many examples of branching processes and we briefly describe some of the most important. In biology, a certain species generates offsprings and the survival of that species after n generations is studied as a branching process. In the same vein, what is the probability that a family name that is inherited by sons only will eventually become extinct? This was the question posed in 1874 by Galton and Watson that gave birth to the theory of branching processes. In physics, branching processes have been studied to understand nuclear chain reactions. A nucleus is split by a neutron and several new free neutrons are generated. Each of these free neutrons again may hit another nucleus producing additional free neutrons and so on. In micro-electronics, the avalanche break-down of a diode is another example of a branching process. In queuing theory, all new arrivals of packets during the service time of a particular packet can be described as a branching process. The process continues as long as the queue lasts. The number of duplicates generated by a flooding process in a communications network is a branching process: a flooded packet is sent on all interfaces of a router except for the incoming interface. The spread of viral information in the Internet via email forwarding can be modeled approximately as a branching process (Iribarren and Moro, 2011). The application of a branching process to compute the hopcount of the shortest path between two arbitrary nodes in a network is discussed in Section 16.8.

12.1 The probability generating function

Since $Y_{k,j}$ are independent, identically distributed random variables with the same distribution as Y and independent of the random variable X_k, the probability generating function $\varphi_{X_{k+1}}(z)$ of X_{k+1} follows from (2.79) and the basic law (12.1) as

$$\varphi_{X_{k+1}}(z) = \varphi_{X_k}(\varphi_Y(z)) \tag{12.2}$$

with $\varphi_{X_0}(z) = E\left[z^{X_0}\right] = f(z)$, where $f(z)$ is a given probability generating function. Iterating the general relation (12.2) gives

$$\varphi_{X_{k+1}}(z) = \varphi_{X_k}(\varphi_Y(z)) = \varphi_{X_{k-1}}(\varphi_Y(\varphi_Y(z))) = \cdots$$

Eventually,

$$\varphi_{X_k}(z) = f\left(\varphi_Y(\varphi_Y(\ldots(\varphi_Y(z))))\right) = f\left(\varphi_Y^{[k\ \text{iterates}]}(z)\right) \qquad (12.3)$$

where the last relation consists of k nested repeated functions $\varphi_Y(.)$, called the iterates of $\varphi_Y(.)$.

The expectation can be derived from the probability generating function by derivation and setting $z = 1$. More elegantly, by taking the expectation of the basic law (12.1) and recalling that X_k and $Y_{k,j}$ are independent, we have with $\mu = E[Y]$ and (2.82)

$$E\left[X_{k+1}\right] = E\left[\sum_{j=1}^{X_k} Y_{k,j}\right] = E\left[E\left[\sum_{j=1}^{X_k} Y_{k,j} \,\Big|\, X_k\right]\right]$$

$$= E\left[\sum_{j=1}^{X_k} E\left[Y_{k,j}\right]\right] = E\left[X_k E\left[Y\right]\right] = \mu E\left[X_k\right]$$

Iteration, starting from a given mean $E[X_0] > 0$ of the initial population, gives

$$E\left[X_k\right] = \mu^k E\left[X_0\right] \qquad (12.4)$$

Using (2.27), the variance of X_{k+1} follows from (12.2) with

$$\varphi'_{X_{k+1}}(z) = \varphi'_{X_k}(\varphi_Y(z))\varphi'_Y(z)$$

$$\varphi''_{X_{k+1}}(z) = \varphi''_{X_k}(\varphi_Y(z))\left(\varphi'_Y(z)\right)^2 + \varphi'_{X_k}(\varphi_Y(z))\varphi''_Y(z)$$

as

$$\begin{aligned} \text{Var}\left[X_{k+1}\right] &= \varphi''_{X_{k+1}}(1) + \varphi'_{X_{k+1}}(1) - \left(\varphi'_{X_{k+1}}(1)\right)^2 \\ &= \varphi''_{X_k}(1)\left(\varphi'_Y(1)\right)^2 + \varphi'_{X_k}(1)\varphi''_Y(1) + \varphi'_{X_k}(1)\varphi'_Y(1) - \left(\varphi'_{X_k}(1)\varphi'_Y(1)\right)^2 \\ &= \varphi''_{X_k}(1)\left(E\left[Y\right]\right)^2 + E\left[X_k\right]\varphi''_Y(1) + E\left[X_k\right]E\left[Y\right] - \left(E\left[X_k\right]E\left[Y\right]\right)^2 \\ &= \mu^2 \text{Var}\left[X_k\right] + E\left[X_k\right]\text{Var}\left[Y\right] \end{aligned}$$

Iteration starting from a given variance Var$[X_0]$ of the initial set of items and employing the expression for the mean (12.4) yields

$$\text{Var}\left[X_1\right] = \mu^2 \text{Var}\left[X_0\right] + E\left[X_0\right]\text{Var}\left[Y\right]$$

$$\begin{aligned} \text{Var}\left[X_2\right] &= \left(E\left[Y\right]\right)^2 \text{Var}\left[X_1\right] + E\left[X_1\right]\text{Var}\left[Y\right] \\ &= \mu^4 \text{Var}\left[X_0\right] + \left(\mu^2 + \mu\right)E\left[X_0\right]\text{Var}\left[Y\right] \end{aligned}$$

$$\begin{aligned} \text{Var}\left[X_3\right] &= \left(E\left[Y\right]\right)^2 \text{Var}\left[X_2\right] + E\left[X_2\right]\text{Var}\left[Y\right] \\ &= \mu^6 \text{Var}\left[X_0\right] + \left(\mu^4 + \mu^3 + \mu^2\right)E\left[X_0\right]\text{Var}\left[Y\right] \end{aligned}$$

from which we deduce

$$\text{Var}\,[X_k] = \mu^{2k}\text{Var}\,[X_0] + \sum_{j=k-1}^{2(k-1)} \mu^j E\,[X_0]\,\text{Var}\,[Y]$$

or

$$\text{Var}\,[X_k] = \mu^{2k}\text{Var}\,[X_0] + E\,[X_0]\,\text{Var}\,[Y]\,\mu^{k-1}\frac{1-\mu^k}{1-\mu} \qquad (12.5)$$

Substitution into the recursion for $\text{Var}[X_k]$ confirms the correctness of (12.5).

The relations for the expectation (12.4) and the variance (12.5) of the number of items in generation k imply that, if the mean production per generation is $E\,[Y] = \mu = 1$, $E\,[X_k] = E\,[X_0]$ and that $\text{Var}[X_k] = \text{Var}[X_0] + kE\,[X_0]\text{Var}[Y]$. In the case that the mean production $E\,[Y] = \mu > 1$ ($E\,[Y] < 1$), the mean population per generation grows (decreases) exponentially in k with rate $\log\mu$ and, similarly for large k, the standard deviation $\sqrt{\text{Var}\,[X_k]}$ grows (decreases) exponentially in k with the same rate $\log\mu$. Hence, the most important factor in the branching process is the mean production $E\,[Y] = \mu$ per generation. The variance terms and $E\,[X_0]$ only play a role as prefactor.

The limiting value of $E\,[X_k]$ in (12.4) for large k leads to three possibilities: (a) a branching process is called *critical* if $\mu = 1$, in which case $\lim_{k\to\infty} E\,[X_k] = E\,[X_0]$, (b) *subcritical* if $\mu < 1$ and then $\lim_{k\to\infty} E\,[X_k] = 0$ and (c) *supercritical* if $\mu > 1$, in which case $\lim_{k\to\infty} E\,[X_k] = \infty$.

Often, the initial set of items consists of only one item. In that case, $X_0 = 1$ and $\varphi_{X_0}(z) = f(z) = z$ and

$$E\,[X_k] = \mu^k$$

$$\text{Var}\,[X_k] = \text{Var}\,[Y]\,\mu^{k-1}\frac{1-\mu^k}{1-\mu}$$

while the explicit nested form of the probability generating function indicates that

$$\varphi_{X_{k+1}}(z) = \varphi_Y(\varphi_{X_k}(z)) \qquad (12.6)$$

This relation is only valid if $f(z) = z$ or, equivalently, only if $X_0 = 1$. In case $f(z) = z$, $\varphi_{X_k}(z)$ is the k-th iterate of $\varphi_Y(z)$. In summary, only if $f(z) = z$ which corresponds to $X_0 = 1$, both the general (12.2) and the specific (12.6) relations hold

$$\varphi_{X_{k+1}}(z) = \varphi_{X_k}(\varphi_Y(z)) = \varphi_Y(\varphi_{X_k}(z))$$

Example 1 Due to the nested structure (12.2), closed form expressions for the k-th generation probability generating function $\varphi_{X_k}(z)$ are rare. Assume that $X_0 = 1$. A simple case that allows explicit computation occurs in a deterministic production of m offspring in each generation for which $\varphi_Y(z) = z^m$. We have from (12.6) that $\varphi_{X_1}(z) = \varphi_Y(z) = z^m$ and

$$\varphi_{X_k}(z) = z^{km}$$

This branching process evolves as an m-ary tree shown in Fig. 18.7. A second example that can be computed exactly is the geometric branching process studied in Section 12.6.

Example 2 We will approximate the mean number C of nodes in a disconnected component of the Erdős-Rényi random graph $G_p(N)$, discussed in Section 15.7, for $p = \frac{\lambda}{N} < p_c$. Below the disconnectivity threshold $p_c \sim \frac{\log N}{N}$, the Erdős-Rényi random graph $G_p(N)$ is almost surely disconnected into several disconnected components or clusters. Starting with a root node, the Erdős-Rényi random graph is grown by attaching the remaining $N-1$ nodes with probability p to the root. To the first child (or neighboring node) of the root node, we attach all remaining $N-2$ (without the root) nodes with probability p, to the second child, all remaining $N-3$ (without root and first child) nodes with probability p. We continue this attachment process for all children of the root, and for all the children of the children, etc. Eventually this growth process ends and the Erdős-Rényi random graph $G_p(N)$ is finally constructed. The growth process shows that the children of the root can interconnect with probability p^3, which is the probability to have a triangle between three nodes in $G_p(N)$. Hence, if p is sufficiently low and $\lambda = O(1)$, the probability to have a loop is at most $p^3 = O(N^{-3})$ and negligibly small. In that case, we can approximate the growth process as a branching tree that starts from the root ($X_0 = 1$) and creates children at each stage k with a binomial production distribution Y, specified by (15.19), which is a sum of $N-1$ Bernoulli random variables each having a connection success with probability p. Thus, $E[Y] = \mu = (N-1)p = \frac{N-1}{N}\lambda$ and the branching process is subcritical if $\mu < 1$, hence, $\lambda < \frac{N}{N-1}$. In that case, the number of nodes in such a branch

$$C = \sum_{k=0}^{\infty} X_k$$

is a good approximation for the number C of nodes in a disconnected component. Using (12.4), we have

$$E[C] = \lim_{N \to \infty} \sum_{k=0}^{N} E[X_k] = \frac{1}{1-\lambda}$$

Notice from (2.60) that $\text{Var}[C]$ requires the knowledge of $\text{Cov}[X_k, X_j]$, which is not easy to compute.

12.2 The limit W of the scaled random variables W_k

First, we compute the conditional expectation, defined in Section 2.6,

$$E\left[X_{k+1}|X_k, X_{k-1}, \ldots, X_0\right] = E\left[X_{k+1}|X_k\right] \qquad \text{(Markov property)}$$

$$= E\left[\sum_{j=1}^{X_k} Y_{k,j} \,\middle|\, X_k\right] = X_k E\left[Y\right] = \mu X_k$$

Thus,

$$E\left[X_{k+1}|X_k\right] = \mu X_k \tag{12.7}$$

which is a random variable. Next, we consider $E\left[X_{k+r}|X_k\right]$ for $r \in \mathbb{N}_0$. We apply the conditional expectation (2.82) to the condition $X = \{X_{k+r-1}, X_{k+r-2}, \ldots, X_k\}$ as

$$E\left[X_{k+r}|X_k\right] = E\left[E_{X_{k+r}}\left[X_{k+r}|X_{k+r-1}, X_{k+r-2}, \ldots, X_k\right]\middle| X_k\right]$$

Using the Markov property,

$$E_{X_{k+r}}\left[X_{k+r}|X_{k+r-1}, X_{k+r-2}, \ldots, X_k\right] = E_{X_{k+r}}\left[X_{k+r}|X_{k+r-1}\right]$$

and letting $k \to k+r-1$ in (12.7) so that $E_{X_{k+r}}\left[X_{k+r}|X_{k+r-1}\right] = \mu X_{k+r-1}$ yields the recursion

$$E\left[X_{k+r}|X_k\right] = \mu E\left[X_{k+r-1}| X_k\right]$$

After iterating $r-1$ times, we have $E\left[X_{k+r}|X_k\right] = \mu^{r-1} E\left[X_{k+1}| X_k\right]$. Finally, invoking (12.7) leads, for all non-negative integers k and r, to

$$E\left[X_{k+r}|X_k\right] = \mu^r X_k \tag{12.8}$$

This property (12.8) suggests us to consider the scaled random variable

$$W_k = \frac{X_k}{\mu^k}$$

because the left-hand side in (12.8) equals $E\left[X_{k+r}|X_k\right] = \mu^{k+r} E\left[W_{k+r}|X_k\right]$, whereas the right-hand side in (12.8) becomes $E\left[X_{k+r}|X_k\right] = \mu^{r+k} W_k$, and combined, we see that

$$E\left[W_{k+r}|X_k\right] = W_k = E\left[W_{k+r}|W_k\right]$$

where the latter follows because the conditional expectation $E\left[X|Y\right]$ is a function of Y.

A sequence $\{S_k\}_{k\geq 0}$ is a *martingale* with respect to the sequence $\{X_k\}_{k\geq 0}$ if, for all $k \geq 0$, it holds that $E\left[S_k\right] < \infty$ and

$$E\left[S_{k+1}|X_k, X_{k-1}, \ldots, X_0\right] = S_k \tag{12.9}$$

Thus, the scaled Markov process $\{W_k\}_{k\geq 1}$ with respect to $\{X_k\}_{k\geq 0}$ or to itself $\{W_k\}_{k\geq 0}$ obeys, for all $r \in \mathbb{N}_0$,

$$E\left[W_{k+r}|W_k, W_{k-1}, \ldots, W_0\right] = E\left[W_{k+r}|W_k\right] = W_k \tag{12.10}$$

while (12.4) shows that $E[W_k] = E[X_0]$ for all k. The stochastic process $\{W_k\}_{k \geq 1}$ is a martingale process, which is a generalization of a fair game with characteristic property that at each step k in the process $E[W_k]$ is a constant (independent of k). The theory of martingales is rich and beautiful (see e.g. Grimmett and Stirzacker (2001)), but only touched here for the determination of the limit random variable W of a branching process.

From (12.5), the variance of the scaled random variables $W_k = \frac{X_k}{\mu^k}$ is

$$\text{Var}[W_k] = \text{Var}[X_0] + E[X_0] \frac{\text{Var}[Y]}{\mu^2 - \mu}\left(1 - \mu^{-k}\right) \qquad (12.11)$$

which geometrically tends, provided $E[Y] = \mu > 1$, to a constant independent of k. The expression for the variance (12.11) indicates that

$$\text{Var}[W] = \lim_{k \to \infty} \text{Var}[W_k] = \text{Var}[X_0] + E[X_0] \frac{\text{Var}[Y]}{\mu^2 - \mu} \qquad (12.12)$$

exists provided $E[Y] = \mu > 1$. We now concentrate[1] on the limit variable $W = \lim_{k \to \infty} W_k$.

Theorem 12.2.1 *If $E[Y] = \mu > 1$ and Var[Y] is finite, the scaled random variables $W_n \to W$ a.s.*

Proof: Consider

$$E\left[(W_{k+n} - W_n)^2\right] = E\left[W_{k+n}^2\right] + E\left[W_n^2\right] - 2E\left[W_{k+n}W_n\right]$$

Using (2.81) with $h(x) = x$ and the martingale property (12.10),

$$E[W_{k+n}W_n] = E[E[W_{k+n}|W_n]W_n] = E[W_n^2]$$

we have, with (2.16), $E[W_{k+n}] = E[W_k] = E[X_0]$ and (12.11),

$$E\left[(W_{k+n} - W_n)^2\right] = \text{Var}[W_{k+n}] - \text{Var}[W_n] = E[X_0]\frac{\text{Var}[Y]\mu^{-n}}{\mu^2 - \mu}\left(1 - \mu^{-k}\right)$$

In the limit $k \to \infty$,

$$E\left[(W - W_n)^2\right] = O(\mu^{-n})$$

which means that the sequence $\{W_n\}_{n \geq 1}$ converges to W in mean square (see Section 6.1.2). Moreover, for $\mu > 1$,

$$\sum_{n=1}^{\infty} E\left[(W - W_n)^2\right] = E\left[\sum_{n=1}^{\infty}(W - W_n)^2\right] = O(1)$$

which means that the series has finite expectation and that $\sum_{n=1}^{\infty}(W - W_n)^2$ is

---

[1] Theorem 12.2.1 is a special and simplified case of the general martingale convergence theorem (proved in Grimmett and Stirzacker (2001, pp. 338-341)): If $\{S_k\}_{k \geq 0}$ is a martingale with $E[S_k^2] < M$ for some finite real number M and all k, then the limit random variable $S = \lim_{k \to \infty} S_k$ exists and S_k converges to S a.s. and in mean square.

finite with probability 1. The convergence of this series implies, for large n, that $(W - W_n)^2 \to 0$ with probability 1 or that $W_n \to W$ a.s. $\qquad\square$

Theorem 12.2.1 indicates that the number of items in generation k is, for large k, well approximated by $X_k \sim W\mu^k$. Hence, an asymptotic analysis of a branching process crucially relies on the properties of the limit random variable W.

The generating function $\varphi_W(z) = E\left[z^W\right]$ of this limit random variable can be deduced as the limit of the sequence of generating functions

$$\varphi_{W_k}(z) = E\left[z^{W_k}\right] = E\left[z^{\frac{X_k}{\mu^k}}\right] = \varphi_{X_k}(z^{\mu^{-k}}) \tag{12.13}$$

Using (12.6) in case[2] $X_0 = 1$ with $z \to z^{\mu^{-k-1}}$

$$E\left[z^{\frac{X_{k+1}}{\mu^{k+1}}}\right] = \varphi_Y\left(E\left[z^{\frac{X_k}{\mu^k\mu}}\right]\right)$$

leads with $W_k = \frac{X_k}{\mu^k}$ to the recursion of the pgf of the scaled random variables W_k,

$$\varphi_{W_{k+1}}(z) = \varphi_Y\left(\varphi_{W_k}\left(z^{\frac{1}{\mu}}\right)\right)$$

In the limit $k \to \infty$ where $W_k \to W$ a.s., we can apply the Continuity Theorem 6.1.3, which results in the functional equation for the pgf of the *continuous* limit random variable W,

$$\varphi_W(z) = \varphi_Y\left(\varphi_W\left(z^{\frac{1}{\mu}}\right)\right) \tag{12.14}$$

Since W is a continuous random variable except at $W = 0$ as explained below (see (12.26)), it is more convenient to define the moment generating function $\chi_W(t) = E\left[e^{-tW}\right]$. Obviously, the relation between the two generating functions is, with $z = e^{-t}$,

$$\chi_W(t) = \varphi_W(e^{-t})$$

With $z = e^{-t}$ in (12.14) the functional equation of $\chi_W(t)$ is, for $t \geq 0$ and $E[W] = E[X_0] = 1$,

$$\chi_W(t) = \varphi_Y\left(\chi_W\left(\frac{t}{\mu}\right)\right) \tag{12.15}$$

The functional equation (12.15) is more convenient than (12.14): $\chi_W(t)$ is convex for all t, while $\varphi_W(z)$ is not convex for all z. In particular, $\varphi_W(z) = \chi_W(-\log z)$ is not analytic at $z = 0$ and appears[3] to have a concave regime near $z \downarrow 0$ where $\chi_W'(-\log z) + \chi_W''(-\log z) < 0$.

Lemma 12.2.2 $\chi_W(t)$ *is the only moment generating function satisfying the functional equation (12.15).*

---

[2] The use of the general equation (12.2) is inadequate.
[3] This fact is observed for both a geometric and Poisson production distribution function.

Proof: Let $\chi_{W^*}(t) = E\left[e^{-tW^*}\right]$ and $\chi_W(t) = E\left[e^{-tW}\right]$ be two probability generating functions that satisfy both (12.15). Then $\chi_{W^*}(t) - \chi_W(t)$ is continuous for $\operatorname{Re} t \geq 0$ and, since $E[W] = E[W^*] = 1$, the Taylor series (2.41) around $t = 0$ is

$$\chi_{W^*}(t) - \chi_W(t) = (-t)E\left[W^* - W\right] + E\left[\sum_{k=2}^{\infty} \frac{(-t)^k}{k!}\left((W^*)^k - W^k\right)\right]$$

$$= tE\left[\sum_{k=1}^{\infty} \frac{(-t)^k}{(k+1)!}\left(W^{k+1} - (W^*)^{k+1}\right)\right]$$

from which we define $\chi_{W^*}(t) - \chi_W(t) = t\,h(t)$ and $h(0) = 0$. Since $|\varphi_Y'(z)| \leq \mu$ for $|z| \leq 1$, equation (5.8) of the Mean Value Theorem implies $|\varphi_Y(a) - \varphi_Y(b)| \leq \mu\,|a - b|$ for any $a, b \in [0, 1]$. Since $|\chi_W(t)| \leq 1$ and $|\chi_{W^*}(t)| \leq 1$ for $\operatorname{Re}(t) \geq 0$, we obtain

$$|t\,h(t)| = \left|\varphi_Y\left(\chi_{W^*}\left(\frac{t}{\mu}\right)\right) - \varphi_Y\left(\chi_W\left(\frac{t}{\mu}\right)\right)\right|$$

$$\leq \mu\left|\chi_{W^*}\left(\frac{t}{\mu}\right) - \chi_W\left(\frac{t}{\mu}\right)\right| = \mu\left|\frac{t}{\mu}\,h\left(\frac{t}{\mu}\right)\right|$$

or

$$|h(t)| \leq \left|h\left(\frac{t}{\mu}\right)\right|$$

After K iterations, we have that $|h(t)| \leq \left|h\left(\frac{t}{\mu^K}\right)\right|$, which holds for any integer K. Hence, for any finite t, $\mu > 1$ and since $h(t)$ is continuous,

$$|h(t)| \leq \lim_{K \to \infty}\left|h\left(\frac{t}{\mu^K}\right)\right| = h(0) = 0$$

which proves the Lemma. $\qquad\square$

Lemma 12.2.2 is important because solving the functional equation, for example by Taylor expansion, is one of the primary tools to determine $\chi_W(t)$. If $\varphi_Y(z)$ is analytic inside a circle with radius $R_Y > 0$ centered at $z = 1$, then the Taylor series around $z_0 = 1$,

$$\varphi_Y(z) = 1 + \sum_{k=0}^{\infty} u_k(z - 1)^k$$

converges for all $|z - 1| < R_Y$. The definition $\chi_W(t) = E\left[e^{-tW}\right]$ implies that the maximum value of $|\chi_W(t)|$ inside and on a circle with radius r around the origin is attained at $\chi_W(-r)$. The functional equation (12.15) then shows that $\chi_W(t)$ is analytic inside a circle around $t = 0$ with radius R_W for which $\chi_W\left(-\frac{R_W}{\mu}\right) < 1 + R_Y$. Since $\chi_W(0) = 1$ and $\chi_W'(0) = -E[W] = -1$, $\chi_W(t)$ is convex and decreasing for real t and $R_Y > 0$, there exists such a non-zero value of R_W. This implies that the Taylor series

$$\chi_W(t) = 1 - t + \sum_{k=2}^{\infty} \omega_k t^k \qquad (12.16)$$

converges around $t = 0$ for $|t| < R_W$. There exists a recursion to compute ω_k for any $k \geq 1$ as shown in Van Mieghem (2005). If $\chi_W(t)$ is not known in closed form, the interest of the Taylor series (12.16) lies in the fast convergence for small values of $|t| < 1$. The recursion for the Taylor coefficients ω_k enables the computation of $\chi_W(t)$ for $|t| < 1$ to any desired degree of accuracy. The functional equation $\chi_W(t) = \varphi_Y \left(\chi_W \left(\frac{t}{\mu} \right) \right)$ extends the t-range to the entire complex plane. For large values of t and in particular for negative real t, $\chi_W(t)$ is best computed from $\chi_W \left(\frac{t}{\mu^{[\log_\mu |t|]+1}} \right)$ after $[\log_\mu |t|] + 1$ functional iteratives of (12.15). Indeed, since $\mu > 1$ such that $\frac{t}{\mu^{[\log_\mu |t|]+1}} < 1$, the Taylor series (12.16) provides an accurate start value $\chi_W \left(\frac{t}{\mu^{[\log_\mu |t|]+1}} \right)$ for this iterative scheme.

12.3 The probability of extinction of a branching process

In many applications the probability that the process will eventually terminate and which parameters influence this extinction probability are of interest. For instance, a nuclear reaction will only lead to an explosion if critical starting conditions are obeyed. The branching process terminates if, for some generation $n > 0$, $X_n = 0$ and, of course, $X_m = 0$ for all $m > n$. Let us denote

$$q_k = \Pr[X_k = 0] = \varphi_{X_k}(0)$$

If we assume that $X_0 = 1$ so that $\varphi_{X_0}(z) = E\left[z^{X_0}\right] = z$, the analysis simplifies because the more specific version (12.6) holds. Hence, only if the initial set consists of a single item $X_0 = 1$,

$$q_{k+1} = \varphi_{X_{k+1}}(0) = \varphi_Y(\varphi_{X_k}(0)) = \varphi_Y(q_k) \qquad (12.17)$$

and with $q_0 = \varphi_{X_0}(0) = 0$ and $q_1 = \varphi_Y(0) = \Pr[Y = 0] \geq 0$. Obviously, if there is no production, $\Pr[Y = 0] = 1$, or always production, $\Pr[Y = 0] = 0$, extinction never occurs. By its definition (2.18), a probability generating function of a non-negative discrete random variable is strict increasing along the positive real z-axis. When excluding the extreme cases such that $0 < \Pr[Y = 0] < 1$, by the strict increase of $\varphi_Y(x)$ for $x = \operatorname{Re} z \geq 0$, we observe that

$$0 = q_0 < q_1 = \varphi_Y(0) < q_2 = \varphi_Y(q_1) < q_3 = \varphi_Y(q_2) < \dots$$

The series $q_0, q_1, q_2, \dots$ is a monotone increasing sequence bounded by 1 because $\varphi_Y(1) = 1$. Hence, the probability of extinction

$$\pi_0 = \lim_{k \to \infty} \Pr[X_k = 0] = \Pr[W = 0]$$

exists and $0 < \pi_0 \leq 1$. The existence of a limiting process and the fact that the probability generating function is analytic for $|z| < 1$ and hence, continuous, which allows us to interchange $\lim_{k \to \infty} \varphi_Y(q_k) = \varphi_Y(\lim_{k \to \infty} q_k)$ in (12.17), yields the equation for the extinction probability π_0,

$$\pi_0 = \varphi_Y(\pi_0) \tag{12.18}$$

Hence, the extinction probability π_0 is a root of $\varphi_Y(x) - x$ in the interval $x \in [0, 1]$. Equation (12.18) also shows that π_0 is zero if and only if $\varphi_Y(0) = \Pr[Y = 0] = 0$: when there is always production, $\Pr[Y > 0] = 1$, there is no extinction.

Since $\varphi_W(0) = \Pr[W = 0] = \pi_0$, equation (12.18) follows more directly from (12.14). Notice, however, that in the functional equation (12.14) the function z^μ is not analytic at $z = 0$ if $\mu \notin \mathbb{N}$, which may cause that $f_W(x)$ is possibly not continuous at $x = 0$, although the limit $\lim_{z \to 0} \varphi_W(0) = \pi_0$ exists. On the other hand, since $\chi_W(t) = \varphi_W(e^{-t})$, the extinction probability is found as

$$\lim_{t \to \infty} \chi_W(t) = \pi_0$$

and the convexity of $\chi_W(t)$ implies that, for any real value of t, $\chi_W(t) \geq \pi_0$.

An alternative, more probabilistic derivation of equation (12.18) is as follows. Applying the law of total probability (2.49) to the definition of the extinction probability

$$\pi_0 = \Pr[X_n = 0 \text{ for some } n > 0]$$

$$= \sum_{j=0}^{\infty} \Pr[X_n = 0 \text{ for some } n > 0 | X_1 = j] \Pr[X_1 = j]$$

Only if $X_0 = 1$, relation (12.6) indicates that $\varphi_{X_1}(z) = \varphi_Y(z)$ which implies that $\Pr[X_1 = j] = \Pr[Y = j]$. In addition, given the first generation consists of j items, the branching process will eventually terminate if and only if each of the j sets of items generated by the first generation eventually dies out. Since each set evolves independently and since the probability that any set generated by a particular ancestor in the first generation becomes extinct is π_0, we arrive at

$$\pi_0 = \sum_{j=0}^{\infty} \pi_0^j \Pr[Y = j] = \varphi_Y(\pi_0)$$

The different viewpoints thus lead to a same result summarized by:

Theorem 12.3.1 *If $X_0 = 1$ and $0 < \Pr[Y = 0] < 1$, the extinction probability π_0 is (a) the smallest positive real root of $x = \varphi_Y(x)$ and (b) $\pi_0 = 1$ if and only if $E[Y] = \mu \leq 1$ and $\Pr[Y = 0] + \Pr[Y = 1] < 1$.*

Proof: (a) Suppose that x_o is the smallest positive real root obeying $\varphi_Y(x_o) = x_o > 0$. Then, $q_1 = \varphi_Y(0) < \varphi_Y(x_o) = x_o$. Assume (induction hypothesis) that $q_n < x_o$. The recursion (12.17) and the strict increase of $\varphi_Y(x)$ then shows that $q_{n+1} = \varphi_Y(q_n) < \varphi_Y(x_o) = x_o$. Hence, the principle of induction demonstrates that $q_n < x_o$ for all (finite) n and, hence, that $\pi_0 \leq x_o$.

(b) First, the condition $\Pr[Y = 0] + \Pr[Y = 1] < 1$ implies that $\Pr[Y > 1] > 0$ and that there exists at least one integer $j > 1$ such that $\Pr[Y = j] > 0$. In that case, for real $x > 0$ but x smaller than the radius R of convergence which is at least $R = 1$, the second derivative $\varphi_Y''(x) = \sum_{j=2}^{\infty} j\,(j-1)\Pr[Y = j]\,x^{j-2}$ is positive, which implies that $\varphi_Y(x)$ is strictly convex in $(0, 1)$. Since $x = 1$ obeys $x = \varphi_Y(x)$ and $\varphi_Y(0) = \Pr[Y = 0] \in (0, 1)$, the strict convex function $y = \varphi_Y(x)$ can only intersect the line $y = x$ in some point $x \in (0, 1)$ if $\varphi_Y(x)$ is below that line near their intersection at $x = 1$ or if $\varphi_Y'(1) = E[Y] > 1$. In the other case, if $E[Y] < 1$, the only intersection is at $x = 1$. □

The two possibilities are drawn in Fig. 12.2.

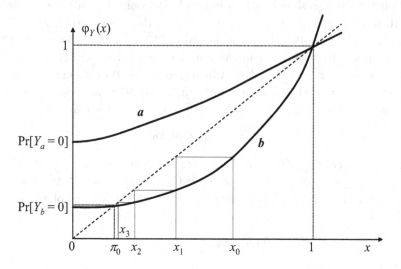

Fig. 12.2. The generating function $\varphi_Y(x)$ along the positive real axis x. The two possible cases are shown: curve a corresponds to $E[Y_a] < 1$ and curve b to $E[Y_b] > 1$. The fast convergence towards the zero π_0 is exemplified by the sequence $x_0 > x_1 = \varphi_Y(x_0) > x_2 = \varphi_Y(x_1) > x_3 = \varphi_Y(x_2)$.

A root equation such as (12.18) also appears in queuing models such as the M/G/1 (Section 14.3) and GI/D/1 (Section 14.4) and reflects the asymptotic behavior as explained in Section 5.7. The extinction probability π_0 can be expressed explicitly as a Lagrange series as demonstrated in Van Mieghem (1996). If Y is a Poisson random variable with $\varphi_Y(z) = e^{\mu(z-1)}$, then the extinction probability is given by the Lagrange series (15.30) with $\pi_0 = 1 - S$. Even when the mean number of descendants/items $E[X_k]$ in the k-th generation grows with k if $E[Y] = \mu > 1$ (see (12.4)), the fundamental Theorem 12.3.1 illustrates, counter intuitively, that such populations may extinguish with positive probability π_0.

If the initial population is $X_0 = m > 1$, then the extinction probability equals π_0^m. Indeed, the extinction probability is, with (12.3),

$$\lim_{k \to \infty} \Pr[X_k = 0] = \lim_{k \to \infty} \varphi_{X_k}(0) = \lim_{k \to \infty} f\left(\varphi_Y^{[k \text{ iterates}]}(0)\right) = f\left(\lim_{k \to \infty} \varphi_Y^{[k \text{ iterates}]}(0)\right)$$

When $X_0 = 1$, the extinction probability π_0 satisfies (12.18) and also, with (12.3), $\pi_0 = \lim_{k \to \infty} \varphi_Y^{[k \text{ iterates}]}(0)$. Hence, for any $X_0 > 0$, the extinction probability can be written in terms of π_0 as $\lim_{k\to\infty} \Pr[X_k = 0] = f(\pi_0)$. In particular, when $X_0 = m$, then $f(z) = z^m$ and the extinction probability equals π_0^m.

The branching process with infinitely many generations $k \to \infty$ can be viewed as an infinite directed tree where each node has a finite degree a.s. The fact that $\pi_0 < 1$ if $E[Y] > 1$ implies that, in infinitely directed trees, there exists an infinitely long path starting from the root with probability $1 - \pi_0$.

Theorem 12.3.2 *The branching process X_k with $X_0 = 1$ obeys for $k \to \infty$*

$$\Pr[X_k = 0] \to \pi_0$$
$$\Pr[X_k = j] \to 0 \qquad \text{for any } j > 0$$

Proof: First, if $E[Y] < 1$, then Theorem 12.3.1 states that $\pi_0 = 1$. For any probability generating function $\varphi(z)$, it holds that $|\varphi(z)| \le 1$ for $|z| \le 1$. Hence, $\varphi_{X_k}(x) \le 1$ for real $x \in [0, 1]$. Moreover, $q_k = \varphi_{X_k}(0) \le \varphi_{X_k}(x)$. In the limit $k \to \infty$, $q_k \to \pi_0 = 1$, which implies that for all $x \in [0, 1]$ it holds that $\varphi_{X_k}(x) \to \pi_0 = 1$. The fact that a probability generating function, a Taylor series around $z = 0$, converges to a constant π_0 for $0 \le x \le 1$ implies that $\Pr[X_k = j] \to 0$ for any $j > 0$ and $\Pr[X_k = 0] \to \pi_0$.

The second case $E[Y] > 1$ possesses an extinction probability $\pi_0 < 1$. For $x \in (\pi_0, 1)$, Fig. 12.2 shows that $\pi_0 < \varphi_Y(x) < x < 1$. By induction using (12.6), we find that $\pi_0 < \varphi_{X_k}(x) < \varphi_{X_{k-1}}(x) < \cdots < 1$ or $\lim_{k\to\infty} \varphi_{X_k}(x) = \pi_0$ for $x \in (\pi_0, 1)$. For $x \in [0, \pi_0)$, the same argument $q_k = \varphi_{X_k}(0) \le \varphi_{X_k}(x) \le \pi_0$ shows that $\lim_{k\to\infty} \varphi_{X_k}(x) = \pi_0$ for $x \in [0, 1)$. This proves the theorem. □

Theorem 12.3.2 states that, regardless of the value of $E[Y]$, the probability that the k-th generation will consists of any finite positive number of items tends to zero if $k \to \infty$. Theorem 12.3.2 is equivalent to the statement that, when $k \to \infty$, $X_k \to \infty$ with probability $1 - \pi_0$. Thus, eventually when $k \to \infty$, the population shows a zero-one law with probability 1: either $X_k \to 0$ with probability π_0 or $X_k \to \infty$ with probability $1 - \pi_0$. The process $\{X_k\}_{k\ge 0}$ does not fluctuate a.s., but either vanishes or explodes when $k \to \infty$.

Theorem 12.3.2 also illustrates that a Markov chain with an infinite number of states behaves differently than a Markov chain with a finite number of states. In particular, Theorem 12.3.2 shows that the infinite Markov chain $\{X_k\}_{k\ge 0}$ has a single absorbing state $X_k = 0$ while all other states j are transient ($\lim_{n\to\infty} P_{ij}^n = 0$ for $1 \le i, j < \infty$). The existence of the steady-state vector π (not all components are zero) does not imply that the branching process with $X_0 = 1$ and $0 < \Pr[Y = 0] < 1$ and infinitely many states is an irreducible Markov chain!

12.4 Conditioning of a supercritical branching process

Let T denote the extinction time of the branching process $\{X_k\}_{k\ge 0}$, which is defined as $T = \inf\{n : X_n = 0\}$, the infimum of the set of values for which $X_n = 0$. Thus, the event $\{T = n\}$ is equivalent to $\{X_{n-1} > 0\} \cap \{X_n = 0\}$. If $T \to \infty$, the process grows unboundedly, while, if $T < \infty$, the process dies out. As defined in Section 12.3, we have that $\Pr[T < \infty] = \Pr[X_n = 0 \text{ for some } n > 0] = \pi_0$ and

$$\Pr[T \le n] = \Pr[X_n = 0] = \varphi_{X_n}(0)$$

We will now study three conditioned branching processes: (a) the "future starving" process $\{F_k\}_{k\geq 0}$, where $F_k = \{X_k | k < T < \infty\}$, which is the branching process $\{X_k\}_{k\geq 0}$ given that extinction occurs in the future after the k-th generation, (b) the "starving" process $\{S_k\}_{k\geq 0}$, where $S_k = \{X_k | T < \infty\}$ and (c) the "up-running" branching process $\{U_k\}_{k\geq 0}$, where $U_k = \{X_k | T = \infty\}$, that never dies out. Thus, $S_k \to 0$, $F_k \to 0$ and $U_k \to \infty$ as $k \to \infty$. The transience of the non-zero states of $\{X_k\}_{k\geq 1}$ as shown in Theorem 12.3.2 implies that $\Pr[F_\infty \cup U_\infty] = 1$ and $\Pr[S_\infty \cup U_\infty] = 1$. These transformations of a branching process $\{X_k\}_{k\geq 0}$ in the supercritical case ($\mu > 1$) with $\pi_0 > 0$, that is equivalent to $\Pr[Y = 0] > 0$, may ease to deduce properties.

Let us consider, for $|z| \leq 1$, the probability generating function

$$\varphi_{F_k}(z) = E\left[z^{X_k} \middle| k < T < \infty\right] = \sum_{j=0}^{\infty} \Pr[X_k = j | k < T < \infty] z^j$$

First, $\Pr[X_k = 0 | k < T < \infty] = 0$ since the event $\{X_k = 0 | k < T < \infty\}$ is impossible for $k > 0$, because it contradicts the definition of T and $X_0 = 1$. In fact $F_k = \{X_k | X_k > 0\}$. Next, with the definition of conditional probability (2.47) for $j \geq 1$,

$$\Pr[X_k = j | k < T < \infty] = \frac{\Pr[\{X_k = j\} \cap \{k < T < \infty\}]}{\Pr[k < T < \infty]}$$

and

$$\Pr[\{X_k = j\} \cap \{k < T < \infty\}] = \Pr[X_k = j \text{ and all subsequent generations die out}]$$
$$= \Pr[X_k = j] \pi_0^j$$

With $\Pr[k < T < \infty] = \Pr[T < \infty] - \Pr[T \leq k] = \pi_0 - \varphi_{X_k}(0)$, we find

$$\varphi_{F_k}(z) = \frac{\sum_{j=1}^{\infty} \Pr[X_k = j] \pi_0^j z^j}{\pi_0 - \varphi_{X_k}(0)} = \frac{\varphi_{X_k}(\pi_0 z) - \varphi_{X_k}(0)}{\pi_0 - \varphi_{X_k}(0)} \quad (12.19)$$

The mean is $E[F_k] = \frac{\pi_0 \varphi'_{X_k}(\pi_0)}{\pi_0 - \varphi_{X_k}(0)}$. Differentiating (12.2) and invoking (12.18) yields $\varphi'_{X_k}(\pi_0) = \varphi'_{X_{k-1}}(\pi_0) \varphi'_Y(\pi_0)$, and after iterating, we obtain $\varphi'_{X_k}(\pi_0) = (\varphi'_Y(\pi_0))^k$. These manipulations illustrate, for iterates of a function f, the importance of a fixed point[4] of that function f. Thus, for $k \geq 0$ and $\Pr[Y = 0] > 0$,

$$E[F_k] = \frac{\pi_0 (\varphi'_Y(\pi_0))^k}{\pi_0 - \varphi_{X_k}(0)}$$

For $\mu > 1$, $\varphi'_Y(\pi_0) < 1$ as shown in the proof of Lemma 12.5.1 below, which shows that $E[F_k] \to 0$ exponentially fast in k. The effective production process Y_F is

---

[4] A fixed point of a mapping $z \to f(z)$ is a point z_0 that is mapped into itself: $f(z_0) = z_0$. A fixed point is invariant for f, similar to the eigenvector x of a matrix A corresponding to an eigenvalue 1 (if x exists).

found for $k = 1$. Since $X_0 = 1$ and $\varphi_{X_1}(0) = \varphi_Y(0) = \Pr[Y = 0] > 0$, we find that

$$\varphi_{F_1}(z) = \varphi_{Y_F}(z) = \frac{\varphi_Y(\pi_0 z) - \Pr[Y = 0]}{\pi_0 - \Pr[Y = 0]}$$

Before proceeding, we derive an interesting functional equation for the limiting probability generating function $\lim_{k \to \infty} \varphi_{F_k}(z) = \varphi_{F_\infty}(z)$. For all k, $\varphi_{F_k}(z)$ is a pgf that is analytic and bounded (i.e. $|\varphi_{F_k}(z)| \le 1$) inside $|z| < 1$. By Vitali's convergence theorem (Titchmarsh, 1964, p. 168), these pgfs converge uniformly inside the unit circle such that $\varphi_{F_\infty}(z)$ exists, is bounded and also analytic for $|z| < 1$. Since also all derivatives of an analytic function converge uniformly (Titchmarsh, 1964, p. 97), we conclude that $\varphi_{F_\infty}(z)$ is a pgf. We rewrite $\varphi_{F_k}(z)$ in (12.19) as

$$\varphi_{F_k}(z) = \frac{\frac{\varphi_{X_k}(\pi_0 z)}{\pi_0} - \frac{\varphi_{X_k}(0)}{\pi_0}}{1 - \frac{\varphi_{X_k}(0)}{\pi_0}} = 1 - \frac{1 - \frac{\varphi_{X_k}(\pi_0 z)}{\pi_0}}{1 - \frac{\varphi_{X_k}(0)}{\pi_0}}$$

Let $f_k(z) = \frac{\varphi_{X_k}(\pi_0 z)}{\pi_0}$ and $f(z) = \frac{\varphi_Y(\pi_0 z)}{\pi_0}$, then $f_k(z) = f(f_{k-1}(z))$, which shows that $f_k(z)$ is the k-th iterate of $f(z)$. Define $G_k(z) = \frac{1 - f_k(z)}{1 - f_k(0)}$ and $g(z) = \frac{1 - f(z)}{1 - z}$. Then, $\varphi_{F_k}(z) = 1 - G_k(z)$ and

$$G_k(f(z)) = \frac{1 - f_k(f(z))}{1 - f_k(0)} = \frac{1 - f_{k+1}(z)}{1 - f_{k+1}(0)} \frac{1 - f_{k+1}(0)}{1 - f_k(0)} = G_{k+1}(z) g(f_k(0))$$

Since $\lim_{k \to \infty} f_k(0) = \frac{1}{\pi_0} \lim_{k \to \infty} \Pr[X_k = 0] = 1$ and $\lim_{z \to 1} g(z) = f'(1) = \varphi_Y'(\pi_0)$, the limit $k \to \infty$ process yields

$$G_\infty(f(z)) = G_\infty(z) \varphi_Y'(\pi_0)$$

Introducing $G_\infty(z) = 1 - \varphi_{F_\infty}(z)$ leads to the limit functional equation

$$\varphi_{F_\infty}(f(z)) = \varphi_Y'(\pi_0) \varphi_{F_\infty}(z) + 1 - \varphi_Y'(\pi_0) \tag{12.20}$$

The interesting point is that this equation holds for $\mu > 1$ as well as for $\mu < 1$. Indeed, in the latter case of a subcritical branching process ($\mu < 1$) is $\pi_0 = 1$. Moreover, $f(z) = \varphi_Y(z)$ and $\varphi_{F_k}(z) = \frac{\varphi_{X_k}(z) - \varphi_{X_k}(0)}{1 - \varphi_{X_k}(0)}$. The case for $\mu = 1$ is more complicated and treated in detail by Athreya and Ney (1972). A noteworthy result in case $\mu = 1$, due to Yaglom, states that, if $\sigma_Y^2 = \text{Var}[Y] < \infty$, then, for $x \ge 0$,

$$\lim_{k \to \infty} \Pr\left[\frac{X_k}{k} > x \,\middle|\, X_k > 0\right] = e^{-\frac{2x}{\sigma_Y^2}}$$

The fact that $0 = \varphi_{F_k}(0) = \Pr[F_k = 0]$ reflects dying out later than in generation k. We can relax the condition of only dying in the future and investigate the process $S_k = \{X_k | T < \infty\} = \{X_k | X_n = 0 \text{ for some } n\}$, which is a branching process that certainly extincts at some time or in some generation, but not necessarily in a future generation. Then,

$$\Pr[S_k = j] = \frac{\Pr[\{X_k = j\} \cap \{T < \infty\}]}{\Pr[T < \infty]} = \frac{\Pr[X_k = j] \pi_0^j}{\pi_0}$$

which implies for $j > 0$ that at k, the process is still living. If $j = 0$, the process S_k is already dead or has just died. The probability generating function is $\varphi_{S_k}(z) = \frac{\varphi_{X_k}(\pi_0 z)}{\pi_0}$, which is thus the k-th iterate of $\frac{\varphi_Y(\pi_0 z)}{\pi_0}$. In contrast to the future dying process, $\varphi_{S_k}(0) = \frac{\Pr[Y = 0]}{\pi_0} > 0$. For example, for a Poisson production process with $\varphi_Y(z) = e^{\mu(z-1)}$ and $\mu > 1$, the corresponding production process conditioned to die is $\varphi_{Y_S}(z) = \varphi_{S_1}(z) = e^{\mu\pi_0(z-1)}$ (where (12.18) is used), which is again a Poisson process but with rate (or mean) $\mu\pi_0 < 1$.

Let us now turn to the "up-running" branching process $\{U_k\}_{k\geq 0}$, where U_k is the number of items among X_k that possess an infinite line of descent. Hence, each of the U_k items will produce an arbitrary number of offsprings and at least one of them must produce infinitely many descendents. The probability generating function of U_k is

$$\varphi_{U_k}(z) = E\left[z^{X_k}\middle| T = \infty\right] = \sum_{j=0}^{\infty} \Pr\left[X_k = j\middle| T = \infty\right] z^j$$

where the conditional probability $\Pr\left[X_k = j\middle| T = \infty\right] = \frac{\Pr[\{X_k=j\}\cap\{T=\infty\}]}{\Pr[T=\infty]}$ requires the determination of the joint probability. Similarly as above, a *contradictio in terminis* causes an impossible event, $\Pr\left[X_k = 0\middle| T = \infty\right] = 0$. By the law (2.49) of total probability,

$$\Pr\left[\{X_k = j\} \cap \{T = \infty\}\right] = \sum_{l=j}^{\infty} \Pr\left[\{X_k = j\} \cap \{T = \infty\}\middle| X_k = l\right] \Pr\left[X_k = l\right]$$

$$= \sum_{l=j}^{\infty} \binom{l}{j} (1 - \pi_0)^j \pi_0^{l-j} \Pr\left[X_k = l\right]$$

Then, for $|z| \leq 1$, and with $\Pr\left[T = \infty\right] = 1 - \pi_0$,

$$\varphi_{U_k}(z) = \frac{1}{1-\pi_0} \sum_{j=1}^{\infty} \sum_{l=j}^{\infty} \binom{l}{j} (1 - \pi_0)^j \pi_0^{l-j} \Pr\left[X_k = l\right] z^j$$

Reversing the j- and l-summation, which is allowed because all terms are non-zero, yields

$$\varphi_{U_k}(z) = \frac{1}{1-\pi_0} \sum_{l=1}^{\infty} \Pr\left[X_k = l\right] \sum_{j=1}^{l} \binom{l}{j} \pi_0^{l-j} (1 - \pi_0)^j z^j$$

$$= \frac{1}{1-\pi_0} \sum_{l=1}^{\infty} \Pr\left[X_k = l\right] \left[(\pi_0 + (1 - \pi_0) z)^l - \pi_0^l\right]$$

$$= \frac{1}{1-\pi_0} \sum_{l=0}^{\infty} \Pr\left[X_k = l\right] \left[(\pi_0 + (1 - \pi_0) z)^l - \pi_0^l\right]$$

$$= \frac{\varphi_{X_k}(\pi_0 + (1 - \pi_0) z) - \varphi_{X_k}(\pi_0)}{1 - \pi_0}$$

which results, with $\varphi_{X_k}(\pi_0) = \pi_0$ as follows after combining (12.2) and (12.18), in

$$\varphi_{U_k}(z) = \frac{\varphi_{X_k}(\pi_0 + (1 - \pi_0) z) - \pi_0}{1 - \pi_0}$$

The mean is $E\left[U_k\right] = \varphi'_{X_k}(1) = \mu^k$. The effective production function of the "up-running" branching process, $\varphi_{Y_U}(z) = \varphi_{U_1}(z)$, has pgf

$$\varphi_{Y_U}(z) = \frac{\varphi_Y(\pi_0 + (1 - \pi_0)z) - \pi_0}{1 - \pi_0}$$

with property $\varphi_{Y_U}(0) = 0$ that, indeed, reflects non-extinction of the process U_k.

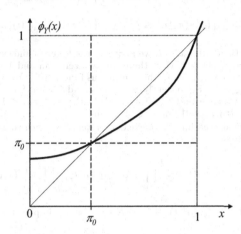

Fig. 12.3. The generating function $\varphi_Y(x)$ and three squares with sizes π_0, $1 - \pi_0$ and 1.

Both $\varphi_{Y_S}(z)$ and $\varphi_{Y_U}(z)$ have an interesting geometrical interpretation in terms of the original, unconditioned production pgf $\varphi_Y(z)$ drawn in Fig. 12.3 together with its two fixed points π_0 and 1. If we rescale the graph of $\varphi_Y(z)$ on the interval $[0, \pi_0]$ to the interval $[0, 1]$, we obtain $\varphi_{Y_S}(z)$, where π_0 coincides with 1 (dying out a.s.). That rescaled part of $\varphi_Y(z)$ corresponds to curve a in Fig. 12.2. Similarly, if the square with opposite corner points (π_0, π_0) and $(1, 1)$ is stretched to the unit square, mapping (π_0, π_0) to $(0, 0)$, the resulting curve is $\varphi_{Y_U}(z)$, where π_0 coincides with 0 (never dying a.s.).

12.5 Asymptotic behavior of W

Since $W \geq 0$ and $\chi_W(t) = E\left[e^{-Wt}\right]$, the Laplace integral (2.38) shows that $\chi'_W(t) \leq 0$ for all real t, thus, that $\chi'_W(t)$ is non-increasing in t. We know that $\chi'_W(0) = -1$. Since $\lim_{t \to \infty} \chi_W(t) = \pi_0$, thus finite, it follows that $\lim_{t \to \infty} \chi'_W(t) = 0$. The following Lemma 12.5.1 is more precise.

Lemma 12.5.1 *For $\mu > 1$ and large t, $|\chi'_W(t)|$ is bounded by*

$$|\chi'_W(1)| t^{-1 + \log_\mu \varphi'_Y(\pi_0)} \leq |\chi'_W(t)| \leq C t^{-1 - \beta}$$

where C is a finite positive real number and $0 < \beta \leq \log_\mu \varphi'_Y(\pi_0)$.

Proof: The derivative of the functional equation (12.15) is $\mu \chi'_W (\mu t) = \varphi'_Y (\chi_W (t)) \chi'_W (t)$. By iteration, we have

$$\mu^K \chi'_W \left(\mu^K t \right) = \chi'_W (t) \prod_{j=0}^{K-1} \varphi'_Y (\chi_W (\mu^j t))$$

Since $\chi_W (t) \in [\pi_0, 1]$ for real $t \geq 0$, then $\varphi'_Y (\pi_0) \leq \varphi'_Y (\chi_W (\mu^j t)) \leq \varphi'_Y (1) = \mu$ for any j. Thus,

$$\left| \chi'_W (t) \right| \left(\varphi'_Y (\pi_0) \right)^K \leq \mu^K \left| \chi'_W \left(\mu^K t \right) \right| \leq \left| \chi'_W (t) \right| \mu^K$$

Let $x = \mu^K t$ and $t = 1$, then, after $x \to t$, we arrive at the lower bound, valid for $t \geq 1$. Theorem 12.3.1 states that if $\mu = \varphi'_Y (1) > 1$, then there are two zeros π_0 and 1 of $f(z) = \varphi_Y (z) - z$ in continuous $z \in [0, 1]$. By Rolle's Theorem applied to the continuous function $f(z) = \varphi_Y(z) - z$, there exists a $\xi \in (\pi_0, 1)$ for which $f'(\xi) = 0$. Equivalently, $\varphi'_Y(\xi) = 1$ and $\xi > \pi_0$. Since $\varphi'_Y (z)$ is monotonously increasing in $z \in (0, 1]$, we have that $\varphi'_Y(0) = \Pr[Y = 1] \leq \varphi'_Y (\pi_0) < 1$. Thus, there exists an non-empty interval $z \in [\pi_0, \xi)$ with $\pi_0 < \xi < 1$ where $\varphi'_Y(z) < 1$. Since $\chi_W (t) \in [\pi_0, 1]$ is continuous and monotone decreasing, there exists an integer K_0 such that $\varphi'_Y (\chi_W (\mu^j t)) < 1$ for $j > K_0$ and any $t > 0$. Hence,

$$\lim_{K \to \infty} \prod_{j=0}^{K-1} \varphi'_Y (\chi_W (\mu^j t)) = \prod_{j=0}^{K_0-1} \varphi'_Y (\chi_W (\mu^j t)) \prod_{j=K_0}^{\infty} \varphi'_Y (\chi_W (\mu^j t)) \to 0$$

and, for any finite $t > 0$, $\mu^K \chi'_W (\mu^K t) \to 0$ for $K \to \infty$ which implies the lemma. $\qquad\square$

Lemma 12.5.1 suggests that

$$\chi'_W (t) = -g_\mu (t) t^{-\beta - 1} \tag{12.21}$$

where $0 < g_\mu (t) \leq C$ for $t \geq 0$.

Lemma 12.5.2 *If $\varphi'_Y (\pi_0) > 0$ and $\mu > 1$, then*

$$F_\mu = \lim_{t \to \infty} g_\mu (t) \tag{12.22}$$

exists, is finite and strictly positive and $g_\mu (t)$ is non-decreasing for $t \geq 0$. Moreover,

$$\beta = -\frac{\log \varphi'_Y (\pi_0)}{\log \mu} \tag{12.23}$$

Proof: We first use (a) the convexity of any mgf $\chi_W (t)$ implying that $\chi''_W (t) \geq 0$ for all t and we then invoke (b) the functional equation (12.15) of $\chi_W (t)$ and (c) verify consistency.

(a) The function $g_\mu (t) = -\chi'_W (t) t^{\beta+1}$ is differentiable, thus continuous, and has for real $t > 0$ only one extremum at $t = \tau$ obeying

$$\tau = \frac{-\chi'_W (\tau)}{\chi''_W (\tau)} (\beta + 1) > 0$$

Since $\chi'_W (0) = -1$, implying that $g_\mu (t) = t^{\beta+1} (1 + O(t))$ as $t \downarrow 0$ or that $g_\mu(t)$ is initially monotone increasing in t, the extremum at $t = \tau$ is a maximum, $\max g_\mu (t) = \frac{\tau^{\beta+2}}{\beta+1} \chi''_W (\tau)$.

(b) Substitution of (12.21) in the derivative of the functional equation (12.15) yields

$$g_\mu (t) = \varphi'_Y \left(\chi_W \left(\frac{t}{\mu} \right) \right) g_\mu \left(\frac{t}{\mu} \right) \mu^\beta \tag{12.24}$$

Since $\varphi_Y' \left(\chi_W \left(\frac{t}{\mu} \right) \right) \geq \varphi_Y' \left(\pi_0 \right) > 0$ (the restriction of this Lemma), there holds with $A = \varphi_Y' \left(\pi_0 \right) \mu^\beta > 0$ for all $t > 0$ that

$$g_\mu (t) \geq A g_\mu \left(\frac{t}{\mu} \right)$$

For $t < \tau$, $g_\mu (t)$ is shown in (a) to be monotone increasing, which requires that $A \leq 1$ for $\mu > 1$. But, since the inequality with $A \leq 1$ holds for all $t > 0$, we must have that $\tau \to \infty$. Hence, $g_\mu(t)$ is continuous and non-decreasing for all $t \geq 0$ with a maximum at infinity, which proves the existence of a unique limit $F_\mu = \max_{t \geq 0} g_\mu (t)$.

(c) We still need to show that $F_\mu > 0$ is finite and that $A = \varphi_Y' \left(\pi_0 \right) \mu^\beta = 1$. When passing to the limit $t \to \infty$ in (12.24) assuming a finite $F_\mu > 0$, we obtain $\mu^{-\beta} = \varphi_Y' \left(\pi_0 \right)$, which determines the exponent $\beta \geq 1$ in (12.23). The fact that $F_\mu > 0$ is finite is directly coupled to the value of β. Indeed, if $F_\mu \to \infty$, then $g_\mu (t) = O \left(t^\varepsilon \right)$ for some real $\varepsilon > 0$ and we propose $\chi_W' (t) = -h_\mu (t) t^{-\beta + \varepsilon - 1}$, where $h_\mu (t) = t^{-\varepsilon} g_\mu (t)$ and $\lim_{t \to \infty} h_\mu (t) > 0$ is finite. This case would lead to a functional equation for h_μ similar as (12.24), from which the exponent $\beta - \varepsilon = \log_\mu \varphi_Y' \left(\pi_0 \right)$ would follow.

If $F_\mu = 0$, the suggestion (12.21) is not correct implying that $\chi_W' (t)$ decreases faster than any power of t^{-1}. The proof of Lemma 12.5.1 indicates that this case can only occur if $\varphi_Y' \left(\pi_0 \right) = 0$.□

The value (12.23) of β is precisely the exponent of the lower bound in Lemma 12.5.1. After integration of (12.21), we have that

$$\chi_W (t) = \pi_0 + \int_t^\infty g_\mu (u) u^{-\beta - 1} du \qquad (12.25)$$

Approximating $g_\mu (u)$ by its limit F_μ for large t, we obtain the asymptotic form

$$\chi_W (t) \sim \pi_0 + \frac{F_\mu}{\beta} t^{-\beta}$$

Beside β and the extinction probability π_0, the parameter F_μ appears as an additional characterizing quantity of a branching process. The behavior of the Laplace transform (2.38) for large t reflects the behavior of the probability density function for small x. Hence, using $\frac{1}{2\pi i} \int_{c-i\infty}^{c+i\infty} \frac{e^{xt}}{t^s} dt = \frac{x^{s-1}}{\Gamma(s)}$ for $\mathrm{Re}\, s > 0$, the probability density function is, for small x,

$$f_W (x) \approx \pi_0 \delta (x) + \frac{F_\mu}{\Gamma (\beta + 1)} x^{\beta - 1} \qquad (12.26)$$

The probability density function $f_W (x)$ is not continuous at $x = 0$ if $\pi_0 > 0$ and reflects the two different regimes: (a) $W = 0$ implying that the branching process extinguishes, $X_k = 0$, from some generation k on and (b) $W > 0$ implying $X_k \sim W\mu^k$ for large k, the number of items per generation grows exponentially with prefactor W. If two sample paths of the same branching process are generated, $X_{k,1} \sim W_1 \mu^k$ and $X_{k,2} \sim W_2 \mu^k$, they may be largely different for large k, because of the random nature of W: although the prefactors W_1 and W_2 both have the same probability density function $f_W (x)$, W_1 can differ substantially from W_2 as illustrated by the pdf $f_W (x)$ in Fig. 12.4.

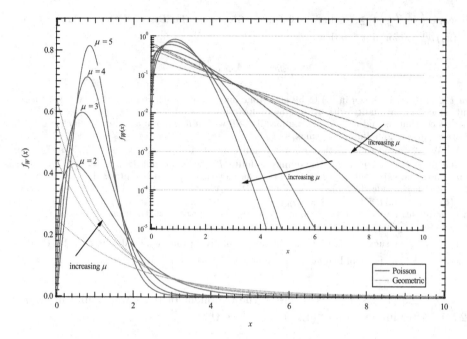

Fig. 12.4. The probability density function $f_W(x)$ of the limit random variable W for both a geometric and a Poisson production process for the same set of values of the average $\mu = E[Y]$.

12.6 A geometric branching process

Consider a pgf $\varphi_Y(z)$ of the form $f(z) = \frac{az+b}{cz+d}$. Beside straightforward iteration of (12.6), a more elegant approach[5] relies on the following property of $f(z)$. For $z = x$, the difference is

$$f(z) - f(x) = \frac{ad - bc}{(d + cz)(d + cx)}(z - x)$$

and, hence, for any two points x_0 and x_1,

$$\frac{f(z) - f(x_0)}{f(z) - f(x_1)} = \left(\frac{cx_1 + d}{cx_0 + d}\right)\left(\frac{z - x_0}{z - x_1}\right)$$

Let us now confine to the two fixed points, x_0 and x_1, of $f(z)$ that are solutions of $f(z) = z$ and let $\xi = \frac{cx_1+d}{cx_0+d}$, then

$$\frac{f(z) - x_0}{f(z) - x_1} = \xi\frac{z - x_0}{z - x_1}$$

---

[5] The linear fractional transformation $f(z)$ is an automorphism of the extended complex plane and basic in the geometric theory of a complex function for which we refer to the book of Sansone and Gerretsen (1960, Vol. 2). Fixed points of an automorphism of the extended plane are the solution of $z = f(z)$, which is a quadratic equation $cz^2 + (d - a)z - b = 0$ and which shows that there are at most two different fixed points.

Now, substitute $z \to f(z)$, then

$$\frac{f(f(z)) - x_0}{f(f(z)) - x_1} = \xi \frac{f(z) - x_0}{f(z) - x_1} = \xi^2 \frac{z - x_0}{z - x_1}$$

Let us denote the iterates of $f(z)$ by $w_n = f_n(z) = f(f_{n-1}(z))$. By iterating, we find that the iterates obey

$$\frac{w_n - x_0}{w_n - x_1} = \xi^n \frac{z - x_0}{z - x_1}$$

or

$$w_n = \frac{x_0(z - x_1) - x_1 \xi^n (z - x_0)}{z - x_1 - \xi^n (z - x_0)} \tag{12.27}$$

Since the probability generating function (3.6) of a geometric random variable Y is of the form $f(z) = \frac{az+b}{cz+d}$, a *geometric branching process* is regarded as a basic reference model in the study of branching processes. The production process in each generation obeys $\Pr[Y = k] = qp^k$ for $k \geq 0$ leading to $\varphi_Y(z) = \frac{q}{1-pz}$, which is slightly different from (3.6). We know that the equation $\varphi_Y(x) = x$ can have two real zeros in $[0,1]$, one at $x_1 = 1$ since $\varphi_Y(z)$ is a probability generating function and another at $x_0 = \frac{q}{p} = \frac{1}{E[Y]} = \frac{1}{\mu} = \pi_0$ such that

$$\xi = \frac{-px_1 + 1}{-px_0 + 1} = \frac{q}{p} = \frac{1}{\mu}$$

The functional equation (12.6) associates $w_n = \varphi_{X_n}(z)$ and after substitution in (12.27) we obtain

$$\varphi_{X_n}(z) = \frac{(\mu^{n-1} - 1) z - \mu^{n-1} + \frac{1}{\mu}}{(\mu^n - 1) z - \mu^n + \frac{1}{\mu}} \tag{12.28}$$

In the case that $E[Y] = \mu \to 1$ or $p = q$, using the rule of de l'Hospital gives

$$\varphi_{X_n}(z) = \frac{(n-1)z - n}{nz - n + 1}$$

From (12.28), the probabilities of extinction at the k-th generation are

$$\Pr[X_k = 0] = \varphi_{X_k}(0) = \frac{1}{\mu} \left(\frac{\mu^k - 1}{\mu^k - \frac{1}{\mu}} \right)$$

If $E[Y] = \mu > 1$, then $\lim_{k \to \infty} \Pr[X_k = 0] = \frac{1}{\mu} = x_0 = \pi_0$ (Theorem 12.3.2) whereas for $E[Y] = \mu \leq 1$, we find that $\lim_{k \to \infty} \Pr[X_k = 0] = 1$. If T_0 is the hitting time defined in Section 9.2.3 as the smallest discrete-time k such that $X_k = 0$, then $\Pr[T_0 \leq k] = \Pr[X_k = 0]$.

The probability generating function of the scaled random variables $W_k = \frac{X_k}{\mu^k}$ follows from (12.28) and (12.13) as

$$\varphi_{W_k}(z) = \frac{(\mu^{k-1} - 1) z^{\mu^{-k}} - \mu^{k-1} + \frac{1}{\mu}}{(\mu^k - 1) z^{\mu^{-k}} - \mu^k + \frac{1}{\mu}}$$

from which $\varphi_{W;Geo}(z) = \lim_{k \to \infty} \varphi_{W_k}(z)$ follows as

$$\varphi_{W;Geo}(z) = \frac{-\frac{1}{\mu} \log z + 1 - \frac{1}{\mu}}{-\log z + 1 - \frac{1}{\mu}} \qquad (12.29)$$

and

$$\chi_{W;Geo}(t) = \frac{\frac{t}{\mu} + 1 - \frac{1}{\mu}}{t + 1 - \frac{1}{\mu}} = 1 + \sum_{k=0}^{\infty} \frac{\mu^{k-1}(-1)^k}{(\mu-1)^{k-1}} t^k \qquad (12.30)$$

Since $\chi_W(t) = E\left[e^{-Wt}\right]$ and using (2.41), all moments are found as $E\left[W^k\right] = \frac{k!\mu^{k-1}}{(\mu-1)^{k-1}}$. Furthermore, with $\pi_0 = \varphi_W(0) = \frac{1}{\mu}$ and from (2.39), the probability density function follows as

$$f_{W;Geo}(x) = \frac{1}{2\pi i} \int_{c-i\infty}^{c+i\infty} \frac{\frac{t}{\mu} + 1 - \frac{1}{\mu}}{t + 1 - \frac{1}{\mu}} e^{xt} dt \qquad (c > 0)$$

By closing the contour for $x > 0$ over the negative $\mathrm{Re}(t)$-plane, we encounter a simple pole at $t = -1 + \frac{1}{\mu} = -(1 - \pi_0) < 0$ (since $\mu > 1$) resulting in

$$f_{W;Geo}(x) = \begin{cases} \left(\frac{1}{\mu} - 1\right)^2 \exp\left(-x\left(1 - \frac{1}{\mu}\right)\right) & x > 0 \\ \frac{1}{\mu}\delta(x) & x = 0 \\ 0 & x < 0 \end{cases} \qquad (12.31)$$

From (12.12) the variance is $\mathrm{Var}[W_{Geo}] = \frac{\mu+1}{\mu-1}$. The limit random variable W_{Geo} of a geometric branching process is exponentially distributed with an atom at $x = 0$ equal to the extinction probability $\pi_0 = \frac{1}{\mu}$. From (12.23), the exponent $\beta_{Geo} = 1$ for any value of $\mu \geq 1$. Comparing (12.31) and the general relation (12.26) for small x indicates that the parameter $F_\mu = \left(\frac{1}{\mu} - 1\right)^2$ for a geometric production process.

The limit random variable W for production processes Y of which all moments exist can be computed via Taylor series expansions. In Van Mieghem (2005), series for both $\chi_{W;Po}(t)$ and $f_{W;Po}(x)$ of a Poisson branching process are presented. Fig. 12.4 illustrates that the probability density function $f_{W;Po}(x)$ of a Poisson branching process is definitely distinct from that of geometric branching process. Since $E[W] = 1$, the variance $\mathrm{Var}[W_{Po}] = \frac{1}{\mu-1}$ of a Poisson limit random variable W_{Po} implies that $f_{W;Po}(x)$ is centered around $x = 1$ more tightly as μ increases.

12.7 Time-dependent branching process

So far, we have only considered simple, discrete-time branching processes, called Galton-Watson processes, in which items are produced per generation without detailing the lifetime of an item. All items are of the same type reproduced by one production process and different generations do not overlap: the tree in Fig. 12.1 is organized per generation. In reality, an item produced at time $t = 0$ possesses a (random) lifetime L with probability distribution $G(t) = \Pr[L \leq t]$ for $t \geq 0$.

When the item dies or fails, it is replaced by Y similar items of age zero, which all live or are operational during a time that is distributed as their parent's lifetime L. Moreover, the lifetimes of all items are independent from each other and from the production process. If $Z(t)$ denotes the number of items living at time t, then $\{Z(t), t \geq 0\}$ is called a Bellman-Harris process, which was first introduced by Harris (1963) as an *age-dependent* branching process. Items in the Bellman-Harris process are still of the same type, but generations can overlap in time, similar to families where children of brothers or sisters can be older than their uncle or aunt.

Let us consider the event $\{Z(t) = k\}$. If $L > t$, then the first item is still living and $Z(t) = 1$, else the item has died. In the latter case, assuming that the item dies at time $0 \leq y \leq t$, it is replaced by Y offspring that produce, in the remaining time $t - y$, a total of k offspring. Using the continuous representation of the law of total probability (2.49), we can write

$$\Pr[Z(t) = k] = \int_0^\infty \Pr[Z(t) = k | L = y] \, d\Pr[L \leq y]$$

$$= \int_0^t \Pr[Z(t) = k | L = y] \, dG(y) + 1_{\{k=1\}} \int_t^\infty d\Pr[L \leq y]$$

$$= 1_{\{k=1\}} (1 - G(t)) + \int_0^t \Pr[Z(t) = k | L = y] \, dG(y)$$

Further, for $0 \leq y \leq t$, the replacement of the first item at time y by Y offspring results in

$$\Pr[Z(t) = k | L = y] = \Pr\left[\sum_{n=1}^Y Z_n(t - y) = k \right]$$

$$= \sum_{j=0}^\infty \Pr\left[\sum_{n=1}^j Z_n(t - y) = k \right] \Pr[Y = j]$$

where Z_n counts all items produced by the n-th descendant of the first item. All random variables $Z_1, Z_2, \ldots, Z_j$ are independent and identically distributed as Z so that $\Pr\left[\sum_{n=1}^j Z_n(t - y) = k \right]$ is the j-th fold convolution of $\Pr[Z(t - y) = k]$. Thus,

$$\Pr[Z(t) = k] = 1_{\{k=1\}} (1 - G(t)) + \int_0^t \sum_{j=0}^\infty \Pr\left[\sum_{n=1}^j Z_n(t - y) = k \right] \Pr[Y = j] \, dG(y)$$

Transforming to probability generating functions by multiplying both sides with s^k and summing over all $k \geq 0$, and using

$$\varphi_{Z(t)}(s) = E\left[s^{Z(t)} \right] = \sum_{k=0}^\infty \Pr[Z(t) = k] s^k$$

with

$$\sum_{k=0}^{\infty} \Pr\left[\sum_{n=1}^{j} Z_n (t-y) = k\right] s^k = E\left[s^{\sum_{n=1}^{j} Z_n(t-y)}\right]$$

$$= \prod_{n=1}^{j} E\left[s^{Z_n(t-y)}\right] = \varphi_{Z(t-y)}^{j}(s)$$

and $\sum_{j=0}^{\infty} \varphi_{Z(t-y)}^{j}(s) \Pr[Y = j] = \varphi_Y\left(\varphi_{Z(t-y)}(s)\right)$, yields the integral equation for the pgf, for $t \geq 0$,

$$\varphi_{Z(t)}(s) = s\left(1 - G(t)\right) + \int_0^t \varphi_Y\left(\varphi_{Z(t-y)}(s)\right) dG(y) \qquad (12.32)$$

which is the basic equation, investigated in detail by Harris (1963) and Athreya and Ney (1972).

Observe that the Bellman-Harris process $\{Z(t), t \geq 0\}$ is not Markovian in general, but only when the lifetime distribution $G(t) = 1 - e^{-\lambda t}$ is an exponential distribution. Also, if the lifetime is deterministic and equal to $L = 1$ so that $G(t) = 1_{\{t \geq 1\}}$, then the basic equation (12.32) reduces to

$$\varphi_{Z(t)}(s) = s1_{\{t<1\}} + \varphi_Y\left(\varphi_{Z(t-1)}(s)\right) 1_{\{t \geq 1\}}$$

which equals, for $t = k + 1$ and $k \in \mathbb{N}$, the basic equation (12.6) of a branching process that started with one item ($X_0 = 1$). In that case, $Z(k) = X_k$, the number of items produced in the k-th generation, studied before. The non-linearity of (12.32) is a characteristic feature of the age-dependence in contrast to the linearity of the Markov processes of age-independent theory.

12.7.1 Relation with renewal theory and mean growth of $Z(t)$

Taking the derivative of (12.32) with respect to s and evaluation of the result for $s = 1$ yields, with $\mu = \varphi_Y'(1)$ and $M(t) = E[Z(t)]$,

$$M(t) = (1 - G(t)) + \mu \int_0^t M(t-y) dG(y) \qquad (12.33)$$

The mean number of living items at time t satisfies an equation of a similar form as the renewal equation (8.9). By conditioning on the lifetime L of the first item (similar as the probabilistic argument in Section 8.1.3), we obtain

$$E[Z(t)|L = y] = \begin{cases} 1 & t < y \\ \mu E[Z(t-y)] & t \geq y \end{cases}$$

or, rewritten as a conditional expectation

$$E[Z(t)|L] = 1_{t<L} + \mu E[Z(t-y)] 1_{t \geq L}$$

After taking expectations and using (2.82), we arrive again at (12.33).

Equation (12.33) is of the form (8.13), whose unique solution (8.14) translates here to

$$M(t) = (1 - G(t)) + \int_0^t (1 - G(t - u)) \, dm_\mu(u) \qquad (12.34)$$

where the generalized renewal function is $m_\mu(t) = \sum_{k=1}^\infty \mu^k G^{(k*)}(t)$. If $\mu = 1$ and, since $M(0) = 1$, we verify that $M(t) = 1$ is a solution of (12.33). In the sequel, we focus on the asymptotic solution. Instead of starting from the solution (12.34), we will transform (12.33) and apply the tools of renewal theory.

Besides $M(0) = 1$, the equation (12.33) for $M(t) = E[Z(t)]$ is not precisely a renewal equation (8.9) when $\mu \neq 1$. Fortunately, we can transform the equation for $M(t)$ into a general renewal equation (8.11) by multiplying both sides by[6] $e^{-\alpha t}$,

$$e^{-\alpha t} M(t) = e^{-\alpha t}(1 - G(t)) + \mu \int_0^t e^{-\alpha(t-y)} M(t - y) e^{-\alpha y} dG(y)$$

Since

$$d\left(\int_0^y e^{-\alpha u} dG(y)\right) = e^{-\alpha y} dG(y)$$

we obtain

$$e^{-\alpha t} M(t) = e^{-\alpha t}(1 - G(t)) + \int_0^t e^{-\alpha(t-y)} M(t - y) \, d\left(\mu \int_0^y e^{-\alpha u} dG(y)\right)$$

which is recast into a general renewal equation (8.11), for $Y(t) = e^{-\alpha t} M(t)$, $h(t) = e^{-\alpha t}(1 - G(t))$ and the distribution

$$F(y) = \mu \int_0^y e^{-\alpha u} dG(u) \qquad (12.35)$$

which is called the *stable age* distribution. In order for $F(y)$ to be a distribution function for which $\lim_{y \to \infty} F(y) = 1$, we must require that α satisfies

$$\mu \int_0^\infty e^{-\alpha u} dG(u) = 1 \qquad (12.36)$$

Since the integral decreases monotonously as a function of α, the solution of (12.36) for α is a unique. If $\mu \geq 1$, there is always a positive solution; if $\mu = 1$, then $\alpha = 0$, whereas for $\mu < 1$, a solution may not exist, but when it exists, $\alpha < 0$.

---

[6] The exponential function is unique. For, after multiplying both sides of (12.33) by $r(t)$ yields

$$r(t) M(t) = r(t)(1 - G(t)) + \mu \int_0^t r(t) M(t - y) dG(y)$$

Reduction to the general renewal equation (8.11) requires that $r(t) = r(t - u) r(u)$. This Cauchy equation is only satisfied by the exponential function: $r(t) = e^{-\alpha t}$ for some real number α.

Since $h(t) = e^{-\alpha t}(1 - G(t))$ is bounded, the solution (8.12) of the general renewal equation is

$$e^{-\alpha t}M(t) = e^{-\alpha t}(1 - G(t)) + \int_0^t e^{-\alpha(t-u)}(1 - G(t-u))\,dm(u)$$

where $m(t) = \sum_{k=1}^{\infty} F^{(k*)}(t)$ is the renewal function. Since $h(t)$ is direct Riemann integrable[7] over $[0, \infty)$, the Key Renewal Theorem 8.2.2 leads to

$$\lim_{t\to\infty} e^{-\alpha t}M(t) = \frac{\int_0^\infty e^{-\alpha u}(1 - G(u))\,du}{\int_0^\infty y\,dF(y)} = \frac{\frac{1}{\alpha} - \int_0^\infty e^{-\alpha u}G(u)\,du}{\mu \int_0^\infty y e^{-\alpha y}\,dG(y)}$$

$$= \frac{\frac{1}{\alpha}\left(1 - \int_0^\infty e^{-\alpha u}\,dG(u)\right)}{\mu \int_0^\infty y e^{-\alpha y}\,dG(y)}$$

Finally, after invoking (12.36), we find asymptotically that the mean number $M(t) = E[Z(t)]$ of living items grows[8] as

$$M(t) \sim \frac{\mu - 1}{\alpha \mu^2 \int_0^\infty y e^{-\alpha y}\,dG(y)} e^{\alpha t} \tag{12.37}$$

This law (12.37) is valid provided there exists a solution for α in the equation (12.36).

12.7.1.1 The Malthusian parameter

The law (12.37) indicates the importance of the constant α, which is a solution of (12.36) and which is called the Malthusian parameter. When $\mu > 1$, (12.37) characterizes the famous exponential growth law of $M(t) = E[Z(t)]$ as postulated by Malthus in 1798 and before by Euler, as mentioned in Haccou *et al.* (2005). When the lifetime L is exponentially distributed, $G(t) = 1 - e^{-\lambda t}$, the Mathusian parameter α is found from (12.36) as $\alpha = (\mu - 1)\lambda$. Because Y and L are independent, the governing equation (12.36) can be written as

$$E\left[Y e^{-\alpha L}\right] = E[Y]E\left[e^{-\alpha L}\right] = 1$$

where $\varphi_L(z) = E\left[e^{-\alpha L}\right]$ the probability generating function of the lifetime L. The Mathusian parameter $\alpha = \varphi_L^{-1}\left(\frac{1}{E[Y]}\right)$ can be regarded as a characteristic rate of the Bellman-Harris branching process, because $1/\alpha$ possesses the same dimension unit (seconds) as that of the lifetime L. The transformation of the integral equation (12.33) for $M(t)$ into a genuine renewal equation (8.11) requires that the adjusted reproduction process creates $Y e^{-\alpha L}$ items, with mean equal to one as in a counting or renewal process, where at each death or failure, only one item is replaced.

---

[7] We exclude so-called lattice distributions. A Δ-lattice distribution, in short lattice distribution, is constant except for jumps at positive integer multiples of the basic length Δ.

[8] Harris (1963) mentions that, when G is a Δ-lattice distribution, $M(t)$ is constant on each interval $(0, \Delta), (\Delta, 2\Delta), \ldots$ and on the interval $(r\Delta, (r+1)\Delta)$, for large r,

$$M(r\Delta + \varepsilon) \sim \frac{(\mu - 1)\Delta}{\mu^2(1 - e^{-\alpha\Delta})\int_0^\infty y e^{-\alpha y}\,dG(y)} e^{\alpha r\Delta}$$

12.7.1.2 Non-existence of the Malthusian parameter

If $\mu < 1$ and a solution of (12.36) for α does not exist, we must resort to the general solution (12.34) to explore the asymptotic behavior of $M(t)$. After integrating by parts, (12.34) becomes

$$M(t) = (1 - G(t))(1 - m_\mu(0)) + (1 - G(0)) m_\mu(t) + \int_0^t m_\mu(u) dG(t - u)$$

Using the bound (8.5) on convolutions shows, for large t, that $G^{(k*)}(t) \to 1$ and $m_\mu(t) \sim \sum_{k=1}^{\infty} \mu^k = \frac{\mu}{1-\mu}$, independent of t. Hence, for large t,

$$\int_0^t m_\mu(u) dG(t - u) \sim \frac{\mu}{1 - \mu} \int_0^t dG(t - u) = \frac{\mu}{1 - \mu} (G(0) - G(t))$$

and, with $G(0) = m_\mu(0) = 0$,

$$M(t) \sim \frac{1 - G(t)}{1 - \mu} \tag{12.38}$$

so that $\lim_{t \to \infty} M(t) = 0$.

Athreya and Ney (1972) specify the distributions $G(t)$ for which the Malthusian parameter α does not exist, which they call the sub-exponential class that obeys

$$\lim_{t \to \infty} \frac{1 - G^{(2*)}(t)}{1 - G(t)} = 2$$

Examples are Pareto-like distributions, $1 - G(t) \sim t^{-k}$ (where $k > 0$), $1 - G(t) \sim \exp(t^{-\beta})$ (where $0 < \beta < 1$) and lognormal distributions. The decay in (12.38) of the mean number $M(t) = E[Z(t)]$ of living items is determined by the tail of the sub-exponential class of distributions.

These sub-exponential lifetimes cause a strongly non-Markovian behavior for $Z(t)$ (even in the case when $\mu > 1$, for which the Malthusian parameter α always exists). Iribarren and Moro (2011) found that the response time to emails is well approximated by a lognormal distribution with parameters (τ, σ). When $\mu < 1$, (12.38) applies and the decay of the lognormal distribution (3.53) gives approximately

$$M(t) \sim \frac{1}{1 - \mu} \frac{\sigma}{\sqrt{2\pi}} \frac{\exp\left(-\frac{(\log t - \tau)^2}{2\sigma^2}\right)}{\log t - \tau}$$

which illustrates that the decay is not exponential and that information propagation prevails for much longer times than expected (based on Markovian behavior).

12.7.2 The age distribution

We now concentrate on the number $Z(x, t)$ of items in a Bellman-Harris process at time t of age smaller than or equal to x. Analogously to (12.32), we can deduce the governing relation for the probability generating function $\varphi_{Z(x,t)}(s) = E\left[s^{Z(x,t)}\right]$ as

$$\varphi_{Z(x,t)}(s) = (1 - G(t))\left(s 1_{\{x \geq t\}} + 1_{\{x < t\}}\right) + \int_0^t \varphi_Y\left(\varphi_{Z(x,t-y)}(s)\right) dG(y)$$

from which the renewal-type of equation for the mean $M(x,t) = E[Z(x,t)]$ follows, by differentation, as

$$M(x,t) = (1 - G(t)) 1_{\{x \geq t\}} + \mu \int_0^t M(x, t - y) dG(y) \tag{12.39}$$

The integral equation (12.39) is very similar to that in (12.33) of $M(t)$, except for the indicator function. Indeed, we have that $Z(t) = \lim_{x \to \infty} Z(x,t)$. The same analysis that led to the Malthusian growth law (12.37) can be followed, where the former $h(t) = e^{-\alpha t}(1 - G(t))$ is now replaced by $h(t) = e^{-\alpha t}(1 - G(t)) 1_{\{x \geq t\}}$. Again excluding lattice distributions, the Key Renewal Theorem 8.2.2 then yields

$$\lim_{t \to \infty} e^{-\alpha t} M(x,t) = \frac{\int_0^\infty e^{-\alpha u}(1 - G(u)) 1_{\{x \geq u\}} du}{\int_0^\infty y dF(y)}$$

$$= \frac{\int_0^x e^{-\alpha u}(1 - G(u)) du}{\int_0^\infty e^{-\alpha u}(1 - G(u)) du} \frac{\int_0^\infty e^{-\alpha u}(1 - G(u)) du}{\int_0^\infty y dF(y)}$$

Hence,

$$\lim_{t \to \infty} e^{-\alpha t} M(x,t) = F_A(x) \lim_{t \to \infty} e^{-\alpha t} M(t)$$

where the limiting distribution (for large t) of the age A equals

$$F_A(x) = \Pr[A \leq x] = \frac{\int_0^x e^{-\alpha u}(1 - G(u)) du}{\int_0^\infty e^{-\alpha u}(1 - G(u)) du} \tag{12.40}$$

The asymptotic law (for large t) of the mean number $M(x,t) = E[Z(x,t)]$ of living items of age smaller than or equal to x is

$$M(x,t) \sim \frac{(\mu - 1) F_A(x)}{\alpha \mu^2 \int_0^\infty y e^{-\alpha y} dG(y)} e^{\alpha t} \tag{12.41}$$

The analogy with the age distribution in renewal theory is worth mentioning. In Section 8.3, we have derived both the residual waiting time $R(t)$ as well as the age $A(t)$ and found that, with $F_\tau(t) = G(t)$,

$$\lim_{t \to \infty} \Pr[A(t) \leq x] = \frac{\int_0^x (1 - G(u)) du}{\int_0^\infty (1 - G(u)) du} \tag{12.42}$$

The limiting age distribution (12.40) in a branching process differs from the limiting age distribution (12.42) in a renewal process by the multiplication effects of produced items. If $\mu > 1$, the Malthusian exponential growth (12.41) indicates that the proportion of younger items is larger than in a renewal process, where a dead item is replaced by precisely one, instead of, on average, μ items in a branching process.

This difference is reflected by the factor $e^{-\alpha u}$ in (12.40). The argument also shows that, when the production of items is deterministic, thus $Y = 1$ or $\varphi_Y(z) = z$, the Bellman-Harris process is a renewal process.

12.8 Problems

(i) Let us consider a simple branching process $\{X_k\}_{k \geq 0}$ with $X_0 = 1$ and a discrete production process Y with $E[Y] = \mu$. The generating function $\varphi_Y(z) = E[z^Y]$ has two fixed points (that obey $z = \varphi_Y(z)$) for real $x \in [0, 1]$, the extinction probability π_0 and $x = 1$. Show that

$$\pi_0 \mu \leq \varphi'_Y(\pi_0) \leq 1 \tag{12.43}$$

if $\log \varphi_Y(z)$ is concave for real $z \in [0, 1]$. (*Hint:* For $\mu > 1$, $\varphi'_Y(\pi_0) < 1$ as shown in the proof of Lemma 12.5.1.)

Notice that for continuous production functions with probability generating function $E[e^{-tY}]$, $\log E[e^{-tY}]$ is always convex for all t (as shown in Section 5.5). Also, $\pi_0 \mu \leq 1$ does not hold in general as shown in the following example. Let $\Pr[Y = 0] = \frac{1}{2}$ and $\Pr[Y = 4] = \frac{1}{2}$, then $E[Y] = \mu = 2$ and $\pi_0 > \Pr[Y = 0] = \frac{1}{2}$, whence, $\pi_0 \mu > 1$.

(ii) Show that the probability of extinction in a time-dependent branching process

$$\pi_0 = \lim_{t \to \infty} \Pr[Z(t) = 0]$$

also satisfies the equation (12.18).

(iii) The propagation of information, where each receiver forwards the message to some of its email contacts, can be approximated by a time-dependent branching process (Iribarren and Moro, 2011). Assume that the mean email production $\mu > 1$. When the lifetime L has a Gamma distribution, with pdf specified by (3.25) with $\alpha = \theta$, verify and discuss that the Malthusian parameter is given by

$$\alpha = \frac{\mu^{\frac{1}{\beta}} - 1}{\theta}$$

The Gamma distribution, with $\beta \simeq 0$ and $\theta \simeq 20$ days, was proposed by Vazquez *et al.* (2007) to model the email response time. For $\beta < 1$, the Gamma function presents a power law with exponential cut-off.

13

General queueing theory

Queueing theory describes basic phenomena such as the waiting time, the through-put, the losses, the number of queueing items, etc. in queueing systems. Following Kleinrock (1975), any system in which arrivals place demands upon a finite-capacity resource can be broadly termed a *queueing system*.

Queuing theory is a relatively new branch of applied mathematics that is generally considered to have been initiated by A. K. Erlang in 1918 with his paper on the design of automatic telephone exchanges, in which the famous Erlang blocking probability, the Erlang B-formula (14.18), was derived (Brockmeyer *et al.*, 1948, p. 139). It was only after the Second World War, however, that queueing theory was boosted mainly by the introduction of computers and the digitalization of the telecommunications infrastructure. For engineers, the two volumes by Kleinrock (1975, 1976) are perhaps the most well-known, while in applied mathematics, apart from the penetrating influence of Feller (1970, 1971), the *Single Server Queue* of Cohen (1969) is regarded as a landmark. Since Cohen's book, which incorporates most of the important work before 1969, a wealth of books and excellent papers have appeared, an evolution that is still continuing today.

13.1 A queueing system

Examples of queueing abound in daily life: queueing situations at a ticket window in the railway station or post office, at the cash points in the supermarket, the waiting room at the airport, train or hospital, etc. In telecommunications, the packets arriving at the input port of a router or switch are buffered in the output queue before transmission to the next hop towards the destination. In general, a queueing system consists of (a) arriving items (packets or customers), (b) a buffer or waiting room, (c) a service center and (d) departures from the system.

The main processes as illustrated in Fig. 13.1 are stochastic in nature. Initially in queueing theory, the main stochastic processes were described in continuous time, while with the introduction of the asynchronous transfer mode (ATM) in the late 1980s, many queueing problems were more effectively treated in discrete time, where the basic time unit or time slot was the minimum service time of one ATM

287

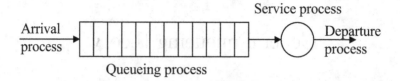

Fig. 13.1. The main processes in a general queueing system.

cell. In the literature, there is unfortunately no widely adopted standard notation for the main random variables, which often troubles the transparency. Let us start defining the main random variables in continuous time.

13.1.1 *The arrival process*

The arrival process is characterized by the arrival time t_n of the n-th packet (customer) and the interarrival time $\tau_n = t_n - t_{n-1}$ between the n-th and $(n-1)$-th packet. If all interarrival times are i.i.d. random variables with distribution $F_\tau(t)$, then

$$\Pr[\tau_n \leq t] = F_\tau(t)$$

As illustrated in Fig. 8.1, we can associate a counting process $\{N(t), t \geq 0\}$ to the arrival process $\{t_n, t \geq 0\}$ by the equivalence $\{N(t) \geq n\} \iff \{t_n \leq t\}$. In other words, *if all interarrival times are i.i.d.*, the number of arriving packets (customers) is a general renewal process with interarrival time distribution specified by $F_\tau(t)$. We mention explicitly the condition of independence, which was initially considered as a natural assumption. In recent measurements, however, arrivals of IP packets are shown not to obey this simple condition of independence, which has lead to the use of complicated self-similar and long-range dependent arrival processes.

In the sequel, we will use the following notation: $N_A(t)$ is the number of arrivals at time t, while $A(t) = \int_0^t N_A(u)du$ is the total number of arrivals in the interval $[0, t]$.

13.1.2 *The service process*

The service process is specified in similar way by the service time x_n of the n-th packet (customer). If the random variables x_n are i.i.d. with distribution $F_x(t)$, then

$$\Pr[x_n \leq t] = F_x(t) \tag{13.1}$$

The service process needs additional specifications. First of all, in a single-server queueing system, only one packet (customer) is served at a time. If there is more than one server, more packets can evidently be served simultaneously. Next, we must detail the *service discipline* or *scheduling rule*, which describes the way a packet is treated. There is a large variety of service disciplines. If all packets are of equal priority, the simplest rule is first-in-first-out (FIFO), which serves the packets in the same order in which they arrive. Other types such as last-in-first-out or a random order are possible, though in telecommunication, FIFO occurs most often. If we have packets of different multimedia flows, all with different quality of service requirements, not all packets have equal priority. For instance, a delay-sensitive packet (of e.g. a voice call) must be served as soon as possible preferably before non-delay-sensitive packets (of e.g. a file transfer). In these cases, packets are extracted from the queue by a certain scheduling rule. The simplest case is a two-priority system with a head-of-the line scheduling rule: high-priority packets are always served before low-priority packets. In the sequel, we confine the presentation to a single-server system with one type of packet and a FIFO discipline. Hence, we omit a discussion of scheduling rules. A next assumption is that of *work conservation*: if there is a packet waiting for service, the server will always serve the packet. Thus, the server is only idle if there are no packets waiting in the buffer and immediately starts service when the first packet is placed in the queue or arrives. In a non-work-conservative system, the server may stay idle, even if there are customers waiting (e.g. a situation where patients have to wait during a coffee break in a hospital). Finally, we assume that the arrival process is independent of the service process. Situations where arriving packets of some type (e.g. control) change the way the remaining packets in the buffer are served, or a service discipline that serves at a rate proportional to the number of waiting packets, are not treated.

The service in a router consists in fetching the packet from the buffer, inspecting the header to determine the correct output port and in placing the packet on the output link for transmission.

In this chapter unless the contrary is explicitly mentioned, we consider a single-server queueing system under a work-conservative, FIFO service discipline in which the arrival and service process are independent.

13.1.3 *The queueing process*

From Fig. 13.1, we observe at least two aspects regarding the queue or buffer: (a) the number of different queues and (b) the number of positions in the queue. In general, a queueing system may have several queues or even a shared queue for different servers. For example, in a router, there is one physical fast memory or buffer in which arriving packets are placed. Depending on the output interfaces, each link driver per output port is a server that extracts the packet destined for its link from the common buffer and transmits the packet on this link. For simplicity, we consider here only one queue with K positions. Often queueing analyses are greatly simplified in the infinite buffer case $K \to \infty$. If the buffer is infinitely long, there is zero loss, as opposed to the finite buffer case in which losses can occur if the queue is full and packets arrive.

So far, the description of the queueing system is complete: we have specified the arrival process, the service process and the physical size of the waiting room or queue. We now turn our attention to desirable quantities that can be deduced from the model specification of the queueing system such as (a) the waiting or queueing time w_n of the n-th packet, (b) the system time $T_n = w_n + x_n$ of the n-th packet, (c) the unfinished work (also called the virtual waiting time or workload) $v(t)$ at time t, (d) the number of packets in the queue $N_Q(t)$ or in the system $N_S(t)$ at time t and (e) the departure time r_n of the n-th packet.

The waiting or queueing time w_n of the n-th packet is only zero if the queue is empty at arrival time t_n. The unfinished work $v(t)$ at time t is the time needed to empty the queueing system or to serve all remaining packets in the system (queue plus server) at time t. Hence, the unfinished work at time t is equal to the sum of the service times of the $N_Q(t)$ buffered packets at time t plus the remaining service time of the packet under service at time t. Precisely at an arrival epoch $t = t_n$ as illustrated in Fig. 13.2, we observe that $v(t_n) = T_n = w_n + x_n$. In addition, $v'(t) = -1$ for all $t \neq t_n$ and $v(t) = \max[T_n - t + t_n, 0]$ for $t_n \leq t < t_{n+1}$.

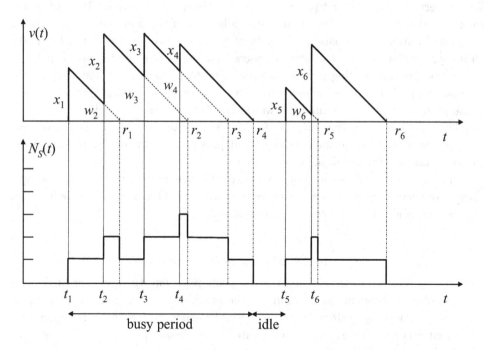

Fig. 13.2. The unfinished work $v(t)$ and the number of packets in the system $N_S(t)$ as function of time. At any new arrival at t_n holds $v(t_n) = w_n + x_n$. The unfinished work $v(t)$ decreases with slope -1 between two arrivals. The waiting times w_n and departure times r_n are also shown. Notice that $w_1 = w_5 = 0$.

The departure times r_n satisfy $r_n = t_n + T_n$. The time during which the server is busy is called a busy period, and likewise, the interval of non-activity is called an idle period.

13.1.4 The Kendall notation for queueing systems

Kendall introduced a notation that is commonly used to describe or classify the type of a queueing system. The general syntax is $A/B/n/K/m$, where A specifies the interarrival process, B the service process, n the number of servers, K the number of positions in the system (queue plus servers) and m restricts the number of allowed arrivals in the queueing system. Examples for both the interarrival distribution A and the service distribution B are M (memoryless or Markovian) for the exponential distribution, G for a general distribution[1] and D for a deterministic distribution. When other letters are used besides these three common assignments, the meaning will be defined. For example, $M/G/1$ stands for a queueing system with exponentially distributed interarrival times, a general service distribution and 1 server. If the two last identifiers K and m are not written, they should be interpreted as infinity[2]. Hence, $M/G/1$ has an infinitely long queue and no restriction on the number of allowed arrivals.

13.1.5 The traffic intensity ρ

An important parameter in any queueing system is the traffic intensity, also called the load or the utilization, defined as the ratio of the mean service time $E[x] = \frac{1}{\mu}$ over the mean interarrival time $E[\tau] = \frac{1}{\lambda}$

$$\rho = \frac{E[x]}{E[\tau]} = \frac{\lambda}{\mu} \tag{13.2}$$

where λ and μ are the mean interarrival and service rate, respectively. Clearly, if $\rho > 1$ or $E[x] > E[\tau]$, which means that the mean service time is longer than the mean interarrival time, then the queue will grow indefinitely long for large t, because packets are arriving faster on average than they can be served. In this case ($\rho > 1$), the queueing system is unstable or will never reach a steady-state. The case where $\rho = 1$ is critical. In practice, therefore, mostly situations where $\rho < 1$ are of interest. If $\rho < 1$, a steady-state can be reached. These considerations are a direct consequence of the law of conservation of packets in the system, but can be proved rigorously by ergodic theory or Markov steady-state theory, which determine when the process is positive recurrent.

13.2 The waiting process: Lindley's approach

From the definition of the waiting time and from Fig. 13.2, a relation between w_{n+1} and w_n is found. Suppose the waiting time for the first packet $w_1 = w$, which is the initialization. If $r_n \leq t_{n+1}$, which means that the n-th packet leaves the queueing

---

[1] Often G is written where GI, general independent process, is meant. We interpret G as a general interarrival process, which can be correlated over time.

[2] If $K \to \infty$, but m is finite, then all identifiers need to be written because $A/B/n/\infty/m$ is obviously different from $A/B/n/m$.

system before the $(n+1)$-th packet arrives, the system is idle and $w_{n+1} = 0$. In all other situations, $r_n > t_{n+1}$, the n-th packet is still in the queueing system while the next $(n+1)$-th packet arrives and $w_{n+1} = t_n + w_n + x_n - t_{n+1}$. Indeed, the waiting time of $(n+1)$-th packet equals the system time $T_n = w_n + x_n$ of the n-th packet which started at t_n minus his own arrival time t_{n+1}. During the interval $[t_n, t_{n+1}]$, the queueing system has processed an amount of the unfinished work equal to $t_{n+1} - t_n = \tau_{n+1}$ time units. Hence, we arrive at the general recursion for the waiting time,

$$w_{n+1} = \max\left[w_n + x_n - \tau_{n+1}, 0\right] \tag{13.3}$$

Let $\sigma_n = x_n - \tau_{n+1}$, then

$$\begin{aligned} w_{n+1} &= \max\left[0, w_n + \sigma_n\right] \\ &= \max\left[0, \max\left[w_{n-1} + \sigma_{n-1}, 0\right] + \sigma_n\right] \\ &= \max\left[0, \sigma_n, w_{n-1} + \sigma_{n-1} + \sigma_n\right] \end{aligned}$$

and, by iteration,

$$w_{n+1} = \max\left[0, \sigma_n, \sigma_{n-1} + \sigma_n, \sigma_{n-2} + \sigma_{n-1} + \sigma_n, \ldots, \sum_{k=1}^{n} \sigma_k + w_1\right] \tag{13.4}$$

A number of observations are in order:

First, *if both the interarrival times τ_{n+1} and the service times x_n are i.i.d. random variables* and mutually independent, then the differences σ_n are i.i.d. random variables. In addition, w_n and σ_n are also independent because (13.4) shows that w_n only depends on σ_k with indices $k < n$. Then, the waiting time process $\{w_n\}_{n\geq 1}$ is a discrete-time Markov process with a continuous state space (the waiting times w_n are positive real numbers) because the general relation (13.3) reveals that, since the random variable σ_n is independent of w_n, the waiting time for the $(n+1)$-th packet is only dependent on the waiting time of the previous n-th packet. This is the Markov property. Since the state space is a continuum, it is not a Markov chain, merely a Markov process.

Second, if there exists a packet m for which $w_m = 0$ (e.g. packet $m = 5$ in Fig. 13.2), which means that the m-th packet finds the system empty, then all packets after the m-th packet are isolated from the effects of those before the m-th. Mathematically, this separation between two busy periods directly follows from (13.3) because $w_{m+1} = \max\left[0, \sigma_m\right]$ leading via iteration for $n \geq m$ to

$$w_{n+1} = \max\left[0, \sigma_n, \sigma_{n-1} + \sigma_n, \ldots, \sum_{k=m+1}^{n} \sigma_k, \sum_{k=m}^{n} \sigma_k\right]$$

In other words, this relation is similar to (13.4) as if the system were started from $k = m$ and $w_m = 0$ instead of $k = 1$ with $w_1 = w$. Any busy period can be regarded as a renewal of the waiting process, independent of the previous busy periods.

Third, again invoking the assumption that σ_n are i.i.d. random variables, then the order in the sequence $\{\sigma_n\}_{n \geq 1}$ is of no importance in (13.4) and we may relabel the random variables in (13.4) as $\sigma_k \to \sigma_{n-k+1}$ to obtain a new random variable

$$\omega_{n+1} = \max\left[0, \sigma_1, \sigma_1 + \sigma_2, \ldots, \sum_{k=1}^{n-1} \sigma_k, \sum_{k=1}^{n} \sigma_k + w_1\right]$$

which is identically distributed as w_{n+1}. The interest of this observation is that, provided $w_1 = 0$ and, hence, $\omega_n = \max_{1 \leq j \leq n} \sum_{k=0}^{n-j} \sigma_k$ where $\sigma_0 = 0$, the sequence $\{\omega_n\}_{n \geq 1}$ can only increase with n, because the maximum cannot decrease[3] if an additional term $\sum_{k=0}^{n} \sigma_k$ is added. Thus, if $w_1 = 0$, the event $\{\omega_{n+1} < x\}$ is always contained in $\{\omega_n < x\}$. In steady-state, which is reached if $n \to \infty$,

$$\lim_{n \to \infty} \{\omega_n < x\} = \cap_{n=1}^{\infty} \{\omega_n < x\} = \{\sup_{j \geq 0} \sum_{k=0}^{j} \sigma_k < x\}$$

which means that the random variable ω_n with same distribution as the waiting time w_n converges to a limit random variable that is the supremum of the terms $\sum_{k=0}^{j} \sigma_k$ in the series. From this relation, it follows that the steady-state distribution $W(x)$ of the waiting time is

$$W(x) = \lim_{n \to \infty} \Pr\left[w_n < x\right] = \lim_{n \to \infty} \Pr\left[\omega_n < x\right] = \Pr\left[\sup_{j \geq 0} \sum_{k=0}^{j} \sigma_k < x\right]$$

if the latter probability exists, i.e. not zero for all x. Lindley has proved that, if $\rho < 1$, the latter corresponds to a proper probability distribution. In other words, the steady-state distribution of the waiting time in a GI/G/1 system[4] exists. Alternatively, the Markov process $\{w_n\}_{n \geq 1}$ is ergodic if $\rho < 1$.

Lindley's proof is as follows. Due to the assumption that σ_n are i.i.d. random variables, the Strong Law of Large Numbers (6.3) is applicable: $\Pr\left[\lim_{n \to \infty} \frac{1}{n} \sum_{k=0}^{n} \sigma_k = E\left[\sigma\right]\right] = 1$ where $E\left[\sigma\right] = E\left[x\right] - E\left[\tau\right] < 0$ (the mean service time is smaller than the mean interarrival time) if $\rho < 1$ while $E\left[\sigma\right] > 0$ if $\rho > 1$. In case $\rho > 1$, there exists a number $\nu > 0$ and $\beta > 1$ such that, for all $n > \nu$, holds $\sum_{k=0}^{n} \sigma_k \geq \frac{1}{\beta} E\left[\sigma\right] n$ with probability 1. For large ν, $\sum_{k=0}^{n} \sigma_k$ can be made larger than any fixed x such that $\Pr\left[\sup_{j \geq 0} \sum_{k=0}^{n} \sigma_k < x\right] = 0$. In case $\rho < 1$, we have for sufficiently large n that $\sum_{k=0}^{n} \sigma_k < 0$. Thus, for any $x > 0$ and $\epsilon > 0$, there exists a number ν (independent of x) such that, for all $n > \nu$,

$$\Pr\left[\sum_{k=0}^{n} \sigma_k < x\right] \geq \Pr\left[\sum_{k=0}^{n} \sigma_k < 0\right] > 1 - \epsilon$$

while, for $n < \nu$, we can always find a number $\xi > 0$ such that for all $x > \xi$,

$$\Pr\left[\sum_{k=0}^{n} \sigma_k < x\right] > 1 - \epsilon$$

---

[3] This observation cannot be made from (13.4) because σ_n, which affects all but the first term in the maximum, can be negative.

[4] Notice that the analysis crucially relies on the independence of the interarrival and service process.

Since $\sup_{j \geq 0} \sum_{k=0}^{j} \sigma_k$ is attained for $j < \nu$ or $j > \nu$ and because both regimes can be bounded by the same lower bound,

$$\Pr\left[\sup_{j \geq 0} \sum_{k=0}^{j} \sigma_k < x\right] > \Pr\left[\sum_{k=0}^{n < \nu} \sigma_k < x\right] 1_{j < \nu} + \Pr\left[\sum_{k=0}^{n > \nu} \sigma_k < x\right] 1_{j > \nu}$$

$$> 1 - \epsilon$$

Clearly, $\lim_{x \to \infty} \Pr\left[\sup_{j \geq 0} \sum_{k=0}^{j} \sigma_k < x\right] = 1$ and $\Pr[w_n < 0] = 0$, thus $\Pr\left[\sup_{j \geq 0} \sum_{k=0}^{j} \sigma_k < x\right]$ is non-decreasing and a proper probability distribution. We omit the considerations for the case $\rho = 1$. $\qquad\square$

We now concentrate on the computation of the steady-state distribution for the waiting time (in the queue) in the case that the load $\rho < 1$ and under the confining assumption that *both the interarrival times τ_{n+1} and the service times x_n are i.i.d. random variables*. We find from (13.3) that

$$\Pr[w_{n+1} < x] = \Pr[w_n + \sigma_n < x] \quad \text{if } x > 0$$
$$\Pr[w_{n+1} < x] = 0 \quad \text{if } x \leq 0$$

With the law of total probability (2.49) and since σ_n can be negative, the right-hand side is

$$\Pr[w_n + \sigma_n < x] = \int_{-\infty}^{\infty} \Pr[w_n < x - s | \sigma_n = s] \frac{d}{ds} \Pr[\sigma_n < s]\, ds$$

Using the independence of w_n and σ_n, and that $w_n \geq 0$, we obtain for $x > 0$,

$$\Pr[w_n + \sigma_n < x] = \int_{-\infty}^{x} \Pr[w_n < x - s]\, d\Pr[\sigma_n < s]$$

The distribution $\Pr[\sigma_n < s] = \Pr[x_n - \tau_{n+1} < s] = C_n(s)$ can be computed (see Problem (v) in Chapter 3) provided the interarrival and service time processes are known. Proceeding to the steady-state by letting $n \to \infty$ amounts to Lindley's integral equation in $W(x) = \lim_{n \to \infty} \Pr[w_n < x]$ with $C(x) = \lim_{n \to \infty} C_n(x)$,

$$W(x) = \int_{-\infty}^{x} W(x - s)dC(s) \quad \text{if } x > 0$$
$$W(x) = 0 \quad \text{if } x \leq 0 \tag{13.5}$$

The integral equation (13.5) is of the Wiener-Hopf type and treated in general by Titchmarsh (1948, Section 11.17) and specifically by Kleinrock (1975, Section 8.2) and Cohen (1969, p. 337). Apart from Lindley's approach, Pollaczek has used variants of the complex integral expression for

$$\max(x, 0) = \frac{x e^{-ax}}{2\pi i} \int_{c-i\infty}^{c+i\infty} \frac{e^{xz}}{z - a} dz \qquad (c > \operatorname{Re}(a) > 0)$$

to treat the complicating non-linear function $\max(x, 0)$ in (13.3). Several other approaches (Kleinrock, 1975, Chapter 8) have been proposed to solve (13.3). We will only discuss the approach due to Beneš, because his approach does not make the confining assumption that both the interarrival times τ_{n+1} and the service times

x_n are i.i.d. random variables. As mentioned before, in Internet traffic, which has been shown to be long-range dependent (i.e. correlated over many time units mainly due to TCP's control loop), the interarrival times can be far from independent.

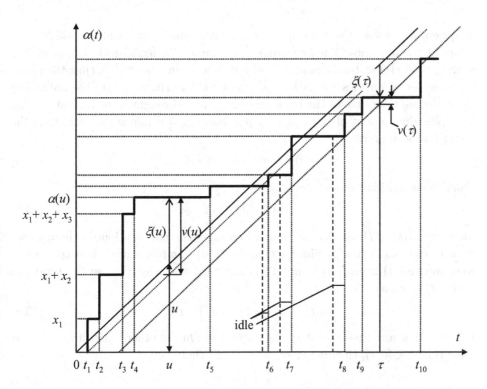

Fig. 13.3. The amount of work arriving to the queueing system $\alpha(t)$ versus time t. At $t = u$, we observe that $\xi(t) = \alpha(t) - t + v(0^-) > 0$. The largest value of $\xi(t) - \xi(u)$ is found for $u = t_1$ because $\xi(t_1) = -t_1$, the only negative value of $\xi(t)$ in $[0, u)$. Graphically, we shift the line at $45°$ so to intersect the point $(t_1, \alpha(t_1))$ to determine $v(u)$. At $t = \tau$, $\xi(\tau) < 0$ and the largest negative value of $\xi(t)$ in $[0, \tau)$ is attained at $t = t_8$. Three of the five idle periods have also been shown.

13.3 The Beneš approach to the unfinished work

Instead of observing the queueing system at a time t, the Beneš approach considers the behavior over a time interval $[0, t)$. Let $\alpha(t)$, $\eta(t)$ and $b(t)$ denote the amount of work arriving to the queueing system in the interval $[0, t)$, the total idle time of the server and the total busy time of the server in the interval $[0, t)$, respectively. The amount of work arriving to the system is expressed in units of time and must be regarded as the time needed to process this work, similarly to the definition of the unfinished work. If the work to process arrives at discrete times, then $\alpha(t)$

increases in jumps, as illustrated in Fig. 13.3,

$$a(t) = \sum_{j=0}^{A(t)} x_j$$

In general, however, the work may arrive continuously over time, possibly with jumps at certain times. The purpose is to determine the unfinished work or virtual waiting time $v(t)$ at time instant t, and not over a time interval $[0, t)$ as the previously defined quantities and $A(t) = \int_0^t N_A(u)du$. Clearly, for $t > 0$, the unfinished work at time t consists of the total amount of work brought in by arrivals during $[0, t)$ plus the amount of work present just before $t = 0$ minus the total time the server has been active,

$$v(t) = v(0^-) + \alpha(t) - b(t) \tag{13.6}$$

From the definitions above,

$$\eta(t) + b(t) = t \tag{13.7}$$

Moreover, $\alpha(t)$, $\eta(t)$ and $b(t)$ are non-decreasing and right continuous (jumps may occur) functions of time t. Since $\eta(t)$ and $b(t)$ are complementary, it is convenient to eliminate $b(t)$ from (13.6) and (13.7) and further concentrate on the total idle time $\eta(t)$ given as

$$\eta(t) = v(t) + t - v(0^-) - \alpha(t) \tag{13.8}$$

If $v(u) > 0$ at any time $u \in [0, t)$, then $\eta(t) = 0$. On the other hand, if $v(u) = 0$ at some time $u \in [0, t)$, then it follows from (13.8) that

$$\eta(u) = u - v(0^-) - \alpha(u) \tag{13.9}$$

Since $\eta(t)$ is non-decreasing in t, the total idle time in the interval $[0, t)$ at moments u when the buffer is empty ($v(u) = 0$) is the largest value for η that can be reached in $[0, t)$,

$$\eta(t) = \sup_{0 < u < t} \left(u - v(0^-) - \alpha(u) \right)$$

and the supremum is needed because $\alpha(t)$ can increase discontinuously (in jumps). Combining the two regimes, we obtain in general that

$$\eta(t) = \max \left[0, \sup_{0 < u < t} \left(u - v(0^-) - \alpha(u) \right) \right] \tag{13.10}$$

Equating the two general expressions (13.8) and (13.10) for the total idle time of the server leads to an new relation for the unfinished work,

$$v(t) = v(0^-) + \alpha(t) - t + \max \left[0, \sup_{0 < u < t} \left(u - v(0^-) - \alpha(u) \right) \right]$$

$$= \max \left[v(0^-) + \alpha(t) - t, \sup_{0 < u < t} \{ \alpha(t) - t - (\alpha(u) - u) \} \right]$$

The quantity $\xi(t) = \alpha(t) - t + v(0^-)$ is recognized as the server overload during $[0, t)$, while $\alpha(t) - t$ is the amount of excess work arriving during the interval $[0, t)$. Thus, $\xi(t) - \xi(u)$ is the amount of excess work during $[u, t)$ or the overload of the server during $[u, t)$ provided $u > 0$ and $\xi(0) = v(0^-)$. Then,

$$v(t) = \max\left[\xi(t),\ \sup_{0 < u < t}\{\xi(t) - \xi(u)\}\right]$$

and, with the convention that $v(0^-) = \sup_{0 < u < 0^-}\{\xi(0) - \xi(u)\} = \xi(0)$,

$$v(t) = \sup_{0 < u < t}\{\xi(t) - \xi(u)\} \tag{13.11}$$

The unfinished work $v(t)$ at time t is equal to the largest value of the overload or excess work during any interval $[u, t) \subset [0, t)$. The relation (13.11) is illustrated and further explained in Fig. 13.3. This general relation (13.11) shows that the unfinished work is the maximum of a stochastic process. Furthermore, if $v(t) = 0$, (13.11) indicates that $\sup_{0 < u < t}\{\xi(t) - \xi(u)\} = 0$. Let u^* denote the value at which $\sup_{0 < u < t}\{\xi(t) - \xi(u)\} = \xi(t) - \xi(u^*) = 0$. But $\xi(u^*)$ is the lowest value of $\xi(t)$ in $[0, t)$ and, unless an arrival occurs during the interval $[t, t + \Delta t]$, $\xi(t + \Delta t) = \xi(t) - \Delta t < \xi(t)$. This argument shows that, as soon as a new idle period begins, $\xi(t)$ attains the minimum value so far.

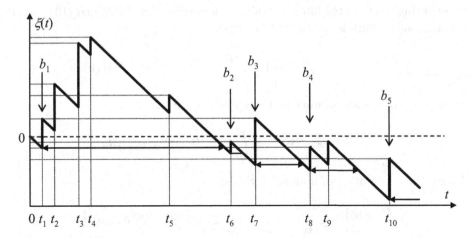

Fig. 13.4. The excess work $\xi(t)$ for the same process as in previous plot. The arrows with b_j denote the start of the j-th busy period. Observe that $\xi(b_j)$ is the minimum so far and that a busy period ends at $t > b_j$ for which $\xi(t) = \xi(b_j)$. The length of a busy period has been represented by a double arrow.

During the idle period as shown in Fig. 13.4, $\xi(t)$ further decreases linearly with slope -1 towards a new minimum $\xi(b_j)$ in $[0, b_j]$ until the beginning of a new busy period, say the j-th at $t = b_j$. Then, for all $b_j < t < b_{j+1}$,

$$v(t) = \xi(t) - \xi(b_j) = \sup_{b_j < u < t}\{\xi(t) - \xi(u)\}$$

In other words, we observe that idle periods decouple the past behavior from future behavior, as deduced earlier from the waiting time analysis in Section 13.2. As illustrated in Fig. 13.4, the series $\{\xi(b_j)\}$, where b_j denotes the start of the j-th busy period, is monotonously decreasing in b_j, i.e. $\xi(b_j) > \xi(b_{j+1})$ for any j.

Let us proceed to compute the distribution of the unfinished work following an idea due to Beneš. Beneš applies the identity[5], valid for all z,

$$e^{-zt} = 1 - z \int_0^t e^{-z\tau} d\tau$$

to the total idle time of the server by putting $\tau = \eta(u)$

$$e^{-zt} = 1 - z \int_{\eta^{-1}(0)}^{\eta^{-1}(t)} e^{-z\eta(u)} d\eta(u)$$

where $\eta^{-1}(t)$ is the inverse function. Note that $\eta(0) = 0$ and that $d\eta(u) = 1_{\{v(u)=0\}} du = \delta(v(u)) du$ where $\delta(x)$ is the Dirac impulse. Let $t \to \eta(t)$, then

$$e^{-z\eta(t)} = 1 - z \int_0^t e^{-z\eta(u)} \delta(v(u)) \, du$$

Substituting (13.9) in the integral, which is only valid if $v(u) = 0$, and (13.8) at the left-hand side, which is generally valid, gives

$$e^{-z\left(v(t)+t-v(0^-)-\alpha(t)\right)} = 1 - z \int_0^t e^{-z\left(u-v(0^-)-\alpha(u)\right)} \delta(v(u)) \, du$$

or, in terms of the excess work $\xi(t) = \alpha(t) - t + v(0^-)$,

$$e^{-zv(t)} = e^{-z\xi(t)} - z \int_0^t e^{-z(\xi(t)-\xi(u))} \delta(v(u)) \, du$$

Taking the expectation of both sides yields,

$$E\left[e^{-zv(t)}\right] = E\left[e^{-z\xi(t)}\right] - z \int_0^t E\left[e^{-z(\xi(t)-\xi(u))} \delta(v(u))\right] du$$

Recall that, with (2.35), with the definition of a generating function (2.38), and further with (2.64),

$$Q = E\left[e^{-z(\xi(t)-\xi(u))} \delta(v(u))\right]$$

$$= \int_{-\infty}^{\infty} \int_{-\infty}^{\infty} e^{-zx} \delta(y) \frac{\partial^2}{\partial x \partial y} \Pr\left[\xi(t) - \xi(u) \le x, v(u) \le y\right] dx dy$$

---

[5] Borovkov (1976, p. 30) proposes another but less simple approach by avoiding the use of the identity ingeniously introduced by Beneš.

and with (2.48), we have

$$Q = \int_{-\infty}^{\infty} \int_{-\infty}^{\infty} e^{-zx} \delta(y) \frac{\partial^2}{\partial x \partial y} \Pr\left[\xi(t) - \xi(u) \le x | v(u) \le y\right] \Pr\left[v(u) \le y\right] dx dy$$

$$= \int_{-\infty}^{\infty} dx e^{-zx} \frac{\partial}{\partial x} \int_{-\infty}^{\infty} \delta(y) \Pr\left[\xi(t) - \xi(u) \le x | v(u) \le y\right] \frac{\partial \Pr\left[v(u) \le y\right]}{\partial y} dy$$

$$= \int_{-\infty}^{\infty} e^{-zx} \frac{d}{dx} \Pr\left[\xi(t) - \xi(u) \le x | v(u) = 0\right] \Pr\left[v(u) = 0\right] dx$$

Combining all leads to

$$\int_{-\infty}^{\infty} e^{-zx} f_{v(t)}(x) dx = \int_{-\infty}^{\infty} e^{-zx} f_{\xi(t)}(x) dx$$
$$- z \int_{-\infty}^{\infty} e^{-zx} \frac{d}{dx} \left(\int_0^t \Pr\left[\xi(t) - \xi(u) \le x | v(u) = 0\right] \Pr\left[v(u) = 0\right] du \right) dx$$

By partial integration, we can remove the factor z at both sides. Indeed, since $\Pr\left[v(t) \le x\right] = 0$ for $x < 0$,

$$\int_{-\infty}^{\infty} e^{-zx} f_{v(t)}(x) dx = z \int_{-\infty}^{\infty} e^{-zx} \Pr\left[v(t) \le x\right] dx$$

Hence, we arrive at

$$\int_{-\infty}^{\infty} e^{-zx} \Pr\left[v(t) \le x\right] dx = \int_{-\infty}^{\infty} e^{-zx} \left[\Pr\left[\xi(t) \le x\right] \right.$$
$$\left. - \frac{d}{dx} \left(\int_0^t \Pr\left[\xi(t) - \xi(u) \le x | v(u) = 0\right] \Pr\left[v(u) = 0\right] du \right) \right] dx$$

which is equivalent to

$$\Pr\left[v(t) \le x\right] = \Pr\left[\xi(t) \le x\right] - \int_0^t \frac{d \Pr\left[\xi(t) - \xi(u) \le x | v(u) = 0\right]}{dx} \Pr\left[v(u) = 0\right] du \quad (13.12)$$

This general relation for the distribution of the unfinished work in terms of the excess work is the Beneš equation. If $v(u) = 0$ for all $u \in [0, t)$, this means that during that interval no work arrives and that $\xi(t) - \xi(u) = u - t$ or that $-t \le \xi(t) - \xi(u) \le 0$ for any $u \in [0, t)$. Thus, if we choose $x \in [-t, 0)$ such that the event $\{\xi(t) - \xi(u) = u - t \le x\}$ is possible, the probabilities appearing in the right-hand side are not identically zero while $\Pr\left[v(t) \le x\right] = 0$. Hence, for $x \in [-t, 0)$, the Beneš equation reduces to

$$\Pr\left[\xi(t) \le x\right] = \int_0^t \frac{d \Pr\left[\xi(t) - \xi(u) \le x | v(u) = 0\right]}{dx} \Pr\left[v(u) = 0\right] du$$

from which the unknown probability of an empty system $\Pr\left[v(u) = 0\right]$ can be found[6]

---

[6] This relation in the unknown function $f(u) = \Pr\left[v(u) = 0\right]$ is a Volterra equation of the first kind (see e.g. Morse and Feshbach (1978, Chapter 8))

$$g(z) = \int_a^z K(z|u) f(u) du$$

These integral equations frequently appear in physics in boundary problems, potential and Green's function theory.

for $t + x \le u \le t$. The Beneš equation translates the problem of finding the time-dependent virtual waiting time or unfinished work in an integral equation that, in principle, can be solved. We further note that in the derivation hardly any assumptions about the queueing system nor the arrival process are made such that the Beneš equation provides the most general description of the unfinished work in any queueing system. Of course, the price for generality is a considerable complexity in the integral equation to be solved. However, we will see examples[7] of its use in ATM.

13.3.1 A constant service rate

If the server operates deterministically as in ATM, for example, the amount of work arriving to the queueing system in the interval $[0, t)$ simplifies to $\alpha(t) = A(t)$, the number of ATM cells in the interval $[0, t)$, because $x_j = x$ is the time to process one ATM cell, which we take as time unit $x = 1$. With this convention, we have that $\xi(t) = A(t) - t$. After substitution of $u = t - y$, the integral I in (13.12) is

$$ I = \int_0^t \frac{d\Pr\left[\xi(t) - \xi(t-y) \le x | v(t-y) = 0\right]}{dx} \Pr\left[v(t-y) = 0\right] dy $$

and the event

$$ \{\xi(t) - \xi(t-y) \le x\} = \{A(t) - A(t-y) \le x + y\} $$

Since $A(t) - A(t-y)$ is a non-negative integer k and thus a discrete random variable, the probability density function is

$$ \frac{d\Pr[\xi(t) - \xi(t-y) \le x | v(t-y) = 0]}{dx} = \Pr[A(t) - A(t-y) = k | v(t-y) = 0]\, 1_{x+y=k} $$

which implies that only values at $y = k - x$ contribute to the integral I. Hence, $0 \le y \le t$ implies that $\lceil x \rceil \le k \le \lfloor x + t \rfloor$, where $\lfloor z \rfloor$ (respectively $\lceil z \rceil$) is the integer equal to or smaller (respectively larger) than z,

$$ I = \sum_{k=\lceil x \rceil}^{\lfloor x+t \rfloor} \Pr\left[A(t) - A(t+x-k) = k | v(t+x-k) = 0\right] \Pr\left[v(t+x-k) = 0\right] $$

Hence, for a discrete queue with time slots equal to the constant service time, the Beneš equation reduces to

$$ \Pr\left[v(t) \le x\right] = \Pr\left[A(t) \le \lfloor x + t \rfloor\right] \tag{13.13} $$

$$ - \sum_{k=\lceil x \rceil}^{\lfloor x+t \rfloor} \Pr\left[A(t) - A(t+x-k) = k | v(t+x-k) = 0\right] \Pr[v(t+x-k) = 0] $$

---

[7] Borovkov (1976) investigates the Beneš method in more detail. He further derives from (13.12) formulae for light and heavy traffic, and the discrete time process.

13.3.2 The steady-state distribution of the virtual waiting time

Let us turn to the steady-state distribution $V(x) = \lim_{t\to\infty} \Pr[v(t) \leq x]$. Since $\alpha(t)$ is the amount of work arriving to the queueing system in the interval $[0, t)$, $\frac{\alpha(t)}{t}$ is the mean amount of work arriving in that interval $[0, t)$. The steady-state stability condition, equivalent to $\rho < 1$, requires that

$$\lim_{t\to\infty} \frac{\alpha(t)}{t} = \rho < 1$$

because the server capacity is 1 unit of work per unit of time. Since $\alpha(t)$ is not decreasing, $\xi(t)$ decreases continuously with slope -1 between two arrivals and increases (possibly discontinuously with jumps) during arrival epochs, as illustrated in Fig. 13.3. In the stable steady-state regime where $\rho < 1$ and $\lim_{t\to\infty} \frac{\alpha(t)}{t} < 1$, we find that

$$\lim_{t\to\infty} \xi(t) = \lim_{t\to\infty} t\left(\frac{\alpha(t)}{t} - 1\right) = -\infty$$

and thus $\lim_{t\to\infty} \Pr[\xi(t) \leq x] = 1$ and $\lim_{t\to\infty} \frac{\xi(t)}{t} = \rho - 1 < 0$. From (13.7), we see that

$$\lim_{t\to\infty} \frac{\eta(t)}{t} = 1 - \lim_{t\to\infty} \frac{b(t)}{t} = 1 - \rho$$

which, with (13.9), suggests that

$$V(0) = \lim_{t\to\infty} \Pr[v(t) = 0] = 1 - \rho \tag{13.14}$$

If the Strong Law of Large Numbers (Theorem 6.2.2) is applicable, which implies that the lengths of the idle periods are independent and identically distributed, this relation $V(0) = 1 - \rho$ is proved to be true by Borovkov (1976, pp. 33–34). Hence, for any stationary single-server system with traffic intensity ρ, the probability of an empty system at an arbitrary time is $1 - \rho$.

Taking the limit $t \to \infty$ in (13.12) then yields

$$V(x) = 1 - \lim_{t\to\infty} \int_0^t \frac{d\Pr[\xi(t) - \xi(t-y) \leq x | v(t-y) = 0]}{dx} \Pr[v(t-y) = 0] dy$$

The tail probability $1 - V(x) = \lim_{t\to\infty} \Pr[v(t) > x] = \Pr[v(t_\infty) > x]$ is

$$\Pr[v(t_\infty) > x] = \int_0^\infty \frac{d\Pr[\xi(t_\infty) - \xi(t_\infty - y) \leq x | v(t_\infty - y) = 0]}{dx} \Pr[v(t_\infty - y) = 0] dy$$

This relation shows that, at a point in the steady-state $t_\infty \to \infty$, the contributions to $V(x)$ are due to arrivals and idle periods in the past. The corresponding steady-state equation for (13.13) is

$$\Pr[v(t_\infty) > x] = \sum_{k=\lceil x \rceil}^{\infty} \Pr\left[v(t_\infty + x - k) = 0 \,\Big|\, \int_{t_\infty + x - k}^{t_\infty} N_A(u)\, du = k\right]$$

$$\times \Pr\left[\int_{t_\infty + x - k}^{t_\infty} N_A(u)\, du = k\right] \tag{13.15}$$

13.4 The counting process

A relation of a same general nature as (13.3) can be deduced for the associated counting process with a single server

$$N_S(r_{k+1}^+) = N_S(r_k^+) - 1 + A(r_{k+1}^+) - A(r_k^+) \qquad (13.16)$$

where $N_S\left(r_k^+\right) = N_S\left(r_k + \epsilon\right)$, which, in fact, means that we observe the system just after a departure at time[8] $r_k^+ = r_k + \epsilon$. The system equation (13.16) states that the number of packets in the system just after the departure time of the $(k+1)$-th packet equals the number of packets in the system just after the departure time of the previous packet k minus the $(k+1)$-th packet, but increased by the number of arrivals in the time interval $(r_k^+, r_{k+1}^+]$, for all $r_k > 0$ and $r_0 = 0$. A number of observations need to be made. First, the departing packet at r_k is *not* included in $N_S\left(r_k^+\right)$. Second, in order to have a departure, there must be at least one packet in the system. Hence, if $N_S\left(r_k^+\right) = 0$, this implies that $A(r_{k+1}^+) - A(r_k^+) > 0$. The max-operator, that was crucial in (13.3), does not appear. Third, we consider the half open interval $(r_k^+, r_{k+1}^+]$ to avoid double counting or any overlap between consecutive intervals at the departure epochs r_k for all $k \geq 0$. In a continuous-time analysis, the boundaries of the interval (open or closed) are not important because the probability that two events happen at precisely the same time is zero almost surely. However, in discrete-time, where all time slots have a unit length and indexed by the integer k, coinciding events are possible.

The extension of (13.16) to multiple servers c requires, in addition to the arrival proccess, the definition of $B\left(t\right)$, the number of served packets in the interval $[0, t]$, where possibly multiple packets are served at the same time. For each time $t \geq 0$, the conservation of packets means that

$$N_S\left(t\right) = A\left(t\right) - B\left(t\right)$$

from which, for any non-negative time u and t,

$$N_S\left(t + u\right) = N_S\left(t\right) + \left(A\left(t + u\right) - A\left(t\right)\right) - \left(B\left(t + u\right) - B\left(t\right)\right)$$

In the particular case where departures are synchronized at certain departure times, observation of the system just after the departure times yields

$$N_S\left(r_{k+1}^+\right) = N_S\left(r_k^+\right) - X\left(r_{k+1}^+\right) + A(r_{k+1}^+) - A(r_k^+)$$

where $X\left(r_{k+1}\right) = B\left(r_{k+1}\right) - B\left(r_k\right)$ is the number of packets that departs from the system at time r_{k+1} due to some synchronization feature. For a single server, where $c = 1$ and, thus $X\left(r_{k+1}\right) = 1$, we find again (13.16), that is completely general because there is no synchronization of departing packets. The argument also shows that multiple server systems are inherently more complicated than the single server case. The dual case occurs for "batch" arrivals, where successive packets arrive at

---

[8] We write $x^+ = x + \epsilon$ and $x^- = x - \epsilon$ where $\epsilon > 0$ is an arbitrarily small, positive real number. The notation x^+ should not be confused with $(x)^+ = \max(x, 0)$.

the system at arrival times t_k for $k \geq 0$. Observation of the system just before the arrival times at $t_k^- = t_k - \epsilon$ leads to

$$N_S\left(t_{k+1}^-\right) = N_S\left(t_k^-\right) + Y\left(t_{k+1}^-\right) - B(t_{k+1}^-) + B(t_k^-)$$

where $Y(t_{k+1}) = A(t_{k+1}) - A(t_k)$ is the number of packets that arrives at the system at time t_{k+1} in "batch".

Whereas the waiting time process is more natural to consider in problems where interarrival times are specified, the counting process has more advantages in a discrete-time analysis. In discrete-time queuing models, the time axis is divided in intervals of equal length, called time slots, and s_k indicates the time of the k-th time slot boundary. In those discrete-time queuing models, it is further assumed that the service time of a packet equals an integer number of time slots and that the service is synchronized, i.e. only starts at time slot boundaries (that occur at time s_k). A major consequence of this assumption is that the service of a packet cannot start earlier than at the beginning of the time slot following the time slot of the packet's arrival. The usual observation epochs of the system in discrete-time are just after the time slot boundaries, that lead to the following system equation:

$$N_S\left(s_{k+1}^+\right) = N_S\left(s_k^+\right) - \mathcal{B}_k + \mathcal{A}_k$$

where $\mathcal{B}_k = B\left(s_{k+1}^+\right) - B\left(s_k^+\right)$ is the number of departing packets and $\mathcal{A}_k = A\left(s_{k+1}^+\right) - A\left(s_k^+\right)$ is the number of arriving packets during time slot k, i.e. during the interval $(s_k, s_{k+1}]$. In case all service times are equal to one time slot, and there are c servers, then we obtain

$$N_S\left(s_{k+1}^+\right) = \left(N_S\left(s_k^+\right) - c\right)^+ + \mathcal{A}_k \tag{13.17}$$

where $(x)^+ \equiv \max(x, 0)$. The $\max(x, 0)$-operator is now needed because we decouple the time dependence relation between departures $\mathcal{B}_k$ and arrivals $\mathcal{A}_k$ by letting $\mathcal{B}_k = c$ for each $k \geq 0$, irrespective of the arrival process. In other words, the number of departures during time slot k is only equal to $\mathcal{B}_k = c$ provided $N_S\left(s_k^+\right) \geq c$, else the number of departures equals $N_S\left(s_k^+\right) < c$. Thus, if $N_S\left(s_k^+\right) < c$, at most $N_S\left(s_k^+\right)$ packets can leave the system at time s_{k+1}, because the arrivals $\mathcal{A}_k$ during the k-th time slot need at least one extra time slot to be processed (which is one of the assumptions of a discrete-time model) and cannot depart in the same time slot during which they have entered the system. Hence, the $\max(x, 0)$-operator precisely "reconciles" the decrease in the number of packets in the system at each time slot by c with the requirement that $N_S(t) \geq 0$: in each time slot, at most c packets can depart from the system.

For any work-conserving system (in continuous-time as well as discrete-time), it holds at any time t that

$$N_Q(t) = (N_S(t) - c)^+ \tag{13.18}$$

because the number of packets in the queue $N_Q(t)$ plus the number of packets currently being served in the c servers equals the total number $N_S(t)$ of packets in the system. In the special discrete-time case where all service times are equal to one time slot, combining (13.17) and (13.18) yields

$$N_S\left(s_{k+1}^+\right) = N_Q\left(s_k^+\right) + \mathcal{A}_k$$

It will be convenient to simplify the notation: $\mathcal{S}_k = N_S(s_k^+)$ denotes the system content (i.e. the number of occupied queue positions including the packets currently being served) at the beginning of time slot k, $\mathcal{Q}_k = N_Q(s_k^+)$ is the queue content at the beginning of time slot k, respectively. Rewriting (13.17) shows that the system content satisfies the continuity, also called balance or system, equation

$$\mathcal{S}_{k+1} = (\mathcal{S}_k - c)^+ + \mathcal{A}_k \tag{13.19}$$

whereas the queue content obeys

$$\mathcal{Q}_{k+1} = (\mathcal{Q}_k - c + \mathcal{A}_k)^+ \tag{13.20}$$

On the other hand, the relation (13.18) between system and queue content implies that $\mathcal{Q}_k = (\mathcal{S}_k - c)^+$ such that (13.19) is rewritten as

$$\mathcal{S}_{k+1} = \mathcal{Q}_k + \mathcal{A}_k \tag{13.21}$$

The number of packets at the beginning of a time slot $k + 1$ in the system is the sum of the number of queued packets at the beginning of the previous time slot k and the newly arrived packets during time slot k.

13.5 Queue observations

It is worthwhile to investigate the relation between observations at various instances of time of the queueing process $\{N_S(t), t \geq 0\}$ which represents the number of packets in the system at time t. As seen before and as illustrated in Fig. 13.5, two time instances seem natural: an inspection at departure times where $N_S(r_n^+)$ describes the number of packets in the system just after departure of the n-th packet and an observation at arrival times where $N_S(t_n^-)$ describes the number of packets in the system just before the n-th packet enters.

Suppose that the n-th packet leaves $N_S(r_n^+) = k \leq j$ packets behind in the system. This implies that precisely k arrivals after the n-th packet have entered the system. Hence, the $(n + k + 1)$-th packet sees, just before entering the system, at most k packets, because during the period $t = r_n$ and $t = t_{n+k+1} \geq r_n$, only departures are possible. Thus, $N_S(t_{n+k+1}^-) \leq k$ and, clearly, for $j \geq k$, the $(n + j + 1)$-th packet observes no more than $N_S(t_{n+j+1}^-) \leq j$. Hence, the following implication holds:

$$\left\{N_S(r_n^+) \leq j\right\} \implies \left\{N_S(t_{n+j+1}^-) \leq j\right\}$$

Consider now the converse. Suppose that the $(n + j + 1)$-th packet sees precisely $k \leq j$ packets in front of it upon arrival: $N_S(t_{n+j+1}^-) = k \leq j$. This implies

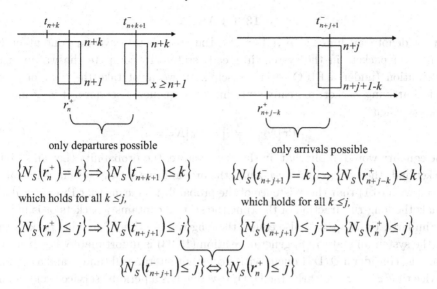

Fig. 13.5. Relation between queue observations at arrival and at departure epochs.

that the $(n + j + 1 - k)$-th packet is the first packet that will leave the system after $t = t_{n+j+1}$ and that the $(n + j - k)$-th packet has already left the system. At its departure at $t = r_{n+j-k} \leq t_{n+j+1}$, it has observed at most k packets behind it, because only arrivals are possible in the interval $[r_{n+j-k}, t_{n+j+1})$. Hence, $N_S(r_{n+j-k}^+) \leq k$ and, setting $k = j$, then $N_S(r_n^+) \leq j$ leads to the implication

$$\{N_S(t_{n+j+1}^-) \leq j\} \implies \{N_S(r_n^+) \leq j\}$$

Combining both implications leads to the equivalence,

$$\{N_S(t_{n+j+1}^-) \leq j\} \iff \{N_S(r_n^+) \leq j\}$$

or, for any sample path (or realization), it holds, for any non-zero integer j, that

$$\Pr\left[N_S(t_{n+j+1}^-) \leq j\right] = \Pr\left[N_S(r_n^+) \leq j\right]$$

In steady-state for $n \to \infty$, with $\lim_{n\to\infty} N_S(t_n^-) = N_{S;A}$ and $\lim_{n\to\infty} N_S(r_n^+) = N_{S;D}$, we find that

$$\Pr\left[N_{S;A} = j\right] = \Pr\left[N_{S;D} = j\right] \tag{13.22}$$

In words, in steady-state, the number of packets in the system observed by arriving packets is equal in probability to the number of packets in the system left behind by departing packets. Of course, we have assumed that the steady-state distribution exists. If one of these distributions exists, the analysis demonstrates that the other must exist. Notice that no assumptions about the distribution or dependence are made and that (13.22) is a general result which only assumes the existence of a steady-state.

13.6 PASTA

Let us denote by $\lim_{t\to\infty} N_S(t) = N_S$ the steady-state system content or the number of packets in the system in steady-state. To compute the waiting time distribution (under a FIFO service discipline), we must take the view of how a typical arriving packet in steady-state finds the queue. Therefore, it is of interest to know when

$$\Pr[N_{S;A} = j] \stackrel{?}{=} \Pr[N_S = j] \tag{13.23}$$

The equality would imply that, in the steady-state, the probability that an arriving packet finds the system in state j equals the probability that the system *is* in state j. Recall with (6.1) that the existence of the probabilities means that $\Pr[N_S = j]$ also equals the long-run fraction of the time the system contains j packets or is in state j. Similarly, $\Pr[N_{S;A} = j]$ also equals the long-run fraction of arriving packets that see the system in state j. In general, relation (13.23) is unfortunately not true. For example, consider a D/D/1 queue with a constant interarrival time τ_c and a constant service time $x_c < \tau_c$. Clearly, the D/D/1 system has a periodic service cycle: a busy period takes x_c time units and the idle period equals $\tau_c - x_c$ time units. Thus, every arriving packet always finds the system empty and concludes $\Pr[N_{S;A} = 0] = 1$, while $\Pr[N_S = 1] = \frac{x_c}{\tau_c}$ and $\Pr[N_S = 0] = \frac{\tau_c - x_c}{\tau_c}$. The waiting time computation of the GI/D/c system in Section 14.4.2 is another counter example.

Relation (13.23) is true for Poisson arrivals and this property is called "*Poisson arrivals see time averages*" (PASTA).

Theorem 13.6.1 (PASTA) *The long-run fraction of time that a process spends in state j is equal to the long-run fraction of Poisson arrivals that find the process in state j,*

$$\Pr[N_{S;A} = j] = \Pr[N_S = j]$$

Proof: See[9] e.g. Wolff (1982). □

The Poisson process has the typical property that future increments are independent of the past and, thus also of the past system history. In certain sense, Poisson arrivals perform a random sampling which is sufficient to characterize the steady-state of the system exactly. The PASTA property also applies to Markov chains. The transitions in continuous-time Markov chains are Poisson processes if self-transitions are allowed (see Section 10.4.1). For any state j, the fraction of Poisson events that see the chain in state j is π_j, which (see Lemma 6.1.2) also equals the fraction of time the chain is in state j.

13.7 Little's Law

Little's Law is perhaps the simplest of the general queueing formulae.

---

[9] Although Wolff's general proof (Wolff, 1982) only contains two pages, it is based on martingales and on axiomatic probability theory.

Theorem 13.7.1 (Little's Law) *In the steady-state, the mean number of packets (customers) in the system $E[N_S]$ (or in the queue $E[N_Q]$) equals the mean arrival rate λ times the mean time spent in the system $E[T]$ (or in the queue $E[w]$),*

$$E[N_S] = \lambda E[T] \tag{13.24}$$
$$E[N_Q] = \lambda E[w]$$

Little's Law holds if two of the three limits

$$\lim_{t\to\infty} \frac{A(t)}{t} = \lambda \tag{13.25}$$

$$\lim_{t\to\infty} \frac{1}{t} \int_0^t N_S(u)\,du = E[N_S] \tag{13.26}$$

$$\lim_{n\to\infty} \frac{1}{n} \sum_{k=1}^n T_k = E[T] \tag{13.27}$$

exist. The arrival rate λ is the rate at which packets enter the system; lost or rejected packets, considered in Section 14.9, are excluded.

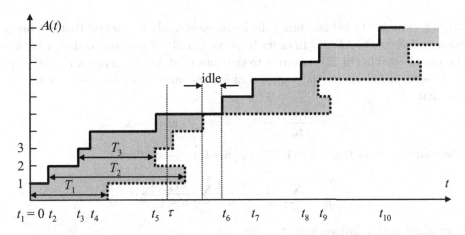

Fig. 13.6. The arrival (bold) and departure (dotted) processes, together with the system time T_k for each packet in the queueing system.

Proof: Recall that $A(t)$ represents the total number of arrivals in time interval $[0, t]$. If $N_S(t) = 0$ or the system is idle at time t, then

$$\int_0^t N_S(u)\,du = \int_0^t \sum_{j=1}^{\infty} 1_{\{u\in[t_j, t_j+T_j)\}}\,du = \sum_{j=1}^{A(t)} \int_0^t 1_{\{u\in[t_j, t_j+T_j)\}}\,du = \sum_{k=1}^{A(t)} T_k$$

The general case where $N_S(t) \geq 0$ is more complicated as Fig. 13.6 shows for $t = \tau$ because not all intervals $[t_j, t_j + T_j)$ for $1 \leq j \leq A(\tau)$ are contained in $[0, \tau)$. Hence, $\sum_{k=1}^{A(\tau)} T_k$ counts too much and is an upper bound for $\int_0^\tau N_S(u)\,du$. If $D(t)$ denotes the number of departures in $[0, t]$, Fig. 13.6 illustrates that the area (in grey) in an interval $[0, t]$, which equals the total number of packets in the system in that interval $\int_0^t (A(u) - D(u))du = \int_0^t N_S(u)du$, can be bounded for any realization (sample path) and any $t \geq 0$ by

$$\sum_{k:T_k+t_k \leq t} T_k \leq \int_0^t N_S(u)du \leq \sum_{k=1}^{A(t)} T_k$$

where the lower bound only counts the packets that have left the system by time t. By dividing by t, we have

$$\frac{A(t)}{t} \sum_{k:T_k+t_k \leq t} \frac{T_k}{A(t)} \leq \frac{1}{t} \int_0^t N_S(u)du \leq \frac{A(t)}{t} \sum_{k=1}^{A(t)} \frac{T_k}{A(t)} \qquad (13.28)$$

Since we assume that the limit (13.25) exists, we have that $A(t) = O(t)$ for $t \to \infty$. From the existence of the limit (13.27), we can thus write

$$\lim_{t \to \infty} \sum_{k=1}^{A(t)} \frac{T_k}{A(t)} = E[T]$$

When $t \to \infty$ in (13.28) and using the limits defined above, we find that the upper bound converges to $\lambda E[T]$. In order to prove (13.24), it remains to show that also the lower bound in (13.28) converges to the same limit $\lambda E[T]$. Since $A(t) = \lambda t + o(t)$ for $t \to \infty$, it follows for the sequence of arrival times t_n that $t_n \to \infty$ if $n \to \infty$ and that

$$\frac{n}{t_n} = \frac{A(t_n)}{t_n} \to \lambda \qquad\qquad \text{as } n \to \infty$$

The convergence of the series (13.27) implies for $n \to \infty$ that

$$\frac{T_n}{n} = \sum_{k=1}^{n} \frac{T_k}{n} - \frac{n-1}{n} \sum_{k=1}^{n-1} \frac{T_k}{n-1} \to 0$$

Combining both relations leads to

$$\frac{T_n}{t_n} = \frac{T_n}{n} \frac{n}{t_n} \to 0 \qquad\qquad \text{as } n \to \infty$$

which implies that, for any $\varepsilon > 0$, there exists a fixed m such that, for all $k > m$, we have that $\frac{T_k}{t_k} < \varepsilon$ or $t_k + T_k < (1+\varepsilon)t_k$. For $t > t_m$, the lower bound in (13.28) is

$$\sum_{k>m:T_k+t_k \leq t} T_k + \sum_{k=1}^{m} T_k = \sum_{k=1}^{A(t/(1+\varepsilon))} T_k$$

or

$$\frac{1}{t} \sum_{k \geq 1:T_k+t_k \leq t} T_k = \frac{1}{1+\varepsilon} \frac{A(t/(1+\varepsilon))}{t/(1+\varepsilon)} \sum_{k=1}^{A(t/(1+\varepsilon))} \frac{T_k}{A(t/(1+\varepsilon))}$$

In the limit $t \to \infty$, we obtain

$$\frac{1}{t} \sum_{k \geq 1 : T_k + t_k \leq t} T_k \to \frac{\lambda}{1 + \varepsilon} E[T]$$

Since ε can be made arbitrarily small, this finally proves (13.24). $\qquad\square$

Although the proof may seem rather technical[10] for what is, after all, an intuitive result, it reveals that no assumptions about the distributions of arrival and service process apart from steady-state convergence are made. There are no probabilistic arguments used. In essence Little's Law is proved by showing that two limits exist for any sample path or realization of the process, which guarantees a very general theorem. Moreover, no assumptions about the service discipline, nor about the dependence between arrival and service process or about the number of servers are made which means that Little's Law also holds for non-FIFO scheduling disciplines, in fact for *any* scheduling discipline! Little's Law connects three essential quantities: once two of them are known the third is determined by (13.24). Fig. 7.4 in Section 7.4 exemplifies a case where the distribution of $N_S(t)$ is analytically computable and where $E[N_S(t)]$ obeys Little's Law only in the limit $t \to \infty$, illustrating the importance of the steady-state or stationary regime for Little's Law to hold.

Little's Law is very important in operations where it relates the mean inventory (similar to $E[N_S]$), the mean flow rate or throughput λ and the mean flow time $E[T]$ in a process flow of products or services. Several examples can be found in Chapter 14, in Anupindi *et al.* (2006) and Bertsekas and Gallager (1992, pp. 157–162).

---

[10] We have chosen for a very general proof. Other proofs (e.g. in Ross (1996) and Gallager (1996)) use arguments from renewal reward theory (Section 8.4) which makes their proofs less general because they require that the system has renewals.

14

Queueing models

This chapter presents some of the simplest and most basic queueing models. Unfortunately, most queueing problems are not available in analytic form and many queueing problems require a specific and sometimes tailor-made solution.

Beside the simple and classical queueing models, we also present two other exact solvable models that have played a key role in the development of Asynchronous Transfer Mode (ATM). In these ATM queueing systems the service discipline is deterministic and only the arrival process is the distinguishing element. The first is the N*D/D/1 queue (Roberts, 1991, Section 6.2) whose solution relies on the Beneš approach. The arrivals consist of N periodic sources each with period of D time slots, but randomly phased with respect to each other. The second model is the fluid flow model of Anick *et al.* (1982), known as the AMS-queue, which considers N on-off sources as input. The solution uses Markov theory. Since the Markov transition probability matrix has a special tri-band diagonal structure, the eigenvector and eigenvalue decomposition can be computed analytically.

We would like to refer to a few other models. Norros (1994) succeeded in deriving the asymptotic probability distribution of the unfinished work for a queue with self-similar input, modeled via a fractal Brownian motion. The resulting asymptotic probability distribution turns out to be a Weibull distribution (3.49). Finally, Neuts (1989) has established a matrix analytic framework and was the founder of the class of Markov Modulated arrival processes and derivatives such as the Batch Markovian Arrival process (BMAP).

14.1 The M/M/1 queue

The M/M/1 queue consists of a Poisson arrival process of packets with i.i.d. exponentially distributed interarrival times, a service process with i.i.d. exponentially distributed service times, one server and an infinitely long queue. The M/M/1 queue is a basic model in queueing theory for several reasons. First, as shown below, the M/M/1 queue can be computed in analytic form, even the transient time behavior. Apart from the computational advantage, the M/M/1 queue possesses the basic feature of queuing systems: the quantities of interest (waiting time, number of packets, etc.) increase monotonously with the traffic intensity ρ.

Packets arrive in the M/M/1 queue with interarrival rate λ and are served with service rate μ. The M/M/1 queue is precisely described by a constant rate birth and death process, analyzed in Section 11.3. Any arrival of a packet to the queueing system can be regarded as a birth. The current state k that reflects the number of packets in the M/M/1 system jumps to state $k+1$ at the arrival of a new packet and the transition rate equals the interarrival rate λ: on average every $\frac{1}{\lambda}$ time units a packet arrives to the system. A packet leaves the M/M/1 system after service, which corresponds to a death: at each departure from the system the current state is decreased by one, with death rate μ equal to the service rate μ: on average every $\frac{1}{\mu}$ time units, a packet is served. In the sequel, we concentrate on the steady-state behavior and refer for the transient behavior to the discussion of the birth and death process in Section 11.3.3.

14.1.1 The system content in steady-state

From the analogy with the constant rate birth and death process as studied in Section 11.3.3, we obtain immediately the steady-state queueing distribution from (11.24) as

$$\Pr\left[N_S = j\right] = (1 - \rho)\,\rho^j \qquad\qquad j \geq 0 \qquad\qquad (14.1)$$

where $N_S = \lim_{t \to \infty} N_S(t)$ is the number of packets in the system in the stationary regime. In other words, $\Pr\left[N_S = j\right]$ is the probability that the M/M/1 system (queue plus server) contains j packets. It has been shown in Section 11.3 that the M/M/1 queue is strongly ergodic (i.e. that an unique equilibrium or steady-state exists) if $\rho = \frac{\lambda}{\mu} < 1$, which is a characteristic of a general queueing system. The probability density function of the system content (14.1) is a geometric distribution reflecting the memoryless property. We observe that the M/M/1 system is empty with probability $\Pr\left[N_S = 0\right] = 1 - \rho$ and of all states, the empty state has the highest probability. Immediately, the chance that there is a packet in the M/M/1 system is precisely equal to the traffic intensity ρ, namely $\Pr\left[N_S > 0\right] = 1 - \Pr\left[N_S = 0\right] = \rho$.

The corresponding probability generating function (2.18) is

$$\varphi_{N_S}(z) = \sum_{k=0}^{\infty} \Pr\left[N_S = k\right] z^k = \frac{1 - \rho}{1 - \rho z}$$

The mean number of packets in the M/M/1 system $E\left[N_S\right] = \varphi'_{N_S}(1)$ equals

$$E\left[N_{S;M/M/1}\right] = \frac{\rho}{1 - \rho}$$

while the variance $\mathrm{Var}[N_S]$ follows from (2.27) as

$$\mathrm{Var}\left[N_{S;M/M/1}\right] = \frac{\rho}{(1 - \rho)^2} \qquad\qquad (14.2)$$

Both the mean and variance of the number of packets in the system diverge as $\rho \to 1$. When the interarrival rate tends to the service rate, the queue grows indefinitely long with indefinitely large variation. From Little's Law (13.24), the mean time spent in the M/M/1 system equals

$$E\left[T_{M/M/1}\right] = \frac{E\left[N_S\right]}{\lambda} = \frac{1}{\mu\left(1-\rho\right)} = \frac{1}{\mu - \lambda} \tag{14.3}$$

where $\rho < 1$ or, equivalently, $\lambda < \mu$. If $\rho = 0$, there is no load in the system and the mean system time attains its minimum equal to the mean service time $\frac{1}{\mu}$. In the other limit $\rho \to 1$, the mean system time grows unboundedly, just as the queue length or number of packets in the system. The behavior of the M/M/1 system in the limit $\rho \to 1$ is characteristic for the mean of quantities $(N_S, w, T, \ldots)$ in many queueing systems: a simple pole at $\rho = 1$.

As a remark, the mean waiting time in the M/M/1 queue follows, after taking expectations from the general relation $T_n - x_n = w_n$ or $E\left[w\right] = E\left[T\right] - \frac{1}{\mu}$, as

$$E\left[w_{M/M/1}\right] = \frac{1}{\mu\left(1-\rho\right)} - \frac{1}{\mu} = \frac{\rho}{\mu\left(1-\rho\right)} \tag{14.4}$$

14.1.2 The virtual waiting time

For the M/M/1 queue, the virtual waiting time $v(t)$ at some time t consists of (a) the residual service time of the packet currently under service, (b) the time needed to serve the $N_Q(t)$ packets in the queue. As mentioned in Section 13.1.3, the virtual waiting time at arrival epochs equals the system time T_n. In other words, if a new packet, say the n-th packet, enters the M/M/1 system at $t = t_n$, the total time (system time) that the packet spends in the M/M/1 system equals $v(t_n) = T_n$. At $t = t_n^-$, the number $N_S(t_n^-)$ does not include the new packet at the last position and the n-th packet "sees" $N_S(t_n^-)$ other packets in the system (queue plus the packet in the server) in front of it. We assume further that the server operates in FIFO (first in, first out) order. Since the service time is exponentially distributed and possesses the memoryless property, the residual or remaining service time of the packet currently under service has the same distribution. In other words, it does not matter how long the packet has already been under service. The more general argument is that the PASTA property, Theorem 13.6.1, applies. The system time of the n-th packet is thus the sum of $N_S(t_n^-) + 1$ exponential i.i.d. random variables. As shown in Section 3.3.1, if $N_S(t_n^-) = k$, the system time has an Erlang distribution given by (3.24) with $n = k + 1$,

$$f_{T_n}(t|N_S(t_n^-) = k) = \frac{\mu(\mu t)^k}{k!}e^{-\mu t}$$

Using the law of total probability (2.49),

$$\Pr\left[T_n \le t\right] = \sum_{k=0}^{\infty} \Pr\left[\left|T_n \le t\right| N_S(t_n^-) = k\right] \Pr\left[N_S(t_n^-) = k\right]$$

the pdf $f_{T_n}(t) = \frac{d}{dt} \Pr[T_n \leq t]$ of the system time T_n of the n-th packet or the virtual waiting time at time $t = t_n$ becomes, after derivation with respect to t,

$$f_{T_n}(t) = \sum_{k=0}^{\infty} f_{T_n}(t|N_S(t_n^-) = k) \Pr[N_S(t_n^-) = k]$$

$$= \mu e^{-\mu t} \sum_{k=0}^{\infty} \frac{(\mu t)^k}{k!} \Pr[N_S(t_n^-) = k]$$

In Section 11.3.3, $s_k(t_n^-) = \Pr[N_S(t_n^-) = k]$ is computed in (11.29) assuming that the system starts with j packets, i.e. $s_k(0) = \delta_{kj}$. In steady-state, where $t_n \to \infty$, it is shown that $s_k(t_n^-) \to (1-\rho)\rho^k$. In most cases, however, a time-dependent solution is not available in closed form. Fortunately, for Poisson arrivals, the PASTA property helps to circumvent this inconvenience. Based on the PASTA property, in steady-state, $\lim_{n\to\infty} \Pr[N_S(t_n^-) = k] = \Pr[N_S = k]$ given by (14.1). The probability density function $f_T(t) = \lim_{n\to\infty} f_{T_n}(t)$ of the steady-state system time T (or the total waiting time of a packet) is

$$f_T(t) = \mu e^{-\mu t} \sum_{k=0}^{\infty} \frac{(\mu t)^k}{k!} (1-\rho)\rho^k$$

or

$$f_T(t) = (1-\rho)\mu e^{-(1-\rho)\mu t} \tag{14.5}$$

In summary, the total time spent in the M/M/1 system in steady-state ($\rho < 1$) has an exponential distribution with mean $\frac{1}{(1-\rho)\mu} = \frac{1}{\mu-\lambda}$, which has been found above in (14.3) by Little's Law. Similarly[1], the waiting time in the M/M/1 queue is

$$f_w(t) = (1-\rho)\delta(t) + (1-\rho)\lambda e^{-(1-\rho)\mu t} \tag{14.6}$$

where the first term with Dirac function reflects a zero queueing time provided the system is empty, which has probability $\Pr[N_S = 0] = 1 - \rho$.

14.1.3 The departure process from the M/M/1 queue

There is a remarkable theorem due to Burke with far-reaching consequences for networks of M/M/1 queues, for which we refer to Walrand (1988).

Theorem 14.1.1 (Burke) *In a steady-state M/M/1 queue, the departure process is a Poisson process with rate λ*

---

[1] The Laplace transform of the waiting time in the queue follows from $T_n = w_n + x_n$ as

$$\varphi_w(z) = \frac{\varphi_T(z)}{\varphi_x(z)} = \frac{(1-\rho)\mu}{z + (1-\rho)\mu} \frac{z+\mu}{\mu}$$

$$= (1-\rho) + \rho \frac{(1-\rho)\mu}{z + (1-\rho)\mu}$$

which, after inverse Laplace transformation, gives (14.6).

Burke's Theorem is equivalent to the statement that the interdeparture times $r_n - r_{n-1}$ in steady-state are i.i.d. exponential random variables with mean $\frac{1}{\lambda}$.

Proof: Let us denote the probability density function of the interdeparture time r by

$$f_r(t) = \frac{d}{dt} \Pr[r \leq t]$$

In steady-state, it holds in general that $\Pr[N_{S;A} = j] = \Pr[N_{S;D} = j]$, as shown in Section 13.5, while the PASTA property (Theorem 13.6.1) states that $\Pr[N_{S;A} = j] = \Pr[N_S = j]$. By combining both, in steady-state in the M/M/1 queue, departing packets see the steady-state system content, i.e. $\Pr[N_{S;D} = j] = \Pr[N_S = j]$. Moreover, in steady-state, the departure process can be decomposed into two different situations after the departure of a packet: (a) the departing packet sees an empty system (which is equivalent to "the system is empty") or (b) the departing packet sees a next packet in the queue (which is equivalent to "the system serves immediately the next packet in the queue"),

$$\Pr[r \leq t] = \Pr[r \leq t|N_S = 0]\Pr[N_S = 0] + \Pr[r \leq t|N_S > 0]\Pr[N_S > 0]$$

In case (a), we must await for the next packet to arrive and to be served. This total time is the sum of an exponential random variable with rate λ and an exponential random variable with rate μ. It is more convenient to compute the Laplace transform as shown in Section 3.3.1,

$$\varphi_{r|N_S=0}(z) = \int_0^\infty e^{-zt}d\left(\Pr[r \leq t|N_S = 0]\right) = \frac{\lambda}{z+\lambda}\frac{\mu}{z+\mu}$$

In case (b), the next packet leaves the M/M/1 queue after an exponential service time with rate μ,

$$\varphi_{r|N_S>0}(z) = \int_0^\infty e^{-zt}d\left(\Pr[r \leq t|N_S > 0]\right) = \frac{\mu}{z+\mu}$$

Hence,

$$\varphi_r(z) = \varphi_{r|N_S=0}(z)\Pr[N_S = 0] + \varphi_{r|N_S>0}(z)\Pr[N_S > 0]$$

$$= \frac{\lambda}{z+\lambda}\frac{\mu}{z+\mu}(1-\rho) + \frac{\mu}{z+\mu}\rho = \frac{\lambda}{z+\lambda}$$

which proves the theorem. ☐

Burke's Theorem states that the steady-state arrival and departure processes of the M/M/1 queue are the same! Consequently, the steady-state departure rate equals the steady-state arrival rate λ.

Burke's Theorem 14.1.1 is more generally proved as a consequence of time-reversibility of the stationary M/M/1 Markov process. In Section 10.7, we have shown that a continuous-time Markov chain is time reversible provided $\pi_i q_{ij} = \pi_j q_{ji}$ for any state i and $j \neq i$. Applying the balance equation of a general random walk, studied in Section 11.2.1, to the M/M/1 or birth and death process yields $\lambda \pi_j = \mu \pi_{j+1}$, for all $j \geq 0$, which illustrates time-reversibility of the M/M/1 queue. Hence, the departure process of a stationary M/M/1 queue is a Poisson process and the departure times up to any time t are independent of the queueing process at time t.

14.2 Variants of the M/M/1 queue

A number of readily obtained variants from the birth-death analogy are worth considering here. Mainly steady-state results are presented.

14.2.1 The M/M/m queue

Instead of one server, we consider the case with m servers. The buffer is still infinitely long and the interarrival process is exponential with interarrival rate λ. The M/M/m queue can model a router with m physically different interfaces (or output ports) with same transmission rate μ towards the same next hop. All packets destined to that next hop can be transmitted over any of the m interfaces. This type of load balancing frequently occurs in the Internet.

As shown in Fig. 14.1, the M/M/m system can still be described by a birth and death process with birth rate $\lambda_k = \lambda$, but with death rate $\mu_k = k\mu$ for $0 \leq k \leq m$ and $\mu_k = m\mu$ if $k \geq m$. Indeed, if there are $k \leq m$ packets in the system, they can all be served and the departure (or death) rate from the system is $k\mu$. If there are more packets, $k > m$, only m of them can be served such that the death rate is limited to the maximum service rate $m\mu$.

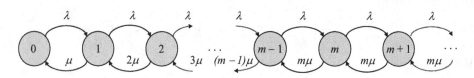

Fig. 14.1. The birth–death process corresponding to the M/M/m queue.

14.2.1.1 System content

From the basic steady-state relations for the birth and death process (11.16) and (11.17), we find

$$\Pr\left[N_S = 0\right] = \frac{1}{1 + \sum_{j=1}^{m-1} \frac{\lambda^j}{j!\mu^j} + \sum_{j=m}^{\infty} \frac{\lambda^j}{m^{j-m}m!\mu^j}}$$

$$= \frac{1}{\sum_{j=0}^{m-1} \frac{\lambda^j}{j!\mu^j} + \frac{\lambda^m}{m!\mu^m} \frac{1}{1 - \frac{\lambda}{m\mu}}} \tag{14.7}$$

and

$$\Pr\left[N_S = j\right] = \frac{\lambda^j}{j!\mu^j}\Pr\left[N_S = 0\right] \qquad j \leq m \qquad (14.8)$$

$$= \frac{m^m}{m!}\left(\frac{\lambda}{m\mu}\right)^j \Pr\left[N_S = 0\right] \qquad j \geq m \qquad (14.9)$$

The traffic intensity is $\rho = \frac{\lambda}{m\mu}$, the ratio between mean interarrival rate and mean (maximum) service rate. Again, $\rho < 1$ corresponds to the stable (ergodic) regime.

For the M/M/m system it is of interest to know what the probability of queueing is. Queueing occurs when an arriving packet finds all servers busy, which happens with probability $\Pr\left[N_S \geq m\right]$, or explicitly,

$$\Pr\left[N_S \geq m\right] = \frac{\Pr\left[N_S = 0\right]}{m!\left(1 - \frac{\lambda}{m\mu}\right)}\frac{\lambda^m}{\mu^m} \qquad (14.10)$$

This probability also corresponds to a situation in classical telephony where no trunk is available for an arriving call. Relation (14.10) is known as the *Erlang C formula*.

14.2.1.2 Waiting (or queueing) time

Instead of computing the virtual waiting time (or system time, unfinished work), we will now concentrate on the waiting time of a packet in the M/M/m queue. The system time can be deduced from the basic relation $T_n = w_n + x_n$, where x_n is an exponential random variable with rate $m\mu$.

A packet only experiences queueing if all servers are occupied. This event has probability $\Pr\left[N_S \geq m\right]$ specified by the Erlang C formula (14.10). Hence, the queueing time w can be decomposed into two cases: (a) an arriving packet does not queue ($w = 0$) and (b) an arriving packet must wait in the M/M/m queue,

$$\Pr[w \leq t] = \Pr[w \leq t|N_S < m]\Pr[N_S < m] + \Pr[w \leq t|N_S \geq m]\Pr[N_S \geq m]$$
$$= \Pr\left[N_S < m\right] + \Pr\left[w \leq t|N_S \geq m\right]\Pr\left[N_S \geq m\right]$$
$$= 1 - \Pr\left[N_S \geq m\right] + \Pr\left[w \leq t|N_S \geq m\right]\Pr\left[N_S \geq m\right] \qquad (14.11)$$

It remains to compute $\Pr\left[w \leq t|N_S \geq m\right]$. The reasoning is analogous to that in the M/M/1 queue. An arriving packet must wait for the departure of $j+1$ packets before it can be served: the packet currently under service and the j packets already in the queue in front of it. Invoking the memoryless property to the packet currently served, w equals the sum of $j + 1$ exponential random variables with rate $m\mu$ because, for the M/M/m queue, the service rate is $m\mu$. Hence (see Section 3.3.1), the distribution for the waiting time w in the queue, provided j packets are in the queue, is an Erlang distribution,

$$f_w(t|N_Q = j) = \frac{m\mu(m\mu t)^j}{j!}e^{-m\mu t}$$

Furthermore, the number of packets in the queue N_Q in steady-state is related to the system content as $N_S = m + N_Q$. Using the law of total probability (2.49), the waiting time in the queue in steady-state is

$$f_w (t | N_S \geq m) = \frac{d}{dt} \Pr [w \leq t | N_S \geq m]$$

$$= \sum_{j=0}^{\infty} f_w(t | N_S = m + j) \Pr [N_S = m + j | N_S \geq m]$$

The conditional probability $\Pr [N_S = m + j | N_S \geq m]$ follows from (2.47) and (14.9) with $\rho = \frac{\lambda}{m\mu}$ as

$$\Pr [N_S = m + j | N_S \geq m] = \frac{\Pr [N_S = m + j]}{\Pr [N_S \geq m]} = \frac{\Pr [N_S = 0]}{\Pr [N_S \geq m]} \frac{m^m}{m!} \left(\frac{\lambda}{m\mu} \right)^{j+m}$$

$$= \left(1 - \frac{\lambda}{m\mu} \right) \left(\frac{\lambda}{m\mu} \right)^{j} = (1 - \rho) \rho^{j}$$

We observe from (14.1) that, if all m servers are busy, the system content of an M/M/m system behaves as that in a M/M/1 system. Combining the above relations yields

$$f_w (t | N_S \geq m) = \sum_{j=0}^{\infty} f_w(t | N_S = m + j) \Pr [N_S = m + j | N_S \geq m]$$

$$= (1 - \rho) \, m\mu e^{-m\mu t} \sum_{j=0}^{\infty} \frac{(m\mu t \rho)^{j}}{j!} = (1 - \rho) \, m\mu e^{-(1-\rho)m\mu t}$$

Thus, the conditional probability density function for the waiting time in the M/M/m queue is also an exponential distribution,

$$\Pr [w \leq t | N_S \geq m] = 1 - e^{-(1-\rho)m\mu t}$$

Substitution in (14.11) finally results in the distribution of the waiting time in the queue of the M/M/m system,

$$F_w(t) = \Pr [w \leq t] = 1 - \Pr [N_S \geq m] \, e^{-(1-\rho)m\mu t} \qquad (14.12)$$

Since $F_w(0) = 1 - \Pr [N_S \geq m] > 0$, while obviously $F_w(0^-) = 0$, there is probability mass at $t = 0$, which is reflected by a Dirac impulse in the probability density function,

$$f_w(t) = (1 - \Pr [N_S \geq m]) \, \delta(t) + (1 - \rho) \, m\mu \Pr [N_S \geq m] \, e^{-(1-\rho)m\mu t} \qquad (14.13)$$

The pdf of the system time $T = w + x$ follows after convolution of (14.13) and $f_x(t) = \mu e^{-\mu t}$ as

$$f_T(t) = (1 - \Pr [N_S \geq m]) \, \mu e^{-\mu t} + \frac{\Pr [N_S \geq m]}{1 - m(1 - \rho)} (1 - \rho) \, m\mu \left(e^{-(1-\rho)m\mu t} - e^{-\mu t} \right) \qquad (14.14)$$

and the mean system time can be computed from (14.14) with (2.34) or directly
from $E[T] = E[w] + E[x]$ as

$$E[T] = \frac{1}{\mu} + \frac{\Pr[N_S \geq m]}{m\mu(1-\rho)} \qquad (14.15)$$

Also, in the single-server case ($m = 1$), (14.13) reduces to the pdf (14.6) of
the M/M/1 queue. Furthermore, Burke's Theorem 14.1.1 can be extended to the
M/M/m queue: the arrival and departure processes of the M/M/m queue are both
Poisson processes with rate λ.

14.2.2 The M/M/m/m queue

The difference with the M/M/m queue is that the number of packets (calls) in
the M/M/m/m queue is limited to m. Hence, when more than m packets (calls)
arrive, they are lost. This situation corresponds with classical telephony where a
conversation is possible if no more than m trunks are occupied, otherwise you hear
a busy tone and the connection cannot be set-up. The limitation to m arrivals is
modeled in the birth and death process by limiting the interarrival rates, $\lambda_k = \lambda$
if $k < m$ and $\lambda_k = 0$ if $k \geq m$. The death rates are the same as in the M/M/m
queue, $\mu_k = k\mu$ for $0 \leq k \leq m$ and $\mu_k = m\mu$ if $k \geq m$.

From the basic steady-state relations for the birth and death process (11.16) and
(11.17), we find

$$\Pr[N_S = 0] = \frac{1}{\sum_{j=0}^{m} \frac{\lambda^j}{j!\mu^j}} \qquad (14.16)$$

$$\Pr[N_S = j] = \frac{\lambda^j}{j!\mu^j} \Pr[N_S = 0] \qquad j \leq m$$

$$= 0 \qquad j > m \qquad (14.17)$$

The quantity of interest in the M/M/m/m system is the probability that all trunks
(servers) are busy, which is known as the *Erlang B formula*,

$$\Pr[N_S = m] = \frac{\frac{\lambda^m}{m!\mu^m}}{\sum_{j=0}^{m} \frac{\lambda^j}{j!\mu^j}} \qquad (14.18)$$

In practice, a telephony exchange is dimensioned (i.e. the number of lines m is
determined) such that the blocking probability $\Pr[N_S = m]$ is below a certain level,
say below 10^{-4}. In summary, the Erlang B formula (14.18) determines the blocking
probability or loss probability (because only m calls or packets are allowed to the
system), while the Erlang C formula (14.10) is the probability that a packet must
wait in the (infinitely long) queue because all servers are busy.

Although the Erlang B formula (14.18) has been derived in the context of the
M/M/m/m queue, it holds under much weaker assumptions, a fact already known to
Erlang, as mentioned by Kelly (1991). Kelly starts his memoir on "loss networks"

with the Erlang B formula, which Erlang obtained from his powerful method of statistical equilibrium. The latter concept is now identified as the steady-state of Markov processes. Kelly further relates the impact of the Erlang B formula from telephony to interacting particle systems and phase transitions in nature (e.g. the famous Ising model). Much effort has been devoted over time to generalize Erlang's results as far as possible. The Erlang B formula (14.18) holds for the M/G/m/m queue as well, thus for an arbitrary service process provided the mean service rate (per server) equals μ. The proof by Gnedenko and Kovalenko (1989, pp. 237–240) is long and complicated, whereas the proof of Ross (1996, Section 5.7.2) is more elegant and is based on the time-reversed Markov chain. As a corollary, Ross demonstrates that the departure process (including both lost and served packets) of the M/G/m/m system is a Poisson process with rate λ.

Example 1 In case $m \to \infty$, (14.16) and (14.7) tend to $\lim_{m \to \infty} \Pr[N_S = 0]$ $= \exp\left(-\frac{\lambda}{\mu}\right)$, while (14.17) and (14.8) become

$$\Pr[N_S = j] = \frac{\lambda^j}{j!\mu^j} \exp\left(-\frac{\lambda}{\mu}\right)$$

This queueing system is denoted as M/M/∞. Thus, the number in the M/M/∞ system (in steady-state) is Poisson distributed with parameter $\frac{\lambda}{\mu}$. Hence, in case $m \to \infty$, the mean number in the system $E[N_S] = \rho$ (as follows from (3.11)) and the mean time in the system follows from Little's theorem (13.24) as $E[T] = \frac{1}{\mu}$. The fact that, if $m \to \infty$, the mean time in the system M/M/∞ equals the mean service time has a consistent explanation: if the number of servers $m \to \infty$, implying that there is an infinite service capacity, it means that there is no waiting room and the only time a packet is in the system is his service time $\frac{1}{\mu}$. The generalization to an M/G/∞ system was analyzed on p. 148.

Example 2 Consider two voice over IP (VoIP) gateways connected by a link with capacity C. Denote the capacity of a voice call by C_{voice} (in bit/s). For example in ISDN, $C_{\text{voice}} = 64$ kb/s. In general, C_{voice} in VoIP depends on the specifics of the codecs used. The arrival rate λ (in events/s) of voice traffic can be expressed in terms of the number a of call attempts per hour and the mean call duration d (in seconds) as[2]

$$\lambda = \frac{a \times d}{3600}$$

The number m of calls that the link can carry simultaneously is

$$m = \frac{C}{C_{\text{voice}}}$$

Since the arrival process of voice calls is well modeled by a Poisson process with

---

[2] Since each call lasts on average d seconds, the number of call events per unit time is $\lambda = \frac{a}{T}$, where the effective time $T = \frac{3600}{d} s$.

exponential holding time, the Erlang B formula (14.18) is applicable to compute the blocking probability ϵ or grade of service (GoS) as

$$\frac{\frac{r^m}{m!}}{\sum_{j=0}^{m} \frac{r^j}{j!}} \leq \epsilon \tag{14.19}$$

where $r = \frac{\lambda}{\mu} = m\rho$ and $\rho = \frac{\lambda}{m\mu}$ is the traffic intensity. This relation (14.19) specifies the probability that admission control will have to refuse a call request between the two VoIP gateways because the link is already transporting m calls. An Internet service provider can make a trade-off between the link capacity C (by hiring more links or a higher capacity link from a network provider) and the blocking probability ϵ or GoS. The latter must be small enough to keep its subscribed customers, but large enough to make profit. A reasonable value for GoS seems $\epsilon = 10^{-4}$. If the Internet service provider hires a 2 Mb/s link and offers its customers VoIP software with codec rate 40 kb/s (G.726 standard), then $m = 50$. Since the left-hand side of (14.19) is strictly increasing in r, solving the equation (14.19) for r yields $r \leq 28.87$ or the traffic intensity equals $\rho = 0.5775$. Furthermore, since $\mu = \frac{C}{m} = 40$ kb/s, we obtain $\lambda = 1.155$ Mb/s. If the mean call duration d (in seconds) is known, the number of call attempts per hour then follows as $a = \frac{4\ 158\ 129}{d}$. If we assume that a telephone call lasts on average 2 minutes or $d = 120$ s, the number of call attempts per hour that the Internet service provider can handle with a GoS of 10^{-4} equals $a = 34\ 651$.

14.2.3 The M/M/1/K queue

In contrast to the basic M/M/1 queue, the M/M/1/K system cannot contain more than K packets (including the packet in the server). Arriving packets that find the system completely occupied (with K packets), are refused service and are to be considered as lost (or marked).

In the basic steady-state relations for the birth and death process (11.16), the appearing summation is limited to K instead of infinity or $\lambda_k = \lambda$ if $k < K$ and $\lambda_k = 0$ if $k \geq K$. Thus, with $\rho = \frac{\lambda}{\mu}$ and

$$\Pr[N_S = 0] = \frac{1}{\sum_{j=0}^{K} \rho^j} = \frac{1 - \rho}{1 - \rho^{K+1}}$$

the pdf of the system content for the M/M/1/K system becomes,

$$\Pr[N_S = j] = \frac{(1 - \rho)\rho^j}{1 - \rho^{K+1}} \qquad\qquad 0 \leq j \leq K \tag{14.20}$$

$$= 0 \qquad\qquad\qquad j > K$$

The probability that j positions in the M/M/1/K system are occupied is proportional to that in the infinite system (14.1) with proportionality factor $\left(1 - \rho^{K+1}\right)^{-1}$.

The probability that the system is completely filled with K packets equals

$$\Pr\left[N_S = K\right] = \frac{(1 - \rho)\,\rho^K}{1 - \rho^{K+1}} \tag{14.21}$$

This probability also equals the loss probability for packets in the M/M/1/K system. Regarding the QoS problem in multimedia based on IP-networks, a first crude estimate of the packet loss in a router with K positions can be derived from (14.21). The estimate is rather crude because the arrival process of packets in the Internet is likely not a Poisson process and the variable length of the packets does not necessarily lead to an exponential service rate.

14.3 The M/G/1 queue

The most general single-server queueing system with Poisson arrivals is the M/G/1 queue. The service time distribution $F_x(t)$ can be any arbitrary distribution. Due to its importance, we will derive the system content and the waiting time distribution in steady-state.

In order to describe the M/G/1 queueing system, special observation points in time must be chosen such that the equations for the evolution of the number of packets in the system are most conveniently deduced. The set of departure times $\{r_n\}$ appears to be a suitable set. Any other set of observation points is likely to lead to a more complex mathematical treatment, mainly because the remaining service time of the packet just under service is a stochastic variable. If the M/G/1 queue is observed at departure times, the evolution of the number of packets in the system $N_S(r_n^+)$ that the departing packets leave behind is a discrete Markov chain, namely the embedded Markov chain of the M/G/1 queue continuous-time process. Section 13.5 has shown that, in steady-state, relation (13.22) tells that the distribution of the number of packets in the system observed by arrivals equals that left behind by departures. In addition, the PASTA Theorem 13.6.1 states that, in steady-state, Poisson arrivals observe the actual distribution of the number of packets in the system. This PASTA property makes that the embedded Markov chain observing the system at departure epoch only, nevertheless provides the steady-state solution since the arrival process is Poissonean.

Let us concentrate in deriving the transition probabilities that specify this embedded Markov chain entirely apart from an initial state distribution. With the notation of Section 13.4, $\mathcal{S}_k = N_S(r_k^+)$ and $\mathcal{A}_k$ denote the number of packets in the system at the discrete-time $r_k^+ = r_k + \epsilon$ and the number of arrivals during a time interval $(r_k, r_{k+1}]$, respectively. The transition probability of the embedded Markov chain is

$$V_{ij} = \Pr\left[\mathcal{S}_{k+1} = j | \mathcal{S}_k = i\right]$$

and the evolution over time follows from (13.19) as

$$\mathcal{S}_{k+1} = (\mathcal{S}_k - 1)^+ + \mathcal{A}_k \tag{14.22}$$

Hence, since $\mathcal{A}_k \geq 0$, we see that $\mathcal{S}_{k+1} \geq (\mathcal{S}_k - 1)^+$ or that $\mathcal{S}_{k+1} < (\mathcal{S}_k - 1)^+$ is impossible. Hence, $V_{ij} = 0$ for $j < i-1$ and $i > 1$ while, for $i > 0$, $V_{ij} = \Pr[i - 1 + \mathcal{A}_k = j | \mathcal{S}_k = i] = \Pr[\mathcal{A}_k = j - i + 1]$. The case for $i = 0$ results in $V_{0j} = \Pr[\mathcal{A}_k = j] = V_{1j}$. Denoting $a_j = \Pr[\mathcal{A}_k = j]$, the transition probability matrix becomes[3]

$$V = \begin{bmatrix} a_0 & a_1 & a_2 & a_3 & \cdots \\ a_0 & a_1 & a_2 & a_3 & \cdots \\ 0 & a_0 & a_1 & a_2 & \cdots \\ 0 & 0 & a_0 & a_1 & \cdots \\ \vdots & \vdots & \vdots & \vdots & \ddots \end{bmatrix}$$

and the corresponding transition graph is sketched in Fig. 14.2.

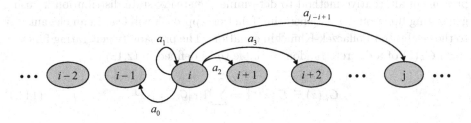

Fig. 14.2. State transition graph for the M/G/1 embedded Markov chain.

The number of Poisson arrivals during a time slot $(r_k, r_{k+1}]$ clearly depends on the length of the service time $x_{k+1} = r_{k+1} - r_k$ that is distributed according to $F_x(t)$, which is independent of a specific packet k. Furthermore, the arrival process is a Poisson process with rate λ and independent of the state of the queueing process, thus $\Pr[\mathcal{A}_k = j] = \Pr[\mathcal{A} = j]$. Hence, using the law of total probability (2.49),

$$\Pr[\mathcal{A} = j] = \int_0^\infty \Pr[\mathcal{A} = j | x = t]\, dF_x(t)$$
$$= \int_0^\infty e^{-\lambda t} \frac{(\lambda t)^j}{j!}\, dF_x(t)$$

If we denote the Laplace transform of the service time by

$$\varphi_x(z) = E\left[e^{-zx}\right] = \int_0^\infty e^{-zt} dF_x(t)$$

then we observe that

$$a_j = \Pr[\mathcal{A} = j] = \frac{\lambda^j}{j!} \int_0^\infty e^{-\lambda t} t^j\, dF_x(t) = \frac{(-\lambda)^j}{j!} \left. \frac{d^j \varphi_x(z)}{dz^j} \right|_{z=\lambda} \qquad (14.23)$$

---

[3] The structure of this transition probability matrix V has been investigated in great depth by Neuts (1989). Moreover, V belongs to the class of matrices whose eigenstructure is explicitly given in Appendix A.6.4.

so that the transition probability matrix V is specified. Since all $a_j > 0$ for all $j > 0$, Fig. 14.2 indicates that all states are reachable from an arbitrary state i as the Markov process evolves over time in at least i steps. These i steps occur in the transition from state i to state 0. This implies that the Markov process is irreducible and the steady-state stability requirement $\rho < 1$ makes the Markov process ergodic. The steady-state vector with components $\pi_j = \Pr[N_{S;D} = j]$ where $\lim_{n \to \infty} N_S(r_n^+) = N_{S;D}$ follows from (9.28) as solution of $\pi = \pi V$, where V is a matrix of infinite dimensions.

14.3.1 The system content in steady-state

Rather than pursuing with the matrix analysis that is explored by Neuts (1989), we present an alternative method to determine the steady-state distribution π_j using generating functions. The generating function approach will lead in an elegant way to the celebrated Pollaczek-Khinchin equation. The probability generating function (pgf) $G_k(z)$ of a discrete random variable $\mathcal{G}_k$ is defined in (2.18) as

$$G_k(z) \triangleq E\left[z^{\mathcal{G}_k}\right] = \sum_{j=0}^{\infty} \Pr[\mathcal{G}_k = j] z^j \qquad (14.24)$$

From (14.22), we have

$$S_{k+1}(z) = E\left[z^{(S_k - 1)^+ + A_k}\right] \qquad (14.25)$$

Anticipating the corresponding result (14.32) derived in Section 14.4 for the GI/D/m system in discrete-time, we observe that the generating function $S_k(z)$ satisfies a formally similar equation when $m = 1$ in (14.32). This correspondence points to a more general framework because, by choosing appropriate observation points, the M/G/1 and GI/D/1 systems (in discrete-time) obey a same equation formally. Since the results deduced for the GI/D/m system are more general because of the m servers instead of 1 here, we content ourselves here to copy the result (14.36) derived below[4] in Section 14.4,

$$S(z) = (1 - A'(1)) \frac{(z - 1) A(z)}{z - A(z)}$$

We further continue to introduce in this general equation the details of the M/G/1 queueing system by specifying $A(z) = \sum_{j=0}^{\infty} \Pr[\mathcal{A} = j] z^j$. With (14.23), we find the Taylor expansion,

$$A(z) = \sum_{j=0}^{\infty} \frac{(-\lambda z)^j}{j!} \left. \frac{d^j \varphi_x(z)}{dz^j} \right|_{z=\lambda} = \varphi_x(\lambda - \lambda z) \qquad (14.26)$$

---

[4] Notice that the notation $A(z)$ here is different from $A(t) = \int_0^t N_A(u)du$ used before.

and the probability generating function of the system content of the steady-state M/G/1 queueing system,

$$S(z) = (1 + \lambda\varphi'_x(0)) \frac{(z-1)\,\varphi_x(\lambda - \lambda z)}{z - \varphi_x(\lambda - \lambda z)}$$

But, since $\varphi'_x(0) = -E\,[x] = -\frac{1}{\mu}$, which is the mean service time, we finally arrive using (13.2) at the famous *Pollaczek-Khinchin equation*,

$$S(z) = (1 - \rho) \frac{(z-1)\,\varphi_x(\lambda - \lambda z)}{z - \varphi_x(\lambda - \lambda z)} \tag{14.27}$$

Let us further investigate what can be concluded from the Pollaczek-Khinchin equation (14.27). First, we can verify by executing the limit $z \to 1$ that $S(1) = 1$, the normalization condition for any probability generating function. More interestingly, the mean number of packets in the M/G/1 system follows after some tedious manipulations using de l'Hospital's rule as

$$E\,[N_S] = S'(1)$$

$$= \rho + \frac{\lambda^2 \varphi''_x(0)}{2(1 - \rho)}$$

From (2.45), $\varphi''_x(0) = E\,[x^2]$,

$$E\,[N_S] = \rho + \frac{\lambda^2 E\,[x^2]}{2(1 - \rho)} \tag{14.28}$$

Hence, the mean number of packets in the M/G/1 system (in steady-state) is proportional to the second moment of the service time distribution. Since $E\,[x^2] = \text{Var}[x] + (E\,[x])^2$, the relation[5] (14.28) shows that, for equal mean service rates, the service process with highest variability leads to the largest mean number of packets in the system. One of the early successes of the Japanese industry was the "just in time" (JIT) principle, which essentially tries to minimize the variability in a manufacturing process. Minimization of variability is also very important in the design of scheduling rules: the less variability, the more efficiently buffer places in a router are used. Since a deterministic server has the lowest variance (namely zero), the M/D/1 queue will occupy on average the lowest number of packets. This design principle was used in ATM, where all service times precisely equal the time needed to serve one ATM cell. The variance $\text{Var}[N_S]$ is computed in (B.19) as an exercise.

---

[5] Sometimes the coefficient of variation for the service time, $C_X = \frac{\sqrt{\text{Var}[X]}}{E[X]}$, is used such that

$$E\,[N_S] = \rho + \frac{\rho^2\,(1 + C_X^2)}{2(1 - \rho)}$$

The mean time spent in the system follows directly from Little's Law (13.24),

$$E[T] = \frac{E[N_S]}{\lambda} = E[x] + \frac{\lambda E[x^2]}{2(1-\rho)}$$

and, since $E[T] = E[x] + E[w]$, the mean waiting time in the queue is

$$E[w] = \frac{\lambda E[x^2]}{2(1-\rho)} \tag{14.29}$$

Observe a general property of "averages" in queueing systems: there is a simple pole at $\rho = 1$. Both the mean number of packets in the system (and in the queue) and the mean waiting time grows unboundedly as $\rho \to 1$.

14.3.2 The waiting time in steady-state

The derivation of the pgf (14.27) of the steady-state system content $S(z)$ has not made any assumption about the service discipline that determines the order in which packets are served. The waiting time (in the queue) and the system time (total time spent in the $M/G/1$ queueing system) will, of course, dependent on the order. As mentioned earlier, a FIFO service discipline is assumed. At each departure time r_k, the number of packets left behind by that k-th packet is precisely $N_S(r_k)$. With FIFO, this implies that during the total time T_k that the k-th packet has spent in the $M/G/1$ queueing system, precisely $N_S(r_k)$ packets have arrived. Similarly, as above in (14.23), we compute the number of Poisson $\mathcal{A}^*$ arrivals during T_k (instead of x_k) and directly find, in steady-state,

$$\Pr[\mathcal{A}^* = j] = \frac{\lambda^j}{j!} \int_0^\infty e^{-\lambda t} t^j \, dF_T(t) = \frac{(-\lambda)^j}{j!} \left. \frac{d^j \varphi_T(z)}{dz^j} \right|_{z=\lambda}$$

and the corresponding pgf,

$$A^*(z) = \sum_{j=0}^\infty \Pr[\mathcal{A}^* = j] \, z^j = \varphi_T(\lambda - \lambda z)$$

where $\varphi_T(z)$ is the Laplace transform of the system time T. Since the number of Poisson arrivals $\mathcal{A}^*$ during the system time T of a packet in steady-state equals the number of packets left behind by that packet, $\Pr[\mathcal{A}^* = j] = \Pr[N_{S;D} = j]$. The PASTA property (Theorem 13.6.1) states that, in steady-state, the observed number of packets in the queue at departure or arrival times is equal in distribution to the actual number of packets in the queue or that $\Pr[N_{S;D} = j] = \Pr[N_S = j]$. By considering the pgfs of both sides, $A^*(z) = S(z)$, such that with (14.27),

$$\varphi_T(\lambda - \lambda z) = (1-\rho) \frac{(z-1)\, \varphi_x(\lambda - \lambda z)}{z - \varphi_x(\lambda - \lambda z)}$$

After a change of variable $s = \lambda - \lambda z$, we end up with the result that the Laplace transform of the total system time in steady-state is a function of the Laplace transform of the service time

$$\varphi_T(s) = (1 - \rho) \frac{s\varphi_x(s)}{s - \lambda + \lambda\varphi_x(s)} \tag{14.30}$$

Since $T_k = w_k + x_k$ and, in steady-state $T = w + x$, we have that $E\left[e^{-sT}\right] = E\left[e^{-sw}e^{-sx}\right] = E\left[e^{-sw}\right]E\left[e^{-sx}\right]$, where the latter follows from independence between x and w. Hence, $\varphi_T(s) = \varphi_x(s)\varphi_w(s)$, from which the Laplace transform of the waiting time in the queue follows as

$$\varphi_w(s) = (1 - \rho) \frac{s}{s - \lambda + \lambda\varphi_x(s)} \tag{14.31}$$

Due to the correspondence with (14.27), these two relations (14.30) and (14.31) are also called the Pollaczek-Khinchin equations for the system time and waiting time respectively. For example, for an exponential service time with mean $\frac{1}{\mu}$, $\varphi_x(s) = \frac{\mu}{s+\mu}$ and (14.30) becomes $\varphi_T(s) = \frac{(1-\rho)\mu}{s+(\mu-\lambda)}$, which is indeed the Laplace transform of the pdf of total system time (14.5) in the M/M/1 queue. The relation (14.31) can be written in terms of the residual service time (8.19), which after Laplace transform becomes $\varphi_{rw}(s) = \frac{1-\varphi_x(s)}{sE[x]}$, as

$$\varphi_w(s) = \frac{1 - \rho}{1 - \rho\varphi_{rw}(s)}$$

It shows that the dominant tail behavior (see Section 5.7) arises from the pole at $\varphi_{rw}(s) = \frac{1}{\rho}$. By formal expansion into a Taylor series (only valid for $|\rho\varphi_{rw}(s)| < 1$), we find

$$\varphi_w(s) = (1 - \rho) \sum_{k=0}^{\infty} \rho^k \varphi_{rw}^k(s)$$

or, after taking the inverse Laplace transform,

$$f_w(t) = \frac{d}{dt} \Pr\left[w \le t\right] = \sum_{k=0}^{\infty} (1 - \rho) \rho^k f_{rw}^{(k*)}(t)$$

The pdf $f_w(t)$ of the waiting time in the queue can be interpreted as a sum of convolved residual service time pdfs $f_{rw}^{(k*)}(t)$ weighted by $(1 - \rho)\rho^k = \Pr\left[N_S = k\right]$, the steady-state probability of the system content in the M/M/1 system (14.1).

14.4 The GI/D/m queue

The analysis of the GI/D/m queue illustrates a discrete-time approach to queueing. Since each of the m servers operate deterministically, which means that per unit time precisely one packet (or an ATM cell or customer) is served, the basic time unit in the analysis, also called a time slot, is equal to that service time. Hence, the arrival process is expressed as a counting process: instead of specifying the interarrival rate, the number of arrivals at each time slot is used.

In the sequel, we confine ourselves to a deterministic server discipline that removes during timeslot k precisely m cells from the queue. Based on the discussion in Section 13.4, we have $\mathcal{B}_k = m$ and $B_k(z) = E\left[z^{\mathcal{B}_k}\right] = z^m$. Substituting (13.19) in (14.24) leads to

$$S_{k+1}(z) = E\left[z^{(\mathcal{S}_k - m)^+ + \mathcal{A}_k}\right] \tag{14.32}$$

and (13.18) shows that $\mathcal{Q}_k = (\mathcal{S}_k - m)^+$. At this point, a further general evaluation of the expression (14.32) is only possible by assuming *independence* between the random variables $\mathcal{A}_k$ and $\mathcal{Q}_k$. From (13.21), it then follows that $S_{k+1}(z) = Q_k(z)A_k(z)$. This crucial assumption facilitates the analysis considerably. For,

$$S_{k+1}(z) = E\left[z^{(\mathcal{S}_k - m)^+} z^{\mathcal{A}_k}\right]$$

$$= E\left[z^{(\mathcal{S}_k - m)^+}\right] E\left[z^{\mathcal{A}_k}\right] \qquad \text{(by independence)}$$

$$= A_k(z) \sum_{j=0}^{\infty} \Pr[(\mathcal{S}_k - m)^+ = j]\, z^j \qquad \text{(by definition (14.24))}$$

The summation can be worked out as

$$B = \sum_{j=0}^{\infty} \Pr[(\mathcal{S}_k - m)^+ = j]\, z^j = \Pr[(\mathcal{S}_k - m)^+ = 0] + \sum_{j=1}^{\infty} \Pr[\mathcal{S}_k = m + j]\, z^j$$

$$= \sum_{j=0}^{m} \Pr[\mathcal{S}_k = j] + \sum_{j=1+m}^{\infty} \Pr[\mathcal{S}_k = j]\, z^{j-m}$$

We rewrite B in terms of the generating function of $S_k(z)$,

$$B = \sum_{j=0}^{m} \Pr[\mathcal{S}_k = j] + z^{-m}\left(\sum_{j=0}^{\infty} \Pr[\mathcal{S}_k = j]\, z^j - \sum_{j=0}^{m} \Pr[\mathcal{S}_k = j]\, z^j\right)$$

$$= z^{-m}\left(S_k(z) - \sum_{j=0}^{m} \Pr[\mathcal{S}_k = j]\, (z^j - z^m)\right)$$

Finally, we obtain a recursion relation for the generating function of the system content in the GI/D/m system at discrete-time k,

$$S_{k+1}(z) = A_k(z)z^{-m}\left(S_k(z) - \sum_{j=0}^{m-1} \Pr[\mathcal{S}_k = j]\, (z^j - z^m)\right) \tag{14.33}$$

In the single-server case $m = 1$, where precisely one cell is served per time slot (provided the queue is not empty), equation (14.33) simplifies with $S_k(0) = \Pr[\mathcal{S}_k = 0]$ to

$$S_{k+1}(z) = A_k(z)z^{-1}\left\{S_k(z) - \Pr[\mathcal{S}_k = 0]\,(1 - z)\right\}$$

$$= A_k(z)\left[\frac{S_k(z) - S_k(0)}{z} + S_k(0)\right] \tag{14.34}$$

Notice that $S_{k+1}(0) = A_k(0)(S'_k(0) + S_k(0))$.

14.4.1 The steady-state of the GI/D/m queue

The steady-state behavior is reached if the system's distributions no longer change in time. With $\lim_{k \to \infty} S_k(z) = S(z)$ and $\lim_{k \to \infty} A_k(z) = A(z)$, (14.33) reduces in steady-state to

$$S(z) = \frac{A(z) \sum_{j=0}^{m-1} \Pr[\mathcal{S} = j] (z^m - z^j)}{z^m - A(z)}$$

Recall that $X_k(z) = z^m$ is the generating function of the service process. At this point, we use the same powerful argument from complex analysis as in Section 11.3.3. Since a generating function of a probability distribution is analytic inside and on the unit circle, the possible zeros of $z^m - A(z)$ inside that unit circle must be precisely cancelled by zeros in the numerator. Clearly, $z = 1$ is a zero of $z^m - A(z)$. On the unit circle, excluding the point $z = 1$ where $A(1) = 1$, we have shown on p. 14 that $|A(z)| < 1$ for all points on the unit circle $|z| = 1$ (except for $z = 1$).

The region around $z = 1$ deserves some closer investigation. From the Taylor expansions

$$A(z) = 1 + \lambda(z - 1) + o(z - 1)$$
$$z^m = 1 + m(z - 1) + o(z - 1)$$

and the fact that $\lambda < m$ because a steady-state requires $\rho < 1$, we substitute $z - 1 = \varepsilon e^{i\theta}$, which describes a circle with radius ε around $z = 1$. Along this circle with arbitrarily small radius ε, we find that

$$|A(z)| = |1 + \lambda \varepsilon e^{i\theta} + o(\varepsilon)| = \sqrt{(1 + \lambda \varepsilon \cos\theta + o(\varepsilon))^2 + (\lambda \varepsilon \sin\theta + o(\varepsilon))^2}$$
$$|z^m| = |1 + m\varepsilon e^{i\theta} + o(\varepsilon)| = \sqrt{(1 + m\varepsilon \cos\theta + o(\varepsilon))^2 + (m\varepsilon \sin\theta + o(\varepsilon))^2}$$

which demonstrates that $|z^m| > |A(z)|$ on this arbitrary small circle if $\cos\theta > 0$ or $-\frac{\pi}{2} < \theta < \frac{\pi}{2}$. Invoking Rouché's Theorem 11.3.1 with $f(z) = z^m$ and $g(z) = -A(z)$ on the contour C, the unit circle including the point $z = 1$ by an arbitrarily small arc ε on the right of $z = 1$ as illustrated in Fig. 14.3, such that $|f(z)| > |g(z)|$ on the contour C, shows that $z^m - A(z)$ has precisely m zeros $\zeta_1, \zeta_2, \ldots, \zeta_m = 1$ inside that contour C.

Since $Q(z) = \frac{S(z)}{A(z)}$ is the generating function of the number of occupied buffer positions, it is also analytic inside the unit circle. Therefore, the zeros $\{\zeta_n\}_{1 \le n \le m-1}$ and $\zeta_m = 1$ are also zeros of $p(z) = \sum_{j=0}^{m-1} \Pr[\mathcal{S} = j] (z^m - z^j)$. This leads to a set of m equations for each $\zeta_n \ne 0$,

$$\sum_{j=0}^{m-1} \Pr[\mathcal{S} = j] (\zeta_n^m - \zeta_n^j) = 0$$

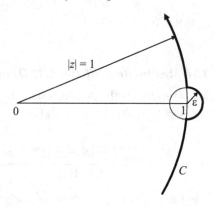

Fig. 14.3. Details of the contour C in the neighborhood of $z = 1$.

which determine the unknown probabilities $\Pr[\mathcal{S} = j]$. Since $p(z)$ is a polynomial of degree m, $p(z)$ is entirely determined by its zeros as

$$p(z) = \alpha(z - 1) \prod_{n=1}^{m-1} (z - \zeta_n)$$

The unknown α is determined from the normalization condition $S(1) = Q(1) = 1$, which is explicitly

$$\lim_{z \to 1} \frac{p(z)}{z^m - A(z)} = 1$$

With de l'Hospital's rule,

$$\alpha^{-1} = \prod_{n=1}^{m-1} (1 - \zeta_n) \lim_{z \to 1} \frac{1}{mz^{m-1} - A'(z)} = \frac{\prod_{n=1}^{m-1}(1 - \zeta_n)}{m - \lambda}$$

Finally, we arrive at the generating function of the buffer content via that of the system content $S(z) = A(z)Q(z)$,

$$Q(z) = \frac{(m - \lambda)(z - 1)}{z^m - A(z)} \prod_{n=1}^{m-1} \frac{z - \zeta_n}{1 - \zeta_n} \tag{14.35}$$

With $A'(1) = \lambda$ the single-server case ($m = 1$) in (14.35) reduces to the well-known result for the pgf of the system and buffer content of a GI/D/1 system respectively as

$$S(z) = (1 - A'(1)) \frac{(z - 1) A(z)}{z - A(z)} \tag{14.36}$$

$$Q(z) = (1 - \lambda) \frac{z - 1}{z - A(z)} \tag{14.37}$$

The probability of an empty buffer, $Q(0) = \Pr[\mathcal{Q} = 0]$, immediately follows from

(14.35). The mean queue length for the single-server ($m = 1$) is obtained as $Q'(1)$, or

$$E[N_Q] = E[Q] = \frac{A''(1)}{2(1 - \lambda)} \tag{14.38}$$

14.4.2 The waiting time in a GI/D/m system

Let $T = w + 1$ denote the steady-state system time of an arbitrary packet, a "test" packet, in units of a timeslot. In addition to the system content S that describes the number of packets in the system at the beginning of a time slot, an additional random variable must be introduced: $\mathcal{F}$ denotes the number of packets that arrive in the same timeslot just before the "test" packet. In the D-server and assuming a FIFO discipline, these $\mathcal{F}$ packets will be served before the "test" packet, possibly in the same time slot. The system time of the "test" packet equals

$$T = \left\lfloor \frac{(S - m)^+ + \mathcal{F}}{m} \right\rfloor + 1$$

where $\lfloor x \rfloor$ denotes the largest integer smaller than or equal to x. Indeed, $(S - m)^+ + \mathcal{F}$ are the number of packets in the system just before the arrival of the "test" packet. At the beginning of a timeslot, at most m packets are served, which explains the integer division. The service time takes precisely one additional time slot. Let us simplify the notation by defining $\mathcal{R} = (S - m)^+ + \mathcal{F}$. From this expression for the system time, we deduce, for each integer $k \geq 1$ (the minimal waiting time in the system equals 1 timeslot), that

$$\Pr[T = k] = \sum_{j=0}^{m-1} \Pr[\mathcal{R} = (k - 1)m + j]$$

such that the generating function of the waiting time $T(z)$ is

$$T(z) = \sum_{k=1}^{\infty} \Pr[T = k] z^k = \sum_{k=1}^{\infty} \sum_{j=0}^{m-1} \Pr[\mathcal{R} = (k - 1)m + j] z^k$$

$$= \sum_{j=0}^{m-1} \sum_{k=0}^{\infty} \Pr[\mathcal{R} = km + j] z^{k+1}$$

Consider

$$T(z^m) = z^m \sum_{j=0}^{m-1} z^{-j} \sum_{k=0}^{\infty} \Pr[\mathcal{R} = km + j] z^{mk+j}$$

and let $n = km + j$,

$$T\left(z^{m}\right) = z^{m} \sum_{j=0}^{m-1} z^{-j} \sum_{k=0}^{\infty} \sum_{n=0}^{\infty} \Pr\left[\mathcal{R} = n\right] z^{n} \delta_{n,mk+j}$$

$$= z^{m} \sum_{j=0}^{m-1} z^{-j} \sum_{n=0}^{\infty} \Pr\left[\mathcal{R} = n\right] z^{n} \sum_{k=0}^{\infty} \delta_{n,mk+j}$$

where the Kronecker delta $\delta_{k,m} = 1$ if $k = m$ else $\delta_{k,m} = 0$. The sum

$$\sum_{k=0}^{\infty} \delta_{n,mk+j} = \sum_{k=-\infty}^{\infty} \delta_{n-j,mk} = 1_{m|n-j}$$

is one if m divides $n - j$ else it is zero. Such expression can be written as

$$\sum_{k=-\infty}^{\infty} \delta_{n-j,mk} = \frac{1 - e^{2\pi i(n-j)}}{m\left(1 - e^{2\pi i(n-j)/m}\right)} = \frac{1}{m} \sum_{k=0}^{m-1} e^{2\pi ik(n-j)/m}$$

Using the latter summation yields

$$T\left(z^{m}\right) = \frac{z^{m}}{m} \sum_{j=0}^{m-1} z^{-j} \sum_{n=0}^{\infty} \Pr\left[\mathcal{R} = n\right] z^{n} \sum_{k=0}^{m-1} e^{2\pi ik(n-j)/m}$$

$$= \frac{z^{m}}{m} \sum_{k=0}^{m-1} \sum_{j=0}^{m-1} \left(ze^{2\pi ik/m}\right)^{-j} \sum_{n=0}^{\infty} \Pr\left[\mathcal{R} = n\right] \left(ze^{2\pi ik/m}\right)^{n}$$

$$= \frac{z^{m}}{m} \sum_{k=0}^{m-1} \frac{1 - z^{-m}}{1 - \left(ze^{2\pi ik/m}\right)^{-1}} R\left(ze^{2\pi ik/m}\right)$$

where we have introduced the generating function $R(z) = \sum_{n=0}^{\infty} \Pr\left[\mathcal{R} = n\right] z^{n}$. Thus, we arrive at

$$T\left(z^{m}\right) = \frac{1}{m} \sum_{k=0}^{m-1} \frac{z^{m} - 1}{1 - \left(ze^{2\pi ik/m}\right)^{-1}} R\left(ze^{2\pi ik/m}\right)$$

The generating function $R(z)$ can be specified further since

$$R(z) = E\left[z^{(\mathcal{S}-m)^{+}+\mathcal{F}}\right] = E\left[z^{\mathcal{F}}\right] E\left[z^{(\mathcal{S}-m)^{+}}\right]$$

where independence of the arrival process and queueing process (GI) has been used. From (14.32), the corresponding steady-state relation is

$$S(z) = E\left[z^{(\mathcal{S}-m)^{+}+\mathcal{A}}\right] = E\left[z^{\mathcal{A}}\right] E\left[z^{(\mathcal{S}-m)^{+}}\right] = A(z)E\left[z^{(\mathcal{S}-m)^{+}}\right]$$

while from $S(z) = A(z)Q(z)$, we observe that $Q(z) = E\left[z^{(\mathcal{S}-m)^{+}}\right]$. Hence,

$$R(z) = F(z)Q(z)$$

where $Q(z)$ is given by (14.35).

We now turn our attention to the determination of $F(z)$, the generating function of the number of packets in front of the "test" packet. The "test" packet has been uniformly chosen out of the total flow of arriving packets at the system. Let us denote by $\mathcal{A}^*$ the number of arriving packets in the same time slot as the "test" packet. The random variable $\mathcal{A}^*$ is not the same as the number of arriving cells $\mathcal{A}$ per time slot. For example, we know that there is at least one arrival in the time slot of the "test" packet, namely, the "test" packet itself, hence, $\Pr[\mathcal{A}^* = 0] = 0$. Furthermore, the larger the number of arriving packets in a time slot, the higher the probability that the "test" packet is chosen out of those packets in this time slot. Hence, $\Pr[\mathcal{A}^* = j]$ is proportional to the number j of arriving packets in a time slot. In addition, $\Pr[\mathcal{A}^* = j]$ is also proportional to $\Pr[\mathcal{A} = j]$, which describes how likely a number j of arriving packets is. Combining both shows that

$$\Pr[\mathcal{A}^* = j] = \alpha j \Pr[\mathcal{A} = j]$$

with the proportionality factor equal to $\alpha = \frac{1}{E[\mathcal{A}]}$ because $\sum_{j=0}^{\infty} \Pr[\mathcal{A}^* = j] = 1$. The "test" packet is uniformly distributed among the arriving packets $\mathcal{A}^*$ in the time slot of the "test" packet (in steady-state). The probability of having precisely k packets in front of the "test" packet given $\mathcal{A}^* = j \geq 1$ equals

$$\Pr[\mathcal{F} = k | \mathcal{A}^* = j] = \frac{1_{k<j}}{j}$$

Indeed, the "test" packet has equal probability $\frac{1}{j}$ of occupying any of the j possible positions. The occupation of a position $k + 1$ implies precisely k cells in front of the "test" packet in a FIFO discipline. Using the law of total probability (2.49),

$$\Pr[\mathcal{F} = k] = \sum_{j=1}^{\infty} \Pr[\mathcal{F} = k | \mathcal{A}^* = j] \Pr[\mathcal{A}^* = j]$$

$$= \sum_{j=k+1}^{\infty} \frac{1}{j} \frac{j \Pr[\mathcal{A} = j]}{E[\mathcal{A}]} = \frac{1}{E[\mathcal{A}]} \sum_{j=k+1}^{\infty} \Pr[\mathcal{A} = j]$$

The generating function $F(z)$ becomes

$$F(z) = \sum_{k=0}^{\infty} \Pr[\mathcal{F} = k] z^k = \frac{1}{E[\mathcal{A}]} \sum_{k=0}^{\infty} \sum_{j=k+1}^{\infty} \Pr[\mathcal{A} = j] z^k$$

$$= \frac{1}{E[\mathcal{A}]} \sum_{j=1}^{\infty} \Pr[\mathcal{A} = j] \sum_{k=1}^{j} z^{k-1} = \frac{1}{E[\mathcal{A}]} \sum_{j=1}^{\infty} \Pr[\mathcal{A} = j] \frac{1 - z^j}{1 - z}$$

$$= \frac{1}{E[\mathcal{A}](z-1)} \left(\sum_{j=1}^{\infty} \Pr[\mathcal{A} = j] z^j - \sum_{j=1}^{\infty} \Pr[\mathcal{A} = j] \right)$$

$$= \frac{z(A(z) - \Pr[\mathcal{A} = 0] - 1 + \Pr[\mathcal{A} = 0])}{E[\mathcal{A}](z-1)}$$

Since $E[\mathcal{A}] = A'(1) = \lambda$, we finally arrive at

$$F(z) = \frac{A(z) - 1}{(z-1)A'(1)}$$

Combining all involved expressions leads to the expression of the generating function of the total time spent in the GI/D/m queue

$$T(z^m) = \frac{m - \lambda}{m\lambda} \sum_{k=0}^{m-1} \frac{z^m - 1}{1 - (ze^{2\pi ik/m})^{-1}} \frac{A(ze^{2\pi ik/m}) - 1}{z^m - A(ze^{2\pi ik/m})} \prod_{n=1}^{m-1} \frac{ze^{2\pi ik/m} - \zeta_n}{1 - \zeta_n}$$

For the single-server case ($m = 1$), the generating function of the system time (queueing time plus service time) considerably simplifies to

$$T(z) = \left(\frac{1-\lambda}{\lambda}\right) z \frac{A(z) - 1}{z - A(z)}$$

from which $E[T]$ and $\mathrm{Var}[T]$ readily follow. The computation of the pdf given the arrival process $A(z)$ is more complex, as illustrated for the M/D/1/K queue in Section 14.5.

14.5 The M/D/1/K queue

Suppose we have a buffer of K cells and an aggregate arrival stream consisting of a large number of sources with none of them dominating the others. This input process is well modeled by a Poisson process with arrival rate λ. Both the input process as well as the buffer content and the output process have been simulated in Fig. 14.4. Observe the effect of variations and maximum number of cells in the queue and input process!

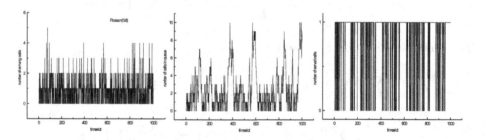

Fig. 14.4. On the left, the Poisson input process with $\lambda = 0.8$ in terms of the number of cells versus the timeslot. In the middle, the buffer occupancy for a buffer with $K = 20$ as function of time. On the right, the M/D/1/K output process in cells served per timeslot.

In the sequel, we concentrate on the buffer occupancy in the M/D/1 queue by applying the general results of the GI/D/m queue.

For a Poisson process, $\Pr[\mathcal{A} = k] = \frac{\lambda^k}{k!}e^{-\lambda}$ and $A(z) = e^{\lambda(z-1)}$. The pgf $Q(z)$ of the buffer content immediately follows from (14.37) as

$$Q(z) = (1 - \lambda)\frac{(1 - z)\,e^{\lambda(1-z)}}{1 - z\,e^{\lambda(1-z)}}$$

The mean queue length is obtained from (14.38) as

$$E\left[N_{Q;M/D/1}\right] = E\left[Q\right] = \frac{\lambda^2}{2\,(1 - \lambda)}$$

and Little's Law (13.24) provides the mean waiting time in the queue

$$E\left[w_{M/D/1}\right] = \frac{E\left[N_{Q;M/D/1}\right]}{\lambda} = \frac{\lambda}{2\,(1 - \lambda)} \tag{14.39}$$

Since $z\,e^{\lambda(1-z)}$ is analytic everywhere, there always exists a neighborhood (depending on λ) around $z = 0$ for which $|z\,e^{\lambda(1-z)}| \leq 1$. Hence, we can use the series expansion for the geometric series to obtain

$$\frac{Q(z)}{1 - \lambda} = (1 - z)\,e^{\lambda(1-z)}\sum_{k=0}^{\infty} z^k\,e^{k\lambda(1-z)} = \sum_{k=0}^{\infty}(1 - z)\,z^k\,e^{(k+1)\lambda(1-z)}$$

Integrating with respect to λ removes the factor $(1 - z)$,

$$\int \frac{Q(z)}{1 - \lambda}\,d\lambda = \sum_{k=0}^{\infty}\frac{z^k}{k + 1}\,e^{(k+1)\lambda}\,e^{-(k+1)\lambda z}$$

$$= \sum_{k=0}^{\infty}\frac{z^k}{k + 1}\,e^{(k+1)\lambda}\sum_{n=0}^{\infty}\frac{(-1)^n(k + 1)^n\lambda^n}{n!}\,z^n$$

$$= \sum_{k=0}^{\infty}\sum_{n=0}^{\infty}\frac{(-1)^n}{n!}\,e^{(k+1)\lambda}\,(k + 1)^{n-1}\lambda^n z^{n+k}$$

After a change in the variable $m = n + k$ with $m \geq 0$, which implies that $k \leq m$ since $n = m - k \geq 0$, we have

$$\int \frac{Q(z)}{1 - \lambda}\,d\lambda = \sum_{m=0}^{\infty}\left(\sum_{k=0}^{m}\frac{(-1)^{m-k}}{(m - k)!}\,e^{(k+1)\lambda}\,(k + 1)^{m-k-1}\lambda^{m-k}\right)z^m$$

$$= \sum_{m=0}^{\infty}\left(\sum_{k=1}^{m+1}\frac{(-1)^{m+1-k}}{(m + 1 - k)!}\,e^{k\lambda}\,k^{m-k}\lambda^{m+1-k}\right)z^m$$

Differentiation with respect to λ gives

$$Q(z) = (1 - \lambda)\sum_{m=0}^{\infty}\left(\sum_{k=1}^{m+1}(-1)^{m+1-k}\,e^{k\lambda}\left[\frac{(k\lambda)^{m+1-k}}{(m + 1 - k)!} + \frac{(k\lambda)^{m-k}}{(m - k)!}\right]\right)z^m$$

from which we finally deduce the probability $q[m] = \Pr\left[Q = m\right]$ that the m-th position in the buffer is occupied[6]

$$q[m] = (1 - \lambda) \sum_{k=1}^{m+1} (-1)^{m+1-k} e^{k\lambda} \left[\frac{(k\lambda)^{m+1-k}}{(m+1-k)!} + \frac{(k\lambda)^{m-k}}{(m-k)!} \right] \qquad (14.40)$$

One observes from the derivation above that the probability $s[m] = \Pr[\mathcal{S} = m]$ that m positions in the system are occupied, is $s[m] = q[m - 1]$ for $m \geq 2$ because the z-transform is $S(z) = (1 - \lambda) \frac{(1-z)}{1 - z\, e^{\lambda(1-z)}}$. This result is a characteristic property of a deterministic server. Next, we rewrite (14.40) as

$$q[m] = (1 - \lambda) \left[\sum_{k=1}^{m+1} e^{k\lambda} \frac{(-k\lambda)^{m+1-k}}{(m+1-k)!} - \sum_{k=1}^{m} e^{k\lambda} \frac{(-k\lambda)^{m-k}}{(m-k)!} \right]$$

$$= (1 - \lambda) \left[e^{(m+1)\lambda} g(-\lambda e^{-\lambda}; m+1) - e^{m\lambda} g(-\lambda e^{-\lambda}; m) \right] \qquad (14.41)$$

where

$$g(x; m) = \sum_{k=0}^{m} (m - k)^k \frac{x^k}{k!} \qquad (14.42)$$

Due to the nature of the differences, we immediately find the cumulative distribution,

$$\sum_{m=1}^{K} q[m] = (1 - \lambda) \left[e^{(K+1)\lambda} g(-\lambda e^{\lambda}; K+1) - e^{\lambda} \right]$$

and since $q[0] = (1 - \lambda)\, e^{\lambda}$, we arrive at

$$\Pr[\mathcal{Q} \leq K] = \sum_{m=0}^{K} q[m] = (1 - \lambda)\, e^{(K+1)\lambda} g(-\lambda e^{\lambda}; K+1) \qquad (14.43)$$

The expressions in (14.41) are numerically only useful for small m because the series is alternating. This problem may be solved by considering a famous result due to Lagrange (Markushevich, 1985, Vol. 2, Chapter 3, Section 14)

$$e^{bz} = 1 + b \sum_{n=1}^{\infty} \frac{(b + na)^{n-1}}{n!} \left(z e^{-az} \right)^n \qquad (14.44)$$

---

[6] Explicitly, we have for the queue content probabilities

$$q[0] = e^{\lambda} (1 - \lambda)$$

$$q[1] = e^{\lambda} (1 - \lambda) (e^{\lambda} - \lambda - 1)$$

$$q[2] = e^{\lambda} (1 - \lambda) (\frac{\lambda^2}{2} - e^{2\lambda} + \lambda - e^{\lambda} - 2\lambda e^{\lambda})$$

while the system content probabilities are

$$s[0] = (1 - \lambda)$$

$$s[1] = (e^{\lambda} - 1) (1 - \lambda)$$

$$s[m] = q[m - 1] \qquad\qquad (m \geq 2)$$

that converges for $|z| \leq a^{-1}$. Differentiation of (14.44) with respect to $w = ze^{-az}$ leads to

$$\frac{e^{(b+a)z}}{1-az} = \sum_{n=0}^{\infty} \frac{(b+a+na)^n}{n!} z^n e^{-naz}$$

because $\frac{d}{dw}e^{bz} = \frac{d}{dz}e^{bz}\frac{dz}{dw} = \frac{be^{bz}}{(1-az)e^{-az}}$. Choosing $a = -1$, $b+a = m$ and $z = -\lambda$, we obtain

$$\frac{e^{-m\lambda}}{1-\lambda} = \sum_{n=0}^{\infty} \frac{(m-n)^n}{n!}(-\lambda e^{-\lambda})^n = g(-\lambda e^{-\lambda}; m) + \sum_{n=m+1}^{\infty} \frac{(n-m)^n}{n!}(\lambda e^{-\lambda})^n$$

and thus

$$g(-\lambda e^{-\lambda}; m) = \frac{e^{-m\lambda}}{1-\lambda} - \sum_{n=m+1}^{\infty} \frac{(n-m)^n}{n!}(\lambda e^{-\lambda})^n$$

where the infinite series consists of merely positive terms. Substitution in (14.43) finally yields

$$\Pr[\mathcal{Q} > K] = (1-\lambda)\,\lambda^{K+1} \sum_{n=1}^{\infty} \frac{n^{n+K+1}}{(n+K+1)!}(\lambda e^{-\lambda})^n \qquad (14.45)$$

In the heavy traffic limit $\rho = \lambda \to 1$, the dominant zero (5.44) of $1 - z\,e^{\lambda(1-z)}$ is approximately equal to $\zeta \approx 1 + \frac{2(1-\lambda)}{\lambda}$ and the resulting tail asymptotic (5.40) for the buffer occupancy pdf is

$$\Pr[\mathcal{Q} > K] \approx -\frac{1-\rho}{1-\rho\zeta}\zeta^{-K-1} \approx e^{-K\log\zeta} \sim e^{-2K\frac{(1-\rho)}{\rho}} \qquad (14.46)$$

14.6 The G/M/1 queue

The G/M/1 infinite queueing system consists of an arrival process with generally distributed interarrival times and of a single processing unit that serves each job or packet independently in an exponential service time with mean $\frac{1}{\mu}$. The G/M/1 queue is the dual to the M/G/1 queue. Mainly due to the "memoryless" service discipline, a nice and elegant analysis is possible. Here, we present a generating function computation of the number of packets in the G/M/1 queueing system at different observation times.

14.6.1 The equation for system content

Whereas the M/G/1 queue was best observed at departure epochs, we analyze the dual G/M/1 queue at the arrival times – strictly speaking just before the k-th arrival at t_k occurs as explained in Section 13.4. Let $\mathcal{R}_k$ denote the number of departing packets during the interarrival time $(t_k, t_{k+1}]$ of the k-th and $(k+1)$-th packet, *assuming* an infinitely large amount of available packets in the system.

The number of packets in the system just before the k-th arrival is denoted by $\mathcal{S}_k = N_S\left(t_k^+\right)$, which satisfies the system equation

$$\mathcal{S}_{k+1} = \left(\mathcal{S}_k + 1 - \mathcal{R}_k\right)^+$$

The random variables $\mathcal{S}_k$ and $\mathcal{R}_k$ are independent as we will demonstrate. Since the packets that leave the system are a subset of those still in the system during the k-th interarrival time, independence between the number of departing $\mathcal{R}_k$ and the number $\mathcal{S}_k$ of those in the system just before a new packet arrives is not immediately obvious. Since the service discipline is a Poisson process with rate μ, a similar reasoning that led to (14.23) can be employed,

$$\Pr\left[\mathcal{R}_k = j\right] = \int_0^\infty \Pr\left[\mathcal{R}_k = j | \mathcal{A} = t\right] dF_{\mathcal{A}}\left(t\right)$$

$$= \int_0^\infty e^{-\mu t} \frac{(\mu t)^j}{j!} dF_{\mathcal{A}}\left(t\right) = \frac{(-\mu)^j}{j!} \left.\frac{d^j \varphi_{\mathcal{A}}\left(z\right)}{dz^j}\right|_{z=\mu}$$

where $\varphi_{\mathcal{A}}\left(z\right) = \int_0^\infty e^{-zt} dF_{\mathcal{A}}\left(t\right)$ is the Laplace transform of the interarrival time $\mathcal{A}$ between two consecutive packets. The corresponding generating function $R_k\left(z\right) = E\left[z^{\mathcal{R}_k}\right]$ is, analogously to (14.26), equal to $R_k\left(z\right) = \varphi_{\mathcal{A}}\left(\mu - \mu z\right)$. Three remarks need to be placed. First, invoking (14.26) implicitly assumes that there is always an infinite amount of packets in the system that can be served. If $\mathcal{R}_k \leq \mathcal{S}_k + 1$, then $\mathcal{R}_k$ denotes the number of departing packets, else, the number of departing packets is, of course, $\mathcal{S}_k + 1$, which underlines the role of the maximum-operator. Thus, $\mathcal{R}_k$ is an auxiliary random variable that allows us to compute the number of departing packets. Second, the service process behaves similarly for any interarrival time such that the time-dependence k has vanished. The underlying assumption here is that the interarrival times are independent for each k. Hence, it is implicitly assumed – similar to the M/G/1 queue analysis – that the arrival process is not correlated over time. Third, the characteristic property of a Poisson service process, in particular the memoryless property of the exponential service distribution, shows that the random variables $\mathcal{S}_k$ and $\mathcal{R}_k$ are independent.

If the mean interarrival time is equal to $\lambda^{-1} = -\varphi'_{\mathcal{A}}\left(0\right)$, then

$$E\left[\mathcal{R}_k\right] = R'_k\left(1\right) = -\varphi'_{\mathcal{A}}\left(0\right)\mu = \frac{\mu}{\lambda} = \rho^{-1}$$

Stability in any queueing system requires that the utilization $\rho < 1$, such that $E\left[\mathcal{R}_k\right] > 1$. In addition, $R_k\left(0\right) = \varphi_{\mathcal{A}}\left(\mu\right) > 0$.

Analogous to Section 14.4, the generating function $S_k\left(z\right) = E\left[z^{\mathcal{S}_k}\right]$ is derived from the system equation as

$$S_{k+1}\left(z\right) = E\left[z^{\max(\mathcal{S}_k+1-\mathcal{R}_k,0)}\right] = \sum_{j=0}^\infty \Pr\left[\max\left(\mathcal{S}_k + 1 - \mathcal{R}_k, 0\right) = j\right] z^j$$

By the law of total probability (2.49), we have

$$\Pr\left[\max\left(\mathcal{S}_k + 1 - \mathcal{R}_k, 0\right) = j\right] = \sum_{l=0}^{\infty} \Pr\left[\max\left(\mathcal{S}_k + 1 - \mathcal{R}_k, 0\right) = j | \mathcal{R}_k = l\right] \Pr\left[\mathcal{R}_k = l\right]$$

The conditional probability is

$$\begin{aligned}
\Pr\left[\max\left(\mathcal{S}_k + 1 - \mathcal{R}_k, 0\right) = j | \mathcal{R}_k = l\right] &= \Pr\left[\max\left(\mathcal{S}_k + 1 - l, 0\right) = j | \mathcal{R}_k = l\right] \\
&= \Pr\left[\max\left(\mathcal{S}_k + 1 - l, 0\right) = j\right]
\end{aligned}$$

where the latter follows by the independence between $\mathcal{S}_k$ and $\mathcal{R}_k$. Next, if $j = 0$, then

$$\Pr\left[\max\left(\mathcal{S}_k + 1 - l, 0\right) = 0\right] = \sum_{r=0}^{l-1} \Pr\left[\mathcal{S}_k = r\right]$$

while, for $j > 0$,

$$\Pr\left[\max\left(\mathcal{S}_k + 1 - l, 0\right) = j\right] = \Pr\left[\mathcal{S}_k = j + l - 1\right]$$

Combining all yields

$$\begin{aligned}
S_{k+1}(z) &= \sum_{j=0}^{\infty} \sum_{l=0}^{\infty} \Pr\left[\max\left(\mathcal{S}_k + 1 - l, 0\right) = j\right] \Pr\left[\mathcal{R}_k = l\right] z^j \\
&= \sum_{l=0}^{\infty} \sum_{r=0}^{l-1} \Pr\left[\mathcal{S}_k = r\right] \Pr\left[\mathcal{R}_k = l\right] + \sum_{j=1}^{\infty} \sum_{l=0}^{\infty} \Pr\left[\mathcal{S}_k = j + l - 1\right] \Pr\left[\mathcal{R}_k = l\right] z^j
\end{aligned}$$

Clearly, the first double sum equals $S_{k+1}(0)$, while the second is

$$\begin{aligned}
B &= \sum_{j=1}^{\infty} \sum_{l=0}^{\infty} \Pr\left[\mathcal{S}_k = j + l - 1\right] \Pr\left[\mathcal{R}_k = l\right] z^j \\
&= \sum_{l=0}^{\infty} \Pr\left[\mathcal{R}_k = l\right] z^{-l} \sum_{j=l}^{\infty} \Pr\left[\mathcal{S}_k = j\right] z^{j+1} \\
&= \sum_{l=0}^{\infty} \Pr\left[\mathcal{R}_k = l\right] z^{-l} \left(\sum_{j=0}^{\infty} \Pr\left[\mathcal{S}_k = j\right] z^{j+1} - \sum_{j=0}^{l-1} \Pr\left[\mathcal{S}_k = j\right] z^{j+1} \right) \\
&= \sum_{l=0}^{\infty} \Pr\left[\mathcal{R}_k = l\right] z^{-l} \sum_{j=0}^{\infty} \Pr\left[\mathcal{S}_k = j\right] z^{j+1} - \sum_{l=0}^{\infty} \Pr\left[\mathcal{R}_k = l\right] z^{-l} \sum_{j=0}^{l-1} \Pr\left[\mathcal{S}_k = j\right] z^{j+1}
\end{aligned}$$

which is rewritten in terms of the generating function $S_k(z)$ and $R_k(z) = E\left[z^{\mathcal{R}_k}\right]$ as

$$B = z R_k\left(z^{-1}\right) S_k(z) - z T_k(z)$$

where

$$T_k(z) = \sum_{l=0}^{\infty} \sum_{j=0}^{l-1} \Pr\left[\mathcal{S}_k = j\right] \Pr\left[\mathcal{R}_k = l\right] z^{j-l}$$

When comparing $S_{k+1}(0)$ above with $T_k(z)$, we recognize that $S_{k+1}(0) = T_k(1)$ such that the recursion for the generating function $S_k(z)$ is deduced

$$S_{k+1}(z) = zS_k(z)R_k(z^{-1}) + V_k(z) \tag{14.47}$$

where $V_k(z) = T_k(1) - zT_k(z)$ and

$$V_k(z) = \sum_{l=0}^{\infty}\sum_{j=0}^{l-2}\Pr[\mathcal{S}_k = j]\Pr[\mathcal{R}_k = l]\left(1 - z^{j+1-l}\right) \tag{14.48}$$

14.6.2 The steady-state of the G/M/1 queue

In the steady-state, where $\lim_{k\to\infty} S_k(z) = S(z)$ and similarly omitting the k-index for the other generating function, we obtain from (14.47)

$$S(z) = \frac{V(z)}{1 - zR(z^{-1})} \tag{14.49}$$

Any probability generating function is bounded in $|z| \leq 1$ and analytic[7] in $|z| < 1$. We assume that $S(z)$ is analytic in $|z| \leq 1$, such that all moments exist. Then,

$$S(z^{-1}) = \frac{zV(z^{-1})}{z - R(z)}$$

must be analytic in $|z| > 1$. Using the same reasoning as in Section 14.4.1, Rouché's Theorem 11.3.1 with $f(z) = z$ and $g(z) = -R(z)$ applied to the contour $\tilde{C}$, similar to that in Fig. 14.3 but with the small circle now not enclosing the point $z = 1$ (because $R'(1) > 1$), shows that $z - R(z)$ has precisely one zero ξ inside the unit circle that is not equal to 1 (since the contour $\tilde{C}$ does not enclose the point $z = 1$). Since $R(0) = \varphi_A(\mu) > 0$, this zero ξ is different from zero. All possible other zeros of $z - R(z)$ lie outside the unit circle. Hence, $1 - zR(z^{-1})$ only has one zero ξ^{-1} outside the unit circle such that $1 - zR(z^{-1}) = (z - \xi^{-1})v(z)$ where all zeros of $v(z)$ lie inside the unit circle. With this preparation, the pgf of steady-state system content is written as

$$S(z) = \frac{1}{z - \xi^{-1}}\frac{V(z)}{v(z)}$$

and the function $\frac{V(z)}{v(z)}$ is analytic in $|z| \leq 1$ (because $S(z)$ is), implying that $V(z)$ has precisely the same zeros as $v(z)$. On the other hand, the definition (14.48) shows that $V(z)$ has no singularities for $|z| > 1$ and is bounded by $T(1) = S(0)$. Thus, $\frac{V(z)}{v(z)}$ is analytic and bounded in the entire complex plane and, by Liouville's theorem (Titchmarsh, 1964, p. 85), it must be a constant c. Since $S(1) = 1$, we arrive at

$$S(z) = \frac{1 - \xi^{-1}}{z - \xi^{-1}} \tag{14.50}$$

---

[7] Only if the radius of convergence of the Taylor series $\varphi_X(z) = \sum_{k=0}^{\infty}\Pr[X = k]z^k$ exceeds unity, the pgf $\varphi_X(z)$ is also analytic on the unit circle $|z| = 1$.

which is recognized from (3.6) as the pgf of a geometric random variable S with parameter $0 < \xi < 1$,

$$\Pr\left[S = j\right] = (1 - \xi)\,\xi^j \tag{14.51}$$

where ξ satisfies $\xi = R(\xi)$, where $R(z) = \varphi_A(\mu - \mu z)$. Notice that, by substituting (14.51) into (14.48), we verify that

$$V(z) = \frac{(1 - \xi)\left(1 - zR\left(z^{-1}\right)\right)}{(1 - z\xi)}$$

as directly follows from (14.49).

In summary, the number of packets in the G/M/1 queueing system is geometrically distributed, just as that of the M/M/1 queue in (14.1), but with mean $E[N_S] = E[S] = \frac{\xi}{1-\xi}$, where ξ is not equal to the traffic load ρ.

14.6.3 The system time in the G/M/1 in steady-state

Since the number S of packets in the system just before arrival time is geometrically distributed as specified in (14.51) and since the service time x is exponentially distributed with parameter μ, the total time that a packet spends in the G/M/1 system in steady-state is

$$T = \sum_{j=1}^{S+1} x_j$$

where all x_j are i.i.d distributed as x. The corresponding pgf follows from (2.79) as

$$\varphi_T(z) = \frac{\mu}{z + \mu} S\left(\frac{\mu}{z + \mu}\right)$$

Using (14.50) yields

$$\varphi_T(z) = \frac{(1 - \xi)\,\mu}{z + \mu\,(1 - \xi)}$$

which shows that the system time T in the G/M/1 queueing system in steady-state is exponentially distributed with mean $\frac{1}{(1-\xi)\mu}$.

14.7 The N*D/D/1 queue

The N*D/D/1 queue is a basic model for constant bit rate sources in ATM, as shown in Fig. 14.5. The input process consists of a superposition of N independent periodic sources with the same period D but randomly phased, i.e. arbitrarily shifted in time with respect to each other. The server operates deterministically and serves one ATM cell per timeslot. The buffer size is assumed to be infinitely long mainly to enable an exact analytic solution. During a time period D measured in time slots or server time units, precisely N cells arrive such that the traffic intensity or load (13.2) equals $\rho = \frac{N}{D}$.

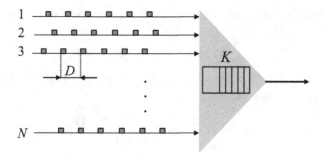

Fig. 14.5. Sketch of an ATM concentrator where N input lines are multiplexed onto a single output line. The N*D/D/1 queue models this ATM basic switching unit accurately.

Whereas the arrivals in the M/D/1 queue are uncorrelated, the successive inter-arrival times in the N*D/D/1 queue are negatively correlated. For the same mean arrival rate, this more regular arrival process results in shorter queues than in the M/D/1 queue, where the higher variability in the arrival process causes longer queues.

Due to the dependence of the arrivals over many timeslots, the solution method is based on the Beneš approach and starts from the complementary distribution (13.15) for the virtual waiting time or unfinished work in steady-state $\Pr\left[v(t_\infty) > x\right]$. Applied to the N*D/D/1 queue the unfinished work equals the number of ATM cells in the system, thus $\Pr\left[v(t_\infty) > x\right] = \Pr\left[N_S > x\right]$. Hence, in the steady-state for $\rho < 1$ or $N < D$,

$$\Pr\left[N_S > x\right] = \sum_{k=\lceil x\rceil}^{\infty} \Pr\left[v(t_\infty + x - k) = 0 \,\bigg|\, \int_{t_\infty + x - k}^{t_\infty} N_A\left(u\right)du = k\right]$$

$$\times \Pr\left[\int_{t_\infty + x - k}^{t_\infty} N_A\left(u\right)du = k\right]$$

The periodic cell trains with period equal to D timeslots at each input line lead to a periodic aggregated arrival stream of the N input lines also with period D. Each cell train transports precisely one cell per period D, which allows us to observe the characteristics of the aggregated arrival process during the time interval $[0, D)$. The computations are most conveniently performed if we choose the steady-state observation point $t_\infty = D$. Each of the N ATM cells arrives uniformly in $[0, D]$ due to the random phasing of each cell train and the probability that it arrives in $[D + x - k, D]$ is $p = \frac{k-x}{D}$. Hence, the number of arrivals in $[D + x - k, D]$ is a sum of Bernoulli random variables, which is binomially distributed,

$$\Pr\left[\int_{D+x-k}^{D} N_A\left(u\right)du = k\right] = \binom{N}{k}\left(\frac{k-x}{D}\right)^k\left(1 - \frac{k-x}{D}\right)^{N-k}$$

The conditional probability is obtained as follows. The unfinished work at time

$D + x - k$ only depends on past arrivals in the interval $[0, D + x - k]$. Given that the number of arrivals in $[D + x - k, D]$ equals k while there are always precisely N in $[0, D]$, the number of arrivals in $[0, D + x - k]$ equals $N - k$ and the corresponding traffic intensity $\rho' = \frac{N-k}{D+x-k} < 1$ since $N < D$ and, thus, $N - k < D - k$ for any k. From Section 13.3.2, we use the *local stationary result*: for any stationary single-server queueing system with traffic intensity ρ, the probability of an empty system at an arbitrary time is $1 - \rho$. If we take a random point in $t \in [0, D + x - k]$, then stationarity implies that $\Pr[v(t) = 0] = 1 - \rho' = \frac{D+x-N}{D+x-k}$. Since $\rho' < 1$, the system is necessarily empty at some instant t^* in $[0, D + x - k)$. As explained in Section 13.2 and Section 13.3, we may consider that the system restarts from t^* on ignoring the past. But the probability $\Pr[v(t) = 0]$ at a random time in the interval $[0, t^*]$ and in $[t^*, D + x - k]$ is the same, which means that $\Pr[v(D + x - k) = 0] = \frac{D+x-N}{D+x-k}$ because we can periodically repeat the system's process in $[0, t^*]$ while omitting any activity in $[t^*, D + x - k]$. With respect to this newly constructed periodic arrival pattern, the point $t = D + x - k$ is arbitrary such that the *local stationary result* is applicable. In summary, we arrive at the overflow probability in a N*D/D/1 system,

$$\Pr[N_S > x] = \sum_{k=\lceil x \rceil}^{N} \frac{D + x - N}{D + x - k} \binom{N}{k} \left(\frac{k - x}{D}\right)^k \left(1 - \frac{k - x}{D}\right)^{N-k} \quad (14.52)$$

Observe that $\Pr[N_S > N] = 0$. Rewriting (14.52) yields

$$\Pr[N_S > x] = \frac{D + x - N}{D^N} \sum_{k=\lceil x \rceil}^{N} \binom{N}{k} (k - x)^k (D + x - k)^{N-k-1}$$

$$= \frac{D + x - N}{D^N} \sum_{j=0}^{N-\lceil x \rceil} \binom{N}{j} (N - j - x)^{N-j} (D + x - N + j)^{j-1}$$

$$= \frac{D + x - N}{D^N} \sum_{j=0}^{N} \binom{N}{j} (D + x - N + j)^{j-1} (N - x - j)^{N-j}$$

$$- \frac{D + x - N}{D^N} \sum_{j=N-\lceil x \rceil+1}^{N} \binom{N}{j} (D + x - N + j)^{j-1} (N - j - x)^{N-j}$$

Applying Abel's identity (Comtet, 1974, p. 128), valid for all u, y, z,

$$(u + y)^n = u \sum_{k=0}^{n} \binom{n}{k} (u - kz)^{k-1} (y + kz)^{n-k} \quad (14.53)$$

with $n = N$, $u = D + x - N$, $y = N - x$ and $z = -1$ gives

$$\Pr[N_S \le x] = \frac{D + x - N}{D^N} \sum_{j=N-\lceil x \rceil+1}^{N} \binom{N}{j} (D + x - N + j)^{j-1} (N - j - x)^{N-j}$$

$$(14.54)$$

demonstrating that, indeed, $\Pr[N_S \le 0] = 0$. For small x, relation (14.54) is convenient, while (14.52) is more suited for large $x \to N$. For example, $\Pr[N_S \le 1] = \frac{D+1-N}{D+1}\left(1 + \frac{1}{D}\right)^N$, while $\Pr[N_S > N - 1] = \left(\frac{1}{D}\right)^N$.

In the heavy traffic regime for $\rho = \frac{N}{D} \to 1$, a Brownian motion approximation (Roberts, 1991) is

$$\Pr[N_S > x] \simeq e^{-2x\left(\frac{x}{N} + \frac{1-\rho}{\rho}\right)} \tag{14.55}$$

Figure 14.6 compares the exact (14.52) overflow probability and the Brownian approximation (14.55) for $\rho = 0.95$. Observe from (14.46) that

$$\Pr[N_S > x] \simeq e^{-\frac{2x^2}{N}} \Pr[\mathcal{Q}_{M/D/1} > x]$$

which shows that, for sufficiently high N, the overflow probability of the N*D/D/1 queue tends to that of the M/D/1 queue. Thus, an arrival process consisting of a superposition of a large number of periodic processes tends to a Poisson arrival process. The decaying factor $e^{-\frac{2x^2}{N}}$ reflects the effect of the negative correlations in the arrival process and shows that a Poisson process overestimates the tail probability in heavy traffic. Comparison of (14.52) and (14.45) for lower loads $\rho = \frac{N}{D}$ illustrates that the Poisson approximation becomes more accurate.

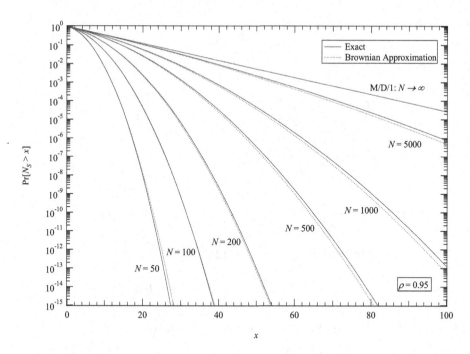

Fig. 14.6. The overflow probability $\Pr[N_S > x]$ in the N*D/D/1 queue for $\rho = 0.95$ and various number of sources N.

14.8 The AMS queue

Many arrival patterns in telecommunication exhibit active or "on" periods succeeded by silent inactive or "off" periods. At the burst or flow level phenomena of the order of time of an on-off period are dominant and the finer details of the individual packet arrivals within an on-period can be ignored. The stream of packets can be regarded as a continuous fluid characterized by the flow arrival rate.

The AMS queue is perhaps the simplest exact solvable queueing model that describes the queueing behavior at the burst or flow level. The AMS queue, named after Anick, Mitra and Sondhi (Anick *et al.*, 1982; Mitra, 1988), considers N homogeneous, independent on-off sources in a continuous fluid flow approach. For each source, both the on- and off-periods are exponentially distributed, which makes the model Markovian by Theorem 10.2.3. In the on-period each source emits a unit amount of information. Hence, at each moment in time when r sources are in the on-period, r packets (units of information) arrive at the buffer. The service time is constant and equal to $c < N$ packets per unit time. If $c > N$, then the buffer is always empty. The time unit is chosen as the mean time of an on-period while the mean time of an off-period is denoted by $\frac{1}{\lambda}$. The buffer size is infinitely long. The traffic intensity then equals $\rho = \frac{N\lambda}{c(1+\lambda)}$ and stability requires that $\rho < 1$. The ratio $\frac{\lambda}{1+\lambda}$ is the long term "on" time fraction of the sources.

Suppose the number of on-sources at time t is i. During the next time interval Δt only two elementary actions can take place: a new source can start with probability $(N-i)\lambda\Delta t$ or a source can turn off with probability $i\Delta t$. Compound events have probabilities $O(\Delta t^2)$. The probability of no change in the arrival process is $1 - [(N-i)\lambda + i]\Delta t$ during which i sources are active and the queue empties at rate $c - i$. The AMS queueing process is a birth-death process where state i describes the number of on-sources and where the birth rate $\lambda_i = (N-i)\lambda$ and the death rate $\mu_i = i$. Let $P_i(t, x)$ where $0 \leq i \leq N$, $t \geq 0$, $x \geq 0$ be the probability that at time t, i sources are on and the buffer content does not exceed x. Then, we have

$$P_i(t + \Delta t, x) = [N - (i-1)]\lambda\,\Delta t\,P_{i-1}(t, x) + (i+1)\,\Delta t\,P_{i+1}(t, x)$$
$$+\,[1 - \{(N-i)\lambda + i\}\,\Delta t]\,P_i(t, x - (i-c)\,\Delta t) + O(\Delta t^2)$$

Passing to the limit $\Delta t \to 0$ yields

$$\frac{\partial P_i(t, x)}{\partial t} + (i-c)\frac{\partial P_i(t, x)}{\partial x} = (N-i+1)\lambda P_{i-1}(t, x) - [(N-i)\lambda + i]P_i(t, x)$$
$$+\,(i+1)\,P_{i+1}(t, x)$$

The time-independent equilibrium probabilities, $\pi_i(x) = \lim_{t\to\infty} P_i(t, x)$, reflect the steady-state where i sources are on and the buffer content does not exceed x. Setting $\frac{\partial P_i(t,x)}{\partial t} = 0$, the steady-state equations become for $0 \leq i \leq N$

$$(i-c)\frac{d\pi_i(x)}{dx} = (N-i+1)\lambda\,\pi_{i-1}(x) - [(N-i)\lambda + i]\,\pi_i(x) + (i+1)\,\pi_{i+1}(x) \quad (14.56)$$

In matrix notation, where $\pi(x)$ is a column vector as opposed to Markov theory where $\pi(x)$ is a row vector,

$$D\frac{d\pi(x)}{dx} = Q\pi(x) \tag{14.57}$$

and $D = \text{diag}[-c, 1 - c, 2 - c, \ldots, N - c]$ and Q is a tri-diagonal $(N + 1) \times (N + 1)$ matrix,

$$Q = \begin{bmatrix} -N\lambda & 1 & 0 & 0 & \cdots & 0 \\ N\lambda & -[(N-1)\lambda + 1] & 2 & 0 & \cdots & 0 \\ 0 & (N-1)\lambda & -[(N-2)\lambda + 2] & 3 & \cdots & 0 \\ \vdots & \vdots & \vdots & \ddots & \vdots & \vdots \\ 0 & 0 & 0 & \cdots & -[\lambda + (N-1)] & N \\ 0 & 0 & 0 & \cdots & \lambda & -N \end{bmatrix}$$

The buffer overflow probability is $\Pr\left[N_S > x\right] = 1 - \|\pi(x)\|_1 = 1 - \sum_{j=0}^{N} \pi_j(x)$, which implies that $\|\pi(\infty)\|_1 = 1$. Moreover, $\lim_{x \to \infty} \frac{d\pi(x)}{dx} = 0$ and $Q\pi(\infty) = 0$ corresponds to the steady-state of the continuous Markov chain (arrival process and service process as a whole). Furthermore,

$$\pi_i(\infty) = \frac{1}{(1+\lambda)^N}\binom{N}{i}\lambda^i \tag{14.58}$$

is the probability that i out of N sources are on simultaneously irrespective of what the buffer level in the system is.

As shown in Section 10.2.2, besides $\pi(x) = e^{D^{-1}Qx}\pi(0)$, the solution of (14.57) can be expressed in terms of the eigenvalues ζ_j, the corresponding right-eigenvector x_j and left-eigenvector y_j of $D^{-1}Q$,

$$\pi(x) = \sum_{j=0}^{N} e^{\zeta_j x} x_j \left(y_j^T \pi(0)\right)$$

where, as shown in Appendix A.6.2.2, the eigenvalues are labeled in increasing order $\zeta_{N-\lceil c\rceil-1} < \cdots < \zeta_1 < \zeta_0 < \zeta_N = 0 < \zeta_{N-1} < \cdots < \zeta_{N-\lceil c\rceil}$. This way of writing distinguishes between underload and overload eigenvalues. Only bounded solutions are allowed. As shown in Appendix A.6.2.2, there are precisely $N - \lceil c\rceil - 1$ negative real eigenvalues such that $j \in [0, N - \lceil c\rceil - 1]$. In addition, $j = N$ that corresponds to the eigenvalue $\zeta_N = 0$ and the $\pi(\infty)$ eigenvector. The general bounded solution of (14.57) is

$$\pi(x) = \pi(\infty) + \sum_{j=0}^{N-\lceil c\rceil-1} a_j e^{\zeta_j x} x_j \tag{14.59}$$

where the scalar coefficients $a_j = y_j^T \pi(0)$ still need to be determined. Rather than determining $y_j^T \pi(0)$ as in Appendix A.6.2.2, a more elegant and physical method is used. The eigenvalue solution in Appendix A.6.2.2 has scaled the eigenvectors by setting the N component equal to 1, hence, $(x_j)_N = 1$. Writing the N-th component in (14.59) gives with (14.58)

$$\pi_N(x) = \frac{\lambda^N}{(1+\lambda)^N} + \sum_{j=0}^{N-[c]-1} a_j e^{\zeta_j x} \tag{14.60}$$

The most convenient choice of x is $x = 0$. If the number of on-source j at any time exceeds the service rate c, then the buffer builds-up and cannot be empty,

$$\pi_j(0) = 0 \qquad \text{for } [c] + 1 \leq j \leq N$$

This observation provides one equation in (14.60) for the coefficients a_k,

$$\sum_{j=0}^{N-[c]-1} a_j = -\frac{\lambda^N}{(1+\lambda)^N}$$

and shows that $N - [c] - 1$ additional equations are needed to determine all coefficients a_j. By differentiating (14.60) m-times and evaluating at $x = 0$, we find these additional equations

$$\frac{d^m \pi_N(x)}{dx^m}\bigg|_{x=0} = \sum_{j=0}^{N-[c]-1} a_j \zeta_j^m$$

which will be determined with the help of the differential equation (14.57). Indeed, for $m = 1$, the differential equation (14.57) gives

$$\frac{d\pi_N(x)}{dx}\bigg|_{x=0} = D^{-1}Q\pi(0)$$

The important observation is that the effect of multiplication by $D^{-1}Q$ decreases the number of zero components in $\pi(0)$ by 1, i.e. $\frac{d\pi_j(x)}{dx}\big|_{x=0} = 0$ for $[c] + 2 \leq j \leq N$. Any additional multiplication by $D^{-1}Q$ has the same effect. Since $\frac{d^m \pi(x)}{dx^m} = (D^{-1}Q)^m \pi(x)$, we thus find, for $0 \leq m \leq N - [c] - 1$ that

$$\frac{d^m \pi_N(x)}{dx^m}\bigg|_{x=0} = 0$$

We write these $N - [c]$ equation in the unknown a_k in matrix form,

$$\begin{bmatrix} 1 & 1 & 1 & \cdots & 1 \\ \zeta_1 & \zeta_2 & \zeta_3 & \cdots & \zeta_{N-[c]-1} \\ \zeta_1^2 & \zeta_2^2 & \zeta_3^2 & \cdots & \zeta_{N-[c]-1}^2 \\ \vdots & \vdots & \vdots & \ddots & \vdots \\ \zeta_1^{N-[c]-2} & \zeta_2^{N-[c]-2} & \zeta_3^{N-[c]-2} & \cdots & \zeta_{N-[c]-1}^{N-[c]-2} \\ \zeta_1^{N-[c]-1} & \zeta_2^{N-[c]-1} & \zeta_3^{N-[c]-1} & \cdots & \zeta_{N-[c]-1}^{N-[c]-1} \end{bmatrix} \cdot \begin{bmatrix} a_0 \\ a_1 \\ a_2 \\ \vdots \\ a_{N-[c]-1} \end{bmatrix} = \begin{bmatrix} \frac{-\lambda^N}{(1+\lambda)^N} \\ 0 \\ 0 \\ \vdots \\ 0 \\ 0 \end{bmatrix}$$

and recognize the matrix, denoted by V, as a Vandermonde matrix (Van Mieghem,

2011, p. 260) with

$$\det\left(V\right) = \prod_{i=0}^{N-[c]-1} \prod_{j=i+1}^{N-[c]-1} \left(\zeta_j - \zeta_i\right)$$

Since all eigenvalues appearing in this Vandermonde matrix are distinct (Appendix A.6.2.2) $\det\left(V\right) \neq 0$ and a unique solution follows for all $0 \leq j \leq N - [c] - 1$ from Cramer's theorem as

$$a_j = -\left(\frac{\lambda}{1+\lambda}\right)^N \prod_{i=0;i\neq j}^{N-[c]-1} \frac{\zeta_j}{\zeta_i - \zeta_j} \qquad (14.61)$$

Together with the exact determination of the eigenvalues ζ_j and corresponding right-eigenvector x_j explicitly given in Appendix A.6.2.2, the coefficients a_j completely solve the AMS queue.

The buffer overflow probability $\Pr\left[N_S > x\right] = 1 - \sum_{j=0}^{N} \pi_j(x)$ becomes with $\sum_{j=0}^{N} \pi_j(\infty) = 1$,

$$\Pr\left[N_S > x\right] = -\sum_{j=0}^{N-[c]-1} a_j e^{\zeta_j x} \sum_{l=0}^{N} (x_j)_l$$

Using the explicit form of the generating function (A.46), where the roots r_1 and r_2 belonging to eigenvalue ζ_k are specified in (A.44) and the residue $k = k_j = c_1$ in (A.45), the buffer overflow probability is

$$\Pr\left[N_S > x\right] = -\sum_{j=0}^{N-[c]-1} a_j e^{\zeta_j x} \left(1 - r_1\right)^{k_j} \left(1 - r_2\right)^{N-k_j} \qquad (14.62)$$

For large x, $\Pr\left[N_S > x\right]$ will be dominated by the exponential with the largest negative eigenvalue ζ_0 (for which $k_0 = N$),

$$\Pr\left[N_S > x\right] \sim -a_0 e^{\zeta_0 x} \sum_{j=0}^{N} (x_N)_j$$

Writing that largest negative eigenvalue (A.49) in terms of the traffic intensity ρ, gives

$$\zeta_0 = -\frac{(1+\lambda)(1-\rho)}{1 - \frac{c}{N}}$$

From (A.51), we have $\sum_{j=0}^{N} (x_N)_j = \left(\frac{N}{c}\right)^N$. Combined with (14.61), the asymptotic formula for the buffer overflow probability becomes

$$\Pr\left[N_S > x\right] \sim \rho^N e^{\zeta_0 x} \prod_{i=1}^{N-[c]-1} \frac{\zeta_i}{\zeta_i - \zeta_0} \qquad (14.63)$$

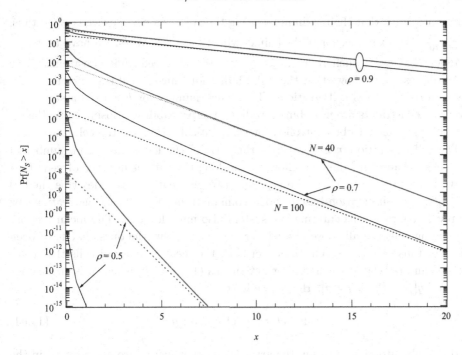

Fig. 14.7. The overflow probability (14.62) in the AMS queue versus the buffer level x for fixed $\lambda = \frac{1}{2}$. For each traffic intensity $\rho = 0.5$, 0.7 and 0.9, the upper curve corresponds to $N = 40$ and the lower to $N = 100$. The asymptotic formula (14.63) is shown as a dotted line.

Fig. 14.7 shows both the exact (14.62) and asymptotic (14.63) overflow probabilities as function of x for various traffic intensities ρ and two size of N. The mean off-period in Fig. 14.7 is twice the mean on-period. For large values of x and large traffic intensities ρ, the asymptotic formula is adequate. For smaller values, clear differences are observed. As mentioned in Section 5.7, the asymptotic regime that nearly coincides with (14.63) refers to the burst scale phenomena while the non-asymptotic regime reflects the smaller scale variations. The AMS queue allows us to analyze the effect of the burstiness of a source by varying λ.

14.9 The cell loss ratio

Due to its importance in ATM and in future time-critical communication services, the QoS loss-performance measure, the cell loss ratio, deserves some attention. In designing a switch for time-critical services with strict delay requirements smaller than D^*, the buffer size K is dimensioned as follows. The order of magnitude of D^* is about 10 ms, the maximum end-to-end delay for high-quality telephony (world wide) advised in ITU-T standards. Let $E[H]$ be the mean number of hops of a path in an ATM network that rarely exceeds 10 hops. The buffer size K is determined such that the maximum waiting time of a cell never exceeds $\frac{D^*}{E[H]}$, thus, $\frac{K}{\mu} \le \frac{D^*}{10}$. For example, for STM-1 links where $C = 155$ Mb/s, we have that $\mu \simeq 366\,800$ ATM cell/s such that $K \le \frac{\mu D^*}{10} \simeq 366$ ATM cell buffer positions. This first-order estimate shows that the ATM buffer for time-critical traffic consists of a few hundreds of ATM cell positions. That small number for K indeed assures that the delay constraints are met, but introduces the probability of losing cells. Hence, the QoS parameter to be controlled for time-critical services is the cell loss ratio.

The cell loss ratio clr is defined as the ratio of the long-time mean number of lost cells because of buffer overflow to the long-time mean number of cells that arrive in steady-state. There are typically two different views to describe the cell loss ratio: a conservation-based and a combinatorial one. The conservation law simply states that cells entering the system also must leave it. The mean number of entering cells are all those offered per time slot minus the ones that have been rejected, thus $(1 - clr)\lambda$. On the other side, the mean number of cells that leave the system are related to the server activity as $(1 - q\,[0])\mu$, where μ is the service rate and $q[j] = \Pr\,[N_Q = j]$. Hence, we have

$$(1 - clr)\,\lambda = (1 - q\,[0])\,\mu \qquad (14.64)$$

In the combinatorial view, only the arrival process is viewed from a position in the buffer and the number of ways in which cells are lost are counted leading to

$$clr = \frac{1}{A'(1)} \sum_{n=0}^{\infty} n \sum_{j=0}^{K} q\,[K - j]\,a\,[j + n] \qquad (14.65)$$

with $A'(1) = \lambda$ and $a[j] = \Pr\,[\mathcal{A} = j]$. Although equation (14.64) is simple, its practical use is limited since the quantities involved are to be known with extremely high accuracy if clr is of the order of 10^{-10}, which in practice means a virtually loss-free service. Therefore, we confine ourselves to the combinatorial result and express (14.65) in terms of a generating function as

$$clr\, A'(1) = \left.\frac{dV(z)}{dz}\right|_{z=1} \qquad (14.66)$$

where

$$V(z) = \sum_{n=0}^{\infty} z^n \sum_{j=0}^{K} q\,[K - j]\,a\,[j + n] = \sum_{n=0}^{\infty} z^n \sum_{j=0}^{K} q\,[K - j]\,z^{-j}a\,[j + n]\,z^j$$

$$= \sum_{j=0}^{K} q\,[K - j]\,z^{-j} \left(\sum_{n=0}^{\infty} a\,[j + n]\,z^{j+n}\right)$$

Rearranging in terms of the generating function for the arrivals $A(z)$ and for the buffer occupancy $Q(z) = \sum_{j=0}^{K} q[j] z^j$, where $q[j] = 0$ for $j > K$, yields

$$V(z) = \sum_{j=0}^{K} q[K-j] z^{-j} \left(A(z) - \sum_{n=0}^{j-1} a[n] z^n \right)$$

$$= A(z) z^{-K} \sum_{j=0}^{K} q[K-j] z^{K-j} - \sum_{j=0}^{K} q[K-j] z^{-j} \sum_{n=0}^{j-1} a[n] z^n$$

$$= z^{-K} A(z)Q(z) - z^{-K} \sum_{j=0}^{K} q[j] z^j \sum_{n=0}^{K-j-1} a[n] z^n \qquad (14.67)$$

In order to express the cell loss ratio entirely in terms of the generating functions $A(z)$ and $Q(z)$, we employ (2.20),

$$\sum_{j=0}^{n} y[j] z^j = \sum_{j=0}^{n} z^j \left(\frac{1}{2\pi i} \int_{C(0)} \frac{Y(\omega)}{\omega^{j+1}} d\omega \right) = \frac{1}{2\pi i} \int_C \frac{Y(\omega)}{\omega - z} \left[1 - \left(\frac{z}{\omega} \right)^{n+1} \right] d\omega$$

$$= Y(z) - \frac{1}{2\pi i} \int_C \frac{Y(\omega)}{\omega - z} \left(\frac{z}{\omega} \right)^{n+1} d\omega \qquad (14.68)$$

where C is a contour enclosing the origin *and* the point z and lying within the convergence region of $Y(z)$. Combining (14.67) and (14.68), we rewrite $V(z)$ as

$$V(z) = z^{-K} A(z)Q(z) - z^{-K} Q(z)A(z) + \frac{1}{2\pi i} \int_C \frac{A(\omega)Q(\omega)}{(\omega - z)\, \omega^K} d\omega$$

$$= \frac{1}{2\pi i} \int_C \frac{A(\omega)Q(\omega)}{(\omega - z)\, \omega^K} d\omega$$

Finally, our expression for the cell loss ratio in a GI/G/1/K system reads

$$clr = \frac{1}{2\pi i A'(1)} \int_C \frac{A(\omega)Q(\omega)}{(\omega - 1)^2\, \omega^K} d\omega \qquad (14.69)$$

where the contour C encloses both the origin and the point $z = 1$ and lies in the convergence region of $A(z)$. Usually, $A(z)$ is known while $Q(z)$ proves to be more complicated to obtain. The product $Q(z)A(z) = S(z)$ is the pgf of the system content.

If $Q(z)$ and $A(z)$ are meromorphic functions[8] and if

$$\lim_{z \to \infty} \left| \frac{A(z)\, Q(z)}{(z-1)^2\, z^{K-1}} \right| = 0,$$

the contour C in (14.69) can be closed over $|\omega| > 1$-plane to get

$$clr = -\frac{1}{A'(1)} \sum_p \mathrm{Res}_{\omega \to p} \frac{A(\omega)Q(\omega)}{\omega^K (\omega - 1)^2} \qquad (14.70)$$

where p are the poles of $A(z)Q(z)$ outside the unit circle. If these conditions are

---

[8] Functions that have as singularities only poles in the complex plane.

met, a non-trivial evaluation of the cell loss ratio can be obtained. If the buffer pgf of the finite system is known, then $Q(z)$ is a polynomial of degree at most K so that the only pole of $\frac{Q(z)}{z^K}$ is zero and $\lim_{z\to\infty} \frac{Q(z)}{z^K} = q(K) \leq 1$ and the above conditions simplify to $\lim_{z\to\infty} \left| \frac{z\,A(z)}{(z-1)^2} \right| = 0$. Executing (14.69) then leads to

$$clr = -\frac{1}{A'(1)} \sum_p \frac{Q(p)}{p^K\,(p-1)^2}\, \mathrm{Res}_{\omega\to p} A(\omega) \tag{14.71}$$

where only the poles p of the arrival process $A(z)$ play a role. For example, if the number of arrivals has a geometric distribution $a[k] = (1-\alpha)\alpha^k$ with $0 \leq \alpha \leq 1$ with generating function (3.6), $A_{\mathrm{geo}}(z) = \frac{1-\alpha}{1-\alpha z}$, then the conditions for (14.71) are satisfied and we obtain,

$$clr_{\mathrm{geo}} = \alpha^K Q\left(\frac{1}{\alpha}\right)$$

An important class excluded from (14.70) consists of entire functions $A(z)$ that possess a Taylor series expansion converging for all complex variables z. The pgf of a Poisson process with parameter λ, $A_{\mathrm{Poisson}}(z) = e^{\lambda(z-1)}$, is an important representative of that class. For a Poisson arrival process, (14.69) is

$$clr_{\mathrm{Poisson}} = \frac{e^{-\lambda}}{2\pi i\lambda} \int_C \frac{e^{\lambda\omega} Q(\omega)}{(\omega-1)^2\,\omega^K} d\omega$$

Deforming the contour to enclose the negative half ω-plane ($\mathrm{Re}(\omega) < c$) yields

$$clr_{\mathrm{Poisson}} = \frac{e^{-\lambda}}{2\pi i\lambda} \int_{c-i\infty}^{c+i\infty} \frac{e^{\lambda\omega} Q(\omega)}{(\omega-1)^2\,\omega^K} d\omega$$

where the real number c exceeds unity. This expression is recognized as an inverse Laplace transform and since the argument of the Laplace transform is a rational function, an exact evaluation is possible leading, however, again to (14.65). Hence, the combinatorial view does not offer much insight immediately which suggests to consider a conservation-based approach. Indeed, it is well known that, owing to the PASTA property (Theorem 13.6.1), an exact expression (Syski, 1986; Bisdikian *et al.*, 1992; Steyaert and Bruneel, 1994) in continuous-time for the cell loss ratio in a M/G/1/K system can be derived, with the result

$$clr_{\mathrm{M/G/1/K;cont}} = (1-\rho)\frac{\Pr[Q > K-1]}{1 - \rho\,\Pr[Q > K-1]} \tag{14.72}$$

where, as usual, the traffic intensity $\rho = \frac{\lambda}{\mu}$ and $\Pr[Q > K-1]$ is the overflow probability in the corresponding infinite system M/G/1. Transforming (14.72) to discrete-time using $clr_{\mathrm{discr}} = \frac{clr_{\mathrm{cont}}}{\rho(1-clr_{\mathrm{cont}})}$ yields

$$clr_{\mathrm{M/G/1/K;discr}} = \frac{1-\rho}{\rho}\frac{\Pr[Q > K-1]}{1 - \Pr[Q > K-1]} \tag{14.73}$$

14.10 Problems

(i) A router processes 80% of the time data packets. On average 3.2 packets are waiting for service. What is the mean waiting time of a packet given that the mean processing time equals $\frac{1}{\mu}$?

(ii) Compute in a M/M/m/m queue the mean number of busy servers.

(iii) Let us model a router by a M/M/1 system with mean service time equal to 0.5 s.

    (a) What is the relation between the mean response time (mean system time) and the arrival rate λ?

    (b) How many jobs/s can be processed for a given mean response time of 2.5 s?

    (c) What is the increase in the mean response time if the arrival rate increases by 10%?

(iv) Assume that a company has a call center with two phone lines for service. During some measurements it was observed that both the lines are busy 10% of the time. On the other hand, the mean call holding time was 10 minutes. Calculate the call blocking probability in the case that the mean call holding time increases from 10 minutes to 15 minutes. Call arrivals are Poissonean with constant rate.

(v) Consider a queueing network with Poisson arrivals consisting of two infinitely long single-server queues in tandem with exponential service times. We assume that the service times of a customer at the first and second queue are mutually independent as well as independent of the arrival process. Let the rate of the Poisson arrival process be λ, and let the mean service rates at queues 1 and 2 be μ_1 and μ_2, respectively. Moreover, assume that $\lambda < \mu_1$ and $\lambda < \mu_2$. Give the probability that in steady-state there are n customers at queue 1 and m customers at queue 2.

(vi) Let us consider the following simple design question: which queue of the M/M/m family is most suitable if the arrival rate is λ and the required service rate is $k\mu$, with $k > 1$. We have the three options illustrated in Fig. 14.8 at our disposal. Since all queues have infinite buffers

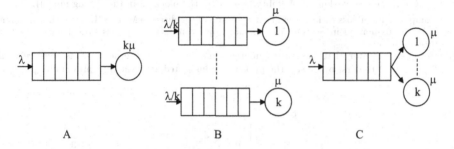

Fig. 14.8. Three different options: (A) one M/M/1 queue with service rate $k\mu$, (B) k M/M/1 queue with service rate μ and (C) one M/M/k queue with service rate $k\mu$.

and the same traffic intensity $\rho = \frac{\lambda}{k\mu}$, and thus the same throughput. The QoS qualifier of interest here is the delay, more precisely, the system time T of a packet. Compare the mean system time and draw conclusions.

(vii) An aeroplane takes exactly 5 minutes to land after the airport's traffic control has sent the signal to land. Aeroplanes arrive at random with an average rate of 6/hour. How long can an aeroplane expect to circle before getting the signal to land? (Only one aeroplane can land at a time.)

(viii) *Mobile communication.* There are two kinds of connection requests arriving at a base station of a mobile telephone network: connection requests generated by new calls (that originate from the same cell as the base station) or handovers (that originate from a different cell, but are transferred to the cell of the base station). The handovers are supposed not to experience blocking. Therefore, the base station has to reject some of the new call connection requests. Every accepted connection request occupies one of the M available channels. These are design rules that are verified by measurements.

During a busy hour, the average measured channel occupation time of a call is 1.64 minutes irrespective of the type of call. Furthermore, the mean number of active calls is 52 and the measured blocking is 2% of the number of *all* the connection requests. The mean interarrival time between two consecutive new call connection requests in the cell is 3 seconds.

(a) Calculate the arrival rate (in calls/minute) for the handover calls.
(b) What is the percentage of new calls that are blocked?

(ix) Let N denote the number of Poisson arrivals with rate λ during the service time x (random variable) of a packet. Assume that the Laplace transform of the service time $\varphi_x(s) = E[e^{-sx}]$ is known.

(a) Show that the pgf of N is given by $\varphi_x(\lambda(1-z))$.
(b) What is the pgf if the service time x is exponential distributed with mean $\frac{1}{\mu}$? Deduce from this the distribution of N.

(x) A single-server queue has exponential inter-arrival and service times with means λ^{-1} and μ^{-1}, respectively. New customers are sensitive to the length of the queue. If there are i customers in the system when a customer arrives, then that customer will join the queue with a probability $(i+1)^{-1}$, otherwise he/she departs and does not return. Find the steady-state probability distribution of this queuing system.

(xi) *The M/M/m/m/s queue (The Engset-formula).* Consider a system with m connections and s customers who all desire to telephone and, hence, need to obtain a connection or line. Each customer can occupy at most one line. The group of s customers consists of two subgroups. When a line has been assigned to a customer, this customer is transferred from the "still demanding subgroup" to the "served group". The number of call attempts decreases with the size of the "served group" whose members all occupy one line. More precisely, the arrival rate in the Engset model is proportional to the size of the "still demanding subgroup" and the number of arrivals is exponential. The holding time of a line is also exponentially distributed with mean $\frac{1}{\mu}$.

(a) Describe the M/M/m/m/s queue as a birth-death process.
(b) Compute the steady-state.
(c) Compute the blocking probability (similar to the blocking in the Erlang model).

(xii) Compare the cell loss ratio of the M/M/1/K and of the discrete M/1/D/K using the dominant pole approximation in Section 5.7. (*Hint:* approximate the cell loss ratio by the overflow probability.)

(xiii) Consider a router that has three identical and independent processors as illustrated in Fig. 14.9. In the first processor, the packers are inspected and sent with probability $p = \frac{1}{2}$ to

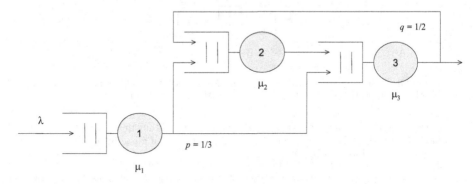

Fig. 14.9. A router with three processors.

the third processor. After the third processor, the packets are sent back with the probability $q = \frac{1}{3}$ to second processor. The service time of packets has an exponential distribution with mean equal to 500 μs. The arrival time of packets is exponentially distributed as well. The mean interarrival time is $\frac{1}{1200}$ s.

(a) What is the mean number of packets in each processor when the system is in the steady-state?
(b) What is the mean delay of packets in the entire system?

(c) What is the joint probability distribution of the number of packets in different processors?

(xiv) A network element in a data communications network is equipped with an output buffer. Packets arrive at the network element according to a Poisson process with mean rate λ. A fraction α of the packets has a deterministic service time of d seconds, while the remaining fraction $(1 - \alpha)$ of the arriving packets has an exponential service time distribution with a mean of h seconds. We shall model the buffer as a M/G/1 queueing system. Compute:

(a) the mean waiting time of the packets;
(b) the mean system time of the packets;
(c) the mean queue length.

(xv) We study the M/M/1/K queue, discussed in Section 14.2.3. Show that:

(a) the mean system size, queue plus buffer, for a traffic intensity $\rho \neq 1$ is given by

$$E\left[N_{S;M/M/1/K}\right]_{\rho \neq 1} = \frac{\rho\left(1 - (K+1)\rho^K + K\rho^{K+1}\right)}{(1-\rho)\left(1 - \rho^{K+1}\right)} \tag{14.74}$$

(b) the mean system size for $\rho = 1$ is given by

$$E\left[N_{S;M/M/1/K}\right]_{\rho=1} = \frac{K}{2}$$

(c) the mean **queue** size is

$$E\left[N_{Q;M/M/1/K}\right] = \frac{\rho\left(\rho - K\rho^K + (K-1)\rho^{K+1}\right)}{(1-\rho)\left(1 - \rho^{K+1}\right)}$$

(*Hint:* Note that $\Pr\left[N_{Q;M/M/1/K} = j\right] = \Pr\left[N_{S;M/M/1/K} = j+1\right]$.)

(d) the mean system time of a job is

$$E\left[T_{S;M/M/1/K}\right] = \frac{1 - \rho^{K+1}}{\lambda(1 - \rho^K)} E\left[N_{S;M/M/1/K}\right] \tag{14.75}$$

(*Hint:* The effective arrival rate into the M/M/1/K queue is affected by the possible rejections due to the limited buffer size.)

(xvi) A small company uses a telephony system with a switch that can handle 10 simultaneous phone calls. The company employs 90 people and the arrival of call requests at the switch can be closely modeled according to a Poisson process. On average, each employee makes 7 phone calls per 8 hours. We assume that the call duration is exponentially distributed with mean 4 minutes. When all the lines are busy, the switch queues the call requests until a line becomes available.

(a) What is the probability that an employee needs to wait for a line?
(b) What is the probability that a request for a telephone line has to wait more than one minute?
(c) The telephony system is updated such that calls that cannot be serviced by the switch are redirected through public telephone lines. What proportion of calls is redirected?

(xvii) The three copy machines in an office building are not very reliable, because, once operational, they break down after four working days on average, independently from each other. Fortunately, the technical department has a technician who needs on average one working day to repair one machine. We view the problem as a queuing system, where the technician is the server and the machines are the customers. The time that a machine is in operation, before it breaks down, as well as the time to repair broken machines are assumed to be exponentially distributed.

(a) Describe this new type of queueing system with the Kendal notation.
(b) Draw the Markov state diagram, where a state is considered as the number of defective machines. The transition rates can be expressed in repair/failure rate per machine and the number of technicians per working machines.
(c) What percentage of time are all machines broken?
(d) How many machines are broken on average?
(e) How long is a machine broken each time on average?

(xviii) Compute Var[N_S] of the M/G/1 queue.

Part III
Network science

15

General characteristics of graphs

The structure or interconnection pattern of a network can be represented by a graph. This chapter mainly focuses on general properties of graphs that are of interest to the modeling of complex networks.

Around 2000, several remarkable, quite universal phenomena observed in many different complex networks gave birth to a new discipline *network science*, which encompasses parts of physics, mathematics, engineering, biology, medical sciences, social sciences, and even finance. The most important universal complex network characteristics are (1) a power-law or scale-free degree distribution (first reported via Internet measurements by Faloutsos *et al.* (1999)), (2) small-world structure proposed by Watts and Strogatz (1998), (3) preferential attachment as a simple driver to explain the power-law degree distribution of scale-free graphs (Barabási and Albert, 1999), (4) clustering and community structure since most complex networks are networks of networks and (5) a high robustness against random failures, but a vulnerability against targeted attacks of mainly the hubs (high-degree nodes).

After the book of Watts (1999), many articles and books on complex networks have followed (see e.g. Strogatz (2001); Barabási (2002); Dorogovtsev and Mendes (2003); Barrat *et al.* (2008); Dehmer and Emmert-Streib (2009); Newman (2010); Cohen and Havlin (2010); Estrada (2012), and references in these books). Those books overview the current state of the art in network science and present many applications to, for example, the Internet, the World Wide Web, and brain, financial, social and biological networks.

15.1 Introduction

Network topologies as drawn in Fig. 15.1 are examples of graphs. A graph G is a data structure consisting of a set of V vertices connected by a set of E edges. In stochastic graph theory and communications networking, the vertices and edges are called nodes and links, respectively. In order to differentiate between the expectation operator $E\left[.\right]$, the set of links is denoted by $\mathcal{L}$ and the number of links by L and similarly, the set of nodes by $\mathcal{N}$ and the number of nodes by N. Thus, the usual notation of a graph $G\left(V, E\right)$ in graph theory is here denoted by $G\left(N, L\right)$.

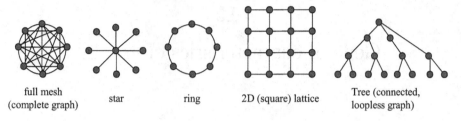

| full mesh (complete graph) | star | ring | 2D (square) lattice | Tree (connected, loopless graph) |

Fig. 15.1. Several types of network topologies or graphs.

The full mesh or complete graph K_N consists of N nodes and $L = L_{\max} = \frac{N(N-1)}{2}$ links, where every node has a link to every other node. The graph that is generated by the statement "any i is directly connected to any j" in a population of N members, is a complete graph K_N. Since in K_N the number of links $L_{\max} = O(N^2)$ for large N, it demonstrates "Metcalfe's law": the value of networking increases quadratically in the number of connected members.

The interconnection pattern of a network with N nodes can be represented by an *adjacency matrix* A consisting of elements a_{ij} that are either one or zero depending on whether there is a link between nodes i and j or not. The adjacency matrix is a real symmetric $N \times N$ matrix when we assume bi-directional transport over links. If there is a link from i to j ($a_{ij} = 1$) then there is a link from j to i ($a_{ji} = 1$) for any $j \neq i$. Moreover, we exclude self-loops ($a_{jj} = 0$) or multiple links between two nodes i and j. More properties of the adjacency matrix of a graph are found in Van Mieghem (2011).

A *walk* from node A to node B with $k - 1$ hops or links is the node list $\mathcal{W}_{A \to B} = n_1 \to n_2 \to \cdots n_{k-1} \to n_k$, where $n_1 = A$ and $n_k = B$. A *path* from node A to node B with $k - 1$ hops or links is the node list $\mathcal{P}_{A \to B} = n_1 \to n_2 \to \cdots n_{k-1} \to n_k$, where $n_1 = A$ and $n_k = B$, and where $n_j \neq n_i$ for each index i and j. Sometimes the shorter notation $\mathcal{P}_{A \to B} = n_1 n_2 \cdots n_{k-1} n_k$ is used. All links $n_i \to n_j$ and the nodes n_j in the path $\mathcal{P}_{A \to B}$ are different, whereas in a walk $\mathcal{W}_{A \to B}$ no restriction on the node list is put. If the starting node A equals the destination node B, that path $\mathcal{P}_{A \to A}$ is called a *cycle* or *loop*. In telecommunications networks, paths and not walks are basic entities in connecting two communicating parties. Two paths between A and B are node(link)-disjoint if they have no nodes(links) in common. The number of links or hops in a path is called the hopcount: if $\mathcal{P}_{A \to B} = n_1 \to n_2 \to \cdots n_{k-1} \to n_k$, then its hopcount equals $H(\mathcal{P}_{A \to B}) = k - 1$ for $k \geq 2$. The minimum hopcount $H_{\min}$ in any connected graph is one. The largest possible hopcount[1] in any graph with N nodes is $H_{\max} = N - 1$. This maximum occurs, for example, in a path graph where the path runs from one extreme node to another or in a ring (see Fig. 15.1) between neighboring nodes where there is a

---

[1] When the graph is disconnected, there may be no path between A and B. In the absense of a path $\mathcal{P}_{A \to B}$, we define $H(\mathcal{P}_{A \to B}) = 0$, though sometimes $H(\mathcal{P}_{A \to B}) \to \infty$ is assumed so that $\frac{1}{H(\mathcal{P}_{A \to B})} = 0$.

one hop and a $(N-1)$-hops path. The *diameter* of a graph equals the hopcount of the longest shortest path.

Apart from the topological structure specified via the adjacency matrix A, the link between nodes i and j is further characterized by a link weight $w(i \rightarrow j)$, most often a positive real number[2] that reflects the importance of that particular link. Often symmetry in both directions, $w(i \rightarrow j) = w(j \rightarrow i)$, is assumed leading to undirected graphs, which are considerably less difficult to analyze than directed graphs. Although this assumption seems rather trivial, we point out that in many real-world complex networks, transport or interaction is, in general, not symmetrical. Via measurements in the Internet, Paxson (1997) found in 1995 that about 50% of the paths from $A \rightarrow B$ were different from those from $B \rightarrow A$. Furthermore, it is often assumed that the link metric $w(i \rightarrow j)$ is independent from $w(k \rightarrow l)$ for all links $(i \rightarrow j)$ different from $(k \rightarrow l)$. In the Internet's intra-domain routing protocol, the Open Shortest Path First (OSPF) protocol, network operators have the freedom[3] to specify the link weight $w(i \rightarrow j) > 0$ on the interfaces of their routers.

15.2 The number of paths with j hops

While the total number of *walks* with j hops in a graph between nodes k and m is given by the entry in $\left(A^j\right)_{km}$, where A^j is the j-th power of the adjacency matrix A, as shown in Van Mieghem (2011), there is no such simple expression for the corresponding total number of *paths* between two nodes.

Let $X_j(A \rightarrow B; N)$ denote the number of paths with j hops between a source node A and a destination node B. The most general expression for the number of paths with j hops between node A and node B is, for $j > 1$ and $N > 2$,

$$X_j(A \rightarrow B; N) = \sum_{k_1 \notin \{A,B\}} \sum_{k_2 \notin \{A,k_1,B\}} \cdots \sum_{k_{j-1} \notin \{A,k_1,\cdots,k_{j-2},B\}} 1_{A \rightarrow k_1} 1_{k_1 \rightarrow k_2} \cdots 1_{k_{j-1} \rightarrow B} \quad (15.1)$$

where 1_x is the indicator function. All j-hop paths between A and B are different, though they can overlap in several segments of the path. The number of paths with one hop equals $X_1(A \rightarrow B; N) = 1_{A \rightarrow B}$. The maximum number of j-hop paths is attained in the complete graph K_N where $1_{k_1 \rightarrow k_2} = 1$ for each link $k_1 \rightarrow k_2$ and equals

$$\max(X_j(A \rightarrow B; N)) = \frac{(N-2)!}{(N-j-1)!} \quad (15.2)$$

---

[2] In quality of service routing, a link is specified by a vector $\vec{w}(i \rightarrow j)$ with positive components, each reflecting a metric (such as delay, jitter, loss, monetary cost, administrative weight, physical distance, available capacity, priority, etc.).

[3] In Cisco's OSPF implementation, it is suggested to use $w(i \rightarrow j) = \frac{10^8}{B(i \rightarrow j)}$, where $B(i \rightarrow j)$ denotes the capacity (in bit/s) of the link between nodes i and j. An approach to optimize the OSPF weights to reflect actual traffic loads is presented by Fortz and Thorup (2000).

The total number of paths M_N between two nodes in the complete graph is

$$M_N = \sum_{j=1}^{N-1} \max(X_j(A \to B; N)) = \sum_{j=1}^{N-1} \frac{(N-2)!}{(N-j-1)!} = (N-2)! \sum_{k=0}^{N-2} \frac{1}{k!}$$

$$= (N-2)!e - R$$

where

$$R = (N-2)! \sum_{j=N-1}^{\infty} \frac{1}{j!} = \sum_{j=0}^{\infty} \frac{(N-2)!}{(N-1+j)!}$$

$$= \frac{1}{N-1} + \frac{1}{(N-1)N} + \frac{1}{(N-1)N(N+1)} + \cdots < \sum_{j=1}^{\infty} \left(\frac{1}{N-1}\right)^j = \frac{1}{N-2}$$

implying that for $N \geq 3$, $R < 1$. But M_N is an integer. Hence, the total number of paths in K_N is exactly equal to

$$M_N = [e(N-2)!] \tag{15.3}$$

where $e = 2.718\,281...$ and $[x]$ denotes the largest integer smaller than or equal to x. Since any graph is a subgraph of the complete graph, the maximum total number of paths between two nodes in any graph is upper bounded by $[e(N-2)!]$.

The other extreme in connected graphs occurs in trees, where there is precisely one path between each node pair, which is considerably less than in the complete graph.

15.3 The degree of a node in a graph

The degree d_j of a node j in a graph $G(N, L)$ equals the number of its neighboring nodes and $0 \leq d_j \leq N - 1$. Clearly, the node j is disconnected from the rest of the graph if and only if $d_j = 0$. Hence, in connected graphs, $1 \leq d_j \leq N - 1$. The basic law for the degree is

$$\sum_{j=1}^{N} d_j = 2L \tag{15.4}$$

since each link belongs to precisely two nodes and, hence, is counted twice. In directed graphs, the in(out)-degree is defined as the number of the in(out)-going links at a node, while the sum of in- and out-degrees equals the degree. The minimum nodal degree in the graph G is denoted by $d_{\min} = \min_{j \in \mathcal{N}} d_j$. The mean degree of a graph is defined as $d_a = \frac{1}{N} \sum_{j=1}^{N} d_j = \frac{2L}{N}$, which is, for a connected graph, bounded by $2 - \frac{2}{N} \leq d_a \leq N - 1$. The lower bound is obtained for any spanning tree, a graph that connects all nodes and that contains no cycles and where $L = L_{\min} = N - 1$. The upper bound is reached in the complete graph K_N with $L_{\max} = \frac{N(N-1)}{2}$. Graphs where $d_{\min} = d_a$ such as K_N and the ring topology in Fig. 15.1 are called regular graphs since any node has precisely d_a links. Regular graphs have many nice and extremal properties (see e.g. Van Mieghem (2011)). The degree vector $d = (d_1, d_2, \ldots, d_N)$ satisfies $d = Au$, where $u = (1, 1, \ldots, 1)$ is the all-one vector and the relation $d = Au$ reflects that the sum of the elements of the i-th row of the adjacency matrix A equals d_i.

Fig. 15.2. A power-law graph with $\tau = 2.4$ and $N = 300$. All nodes are drawn on a circle.

Sometimes networks are classified either as dense if d_a is high, or as sparse if d_a is small. For instance, the Internet is sparse with mean degree $d_a \approx 3$, although some backbone routers may have a much higher degree, exceeding 100. The distribution of the degree D_{Internet} of an arbitrary node in the Internet is found to follow approximately a power law (Siganos *et al.*, 2003),

$$\Pr\left[D_{\text{Internet}} = k\right] \approx \frac{k^{-\tau}}{\zeta\left(\tau\right)} \qquad (15.5)$$

with[4] $\tau \in (2.2, 2.5)$ and $\zeta\left(s\right) = \sum_{k=1}^{\infty} k^{-s}$ for $\text{Re}(s) > 1$ is the Riemann Zeta function (Titchmarsh and Heath-Brown, 1986). A graph of this class is called a power-law graph. Figures 15.2 and 15.3 show two instances of a power-law graph.

---

[4] A more general expression than (15.5) is $\Pr\left[d_j = k\right] = ck^{\alpha}g(k)$, where c is a normalization constant and where $g(k)$ is a slowly varying function (Feller, 1971, pp. 275-284) with basic property that $\lim_{t \to \infty} \frac{g(tx)}{g(t)} = 1$, for every $x > 0$.

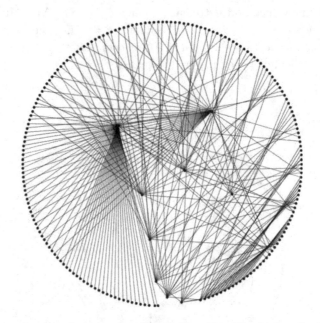

Fig. 15.3. A power-law graph with $\tau = 2.4$ and $N = 200$. The higher degree nodes are put inside the circle.

15.4 The origin of power-law degree distributions in complex networks

Besides the Internet's topology, the web graph consisting of websites and hyperlinks also features a power law for the in-degree. David Aldous has given the following argument why a power law of the in-degree of the web graph is natural. As shown, one of the underlying mechanisms, that cause a particular property of the network to be power-law distributed, is exponential growth over time.

To a good approximation, experiments show that the number of websites is growing exponentially at rate $\rho > 0$. At observation time t, the number of websites is approximately equal to $N(t) \approx e^{\rho t}$ and the counting process $N(t)$ has started at $t = 0$ with $N(0) = 1$ website. Referring to Fig. 8.1, there are precisely n websites at time W_n and the basic equivalence $\{N(\tau) \geq n\} \iff \{W_n \leq \tau\}$ of renewal theory (see Chapter 8) will now be used to estimate the probability that the lifetime $T_U = t - W_U$ of a uniformly and independently chosen website U out of the $N(t)$ websites at the observation moment t is smaller than τ. Given $N(t)$ websites at time t, application of the law of total probability (2.49) yields

$$\Pr[W_U \leq \tau] = \sum_{k=1}^{N(t)} \Pr[W_U \leq \tau \,|\, U = k] \Pr[U = k]$$

The basic equivalence implies that $\Pr[W_U \leq \tau \,|\, U = k] = \Pr[N(\tau) \geq U \,|\, U = k] =$

$\Pr\left[N(\tau) \geq k\right]$ and with $\Pr\left[U = k\right] = \frac{1}{N(t)}$, we obtain, for $0 \leq \tau \leq t$,

$$\Pr\left[W_U \leq \tau\right] = \frac{1}{N(t)} \sum_{k=1}^{N(t)} \Pr\left[N(\tau) \geq k\right] = \frac{1}{N(t)} \sum_{k=1}^{N(\tau)} 1 = \frac{N(\tau)}{N(t)}$$

This result also follows from (2.1): the website U is uniformly chosen out of the $N(\tau)$ website creation events before time τ, while $N(t)$ is the total number of created websites in the interval $[0, t]$. Given $N(t) \approx e^{\rho t}$ and $T = T_U = t - W_U$, the lifetime T of a random website satisfies $\Pr\left[T \geq \tau\right] \approx e^{-\rho \tau}$ for $0 \leq \tau \leq t$.

Let $l(u)$ denote the number of links into a website at time u after its creation. At observation time t, the distribution of the number of links or the in-degree D_{in} into a random website is, again by the law of total probability,

$$\Pr\left[D_{in} > k\right] = \int_0^t \Pr\left[D_{in} > k | T = u\right] \frac{d\Pr\left[T \leq u\right]}{du} du$$

The above estimate $\Pr\left[T \geq u\right] \approx e^{-\rho u}$ gives

$$\Pr\left[D_{in} > k\right] \approx \rho \int_0^t e^{-\rho u} \Pr\left[D_{in} > k | T = u\right] du$$

$$= \rho \int_0^t e^{-\rho u} 1_{\{l(u) > k\}} du = \rho \int_{l^{-1}(k)}^t e^{-\rho u} du = e^{-\rho l^{-1}(k)} - e^{-\rho t}$$

because, by definition of $l(u)$, the conditional probability $\Pr\left[D_{in} > k | T = u\right] = 1$, if $l(u) > k$, and zero otherwise. Only if l increases exponentially fast as $l(u) \sim e^{\beta u}$ for some $\beta < \rho$ so that the inverse function $l^{-1}(u) = \frac{1}{\beta} \ln u$, a power-law behavior of the in-degree

$$\Pr\left[D_{in} > k\right] \approx k^{-\frac{\rho}{\beta}}$$

arises for sufficiently large time t. For a polynomial growth $l(u) \sim u^{\beta}$ and large t,

$$\Pr\left[D_{in} > k\right] \approx e^{-\rho k^{\frac{1}{\beta}}}$$

The large difference in the decrease of $\Pr\left[D_{in} > k\right]$ with k between both examples illustrates the importance of the growth law of the number of links $l(u)$ into a website at time u since the website's creation.

The argument shows that a power law arises as a natural consequence of exponential growth. An exponential growth possesses the property that $\frac{dl(u)}{du} = \beta l(u)$ which is established by preferential attachment. Preferential attachment means that new links are on average added to sites proportional to their size. The more links a site has, the larger the probability that a new link attaches to this site. For example, popular websites are more often linked to than small or less popular websites. Since many aspects of the Internet, such as the number of IP packets, number of users, number of websites, number of routers, etc., are currently still growing approximately exponentially, the observed power laws are more or less as expected.

15.5 Connectivity and robustness

A graph G is connected if there is a path between each pair of nodes and disconnected otherwise. Most well-functioning complex networks, such as telecommunication networks, are connected most of the time. Moreover, it is desirable that the network is robust: it should still operate if some of the links between routers or switches are broken or temporarily blocked by other calls. Hence, the network should possess a redundancy of links. The minimum number of links to connect all nodes in the network equals $N - 1$. This minimum configuration is called redundancy level 1. In general, a redundancy level of D is defined by Baran (2002) as the link-to-node ratio in an infinite D-lattice[5]. A redundancy level of at least 3 is regarded as a highly robust network. A consequence of this insight has been employed in the design of the early Internet (Arpanet): it would be theoretically possible to build extremely reliable communication networks out of unreliable links by the proper use of redundancy. Another more timely application of the same principle is the design of reliable ad-hoc and sensor networks.

There exist interesting results from graph theory that help to dimension a reliable telecommunication network. Instead of the redundancy level, the edge and vertex connectivity seem more natural quantifiers from which robustness can be derived. The edge connectivity $\lambda(G)$ of a connected graph G is the smallest number of edges (links) whose removal disconnects G. The vertex connectivity $\kappa(G)$ of a connected graph, different[6] from the complete graph K_N, is the smallest number of vertices (nodes) whose removal disconnects G.

These definitions are illustrated in Fig. 15.4. For any connected graph G, it holds that

$$\kappa(G) \leq \lambda(G) \leq d_{\min}(G) \qquad (15.6)$$

In particular, equality occurs in (15.6) if G is the complete graph K_N, because then $\kappa(K_N) = \lambda(K_N) = d_{\min}(K_N) = N - 1$. Due to the importance of the inequality[7] (15.6), it deserves more discussion. Let us concentrate on a connected graph G

---

[5] A D-lattice is a graph where each nodal position corresponds to a point with integer coordinates within a D-dimensional hyper-cube with size Z. Apart from the border nodes, each node has a same degree equal to $2D$. The number of nodes equals $N = Z^D$. From (15.4), the link-to-node ratio follows as

$$\frac{L}{N} = \frac{1}{2N} \sum_{j=1}^{N} d_j = D - r$$

where the correction $r = O\left(N^{\frac{1}{D}-1}\right)$ is due to the border nodes. For an infinite D-lattice, where the limit $Z \to \infty$ (which implies $N \to \infty$), we obtain

$$\lim_{Z \to \infty} \frac{L}{N} = D$$

[6] The complete graph K_N cannot be disconnected by removing nodes and we define $\kappa(K_N) = N - 1$ for $N \geq 3$.

[7] The edge and vertex connectivity can be related the second smallest eigenvalue of the Laplacian (see Van Mieghem (2011)), also called the algebraic connectivity.

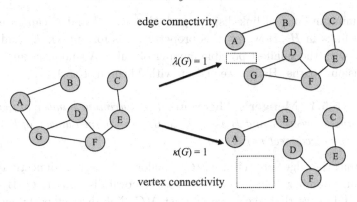

Fig. 15.4. An example of the edge and the vertex connectivity of a graph.

that is not a complete graph. Since $d_{\min}(G)$ is the minimum degree of a node, say n, in G, by removing all links incident to node n, G becomes disconnected. By definition, since $\lambda(G)$ is the minimum number of links that leads to disconnectivity, it follows that $\lambda(G) \le d_{\min}(G)$ and $\lambda(G) \le N-2$ because G is not a complete graph and consequently the minimum nodal degree is at most $N-2$. Furthermore, the definition of $\lambda(G)$ implies that there exists a set S of $\lambda(G)$ links whose removal splits the graph G into two connected subgraphs G_1 and G_2, as illustrated in Fig. 15.5. Any link of that set S connects a node in G_1 to a node in G_2. Indeed, adding an arbitrary link of that set S makes G again connected. But G can be disconnected into the same two connected subgraphs by removing nodes (with their incident links) in G_1 and/or G_2. Since possible disconnectivity inside either G_1 or G_2 can occur before $\lambda(G)$ nodes are removed, it follows that $\kappa(G)$ cannot exceed $\lambda(G)$, which establishes the inequality (15.6).

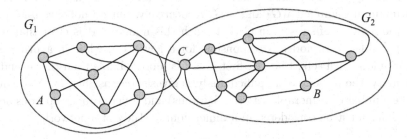

Fig. 15.5. A graph G with $N = 16$ nodes and $L = 32$ links. Two connected subgraphs G_1 and G_2 are shown. The graph's connectivity parameters are $\kappa(G) = 1$ (removal of node C), $\lambda(G) = 2$ (removal of links from C to G_1), $d_{\min}(G) = 3$ and $d_a = \frac{2L}{N} = 4$.

Let us proceed to find the number of link-disjoint paths between A and B in a connected graph G. Suppose that H is *a* set of links whose removal separates A from B. Thus, the removal of all links in the set H destroys all paths from A to B.

The maximum number of link-disjoint paths between A and B cannot exceed the number of links in H. However, this property holds for any set H, and thus also for the set with the smallest possible number of links. A similar argument applies to node-disjoint paths. Hence, we end up with Theorem 15.5.1:

Theorem 15.5.1 (Menger's Theorem) *The maximum number of link- (node)-disjoint paths between A and B is equal to the minimum number of links (nodes) separating (or disconnecting) A and B.*

Recall that the edge connectivity $\lambda(G)$ (analogously vertex connectivity $\kappa(G)$) is the minimum number of links (nodes) whose removal disconnects G. By Menger's Theorem, it follows that there are at least $\lambda(G)$ link-disjoint paths and at least $\kappa(G)$ node-disjoint paths between any pair of nodes in G.

In order to dimension the graph G of a robust complex network, the goal is to maximize both $\kappa(G)$ and $\lambda(G)$. Of course, the most reliable graph is the complete graph, which leads, however, also to the most expensive interconnection. Usually, since the cost of digging and of installing/connecting the fibres is around 70% of the total cost in a telecommunications network, the total number of links L is minimized. Since the minimum cannot exceed the average, we have that $d_{\min} \leq d_a = \frac{2L}{N}$. From (15.6), it follows that the best possible reliability is achieved if the network topology is designed such that

$$\kappa(G) = \frac{2L}{N}$$

The optimum implies that $d_{\min}(G) = d_a = \frac{2L}{N}$ or that each node has the same degree $d_j = d_a$. Hence, a best possible reliable graph is a regular graph ($d_j = d_a$), but not every regular graph necessarily obeys $\kappa(G) = d_a$. Furthermore, two different graphs with the same parameters N, L, $\kappa(G)$, $\lambda(G)$ and $d_{\min}(G)$ are not necessarily equally reliable. Indeed, the edge and vertex disconnectivity only give a minimum number $\lambda(G)$ and $\kappa(G)$ respectively, but do not give information about the number of subsets in G that reach this number. It is clear that if only one vulnerable set of nodes is responsible for a low $\kappa(G)$, while in another graph there are more such sets, that the first graph is more reliable than the second one. In summary, the presented simplified analysis gives some insights, but more details (e.g. the number of vulnerable sets or subgraphs) and other graph properties besides connectivity must be considered in a dimensioning study of a network.

15.6 Graph metrics

In Section 15.1, we have seen that an unweighted graph G, consisting of a set $\mathcal{N}$ of N nodes and a set $\mathcal{L}$ of L links, can be represented by an $N \times N$ adjacency matrix A, that completely specifies the graph's topology. Besides the topological structure, real-world complex networks execute *functions* over the underlying network, such as transport of information from source to destination in communication and social networks, neuron transport in the brain, money flows in economic networks, reaction processes in metabolic networks, epidemic spread of viruses in contact networks, etc. Thus, each network performs a function or service by using the underlying topology. The function is usually a dynamic process acting on the graph (when the topology is fixed) or even interacting with the graph (when the topology can be changed by the processes, such as in biological or epidemic processes). Examples of dynamic processes on networks are the protocols in telecommunication networks (Van Mieghem, 2010a), shortest path transport (Chapter 16), epidemic spread in networks (Chapter 17), multicast and anycast delivery of information (Chapter 18 and 19). In the remainder of this section, we confine ourselves to *graph* metrics, a subclass of *network* metrics, that can be computed from the knowledge of the graph's topology. We will not discuss *service* metrics (such as robustness, availability, reliability, dependability, etc.) that characterize the service or function or dynamic process on the network.

An important challenge in the modeling of a network is to characterize and determine the class of graphs that represents best the global and local structure of the network. Most of the valuable networks like the Internet, road infrastructures, neural networks in the human brain, social networks, etc. are large and changing over time. In order to classify graphs a set of distinguishing properties, called graph metrics, needs to be chosen. These metrics are in general a function of the graph's structure $G(N, L)$ and can all be computed based on the adjacency matrix A. Although graph metrics can be easily defined, skilful application of graph metrics proves to be far more complex than realized at first glance, as illustrated in Van Mieghem *et al.* (2010a).

Graph metrics are invariant under node relabeling and can be divided into topology metrics and spectral metrics, such as the spectral radius (largest eigenvalue of the adjacency matrix), the algebraic connectivity (second smallest eigenvalue of the Laplacian), eigenvector centrality, etc. are derived from the eigenvalue decomposition of graph matrices and are discussed in our book on *Graph Spectra*. Here, we confine ourselves to topology metrics, that can be computed directly from the adjacency matrix (without resorting to eigenvalues or eigenvectors).

Normally, a graph metric is a real number, but sometimes a random variable. We have already introduced the degree D of an arbitrary node in a graph, which is a random variable. Also the hopcount H_N of the shortest path between two arbitrary nodes in a graph, studied in Chapter 16, is a random variable. The natural transformation from a random variable towards a real number is by choosing an operator acting on (a function of) that random variable, such as the mean $E[.]$ or the probability $\Pr[.]$. For example, the most common degree related graph metrics are the mean degree, the variance of the degree and the linear degree correlation coefficient which measures the assortativity of a graph. In the sequel, we will list a subset of the published graph metrics and we refer to e.g. Da F. Costa *et al.* (2007); Newman (2010); Estrada (2012); Van Mieghem (2011) for a discussion of additional graph metrics (such as the rich-club coefficient, persistence, Estrada index, entropy, graph energy, effective graph resistance, eigenvector centrality, etc.).

15.6.1 Closeness

The closeness Cl_i of a node i is the reciprocal of the total hopcount of all shortest paths starting at this node i to all the other nodes in G,

$$Cl_i = \frac{1}{\sum_{j \in \mathcal{N} \setminus \{i\}} H\left(\mathcal{P}_{i \to j}^*\right)} \tag{15.7}$$

where $\mathcal{P}_{A \to B}^*$ (with asterisk $*$) denotes a *shortest* path from A to B. If the hopcount is defined as $H\left(\mathcal{P}_{A \to B}\right) \to \infty$ when the path $\mathcal{P}_{A \to B}$ does not exist, then often a modified definition of closeness appears,

$$\widetilde{Cl}_i = \sum_{j \in \mathcal{N} \setminus \{i\}} \frac{1}{H\left(\mathcal{P}_{i \to j}^*\right)}$$

The closeness is often regarded as a measure to quantify the node's participation in a network. Nodes with low closeness have short hopcount paths to other nodes and will receive information sooner and disseminate information faster. The closeness has been used in biology to identify central metabolites in metabolic networks.

15.6.2 Eccentricity, diameter, radius and girth

The eccentricity ε_i of a node i is defined as the longest hopcount of the shortest path between the node i and any other node j in G,

$$\varepsilon_i = \max_{j \in \mathcal{N}} \left(H\left(\mathcal{P}_{i \to j}^*\right)\right) \tag{15.8}$$

The eccentricity of a graph ε is the mean eccentricity over all the nodes in G. The graph eccentricity is related to the flooding time (Section 16.6), which is the minimum time needed to inform the last node in a network. Intuitively, important nodes should be easily reachable by other nodes in a graph and have low eccentricity.

The diameter ρ of a graph G is the maximum node eccentricity over all the nodes

$$\rho = \max_{i \in \mathcal{N}}(\varepsilon_i) \tag{15.9}$$

The diameter ρ is thus equal to the hopcount of the *longest shortest* path in a graph. The diameter gives an indication how extended a graph is. Most real-world networks have a small diameter and possess the "small-world graph" property (see p. 394), a term introduced by Watts and Strogatz (1998).

The radius R of a graph is the minimum node eccentricity over all the nodes

$$R = \min_{i \in \mathcal{N}}(\varepsilon_i) \tag{15.10}$$

The girth γ of a graph is the hopcount of the shortest cycle (closed path) contained in the graph.

15.6.3 Expansion, resilience and distortion

The expansion $e_G(h)$ of a graph reflects the mean number of nodes that can be reached in h (or less) hops from a node v,

$$e_G(h) = \frac{1}{N^2} \sum_{v \in \mathcal{N}} |C_v(h)| \tag{15.11}$$

where $C_v(h)$ is the set of nodes that can be reached in h hops from a node v and $|C_v(h)|$ denotes the number of elements in the set $C_v(h)$. Thus, $C_v(h)$ is the subgraph of G that contains all the nodes of the graph G, including node v, that can be reached in at most h hops from node v. We can interpret $C_v(h)$ geometrically as the subgraph of G enclosed by a ball with radius h centered at node v.

The resilience $r_G(h)$ measures the connectivity or robustness of a graph. Let $m = |C_v(h)|$ denote the number of nodes in a ball with radius h centered at node v, and define $l(v, m)$ as the number of links that need to be removed to split $C_v(h)$ into two sets with roughly equal numbers of nodes (around $m/2$). The resilience $r_G(h)$ of a graph is

$$r_G(h) = \frac{1}{L} \sum_{v \in \mathcal{N}} l(v, |C_v(h)|) \tag{15.12}$$

The distortion measures how closely a graph resembles a tree or how "tree-like" a graph is. The distortion $t_G(h)$ is defined as

$$t_G(h) = \frac{1}{N} \sum_{v \in \mathcal{N}} w(C_v(h)) \tag{15.13}$$

where $w(G)$ is the weight of the minimum spanning tree in G with unit link weight $w(i \to j) = 1$ for each link of G.

Consider any spanning tree T of a graph G. Denote by $E[H_T]$ the mean hopcount on T between any two nodes i and j that share a link in G, i.e. $a_{ij} = 1$. The distortion measures the extra hops required to travel from one side of a link in G to the other, if traveling is restricted to T.

15.6.4 Betweenness

Consider a flow with a unit amount of traffic between each pair of nodes in the graph G. Each flow between a node pair follows the shortest path between that node pair. The betweenness B_l (B_n) of a link l (and node n) is defined as the number of shortest paths between all possible pairs of nodes in G that traverse the link l (node n). If $H_{i \to j} = H(\mathcal{P}^*_{i \to j})$ denotes the number of hops in the shortest path[8] from $i \rightarrow j$, then the total number of hops H_G in the $\binom{N}{2}$ shortest paths in G is $H_G = \sum_{i=1}^{N} \sum_{j=i+1}^{N} H_{i \rightarrow j}$. This number is also equal to $H_G = \sum_{l=1}^{L} B_l$, where B_l is the betweenness of a link l in G. Taking the expectation of both relations gives the mean betweenness of a link in terms of the mean hopcount

$$E[B_l] = \frac{\binom{N}{2}}{L} E[H_N] \geq E[H_N]$$

with equality only for the complete graph.

The node betweenness is a measure of the centrality or influence of nodes in complex networks. In communication networks, the betweenness measures the maximum number of units of traffic that crosses a node or link. This potential traffic will be affected when the node or link fails.

15.6.5 Clustering coefficient

The clustering coefficient $c_G(v)$ characterizes the density of connections in the environment of a node v and is defined as the ratio of the number of links y connecting the d_v neighbors of v over the total possible $\frac{d_v(d_v - 1)}{2}$,

$$c_G(v) = \frac{2y}{d_v(d_v - 1)} \tag{15.14}$$

When the degree d_w of node w equals one, then the clustering coefficient is $c_G(w) = 0$. The clustering coefficient of the graph G is defined as the average over all nodes,

$$C_G = \frac{1}{N} \sum_{v=1}^{N} c_G(v)$$

The largest possible $C_G = 1$ is attained in the complete graph, while the lowest possible $C_G = 0$ occurs in trees.

Another graph theoretical definition of the graph's clustering coefficient (Newman *et al.*, 2001) is three times the number $\blacktriangle_G$ of triangles divided by the number of connected triples,

$$\widetilde{C}_G = \frac{3\blacktriangle_G}{N_2} = \frac{W_3}{2d^T d} = \frac{\text{trace}\left(A^3\right)}{2 \sum_{j=1}^{N} d_j^2}$$

where $N_k = u^T A^k u = d^T A^{k-2} d$ is the total number of walks with length k and $W_k = \text{trace}(A^k)$ is the number of closed walks with length k. Moreover (Van Mieghem, 2011), $W_3 = 6\blacktriangle_G$ and the number of connected triples equals the total number $N_2 = d^T d$ of walks of length 2, connecting three nodes. The factor of 3 in the numerator accounts for the fact that each triangle contributes to three connected triples of nodes, one for each of its three vertices. With this factor of 3, the value of $\widetilde{C}_G$ lies strictly in the range from zero to one.

---

[8] If there exist multiple shortest paths between a same node pair (i, j), the betweenness may differ depending on which shortest path is chosen. In order to circumvent this arbitrariness, all m shortest paths between (i, j) should be considered, but each one weighted by $1/m$.

15.6.6 Assortativity

"Mixing" in complex networks refers to the tendency of network nodes to connect preferentially to other nodes with either similar or opposite properties. Mixing is computed via the linear correlation coefficient (2.61) between the properties, such as the degree, of nodes in a network. When $\rho_D > 0$, the graph possesses *assortative* mixing, a preference of high-degree nodes to connect to other high-degree nodes and, when $\rho_D < 0$, the graph features *disassortative* mixing, where high-degree nodes are connected to low-degree nodes.

Denote by D_i and D_j the node degree of two *connected* nodes i and j in an undirected graph with N nodes. In fact, we are interested in the degree of nodes at both sides of a link, without taking the link that we are looking at into consideration. As Newman (2003) points out, we need to consider the number of excess links at both sides, and, hence the degree $D_{l+} = D_i - 1$ and $D_{l-} = D_j - 1$, where the link l has a start at node $l^+ = i$ and an end at node $l^- = j$. The linear correlation coefficient (2.61) of those excess degrees is

$$\rho\left(D_{l+}, D_{l-}\right) = \frac{E\left[\left(D_{l+} - E\left[D_{l+}\right]\right)\left(D_{l-} - E\left[D_{l-}\right]\right)\right]}{\sqrt{E\left[\left(D_{l+} - E\left[D_{l+}\right]\right)^2\right] E\left[\left(D_{l-} - E\left[D_{l-}\right]\right)^2\right]}}$$

Since $D_{l+} - E\left[D_{l+}\right] = D_i - E\left[D_i\right]$, subtracting one link everywhere does not change the linear correlation coefficient, provided $D_i > 0$ (and similarly that $D_j > 0$), which is the case if there are no isolated nodes. Removing isolated nodes from the graph does not alter the linear degree correlation coefficient. Hence, we can assume that the graph has no zero-degree nodes. In summary, the linear degree correlation coefficient is, with $E\left[D_i\right] = \mu_D$,

$$\rho\left(D_{l+}, D_{l-}\right) = \rho\left(D_i, D_j\right) = \frac{E\left[D_i D_j\right] - \mu_D^2}{E\left[D^2\right] - \mu_D^2} \tag{15.15}$$

which can be written (Van Mieghem *et al.*, 2010c; Van Mieghem, 2011) in terms of the total number $N_k = u^T A^k u$ of walks of length k,

$$\rho_D = \rho\left(D_i, D_j\right) = \frac{N_1 N_3 - N_2^2}{N_1 \sum_{i=1}^{N} d_i^3 - N_2^2} \tag{15.16}$$

The crucial understanding of (dis)assortativity lies in the total number N_3 of walks with three hops compared to those with two hops, N_2, and one hop, $N_1 = 2L$. Denoting a link by $l = i \sim j$, the degree correlation (15.16) can be rewritten (Van Mieghem, 2011) as

$$\rho_D = 1 - \frac{\sum_{i \sim j} \left(d_i - d_j\right)^2}{\sum_{i=1}^{N} d_i^3 - \frac{1}{2L}\left(\sum_{i=1}^{N} d_i^2\right)^2} \tag{15.17}$$

Degree-preserving rewiring (Van Mieghem, 2011), a tool to alter the graph's assortativity without modifying the degree of a node, only affects $\sum_{i \sim j} (d_i - d_j)^2$ in (15.17).

Highly connected nodes tend to be connected with other high-degree nodes in social networks (assortative mixing). On the other hand, technological and biological networks typically show disassortative mixing, where high-degree nodes tend to attach to low-degree nodes. van der Hofstad and Litvak (2013) show peculiar behavior of the Pearson's linear correlation coefficient (2.61) in some graphs or degree sequences for large N and they suggest to replace the measure of (dis)assortativity by a rank correlation, such as Spearman's correlation coefficient.

15.6.7 Coreness

The k-core is the subgraph obtained from the original graph G by the recursive removal of all nodes of degree less than or equal to k. Hence, in a k-core subgraph all nodes have at least degree k.

The node coreness k_i of a given node n_i is the maximum k such that this node is present in the k-core subgraph, but removed from the $(k+1)$-core subgraph. The coreness can be regarded as an indicator of node centrality, since it measures how *deep* within the network a node is located.

15.6.8 Modularity

The modularity, proposed by Newman and Girvan (2004), is a measure of the quality of a particular division of the network. The modularity is proportional to the number of links falling within clusters or groups minus the expected number in an equivalent network with links placed at random. Thus, if the number of links within a group is no better than random, the modularity is zero. A modularity approaching one reflects networks with strong community structure: a dense intragroup and a sparse inter-group connection pattern.

If links are placed at random, then the expected number of links between node i and node j equals $\frac{d_i d_j}{2L}$. The modularity $m \in [-1, 1]$ is defined by Newman (2006) as

$$m = \frac{1}{2L} \sum_{i=1}^{N} \sum_{j=1}^{N} \left(a_{ij} - \frac{d_i d_j}{2L} \right) 1_{\{i \text{ and } j \text{ belong to the same cluster}\}} \qquad (15.18)$$

A graph-theoretic study of the modularity is presented in (Van Mieghem, 2011, p.96-108), maximum modular graphs are determined in Trajanovski *et al.* (2012), while graphs with a prescribed modularity are constructed in Trajanovski *et al.* (2013).

15.7 Random graphs

Besides the regular topologies in Fig. 15.1, the class of random graphs constitutes an attractive set of topologies to analyze network performance. Random graphs are constructed by a probabilistic rule and each realization is a specific graph that belongs to that particular random graph class. For example, if a complex network changes over time, then each topology change can be regarded as a realization of the graph class to which that complex network belongs. The interest in random graphs is fueled by the fact that the topology of complex networks is inaccurately known and also that good models[9] are lacking. Many complex networks can be regarded as growing and changing organisms, and random graphs can be a first-order description. Due to their relative simplicity, random graphs are an elegant vehicle to analyze the performance of, for example, routing algorithms or to study structural growth, topological properties and scaling laws of networks.

This section will mainly focus on the simplest type of random graphs, the Erdős-Rényi random graph, that is analytically tractable in many cases. Subsection 15.7.2 introduces the Barabási-Albert random graph, that is more representative for complex networks, but unfortunately, less easy in analytic computations.

15.7.1 Erdős-Rényi random graphs

The theory of random graphs originated from a series of papers by Erdős and Rényi in the late 1950s. There exists an astonishingly large amount of literature on random graphs. The standard work on random graphs is the book by Bollobas (2001). We also mention the work of Janson *et al.* (1993) on evolutionary processes in random graphs.

The two most frequently occurring models for random graphs are the Erdős-Rényi (ER) random graphs $G_p(N)$ and $G_r(N, L)$. The class of random graphs denoted by $G_p(N)$ consists of all graphs with N nodes in which the links are chosen independently and with probability p. In the class $G_p(N)$ the total number of links is not deterministic, but on average equal to $E[L] = p L_{\max}$, where $L_{\max} = \binom{N}{2}$. Since $p = \frac{E[L]}{L_{\max}}$ we also call p the link density of $G_p(N)$. An instance of the class $G_{0.013}(300)$ is drawn in Fig. 15.6. Related to $G_p(N)$ are the *geometric* random graphs $G_{\{p_{ij}\}}(N)$ where the links are still chosen independently but where the probability of $i \rightarrow j$ being an edge is p_{ij}. An example of $G_{\{p_{ij}\}}(N)$ is the Waxman graph (Waxman, 1988; Van Mieghem, 2001) with $p_{ij} = \exp(-a|\vec{r_i} - \vec{r_j}|)$ and where the vector $\vec{r_i}$ represents the position of node i and a is a real, non-negative number. Geometric random graphs are good models for ad-hoc wireless networks where the probability $p_{ij} = f(r_{ij})$ that there is a wireless link between nodes i and j is specified by the radio propagation that is briefly explained at the end of Section 3.5.

---

[9] A detailed discussion on difficulties in modeling or simulating the Internet is presented by Floyd and Paxson (2001).

The class $G_r(N, L)$ is the set of random graphs with N nodes and L links. In total, we can construct $\binom{L_{\max}}{L}$ different graphs, which corresponds to the number of ways we can distribute a set of L ones in the $L_{\max}$ possible places in the upper triangular part above the diagonal of the adjacency matrix A. Each of the possible $L_{\max}$ links has equal probability to belong to a random graph of the class $G_r(N, L)$. The probability that an element in the adjacency matrix A is $a_{ij} = 1$ equals $p = \frac{L}{L_{\max}}$. As opposed to the class $G_p(N)$, the number of non-zero elements in A in each random graph of $G_r(N, L)$ is precisely $2L$, which induces weak dependence between links in $G_r(N, L)$. This dependence explains why stochastic computations are easier in $G_p(N)$ than in $G_r(N, L)$, while combinatorial derivations are easier in $G_r(N, L)$.

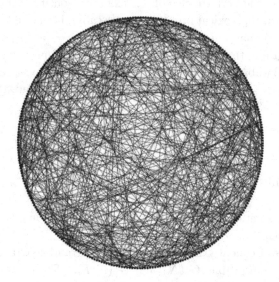

Fig. 15.6. A connected random graph $G_p(N)$ with $N = 300$ and $p = 0.013$ drawn on a circle.

Since the degree d_j of a node j is the number of links incident with that node, $d_j = \sum_{k=1}^{N} a_{kj}$ with $a_{jj} = 0$, and since the definition of $G_p(N)$ implies that the adjacency matrix elements a_{kj} are $N - 1$ independent Bernoulli random variables with mean p, the probability density function of the degree of an arbitrary node in $G_p(N)$ follows from (3.3) as

$$\Pr\left[D_{G_p(N)} = k\right] = \binom{N-1}{k} p^k (1-p)^{N-1-k} \tag{15.19}$$

The mean number of paths with j hops between two arbitrary nodes in $G_p(N)$ follows from (15.1) and (2.13) as

$$E[X_j] = \frac{(N-2)!}{(N-j-1)!} p^j \tag{15.20}$$

for $1 \leq j \leq N - 1$. The mean total number of paths between two arbitrary nodes A and B equals

$$E \left[\sum_{j=1}^{N-1} X_j \right] = (N - 2)! p^{N-1} \sum_{l=0}^{N-2} \frac{p^{-l}}{l!} \leq (N - 2)! p^{N-1} e^{\frac{1}{p}}$$

where the latter bound is closely approached for large N. Moreover, when the random graph reduces to the complete graph ($p = 1$), we again obtain (15.3).

Some constructed overlay networks such as Gnutella and mobile wireless ad-hoc networks seem reasonably well modeled by $G_p(N)$. However, the class $G_p(N)$ does not describe the Internet topology well, and the degree distribution especially deviates significantly. The degree distribution (15.19) in $G_p(N)$ is binomially distributed, while that of the Internet is close to a power law (15.5). Hence, in many complex networks, there are discrepancies between measurements and properties of the ER random graph $G_p(N)$, that resulted in new graph models.

15.7.2 Barabási-Albert random graphs

A second popular random graph model, due to Barabási and Albert (1999), is coined the BA random graph. Based on preferential attachment explained in Section 15.4, a BA graph is constructed by sequentially adding nodes to an initial, small graph G_0 consisting of N_0 nodes and L_0 links, whose structure is assumed to barely affect the final graph. We limit the discussion to undirected graphs. Each new node is connected to m already existing nodes with probability proportional to the degree of the existing nodes. Thus, the probability of connecting the n-th node at time t_n to an existing node i with degree d_i equals $\xi_n d_i$, where $\xi_n^{-1} = \sum_{j=1}^{n-1} d_j = 2L(t)$ and $L(t)$ is the number of links at the connection time $t = t_n$. Since at each time step, a new node is connected to m existing nodes, the number of links equals $L(t) = mt + L_0$ for $t \geq 0$. This BA growth rule results in a power-law graph with exponent $\tau = 3$ in (15.5). The approximate argument (by ignoring the difference between discrete and continuous time), due to Barabási and Albert (1999), starts from the governing equation for the degree of a node i, created at time $t_i < t$,

$$\frac{\partial d_i(t)}{\partial t} = m \frac{d_i}{2L(t)} = m \frac{d_i}{2mt + 2L_0}$$

whose solution is

$$\ln \frac{d_i(t)}{d_i(t_i)} = \int_{t_i}^{t} \frac{m \, du}{2mu + 2L_0} = \frac{1}{2} \ln \frac{2mt + 2L_0}{2mt_i + 2L_0}$$

Since L_0 is small compared to $2mt_i$, we obtain that $d_i(t) \approx d_i(t_i) \left(\frac{t}{t_i} \right)^{1/2}$. Furthermore, $d_i(t_i) = m$ for $t_i > 0$ so that

$$\Pr[d_i(t) \leq k] \simeq \Pr \left[m \sqrt{\frac{t}{t_i}} \leq k \right] = \Pr \left[\frac{m^2}{k^2} t \leq t_i \right]$$

The probability that node i is added at time $t_i \leq x$ equals $\Pr[t_i \leq x] = \frac{x}{N}$, where the total number of nodes at time t equals $N = N_0 + t$, because at each time step a node is connected. In other words, t_i is approximately a uniform random variable. Hence,

$$\Pr[d_i(t) \leq k] \simeq 1 - \frac{\frac{m^2}{k^2}t}{t + N_0}$$

and, using $\Pr[d_i(t) = k] = \Pr[d_i(t) \leq k] - \Pr[d_i(t) \leq k - 1]$,

$$\Pr[d_i(t) = k] \simeq \frac{m^2 t}{t + N_0}\left(\frac{1}{(k-1)^2} - \frac{1}{k^2}\right) \approx \frac{2m^2}{k^3}$$

where we have assumed that the final time t is large and the initial number of nodes N_0 is small. These approximate arguments show that the degree distribution in BA random graphs has a power-law exponent $\tau = 3$. Unfortunately, many complex networks possess a different power-law exponent. Moreover, Bollobas and Riordan (2001) criticize the model proposed by Barabási and Albert (1999). First, the impact of the initial graph G_0 can be large. For example, if $m = 1$ and G_0 is disconnected, then the final graph will be a disconnected tree. Second, the above mathematical analysis is intuitive, however, a rigorous derivation of the degree distribution is found via the Price citation graph defined in Problem (xiv). A Price citation graph can be constructed with a power-law degree exponent $\tau > 2$.

Since the introduction of the BA graph, many variants of the preferential attachment growth model have been proposed and studied (Cohen and Havlin, 2010; van der Hofstad, 2013). For example, if the attachment rule is proportional to the degree plus a constant (as in the Price citation graph), any power-law exponent $\tau > 2$ can be designed (van der Hofstad, 2013, Chapter 8). The configuration model, on the other hand, is a static model, proposed to create a graph with a given a degree sequence $d_1, d_2, \ldots, d_N$.

15.7.3 The number $C(N, L)$ of connected ER random graphs in the class $G_r(N, L)$

From the point of view of telecommunications networks, by far the most interesting graphs are those with connected topology. This limitation restricts the value of the link density p from below by a critical threshold p_c. For large N, the critical threshold is $p_c \sim \frac{\log N}{N}$, as shown in Section 15.7.5. In the theory of random graphs, the problem to determine the number of connected random graphs $C(N, L)$ in the class $G_r(N, L)$ has been intensively studied. Gilbert (1956) has presented an exact recursion formula for $C(N, L)$ via the technique of enumeration. Erdős and Rényi (1959, 1960) have determined the asymptotic behavior of random graphs via the probabilistic method, largely introduced by Erdős himself. Since the analysis[10] of

---

[10] A different, less admissible approach is found in Goulden and Jackson (1983).

Gilbert is both exact and simple, we will review his results here and those of Erdős and Rényi in the next section.

Consider a particular random graph J of the class of random graphs $G_r(N+1, L)$, which is constructed from the class $G_r(N, L)$ by adding one node labelled $N + 1$. Suppose that the node labelled $N+1$ in the random graph J belongs to a connected component K that possesses v other nodes and some number μ of links. The remaining part of J has $N - v$ nodes and $L - \mu$ links. There are $\binom{N}{v}$ ways in which the v nodes of K can be chosen out of the N nodes in $G_r(N, L)$. On the other hand, there are $C(v + 1, \mu)$ ways of picking a connected random graph K while there are $\binom{\binom{N-v}{2}}{L-\mu}$ ways of constructing the remaining part of J. Hence, since the number of ways we can construct a graph J equals $\binom{\binom{N+1}{2}}{L}$, we obtain Gilbert's recursion formula,

$$\binom{\binom{N+1}{2}}{L} = \sum_{v=0}^{N} \binom{N}{v} \sum_{\mu=v}^{\binom{v+1}{2}} C(v+1, \mu) \binom{\binom{N-v}{2}}{L-\mu} \qquad (15.21)$$

Gilbert (1956) further derives the generating function for $C(N, L)$ as

$$\sum_{N=1}^{\infty} \sum_{L=1}^{\infty} \frac{C(N, L)}{N!} x^N y^L = \log\left(1 + \sum_{k=1}^{\infty} \frac{(1+y)^{\binom{k}{2}} x^k}{k!}\right) \qquad (15.22)$$

which converges for $-2 \leq y \leq 0$ and all x. So far, no other explicit formulae for $C(N, L)$ exist.

In 1889, Caley proved that $C(N, N - 1) = N^{N-2}$ in the special case of a tree where $L = N - 1$. In the other extreme of a full mesh where $L = L_{\max}$, we have $C(N, L_{\max}) = 1$. Actually, when $\binom{N}{2} - (N - 1) < L \leq L_{\max}$, the graph is always connected because there is at least one node that is connected to all the other nodes. This means that $C(N, L) = \binom{\binom{N}{2}}{L}$ when $\binom{N}{2} - (N - 1) < L \leq L_{\max}$. In all cases where $L < N - 1$, the random graphs are necessarily disconnected, leading to $C(N, L) = 0$.

For computational purposes, we rewrite (15.21) as

$$\binom{\binom{N+1}{2}}{L} = \sum_{\mu=N}^{\binom{N+1}{2}} C(N+1, \mu) \binom{\binom{0}{2}}{L-\mu} + \sum_{v=0}^{N-1} \binom{N}{v} \sum_{\mu=v}^{\binom{v+1}{2}} C(v+1, \mu) \binom{\binom{N-v}{2}}{L-\mu}$$

Since $\binom{0}{L-\mu} = \delta_{\mu,L}$, we arrive after a substitution of $N \to N - 1$ at the recursion formula,

$$C(N, L) = \binom{\binom{N}{2}}{L} - \sum_{v=0}^{N-2} \binom{N-1}{v} \sum_{\mu=v}^{\binom{v+1}{2}} C(v+1, \mu) \binom{\binom{N-1-v}{2}}{L-\mu} \qquad (15.23)$$

Below we list a few values:

$C(2,1) = 1$			
$C(3,2) = 3$	$C(3,3) = 1$		
$C(4,3) = 16$	$C(4,4) = 15$	$C(4,5) = 6$	$C(4,6) = 1$
$C(5,4) = 125$	$C(5,5) = 222$	$C(5,6) = 205$	$C(5,7) = 120$
$C(5,8) = 45$	$C(5,9) = 10$	$C(5,10) = 1$	
$C(6,5) = 1296$	$C(6,6) = 3660$	$C(6,7) = 5700$	$C(6,8) = 6165$
$C(6,9) = 4945$	$C(6,10) = 2997$	$C(6,11) = 1365$	$C(6,12) = 455$
$C(6,13) = 105$	$C(6,14) = 15$	$C(6,15) = 1$	
$C(7,6) = 16807$	$C(7,7) = 68295$	$C(7,8) = 156555$	$C(7,9) = 258125$
$C(7,10) = 331506$	$C(7,11) = 343140$	$C(7,12) = 290745$	$C(7,13) = 202755$
$C(7,14) = 116175$	$C(7,15) = 54257$	$C(7,16) = 20349$	$C(7,17) = 5985$
$C(7,18) = 1330$	$C(7,19) = 210$	$C(7,20) = 21$	$C(7,21) = 1$

15.7.4 The Erdős and Rényi asymptotic analysis

In a classical paper, Erdős and Rényi (1959) proved that

$$\lim_{N \to \infty} \Pr\left[G_r\left(N, \left[\frac{1}{2}N \log N + xN \right] \right) \text{ is connected} \right] = e^{-e^{-2x}} \tag{15.24}$$

Ignoring the integral part [.] operator and eliminating x using the number of links $L = \frac{1}{2}N \log N + xN$ gives, for large N,

$$\Pr[G_r(N, L) = \text{connected}] \sim e^{-N e^{-\frac{2L}{N}}} \tag{15.25}$$

which should be compared with the exact result,

$$\Pr[G_r(N, L) = \text{connected}] = \frac{C(N, L)}{\binom{\binom{N}{2}}{L}} \tag{15.26}$$

In contrast to the unattractive computation of the exact $C(N, L)$ via recursion (15.23), the Erdős and Rényi asymptotic expression (15.25) is simple. The accuracy for relatively small N is shown in Fig. 15.7.

The key observation of Erdős and Rényi (1959) is that a phase transition in random graphs with N nodes occurs when the number of links L is around $L_c = \frac{1}{2}N \log N$. Phase transitions are well-known phenomena in physics. For example, at a certain temperature, most materials possess a solid-liquid transition and at a higher temperature a second liquid-gas transition. Below that critical temperature, most properties of the material are completely different than above that temperature. Some materials are superconductive below a certain critical temperature T_c, but normally conductive (or even isolating) above T_c. Erdős and Rényi concentrated on the property A_k that a random graph $G_r(N, L)$ with $L_x = \left[\frac{1}{2}N \log N + xN \right]$ consists of $N - k$ connected nodes and k isolated nodes for fixed k. If A_k^c means the absence of property A_k, they proved that, for all fixed k, $\Pr[A_k^c] \to 0$ if $N \to \infty$, which means that for a large number of nodes N, almost all random graphs $G_r(N, L_x)$ possess property A_k. This result is equivalent to a result proved

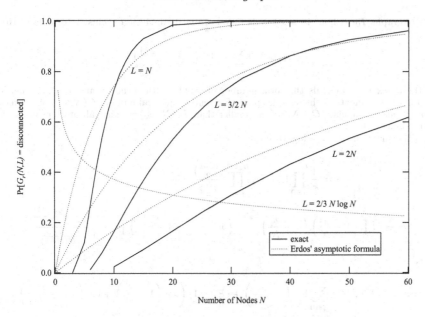

Fig. 15.7. The probability that a random graph $G_r(N, L)$ is *disconnected*: a comparison between the exact result (15.26) and Erdős' asymptotic formula (15.25) for $L = N$, $L = \frac{3}{2}N$, $L = 2N$ and $L = \frac{2}{3}N \log N$.

in Section 15.7.5 that the class of random graphs $G_p(N)$ is almost surely disconnected if the link density p is below $p_c \sim \frac{\log N}{N}$ and connected for $p > p_c$. In view of the analogy with physics, it is not surprising that corresponding sharp transitions also are observed for other properties than just A_k.

In the sequel, we will demonstrate:

Theorem 15.7.1 *For the random graph* $G_r(N, L_x)$ *with* $L_x = \left[\frac{1}{2}N \log N + xN\right]$ *links, the probability that the largest connected component, called the giant component* $GC(N, L_x)$, *has* $N - k$ *nodes is, for large* N, *Poisson distributed with mean* e^{-2x},

$$\lim_{N \to \infty} \Pr\left[\text{number of nodes in } GC(N, L_x) = N - k\right] = \frac{(e^{-2x})^k e^{-e^{-2x}}}{k!} \quad (15.27)$$

If $k = 0$, then all nodes belong to the giant component and the graph is completely connected in which case (15.27) leads to (15.25).

Proof: The total number of graphs $G_r(N, L_x)$ with $k \geq 1$ isolated nodes equals $\binom{N}{k}\binom{\binom{N-k}{2}}{L_x}$, the number of ways in which k isolated nodes can be chosen out of the total of N nodes multiplied by the number of graphs that can be constructed with $N-k$ nodes and L_x links. Observe that this total number also includes those graphs where not all the $N - k$ nodes are necessarily connected. In other words, this total number includes the graphs that do not possess property A_k. The total number of graphs T_0 without isolated node follows from the inclusion-exclusion formula (2.10) as

$$T_0\left(N, L_x\right) = \sum_{k=0}^{N} (-1)^k \binom{N}{k} \binom{\binom{N-k}{2}}{L_x}$$

where the index $k = 0$ equals the total number of graphs with N nodes and L_x links, i.e. the total number of elements in the sample space. Evidently, the total number $C(N, L_x)$ of connected random graphs of the class $G_r\left(N, L_x\right)$ is smaller than $T_0\left(N, L_x\right)$ because all of them must obey property A_0 as well.

It is convenient to take the logarithm of

$$t_k = \frac{\binom{N}{k}\binom{\binom{N-k}{2}}{L_x}}{\binom{\binom{N}{2}}{L_x}} = \frac{1}{k!} \prod_{j=0}^{k-1} (N-j) \prod_{j=0}^{L_x-1} \frac{\binom{N-k}{2} - j}{\binom{N}{2} - j}$$

$$= \frac{N^k}{k!} \prod_{j=0}^{k-1} \left(1 - \frac{j}{N}\right) \left(1 - \frac{k}{N}\right)^{L_x-1} \left(1 - \frac{k}{N-1}\right)^{L_x-1} \prod_{j=0}^{L_x-1} \frac{1 - \frac{2j}{(N-k)(N-1-k)}}{1 - \frac{2j}{N(N-1)}}$$

which is

$$\log\left(k! t_k\right) = k \log N + \sum_{j=0}^{k-1} \log\left(1 - \frac{j}{N}\right) + (E_x - 1)\left(\log\left(1 - \frac{k}{N}\right) + \log\left(1 - \frac{k}{N-1}\right)\right)$$

$$+ \sum_{j=0}^{E_x-1} \log\left(1 - \frac{2j}{(N-k)(N-1-k)}\right) - \log\left(1 - \frac{2j}{N(N-1)}\right)$$

For large N and using the expansion $\log(1-z) = -z + O(z^2)$, we have for fixed k with

$$\log\left(1 - \frac{2j}{(N-k)(N-1-k)}\right) = \log\left(1 - \frac{2j}{N(N-1)} + O(N^{-3})\right)$$

that

$$\log\left(k! t_k\right) = k \log N + O(N^{-1}) - L_x \frac{2k}{N} + O(L_x^2 N^{-3})$$

In order to have a finite limit $\lim_{N \to \infty} \log\left(k! t_k\right) = c \in \mathbb{R}$, we must require that $k \log N - L_x \frac{2k}{N} = c$ which implies that $L_x = \frac{N}{2} \log N - \frac{Nc}{2k}$. For this scaling the order term $O(L_x^2 N^{-3})$ indeed vanishes if $N \to \infty$. By choosing $x = -\frac{c}{2k}$, we arrive at the correct scaling of $L_x = \frac{1}{2} N \log N + xN$ postulated above and $c = -2kx$. Hence, with

$$\lim_{N \to \infty} \frac{\binom{N}{k}\binom{\binom{N-k}{2}}{L_x}}{\binom{\binom{N}{2}}{L_x}} = \frac{(e^{-2x})^k}{k!}$$

we obtain

$$\lim_{N \to \infty} \frac{T_0\left(N, L_x\right)}{\binom{\binom{N}{2}}{L_x}} = \sum_{k=0}^{\infty} (-1)^k \frac{(e^{-2x})^k}{k!} = e^{-e^{-2x}}$$

But, if $N \to \infty$, the difference

$$\frac{T_0\left(N, L_x\right) - C(N, L_x)}{\binom{\binom{N}{2}}{L_x}} \leq \Pr\left[A_0^c\right] \to 0$$

which demonstrates (15.27) for $k = 0$. The remaining case for $k > 0$ in (15.27) follows from the observation that the number of graphs in $G_r(N, L_x)$ with property A_k is equal to $\binom{N}{k}$ multiplied by the number of connected graphs with $N - k$ nodes and L_x links, which is approximately

$$\frac{\binom{N}{k}\binom{\binom{N-k}{2}}{L_x}}{\binom{\binom{N}{2}}{L_x}} \frac{T_0(N-k, L_x)}{\binom{\binom{N-k}{2}}{L_x}} \to \frac{(e^{-2x})^k}{k!} e^{-e^{-2x}}$$

where the limit gives the correct result because the small difference between the total number and that without property A_k tends to zero. □

15.7.5 Connectivity and degree in ER random graphs

There is an interesting relation between the connectivity of a graph, a global property, and the degree D of an arbitrary node, a local property. The implication $\{G$ is connected$\} \implies \{D_{\min} \geq 1\}$ where $D_{\min} = \min_{\text{all nodes } \in G} D$ is always true. The opposite implication is not always true, because a network can consist of separate, disconnected clusters containing nodes each with minimum degree larger than 1. A random graph can be generated from a set of N labelled nodes by randomly assigning a link with probability p to each pair of nodes. During this construction process, initially separate clusters originate, but at a certain moment, one of those clusters starts dominating (and swallowing) the other clusters. This largest cluster becomes the *giant* component. For large N and a certain p_N which depends on N, the implication $\{D_{\min} \geq 1\} \implies \{G_p(N)$ is connected$\}$ is almost surely (a.s.) correct. A rigorous mathematical proof is fairly complex and omitted. Thus, for large random graphs $G_p(N)$ holds the equivalence $\{G_p(N)$ is connected$\} \iff \{D_{\min} \geq 1\}$ almost surely such that

$$\Pr[G_p(N) \text{ is connected}] = \Pr[D_{\min} \geq 1] + o(1),$$

meaning that the difference between left- and right-hand sides vanishes as $N \to \infty$.

The basic law (15.4) couples the degree of nodes. But, for large N and $p < 1$ in $G_p(N)$, this dependence is negligibly weak. Assuming independence, from (3.39) and (15.19), we have that

$$\Pr[D_{\min} \geq 1] \simeq \left(\Pr[D_{G_p(N)} \geq 1]\right)^N = \left(1 - \Pr[D_{G_p(N)} = 0]\right)^N = \left(1 - (1-p)^{N-1}\right)^N$$

which shows that $\Pr[D_{\min} \geq 1]$ rapidly tends to one for fixed $0 < p < 1$ and large N. Therefore, the asymptotic behavior of $\Pr[G_p(n)$ is connected$]$ requires an investigation of the influence of p as a function of N,

$$\Pr[G_p(N) \text{ is connected}] \simeq \exp\left(N \log\left(1 - (1 - p_N)^{N-1}\right)\right)$$

Invoking the Taylor series of $\log(1 - x) = -\sum_{k=1}^{\infty} \frac{x^k}{k}$ for $|x| < 1$ yields

$$\log\left(1 - (1 - p_N)^{N-1}\right) = -(1 - p_N)^{N-1} - \sum_{j=2}^{\infty} \frac{\left((1 - p_N)^{(N-1)}\right)^j}{j}$$

and

$$\Pr\left[G_p\left(N\right) \text{ is connected}\right] \simeq e^{-N(1-p_N)^{N-1}} \exp\left(-N\sum_{j=2}^{\infty} \frac{(1-p_N)^{(N-1)j}}{j}\right)$$

If we denote $c_N \triangleq N \cdot (1-p_N)^{N-1}$, then

$$N\sum_{j=2}^{\infty} \frac{(1-p_N)^{(N-1)j}}{j} = \sum_{j=2}^{\infty} \frac{c_N^j}{jN^{j-1}}$$

can be made arbitrarily small for large N provided we choose $c_N = cN^\beta$ with $\beta < \frac{1}{2}$, where c is a positive real and constant number. Thus, for large N, we have that

$$\Pr\left[G_p\left(N\right) \text{ is connected}\right] = e^{-cN^\beta}\left(1 + O\left(N^{2\beta-1}\right)\right)$$

which tends to 0 for $0 < \beta < \frac{1}{2}$ and to 1 for $\beta < 0$. Hence, the critical exponent where a sharp transition occurs is $\beta = 0$. In that case, $c_N = c$ and

$$p_N = 1 - \exp\left(\frac{\log\frac{c}{N}}{N-1}\right) = \frac{\log N}{N} + O\left(\frac{\log c}{N}\right)$$

In summary, for large N and writing $p = a(\log N)/N$ for some $a > 0$,

$$\Pr\left[G_p\left(N\right) \text{ is connected}\right] \longrightarrow \begin{cases} 0 & \text{if } a < 1 \\ 1 & \text{if } a > 1 \end{cases} \tag{15.28}$$

with a transition region (corresponding to $a = 1$) around $p_c \sim \frac{\log N}{N}$ with a width of $O(1/N)$. Notice the agreement with (15.24) where $p_x = \frac{L_x}{L_{\max}} \simeq \frac{\log N}{N} + \frac{x}{N}$: for large $x < 0$, $e^{-e^{-2x}} \to 0$, while for large $x > 0$, $e^{-e^{-2x}} \to 1$ and the width of the transition region for the link density p is $O(1/N)$.

15.7.6 Size of the giant component

15.7.6.1 In ER random graphs

Let $S = \Pr\left[n \in C\right]$ denote the probability that a node n in $G_p\left(N\right)$ belongs to the giant component C. If $n \notin C$, then none of the neighbors of node n belongs to the giant component. The number of neighbors of a node n equals the degree d_n of a node such that

$$\Pr\left[n \notin C\right] = \Pr\left[\text{all neighbors of } n \notin C\right]$$
$$= \sum_{k\geq 0} \Pr\left[\text{all } k \text{ neighbors of } n \notin C | d_n = k\right] \Pr\left[d_n = k\right]$$

Since in $G_p(N)$ all neighbors of n are independent[11], the conditional probability becomes, with $1 - S = \Pr[n \notin C]$,

$$\Pr[\text{all } k \text{ neighbors of } n \notin C | d_n = k] = (\Pr[n \notin C])^k = (1 - S)^k$$

provided that we assume that N is large. Indeed, when $k = N - 1$ so that node n is connected to all other nodes in the graph, then the left-hand side is zero, whereas the right-hand side is positive if N is finite. Hence, in the sequel, we confine ourselves to an asymptotic analysis ($N \to \infty$). Since the above holds for any node in $n \in G_p(N)$ with large N, we can write the random variable $D_{G_p(N)}$ instead of an instance d_n and obtain

$$1 - S = \sum_{k=0}^{\infty} (1 - S)^k \Pr[D_{G_p(N)} = k] = \varphi_{D_{G_p(N)}}(1 - S)$$

where $\varphi_{D_{G_p(N)}}(u) = E\left[u^{D_{G_p(N)}}\right]$ is the generating function of the degree of an arbitrary node in $G_p(N)$. The fraction S of nodes in the giant component in the random graph thus satisfies the functional equation (12.18) of the extinction probability in a branching process with $\pi_0 = 1 - S$. For large N and constant mean degree, the binomial degree distribution in $G_p(N)$ is approximately Poisson distributed with mean degree $\mu_{G_p(N)} = p(N - 1)$ by the law of rare events (Section 3.1.4) and $\varphi_{D_{G_p(N)}}(u) \simeq e^{\mu_{G_p(N)}(u-1)}$ such that

$$S = 1 - e^{-\mu_{G_p(N)}S} \tag{15.29}$$

The mean size of the giant component is NS. From the inequality (5.11), we find for any $S \neq 0$ that

$$\mu_{G_p(N)}S > 1 - e^{-\mu_{G_p(N)}S}$$

illustrating that, for a mean degree $\mu_{G_p(N)} \leq 1$, the only solution of (15.29) is the trivial solution $S = 0$, whereas for $\mu_{G_p(N)} > 1$ there is a non-zero solution for the size of the giant component. Hence, if the link density in the ER random graph $G_p(N)$ satisfies $p \leq \frac{1}{N-1}$, there is no giant component. The solution can be expressed as a Lagrange series using (5.43),

$$S\left(\mu_{G_p(N)}\right) = 1 - e^{-\mu_{G_p(N)}} \sum_{n=0}^{\infty} \frac{(n+1)^n}{(n+1)!} \left(\mu_{G_p(N)}e^{-\mu_{G_p(N)}}\right)^n \tag{15.30}$$

which shows that $S\left(\mu_{G_p(N)}\right)$ is a continuous function of $\mu_{G_p(N)}$. From (5.44), we deduce that $S(1) = 0$. By reversing (15.29), the mean degree $\mu_{G_p(N)}$ in the ER

---

[11] This argument is not valid, for example, for a two-dimensional lattice $\mathbb{Z}_p^2$ in which each link between adjacent nodes at integer value coordinates in the plane exists with probability p. The critical link density for connectivity in $\mathbb{Z}_p^2$ is $p_c = \frac{1}{2}$, a famous result proved in the theory of percolation (see e.g., Grimmett (1989)). The argument is valid when the graph is locally treelike.

random graph $G_p(N)$ for large N can be expressed in terms of the fraction S of nodes in the giant component,

$$\mu_{G_p(N)}(S) = -\frac{\log(1-S)}{S} \qquad (15.31)$$

The determination of the size of the giant component, rather than the fraction S, proves to be more difficult and we content ourselves to merely state the result:

Theorem 15.7.2 *The size $|C|$ of the largest component in the ER random graph $G_p(N)$ is, for large N and mean degree $\mu_{G_p(N)} = p(N-1) > 0$:*

(i) $\mu_{G_p(N)} < 1$ *(subcritical regime): for some positive real number α depending on μ, it holds that*

$$\lim_{N \to \infty} \Pr\left[|C| \le \alpha \log N\right] = 1$$

(ii) $\mu_{G_p(N)} > 1$ *(supercritical regime): for some positive real number α' depending on μ_{rg} and any $\delta > 0$, it holds that*

$$\lim_{N \to \infty} \Pr\left[\left\{\left|\frac{|C|}{N} - S\left(\mu_{G_p(N)}\right)\right| \le \delta\right\} \cap \left\{|C_2| \le \alpha' \log N\right\}\right] = 1$$

where $1 - S(\mu)$ is the extinction probability of a branching process with Poissonean production function, $S(\mu)$ satisfies (15.29) and C_2 is the second largest component in $G_p(N)$.

(iii) $\mu_{G_p(N)} = 1$ *(critical regime): there exists a constant κ and for all $\alpha > 0$, it holds that*

$$\lim_{N \to \infty} \Pr\left[|C| \ge \alpha N^{\frac{2}{3}}\right] \le \frac{\kappa}{\alpha^2}$$

Proof: See e.g. Draief and Massoulié (2010, p. 21-33)   □

15.7.6.2 In any graph

The analysis of the giant component can be generalized to other graphs than the ER random graph. Following Newman *et al.* (2001), we consider two points of view: statistics obtained by choosing a random node in the graph G and by following an arbitrary link towards one of its end points. Given the pgf $\varphi_D(z) = E\left[z^D\right]$ of the degree D of an arbitrary node in G, the pgf $\varphi_{D_{l^+}}(z)$ of the degree found at an end point l^+ of an arbitrarily chosen link l is specified in (B.33). We will now concentrate on the distribution of the sizes of connected components in the graph, when the link density $p = L/\binom{N}{2}$ is so low that the graph's structure is treelike and that the occurrence of loops is negligibly small. Moreover, we assume that the size N of the graph is large. Let $\varphi_{C_{l^+}}(z) = E\left[z^{C_{l^+}}\right]$ be the pgf of the size C_{l^+} of components that are reached by following a random link l towards one of its endnodes, say l^+. Our main assumption of tree-like structure *excludes* the giant component: we assume such a low link density p that the formation of the giant component has not yet occurred. Each component is treelike in structure, consisting of the single node l^+, reached by following our initial, random link l, plus any integer number of other treelike clusters, with the same size distribution, joined to node l^+ by single links. By the law of total probability (2.49), we have

$$\varphi_{C_{l^+}}(z) = \sum_{n=0}^{N-1} E\left[z^{C_{l^+}} \mid D_{l^+} - 1 = n\right] \Pr\left[D_{l^+} - 1 = n\right]$$

Due to the tree structure at node l^+, each of the n branches at l^+ (except for link l) contains again components, independent and non-overlapping, and identically distributed in size, so that

$$E\left[z^{C_{l^+}} \mid D_{l^+} - 1 = n\right] = E\left[z^{1+\sum_{j=1}^{n} C_{j^+}} \mid D_{l^+} - 1 = n\right] = z\left(\varphi_{C_{l^+}}(z)\right)^n$$

where the factor z refers to the fact that node l^+ must be counted in each possible tree configuration rooted at l^+. Combining the above, we arrive at the functional or "self-consistent" equation[12]

$$\varphi_{C_{l^+}}(z) = z\varphi_{(D_{l^+}-1)}\left(\varphi_{C_{l^+}}(z)\right) \tag{15.32}$$

that is typical for branching theory (see Chapter 12) and where $\varphi_{(D_{l^+}-1)}(z) = E\left[z^{D_{l^+}-1}\right] = \frac{\varphi'_D(z)}{\varphi'_D(1)}$ follows from (B.33).

If we choose a random node n in G, then we have one such component at the end l^+ of each outgoing link l incident to $n = l^-$, and hence the pgf $\varphi_{C_n}(z) = E\left[z^{C_n}\right]$ for the size of the whole component is

$$\varphi_{C_n}(z) = z\varphi_D\left(\varphi_{C_{l^+}}(z)\right) \tag{15.33}$$

By solving $\varphi_{C_{l^+}}(z)$ from (15.32) and substituting into (15.33), we can find the probability $\Pr\left[C_n = k\right]$ that the connected component that contains the node n has size equal to k. In most cases, an analytic solution is not possible, but an analysis of the mean size of connected component $s = E\left[C_n\right] = \varphi'_{C_n}(1)$ is more amenable. Indeed, differentiating (15.33) with respect to z,

$$\varphi'_{C_n}(z) = \varphi_D\left(\varphi_{C_{l^+}}(z)\right) + z\varphi'_D\left(\varphi_{C_{l^+}}(z)\right)\varphi'_{C_{l^+}}(z) \tag{15.34}$$

and evaluating at $z = 1$, yields

$$s = E\left[C_n\right] = 1 + E\left[D\right]E\left[C_{l^+}\right]$$

Similarly, from (15.32), we find

$$E\left[C_{l^+}\right] = 1 + E\left[D_{l^+} - 1\right]E\left[C_{l^+}\right]$$

---

[12] In stead of relying on the law of total probability, Newman *et al.* (2001) deduce the Dyson-like equation (15.32) by using Feynman-diagrams in many-particle systems (see e.g. Fetter and Walecka (1971)): a cluster C can be decomposed as a single node, a single node plus a cluster C_1, a single node plus two independent clusters C_1 and C_2, a single node plus three independent clusters etc, where each cluster C_j has the same distribution as that of C.

from which

$$E\left[C_{l+}\right] = \frac{1}{1 - E\left[D_{l+} - 1\right]}$$

and $E\left[D_{l+} - 1\right] = \frac{E[D^2]}{E[D]} - 1$, also called the branching factor of a network, follows from (B.33). Substituted into the above expression for $E\left[C_n\right]$ gives the mean size of connected clusters below the critical link density p_c

$$s = E\left[C_n\right] = 1 + \frac{\left(E\left[D\right]\right)^2}{2E\left[D\right] - E\left[D^2\right]} \tag{15.35}$$

We observe that both s and $E\left[C_{l+}\right]$ diverge when the branching factor $E\left[D_{l+} - 1\right] \to 1$ or, when $E\left[D^2\right] - 2E\left[D\right] = E\left[D\left(D - 2\right)\right] = 0$. This divergence marks the phase transition at which a giant component first appears and was also found by Molloy and Reed (1995). Since $E\left[D\left(D - 2\right)\right]$ increases monotonically as links are added to the graph, it follows that the giant component exists if and only if this sum is positive.

The generating function approach of Newman *et al.* (2001) can be extended to the situation when there is a giant component in the graph, but then, by definition, $\varphi_{C_n}\left(z\right)$ generates the probability distribution of the sizes of components *excluding* the giant component. In contrast to the basic property of a pgf $\varphi_X\left(z\right) = E\left[z^X\right]$ that $\varphi_X\left(1\right) = 1$ for any random variable X, this exclusion means that then $\varphi_{C_n}\left(1\right) = 1 - S$, where S is the fraction of the graph occupied by the giant component, so that we obtain from (15.33),

$$1 - S = \varphi_D\left(u\right) \tag{15.36}$$

where $u = \varphi_{C_{l+}}\left(1\right)$ follows from (15.32) as

$$u = \varphi_{\left(D_{l+} - 1\right)}\left(u\right) \tag{15.37}$$

Hence, $1 - u$ is the fraction of nodes belonging to the giant component observed by the endnode l^+ of a random link l. The mean size s of the components (excluding a possible giant component) can now be deduced from (15.34) as

$$s = \frac{\varphi'_{C_n}\left(1\right)}{\varphi_{C_n}\left(1\right)} = \frac{\varphi_D\left(u\right) + \varphi'_D\left(u\right)\varphi'_{C_{l+}}\left(1\right)}{1 - S}$$

and

$$\varphi'_{C_{l+}}\left(1\right) = \varphi_{\left(D_{l+} - 1\right)}\left(u\right) + \varphi'_{\left(D_{l+} - 1\right)}\left(u\right)\varphi'_{C_{l+}}\left(1\right)$$

resulting, with (15.36) and (15.37), in

$$s = 1 + \frac{u\varphi'_D\left(u\right)}{\left(1 - S\right)\left(1 - \varphi'_{\left(D_{l+} - 1\right)}\left(u\right)\right)}$$

Finally, after combining (B.33) and (15.37) to $u = \varphi_{(D_{l+}-1)}(u) = \frac{\varphi'_D(u)}{E[D]}$, we arrive at the mean size of (non-giant) components

$$s = 1 + \frac{E[D]\,u^2}{(1-S)\left(1 - \varphi'_{(D_{l+}-1)}(u)\right)} \tag{15.38}$$

which reduce for $S = 0$ and $u = 1$ to (15.35).

Newman *et al.* (2001) observe that, when the degree D has a Poisson distribution (which occurs as a limiting case for ER random graphs for fixed $\mu_{G_p(N)}$ and large N), then $\varphi_{(D_{l+}-1)}(z) = \varphi_D(z) = e^{\mu_{G_p(N)}(z-1)}$, so that the distribution of outgoing links at a node is the same as that by following a randomly chosen link. Hence, when D is Poissonean, the coupled equations (15.36) and (15.37) reduce to (15.29). The mean size (15.38) excluding the giant component in the ER random graph $G_p(N)$ becomes

$$s_{G_p(N)} = \frac{1}{1 - \mu_{G_p(N)}(1-S)}$$

15.8 Interdependent networks

Buldyrev *et al.* (2010) announced that cascading failures in interdependent networks behave dramatically different from those in single networks. An interdependent network, depicted in Fig. 15.8, is a network consisting of different types of networks, that depend upon each other for their functioning. For example, many human-made infrastructural networks are interdependent networks: transport networks like car, train, airplane and ship networks depend on each other as well as on energy and oil supply networks. Smart grids or electricity networks depend upon computer networks, where each node (computer) controls one or more power generators, while each computer needs electricity to function. Our human body consists of a cardiovascular network, a brain network, a respiratory system, a nervous network and many more organs that are all interdependent. In fact, single systems or networks occur very rarely in nature and most networks are interdependent.

Let us start by considering an interdependent network as in Fig. 15.8 consisting of two networks A and B, each represented by a graph $G_A(N_A, L_A)$ and $G_B(N_B, L_B)$ and topology G_A is independent of topology G_B. The interdependency link at a node $n_A \in G_A$ to a node $n_B \in G_B$ means that node n_A depends for its operation on node n_B and vice-versa. Thus, interdependency links are bidirected. The adjacency matrix of a general interdependent network consisting of m different networks has the block structure

$$A = \begin{bmatrix} A_1 & B_{12} & \cdots & B_{1m} \\ B_{12}^T & A_2 & \cdots & B_{2m} \\ \vdots & \vdots & \ddots & \vdots \\ B_{1m}^T & B_{2m}^T & \cdots & A_m \end{bmatrix} \tag{15.39}$$

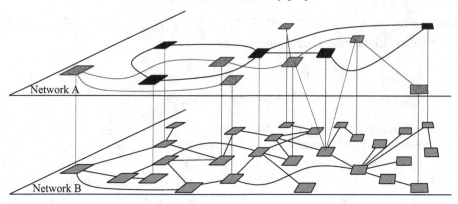

Fig. 15.8. A schematic representation of two interdependent networks A and B. Each layer represents a single network in which nodes are joined by links (in full line). Some of the nodes in network A are interconnected to nodes in network B by dependency links (in thin lines).

where A_j is a $N_j \times N_j$ symmetric adjacency matrix ($1 \leq j \leq m$) of the graph G_j and B_{ij} is the $N_i \times N_j$ zero-one interdependency matrix. Here in Fig. 15.8 with $m = 2$ networks, $G_1 = G_A$, $G_2 = G_B$ and B_{12} is an $N_1 \times N_2$ zero-one matrix in which each element $(B_{12})_{kl}$ represents an interdependency link between node k in G_A and node l in G_B. Instead of fully interdependent networks, partially interdependent networks, where only a fraction q_A of nodes in G_A are dependent on a node in G_B, are investigated in Son *et al.* (2012). Since network A and network B are different in nature (e.g. a computer control network and a power grid), the interdependency links in the matrix B_{12} are usually differently weighted than the links in the adjacency matrices A_j, as illustrated in Wang *et al.* (2013). In the sequel, we only consider fully interdependent networks.

15.8.1 Giant component of interdependent networks

If node n_A fails for whatever reason, then node n_B stops functioning as well. Buldyrev *et al.* (2010) investigate the effect of removing a fraction of the nodes in G_A and all links incident to these nodes. Due to the interdependence, all nodes in G_B connected by interdependency links to removed nodes in G_A must be removed as well. Only nodes belonging to the giant component of G_A and of G_B remain potentially functional. The removal of nodes in G_B can only decrease the size of the giant component. Nodes in G_B that no longer belong to the decreased giant component cease to function and may trigger additional removals of nodes in G_A. Hence, a cascade in removing nodes from G_A and G_B may occur and both networks become increasingly more fragmented in disconnected components (or clusters). For single networks, we know (Section 15.7.6.1) that there is a critical link density p_c below which there is no giant component. The aim here is to compute the size of the mutual giant component of G_A and G_B, which can be regarded as the operational part in an interdependent network. We limit ourselves to an interdependent network with $m = 2$ in (15.39), where $B_{12} = I$ and $N_A = N_B$. Thus, each node in G_A depends only on one node in G_B and vice versa.

Due to the one-to-one mapping of nodes in G_A on nodes in G_B, Son et al. (2012) merge each node in G_A with its partner in G_B to one set $\mathcal{N}$ of nodes, each with two sets of links, A-links belonging to G_A and B-links to G_B. A mutual cluster is a set $\mathcal{C} \subset \mathcal{N}$ in which each pair of nodes, $i, j \in \mathcal{C}$, are connected by (at least) two paths, one path consisting of A-links only in G_A and the other path with B-links only in G_B. The probability that node n belongs to the mutual giant component of the interdependent network then equals the probability that node n belongs to the mutual cluster, $S = \Pr[n \in \mathcal{C}]$. Since node n is a member of two independent networks G_A and G_B,

$$S = \Pr\left[\{n \in C_A\} \cap \{n \in C_B\}\right] = \Pr[n \in C_A]\Pr[n \in C_B] = S_A S_B$$

where C_A denotes the giant component in G_A (and similarly for C_B), we obtain from (15.36) that

$$S = \left(1 - \varphi_{D_{G_A}}(u_A)\right)\left(1 - \varphi_{D_{G_B}}(u_B)\right) \tag{15.40}$$

where $1 - u_A$ (and analogously $1 - u_B$) is the probability that a node reached by following a random A-link belongs to the mutual cluster $\mathcal{C}$, because $u = \varphi_{C_{l+}}(1)$ in (15.37) excludes the contribution of the giant component. Since the partner node n_B of n_A must also belong to the mutual cluster via B-links and since G_A and G_B are independent, we find from (15.37) that $1 - u_A = S_B\left(1 - \varphi_{(D_{l+}-1)}(u_A)\right)$. Explicitly,

$$1 - u_A = \left(1 - \varphi_{D_{G_B}}(u_B)\right)\left(1 - \frac{\varphi'_{D_{G_A}}(u_A)}{\varphi'_{D_{G_A}}(1)}\right) \tag{15.41}$$

Indeed, the end node l_A^+ of a randomly chosen link l_A of G_A can only belong to the mutual cluster $\mathcal{C}$ if its partner node in G_B also belongs to $\mathcal{C}$ and vice versa. Similarly for $1 - u_B = S_A\left(1 - \varphi_{(D_{l+}-1)}(u_B)\right)$, we have

$$1 - u_B = \left(1 - \varphi_{D_{G_A}}(u_A)\right)\left(1 - \frac{\varphi'_{D_{G_B}}(u_B)}{\varphi'_{D_{G_B}}(1)}\right) \tag{15.42}$$

The approach is readily generalized to M interdependent networks and the corresponding relations are

$$\begin{cases} S = \prod_{j=1}^{M}\left(1 - \varphi_{D_{G_j}}(u_j)\right) \\ 1 - u_j = \frac{S}{\left(1 - \varphi_{D_{G_j}}(u_j)\right)}\left(1 - \frac{\varphi'_{D_{G_j}}(u_j)}{\varphi'_{D_{G_j}}(1)}\right) \end{cases} \tag{15.43}$$

Let us consider the case where G_A and G_B are two large ER random graphs with (approximate) Poisson pgf $\varphi_{DG_A}(z) = e^{\mu_A(z-1)}$ and $\varphi_{DG_B}(z) = e^{\mu_B(z-1)}$. The fraction of nodes in the giant mutual clusters follows from (15.40) as

$$S = \left(1 - e^{\mu_A(u_A-1)}\right)\left(1 - e^{\mu_B(u_B-1)}\right)$$

while (15.41) and (15.42) show that $u_A = u_B = 1 - S$. Thus, we arrive at the functional equation for the fraction S of nodes in the giant mutual component of two ER random graphs

$$S = \left(1 - e^{-\mu_A S}\right)\left(1 - e^{-\mu_B S}\right) \tag{15.44}$$

In the particular case when the mean degree $\mu_j = \mu$ for $1 \leq j \leq M$ interdependent

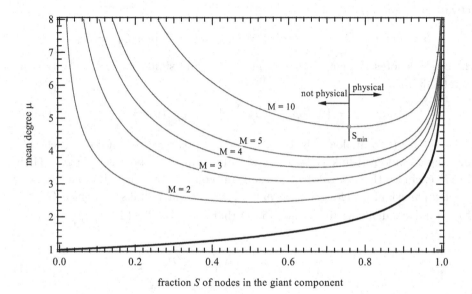

Fig. 15.9. The mean degree μ versus the fraction S of nodes in the giant component for various numbers M of interdependent ER random graphs. Usually, S versus μ is plotted, which necessitates to solve S in $S = \left(1 - e^{-\mu S}\right)^M$ numerically. When, starting from $S = 1$, the minimum degree $\mu_{\min}$ at $S_{\min}$ is reached for $M > 1$, the "physical" curve $\mu(S)$ jumps to the point $\mu(0) = \mu_{\min}$ at $S = 0$.

ER random graphs, (15.43) reduces to $S = \left(1 - e^{-\mu S}\right)^M$, from which, as in Section 15.7.6.1, the mean degree μ is written explicitly as a function of S,

$$\mu(S) = -\frac{\log\left(1 - \sqrt[M]{S}\right)}{S} \tag{15.45}$$

While the mean degree $\mu(S) = p(N-1) \to \infty$ if $S \to 1$ is meaningful for large ER graphs ($N \to \infty$), the limit $\lim_{S \to 0} \mu(S) = \infty$, provided $M > 1$, is not. Fig. 15.9 illustrates that, for $M > 1$, there are two possible values of S corresponding to one μ above a minimum degree $\mu_{\min}$, which is reached at a unique minimum fraction $S_{\min}$. Moreover, Fig. 15.9 shows that the mean degree $\mu(S)$ is decreasing for small S values until $S_{\min}$. The region $[0, S_{\min}]$ is not physical, because the mean degree $\mu(S)$ must be increasing with the fraction S of nodes in the mutual giant component. Hence, Fig. 15.9 demonstrates the curious property that, for $M > 1$, the function $\mu(S)$ possesses a discontinuity at $S_{\min}$ with minimum degree $\mu_{\min}$, whereas for $M = 1$, there is no such discontinuity in $[0, 1]$. In other words, by gradually lowering the mean degree μ or the link density $p = \frac{\mu}{N-1}$ starting from $S = 1$, the fraction S of nodes in the mutual giant component $(M > 1)$ suddenly ceases to exist at $S_{\min}$, which points to the existence of a so-called *first-order or abrupt* phase transition. Single networks $(M = 1)$ feature a *second-order or smooth* phase transition without a discontinuity in the S interval $[0, 1]$. The abrupt phase transition is the distinguishing factor of interdependent networks, which results in a more dramatic breakdown of interdependent networks due to link failures in comparison to single networks. If $S > S_{\min}$, then the interdependent network still has parts that are operational, while if $S < S_{\min}$, the interdependent network collapses totally and abrupt.

15.9 Problems

(i) An extremely regular graph is a d-lattice where each nodal position corresponds to a point with integer coordinates within a d-dimensional hyper-cube with size Z. Apart from border nodes, each node has a constant degree (number of neighbors), precisely equal to $2d$. Assuming that all link metrics are equal to one, compute the probability generating function of the hopcount of the shortest path between two uniformly chosen points.

(ii) Compute $\Pr\left[c_{G_p(N)} \le x\right]$ and $E\left[c_{G_p(N)}\right]$, where $c_{G_p(N)}$ denote the clustering coefficient of the random graph $G_p(N)$.

(iii) Derive the mean number of trees of size k in the random graph $G_p(N)$ by using Caley's famous formula (Van Mieghem, 2011, p. 77): the total number of spanning trees in G equals N^{N-2}.

(iv) Deduce a recursion relation for the total number of trees that can be constructed on N nodes. From that recursion, Caley's formula can be found.

(v) Compute the degree distribution of an arbitrary neighbor of a node j with degree k in the Erdős-Rényi random graph $G_p(N)$.

(vi) *Independence in graphs.* In most cases, dependence among random variables prevents probabilistic calculus. Since a graph is connected set of node and links, most properties are dependent, even in the Erdős-Rényi random graph $G_p(N)$. Nevertheless, by ignoring weak dependence, one often obtains reasonably accurate results. For example, in Section 15.7.5, the implicit assumption of independence that appears in the application of (3.39) leads to correct, asymptotic results. Discuss the combination of assuming independence of the degree among the nodes in $G_p(N)$ and the basis law (15.4) of the degree.

(vii) The linear correlation coefficient of the degree D_i and D_j of two random nodes i and j in $G_p(N)$ for $0 < p < 1$ is

$$\rho(D_i, D_j) = \frac{\text{Cov}[D_i, D_j]}{\sqrt{\text{Var}[D_i]}\sqrt{\text{Var}[D_j]}} = \frac{1}{N-1} \tag{15.46}$$

Prove this property of the Erdős-Rényi random graph $G_p(N)$, which indicates that, if N is large enough, D_i and D_j are almost independent for $i \neq j$. Hence, all random variables of the set $\{D_i\}_{1 \le i \le N}$ can be regarded as almost i.i.d. binomially distributed for large N.

(viii) Let $D_{\min}(p)$ denote the minimum degree in the Erdős-Rényi random graph $G_p(N)$. Show that

$$D_{\min}(p) = N - 1 - D_{\max}(1 - p) \qquad (15.47)$$

(ix) When does the adjacency matrix of a graph possess a zero eigenvalue? Discuss its physical meaning.

(x) When does the adjacency matrix A of a graph possess an eigenvalue $\lambda(A) = -1$?

(xi) Consider a circular area of radius R in which N wireless nodes are placed independently and uniformly with density δ nodes/m^2. The transmission radius of each node is r. Show that the asymptotic, critical transmission radius r_c for each node in the ad hoc network to be connected is

$$r \ge r_c = R\sqrt{\frac{\log N}{N}}$$

for large N.

(xii) *Watts-Strogatz Small World Graph.* Watts and Strogatz (see e.g. Watts (1999)) have proposed a class of graphs that are created by rewiring the links. The construction of a Watts-

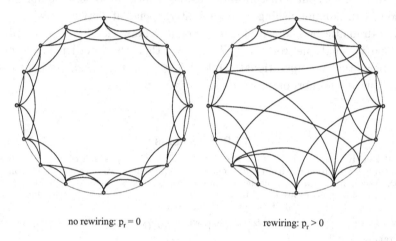

no rewiring: $p_r = 0$ rewiring: $p_r > 0$

Fig. 15.10. If there is no rewiring (left), the Watts-Strogatz model is a regular graph drawn on a circle where each node has precisely 2κ links, κ to left-hand side neighbors and κ to right-hand side neighbors. The resulting graph has an adjacency matrix of the form of a circulant matrix. When there is rewiring (right), each end-point of a link, corresponding a node n, has probability p to reconnected uniformly to any other node that is not a direct neighbor of n. When the rewiring probability tends to $p \to 1$, the Watts-Strogatz graph tends to a random graph.

Strogatz graph starts with a regular graph with N nodes, positioned on a circle as illustrated in Fig. 15.10, where the degree of each node is precisely 2κ. The basic law of the degree (15.4) shows that this regular graph has $L = \kappa N$ links. Then, rewiring is introduced. Each end point of a link has probability p_r to be rewired. The nodes are sequentially visited in one direction (e.g. clock-wise sense). Each end point of a link at a node n that connects neighbors in the visiting direction are replaced with probability p_r to a uniformly chosen other node that is not a direct neighbor of node n (to prevent multiple links). After rewiring (right-hand side of Fig. 15.10), the graph still has $L = \kappa N$ links, but the degree per node is not necessarily equal to 2κ. The number of "forwards pointing links" is always equal to κ. The number of "backwards pointing links" of a node n are possibly reduced by previous neighbors (as those links are, for previous neighbors, "forwards pointing links" that can be rewired), but they are possibly increased by previously visited non-neighbors that rewire their "forwards pointing links" to node n. Hence, the minimum degree of each node is at least κ, while the maximum degree can be $N - 1$.

An interesting property of the Watts-Strogatz graph is that the mean hopcount varies from $E[H_N] = O(N)$ when $p_r = 0$ ("big world") to $E[H_N] = O(\log N)$ when p_r large enough ("small world"). Hence, the existence of a few "long links" significantly reduces the mean hopcount.

(a) Compute the mean hopcount in case $p_r = 0$.

(b) The degree distribution when $p_r > 0$: We allow that a link at a node is rewired to itself, but not to other direct neighbors of that node. In general, this computation is difficult because, in the process of rewiring each node n in clock-wise sense starting at node 1, the possible number of nodes to which can be rewired is $N - d_n + 1$, where d_n is the degree of that node n at his rewiring time (i.e. at time step n). This implies that the full history from node 1 up to node n is needed to determine the random variable d_n. However, if the degree 2κ is sufficiently small and N is large, we may neglect a multiple link rewiring such that the available number of nodes for rewiring at each step is constant and equal to N. Compute, under this assumption, the degree distribution.

(c) Under the assumptions of (b), i.e. for large N and fixed κ and ignoring multiple link rewiring, show that the pgf of the degree in the Watts-Strogatz graph equals

$$\varphi_{D|p_r}(z) = E\left[z^{D|p_r}\right] = z^\kappa \left(1 + (1 - p_r)(z - 1)\right)^\kappa e^{\kappa p_r (z-1)} \qquad (15.48)$$

(d) Compare the degree distribution of the Erdős-Rényi random graph $G_p(N)$ and the Watts-Strogatz graph for $p_r \to 1$. Use the solution of (b) or (15.48) for large N and fixed κ.

(xiii) Show that the clustering coefficient $c_{WS}(p_r)$ of the Watts-Strogatz graph, without rewiring $p_r = 0$ and defined in (15.14), equals

$$c_{WS}(0) = \frac{3(\kappa - 1)}{2(2\kappa - 1)} \qquad (15.49)$$

Remark. In case $p_r > 0$, a rough argument is that a link between neighbors of a node in the case $p_r = 0$ is still a valid link when $p_r > 0$ provided the two neighbors are still neighbors of that node and they are still connected. Each of those three independent events has roughly the same probability of $1 - p_r$ (no rewiring), such that the initial number of links between neighbors $y(0)$ is maintain roughly with probability $(1 - p_r)^3$, or $E[y(p_r)] \approx y(0)(1 - p_r)^3$. The event that more neighbors are rewired to the node is neglected for large N and fixed κ because a rewiring of links into a node has probability $O\left(\frac{1}{N}\right)$. When the mean (over all nodes) of the clustering coefficient is approximated by

$$E[c_{WS}(p_r)] = E\left[\frac{y(p_r)}{\binom{D|p_r}{2}}\right] \approx \frac{E[y(p_r)]}{E\left[\binom{D|p_r}{2}\right]}$$

we obtain, using (15.48) and (15.49),

$$E[c_{WS}(p_r)] \approx \frac{3(\kappa - 1)(1 - p_r)^3}{2\left(2\kappa - \frac{1}{2}\right)} \approx c_{WS}(0)(1 - p_r)^3$$

Simulations, for large N, show that this rough estimate is a reasonable approximation. The mean of the clustering coefficient of the random graph $G_p(N)$ is $E\left[c_{G_p(N)}\right] = p = \frac{E[D]}{N-1}$ (see (B.21)). For large N and constant degree $E[D] = 2\kappa$, we thus observe that both $E[c_{WS}(1)] \to 0$ and $E\left[c_{G_{\frac{2\kappa}{N}}(N)}\right] \to 0$.

(xiv) *Price's citation graph.* Newman (2010, Section 14.1) describes a preferential attachment model, due to Price. Papers (nodes in the citation graph) are generated continuously. The mean number of citations per paper equals c and citations correspond to links and their number is equal to the out-degree of a node in the citation graph. Thus, papers can only cite older papers so that the resulting graph is directed. Price assumes that a paper is cited by (newer) papers randomly and proportional to its number of citations at that time (i.e. in-degree) plus a constant a. That constant a is mainly needed to enable a paper to receive citations, because, in the beginning a new paper does not have citations and the proportionality rule would also result in zero citations. Hence, a can be interpreted as an initial number of "free" citations.

(a) Derive the governing equations for the probability $p_q(n)$ that the in-degree equals q when the citation network contains n nodes (papers).

(b) Solve those equations asymptotically for large n and deduce, for large q and n, that $p_q(n) \sim bq^{-2-a/c}$, where b is a normalization constant. Hence, the Price's citation graph has a power-law degree distribution with exponent $2 + \frac{a}{c} > 2$.

(c) Compare Price's citation graph with the BA random graph in Section 15.7.2 and show that the BA random graph can be considered as a special case of the Price citation graph with $a = c$.

(xv) *Chromatic number of a graph.* Let $\mathcal{N}$ denote the set of nodes of a graph G, and let $\mathcal{S}$ be the set of colors. A node coloring of a graph $G(\mathcal{N}, \mathcal{L})$ is a map $c : \mathcal{N} \to \mathcal{S}$ such that $c(v) \neq c(w)$ whenever v and w are adjacent. A set of nodes is independent if each pair of nodes is not adjacent. The elements of the set $\mathcal{S} = \{1, \cdots, k\}$ are called the k available colors. What is the minimal value of k for which the set of nodes can be partitioned into k classes, $\mathcal{N}_1, \mathcal{N}_2, \ldots, \mathcal{N}_k$ and $\mathcal{N} = \cup_{j=1}^{k} \mathcal{N}_j$, such that no link joins two nodes of a same class. That smallest number k of colors in $\mathcal{S}$ is called the chromatic number of G, denoted by $\chi(G)$. In general, it is difficult to determine the chromatic number of a graph, but for the class of Erdős-Rényi graph $G_p(N)$ we can deduce an interesting lower bound. Show that, for every constant $p \in (0,1)$ and every $\epsilon > 0$, when N tends to infinity, almost every Erdős-Rényi graph $G \in G_p(N)$ has chromatic number

$$\chi(G) > \frac{\log(q^{-1})}{2 + \epsilon} \frac{N}{\log N} \qquad (15.50)$$

where $q = 1 - p$. Bollobas (2001, Section 11.4) presents sharper asymptotic results. (*Hint:* Upper bound the probability that the largest set of independent nodes in $G_p(N)$ is larger than k and determine k in such a way that the upper bound tends to zero for large N.)

(xvi) *Friendship paradox.* Individuals use the number of friends that their friends have as one basis for determining whether they, themselves, have an adequate number of friends. Feld (1991) shows that, if individuals compare themselves with their friends, it is likely that most of them will feel relatively inadequate, because most people have fewer friends than their friends have. This disproportionate experiencing of friends with many friends is similar to "class size paradoxes" such as the tendencies for college students to experience the mean class size as larger than it actually is and for people to experience beaches and parks as more crowded than they usually are.

(a) Show that the mean number of friends of friends is always greater than or equal to the mean number of friends of an individual.

(b) Consider in a random graph $G_p(N)$ the sum of the degree of the neighbors of an arbitrary node and demonstrate that Wald's identity (2.78) is not valid to compute the mean.

(xvii) *Degree distribution after adding or removing links.* Consider an undirected graph $G_0(N, L)$ with N nodes and L links. Let us denote the degree of a randomly chosen node in G_0 by D_{G_0}. We assume that the degree distribution $\Pr[D_{G_0} = j]$ is known. After the addition of m (undirected) links in a uniform and random way to G_0, the resulting graph G has a degree distribution $\Pr[D = k]$.

(a) Compute the explicit expression for $\Pr[D = k]$ in terms of $\Pr[D_{G_0} = k]$.

(b) Instead of adding links, m arbitrary links are removed from G_0. Given $\Pr[D_{G_0} = k]$, compute $\Pr[D = k]$ and $\varphi_D(z) = E[z^D]$. Also, discuss how the degree distribution changes when n arbitrary nodes, instead of m arbitrary links, are removed in a graph G_0.

(xviii) *Degree distribution of the end node of an arbitrarily chosen link.* Let l denote an arbitrarily chosen link in the graph G and $l^+ = i$ and $l^- = j$ are the two end nodes of link l.

(a) Compute $\Pr[D_{l^+} = k]$ and the corresponding pgf $\varphi_{D_{l^+}}(z)$, given the degree distribution $\Pr[D = k]$ of a node in G.

(b) Compute the pgf of all the neighbors of the neighbors of a node, i.e. all second-nearest neighbors of an arbitrary node in G, by assuming that all those neighbors are different.

16

The shortest path problem

The shortest path problem asks for the computation of the path from a source to a destination node that minimizes the sum of the positive weights[1] of its constituent links. The related shortest path tree (SPT) is the union of the shortest paths from a source node to a set of m other nodes in the graph with N nodes. If $m = N - 1$, the SPT connects all nodes and is termed a spanning tree. The SPT belongs to the fundamentals of graph theory and has many applications. Moreover, powerful shortest path algorithms like that of Dijkstra exist.

Routers in the Internet forward IP packets to the next hop router, which is found by routing protocols (such as OSPF and BGP). Intra-domain routing as OSPF is based on the Dijkstra *shortest path* algorithm, while inter-domain routing with BGP is policy-based, which implies that BGP does not minimize a length criterion. Nevertheless, end-to-end paths in the Internet are shortest paths in roughly 70% of the cases. Therefore, we consider the shortest path between two *arbitrary* nodes because (a) the IP address does not reflect a precise geographical location and (b) uniformly distributed world wide communication, especially, on the web seems natural since the information stored in servers can be located in places unexpected and unknown to browsing users. The Internet type of communication is different from classical telephony because (a) telephone numbers have a direct binding with a physical location and (b) the intensity of average human interaction rapidly decreases with distance. We study the *hopcount* H_N in a graph with N nodes because it is simple to measure via the trace-route utility, it is an integer, dimensionless, and the quality of service (QoS) measures (such as packet delay, jitter and packet loss) depend on the hopcount, the number of traversed routers.

In this chapter, the influence of the link weight structure on the properties of the SPT will be analyzed. Starting from one of the simplest possible graph models, the complete graph with i.i.d. exponential link weights, the characteristics of the shortest path will be derived and compared to Internet measurements. Next, since most complex networks are sparse, the shortest path problem in large, sparse networks is investigated. Finally, we study the minimum spanning tree problem.

---

[1] A zero link weight is regarded as the coincidence of two nodes (which we exclude), while an infinite link weight means the absence of a link.

The link weights seriously impact the path properties in QoS routing (Kuipers and Van Mieghem, 2003). In addition, from a traffic engineering perspective, an ISP may want to tune the weight of each link such that the resulting shortest paths between a particular set of in- and egresses follow the desirable routes in its network. Thus, apart from the topology of the graph, the link weight structure clearly plays an important role. Often, as in the Internet or other large infrastructures, both the topology and the link weight structure are not accurately known. This uncertainty about the precise structure leads us to consider both the underlying graph and each of the link weights as random variables.

16.1 The shortest path and the link weight structure

Since the shortest path is mainly sensitive to the smaller, positive link weights, the probability distribution of the link weights around zero will dominantly influence the properties of the resulting shortest path. A *regular* link weight distribution $F_w(x) = \Pr[w \le x]$ has a Taylor series expansion around $x = 0$,

$$F_w(x) = f_w(0)x + O(x^2)$$

since $F_w(0) = 0$ and $F_w'(0) = f_w(0)$ exists. A regular link weight distribution is thus linear in x around zero. The factor $f_w(0)$ only scales all link weights, but does not influence the shortest path. The simplest distribution of the link weight w with a distinct different behavior for small values is the polynomial distribution

$$F_w(x) = x^\alpha 1_{x \in [0,1]} + 1_{x \in (1,\infty)}, \quad \alpha > 0 \tag{16.1}$$

The corresponding density is $f_w(x) = \alpha x^{\alpha - 1} 1_{x \in [0,1]}$. The exponent

$$\alpha = \lim_{x \downarrow 0} \frac{\log F_w(x)}{\log x}$$

is called the *extreme value index* of the probability distribution of w and $\alpha = 1$ for regular distributions. By varying the exponent α over all non-negative real values, any extreme value index can be attained and a large class of corresponding SPTs, in short α-trees, can be generated.

Figure 16.1 illustrates schematically the probability distribution of the link weights around zero, thus for $x \in (0, \epsilon]$, where $\epsilon > 0$ is an arbitrarily small, positive real number. The larger link weights in the network will hardly appear in a shortest path provided the network possesses enough links. These larger link weights are drawn in Fig. 16.1 from the double dotted line to the right. The nice advantage that only small link weights dominantly influence the property of the resulting shortest path tree implies that the remainder of the link weight distribution (denoted by the arrow with larger scale in Fig. 16.1) only plays a second-order role. To some extent, it also explains the success of the simple SPT model based on the complete graph K_N with i.i.d. exponential link weights, which we derive in Section 16.2. A link

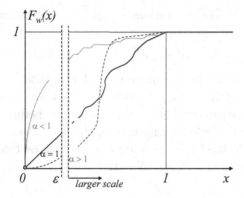

Fig. 16.1. A schematic drawing of the distribution of the link weights for the three different α-regimes. The shortest path problem is mainly sensitive to the small region around zero. The scaling invariant property of the shortest path allows us to divide all link weights by the largest possible such that $F_w(1) = 1$ for all link weight distributions.

weight structure effectively thins the complete graph K_N – any other graph is a subgraph of K_N – to the extent that a specific shortest path tree can be constructed.

Finally, we assume the independence of link weights, which we deem a reasonable assumption in large networks, such as the Internet with its many independent autonomous systems (ASs). Apart from Section 16.9, we will mainly consider the case for $\alpha = 1$, which allows an exact analysis.

16.2 The shortest path tree in K_N with exponential link weights

16.2.1 The Markov discovery process

Let us consider the shortest path problem in the complete graph K_N, where each node in the graph is connected to each other node. The problem of finding the shortest path between two nodes A and B in K_N with exponentially distributed link weights with mean 1 can be rephrased in terms of a Markov discovery process. The discovery process evolves as a function of time and stops at a random time T when node B is found. The process is shown in Fig. 16.2.

The evolution of the discovery process can be described by a continuous-time Markov chain $X(t)$, where $X(t)$ denotes the number of discovered nodes at time t, because the characteristics of a Markov chain (Theorem 10.2.3) are based on the exponential distribution and the memoryless property. Of particular interest here is the property (see Section 3.4.1) that the minimum of n independent exponential variables each with parameter α_i is again an exponential variable with parameter $\sum_{i=1}^{n} \alpha_i$.

The discovery process starts at time $t = T_0$ with the source node A and for the initial distribution of the Markov chain, we have $\Pr[X(T_0) = 1] = 1$. The state space of the continuous Markov chain is the set S_N consisting of all positive integers (nodes) n with $n \leq N$. For the complete graph K_N, the transition rates are given by

$$\lambda_n = n(N - n), \qquad\qquad n \in S_N \qquad\qquad (16.2)$$

Indeed, initially there is only the source node A with label[2] 0, hence $n = 1$. From this first node A precisely $N - 1$ new nodes can be reached in the complete graph K_N. Alternatively one can say that $N-1$ nodes are competing with each other each with exponentially distributed strength to be discovered and the winner amongst them, say node C with label 1, is the one reached in shortest time which corresponds to an exponential variable with rate $N - 1$.

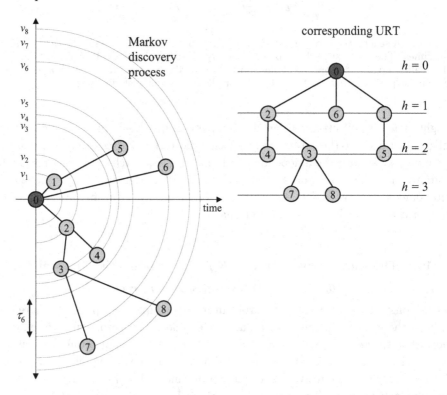

Fig. 16.2. On the left, the Markov discovery process as function of time in a graph with $N = 9$ nodes. The circles centered at the discovering node A with label 0 present equi-time lines and v_k is the discovery time of the k-th node, while $\tau_k = v_k - v_{k-1}$ is the k-th interattachment time. The set of discovered nodes redrawn per level are shown on the right, where a level gives the number of hops h from the source node A. The tree is a uniform recursive tree (URT).

After having reached C from A at hitting time v_1, two nodes $n = 2$ are found and the discovery process restarts from both A (label 0) and C (label 1). Although

---

[2] When continuous measures such as time and weight of a path are computed, the source node is most conveniently labeled by zero, whereas in counting processes, such as the number of hops of a path, the source node is labeled by one.

at time v_1 we were already progressed a certain distance towards each of the $N-2$ other, not yet discovered, nodes, the memoryless property of the exponential distribution tells us that the remaining distance to these $N-2$ nodes is again exponentially distributed with the same parameter 1. Hence, this allows us to restart the process from A and C by erasing the previously partial distance to any other not yet discovered node as if we ignore that it were ever travelled. From the discovery time v_1 of the first node on, the discovery process has double strength to reach precisely $N-2$ new nodes. Hence, the next winner, say node D labeled by 2, is reached at v_2 in the minimum time out of $2(N-2)$ traveling times. This node D has equal probability to be attached to A or C because of symmetry. When D is attached to A as in Fig. 16.2 (the argument below holds similarly for attachment to C), symmetry appears to be broken, because D and C have only one link used, whereas A already has two links used. However, since we are interested in the shortest path problem and since the direct link from A to D is shorter than the path $A \to C \to D$, we exclude the latter in the discovery process, hereby establishing again the full symmetry in the Markov chain. This exclusion also means that the Markov chain maintains single paths from A to each newly discovered node and this path is also the shortest path. Hence, there are no cycles possible. Furthermore, similar to Dijkstra's shortest path algorithm, each newly reached node is withdrawn from the next competition round, which guarantees that the Markov chain eventually terminates. Besides terminating by extinction of all available nodes, after each transition when a new node is discovered, the Markov chain stops with probability equal to $\frac{1}{N-n}$, since each of the n already discovered nodes has precisely 1 possibility out of the remaining $N-n$ to reach B and only one of them is the discoverer. The stopping time T is defined as the infimum for $t \geq 0$ at which the destination node B is discovered.

In summary, the described Markov discovery process on K_N with i.i.d. exponential link weights is a pure birth process (see Section 11.3.2) with birth rate $\lambda_n = n(N-n)$, that models exactly the shortest path tree for all values of N.

16.2.2 The uniform recursive tree

A uniform recursive tree (URT) of size N is a random tree rooted at A. At each stage a new node is attached uniformly to one of the existing nodes until the total number of nodes is equal to N. The hopcount h_N (equivalent to the depth or distance) is the smallest number of links between the root A and a destination chosen uniformly from all nodes $\{1, 2, \ldots, N\}$.

Denote by $\left\{ X_N^{(k)} \right\}$ the k-th level set of a tree T, which is the set of nodes in the tree T at hopcount k from the root A in a graph with N nodes, and by $X_N^{(k)}$ the number of elements in the set $\left\{ X_N^{(k)} \right\}$. Then, we have $X_N^{(0)} = 1$ because the zeroth level can only contain the root node A itself. For all $k > 0$, it holds that

$0 \leq X_N^{(k)} \leq N - 1$ and that

$$\sum_{k=0}^{N-1} X_N^{(k)} = N \tag{16.3}$$

Another consequence of the definition is that, if $X_N^{(n)} = 0$ for some level $n < N - 1$, then all $X_N^{(j)} = 0$ for levels $j > n$. In such a case, the longest possible shortest path in the tree has a hopcount of n. The level set

$$L_N = \left\{ 1, X_N^{(1)}, X_N^{(2)}, \ldots, X_N^{(N-1)} \right\}$$

of a tree T is defined as the set containing the number of nodes $X_N^{(k)}$ at each level k. An example of a URT organized per level k is drawn on the right in Fig. 16.2 and in Fig. 16.3. A basic theorem for URTs proved in van der Hofstad *et al.* (2002b), is the following, where it is assumed that $X_N^{(k)} = 0$ for $k < 0$:

Theorem 16.2.1 *Let* $\{Y_N^{(k)}\}_{k,N \geq 0}$ *and* $\{Z_N^{(k)}\}_{k,N \geq 0}$ *be two independent copies of the vector of level sets of two sequences of independent URTs. Then*

$$\{X_N^{(k)}\}_{k \geq 0} \overset{d}{=} \{Y_{N_1}^{(k-1)} + Z_{N-N_1}^{(k)}\}_{k \geq 0}, \tag{16.4}$$

where on the right-hand side the random variable N_1 *is uniformly distributed over the set* $\{1, 2, \ldots, N - 1\}$.

Theorem 16.2.1 also implies that a subtree rooted at a direct child of the root is a URT. For example, in Fig. 16.3, the tree rooted at node 5 is a URT of size 13 as well as the original tree without the tree rooted at node 5. By applying Theorem 16.2.1 to the URT subtree, any subtree rooted at a member of a URT is also a URT. Theorem 16.2.1 is further illustrated in Fig. 19.3 and applied in Section 19.4.1.

An arbitrary URT U consisting of N nodes and with the root labeled by 1 can be represented as

$$U = (n_2 \longleftarrow 2)(n_3 \longleftarrow 3) \ldots (n_N \longleftarrow N) \tag{16.5}$$

where $(n_j \longleftarrow j)$ means that the j-th node is attached to node $n_j \in [1, j - 1]$ and $n_2 = 1$. Hence, n_j is the predecessor of j and the predecessor relation is indicated by the arrow "$\longleftarrow$". Moreover, n_j is a discrete uniform random variable on $[1, j-1]$ and all $n_2, n_3, \ldots, n_N$ are independent.

Theorem 16.2.2 *The total number of URTs with N nodes is* $(N - 1)!$

Proof: (a) Let the nodes be labeled in the order of attachment to the URT and assign label 1 for the root. The URT growth law indicates that node 2 can only be attached in one way, node 3 in two ways, namely to node 1 and node 2 with equal probability. The k-th node can be attached at $k - 1$ possible nodes. Each of these possible constructions leads to a URT.

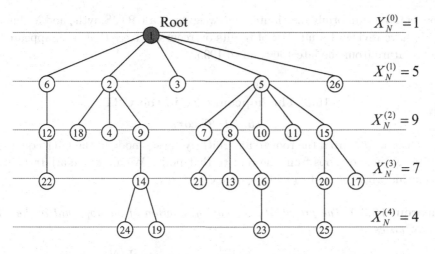

Fig. 16.3. An instance of a uniform recursive tree with $N = 26$ nodes organized per level $0 \leq k \leq 4$. The node number (inside the circle) indicates the order in which the nodes were attached to the tree.

(b) By summing over all allowable configurations in (16.5), we obtain

$$\sum_{n_2=1}^{1} \sum_{n_3=1}^{2} \cdots \sum_{n_N=1}^{N-1} 1 = (N-1)!$$

and this proves the theorem.   □

In general, Cayley's Theorem (Van Mieghem, 2011) states that there are N^{N-2} labeled trees possible. The URT is a subset of the set of all possible labeled trees. Not all labeled trees are URTs, because the nodes that are further away from the root must have larger labels.

The shortest path tree from the source or root A to other nodes in the complete graph is the tree associated with the Markov discovery process, where the number of nodes $X(t)$ at time t is constructed as follows. Just as in the discovery process, the associated tree starts at the root A. We now investigate the embedded Markov chain (Section 10.4) of the continuous-time discovery process. After each transition in the continuous-time Markov chain, $X(t) \to X(t) + 1$, an edge of *unit* length is attached *randomly* to one of the n already discovered nodes in the associated tree because a new edge is equally likely to be attached to any of the n discovering nodes. Hence, the construction of the tree associated with the Markov discovery process, as the right-hand side of Fig. 16.2, demonstrates that the shortest path tree in the complete graph K_N with exponential link weights is a *uniform recursive tree*. This property of the shortest path tree in K_N with exponential link weights is an important motivation to study the URT. More generally, in van der Hofstad *et al.* (2001), we have proved that, for a fixed link density p and sufficiently large N, the shortest path tree in the *class RGU*, the class of ER random graphs $G_p(N)$ with exponential or uniformly distributed link weights, is a URT. Smythe and Mahmoud (1995) have reviewed a number of results on recursive trees that have appeared in the literature from the late 1960s up to 1995.

16.3 The hopcount h_N in the URT

16.3.1 Theory

The hopcount h_N from the root to an arbitrary chosen node in the URT equals the number of links or hops from the root to that node. We allow the arbitrary node to coincide with the root in which case $h_N = 0$.

Theorem 16.3.1 *The probability generating function of the hopcount in the URT with N nodes is*

$$\varphi_{h_N}(z) = E\left[z^{h_N}\right] = \frac{1}{N!}\prod_{k=1}^{N-1}(z+k) = \frac{\Gamma(N+z)}{\Gamma(N+1)\Gamma(z+1)} \qquad (16.6)$$

Proof: Since the number of nodes at hopcount k from the root (or at level k) is $X_N^{(k)}$, a node uniformly chosen out of N nodes in the URT has probability $\frac{E\left[X_N^{(k)}\right]}{N}$ of having hopcount k,

$$\Pr[h_N = k] = \frac{E\left[X_N^{(k)}\right]}{N} \qquad (16.7)$$

If the size of the URT grows from n to $n+1$ nodes, each node at hopcount $k-1$ from the root can generate a node at hopcount k with probability $1/n$. Hence, for $k \geq 1$, the basic recursion for the expectation of these level sets is

$$E\left[X_N^{(k)}\right] = \sum_{n=k}^{N-1} \frac{E\left[X_n^{(k-1)}\right]}{n} \qquad (16.8)$$

With (16.7), a recursion for $\Pr[h_N = k]$ follows for $k \geq 1$ as

$$\Pr[h_N = k] = \frac{1}{N}\sum_{n=k}^{N-1}\Pr[h_n = k-1]$$

The probability generating function of h_N equals

$$\varphi_{h_N}(z) = E\left[z^{h_N}\right] = \Pr[h_N = 0] + \sum_{k=1}^{N-1}\Pr[h_N = k]z^k$$

$$= \frac{1}{N} + \frac{1}{N}\sum_{k=1}^{N-1}\sum_{n=k}^{N-1}\Pr[h_n = k-1]z^k$$

$$= \frac{1}{N} + \frac{1}{N}\sum_{n=1}^{N-1}\sum_{k=1}^{n}\Pr[h_n = k-1]z^k$$

Since

$$\sum_{k=1}^{n} \Pr[h_n = k-1]z^k = \sum_{k=0}^{n-1} \Pr[h_n = k]z^{k+1}$$

we find, with the definition $\varphi_{h_m}(z) = \sum_{k=0}^{m-1} \Pr[h_m = k] z^k$, the relation

$$\varphi_{h_N}(z) = \frac{1}{N} + \frac{z}{N} \sum_{n=1}^{N-1} \varphi_{h_n}(z)$$

Taking the difference between $(N+1)\varphi_{h_{N+1}}(z)$ and $N\varphi_{h_N}(z)$ results in the recursion

$$(N+1)\varphi_{h_{N+1}}(z) = (N+z)\varphi_{h_N}(z)$$

Iterating this recursion starting from $\varphi_{h_1}(z) = E\left[z^{h_1}\right] = E\left[z^0\right] = 1$ leads to (16.6). $\qquad\square$

Corollary 16.3.2 *The probability density function of the hopcount in the URT with N nodes is*

$$\Pr[h_N = k] = \frac{(-1)^{N-(k+1)} S_N^{(k+1)}}{N!} \qquad (16.9)$$

Proof: The probability generating function $\varphi_{h_N}(z) = \sum_{k=0}^{N-1} \Pr[h_N = k]z^k$ in (16.6) is related to the generating function of the Stirling numbers $S_N^{(k)}$ of the first kind (Abramowitz and Stegun, 1968, 24.1.3),

$$m!\binom{x}{m} = \frac{\Gamma(x+1)}{\Gamma(x+1-m)} = \prod_{k=0}^{m-1}(x-k) = \sum_{k=0}^{m} S_m^{(k)} x^k$$

Indeed, for $x = -z$ and using the functional equation of the Gamma function $\Gamma(z+1) = z\Gamma(z)$, we have that $\prod_{k=0}^{m-1}(x-k) = (-1)^m \prod_{k=0}^{m-1}(z+k) = (-1)^m \frac{\Gamma(z+m)}{\Gamma(z)}$ such that

$$\frac{\Gamma(z+m)}{\Gamma(z)} = \sum_{k=0}^{m} (-1)^{m-k} S_m^{(k)} z^k$$

where $S_m^{(0)} = \delta_{m0}$. Hence, for $N > 0$, we have that $S_N^{(0)} = 0$ and

$$\frac{\Gamma(N+z)}{\Gamma(N+1)\Gamma(z+1)} = \frac{1}{N!} \sum_{k=1}^{N} (-1)^{N-k} S_N^{(k)} z^{k-1} = \frac{1}{N!} \sum_{k=0}^{N-1} (-1)^{N-(k+1)} S_N^{(k+1)} z^k$$

such that the probability that a uniformly chosen node in the URT has hopcount k equals (16.9). $\qquad\square$

The explicit form of the generating function shows that the mean hopcount h_N in a URT of size N equals

$$E[h_N] = \varphi'_{h_N}(1) = \frac{d}{dz}\log\varphi_{h_N}(z)\bigg|_{z=1} = \sum_{l=2}^{N}\frac{1}{l} \qquad (16.10)$$

$$= \psi(N+1) + \gamma - 1$$

where $\psi(z) = \frac{\Gamma'(z)}{\Gamma(z)}$ is the digamma function (Abramowitz and Stegun, 1968, Section 6.3) and the Euler constant is $\gamma = 0.577\,215\ldots$. Similarly, the variance (2.27) follows from the logarithm of the generating function $L_{h_N}(z) = \log\Gamma(N+z) - \log\Gamma(N+1) - \log\Gamma(z+1)$ as

$$\mathrm{Var}[h_N] = \psi'(N+1) - \psi'(2) + \psi(N+1) + \gamma - 1$$

$$= \psi(N+1) + \gamma - \frac{\pi^2}{6} + \psi'(N+1)$$

Using the asymptotic formula for the digamma function

$$\psi(z) \sim \log z - \frac{1}{2z} + O\left(\frac{1}{z^2}\right)$$

leads to

$$E[h_N] = \log N + \gamma - 1 + O\left(\frac{1}{N}\right) \qquad (16.11)$$

$$\mathrm{Var}[h_N] = \log N + \gamma - \frac{\pi^2}{6} + O\left(\frac{1}{N}\right) \qquad (16.12)$$

For large N, we apply an asymptotic formula of the Gamma function (Abramowitz and Stegun, 1968, Section 6.1.47) to the generating function of the hopcount (16.6),

$$\varphi_{h_N}(z) = \frac{N^{z-1}}{\Gamma(z+1)}\left(1 + O\left(\frac{1}{N}\right)\right)$$

Introducing the Taylor series of $\frac{1}{\Gamma(z)} = \sum_{k=1}^{\infty}c_k z^k$, where the coefficients c_k are listed in Abramowitz and Stegun (1968, Section 6.1.34), in particular, $c_1 = 1$ and $c_2 = \gamma$, we obtain with $N^z = e^{z\log N}$,

$$\varphi_{h_N}(z) = \frac{1}{N}\sum_{k=1}^{\infty}c_k z^{k-1}\sum_{k=0}^{\infty}\frac{\log^k N}{k!}z^k\left(1 + O\left(\frac{1}{N}\right)\right)$$

$$= \frac{1 + O\left(\frac{1}{N}\right)}{N}\sum_{k=0}^{\infty}\sum_{m=0}^{k}c_{m+1}\frac{\log^{k-m}N}{(k-m)!}z^k$$

With the definition (2.18) of the probability generating function, we conclude that the asymptotic form of the probability density function (16.9) of the hopcount in the URT is

$$\Pr[h_N = k] = \frac{1 + O\left(\frac{1}{N}\right)}{N}\sum_{m=0}^{k}c_{m+1}\frac{\log^{k-m}N}{(k-m)!} \qquad (16.13)$$

Since the coefficients c_k are rapidly decreasing, approximating the sum in (16.13) by its first term ($m = 0$) yields to first order in N,

$$\Pr[h_N = k] \sim \frac{(\log N)^k}{N k!} \qquad (16.14)$$

which is recognized as a Poisson distribution (3.9) with mean $\log N$. Hence, for large N and to first order, the mean and variance of the hopcount in the URT are approximately $E[h_N] \sim \mathrm{Var}[h_N] \sim \log N$. The accuracy of the Poisson approximation can be estimated by comparison with the mean (16.11) and the variance (16.12) found above up to second order in N. For example, if the URT has $N = 10^4$ nodes, the Poisson approximation yields $E[h_N] = \mathrm{Var}[h_N] = 9.21034$, while the mean (16.11) is $E[h_N] = 8.78756$ accurate up to 10^{-4} and the variance (16.12) is $\mathrm{Var}[h_N] = 8.14262$. The exact results are $E[h_N] = 8.78761$ and $\mathrm{Var}[h_N] = 8.14277$.

16.3.2 Application of the URT to the hopcount in the Internet

In trace-route measurements explained in Van Mieghem (2010a), we are interested in the hopcount H_N denoted with capital H, which equals h_N in the URT excluding the event $h_N = 0$. In other words, the source and the destination are different nodes in the graph. Since from (16.9) $\Pr[h_N = 0] = \frac{(-1)^{N-1} S_N^{(1)}}{N!} = \frac{1}{N}$ we obtain, for $1 \le k \le N - 1$,

$$\Pr[H_N = k] = \Pr[h_N = k | h_N \ne 0] = \frac{\Pr[h_N = k, h_N \ne 0]}{\Pr[h_N \ne 0]}$$

$$= \frac{N}{N-1} \Pr[h_N = k]$$

Using (16.9), we find

$$\Pr[H_N = k] = \frac{N}{N-1} \frac{(-1)^{N-(k+1)} S_N^{(k+1)}}{N!} \qquad (16.15)$$

with corresponding generating function,

$$\varphi_{H_N}(z) = \sum_{k=1}^{N-1} \Pr[H_N = k] \, z^k$$

$$= \frac{N}{N-1} \sum_{k=0}^{N-1} \Pr[h_N = k] \, z^k - \frac{N}{N-1} \Pr[h_N = 0]$$

$$= \frac{N}{N-1} \left(\varphi_{h_N}(z) - \frac{1}{N} \right) \qquad (16.16)$$

The mean hopcount $E[H_N] = E[h_N | h_N \ne 0]$ is

$$E[H_N] = \frac{N}{N-1} \sum_{l=2}^{N} \frac{1}{l} \qquad (16.17)$$

Hence, for large N and in practice, we find that

$$\Pr[H_N = k] = \Pr[h_N = k] + O\left(\frac{1}{N}\right)$$

which allows us to use the previously derived expressions (16.13), (16.11) and (16.12).

The histogram of the number of traversed routers in the Internet measured between two arbitrary communicating parties seems reasonably well modeled by the pdf (16.13). Figure 16.4 shows both the histogram of the hopcount deduced from paths in the Internet measured via the trace-route utility and the fit with (16.13). From the fit, we find a rather high number of nodes $e^{12.6} \approx 3\ 10^5 \le N \le e^{13.5} \approx 7$

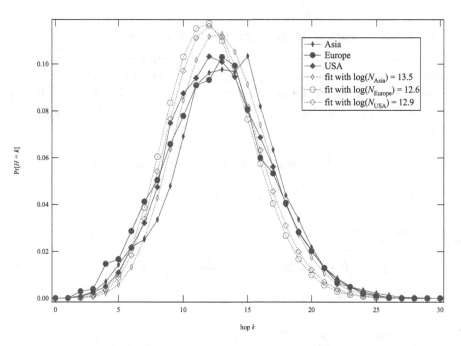

Fig. 16.4. The histograms of the hopcount derived from the trace-route measurement in three continents from CAIDA in 2004 are fitted by the pdf (16.13) of the hopcount in the URT.

10^5, which points to the approximate nature of modeling the Internet hopcount by that deduced from a URT. The relation between Internet measurements and the properties of the URT is further analyzed in a series of articles (Van Mieghem *et al.*, 2000; van der Hofstad *et al.*, 2001; Van Mieghem *et al.*, 2001b; Janic *et al.*, 2002; van der Hofstad *et al.*, 2002b). A more accurate, though more complex model of the hopcount in the Internet is discussed in Section 16.8.3.

16.4 The weight of the shortest path

The weight – sometimes also called the length – of the shortest path is defined as the sum of the link weights that constitute the shortest path. In Section 16.2.1, the shortest path tree in the complete graph with exponential link weights was shown to be a URT. In this section, we confine ourselves to the same type of graph and require that the source node A (or root) is different from the destination node B.

By Theorem 10.2.3 of a continuous-time Markov chain, the discovery time of the k-th node from node A equals $v_k = \sum_{n=1}^{k} \tau_n$, where $\tau_1, \tau_2, \ldots, \tau_k$ are independent, exponentially distributed random variables with parameter $\lambda_n = n(N - n)$ with $1 \leq n \leq k$. We call τ_j the interattachement time between the discovery or the attachment to the URT of the $(j - 1)$-th and j-th nodes in the graph. The weight of an arbitrary shortest path can thus be written as

$$W_N = \sum_{k=1}^{N-1} v_k 1_{\{\text{node } k \text{ is the end node of the shortest path}\}} \tag{16.18}$$

where it is understood that the end node is different from the root. The Laplace transform of v_k is

$$E\left[e^{-zv_k}\right] = \int_0^\infty e^{-zt} \frac{d}{dt} \Pr\left[v_k \leq t\right]$$

For a sum of independent exponential random variables, using the probability generating function (3.16), we have

$$E\left[e^{-zv_k}\right] = E\left[\exp\left(-z\sum_{n=1}^{k} \tau_n\right)\right] = \prod_{n=1}^{k} E\left[e^{-z\tau_n}\right] = \prod_{n=1}^{k} \frac{n(N - n)}{z + n(N - n)} \tag{16.19}$$

The probability generating function[3] $\varphi_{W_N}(z) = E\left[e^{-zW_N}\right]$ of the weight W_N of the shortest path follows from the law of total probability (2.49) as

$$E\left[e^{-zW_N}\right] = \sum_{k=1}^{N-1} E\left[e^{-zW_N}|\text{node } k \text{ is the end node}\right] \Pr\left[\text{node } k \text{ is the end node}\right]$$

$$= \frac{1}{N-1} \sum_{k=1}^{N-1} E\left[e^{-zv_k}\right]$$

because any node apart from the root A but including the destination node B has equal probability to be the k-th attached node. Introducing (16.19) finally yields

$$\varphi_{W_N}(z) = \frac{1}{N-1} \sum_{k=1}^{N-1} \prod_{n=1}^{k} \frac{n(N - n)}{z + n(N - n)} \tag{16.20}$$

---

[3] If the link weights have mean $\frac{1}{a}$ (instead of 1), then W_N is multiplied by a as explained in Sections 16.2.1 and 3.4.1. The weight of the scaled shortest path $W_{N,a}$ has pgf

$$\varphi_{W_{N,a}}(z) = E\left[e^{-zaW_N}\right] = \varphi_{W_N}(az)$$

The mean weight is obtained from (2.45) as

$$E\left[W_N\right] = -\left.\frac{d\varphi_{W_N}(z)}{dz}\right|_{z=0} = -\frac{1}{N-1}\sum_{k=1}^{N-1}\left.\frac{d}{dz}\prod_{n=1}^{k}\frac{n(N-n)}{z+n(N-n)}\right|_{z=0}$$

Using the logarithmic derivative of the product,

$$\left.\frac{d}{dz}\prod_{n=1}^{k}\frac{n(N-n)}{z+n(N-n)}\right|_{z=0} = \left.\prod_{n=1}^{k}\frac{n(N-n)}{z+n(N-n)}\frac{d}{dz}\left(\sum_{n=1}^{k}\log\frac{n(N-n)}{z+n(N-n)}\right)\right|_{z=0}$$

$$= -\sum_{n=1}^{k}\frac{1}{n(N-n)}$$

gives[4]

$$(N-1)\,E\left[W_N\right] = \sum_{k=1}^{N-1}\sum_{n=1}^{k}\frac{1}{n(N-n)} = \sum_{n=1}^{N-1}\frac{1}{n(N-n)}\sum_{k=n}^{N-1}1 = \sum_{n=1}^{N-1}\frac{N-n}{n(N-n)}$$

The mean weight is

$$E\left[W_N\right] = \frac{1}{N-1}\sum_{n=1}^{N-1}\frac{1}{n} = \frac{\psi(N)+\gamma}{N-1} \tag{16.21}$$

For large N,

$$E\left[W_N\right] = \frac{\log N + \gamma}{N} + O\left(\frac{1}{N^2}\right)$$

Curiously, the probability that the shortest path consists of the direct link between source and destination is, with (16.15), (16.21) and $S_N^{(2)} = (-1)^N(N-1)!\sum_{k=1}^{N-1}\frac{1}{k}$,

$$\Pr[H_N = 1] = \frac{1}{N-1}\sum_{k=1}^{N-1}\frac{1}{k} = E\left[W_N\right]$$

Similarly, the variance is computed (see Problem (ii) in Section 16.10) as,

$$\mathrm{Var}\left[W_N\right] = \frac{3}{N(N-1)}\sum_{n=1}^{N-1}\frac{1}{n^2} - \frac{\left(\sum_{n=1}^{N-1}\frac{1}{n}\right)^2}{(N-1)^2 N} \tag{16.22}$$

---

[4] The definition (16.18) directly yields

$$E\left[W_N\right] = \sum_{k=1}^{N-1} E\left[v_k 1_{\{\text{node } k \text{ is the end node of the shortest path}\}}\right]$$

Since v_k and the choice of the end node are independent,

$$E\left[v_k 1_{\{\text{node } k \text{ is the end node}\}}\right] = E\left[v_k\right] E\left[1_{\{\text{node } k \text{ is the end node}\}}\right]$$

$$= E\left[v_k\right] \Pr\left[\text{node } k \text{ is the end node}\right]$$

$$= \frac{E\left[v_k\right]}{N-1} = \frac{1}{N-1}\sum_{n=1}^{k}\frac{1}{n(N-n)}$$

and for large N,

$$\text{Var}\left[W_N\right] = \frac{\pi^2}{2N^2} + O\left(\frac{\log^2 N}{N^3}\right)$$

By inverse Laplace transform of (16.20), the distribution $\Pr\left[W_N \leq t\right]$ can be computed. A more general analysis is presented in Section 19.6.1, where the weight of the shortest path from an arbitrary node to the nearest (in weight) of a group of m nodes is computed. The asymptotic distribution for the weight of the shortest path follows from (19.24) for $m = 1$ as

$$\lim_{N\to\infty} \Pr\left[NW_N - \log N \leq x\right] = e^{-x}e^{e^{-x}} \int_{e^{-x}}^{\infty} \frac{e^{-u}}{u^2}\,du = 1 - e^{-x}e^{e^{-x}} \int_{e^{-x}}^{\infty} \frac{e^{-u}}{u}\,du \tag{16.23}$$

An analysis related to Section 19.6.1 is presented in Section 16.6.1, where we study the flooding time. The interest of such an asymptotic analysis is that it often leads to tractable solutions that are physically more appealing to interpret. Moreover, it turns out that results for finite, not too small N are reasonably approximated by the asymptotic law.

Finally, as derived in Van Mieghem (2010b), the weight w^* of an arbitrary link in a URT consisting of N nodes possesses the pgf

$$E\left[e^{-zw^*}\right] = \frac{1}{N-1} \sum_{j=1}^{N-1} \frac{1}{j} \sum_{k=1}^{j} \prod_{n=k}^{j} \frac{n\left(N-n\right)}{z + n(N-n)} \tag{16.24}$$

from which the mean and variance can be deduced as

$$E\left[w^*\right] = \frac{1}{N} \sum_{n=1}^{N} \frac{1}{n^2}$$

$$\text{Var}\left[w^*\right] = \frac{4}{N(N+1)} \sum_{k=1}^{N} \frac{1}{k^3} - \frac{1}{N^2} \left(\sum_{n=1}^{N} \frac{1}{n^2}\right)^2$$

The weight w^* of an arbitrary link in an infinitely large URT possesses, for $|x| < 1$, the asymptotic pgf

$$\lim_{N\to\infty} E\left[e^{-Nxw^*}\right] = \frac{1}{(x+1)^2} + x \sum_{j=1}^{\infty} \frac{1}{(x+1+j)^2 (x+j)} \tag{16.25}$$

A generalization of the Riemann Zeta function is the Hurwitz Zeta function, defined for $\text{Re}(s) > 1$ and $\text{Re}(a) \geq 0$ as

$$\zeta(s, a) = \sum_{k=1}^{\infty} \frac{1}{(a+k)^s} \tag{16.26}$$

which shows that $\zeta(s, 0) = \zeta(s)$. The asymptotic pgf can be written in terms of the Hurwitz Zeta function as

$$\lim_{N\to\infty} E\left[e^{-Nxw^*}\right] = 1 - x\zeta(2, x) \tag{16.27}$$

from which all moments are deduced as

$$\lim_{N\to\infty} E\left[(Nw^*)^k\right] = kk!\zeta(k+1)$$

The corresponding asymptotic pdf, for $t \geq 0$, is

$$\lim_{N\to\infty} f_{Nw^*}(t) = \frac{(t-1+e^{-t})\,e^{-t}}{(1-e^{-t})^2} \qquad (16.28)$$

Items (e.g. packets, information) in real-world networks are most often transported along shortest paths. Inferring the whole network topology (see e.g. Van Mieghem and Wang (2009); Wang and Van Mieghem (2010)) via measurements based on transport is inherently biased, because mainly shortest path links are observed. The distribution (16.28) of an arbitrary – in the sense of uniformly chosen at random – link weight w^* in a shortest path tree is significantly different than that of an arbitrary link in the complete graph, that has an exponential distribution with mean 1.

16.5 Joint distribution of the hopcount and the weight

Since W_N equals the sum of the link weights of the shortest path from the root to an arbitrary node and since $H_N = h_N | h_N > 0$ is the number of links in that shortest path (where the arbitrary destination node is different from the root), one may wonder whether there is a relation between them.

Theorem 16.5.1 *The joint probability generating function $E[s^{H_N}e^{-tW_N}]$ in the URT is*

$$\varphi_{H_N W_N}(s,t) = E[s^{H_N}e^{-tW_N}] = \frac{1}{N-1}\sum_{k=1}^{N-1}\prod_{n=1}^{k}\frac{n(N-n)}{t+n(N-n)}\frac{\Gamma(k+s)}{k!\Gamma(s)} \qquad (16.29)$$

Proof: The hopcount and the weight of the shortest path between two random nodes is in distribution equal to the same quantities between node 1 and a randomly chosen node from the set $\{2, 3, \ldots, N\}$. We denote the label of this random node by Z, which consequently has a uniform distribution over the above-mentioned discrete set of size $N-1$. Conditioning on the end node Z thus gives by the law of total probability (2.49),

$$\varphi_{H_N W_N}(s,t) = \frac{1}{N-1}\sum_{k=1}^{N-1} E[s^{H_N}e^{-tW_N}|Z = k+1]$$

As shown in Section 16.4, the weight or discovery time of the k-th node (with label $k+1$) equals $v_k = \sum_{n=1}^{k}\tau_n$, where $\tau_1, \tau_2, \ldots, \tau_k$ are independent, exponentially distributed random variables with parameter $\lambda_n = n(N-n)$ with $1 \leq n \leq k$. The pgf $E\left[e^{-tv_k}\right]$ is given in (16.19). After the discovery of the k-th node, the number of hops h_{k+1} of this node to the root (with label 1) is determined by attaching this node independently to a URT of size k. After attachment of the k-th node, the size of the URT is $k+1$. Thus[5], $E\left[s^{h_{k+1}}\right] = E\left[s^{1+h_k}\right] = sE\left[s^{h_k}\right] =$

---

[5] Alternatively, we may repeat the argument in the proof of Theorem 16.3.1 based on the level sets and (16.7), such that

$$E\left[s^{h_{k+1}}\right] = \sum_{l=0}^{k-1} s^{l+1}\frac{E[X_n^{(l)}]}{k}$$

$s\frac{\Gamma(k+s)}{k!\Gamma(s+1)} = \frac{\Gamma(k+s)}{k!\Gamma(s)}$, where (16.6) is used. For a specific node with label $k+1$, both v_k and h_{k+1} are independent. From this construction, we find

$$E[s^{H_N}e^{-tW_N}|Z=k+1] = E\left[e^{-tv_k}\right]E\left[s^{h_{k+1}}\right]$$

$$= \prod_{n=1}^{k}\frac{n(N-n)}{t+n(N-n)}\frac{\Gamma(k+s)}{k!\Gamma(s)}, \quad k=1,2,\ldots,N-1,$$

and, hence, (16.29) follows. $\qquad\square$

Obviously, putting $s=1$ in (16.29) yields (16.20). On the other hand, for $t=0$ and using the identity, proved in (van der Hofstad *et al.*, 2002b, Lemma A.4),

$$\sum_{j=m}^{n}\frac{\Gamma(a+j)}{\Gamma(b+j)} = \frac{1}{(1+a-b)}\left(\frac{\Gamma(a+n+1)}{\Gamma(b+n)} - \frac{\Gamma(a+m)}{\Gamma(b+m-1)}\right) \tag{16.30}$$

we find that

$$\sum_{k=1}^{N-1}\frac{\Gamma(k+s)}{k!\Gamma(s)} = \frac{1}{\Gamma(s+1)}\left\{\frac{\Gamma(s+N)}{(N-1)!} - \Gamma(s+1)\right\}$$

and

$$\varphi_{H_N W_N}(s,0) = E[s^{H_N}] = \frac{1}{N-1}\left\{\frac{\Gamma(s+N)!}{\Gamma(s+1)(N-1)} - 1\right\}$$

which is, indeed, (16.16).

As shown in Hooghiemstra and Van Mieghem (2008), the expectation of the product $H_N W_N$ is

$$E[H_N W_N] = \frac{1}{N-1}\left(\left(\sum_{k=1}^{N-1}\frac{1}{k}\right)^2 - \sum_{k=1}^{N-1}\frac{1}{k} + \sum_{k=1}^{N-1}\frac{1}{k^2}\right) \tag{16.31}$$

For large N, we observe that $E[H_N W_N] = \frac{\log^2 N - \log N + O(1)}{N}$. The asymptotics of the correlation coefficient $\rho(W_N, H_N)$, derived in Hooghiemstra and Van Mieghem (2008), is

$$\rho(W_N, H_N) = \frac{\pi\sqrt{2}}{6\sqrt{\ln N}} + o\left((\ln N)^{-1}\right), \tag{16.32}$$

which clearly tends to zero for $N \to \infty$. Moreover, Hooghiemstra and Van Mieghem (2008) show that W_N and H_N are asymptotically independent, and this matches nicely with one of our earlier findings in Hooghiemstra and Van Mieghem (2001), that the hopcount and the end-to-end delay of an Internet path are seemingly uncorrelated.

Using the basic recursion (16.8) for the level sets in the URT leads to $E\left[s^{h_{k+1}}\right] = \frac{\Gamma(k+s)}{k!\Gamma(s)}$.

16.6 The flooding time T_N

The most commonly used process that informs each node (router) about changes in the network topology is called *flooding*: the source node initiates the flooding process by sending the packet with topology information to all adjacent neighbors and every router forwards the packet on all interfaces except for the incoming one and duplicate packets are discarded. Flooding is particularly simple and robust since it progresses, in fact, along all possible paths from the emitting node to the receiving node. Hence, a flooded packet reaches a node in the network in the shortest possible time (if overheads in routers are ignored). Therefore, an interesting problem lies in the determination of the flooding time T_N, which is the minimum time needed to inform all nodes in a network with N nodes. Only after a time T_N, all topology databases at each router in the network are again synchronized, i.e. all routers possess the same topology information. The flooding time T_N is defined as the minimum time needed to reach all $N - 1$ remaining nodes from a source node over their respective shortest paths.

We will here consider the flooding time T_N in the complete graph containing N nodes and with independent, exponentially distributed link weights with mean 1. The generalization to the random graph $G_p(N)$ with i.i.d. exponential (or uniform[6]) distributed link weight is treated in van der Hofstad *et al.* (2002a).

The flooding time T_N equals the absorption time, starting from state $n = 1$ of the birth-process with rates (16.2). The probability generating function follows directly from (16.19) with $k = N - 1$,

$$\varphi_{T_N}(x) = E[e^{-xT_N}] = \int_0^\infty e^{-xt} f_{T_N}(t)\, dt = \prod_{n=1}^{N-1} \frac{n(N-n)}{n(N-n)+x} \qquad (16.33)$$

The mean flooding time equals

$$E[T_N] = \sum_{n=1}^{N-1} E[\tau_n] = \sum_{n=1}^{N-1} \frac{1}{n(N-n)} = \frac{2}{N} \sum_{n=1}^{N-1} \frac{1}{n} = \frac{2}{N}(\psi(N)+\gamma) \qquad (16.34)$$

Using the asymptotic expansion (Abramowitz and Stegun, 1968, Section 6.3.18) of the digamma function, we conclude that

$$E[T_N] \sim \frac{2 \log N}{N}$$

which demonstrates that the mean flooding time in the complete graph with exponential link weights with mean 1 decreases to zero when $N \to \infty$. Also, the mean flooding time is about twice as long as the mean weight of an arbitrary shortest

---

[6] Both the exponential and uniform distribution are regular distributions with extreme value index $\alpha = 1$. This means that the small link weights that are most likely included in the shortest path are almost identically distributed for all regular distributions with same $f_w(0)$.

path (16.21). The variance of T_N equals

$$\text{Var}[T_N] = \sum_{n=1}^{N-1} \text{Var}\,[\tau_n] = \sum_{n=1}^{N-1} \frac{1}{n^2(N-n)^2} = \frac{2}{N^2} \sum_{n=1}^{N-1} \frac{1}{n^2} + \frac{4}{N^3} \sum_{n=1}^{N-1} \frac{1}{n} \quad (16.35)$$

For large N, we have that $\text{Var}[T_N] = \frac{\pi^2}{3N^2} + O\left(\frac{\log N}{N^3}\right)$.

16.6.1 The asymptotic law for flooding time T_N

The exact expression $f_{T_N}(t)$ for probability density function of the flooding time T_N derived in van der Hofstad *et al.* (2002a), does not provide much insight. Because we are interested in the flooding time in *large* networks, we investigate the asymptotic distribution of T_N, for N large. We rewrite (16.33) as

$$\varphi_{T_N}(x) = \frac{[(N-1)!]^2}{\prod_{n=1}^{N-1}\left[x + \frac{N^2}{4} - \left(n - \frac{N}{2}\right)^2\right]} \quad (16.36)$$

For $N = 2M$, using $\frac{\Gamma(z+m)}{\Gamma(z+1)} = \prod_{n=1}^{m-1}(n+z)$, we deduce that

$$\varphi_{T_{2M}}(x) = \left(\frac{\Gamma(2M)\Gamma(1 + \sqrt{x + M^2} - M)}{\Gamma(M + \sqrt{x + M^2})}\right)^2 \quad (16.37)$$

For large M, there holds $\sqrt{x + M^2} \sim M + \frac{x}{2M}$, provided $|x| < 2M$. After substitution of $x = 2My$ in (16.37), with $|y| < 1$, we obtain

$$\varphi_{T_{2M}}(2My) \sim \Gamma^2(1+y)\frac{\Gamma^2(2M)}{\Gamma^2(2M+y)} \sim \Gamma^2(1+y)(2M)^{-2y}$$

from which follows the asymptotic relation

$$\lim_{N\to\infty} N^{2y}\varphi_{T_N}(Ny) = \Gamma^2(1+y), \qquad |y| < 1 \quad (16.38)$$

Equivalently, we have for $|y| < 1$,

$$\lim_{N\to\infty} E[e^{-y(NT_N - 2\log N)}] = \lim_{N\to\infty} \frac{1}{N}\int_{-\infty}^{\infty} e^{-yt} f_{T_N}\left(\frac{t + 2\log N}{N}\right) dt$$
$$= \Gamma^2(1+y)$$

This limit demonstrates that the probability distribution function of the random variable $NT_N - 2\log N$ converges to a probability distribution with Laplace transform $\Gamma^2(1+y)$. Let us define the normalized density function

$$g_N(t) = \frac{1}{N} f_{T_N}\left(\frac{t + 2\log N}{N}\right) \quad (16.39)$$

We can prove convergence in *density*, i.e. $\lim_{N\to\infty} g_N(t) = g(t)$ and that the latter exists. By the inversion theorem for Laplace transforms we obtain for $t \in \mathbb{R}$,

$$\lim_{N\to\infty} g_N(t) = \lim_{N\to\infty} \frac{1}{2\pi i}\int_{c-i\infty}^{c+i\infty} e^{yt} N^{2y}\varphi_{T_N}(Ny)dy$$

where $0 < c < 1$. Since $\Gamma(z)$ is analytic over the entire complex plane except for simple poles at the points $z = -n$ for $n = 0, 1, 2, ...$, we find that $N^{2y}\varphi_{T_N}(Ny)$ is analytic whenever the real part of y is non-negative. Evaluation along the line $\mathrm{Re}(y) = c = 0$ then gives

$$\lim_{N\to\infty} g_N(t) = \lim_{N\to\infty} \frac{1}{2\pi} \int_{-\infty}^{\infty} e^{itu} N^{2iu} \varphi_{T_N}(iNu)du$$

As dominating function we take

$$\left|e^{itu} N^{2iu} \varphi_{T_N}(iNu)\right| = \left|\varphi_{T_N}(iNu)\right| \leq \frac{1+u^2}{u^4}$$

when $|u| > 1$, and $|\varphi_{T_N}(iNu)| \leq 1$, for $|u| \leq 1$. This follows from the first equality in (16.36), using only the factors in the product with $n = 1$ and $n = N - 1$, and bounding the other factors using

$$\frac{n(N-n)}{|n(N-n) + iNu|} \leq 1$$

The Dominated Convergence Theorem 6.1.4 allows us to interchange the limit and integration operators such that

$$\lim_{N\to\infty} g_N(t) = \frac{1}{2\pi} \int_{-\infty}^{\infty} e^{itu} \lim_{N\to\infty} N^{2iu} \varphi_{T_N}(iNu)du = \frac{1}{2\pi i} \int_{-i\infty}^{i\infty} e^{ty} \lim_{N\to\infty} N^{2y}\varphi_{T_N}(Ny)dy$$

$$= \frac{1}{2\pi i} \int_{-i\infty}^{i\infty} e^{ty}\Gamma^2(1+y)dy \qquad (16.40)$$

The right-hand side of (16.38) is a perfect square, which indicates that the limit distribution is a two-fold convolution. It is a special case of a product (2.69) of Mellin transforms. Now, the Mellin transform (Titchmarsh, 1948) of the exponential function is

$$e^{-t} = \frac{1}{2\pi i} \int_{c-i\infty}^{c+i\infty} t^{-y}\Gamma(y)\,dy, \qquad c > 0 \qquad (16.41)$$

and thus with $t = e^{-u}$,

$$\frac{d}{du}\left(e^{-e^{-u}}\right) = \frac{1}{2\pi i} \int_{c-i\infty}^{c+i\infty} e^{yu}\Gamma(y+1)\,dy$$

which shows that (16.40) is the two-fold convolution of the probability density function $\frac{d}{dt}\Lambda(t)$, where $\Lambda(t) = e^{-e^{-t}}$ is the Gumbel distribution (3.44). Furthermore, the two-fold convolution is given by

$$\frac{d}{dt}\left(\Lambda^{(2*)}(t)\right) = e^{-t} \int_{-\infty}^{\infty} e^{-e^{-u}} e^{-e^{-(t-u)}}\,du$$

$$= e^{-t} \int_{-\infty}^{\infty} \exp\left[-2e^{-t/2}\cosh\left(\frac{t}{2} - u\right)\right]\,du$$

$$= 2e^{-t} \int_0^{\infty} \exp\left[-2e^{-t/2}\cosh(u)\right]\,du = 2e^{-t}K_0\left(2e^{-t/2}\right)$$

where $K_\nu(x)$ denotes the modified Bessel function (Abramowitz and Stegun, 1968, Section 9.6) of order ν.

In summary,

$$\lim_{N\to\infty} g_N(t) = g(t) = \frac{d}{dt}\left(\Lambda^{(2*)}(t)\right) = 2e^{-t}K_0\left(2e^{-t/2}\right) \qquad (16.42)$$

and the corresponding distribution function is

$$\lim_{N\to\infty} \Pr[NT_N - 2\log N \le z] = 2\int_{-\infty}^{z} e^{-t}K_0(2e^{-t/2})dt = 2e^{-z/2}K_1\left(2e^{-z/2}\right) \qquad (16.43)$$

The right-hand side of (16.42) is maximal for $t = 0.506\ 357$, which is slightly smaller than $\gamma = 0.577\ 216$, but still in accordance with $E[T_N]$ given by (16.34). The asymmetry shows that $\{NT_N \ge 2\log N + z\}$ is much more likely than the

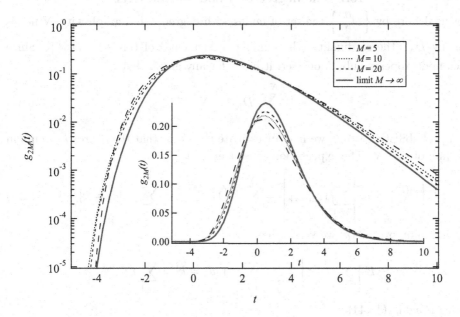

Fig. 16.5. The scaled density $g_N(t)$ for three values of $N = 2M$ (dotted lines) and the asymptotic result (full line) on a log-lin scale. The insert is drawn on a lin-lin scale.

event $\{NT_N \le 2\log N - z\}$, which confirms the intuition that the flooding time can be much longer than the mean $E[T_N]$, but not so much shorter than $E[T_N]$. Figure 16.5 illustrates the convergence of $g_N(t)$ to the limit in (16.42).

Consider the scaled pgf of the weight of the shortest path in (19.23) with $m = 1$, which equals

$$\lim_{N\to\infty} E\left[e^{-y(NW_N - \log N)}\right] = \frac{\pi y}{\sin \pi y}\Gamma(1 + y)$$

and which shows that $NW_N - \log N$ is the sum of a pure Gummel random variable, whose pgf is $\Gamma(1+y)$ and mean γ and a Fermi-Dirac random variable with mean zero (see p. 536) that causes a symmetric broadening similar to the addition of noise.

When comparing (16.38) with the above scaled random variable $NW_N - \log N$, we observe that, for large N, the random variable $NT_N - 2\log N$ is smaller than the sum of $NW_{N;1} - \log N + NW_{N;2} - \log N$, where both $NW_{N;j} - \log N$ are i.i.d. random variables. Intuitively, we can say that the flooding time is smaller than the time to travel from a left-hand corner of the graph to the center and from the center to a right-hand corner of the graph.

The asymptotic distribution (16.43) is a beautiful example of a sum of N independent random variables that clearly does not converge to a Gaussian and, hence, does not obey the (extended) Central Limit Theorem 6.3.1.

16.7 The degree of a node in the URT

Let us denote by $\left\{ D_N^{(k)} \right\}$ the set of nodes with degree k in a graph with N nodes and by $D_N^{(k)}$ the cardinality (the number of elements) of this set $\left\{ D_N^{(k)} \right\}$. Since each node appears only in one set, it holds for any graph that

$$\sum_{k=1}^{N-1} D_N^{(k)} = N \tag{16.44}$$

In a probabilistic setting, we may investigate the event that the degree k occurs in a graph of size N. The expectation of that event is

$$E\left[D_N^{(k)} \right] = E\left[\sum_{j=1}^{N} 1_{\{d_j = k\}} \right] = \sum_{j=1}^{N} E\left[1_{\{d_j = k\}} \right] = \sum_{j=1}^{N} \Pr\left[d_j = k \right] \tag{16.45}$$

By summing over all k, we verify that

$$E\left[\sum_{k=1}^{N-1} D_N^{(k)} \right] = \sum_{j=1}^{N} \sum_{k=1}^{N-1} \Pr\left[d_j = k \right] = \sum_{j=1}^{N} 1 = N$$

which is again (16.44).

16.7.1 The degree of the root of a URT

In the complete graph K_N with i.i.d. exponential link weights, any node n possesses equal properties in probability because of symmetry. If we denote by d_n the degree of node n in the shortest path tree rooted at that node n, the symmetry implies that $\Pr\left[d_n = k \right] = \Pr\left[d_i = k \right]$ for any node n and i. In fact, we consider here the degree of a URT as an overlay tree in a complete graph. Concentrating on a node with label n, we obtain from (16.45)

$$E\left[D_N^{(k)} \right] = N \Pr\left[d_n = k \right] = N \Pr\left[X_N^{(1)} = k \right]$$

The latter follows from the fact that the degree of a node is equal to the number of its direct neighbors, the nodes at level 1, $X_N^{(1)}$. By definition of the URT, the second node surely belongs to the level set 1, while node 3 has equal probability to be attached to the root or to node 2. In general, when attaching a node j to a URT of size $j - 1$, the probability that node j is attached to the root equals $\frac{1}{j-1}$. Thus, the number of nodes at level 1 in the URT (constructed upon the complete graph) is in distribution equal to the sum of $N - 1$ independent Bernoulli random variables each with different mean $\frac{1}{j-1}$,

$$X_N^{(1)} \overset{d}{=} \sum_{j=2}^{N} \text{Bernoulli}\left(\frac{1}{j-1}\right) = \sum_{j=1}^{N-1} \text{Bernoulli}\left(\frac{1}{j}\right)$$

because each node in the complete graph is connected to $N - 1$ neighbors. The generating function is

$$E\left[z^{X_N^{(1)}}\right] = E\left[z^{\sum_{j=1}^{N-1} \text{Bernoulli}\left(\frac{1}{j}\right)}\right] = \prod_{j=1}^{N-1} E\left[z^{\text{Bernoulli}\left(\frac{1}{j}\right)}\right]$$

Using the probability generating function (3.1) of a Bernoulli random variable, $E\left[z^{\text{Bernoulli}\left(\frac{1}{j}\right)}\right] = 1 - \frac{1}{j} + \frac{z}{j}$, yields

$$E\left[z^{X_N^{(1)}}\right] = \prod_{j=1}^{N-1} \left(\frac{z + j - 1}{j}\right) = \frac{\Gamma(z + N - 1)}{\Gamma(z)\Gamma(N)}$$

Compared to the generating function (16.6) of the hopcount h_N, we recognize that

$$E\left[z^{X_N^{(1)}}\right] = z\varphi_{N-1}(z) = \sum_{k=0}^{N-2} \Pr[h_{N-1} = k]z^{k+1} = \sum_{k=1}^{N-1} \Pr[h_{N-1} = k - 1]z^k$$

from which we deduce, for $1 \leq k \leq N - 1$,

$$\Pr\left[X_N^{(1)} = k\right] = \Pr[h_{N-1} = k - 1]$$

Using (16.7), we arrive at the curious result

$$\Pr\left[X_N^{(1)} = k\right] = \frac{E\left[X_{N-1}^{(k-1)}\right]}{N - 1} \qquad \text{for } 1 \leq k \leq N - 1.$$

The probability that the number of level 1 nodes in the shortest path tree in the complete graph with i.i.d. exponential link weights is k equals the mean number of nodes on level $k - 1$ in a URT of size $N - 1$ divided by that size $N - 1$. In other words, the "horizontal" distribution at level 1 is related to the "vertical" distribution of the size of the level sets. In summary[7], in the complete graph with i.i.d. exponential link weights, the probability that an arbitrary node n as root of a shortest path tree has degree k is

$$\Pr\left[d_n = k\right] = \Pr[h_{N-1} = k - 1] = \frac{(-1)^{N-1-k} S_{N-1}^{(k)}}{(N - 1)!} \tag{16.46}$$

---

[7] This result is due to Remco van der Hofstad (private communication).

The degree of an arbitrary node in the union of all shortest paths trees in the complete graph K_N with i.i.d. exponential link weights is also given by (16.46) because in that union each node n is once a root and further plays, by symmetry, the role of the j-th attached node in the URT rooted at any other node in K_N.

16.7.2 The degree of an arbitrary node in a URT

An arbitrary node can be either the root of the URT or the k-th ($2 \leq k \leq N$) discovered node in the Markov discovering process. The degree of the k-th discovered node, which is denoted by $d_{(k)}$ is, as shown in Section 16.7.1, the sum of independent Bernoulli random variables

$$d_{(1)} = \sum_{j=1}^{N-1} \text{Bernoulli}\left(\frac{1}{j}\right)$$

$$d_{(k)} = 1 + \sum_{j=k}^{N-1} \text{Bernoulli}\left(\frac{1}{j}\right), \quad 2 \leq k \leq N$$

The existence of the k-th discovered node with $k > 1$ implies that it has surely one and only one parent node, accounting for the in-degree of 1. An arbitrarily chosen node n in the SPT – including the root – can be the k-th attached node in the URT with equal probability. Similar to $\varphi_{W_N}(z)$, the generating function $\varphi_{D_{\mathrm{URT}}}(z) = E[z^{D_{\mathrm{URT}}}]$ of the degree D_{URT} of an arbitrary node n in the URT follows from the law of total probability (2.49) as

$$\varphi_{D_{\mathrm{URT}}}(z) = \sum_{k=1}^{N} E[z^{d_{(k)}}] \Pr\left[n \text{ is the } k\text{-th attached node in URT}\right]$$

$$= \frac{1}{N} \prod_{j=1}^{N-1}\left(\frac{z+j-1}{j}\right) + \frac{z}{N}\sum_{k=2}^{N}\prod_{j=k}^{N-1}\left(\frac{z+j-1}{j}\right)$$

$$= \frac{\Gamma(z+N-1)}{N!}\left(\frac{1}{\Gamma(z)} + z\sum_{k=2}^{N}\frac{(k-1)!}{\Gamma(z+k-1)}\right)$$

where $E\left[z^{\sum_{j=k}^{N-1}\text{Bernoulli}\left(\frac{1}{j}\right)}\right] = \prod_{j=k}^{N-1}\left(\frac{z+j-1}{j}\right) = \frac{(k-1)!\,\Gamma(z+N-1)}{(N-1)!\,\Gamma(z+k-1)}$ has been used. Using the identity (16.30), the k-sum simplifies to

$$\sum_{k=2}^{N}\frac{\Gamma(k)}{\Gamma(z+k-1)} = \frac{1}{(2-z)}\left(\frac{\Gamma(1+N)}{\Gamma(z+N-1)} - \frac{1}{\Gamma(z)}\right)$$

such that we arrive at

$$\varphi_{D_{\mathrm{URT}}}(z) = \frac{2\Gamma(z+N-1)}{N!\,\Gamma(z)}\frac{(z-1)}{(z-2)} - \frac{z}{(z-2)}$$

$$= \frac{2}{N!}\frac{\prod_{j=0}^{N-1}(z+j-1)}{(z-2)} - \frac{z}{(z-2)} \tag{16.47}$$

We expand $\varphi_{D_{\text{URT}}}(z)$ in a Taylor series around $z = 0$ by using,

$$\frac{\Gamma(z+N)}{\Gamma(z)} = \sum_{k=0}^{N-1} (-1)^{N-1-k} S_N^{(k+1)} z^{k+1} \tag{16.48}$$

that follows from (16.6) and (16.9) and $\frac{(z-1)}{(z-2)} = 1 + \frac{1}{z-2} = 1 - \frac{1}{2}\sum_{k=0}^{\infty}\frac{z^k}{2^k}$, and obtain

$$\varphi_{D_{\text{URT}}}(z) = \frac{2}{N!}\sum_{k=0}^{N-2}(-1)^{N-k}S_{N-1}^{(k+1)}z^{k+1}\left(1 - \frac{1}{2}\sum_{k=0}^{\infty}\frac{z^k}{2^k}\right) + \sum_{k=0}^{\infty}\frac{z^{k+1}}{2^{k+1}}$$

We use the fact that $S_N^{(k)} = 0$ if $k > N$ to apply the Cauchy product rule for power series in

$$\sum_{k=0}^{N-2}(-1)^{N-k}S_{N-1}^{(k+1)}z^{k+1}\sum_{k=0}^{\infty}\frac{z^k}{2^k} = \sum_{k=0}^{\infty}(-1)^{N-k}S_{N-1}^{(k+1)}z^{k+1}\sum_{k=0}^{\infty}\frac{z^k}{2^k}$$

$$= (-1)^N\sum_{j=0}^{\infty}\left(\sum_{k=0}^{j}S_{N-1}^{(k+1)}(-2)^k\right)\frac{z^{j+1}}{2^j}$$

such that

$$\varphi_{D_{\text{URT}}}(z) = \sum_{j=1}^{N-1}\left(\frac{1}{2^j} + \frac{2(-1)^{N-j-1}S_{N-1}^{(j)}}{N!} - \frac{(-1)^N}{2^{j-1}N!}\sum_{k=0}^{j-1}S_{N-1}^{(k+1)}(-2)^k\right)z^j$$

$$+ \sum_{j=N}^{\infty}\left(1 - \frac{2(-1)^N}{N!}\sum_{k=0}^{j-1}S_{N-1}^{(k+1)}(-2)^k\right)\frac{z^j}{2^j}$$

Since $S_N^{(k)} = 0$ if $k > N$, the coefficients in the last sum equal, using (16.48),

$$1 - \frac{2(-1)^N}{N!}\sum_{k=0}^{j-1}S_{N-1}^{(k+1)}(-2)^k = 1 - \frac{1}{N!}\sum_{k=0}^{N-2}(-1)^{N-2-k}S_{N-1}^{(k+1)}2^{k+1}$$

$$= 1 - \frac{1}{N!}\frac{\Gamma(1+N)}{\Gamma(2)} = 0$$

which is consistent with the fact that $\Pr[D_{\text{URT}} = j] = 0$ for $j > N - 1$. Finally, from (2.18), the probability that an arbitrary node in the URT has degree equal to j is

$$\Pr[D_{\text{URT}} = j] = \frac{1}{2^j} + \frac{2(-1)^{N-j-1}S_{N-1}^{(j)}}{N!} + \frac{(-1)^N}{2^j N!}\sum_{k=1}^{j}S_{N-1}^{(k)}(-2)^k$$

or

$$\Pr[D_{\text{URT}} = j] = \frac{1}{2^j} + \frac{(-1)^{N-j-1}S_{N-1}^{(j)}}{N!} + \frac{(-1)^N}{2^j N!}\sum_{k=1}^{j-1}S_{N-1}^{(k)}(-2)^k \tag{16.49}$$

For large N and using the asymptotics of the Stirling numbers $S_N^{(k)}$ of the first kind (Abramowitz and Stegun, 1968, Section 24.1.3.III), the asymptotic law is

$$\Pr\left[D_{\mathrm{URT}} = j\right] = \frac{E\left[D_N^{(j)}\right]}{N} = \frac{1}{2^j} + O\left(\frac{\log^{j-1} N}{N^2}\right) \tag{16.50}$$

The ratio of the mean number of nodes with degree j over the total number of nodes, which equals the probability that an arbitrary node in a URT of size N has degree j, decreases exponentially fast with rate $\ln 2$.

We may verify from (16.47) that $E[D_{\mathrm{URT}}] = 2\left(1 - \frac{1}{N}\right)$, which, of course, follows from the basic law (15.4) that directly gives the mean degree as $E[D] = \frac{2L}{N}$ where any tree has $L = N - 1$ links. The variance is computed from (2.27), and $\varphi''_{D_{\mathrm{URT}}}(1)$ elegantly follows from Cauchy's integral as

$$\varphi''_{D_{\mathrm{URT}}}(1) = \frac{2!}{2\pi i} \int_{C(1)} \frac{\varphi_{D_{\mathrm{URT}}}(z)}{(z-1)^3} dz$$

$$= \frac{4}{N!} \frac{1}{2\pi i} \int_{C(1)} \frac{\prod_{j=1}^{N-1}(z+j-1)}{(z-1)^2(z-2)} dz - \frac{4}{(z-2)^3}\bigg|_{z=1}$$

$$= \frac{4}{N!} \frac{d}{dz} \frac{\prod_{j=1}^{N-1}(z+j-1)}{(z-2)}\bigg|_{z=1} + 4$$

Using the logarithmic derivative, $f'(x) = f(x)\frac{d\log f(x)}{dx}$, yields

$$\frac{d}{dz} \frac{\prod_{j=1}^{N-1}(z+j-1)}{(z-2)}\bigg|_{z=1} = \frac{\prod_{j=1}^{N-1}(z+j-1)}{(z-2)}\left(\sum_{j=1}^{N-1} \frac{1}{z+j-1} - \frac{1}{z-2}\right)\bigg|_{z=1}$$

$$= -(N-1)!\left(\sum_{j=1}^{N-1} \frac{1}{j} + 1\right)$$

Thus, $\varphi''_{D_{\mathrm{URT}}}(1) = 4\left(1 - \frac{1}{N}\right) - \frac{4}{N}\sum_{j=1}^{N-1}\frac{1}{j}$ and the variance is

$$\mathrm{Var}\left[D_{\mathrm{URT}}\right] = 2 + \frac{2}{N} - \frac{4}{N^2} - \frac{4}{N}\sum_{j=1}^{N-1}\frac{1}{j} \tag{16.51}$$

which increases from 0 at $N = 2$ to $\mathrm{Var}[D_{\mathrm{URT}}] = 2$ for $N \to \infty$.

The law (16.50) is observed in Fig. 16.6, which plots the histogram of the degree D_U in the graph G_U. The graph G_U is obtained from the union of trace-routes from each RIPE measurement box to any other box positioned mainly in the European part of the Internet. For about 50 measurement boxes in 2003, the correspondence is striking because the slope of the fit on a log-lin scale equals -0.668 while the law (16.50) gives $-\ln 2 = -0.693$. Ignoring in Fig. 16.6 the leave nodes with $k = 1$ suggests that the graph G_U is URT-like. For 72 measurement boxes in 2004 which obviously results in a larger graph G_U, deviations from the URT law

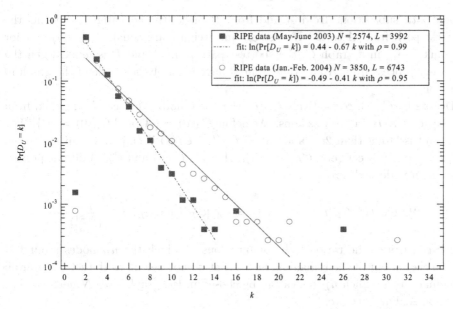

Fig. 16.6. The histogram of the degree D_U derived from the graph G_U formed by the union of paths measured via trace-route in the Internet. Both measurements in 2003 and 2004 are fitted on a log-lin plot and the correlation coefficient ρ quantifies the quality of the fit.

(16.50) are observed. If measurements between a larger full mesh of boxes were possible and if the measurement boxes were more homogeneously spread over the Internet, a power-law behavior is likely to be expected as mentioned in Section 15.3. However, these earlier reported trace-route measurements that lead to power-law degrees have been performed from a relatively small number of sources to a large number of destinations. These results question the observability of the Internet: how accurate are Internet properties such as hopcount and degree that are derived from incomplete measurements, i.e. from a selected small subset of nodes at which measurement boxes are placed?

16.8 The hopcount in a large, sparse graph

In this section, we investigate the hopcount of the shortest path in a sparse, but connected graph. Most real-world networks are large and relatively sparse (i.e. with low mean degree). First, we consider the case where all links have unit weight. Section 16.8.3 generalizes the method to i.i.d. link weight structures.

16.8.1 Bi-directional search

The basic idea of a bi-directional search to find the shortest path is by starting the discovery process (e.g. using Dijkstra's algorithm) from A and B simultaneously.

When both subsections from A and from B meet, the concatenation forms the shortest path from A to B. In case all link weights are equal, $w(i \to j) = 1$ for any link $i \to j$ in a graph G, the shortest path from A and B is found when the discovery process from A and that from B have precisely one node of the graph in common.

Denote by $C_A(l)$, respectively $C_B(l)$, the set of nodes that can be reached from A, respectively B, in l or less hops. We define $C_A(0) = \{A\}$ and $C_B(0) = \{B\}$. The hopcount is larger than $2l$ if and only if $C_A(l) \cap C_B(l)$ is empty. Conditionally on $|C_A(l)| = n_A$, respectively $|C_B(l)| = n_B$, the sets $C_A(l)$ and $C_B(l)$ do not possess a common node with probability

$$\Pr\left[C_A(l) \cap C_B(l) = \varnothing \,\big|\, |C_A(l)| = n_A, |C_B(l)| = n_B\right] = \frac{\binom{N-1-n_A}{n_B}}{\binom{N-1}{n_B}}$$

which consists of the ratio of all combinations in which the n_B nodes around B can be chosen out of the remaining nodes that do not belong to the set C_A over all combinations in which n_B nodes can be chosen in the graph with N nodes except for node A. Furthermore,

$$\frac{\binom{N-1-n_A}{n_B}}{\binom{N-1}{n_B}} = \frac{(N-n_A-1)(N-n_A-2)\cdots(N-n_A-n_B)}{(N-1)(N-2)\cdots(N-n_B)}$$

$$= \frac{(1 - \frac{n_A+1}{N})(1 - \frac{n_A+2}{N})\cdots(1 - \frac{n_A+n_B}{N})}{(1 - \frac{1}{N})(1 - \frac{2}{N})\cdots(1 - \frac{n_B}{N})}$$

For large N, we apply the Taylor series around $x = 0$ of $\log(1-x) = -x - \sum_{j=2}^{\infty} \frac{x^j}{j}$,

$$\log \frac{\binom{N-1-n_A}{n_B}}{\binom{N-1}{n_B}} = \sum_{k=1}^{n_B} \log\left(1 - \frac{n_A+k}{N}\right) - \log\left(1 - \frac{k}{N}\right)$$

$$= -\sum_{k=1}^{n_B}\left(\frac{n_A+k}{N} - \frac{k}{N}\right) - \sum_{j=2}^{\infty} \frac{1}{jN^j}\left(\sum_{k=1}^{n_B}(n_A+k)^j - k^j\right)$$

$$= -\frac{n_A n_B}{N} - \left(\frac{n_A n_B}{N}\right)^2 \left(\frac{1}{2n_A} + \frac{1}{2n_B} + \frac{1}{2n_A n_B}\right) - R$$

where the remainder is

$$R = \sum_{j=3}^{\infty} \frac{1}{jN^j} \sum_{m=0}^{j-1} \binom{j}{m} n_A^{j-m} \sum_{k=1}^{n_B} k^m = O\left(\left(\frac{n_A n_B}{N}\right)^3\right)$$

After exponentiation

$$\Pr\left[H_N > 2l \,\big|\, |C_A(l)| = n_A, |C_B(l)| = n_B\right] = e^{-\frac{n_A n_B}{N}}\left(1 + O\left(\left(\frac{n_A n_B}{N}\right)^2\right)\right)$$

By the law of total probability (2.50) and up to $O\left(\frac{E\left[|C_A(l)|^2 |C_B(l)|^2\right]}{N^2}\right)$ for large N, we obtain

$$\Pr\left[H_N > 2l\right] \approx E\left[\exp\left(-\frac{|C_A(l)|\,|C_B(l)|}{N}\right)\right] \tag{16.52}$$

This probability (16.52) holds for any large graph with a unit link weight structure provided $E\left[|C_A(l)|^2\,|C_B(l)|^2\right] = o(N^2)$. Formula (16.52) becomes increasingly accurate for decreasing $|C_A(l)|$ and $|C_B(l)|$, and so for sparser large graphs.

When $n_A n_B = O(N)$ or $\frac{n_A n_B}{N} = O(1)$, the approximation (16.52) is no longer valid and the above Taylor expansion requires all terms to be considered, indicating that $-\log\frac{\binom{N-1-n_A}{n_B}}{\binom{N-1}{n_B}}$ grows large. Hence, when $n_A n_B = O(N)$,

$$\Pr\left[C_A(l) \cap C_B(l) = \varnothing \,||\, C_A(l)| = n_A, |C_B(l)| = n_B\right] \to 0$$

meaning that two arbitrary clusters both of size $O\left(\sqrt{N}\right)$ have at least one node in common, with high probability.

16.8.2 Sparse large graphs and a branching process

In order to proceed, the number of nodes in the sets $C_A(l)$ and $C_B(l)$ needs to be determined, which is difficult in general. Therefore, we concentrate here on a special class of graphs in which the discovery process from A and B is reasonably well modeled by a branching process (Chapter 12). A branching process evolves from a given set $C_A(l-1)$ in the next l-th discovery cycle (or generation) to the set $C_A(l)$ by including only new nodes, not those previously discovered. The application of a branching process implies that the newly discovered nodes do not possess links to any previously discovered node of $C_A(l-1)$ except for its parent node in $C_A(l-1)$. Hence, only for large and sparse graphs or tree-like graphs, this assumption can be justified, provided that the number of links that point backwards to early discovered nodes in $C_A(l-1)$ is negligibly small.

Assuming that a branching process models the discovery process well, we will compute the number of nodes that can be reached from A and similarly from B in l hops from a branching process with production Y specified by the degree distribution of the nodes in the graph. The additional number of nodes X_l discovered during the l-th cycle of a branching process that are included in the set $C_A(l)$ is described by the basic law (12.1). Thus, $|C_A(l)| = \sum_{k=0}^{l} X_k$ with $X_0 = 1$ (namely node A). In terms of the scaled random variable $W_k = \frac{X_k}{\mu^k}$ with unit mean $E[W_k] = 1$,

$$|C_A(l)| = \sum_{k=0}^{l} W_k \mu^k$$

and where $\mu = E[Y] - 1 > 1$ denotes the mean degree minus 1, i.e. the outdegree, in the graph. Only the root has $E[Y]$ equal to the mean degree. Immediately, the mean size of the set of nodes reached from A in l hops is with $E[W_k] = 1$,

$$E[|C_A(l)|] = \sum_{k=0}^{l} \mu^k = \frac{\mu^{l+1} - 1}{\mu - 1}$$

which equally holds for $E[|C_B(l)|]$.

Applying Jensen's inequality (5.7) to (16.52) yields

$$\exp\left(-\frac{E[C_A(l)]\,E[C_B(l)]}{N}\right) \le E\left[\exp\left(-\frac{C_A(l)C_B(l)}{N}\right)\right]$$

such that

$$\Pr[H_N > k] \ge \exp\left(-\frac{\mu^2}{N(\mu - 1)^2}\mu^k\right)$$

With the tail probability expression (2.37) for the mean, we arrive at the lower bound for the expected hopcount in large graphs,

$$E[H_N] = \sum_{k=0}^{\infty} \Pr[H_N > k] \ge \sum_{k=0}^{\infty} \exp\left(-\frac{\mu^2}{N(\mu - 1)^2}\mu^k\right)$$

The sum $S_1(t) = \sum_{k=0}^{\infty} \exp\left(-t\mu^k\right)$ can be evaluated exactly[8] as

$$S_1(t) = \frac{1}{2} - \frac{\log t + \gamma}{\log \mu} + 2\sum_{k=1}^{\infty} \frac{\cos\left[\frac{2k\pi}{\log \mu}\log\left(\frac{\mu}{t}\right) + \arg\Gamma\left(\frac{2k\pi i}{\log \mu}\right)\right]}{\sqrt{\log \mu}\sqrt{2k\sinh\frac{2k\pi^2}{\log \mu}}} + \sum_{k=1}^{\infty}\left(1 - e^{-t\mu^{-k}}\right)$$

Furthermore,

$$\left|\sum_{k=1}^{\infty} \frac{\cos\left[\frac{2k\pi}{\log \mu}\log\left(\frac{\mu}{t}\right) + \arg\Gamma\left(\frac{2k\pi i}{\log \mu}\right)\right]}{\sqrt{2k\sinh\frac{2k\pi^2}{\log \mu}}}\right| \le \sum_{k=1}^{\infty} \frac{1}{\sqrt{2k\sinh\frac{2k\pi^2}{\log \mu}}} = b(\mu)$$

and the function $T(\mu) = \frac{2b(\mu)}{\sqrt{\log \mu}}$ is increasing, but for $1 < \mu \le 5$ its maximum value

---

[8] From $\frac{\Gamma(s)}{\mu^{ps}} = \int_0^{\infty} t^{s-1}\,e^{-\mu^p t}\,dt$ for $\sigma = Re(s) > 0$ and $Re(p) \ge 0$, we find that

$$\Gamma(s)\sum_{k=0}^{\infty}\frac{1}{\mu^{ks}} = \int_0^{\infty} t^{s-1}\sum_{k=0}^{\infty} e^{-\mu^k t}\,dt$$

or

$$\int_0^{\infty} t^{s-1}\,S_1(t)\,dt = \Gamma(s)\frac{\mu^s}{\mu^s - 1}$$

By Mellin inversion, for $c > 0$,

$$S_1(t) = \frac{1}{2\pi i}\int_{c-i\infty}^{c+i\infty} \frac{\Gamma(s)}{\mu^s - 1}\left(\frac{\mu}{t}\right)^s\,ds$$

By moving the line of integration to the left, we encounter a double pole at $s = 0$ from $\Gamma(s)$ and $\frac{1}{\mu^s-1}$ and simple poles at $s = \frac{2k\pi i}{\log \mu}$ from $\frac{1}{\mu^s-1}$. Invoking Cauchy's residue theorem leads to the result.

$T(5)$ is smaller than 0.0035. Since $t = \frac{\mu^2}{N(\mu-1)^2}$ is small and $\mu > 1$, we approximate

$$S_1(t) \approx \frac{1}{2} - \frac{\log t + \gamma}{\log \mu} \tag{16.53}$$

and arrive, for large N, at

$$E[H_N] \geq \frac{\log N}{\log \mu} + \frac{1}{2} - \frac{\log \frac{\mu^2}{(\mu-1)^2} + \gamma}{\log \mu} \approx \frac{\log N}{\log \mu}$$

This shows that in large, sparse graphs for which the discovery process is well modeled by a branching process, it holds that $E[H_N]$ scales as $\frac{\log N}{\log \mu}$ where $\mu = E[Y] - 1 > 1$ is the mean degree minus 1 in the graph.

We can refine the above analysis. Let us now assume that the convergence of $W_k \to W$ is sufficiently fast for large N and that $W > 0$ such that,

$$|C_A(l)| \sim W_A \sum_{k=0}^{l} \mu^k = W_A \frac{\mu^{l+1} - 1}{\mu - 1} \approx W_A \frac{\mu^{l+1}}{\mu - 1}$$

is a good approximation (and similarly for $|C_B(l)|$). The verification of this approximation is difficult in general. Theorem 12.3.2 states that $\Pr[W = 0] = \pi_0$ and equivalently $\Pr[W > 0] = 1 - \pi_0$, where the extinction probability π_0 obeys the equation (12.18). Using this approximation, we find from (16.52)

$$\Pr[H_N > 2l] \approx E\left[\exp\left(-\frac{W_A W_B}{N} \frac{\mu^{2l+2}}{(\mu-1)^2} \right) \middle| W_A, W_B > 0 \right]$$

where the condition on $W > 0$ is required else there are no clusters $C_A(l)$ and $C_B(l)$ nor a path. Since the same asymptotics also holds for odd values of the hopcount, we finally arrive, for $k \geq 1$ and large N, at

$$\Pr[H_N > k] \approx E\left[\exp\left(-Z\mu^k \right) \middle| W_A, W_B > 0 \right]$$

where the random variable

$$Z = \frac{W_A W_B}{N} \frac{\mu^2}{(\mu - 1)^2}$$

and $W_A \overset{d}{=} W_B \overset{d}{=} W$. A more explicit computation of $\Pr[H_N > k]$ requires the knowledge of the limit random variable W, which strongly depends on the nodal degree Y.

The mean hopcount $E[H_N]$ is found similarly as in the analysis above by using

(16.53) with $t = Z$,

$$E[H_N] \approx E[S_1(Z)|W_A, W_B > 0]$$

$$= E\left[\frac{1}{2} - \frac{2\log W - \log N + 2\log\frac{\mu}{(\mu-1)} + \gamma}{\log\mu}\middle|W > 0\right]$$

$$= \frac{1}{2} + \frac{\log N - 2\log\frac{\mu}{(\mu-1)} - \gamma}{\log\mu} - 2\frac{E[\log W|W > 0]}{\log\mu}$$

In sparse graphs with mean degree $E[Y]$ equal to μ and for a large number of nodes N, the mean hopcount is well approximated[9] by

$$E[H_N] = \frac{\log N}{\log\mu} + \frac{1}{2} - \frac{\gamma - 2\log\left(1 - \frac{1}{\mu}\right)}{\log\mu} - 2\frac{E[\log W|W > 0]}{\log\mu} \qquad (16.54)$$

This expression (16.54) for the mean hopcount – which is more refined than the often used estimate $E[H_N] \approx \frac{\log N}{\log\mu}$ – contains the curious mean $E[\log W|W > 0]$, where W is the limit random variable of the branching process produced by the graph's degree distribution Y.

Application to $\mathbf{G_p(N)}$ The above analysis holds for fixed $E[Y] = p(N-1)$ such that, for large N, we require that $p = \frac{\mu}{N}$, where μ is approximately equal to the mean degree. Since the binomial distribution (15.19) for the degree in the Erdős-Rényi random graph $G_p(N)$ is very well approximated by the Poisson distribution $\Pr[D_{G_p(N)} = k] \approx \frac{\mu^k}{k!}e^{-\mu}$ for large N and constant μ, formula (16.54) requires the computation of $E[\log W|W > 0]$ in a Poisson branching process, which is presented in Hooghiemstra and Van Mieghem (2005) but here summarized in Fig. 16.7. The numerical evaluation of the mean hopcount (16.54) in a random graph of the class $G_p(N)$ for small mean degree μ and large N shows that (16.54) is more accurate than only its first term $\log_\mu N$.

At the other end of the scale for a constant link density $p < 1$, which corresponds to a mean degree $E[Y] = p(N-1)$, the above analysis no longer applies for such large values of the mean degree $E[Y]$. Fortunately, in that case, an exact asymptotic analysis is possible (see Problem (iii)):

$$\Pr[H_N = 1] = p$$
$$\Pr[H_N = 2] = (1 - p)\left(1 - (1 - p^2)^{N-2}\right) \qquad (16.55)$$

Values of H_N higher than 2 are extremely unlikely since $\Pr[H_N > 2] = (1 - p)\left[1 - p^2\right]^{N-2}$ tends to zero rapidly for sufficiently large N and constant p. Hence,

$$E[H_N] \simeq \Pr[H_N = 1] + 2\Pr[H_N = 2] \simeq 2 - p$$

---

[9] A more rigorous derivation that stochastically couples the graph's growth specified by a certain degree distribution to a corresponding branching process is found in van der Hofstad *et al.* (2005). In particular, the analysis is shown to be valid for any randomly constructed graph with a finite variance of the degree. More details on the result for the average hopcount are presented in Hooghiemstra and Van Mieghem (2005).

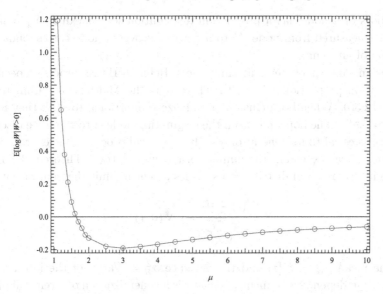

Fig. 16.7. The quantity $E[\log W \,|\, W > 0]$ of a Poisson branching process versus the average degree μ.

and, similarly, we find $\text{Var}[H_N] \simeq p(1-p)$. This asymptotic analysis even holds for a larger link density regime $p = cN^{-\frac{1}{2}+\epsilon}$ with $\epsilon > 0$ because

$$\lim_{N\to\infty} \Pr[H_N > 2] = \lim_{N\to\infty} \left(1 - cN^{-\frac{1}{2}+\epsilon}\right)\left[1 - cN^{-1+2\epsilon}\right]^{N-2} = 0$$

but for $\epsilon = 0$, it holds that $\lim_{N\to\infty} \Pr[H_N > 2] = e^{-c} > 0$.

In summary, if the link density p scales as $p = cN^{-\alpha}$ with $\alpha \in [0, \frac{1}{2})$, the mean hopcount $E[H_N] \simeq 2 - p$ is very small. If $p = \frac{\lambda}{N^{1-\epsilon}}$, equation (16.54) shows that $E[H_N] \approx \log_\lambda N$. The regime in between for $\alpha \in [\frac{1}{2}, 1)$ needs other analysis techniques.

16.8.3 The hopcount in certain large graphs with i.i.d. link weights

The setting of Section 16.8.1 with unit link weights is naturally generalized by considering link weights that are identically and independently distributed with probability distribution function G. We assume again that the underlying graph is sufficiently sparse to allow us to model the hopcount distribution of the shortest path by a branching process, now a time-dependent branching process studied in Section 12.7.

Consider again a bi-directional search starting from two arbitrary nodes A and B. The discovery processes from A and B grow two separate clusters of discovered nodes that "meet" each other when their size is proportional to $\sqrt{N}$ as shown in Section 16.8.1. The discovery process is modeled by a Bellman-Harris branching process: the weight of a link corresponds to the lifetime of an item and the number of neighbors of a node has a production distribution with mean $E[Y] = \mu$. The hopcount (measured from node A) to a newly discovered node corresponds to the generation of an item.

The mean number of items at time t in a Bellman-Harris process, specified in (12.37), grows proportional to $e^{-\alpha t}$, where α is the Malthusian parameter that satisfies (12.36). When both clusters are of size proportional to $\sqrt{N}$, they meet at time $t_N \sim \frac{\log N}{2\alpha}$. The hopcount from the originating node A to the node discovered at time t_N is equal to its generation g_N. In a renewal process, the generation of an item born at time t_N equals the number of renewals $N(t_N)$. Theorem 8.2.3 states that asymptotic renewal distribution satisfies a central limit theorem for large t,

$$\frac{N(t) - \frac{t}{\mu_R}}{\sigma_R \sqrt{\frac{t}{\mu_R^3}}} \xrightarrow{d} N(0,1)$$

where the mean $\mu_R = E[\tau]$ and the variance $\sigma_R^2 = \text{Var}[\tau]$ of the interarrival τ. Since the time-dependent branching process is a generalization of a renewal process, as shown in Section 12.7, the generation g_N will satisfy a similar central limit theorem, but with a slightly adjusted underlying distribution function, $F(y) = \mu \int_0^y e^{-\alpha u} dG(u)$, called the *stable age* distribution (12.35) of the corresponding link weight distribution G. Thus,

$$\frac{g_N - \frac{t_N}{\mu_{BP}}}{\sigma_{BP} \sqrt{\frac{t_N}{\mu_{BP}^3}}} \xrightarrow{d} N(0,1)$$

where μ_{BP} and σ_{BP}^2 are the mean and variance of the stable age distribution $F(y)$. The hopcount H_N in the graph consists of the sum of the two random variables $g_{N,A}$ and $g_{N,B}$, which represent the hopcount (or generation) to the respective starting nodes A and B. For sufficiently large N (and consequently t_N), both random variables $g_{N,A}$ and $g_{N,B}$ are asymptotically independent so that

$$H_N - \frac{2t_N}{\mu_{BP}} = \left(g_{N,A} - \frac{t_N}{\mu_{BP}}\right) + \left(g_{N,B} - \frac{t_N}{\mu_{BP}}\right)$$

$$\sim N\left(0, \sigma_{BP}^2 \frac{t_N}{\mu_{BP}^3}\right) + N\left(0, \sigma_{BP}^2 \frac{t_N}{\mu_{BP}^3}\right)$$

Since the sum of two Gaussian random variables is again a Gaussian random variable (Section 3.2.3), we have

$$H_N - \frac{2t_N}{\mu_{BP}} \sim N\left(0, \sigma_{BP}^2 \frac{2t_N}{\mu_{BP}^3}\right)$$

Finally, after substituting $2t_N \sim \frac{\log N}{\alpha}$ for large N, we arrive at the central limit theorem for the hopcount in a certain graph type with i.i.d. link weights, distributed as G,

$$\frac{H_N - \frac{\log N}{\alpha \mu_{BP}}}{\sigma_{BP} \sqrt{\frac{\log N}{\alpha \mu_{BP}^3}}} \xrightarrow{d} N(0,1) \tag{16.56}$$

The presented arguments are intuitive. A rigorous proof of (16.56) is found in Bhamidi *et al.* (2012), who also prove a limit law for the weight of the shortest path.

16.9 The minimum spanning tree

From an algorithmic point of view, the shortest path problem is closely related to the computation of the minimum spanning tree (MST). The Dijkstra shortest path algorithm is similar to Prim's minimum spanning tree algorithm (Cormen *et al.*, 1991). In this section, we compute the mean weight of the MST in a complete graph K_N with a general link weight structure.

16.9.1 The Kruskal growth process of the MST of K_N

Since the link weights in the underlying complete graph K_N are chosen independently and assigned randomly to links in K_N, the resulting graph is probabilistically the same if we first order the set of link weights and assign them in increasing order randomly to links in the complete graph. In the latter construction process, only the order statistics or the ranking of the link weights suffice to construct the graph because the precise link weight can be unambiguously associated to the rank of a link. This observation immediately favors the Kruskal algorithm for the MST over Prim's algorithm. Although the Prim algorithm leads to the same MST, it gives a more complicated, long-memory growth process, where the attachment of each new node depends stochastically on the whole growth history so far. Pietronero and Schneider (1990) illustrate that in our approach Prim, in contrast with Kruskal, leads to a very complicated stochastic process for the construction of the MST.

The Kruskal growth process described here is closely related to a growth process of the ER random graph $G_r(N, L)$ with N nodes and L links. The construction or growth of $G_r(N, L)$ starts from N individual nodes and in each step an arbitrary, not yet connected random pair is connected. The only difference with Kruskal's algorithm for the MST is that, in Kruskal, links generating loops are forbidden. Those forbidden links are the links that connect nodes within the same connected component or "cluster". As a result, the internal wiring of the clusters differs, but the cluster size statistics (counted in nodes, not links) is exactly the same as in the corresponding random graph. The metacode of the Kruskal growth process for the construction of the MST is shown in Fig. 16.8.

The growth process of the random graph $G_p(N)$, which is asymptotically equal to that of $G_r(N, L)$, is quantified in Section 15.7.6.1 for large N. The fraction of

---

KRUSKALGROWTHMST

1. start with N disconnected nodes
2. **repeat until** all nodes are connected
3.   randomly select a node pair (i, j)
4.   **if** a path $\mathcal{P}_{i \to j}$ does not exist
5.   **then** connect i to j

Fig. 16.8. Kruskal growth process.

nodes S in the giant component of $G_p(N)$ is related to the mean degree or to the link density p because $\mu_{G_p(N)} = p(N-1)$ in $G_p(N)$ by (15.29). For large N, the size of the giant cluster in the forest is thus determined as a function of the number of added links that increase $\mu_{G_p(N)}$.

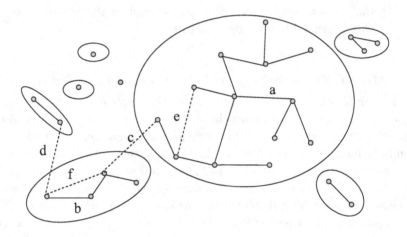

Fig. 16.9. Component structure during the Kruskal growth process.

We will now transform the mean degree $\mu_{G_p(N)}$ in the ER random graph $G_p(N)$ to the mean degree μ_{MST} in the corresponding stage in Kruskal growth process of the MST. In early stages of the growth each selected link will be added with high probability such that $\mu_{\text{MST}} = \mu_{G_p(N)}$ almost surely. After some time the probability that a selected link is forbidden increases, and thus $\mu_{G_p(N)}$ exceeds μ_{MST}. In the end, when connectivity of all N nodes is reached, $\mu_{\text{MST}} = 2$ (since it is a tree) while $\mu_{G_p(N)} = O(\log N)$, as follows from (15.28) and the critical threshold $p_c \sim \frac{\log N}{N}$.

Consider now an intermediate stage of the growth as illustrated in Fig. 16.9. Assume there is a giant component of mean size NS and $n_l = N(1-S)/s_l$ small components of average size s_l each. Then we can distinguish six types of links labelled a-f in Fig. 16.9. Types a and b are links that have been chosen earlier in the giant component (a) and in the small components (b) respectively. Types c and d are eligible links between the giant component and a small component

(c) and between small components (d) respectively. Types e and f are forbidden links connecting nodes within the giant component (e), respectively within a small component (f). For large N, we can enumerate the mean number of links L_y of each type y:

$$
\begin{aligned}
L_a + L_b &= \tfrac{1}{2}\mu_{MST}N & L_e &= \tfrac{1}{2}(SN)^2 - SN \\
L_c &= SN \cdot (1 - S)N & L_f &= \tfrac{1}{2}n_l s_l(s_l - 1) - n_l(s_l - 1) \\
L_d &= \tfrac{1}{2}n_l^2 \cdot s_l^2 &&
\end{aligned}
$$

To highest order in $O(N^2)$, we have

$$
L_c = N^2 S(1 - S), \qquad L_d = \frac{1}{2}N^2(1 - S)^2, \qquad L_e = \frac{1}{2}N^2 S^2
$$

The probability that a randomly selected link is eligible is $q = \frac{L_c + L_d}{L_c + L_d + L_e + L_f}$ or, to order $O\left(N^2\right)$,

$$
q = 1 - S^2 \tag{16.57}
$$

In contrast with the growth of the random graph $G_p(N)$, where at each stage a link is added with probability p, in the Kruskal growth of the MST, we only successfully add one link (with probability 1) per $\frac{1}{q}$ stages on average. Thus the mean number of links added in the random graph corresponding to one link in the MST is $\frac{1}{q} = \frac{1}{1-S^2}$. This provides an asymptotic mapping between μ and μ_{MST} in the form of a differential equation,

$$
\frac{d\mu_{G_p(N)}}{d\mu_{MST}} = \frac{1}{1 - S^2}
$$

By using (15.31), we find

$$
\frac{d\mu_{MST}}{dS} = \frac{d\mu_{MST}}{d\mu_{G_p(N)}} \frac{d\mu_{G_p(N)}}{dS} = \frac{(1 + S)(S + (1 - S)\log(1 - S))}{S^2}
$$

Integration with the initial condition $\mu_{MST} = 2$ at $S = 1$, finally gives the mean degree μ_{MST} in the MST as function of the fraction S of nodes in the giant component

$$
\mu_{MST}(S) = 2S - \frac{(1 - S)^2}{S}\log(1 - S) \tag{16.58}
$$

As shown in Fig. 16.10, the asymptotic result (16.58) agrees well with the simulation (even for a single sample), except in a small region around the transition $\mu_{MST} = 1$ and for relatively small N.

The key observation is that all transition probabilities in the Kruskal growth process asymptotically depend on merely one parameter S, the fraction of nodes in the giant component, and S is called an *order parameter* in statistical physics. In general, the expectation of an order parameter distinguishes the qualitatively different regimes (states) below and above the phase transition. In higher dimensions, fluctuations of the order parameter around the mean can be neglected and the mean value can be computed from a self-consistent mean-field theory. In our

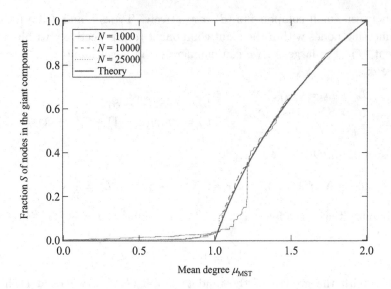

Fig. 16.10. Size of the giant component (divided by N) as a function of the mean degree μ_{MST}. Each simulation for a different number of nodes N consists of one MST sample.

problem, the underlying complete (or random) graph topology makes the problem effectively infinite-dimensional. The argument leading to (15.29) is essentially a mean-field argument.

16.9.2 Mean weight of the minimum spanning tree

By definition, the weight of the MST is

$$W_{\mathrm{MST}} = \sum_{j=1}^{L} w_{(j)} 1_{j \in \mathrm{MST}} \qquad (16.59)$$

where $w_{(j)}$ is the j-th smallest link weight. The mean MST weight is

$$E\left[W_{\mathrm{MST}}\right] = \sum_{j=1}^{L} E\left[w_{(j)} 1_{j \in \mathrm{MST}}\right]$$

The random variables $w_{(j)}$ and $1_{j \in \mathrm{MST}}$ are independent because the j-th smallest link weight $w_{(j)}$ only depends on the link weight distribution and the number of links L, while the appearance of the j-th link in the MST only depends on the graph's topology, as shown in Section 16.9.1. Hence,

$$E\left[w_{(j)} 1_{j \in \mathrm{MST}}\right] = E\left[w_{(j)}\right] E\left[1_{j \in \mathrm{MST}}\right] = E\left[w_{(j)}\right] \Pr\left[j \in \mathrm{MST}\right]$$

such that the mean weight of the MST is

$$E\left[W_{\text{MST}}\right] = \sum_{j=1}^{L} E\left[w_{(j)}\right] \Pr\left[j \in \text{MST}\right] \tag{16.60}$$

In general for independent link weights with probability density function $f_w(x)$ and distribution function $F_w(x) = \Pr\left[w \le x\right]$, the probability density function of the j-th order statistic follows from (3.43) as

$$f_{w_{(j)}}(x) = \frac{j f_w(x)}{F_w(x)} \binom{L}{j} (F_w(x))^j (1 - F_w(x))^{L-j} \tag{16.61}$$

The factor $\binom{L}{j}(F_w(x))^j (1 - F_w(x))^{L-j}$ is a binomial distribution with mean $\mu = F_w(x) L$ and variance $\sigma^2 = L F_w(x)(1 - F_w(x))$ that, by the Central Limit Theory 6.3.1, tends for large L to a Gaussian $\frac{1}{\sigma\sqrt{2\pi}} e^{-\frac{(j-\mu)^2}{2\sigma^2}}$, which peaks at $j = \mu$. For large N and fixed $\frac{j}{L}$, we have[10] $x_j = E\left[w_{(j)}\right] \simeq F_w^{-1}(\frac{j}{L})$.

We found before in (16.57) that the link ranked j appears in the MST with probability

$$\Pr\left[j \in \text{MST}\right] = 1 - S_j^2$$

where S_j is the fraction of nodes in the giant component during the construction process of the random graph at the stage where the number of links precisely equals j. Since links are added independently, that stage in fact establishes the random graph $G_r(N, L = j)$. Our graph under consideration is the complete graph K_N such that we add in total $L = \binom{N}{2}$ links. With (15.31) and $\mu_{G_p(N)} = \frac{2L}{N}$, it follows that

$$\frac{2j}{N} = -\frac{\log(1 - S_j)}{S_j} \tag{16.62}$$

Hence, for large N,

$$E\left[W_{\text{MST}}\right] \simeq \sum_{j=1}^{L} F_w^{-1}\left(\frac{j}{L}\right)(1 - S_j^2)$$

---

[10] In general, it holds that $w_{(k)} = F_w^{-1}(U_{(k)})$ and

$$E\left[w_{(k)}\right] = E\left[F_w^{-1}(U_{(k)})\right] \ne F_w^{-1}(E\left[U_{(k)}\right])$$

but, for a large number of order statistics L, the Central Limit Theorem 6.3.1 leads to

$$E\left[w_{(k)}\right] \simeq F_w^{-1}\left(\frac{j}{L}\right) \simeq F_w^{-1}(E\left[U_{(k)}\right])$$

because for a uniform random variable U on [0,1] the average weight of the j-th smallest link is exactly

$$E\left[w_{(i)}\right] = \frac{j}{L+1} \simeq \frac{j}{L}$$

We now approximate the sum by an integral,

$$E\left[W_{\mathrm{MST}}\right] \simeq \int_1^L F_w^{-1}\left(\frac{u}{L}\right)\left(1 - S_u^2\right) du$$

Substituting $x = \frac{2u}{N}$ (which is the mean degree in any graph $G\left(N, u\right)$) yields for large N where $L \simeq \frac{N^2}{2}$,

$$E\left[W_{\mathrm{MST}}\right] \simeq \frac{N}{2} \int_{\frac{2}{N}}^{N-1} F_w^{-1}\left(\frac{x}{N}\right)\left(1 - S_{\frac{N}{2}x}^2\right) dx \simeq \frac{N}{2} \int_0^N F_w^{-1}\left(\frac{x}{N}\right)\left(1 - S^2(x)\right) dx$$

It is known (Janson *et al.*, 1993) that, if the number of links in the growth process of the random graph is below $\frac{N}{2}$, with high probability (and ignoring a small onset region just below $\frac{N}{2}$), there is no giant component such that $S\left(x\right) = 0$ for $x \in [0, 1]$. Thus, we arrive at the general formula valid for large N,

$$E\left[W_{\mathrm{MST}}\right] \simeq \frac{N}{2} \int_0^1 F_w^{-1}\left(\frac{x}{N}\right) dx + \frac{N}{2} \int_1^N F_w^{-1}\left(\frac{x}{N}\right)\left(1 - S^2(x)\right) dx \quad (16.63)$$

The first term is the contribution from the smallest $N/2$ links in the graph, which are included in the MST almost surely. The remaining part comes from the more expensive links in the graph, which are included with diminishing probability since $1 - S^2(x)$ decreases exponentially for large x as can be deduced from (15.30). The rapid decrease of $1 - S^2(x)$ makes only relatively small values of the argument $F_w^{-1}\left(\frac{x}{N}\right)$ contribute to the second integral.

At this point, the specifics of the link weight distribution needs to be introduced. The Taylor expansion of $\frac{N}{2} F_w^{-1}\left(\frac{x}{N}\right)$ for large N to first order is

$$\frac{N}{2} F_w^{-1}\left(\frac{x}{N}\right) = \frac{N}{2} F_w^{-1}(0) + \frac{x}{2 f_w(0)} + O\left(\frac{1}{N}\right) = \frac{x}{2 f_w(0)} + O\left(\frac{1}{N}\right)$$

since we require that link weights are positive such that $F_w^{-1}(0) = 0$. This expansion is only useful provided f_w is regular, i.e. $f_w(0)$ is neither zero nor infinity. These cases occur, for example, for polynomial link weights with $f_w(x) = \alpha x^{\alpha-1}$ and $\alpha \neq 1$. For polynomial link weights, however, it holds that $\frac{N}{2} F_w^{-1}\left(\frac{x}{N}\right) = \frac{N^{1-\frac{1}{\alpha}}}{2} x^{\frac{1}{\alpha}}$. Formally, this latter expression reduces to the first-order Taylor approach for $\alpha = 1$, apart from the constant factor $\frac{1}{f_w(0)}$. Therefore, we will first compute $E\left[W_{\mathrm{MST}}\right]$ for polynomial link weights and then return to the case in which the Taylor expansion is useful.

16.9.2.1 Polynomial link weights

The mean weight of the MST for polynomial link weights follows[11] from (16.63) for large N as

$$E\left[W_{\mathrm{MST}}(\alpha)\right] \simeq \frac{N^{1-\frac{1}{\alpha}}}{2}\left(\frac{1}{\frac{1}{\alpha}+1} + \int_1^N x^{\frac{1}{\alpha}}\left(1 - S^2(x)\right) dx\right)$$

Let $y = S(x)$ and use (15.31), then $x = S^{-1}(y) = -\frac{\log(1-y)}{y}$ and the differential thus is $dx = -\frac{d}{dy}\left(\frac{\log(1-y)}{y}\right) dy$, while $y = S(1) = 0$ and $y = S(N) = 1$, such that

$$I = \int_1^N x^{\frac{1}{\alpha}}\left(1 - S^2(x)\right) dx$$

$$= \int_0^1 \left(-\frac{\log(1-y)}{y}\right)^{\frac{1}{\alpha}}\left(1 - y^2\right)\frac{d}{dy}\left(-\frac{\log(1-y)}{y}\right) dy$$

After partial integration, we have

$$I = -\frac{1}{\frac{1}{\alpha}+1} + \frac{2}{\frac{1}{\alpha}+1}\int_0^\infty x^{\frac{1}{\alpha}+1}\frac{e^{-x}}{(1-e^{-x})^{\frac{1}{\alpha}}} dx$$

Finally, for large N, we end up with

$$E\left[W_{\mathrm{MST}}(\alpha)\right] \simeq N^{1-\frac{1}{\alpha}}\left(\frac{1}{\frac{1}{\alpha}+1}\int_0^\infty x^{\frac{1}{\alpha}+1}\frac{e^{-x}}{(1-e^{-x})^{\frac{1}{\alpha}}} dx\right) \tag{16.64}$$

If $\alpha < 1$, then $E[W_{\mathrm{MST}}(\alpha)] \to 0$ for $N \to \infty$, while for $\alpha > 1$, $E[W_{\mathrm{MST}}(\alpha)] \to \infty$. In particular, $\lim_{\alpha\to\infty} E[W_{\mathrm{MST}}(\alpha)] = N - 1$. Only for $\alpha = 1$, $E[W_{MST}(1)]$ is finite for large N. More precisely,

$$\lim_{N\to\infty} E[W_{\mathrm{MST}}(1)] = \zeta(3) = 1.202\ldots \tag{16.65}$$

where we have used (Abramowitz and Stegun, 1968, Section 23.2.7) the integral of the Riemann Zeta function $\Gamma(s)\zeta(s) = \int_0^\infty \frac{u^{s-1}}{e^u-1} du$, which is convergent for $\mathrm{Re}(s) > 1$. This particular case for $\alpha = 1$ has been proved earlier by Frieze (1985) based on a different method.

---

[11] Since the average of the k-th smallest link weight can be computed from (3.43) as

$$E\left[w_{(k)}\right] = \frac{L!}{\Gamma\left(L+1+\frac{1}{\alpha}\right)}\frac{\Gamma\left(k+\frac{1}{\alpha}\right)}{\Gamma(k)}$$

the exact formula (16.60) reduces to

$$E\left[W_{\mathrm{MST}}(\alpha)\right] = \frac{L!}{\Gamma\left(L+1+\frac{1}{\alpha}\right)}\sum_{j=1}^L \frac{\Gamma\left(j+\frac{1}{\alpha}\right)}{\Gamma(j)}\left(1 - S_j^2\right)$$

Analogously to the above manipulations, after convertion to an integral, substituting $x = \frac{2u}{N}$ and using (Abramowitz and Stegun, 1968, Section 6.1.47), for large z, that $\frac{\Gamma\left(z+\frac{1}{\alpha}\right)}{\Gamma(z)} = z^{\frac{1}{\alpha}}\left(1 + O\left(\frac{1}{z}\right)\right)$, we arrive at the same formula.

16.9.2.2 MST of an ER random graph

We now return to the Taylor series valid for link weights where $0 < f_w(0) < \infty$. The above result for $\alpha = 1$ immediately yields

$$\lim_{N \to \infty} E[W_{\mathrm{MST}}] = \frac{\zeta(3)}{f_w(0)} \tag{16.66}$$

This result is for the complete graph K_N. A random graph $G_p(N)$ with $p < 1$ and weight density $f_w(x)$ is equivalent to K_N with a fraction $1 - p$ of infinite link weights. Thus the effective link weight distribution is $pf_w(x) + (1-p)\delta_{w,\infty}$, and we can simply replace $f_w(0)$ by $pf_w(0)$ in the expression (16.66) to obtain the mean weight of the MST in the random graph $G_p(N)$.

16.10 Problems

(i) *Comparison of simulations with exact results.* Many of the theoretical results are easily verified by simulations. Consider the following standard simulation: (a) Construct a graph of a certain class, e.g. an instance of the random graphs $G_p(N)$ with exponentially distributed link weights; (b) Determine in that graph a desired property, e.g. the hopcount of the shortest path between two different arbitrary nodes; (c) Store the hopcount in a histogram; and (d) repeat the sequence (a)-(c) n times with each time a different graph instance in (a). Estimate the relative error of the simulated hopcount in $G_p(N)$ with $p = 1$ for $n = 10^4, 10^5$ and 10^6.

(ii) Given the probability generating function (16.20) of the weight of the shortest path in a complete graph with independent exponential link weights, compute the variance of W_N.

(iii) Derive (16.55) in $G_p(N)$ with unit link weights.

(iv) In a communication network often two paths are computed for each important flow to guarantee sufficient reliability. Apart from the shortest path between a source A and a destination B, a second path between A and B is chosen that does not travel over any intermediate router of the shortest path. We call such a path node-disjoint to the shortest path. Derive a good approximation for the distribution of the hopcount of the shortest node-disjoint path to the shortest path in the complete graph with exponential link weights with mean 1.

(v) In the complete graph with exponentially distributed link weights, a fraction p_R of the links is defined as red links while the remaining links with fraction $p_B = 1 - p_R$ are called blue links. Assume that the red links are uniformly distributed.

    (a) Compute the probability distribution of the number of red/blue links that appear in the shortest path.

    (b) Discuss potential applications.

(vi) Hot potato routing is a stochastic variant of packet routing that incurs a minimum routing protocol overhead. The principle is simple: when a packet arrives at a node, the node forwards that packet to the next hop unless the packet is destined for itself. The next hop is chosen uniformly out of the neighbors of that node. This type of routing is also called a random walk on a graph (Van Mieghem, 2011, p. 63-65).

    (a) Derive the hopcount distribution of hot potato routing in a given graph G.

    (b) Apply the result to the complete graph K_N.

    (c) The efficiency of this random walk or hot potato routing strategy can be defined as the probability that the random walk from a source to a destination precisely follows the shortest path between that source-destination pair. Derive this probability, assuming that there is only one shortest path between source and destination.

    (d) A more efficient variant of hot potato routing stores a list of previous hops to prevent that a next hop again visits an unsuccessful previous hop. Derive a general formula for the hopcount of this variant with memory and apply the result to the complete graph.

(vii) The probability generating function of the hopcount in the URT with N nodes is given by

(16.6), which can be rewritten as

$$\varphi_{h_N}(z) = \prod_{j=2}^{N} \left(\frac{j-1}{j} + \frac{z}{j} \right)$$

From (3.1), we recognize that $E[z^{V_j}] = \frac{j-1}{j} + \frac{z}{j}$ is the generating function of a Bernoulli random variable V_j with mean $E[V_j] = \frac{1}{j}$. In view of (2.74), we find that the hopcount h_N in the URT is the sum of $N-1$ independent Bernoulli random variables V_j,

$$h_N = \sum_{j=2}^{N} V_j \qquad (16.67)$$

(a) Derive from this observation (16.67) the mean $E[h_N]$ and the variance $\text{Var}[h_N]$.
(b) What is the meaning of equation (16.67)?

(viii) Let $T_j^{(N)}$ denote the subtree of the complete uniform recursive tree of N nodes rooted at the j-th attached node, that is different from the root of the URT. Show that the number $\left| T_j^{(N)} \right|$ of nodes in that subtree has distribution

$$\Pr\left[|T_j^{(N)}| = k \right] = \frac{(j-1)(N-j)!(N-k-1)!}{(N-1)!(N-j-k+1)!} \qquad (16.68)$$

(ix) A reasoning to derive $\Pr\left[|T_j^{(N)}| = k \right]$ is presented. After the attachment of the j-th node, we split the URT into two parts, the set A containing node j and the set B containing all other nodes. We first examine the probability that a new node $k > j$ is attached to the set A, the subtree $T_j^{(N)}$ rooted at node j. The probability that the $(j+1)$-th node is attached to $T_j^{(N)}$ is

$$p_{j+1} = \frac{1}{j}$$

since, in the URT, a node is uniformly attached one of existing nodes. For the $(j+2)$-th node, using the law of total probability (2.49), we have

$$p_{j+2} = \Pr\left[n_{j+2} \in T_j^{(N)} \middle| n_{j+1} \in T_j^{(N)} \right] \Pr\left[n_{j+1} \in T_j^{(N)} \right]$$
$$+ \Pr\left[n_{j+2} \in T_j^{(N)} \middle| n_{j+1} \notin T_j^{(N)} \right] \Pr\left[n_{j+1} \notin T_j^{(N)} \right]$$
$$= \frac{2}{j+1}\frac{1}{j} + \frac{1}{j+1}\left(1 - \frac{1}{j} \right) = \frac{1}{j}$$

We prove that $p_{j+q} = \frac{1}{j}$ for $q \geq 1$ by induction. It holds for $q = 1, 2$. Assume that it holds for p_{j+q}, then

$$p_{j+q+1} = \sum_{k=0}^{q} \binom{q}{k} \left(\frac{1}{j} \right)^k \left(1 - \frac{1}{j} \right)^{q-k} \cdot \frac{1+k}{j+q}$$
$$= \frac{1}{j+q} + \frac{q \cdot \frac{1}{j}}{j+q} = \frac{1}{j}$$

Hence, the nodes $j + q$ will always have probability $\frac{1}{j}$ to be attached to the subtree $T_j^{(N)}$ rooted at node j. The number of nodes in the subtree $T_j^{(N)}$ is the sum of $N - j$ Bernoulli random variables, thus

$$\Pr\left[|T_j^{(N)}| = k \right] = 1 + \binom{N-j}{k-1} \left(\frac{1}{j} \right)^{k-1} \left(1 - \frac{1}{j} \right)^{N-1-k+1}$$

which is different from (16.68). Where is the error?

(x) The link betweenness B_l of a link l is equal to the number of shortest paths between all possible pairs of nodes in G that traverse the link l. Given a uniform recursive tree with N nodes and unit link weight, show that the probability that an arbitrary link in the URT has a link betweenness equal to $k(N-k)$ is, for $k < \left[\frac{N}{2}\right]$,

$$\Pr\left[B_l = k(N-k)\right] = \frac{N}{(N-1)\,k(N-k)}\left(\frac{N-k}{k+1} + \frac{k}{N-k+1}\right) \qquad (16.69)$$

(*Hint:* Use the expression (16.68) for $\Pr\left[|\mathcal{T}_j^{(N)}| = k\right]$.)

(xi) A network of N routers is connected as a full mesh K_N. Each link is equipped with an independent identically distributed (i.i.d.) exponential (mean 1) link weight. An arbitrary router A is communicating with the nearest (according to the weight of the shortest path) m routers along the shortest paths and each shortest path has hopcount h_i, $1 \le i \le m$. What is the mean hopcount from A to these m destinations $E[h] = \frac{1}{m}\sum i = 1^m h_i$?

Notes

There is a remarkable asymptotic relation between the shortest path and the random assignment problem. The random assignment problem, as explained in Aldous (2001), is the stochastic variant of the following task: choose an assignment of N jobs to N machines with the objective to minimize the total cost of performing the N jobs, given the $N \times N$ matrix C, where the element c_{ij} equals the cost of performing job i on machine j. The assignment problem thus consists of determining the permutation π that minimizes the sum $\sum_{j=1}^{N} c_{j,\pi(j)}$. Probability enters in the most simple setting when the elements c_{ij} are i.i.d. exponentially random variables with mean 1. The corresponding random assignment problem (RAP) investigates the properties of the random variable $R_N = \min_\pi \sum_{j=1}^{N} c_{j,\pi(j)}$.

The RAP has a long history of which parts are overviewed in Aldous (2001) and Wästlund (2006). Here, we only illustrate the remarkable similarity with the shortest path tree problem. A basic result and analog of (18.20) is

$$E[R_N] = \sum_{k=1}^{N} \frac{1}{k^2} \to \zeta(2) \qquad (16.70)$$

which was asymptotically proved by Aldous (2001), and for finite N, independently, by Linusson and Wästlund (2004) and by Nair *et al.* (2005). Earlier, Parisi (1998) conjectured (16.70) based on simulations and Coppersmith and Sorkin (1999) extended the conjecture to partial assignments. Moreover, Aldous (2001) showed that a random cost $c^* = c(1, \pi(1))$ in RAP converges, for large N, to the pdf given in the right-hand side of (16.28), the density of the scaled weight of an arbitrary link in the URT. In addition, Aldous showed that

$$\lim_{N \to \infty} \Pr\left[c^* \text{ is } k\text{-th smallest of the entries } c_{11}, c_{12}, \ldots, c_{1N}\right] = 2^{-k}$$

which is the asymptotic analogon of the probability (16.50) that the degree of a node in the URT is k. The overwhelming similarity between these two different problems is striking and may be worth exploring in more depth.

The union of all shortest path trees in a graph shows a remarkable phase transition towards the minimum spanning tree (MST) when the extreme value index α of the link weights is smaller than a critical value α_c. Van Mieghem and Magdalena (2005) and Van Mieghem and Wang (2009) show that the critical value α_c scales as $N^{-\beta}$, where $0.4 < \beta < 0.7$ depends on the topology of the underlying graph. Tuning the link weights thus separates two phases akin to normal conduction ($\alpha > \alpha_c$) and to superconductivity ($\alpha < \alpha_c$), where all traffic flows over the MST backbone.

Recently, Addario-Berry *et al.* (2013) discovered a deep connection between the MST of the complete graph with i.i.d. link weights and the ER random graph, whose relation is explained in Section 16.9.1, around the critical percolation threshold p_c. Theorem 15.7.2 demonstrates that at $p = \frac{1}{N}$, the size of the clusters is $O\left(N^{2/3}\right)$ and their mutual distance is $O\left(N^{1/3}\right)$, which also holds for the entire MST. In particular, earlier Addario-Berry *et al.* (2009) demonstrated that the mean diameter of the MST on the complete graph with i.i.d. link weights is of order $O\left(N^{1/3}\right)$, provided each weight is different. Addario-Berry *et al.* (2013) present a construction of a scaling limit of the MST, which is a continuous type of tree, based on a scaling $N^{1/3}$ akin to ER random graphs at the percolation threshold. Using the theory of the topology of spaces, properties of that scaling limit of the MST are demonstrated. They show that the diameter of the MST, scaled by $N^{1/3}$, converges to a limit random variable.

17

Epidemics in networks

The spread of information in communications and on-line social networks is similarly described as the virus spread in a biological population. In epidemiology (see e.g. Bailey (1975); Anderson and May (1991); Daley and Gani (1999); Diekmann *et al.* (2012)), two simple "compartment" models form the basis on which many variants exist: the Susceptible-Infected-Susceptible (SIS) and the Susceptible-Infected-Removed (SIR) models. Both models start with infected items that can infect their direct, healthy neighbors that are susceptible to the disease. In the SIS model, each infected item can cure and become healthy, but susceptible again after recovering from the disease, while in the SIR model, the infected item that recovers (or dies) is removed from the population and from the infection process. Just as in queuing theory where the Kendall notation (Section 13.1.4) specifies the queueing system, epidemic models mainly differ by the number and type of compartments. Typical compartment names are healthy, but susceptible S, infected I, recovered or removed R, exposed to the virus E, alert A, and so on. Before concentrating on epidemic spread on networks, we briefly sketch important modelling aspects in classical epidemiology in Section 17.1.

This chapter mainly focuses on the SIS model, in continuous-time, applied to networks: the epidemic[1] spreads over the links that interconnect the nodes (computers or living beings) of a network. This continuous-time SIS epidemic process on a graph allows us to study the effect of the network topology on the process that runs over that topology and can be regarded as one of the simplest, though practically meaningful examples of the coupling between the function (or service) of a network and its underlying topology. The SIS model approximately describes epidemics such as the flu among humans and malware in computer networks. In spite of the simplicity of the SIS model, its analysis on networks is difficult, as this chapter will witness.

Before the advent of network science around 2000, earlier epidemic modeling hardly considered the specific details of the underlying contact network. Some

---

[1] The word "epidemic" is derived from the Greek $\epsilon\pi\iota$ (upon) and $\delta\eta\mu\sigma\varsigma$ (people), meaning "something living among people". The word $\epsilon\pi\iota\delta\eta\mu\iota\alpha$ (epidemia) means "stay, residence, sojourn".

examples of such network-unaware epidemic models are the Yule process (Section 11.3.2.3) and the linear birth and death process (Section 11.3.4).

17.1 Classical epidemiology

Perhaps the first classical paper in epidemiology is that of Kermack and McKendrick (1927), which not only introduces what are now known as the SIR equations, but also the concept of a "per capita force of infectivity". The legacy of Kermack and McKendrick is thoroughly discussed in Breda *et al.* (2012). Here and following Breda *et al.* (2012), we review the simplest variant of an epidemic in a fixed (or closed) population to illustrate the main characteristics of modelling in classical epidemiology.

Let $S(t)$ denote the density (number per unit area) of susceptibles in a population at time t. The force $F(t)$ of infection is the probability per unit of time that a susceptible becomes infected. After infection, an individual recovers and becomes immune to the infection, which reflects the SIR model. The governing equation for the density of susceptibles in a closed population

$$\frac{dS(t)}{dt} = -F(t) S(t) \qquad (17.1)$$

states that the change in S equals minus the incidence $F(t) S(t)$, the number of new cases per unit of time and area. After integrating (17.1), we obtain

$$S(t) = S(t_0) \exp\left(-\int_{t_0}^{t} F(u) \, du\right) \qquad (17.2)$$

where at time t_0, the infection process started (which we further take as $t_0 = -\infty$). The key modeling step is to include how the current force of infection depends on past incidence as

$$F(t) = \int_0^\infty F(t-u) S(t-u) \alpha(u) \, du \qquad (17.3)$$

where $\alpha(t) \geq 0$ is the expected contribution to the force of infection by an individual that was infected t units of time ago. Breda *et al.* (2012) mention how α is related to current compartimental models, for example, $\alpha_{\text{SIR}}(t) = \beta e^{-\delta t}$. Generally, α depends on the contact intensity and the infectiousness, which is the probability of infection transmission during a contact with a susceptible. The sequel consists of deriving conclusions, based on (17.2) and (17.3).

After substituting (17.2) into (17.3), the resulting equation is unfortunately not of a convolution type. However, by introducing the cumulative force of infection,

$$y(t) = \int_{-\infty}^{t} F(u) \, du$$

we obtain a scalar non-linear renewal equation after integrating (17.3) from $-\infty$ to

t, reversing the order of integration and using (17.1) followed by (17.2),

$$y\left(t\right) = S\left(-\infty\right) \int_0^\infty \left(1 - e^{-y(t-u)}\right) \alpha\left(u\right) du$$

Since $y \geq 0$ and $1 - e^{-y} \leq 1$, the limit $y\left(\infty\right) = \lim_{t\to\infty} y\left(t\right)$ exists and satisfies[2]

$$y\left(\infty\right) = R_0 \left(1 - e^{-y(\infty)}\right) \tag{17.4}$$

where the basic reproduction ratio

$$R_0 = S\left(-\infty\right) \int_0^\infty \alpha\left(u\right) du \tag{17.5}$$

can be interpreted as the expected number of secondary cases caused by a primary case introduced in a population with initial susceptible density $S\left(-\infty\right)$. Moreover, (17.2) together with (17.4) shows that

$$1 - e^{-y(\infty)} = 1 - \frac{S\left(\infty\right)}{S\left(-\infty\right)} = \frac{y\left(\infty\right)}{R_0}$$

equals the final size of the epidemic or the fraction of the population that is infected, sooner or later, during an outbreak. Apart from the trivial $y\left(\infty\right) = 0$ solution, the equation (17.4) only has a unique, strict positive solution when $R_0 > 1$, as shown in Section 15.7.6.1. Hence, we have established the famous Kermack–McKendrick threshold theorem: (a) if $R_0 > 1$, an infective agent causes an outbreak with the final size given by $\frac{y(\infty)}{R_0}$ and (b) if $R_0 \leq 1$, the final size of the epidemics is negligibly small.

In summary, the Kermack–McKendrick approach is characterized by a few features[3]: (a) a population is divided into compartments (here, the susceptible S compartment and, implicitly, the others like I and R combined); (b) a differential equation per compartment (as (17.1)) describes the density changes (or mass flow) according to the infection rule for that compartment; and (c) a fundamental result is the observation of threshold behavior that appears to be a general characteristic in epidemics, as will become clear in the sequel.

Although Kermack and McKendrick (1927) described an infectious process masterly in great generality (due to the free choice of $\alpha\left(t\right)$ in (17.3)), a number of implicit assumptions are made. First, a deterministic approach is used, assuming *fractions* of a large population. When we are concerned with a small *number* of infected individuals, the initial stages should be described probabilistically by a branching process (Chapter 12). The condition $R_0 > 1$ then refers to a supercritical branching process. Second, classical epidemology assumes *random and homogeneous mixing*, where each member in a compartment is treated similarly and

---

[2] Observe that equation (17.4) is an instance of the functional equation (12.18) of the extinction probability in a branching process.

[3] Breda *et al.* (2012) further present various generalizations of the Kermack–McKendrick model by including open populations where births and deaths occur, aging in which $S\left(t\right)$ now also depends on the age of an individual, partial or waning immunity, and extensions to other compartimental models like SIS.

indistinguishably from the others in that same compartment. In reality, however, individuals have their own social contact network over which diseases propagate and that contact network usually differs from the contact network of the other members in a group or compartment. Karrer and Newman (2010) even declare that "the incorporation of more realistic mixing patterns into epidemiological modeling has given rise to the field of *network epidemiology*", which is the main topic of this chapter.

In classical epidemiology, the basic reproduction ratio R_0 is the key parameter to assess the threshold behavior of an outbreak. A table with R_0 values in Keeling and Rohani (2008) shows that $1.1 \leq R_0 \leq 18$ for known, different infectious diseases. The basic reproduction ratio R_0 bears some resemblance to the epidemic threshold in network epidemiology (see Section 17.2.1). However, the basic reproduction ratio R_0 does not contain any information about the underlying contact network. Since the network structure is rich and complex, the concept of R_0 looses its easy interpretation. Diekmann *et al.* (2012, p. 249-250) illustrate the weakness of R_0 by discussing a line and square lattice topology and they conclude that network and percolation theory needs to be consulted to compute the epidemic threshold as shown in this chapter. In tree-like networks, where the infection only can follow a single path between nodes, branching theory[4] is adequate and the basic reproduction ratio R_0 can be applied.

17.2 The continuous-time SIS Markov process

In the sequel, we consider virus spread in an undirected graph $G(\mathcal{N}, \mathcal{L})$, where $\mathcal{N}$ is the set with N nodes and $\mathcal{L}$ the set with L links. The graph G is characterized by a symmetric adjacency matrix A. The graph G is fixed and does not change over time. The viral state of a node i at time t is specified by a Bernoulli random variable $X_i(t) \in \{0, 1\}$: $X_i(t) = 0$ for a healthy node and $X_i(t) = 1$ for an infected node. A node i at time t can be in one of the two states: *infected*, with probability $w_i(t) = \Pr[X_i(t) = 1]$, or *healthy*, with probability $1 - w_i(t)$, but susceptible to the virus. We assume that the curing process per node i is a Poisson process with rate δ and that the infection process per link is a Poisson process with rate β. The effective infection rate is $\tau = \frac{\beta}{\delta}$. Obviously, only when a node is infected, can it infect its direct neighbors, which are still healthy. All curing and infection Poisson processes are independent. This is the general continuous-time description of the simplest type of a Susceptible-Infected-Susceptible (SIS) virus process on a network.

It is convenient to slightly generalize the SIS process by adding a nodal component to the infection. We assume that each node i can be infected spontaneously with a rate ε. Hence, besides receiving the infection over links from infected neigh-

---

[4] Generating function approaches to epidemics as proposed by Newman (2002) apply when branching theory (Chapter 12) is valid, since generating functions belong to the basic tools of analysis in branching processes.

bors with rate β, the node i can also itself produce a virus with rate ε. Again, all involved Poisson processes are independent. The motivation to consider a nodal infection component stems from the analogy of epidemics with information spread in social networks, where nodes can generate information, which is spread over links to neighbors. This generalization is here called the ε−SIS model, which clearly reduces to the "classical" SIS model when $\varepsilon = 0$. Apart from the greater flexibility of the ε−SIS epidemic to model practical cases of information diffusion, there is a second, more fundamental motivation to consider this generalization of the "classical" SIS model: the ε−SIS model, as shown in Section 17.2.1, possesses a non-trivial steady-state, while the steady-state of the "classical" SIS in any finite network is the overall-healthy state (which is the absorbing state).

A "physical" description of the ε−SIS epidemic process is as follows. Let I denote the set of infected nodes in the graph G and let a_{ij} be the element of the adjacency matrix A. Then, the Markov transitions

$$\begin{cases} \text{for } j \notin I: & I \mapsto I \cup \{j\} & \text{at rate } \beta \sum_{k \in I} a_{kj} + \varepsilon \\ \text{for } i \in I: & I \mapsto I \backslash \{i\} & \text{at rate } \delta \end{cases} \qquad (17.6)$$

detail the dynamics between the infected subgraph I and its complement $I^c = G \backslash I$. Computationally, enumerating the subgraphs I in G leads to another description, explained in Section 17.2.1.

17.2.1 *The ε−SIS 2^N-state Markov chain*

Each node in the network can be in one of two possible states, either zero (healthy, but susceptible) or one (infected). All possible states, in which the set of N nodes of the graph G can be, thus equals 2^N. A state of the Markov chain is represented by the N-digits binary number $x_N x_{N-1}...x_2 x_1$ with[5] $x_k \in \{0,1\}$, where $1 \le k \le N$ refers to the node with label k in G. For example, state 0 is the all-healthy state $00...000$, state 3 is $00...011$ and state $2^N - 1$ is the all-infected state $11...11$. Here[6], we slightly modify the representation for an integer i as

$$i = \sum_{k=1}^{N} x_k(i) \, 2^{k-1}$$

---

[5] Each x_k is called a bit or binary digit and a 1 bit means that $x_k = 1$, while a 0 bit means that $x_k = 0$.

[6] The representation of an integer number n in a binary base is

$$n = \sum_{k=0}^{\log_2(n)} c_k(n) \, 2^k$$

where the binary k-th digit $c_k(n) = \frac{1}{2}\left(1 - (-1)^{\left\lfloor \frac{n}{2^k} \right\rfloor}\right)$ is either 0 or 1 and where $\lfloor x \rfloor$ denotes the integral part of the real number x.

because the binary k-th digit $x_k(i)$ represents the infectious state of a node k in the network, while being in state i, and node labels in a graph G range from 1 to N, rather than from 0 to $N-1$.

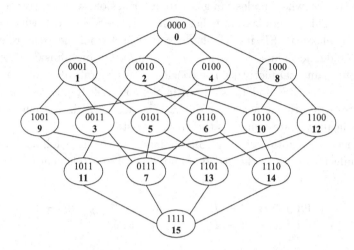

Fig. 17.1. The state diagram in a graph with $N = 4$ nodes and the binary numbering of the states. State 0 is the overall-healthy, virus-free network state, whereas in state 15, all nodes are infected.

The number of the states with j infected nodes is $\binom{N}{j}$. Fig. 17.1 shows an example of the Markov state graph with $N = 4$ nodes. A state with j infected nodes needs to make at least j transitions to reach the overall-healthy state 0. The total number of transitions from any state to state 0 is at least $\sum_{j=1}^{N} j\binom{N}{j} = N2^{N-1}$, so that the mean number of transitions from an arbitrary state to state 0 is at least $\frac{N}{2}$. Likewise, the corresponding variance equals $\sum_{j=1}^{N} j^2 \binom{N}{j} 2^{-N} - \left(\frac{N}{2}\right)^2 = \frac{N}{4}$. Hence, for large N, it requires for a randomly infected network, in the best possible case[7], on average $\frac{N}{2} + O\left(\sqrt{N}\right)$ hops to eradicate the virus.

In any continuous-time Markov process, there is only one event possible during an arbitrary small time interval (see Section 10.2), which implies for the $\varepsilon-$SIS process that a transition out of state i can only end in a state j, whose binary representation $x_N x_{N-1}...x_2 x_1$ has either one more 1 bit or one less than that of state i. If the state j has less 1 bits, then precisely one of the 1 bits (corresponding with one node in the network) of state i has been cured with rate δ. If the state j has precisely one additional 1 bit, then one of the nodes, say m, with $x_m(i) = 0$ has become infected (so that $x_m(j) = 1$) with rate $\varepsilon + \beta \sum_{k=1}^{N} a_{mk} x_k(i)$: the strength of infection is due to all neighbors of node m that are infected, i.e. $x_k(i) = 1$, augmented with the nodal self-infection rate ε of node m. The defined virus infection process is a continuous-time Markov chain with 2^N states specified by the infinitesimal

---

[7] Only curing transitions occur and all transition links are equally weighted.

generator Q with elements

$$
q_{ij} = \begin{cases}
\delta & \text{if } \begin{cases} j = i - 2^{m-1}; m = 1, 2...N \\ \text{and } x_m(i) = 1 \end{cases} \\
\varepsilon + \beta \sum_{k=1}^{N} a_{mk} x_k(i) & \text{if } \begin{cases} j = i + 2^{m-1}; m = 1, 2...N \\ \text{and } x_m(i) = 0 \end{cases} \\
-\sum_{k=1; k \neq j}^{N} q_{kj} & \text{if } i = j \\
0 & \text{otherwise}
\end{cases} \tag{17.7}
$$

where $i = \sum_{k=1}^{N} x_k(i) 2^{k-1}$. For example, if $i = 0$, then all $x_k = 0$ and the transition probability rates out of the network overall-healthy state 0 are $q_{0j} = \varepsilon$ for $j = 2^{m-1}$ with $1 \leq m \leq N$ (ranging over all network states with 1 infected node), while $q_{00} = -N\varepsilon$ and $q_{0j} = 0$ for all other j. Hence, for $\varepsilon > 0$, the first row in the infinitesimal generator Q is non-zero, whereas it is zero in the "classical" SIS model ($\varepsilon = 0$), corresponding to the absorbing state. When $\beta = 0$, there are no link-based infections, only nodal infections. Thus, locally, the infection process per node is a two-state continuous-time Markov process with self-infection rate ε and curing rate δ, from which we know (Section 10.6) that the steady-state infection probability for each node $k \in \mathcal{N}$ equals $\lim_{t \to \infty} \Pr[X_k(t) = 1] = \frac{\varepsilon}{\varepsilon + \delta}$. The infinitesimal generator Q, defined in (17.7), exhibits a recursive structure, which is studied in Van Mieghem and Cator (2012).

The probability state vector $s(t)$, with components

$$
s_i(t) = \Pr[X_1(t) = x_1(i), X_2(t) = x_2(i), ..., X_N(t) = x_N(i)]
$$

and normalization $\sum_{i=0}^{2^N - 1} s_i(t) = 1$, satisfies (10.19) in transposed form

$$
s^T(t) = s^T(0) e^{Qt}
$$

The definition of $s_i(t)$ as a joint probability distribution shows that, if we sum over all the states of all nodes except for the node j, we obtain the probability that a node j is either healthy $x_j = 0$ or infected $x_j = 1$,

$$
\Pr[X_j(t) = x_j] = \sum_{i=0; i \neq j}^{2^N - 1} s_i(t)
$$

where, in the index $i = \sum_{k=1}^{N} x_k(i) 2^{k-1}$ in the sum above, every $x_k(i)$ with $k \neq j$ takes both values from the set $\{0, 1\}$, while for $k = j$, $x_k(i) = x_j$. Defining the nodal viral infection probability as

$$
w_j(t) = \Pr[X_j(t) = 1] = E[X_j(t)] \tag{17.8}
$$

then the relation between the vectors $s(t)$ and $w(t)$ is[8]

$$
w^T(t) = s^T(t) M
$$

---

[8] The joint probabilities $\Pr[X_i = 1, X_j = 1]$ are computed similarly in Van Mieghem and Cator (2012).

where the $2^N \times N$ matrix M contains the states in binary notation, but bit-reversed:

$$M = \begin{bmatrix} 0 & 0 & 0 & \cdots & 0 \\ 1 & 0 & 0 & \cdots & 0 \\ 0 & 1 & 0 & \cdots & 0 \\ 1 & 1 & 0 & \cdots & 0 \\ 0 & 0 & 1 & \cdots & 0 \\ \vdots & \vdots & \vdots & \vdots & \vdots \\ 1 & 1 & 1 & \cdots & 1 \end{bmatrix}$$

Thus, the element $m_{ij} = x_{j-1}(i-1)$. The average fraction of infected nodes in G at time t equals $y(t) = E\left[\frac{1}{N}\sum_{j=1}^{N} X_j(t)\right] = \frac{1}{N}u^T w(t)$ and

$$y(t) = \frac{1}{N}s^T(0)e^{Qt}Mu \qquad (17.9)$$

where $u = (1, 1, \ldots, 1)$ is the all-one vector. We denote the steady-state average fraction of infected nodes for an effective infection rate τ by $y_\infty(\tau)$. For each graph, the steady-state average fraction of infected nodes in G at $\beta = \tau = 0$ equals $y_\infty(0) = \frac{\varepsilon}{\varepsilon+\delta}$ and, obviously, $1 \geq y_\infty(\tau) \geq \frac{\varepsilon}{\varepsilon+\delta}$ for $\tau \geq 0$.

Since the sum of the rows in any infinitesimal generator Q is zero, $Qu = 0$, the $2^N \times 1$ all-one vector u is the right-eigenvector belonging to the largest eigenvalue $\mu = 0$. The steady-state vector π in the $\varepsilon-$SIS model obeys (10.21), written here[9] as $\pi^T Q = 0$, so that π equals the left-eigenvector, normalized as $\sum_{n=0}^{2^N-1} \pi_n = 1$, belonging to the zero eigenvalue.

Fig. 17.2 shows the average number $Ny(t)$ of infected nodes as a function of time t in the Erdős-Rényi random graph $G_p(N)$ with $N = 64$ nodes and link density $p = 2p_c = 2\frac{\log N}{N} \simeq 0.13$ (see Section 15.7.5) for an effective infection rate τ above the epidemic threshold. The height of the apparent flat region for $t \geq 10$, which corresponds to the metastable state defined below, depends on the initial condition in agreement with (17.9). The exponential time scale in Fig. 17.2 obscures somehow the fact that $Ny(t) \sim e^{\zeta t}$ for large $t \geq 10$, where $\zeta < 0$ is the second largest eigenvalue of the infinitesimal generator Q. Thus, Fig. 17.2 depicts two regions: for small times (here $t \leq 10$), the SIS process tends exponentially fast to the metastable state, after which it decays exponentially in t (though with very small rate ζ) towards the absorbing state.

17.3 The governing $\varepsilon-$SIS equations

Instead of starting from the Markov state space as in Section 17.2.1, we now present another, more physical or intuitive approach that provides more insight.

---

[9] Here, we write a vector as a column vector, as usual in linear algebra, but as opposed to Markov theory.

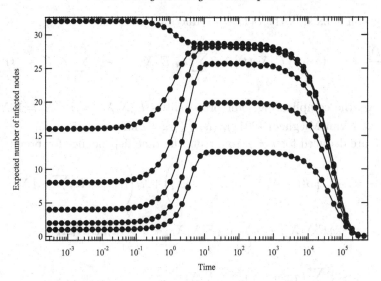

Fig. 17.2. The average number of infected nodes in the Erdős-Rényi random graph $G_{0.13}(64)$ with spectral radius $\lambda_1 = 9.671$ as a function of time for different numbers of initially infected nodes, with $\delta = 1$ and $\tau = \beta = 0.23$. The time is measured in units of $1/\delta$.

Since X_i is a Bernoulli random variable with the useful property that $E\left[X_i\right] = \Pr\left[X_i = 1\right]$, the exact SIS governing equation for node i equals

$$
\frac{dE\left[X_i\left(t\right)\right]}{dt} = E\left[-\delta X_i\left(t\right) + \left(1 - X_i\left(t\right)\right)\left\{\beta\sum_{k=1}^{N}a_{ki}X_k\left(t\right) + \varepsilon\right\}\right]
$$
$$
= E\left[\varepsilon - \left(\delta + \varepsilon\right)X_i\left(t\right) + \beta\sum_{k=1}^{N}a_{ki}X_k\left(t\right) - \beta\sum_{k=1}^{N}a_{ki}X_i\left(t\right)X_k\left(t\right)\right]
$$
$$(17.10)$$

The SIS governing equation (17.10) states that the change over time of the probability of infection $E\left[X_i\left(t\right)\right] = \Pr\left[X_i\left(t\right) = 1\right]$ of node i equals the average of two competing random variables: (a) if the node i is infected (X_i), then $\frac{dE[X_i]}{dt}$ decreases with rate equal to the curing rate δ and (b) if the node is healthy $(1 - X_i)$, it can be infected with infection rate β from each infected neighbor plus its own self-infection with rate ε. The total number of infected neighbors of node i is $\sum_{j=1}^{N}a_{ij}X_j$, where a_{ij} is the adjacency matrix element and the explicit reference to the underlying network over which the epidemic spreads. The differential equation (17.10) also holds for asymmetric adjacency matrices. In the sequel, we confine ourselves to undirected graphs such that $A = A^T$.

If the graph is fixed[10], then (17.10) reduces to

$$\frac{dE\left[X_i\left(t\right)\right]}{dt} = \varepsilon - \left(\delta + \varepsilon\right) E\left[X_i\left(t\right)\right] + \beta \sum_{k=1}^{N} a_{ki} E\left[X_k\left(t\right)\right] - \beta \sum_{k=1}^{N} a_{ki} E\left[X_i\left(t\right) X_k\left(t\right)\right]$$

$$(17.11)$$

which shows the complicating joint probabilities $E\left[X_i X_j\right] = \Pr\left[X_i = 1, X_j = 1\right]$. In Cator and Van Mieghem (2012), the $\binom{N}{2}$ governing equations for $\frac{dE[X_i X_j]}{dt}$ in case $\varepsilon = 0$ are deduced for $i \neq j$, (omitting the time-dependence for brevity)

$$\frac{dE\left[X_i X_j\right]}{dt} = E\left[X_j \left(\beta(1 - X_i) \sum_{k=1}^{N} a_{ik} X_k - \delta X_i\right) + X_i \left(\beta(1 - X_j) \sum_{k=1}^{N} a_{jk} X_k - \delta X_j\right)\right]$$

$$= -2\delta E[X_i X_j] + \beta \sum_{k=1}^{N} a_{ik} E[X_j X_k] + \beta \sum_{k=1}^{N} a_{jk} E[X_i X_k]$$

$$- \beta \sum_{k=1}^{N} (a_{ik} + a_{jk}) E[X_i X_j X_k] \qquad (17.12)$$

and when $i = j$, we find again (17.10). The first line expresses formally that $\frac{dE[X_i X_j]}{dt} = E\left[X_j \frac{dX_i}{dt} + X_i \frac{dX_j}{dt}\right]$. In a time-dependent Markov process, we can ignore the occurrence of multiple events in the arbitrarily small time interval $[t, t + dt]$. The right-hand side of the first equation states that the change in the joint expectation is due to an infection or/and curing in either node i or node j (not in both due to the Markov property). The equation (17.12) involves terms as $E\left[X_i\left(t\right) X_j\left(t\right) X_k\left(t\right)\right]$, which, in turn need to be determined.

In summary, by translating the SIS epidemic process directly in differential equations, we find that the N equations for $E\left[X_i\left(t\right)\right]$ in (17.10) require the knowledge of the joint expectations $E\left[X_k\left(t\right) X_l\left(t\right)\right]$, whose $\binom{N}{2}$ differential equations require the knowledge of $E\left[X_k\left(t\right) X_l\left(t\right) X_m\left(t\right)\right]$, whose $\binom{N}{3}$ differential equations involve joint fourth expectations and so on. Continuing in this way, $\sum_{k=1}^{N} \binom{N}{k} = 2^N - 1$ equations are needed, plus the conservation of probability equation, $\Pr\left[X_i\left(t\right) = 1\right] + \Pr\left[X_i\left(t\right) = 0\right] = 1$. All equations for higher-order joint probabilities can be computed from (17.10). Indeed, interchanging the derivative and expectation operator yields

$$\frac{dX_j}{dt} = -\delta X_j + (1 - X_j) \left\{\beta \sum_{k=1}^{N} a_{kj} X_k + \varepsilon\right\} \qquad (17.13)$$

Strictly speaking, the derivative of an indicator does not exist, but we agree to

---

[10] In adaptive networks as studied in Guo *et al.* (2013), the topology can be changed depending on the state of the infection. If people recognize that their friend is highly infectious, they avoid contact until the friend has recovered, whereafter the link (contact) is re-established. Hence, in adaptive networks, the adjacency matrix elements depend on the set $\{X_k\}_{1 \leq k \leq N}$ and $E\left[a_{ij} X_k\right]$ cannot be simplified without additional knowledge of how the topology interacts with the epidemics.

formally define it by the random variable equation (17.13). Next, making the same reversal of operators,

$$\frac{d}{dt} E\left[\prod_{j=1}^{n} X_j\right] \overset{\text{formally}}{=} E\left[\frac{d}{dt}\prod_{j=1}^{n} X_j\right] = E\left[\sum_{m=1}^{n} \frac{dX_m}{dt} \prod_{j=1; j\neq m}^{n} X_j\right]$$

substituting (17.13) and executing the expectation $E\left[.\right]$ returns the correct result[11],

$$\frac{d}{dt} E\left[\prod_{j=1}^{n} X_j\right] = \varepsilon \sum_{m=1}^{n} E\left[\prod_{j=1; j\neq m}^{n} X_j\right] + \beta \sum_{m=1}^{n}\sum_{k=1}^{N} a_{km} E\left[X_k \prod_{j=1; j\neq m}^{n} X_j\right]$$
$$- (\delta + \varepsilon)\, nE\left[\prod_{j=1}^{n} X_j\right] - \beta \sum_{k=1}^{N}\left(\sum_{m=1}^{n} a_{km}\right) E\left[X_k \prod_{j=1}^{n} X_j\right]$$

$$(17.14)$$

For each combination of n out of N states, such a differential equation for the joint probability

$$E\left[\prod_{j=1}^{n} X_j\right] = \Pr\left[X_1 = 1, X_2 = 1, \ldots, X_n = 1\right]$$

can be written. The expectation in the last summation in (17.14) contains, except when $X_l^2 = X_l$ occurs, a product of $n+1$ different random variables X_j, for which a new differential equation is needed as outlined above. The approach in Section 17.2.1 directly writes the 2^N equations using the Markov transition probability graph. When the exact solution is required, the Markov matrix method is recommended, whereas approximations and physical insight are more elegantly derived from (17.10) as shown below.

17.3.1 The steady-state of the ε−SIS epidemic process

As in Section 10.3, we define the steady-state random variable as

$$X_i = \lim_{t\to\infty} X_i\,(t)$$

which obeys $\lim_{t\to\infty} \frac{dE[X_i(t)]}{dt} = 0$. The fraction of infected nodes in the steady-state is denoted by

$$S_\infty = \frac{1}{N}\sum_{i=1}^{N} X_i \qquad (17.15)$$

Equation (17.10) reduces in the steady-state to

$$\varepsilon = (\delta + \varepsilon)\,E\left[X_i\right] - \beta \sum_{k=1}^{N} a_{ki}E\left[X_k\right] + \beta \sum_{k=1}^{N} a_{ki}E\left[X_i X_k\right] \qquad (17.16)$$

---

[11] The formal method can be made mathematically rigorous (using the framework of stochastic differential equations).

which demonstrates that $X_i \neq 0$ for *all* $1 \leq i \leq N$ when $\varepsilon > 0$. In other words, the Markovian ε–SIS epidemic process has a unique, well-defined steady-state solution, that is different from the absorbing state when $\varepsilon = 0$, in which $X_i = 0$ for all $1 \leq i \leq N$. These observations were already made in Section 17.2.1 based on the Markov graph and infinitesimal generator. The complicating absorbing state in the "classical" SIS model ($\varepsilon = 0$) is one of the reasons to introduce and to study the ε–SIS model. Since any SIS epidemic process ($\varepsilon = 0$) on any finite graph, eventually stops when the absorbing state is reached, the interesting question is how long it takes to converge to the absorbing state. This question is answered in Section 17.3.3, after we have deduced general expressions for the average steady-state fraction of infected nodes.

17.3.2 The average fraction of infected nodes in the steady-state

Summing the SIS governing equation (17.10) over all nodes i, yields

$$\frac{dE\left[\sum_{i=1}^{N} X_i\right]}{dt} = E\left[N\varepsilon - (\delta + \varepsilon)\sum_{i=1}^{N} X_i + \beta\sum_{k=1}^{N}\sum_{i=1}^{N} a_{ki}X_k - \beta\sum_{k=1}^{N}\sum_{i=1}^{N} a_{ki}X_iX_k\right]$$

and, after letting $t^* = \delta t$ with $\varepsilon^* = \frac{\varepsilon}{\delta}$ and $\tau = \frac{\beta}{\delta}$ and using that the degree $d_k = \sum_{i=1}^{N} a_{ki}$,

$$\frac{dE\left[\sum_{i=1}^{N} X_i\right]}{dt^*} = E\left[N\varepsilon^* - (1 + \varepsilon^*)\sum_{i=1}^{N} X_i + \tau\sum_{k=1}^{N} d_k X_k - \tau\sum_{k=1}^{N}\sum_{i=1}^{N} a_{ki}X_iX_k\right]$$

We consider the case where $\frac{d}{dt}E\left[\sum_{i=1}^{N} X_i\right] = 0$, and concentrate further on the steady-state ($t \to \infty$), although a zero derivative also determines an extremal value of $E\left[\sum_{i=1}^{N} X_i\right]$ (possibly attained at some finite time). The average steady-state fraction of infected nodes, also known as the order parameter or the prevalence, is denoted as

$$y_\infty(\tau) = E\left[S_\infty\right] = \frac{1}{N}\sum_{i=1}^{N} E\left[X_i\right]$$

When introducing this definition, we obtain

$$y_\infty(\tau) = \frac{\varepsilon^*}{1 + \varepsilon^*} + \frac{\tau}{(1 + \varepsilon^*)N}E\left[\sum_{k=1}^{N} d_k X_k - \sum_{k=1}^{N}\sum_{i=1}^{N} a_{ki}X_iX_k\right]$$

If we define the vector $X_\infty = (X_1, X_2, \ldots, X_N)$ of random variables and the degree vector D with i-th vector component the degree d_i of node i, then

$$y_\infty(\tau) = \frac{\varepsilon^*}{1 + \varepsilon^*} + \frac{\tau}{(1 + \varepsilon^*)N}E\left[D^T X_\infty - X_\infty^T A X_\infty\right] \tag{17.17}$$

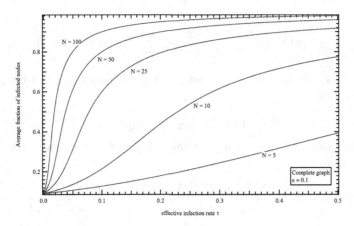

Fig. 17.3. The average steady-state fraction $y_\infty(\tau)$ of infected nodes in the complete graph versus the effective infection rate τ for $\varepsilon^* = 10^{-1}$ and for various sizes N.

where $X_\infty^T A X_\infty$ is the sum over all links with both end nodes infected. The total number of links, possibly counted twice, with at least one infected end node is $D^T X_\infty$. Hence, $D^T X_\infty \geq X_\infty^T A X_\infty$ from which the lower bound $y_\infty(\tau) \geq \frac{\varepsilon^*}{1+\varepsilon^*} = y_\infty(0)$ follows from (17.17). The relation (17.17) expresses the average steady-state fraction of infected nodes in terms of the number of infected link interactions. A typical shape of $y_\infty(\tau)$ as a function of the effective infection rate τ is plotted in Fig. 17.3.

If we write the degree vector as $D = \Delta u$, where $\Delta = \text{diag}(d_1, d_2, \ldots, d_N)$, then

$$D^T X_\infty - X_\infty^T A X_\infty = u^T \Delta X_\infty + X_\infty^T \Delta X_\infty - X_\infty^T \Delta X_\infty - X_\infty^T A X_\infty$$

$$= (u - X_\infty)^T \Delta X_\infty + X_\infty^T (\Delta - A) X_\infty$$

Now, $u - X_\infty$ is the (random) vector with non-infected nodes and

$$(u - X_\infty)^T \Delta X_\infty = \sum_{j=1}^N (1 - X_j) d_j X_j = \sum_{j=1}^N (X_j - X_j^2) d_j = 0$$

because $X_j = X_j^2$ as $X_j \in \{0, 1\}$. Introducing the Laplacian $Q = \Delta - A$ of the graph G, the steady-state average fraction of infected nodes y_∞ is expressed as a quadratic form of the Laplacian Q as

$$y_\infty(\tau) = \frac{\varepsilon^*}{1 + \varepsilon^*} + \frac{\tau E\left[X_\infty^T Q X_\infty\right]}{(1 + \varepsilon^*) N} \tag{17.18}$$

Introducing the basic Laplacian property $z^T Q z = \sum_{l \in \mathcal{L}} (z_{l+} - z_{l-})^2$, where the link l connects the node l^+ and node l^-, yields

$$E\left[X_\infty^T Q X_\infty\right] = 2 \sum_{l \in \mathcal{L}} E\left[X_{l+}(1 - X_{l-})\right]$$

and

$$y_\infty(\tau) = \frac{\varepsilon^*}{1+\varepsilon^*} + \frac{2\tau}{(1+\varepsilon^*)\,N} \sum_{l \in \mathcal{L}} E\left[X_{l^+}\left(1 - X_{l^-}\right)\right] \qquad (17.19)$$

illustrating that only links with one end infected contribute to the average fraction of infected nodes.

17.3.2.1 Regular graphs

For a regular graph, where each node j has a degree $d_j = r$, the analysis leading to (17.17) simplifies to

$$y_{\infty;\text{regular}}(\tau) = \frac{\frac{1}{N}\sum_{k=1}^{N}\sum_{i=1}^{N} a_{ki}E\left[X_i X_k\right] - \frac{\varepsilon^*}{\tau}}{r - \frac{(1+\varepsilon^*)}{\tau}} = \frac{\frac{1}{N}E\left[X_\infty^T A X_\infty\right] - \frac{\varepsilon^*}{\tau}}{r - \frac{(1+\varepsilon^*)}{\tau}} \qquad (17.20)$$

In case $\varepsilon^* = 0$, the average fraction of infected nodes in a regular graph equals

$$y_{\infty;\text{regular}}(\tau) = \frac{\frac{1}{N}E\left[X_\infty^T A X_\infty\right]}{r - \frac{1}{\tau}} \qquad (17.21)$$

Since $E\left[X_\infty^T A X_\infty\right] \geq 0$ and the fraction $y_\infty(\tau)$ of infected nodes is also non-negative, relation (17.21) illustrates the threshold behavior (see Section 17.1): if $\tau < \frac{1}{r}$, then $y_{\infty;r}(\tau) = 0$ and thus $E\left[X_\infty^T A X_\infty\right] = 0$, whereas, if $\tau > \frac{1}{r}$, then $y_{\infty;r}(\tau) \geq 0$. These two τ-regimes when $\varepsilon^* = 0$, separated by a threshold point $\tau^* = \frac{1}{r}$, are a general characteristic of the classical SIS-process ($\varepsilon^* = 0$) in any graph as shown in Section 17.3.3, though not so directly apparent as here for a regular graph.

Example The adjacency matrix of the complete graph K_N is $A = J - I = u.u^T - I$ so that, using the definition (17.15) of S_∞ and $X_j^2 = X_j$,

$$E\left[X_\infty^T A X_\infty\right] = E\left[X_\infty^T u.u^T X_\infty - X_\infty^T X_\infty\right] = E\left[\left(\sum_{j=1}^{N} X_j\right)^2 - \sum_{j=1}^{N} X_j^2\right]$$

$$= N^2 E\left[S_\infty^2\right] - N E\left[S_\infty\right]$$

In terms of the variance of $S_\infty = \frac{1}{N}\sum_{j=1}^{N} X_j$ and $y_{\infty;K_N}(\tau) = E\left[S_\infty\right]$,

$$E\left[X_\infty^T A X_\infty\right] = N^2 \text{Var}\left[S_\infty\right] + N^2 y_{\infty;K_N}^2(\tau) - N y_{\infty;K_N}(\tau)$$

and substituted into (17.21), yields, after reworking,

$$y_{\infty;K_N}^2(\tau) - \left(1 - \frac{1}{N\tau}\right) y_{\infty;K_N}(\tau) + \text{Var}\left[S_\infty\right] = 0$$

Solving the quadratic equation, taking into account that $y_{\infty;K_N}(0) = 0$ for $\varepsilon = 0$, yields

$$y_{\infty;K_N}(\tau) = \frac{1}{2}\left(1 - \frac{1}{N\tau}\right) + \frac{1}{2}\sqrt{\left(1 - \frac{1}{N\tau}\right)^2 - 4\text{Var}\left[S_\infty\right]}$$

where a non-negative discriminant imposes that $\text{Var}[S_\infty] \leq \frac{1}{4}\left(1 - \frac{1}{N\tau}\right)^2$ or, in terms of the variance,

$$\tau \geq \frac{1}{N\left(1 - 2\sqrt{\text{Var}[S_\infty]}\right)} \geq \frac{1}{N}\left(1 + 2\sqrt{\text{Var}[S_\infty]}\right) \qquad (17.22)$$

which should be compared with the exact epidemic threshold in (17.101). Curiously, at $\tau = \frac{1}{N} <$ τ^*, $\mathrm{Var}[S_\infty] = 0$ as well as $y_{\infty;K_N}(\tau) = E[S_\infty] = 0$. Using the expansion $(1 + x)^\alpha = \sum_{k=0}^\infty \binom{\alpha}{k} x^k$ valid for $|x| < 1$, we obtain with $x = \frac{4\mathrm{Var}[S_N]}{\left(1 - \frac{1}{N\tau}\right)^2} \in [0, 1]$, that

$$
\begin{aligned}
y_{\infty;K_N}(\tau) &= \frac{1}{2}\left(1 - \frac{1}{N\tau}\right)\left(1 + \sqrt{1 - x}\right) \\
&= \left(1 - \frac{1}{N\tau}\right)\left(1 - \frac{x}{4} - \frac{x^2}{16} + O(x^3)\right) \\
&= 1 - \frac{1}{N\tau} - \frac{\mathrm{Var}[S_\infty]}{1 - \frac{1}{N\tau}} - \frac{(\mathrm{Var}[S_\infty])^2}{\left(1 - \frac{1}{N\tau}\right)^3} + O(x^3)
\end{aligned}
$$

This expression, from which $y_{\infty;K_N}(\tau) < 1 - \frac{1}{N\tau}$ for $\tau > \frac{1}{N}$, should be compared with the mean-field approximation $y^{(1)}_{\infty;K_N}(\tau) = 1 - \frac{1}{(N-1)\tau} \geq y_{\infty;K_N}(\tau)$ derived later in (17.39).

In summary, the exact analysis of the regular graph (also for $\varepsilon = 0$) reveals the physical aspects of the metastable state, further elaborated in Section 17.3.3 where the threshold point will be called the epidemic threshold.

17.3.3 General properties of the SIS epidemic process

Since $0 \leq \sum_{k=1}^N a_{ki} X_i(t) X_k(t)$, we deduce from (17.10) the upper bound

$$
\frac{dE[X_i(t)]}{dt} \leq \varepsilon - (\delta + \varepsilon) E[X_i(t)] + \beta \sum_{k=1}^N a_{ki} E[X_k(t)]
$$

With (17.8) and denoting the vector $W = (w_1, w_2, \ldots, w_N)$, the set of all nodes becomes the matrix inequality

$$
\frac{dW(t)}{dt} \leq \varepsilon u + (\beta A - (\delta + \varepsilon) I) W(t) \tag{17.23}
$$

where $u = (1, 1, \ldots, 1)$ is the all-one vector. The solution of (17.23) is

$$
\begin{aligned}
W(t) &\leq e^{(\beta A - (\delta + \varepsilon)I)t} W(0) + \int_0^t e^{(\beta A - (\delta + \varepsilon)I)(t-s)} \varepsilon u \, ds \\
&= e^{(\beta A - (\delta + \varepsilon)I)t} W(0) + \varepsilon \frac{I - e^{(\beta A - (\delta + \varepsilon)I)t}}{\beta A - (\delta + \varepsilon) I} u
\end{aligned}
$$

Thus[12],

$$
W(t) \leq e^{(\tau A - (1 + \varepsilon^*)I)t^*} W(0) + \varepsilon^* \frac{I - e^{(\tau A - (1 + \varepsilon^*)I)t^*}}{\tau A - (1 + \varepsilon^*) I} u \tag{17.24}
$$

where $\varepsilon^* = \frac{\varepsilon}{\delta}$, the effective infection rate is $\tau = \frac{\beta}{\delta}$ and the normalized time $t^* = \delta t$ is measured in units of the curing rate δ. The upper bound is dominated by the fastest growth in t^*, which is due to the largest eigenvalue of $\tau A - (1 + \varepsilon^*) I$. The

---

[12] We remark that a symmetric matrix A commutes with e^{At} because $A = X\Lambda X^T$, from which it follows that $e^{At} = X\mathrm{diag}(e^{\lambda_i t}) X^T$ and $Ae^{At} = e^{At} A$. Hence, the inverse matrix $(\beta A - (\delta + \varepsilon) I)^{-1}$ commutes with $e^{(\beta A - (\delta + \varepsilon)I)t}$ and can thus be written in fractional form.

exponential factor is dominated by $\tau\lambda_1 - (1 + \varepsilon^*)$, where λ_1 is the real, largest eigenvalue of the non-negative matrix A (by the Perron-Frobenius Theorem, Appendix A.5.1). When $\tau\lambda_1 - (1 + \varepsilon^*) \leq 0$ or $\tau \leq \frac{1+\varepsilon^*}{\lambda_1}$, then $w_i(t) = E[X_i(t)]$ decreases exponentially in t^* towards $\varepsilon^* \left\{ (\tau A - (1 + \varepsilon^*) I)^{-1} u \right\}_i$, which is positive for $\varepsilon^* > 0$. When there is no self-infection, $\varepsilon = 0$, the epidemic will quickly die out. The argument shows that, for $\tau \leq \frac{1+\varepsilon^*}{\lambda_1}$, the ε–SIS epidemic process tends exponentially fast to a steady-state. For effective infection rates $\tau > \frac{1+\varepsilon^*}{\lambda_1}$, we cannot conclude an exponentially fast tendency towards a steady-state.

In the sequel, we confine ourselves to the "classical" SIS process where $\varepsilon = 0$. The epidemic threshold τ_c is defined as the border between exponential die-out and a non-zero fraction of infected nodes in the metastable state. From (17.24) and the subsequent arguments above, we conclude with:

Theorem 17.3.1 *In any finite sized network and for $\varepsilon = 0$, the exact SIS epidemic threshold*

$$\tau_c \geq \tau_c^{(1)} = \frac{1}{\lambda_1} \tag{17.25}$$

where λ_1 is the spectral radius of the adjacency matrix A.

The lower bound $\tau_c^{(1)} = \frac{1}{\lambda_1}$ is found in the first-order mean-field approximation (Lemma 17.4.6), denoted by the superscript $^{(1)}$, and is of great practical use: if the effective infection rate τ can be controlled such that $\tau \leq \tau_c^{(1)}$, then the network is safeguarded from long-term, massive infection. The first-order mean-field epidemic threshold $\tau_c^{(1)} = \frac{1}{\lambda_1}$ can thus be regarded as a viral graph metric. In Cator and Van Mieghem (2012), we show that (i) higher-order mean-field approximations lead to better lower bounds of the "exact" threshold, i.e. $\tau_c \geq \cdots \geq \tau_c^{(m)} \cdots \geq \tau_c^{(2)} = \frac{1}{\lambda_1(H)} \geq \tau_c^{(1)} = \frac{1}{\lambda_1(A)}$, where H is a $N^2 \times N^2$ matrix involving features of the topology, and that (ii) computationally, higher-order mean-field models are not efficient to determine τ_c. Recently, we proved in Van Mieghem (2014) that:

Theorem 17.3.2 *For large N, the SIS epidemic threshold τ_c in any graph is upper bounded by*

$$\tau_c \leq \frac{1}{d_{\min}} \left(1 + O\left(\frac{1}{\sqrt{N}} \right) \right) \tag{17.26}$$

Proof: We start from (17.19) for $\varepsilon = 0$, which we rewrite with $Ny(\tau) = \sum_{i=1}^{N} \Pr[X_i = 1]$ and

$$2\sum_{l\in\mathcal{L}} E[X_{l+}(1 - X_{l-})] = \sum_{i=1}^{N}\sum_{j=1}^{N} a_{ij}\Pr[X_i = 1, X_j = 0] = \sum_{i=1}^{N}\Pr[X_i = 1]\sum_{j=1}^{N} a_{ij}\Pr[X_j = 0|X_i = 1]$$

as

$$\tau^{-1} = \frac{\sum_{i=1}^{N} \Pr[X_i = 1] \sum_{j=1}^{N} a_{ij}\Pr[X_j = 0|X_i = 1]}{\sum_{i=1}^{N} \Pr[X_i = 1]}$$

The inequality (5.4) leads to

$$\min_{1\le i\le N}\sum_{j=1}^{N}a_{ij}\Pr\left[X_j=0|X_i=1\right]\le\tau^{-1}\le\max_{1\le i\le N}\sum_{j=1}^{N}a_{ij}\Pr\left[X_j=0|X_i=1\right]\le d_{\max}$$

Using the degree $d_i=\sum_{j=1}^{N}a_{ij}$, we proceed with the lower bound,

$$\tau^{-1}\ge\min_{1\le i\le N}\sum_{j=1}^{N}a_{ij}\Pr\left[X_j=0|X_i=1\right]\ge\min_{1\le i\le N}\left(\min_{(k,l)\in\mathcal{L}}\Pr\left[X_k=0|X_l=1\right]d_i\right)$$

$$=\min_{(k,l)\in\mathcal{L}}\Pr\left[X_k=0|X_l=1\right]d_{\min}=d_{\min}\left(1-\max_{(k,l)\in\mathcal{L}}\Pr\left[X_k=1|X_l=1\right]\right)$$

We define the epidemic threshold τ_c as that value of τ when the prevalence or order parameter $y_\infty(\tau)=\frac{1}{N}\sum_{i=1}^{N}\Pr\left[X_i=1\right]$ approaches zero from above, denoted as $y\downarrow0$, so that

$$\tau_c^{-1}=\lim_{y_\infty\downarrow0}\frac{E\left[X_\infty^T Q X_\infty\right]}{Ny_\infty} \tag{17.27}$$

and

$$\tau_c^{-1}\ge d_{\min}\left(1-\lim_{y_\infty\downarrow0}\max_{(k,l)\in\mathcal{L}}\Pr\left[X_k=1|X_l=1\right]\right) \tag{17.28}$$

Definition (17.27) is increasingly accurate for large N. In any graph G, the conditional probability $\varepsilon_G=\lim_{y_\infty\downarrow0}\max_{(k,l)\in\mathcal{L}}\Pr\left[X_k=1|X_l=1\right]$ can be upper bounded by $\varepsilon_G\le\varepsilon_{K_N}$, because the infection probability ε_G on a link (k,l) in the graph G is largest in the complete graph. Using (17.101) implies that $\varepsilon_{K_N}=O\left(\frac{1}{\sqrt{N}}\right)$ for large N and we arrive at (17.26). $\qquad\square$

We know already that this SIS process possesses an absorbing state, which will eventually be reached irrespective of the effective infection rate τ. However, when $\tau\le\tau_c^{(1)}=\frac{1}{\lambda_1}$, the absorbing state is reached at least exponentially fast. For $\tau>\tau_c^{(1)}$, an entirely different behavior is observed, where the epidemic process needs a huge time to reach the absorbing state. The Markov graph in Fig. 17.1 illustrates that, for large τ, the process (without self-infection, $\varepsilon=0$) resides in the states with many infected nodes. Intuitively, we can understand that, only with very low probability, the absorbing state is hit out of 2^N states. Due to the huge time to absorption, the epidemic lives a long time before reaching the steady-state. Therefore, this regime is called the metastable state to which the epidemic rapidly converges, but only very slowly leaves. These two different regimes, separated by the epidemic threshold, are now investigated. We first prove:

Theorem 17.3.3 *For any initial vector $X(0)=(X_1(0),X_2(0),...,X_N(0))$ and all times $t\ge0$, the probability that a SIS Markov process ($\varepsilon=0$) has not yet entered the absorbing state (where $X_k(t)=0$ for all nodes $1\le k\le N$) is upper bounded as*

$$\Pr\left[X(t)\ne0\right]\le\sqrt{N\sum_{j=0}^{N}X_j(0)e^{(\beta\lambda_1-\delta)t}}$$

Proof: Invoking the Boole inequality (2.9) yields

$$\Pr\left[\cup_{k=1}^{N}\{X_k(t) \neq 0\}\right] \leq \sum_{k=1}^{N}\Pr[X_k(t) \neq 0] = \sum_{k=1}^{N}\Pr[X_k(t) = 1] = \sum_{k=1}^{N}E[X_k(t)] = u^T W(t)$$

and, thus, $\Pr[X(t) \neq 0] \leq u^T W(t)$. We introduce the upper bound (17.24) for $W(t)$ in case $\varepsilon = 0$,

$$\Pr[X(t) \neq 0] \leq u^T W(t) \leq u^T e^{(\beta A - \delta)t} W(0)$$

Using the Cauchy-Schwarz inequality (A.28), we have that

$$u^T e^{(\beta A - \delta)t} W(0) = \left|u^T e^{(\beta A - \delta)t} W(0)\right| \leq \|u\|_2 \left\|e^{(\beta A - \delta)t} W(0)\right\|_2$$

From (A.37) and since A is symmetric, it follows that

$$\left\|e^{(\beta A - \delta)t} W(0)\right\|_2 \leq e^{(\beta \lambda_1 - \delta)t} \|W(0)\|_2$$

Hence,

$$\Pr[X(t) \neq 0] \leq \|u\|_2\, e^{(\beta \lambda_1 - \delta)t} \|W(0)\|_2 = \sqrt{N} e^{(\beta \lambda_1 - \delta)t} \sqrt{W^T(0) W(0)}$$

where the scalar product is

$$W^T(0) W(0) = E\left[\sum_{j=0}^{N}X_j^2(0)\right] = E\left[\sum_{j=0}^{N}X_j(0)\right] = \sum_{j=0}^{N}X_j(0)$$

which proves Theorem 17.3.3. □

Let T denote the time to reach the absorbing state, in short, the time to absorption. A consequence of Theorem 17.3.3 is:

Corollary 17.3.4 *The mean time for the SIS Markov process without self-infection* ($\varepsilon = 0$) *to hit the absorbing state when the effective infection rate* $\tau < \frac{1}{\lambda_1}$ *is not larger than*

$$E[T] \leq \frac{1}{\delta} \frac{\log N + 1}{1 - \tau \lambda_1} \tag{17.29}$$

Proof: For an arbitrary initial condition $X(0)$, the mean time for the SIS Markov process to enter the absorbing state equals in terms of the tail probability (2.36)

$$E[T] = \int_0^{\infty} \Pr[T > t]\, dt$$

Since the event $\{T > t\}$ is equivalent to $\{X(t) \neq 0\}$, we have $E[T] = \int_0^{\infty} \Pr[X(t) \neq 0]\, dt$. Invoking Theorem 17.3.3 in the worst case, where $X_k(0) = 1$ for each node k, yields

$$E[T] \leq \int_0^{\infty} \min\left(1, N e^{(\beta \lambda_1 - \delta)t}\right) dt$$

Only when $\beta \lambda_1 - \delta < 0$ or $\tau < \frac{1}{\lambda_1} = \tau_c^{(1)}$, can the second argument in the minimum be smaller than 1. Let $r = \frac{\log N}{\delta - \beta \lambda_1}$ denote the smallest time t for which $N e^{(\beta \lambda_1 - \delta)t} = 1$, then

$$E[T] \leq r + \int_r^{\infty} N e^{(\beta \lambda_1 - \delta)t} dt = r + \frac{N e^{(\beta \lambda_1 - \delta)r}}{\delta - \beta \lambda_1}$$

Since $N e^{(\beta \lambda_1 - \delta)r} = 1$, we arrive at (17.29). □

The regime where $\tau \geq \tau_c^{(1)} = \frac{1}{\lambda_1}$ involves the isoperimetric constant of a graph G, defined as

$$\eta \leq \eta_m = \min_{\mathcal{N}_A} \left\{ \frac{\partial A}{N_A} \,\middle|\, N_A = m \right\}$$

where ∂A is the number of links between the set $\mathcal{N}_A$ and its complement $\mathcal{N} \backslash \mathcal{N}_A$ and $m \leq \frac{N}{2}$. The isoperimetric constant naturally emerges from the "physical" description (17.6) of the ε-SIS process in terms of the infected subgraph I.

Theorem 17.3.5 *The probability that the time T to absorption exceeds $\frac{s}{2m}$ equals*

$$\Pr\left[T \geq \frac{s}{2m} \right] \geq \frac{1-x}{1-x^m} \left(\frac{1-x^{m-1}}{1-x^m} \right)^s \left(1 + o\left(s^{-1}\right) \right)$$

where $s \in \mathbb{N}$, but large, $m < N$ and $\tau \eta_m \geq \frac{1}{x}$.

Proof: see Draief and Massoulié (2010, Section 8.3.2). □

The isoperimetric constant is bounded by Mohar's bound (Van Mieghem, 2011, p. 95)

$$\eta \leq \sqrt{\mu_{N-1} \left(2 d_{\max} - \mu_{N-1} \right)}$$

where μ_{N-1} is the algebraic connectivity of the graph. The requirement $\tau \eta_m \geq \frac{1}{x}$ can be replaced by $x \geq \frac{1}{\tau \sqrt{\mu_{N-1}(2d_{\max} - \mu_{N-1})}}$, which is easier to compute than η_m. Draief and Massoulié (2010, p. 99) also show that, for $\tau \geq \frac{1}{x\eta_m}$, the mean time to absorption is $E\left[T\right] = O\left(e^{bN^a}\right)$ for some constants $a, b > 0$. The decay rate ζ of the Markovian SIS epidemic process towards the absorbing state in the complete graph K_N for effective infection rates $\tau > \tau_c > \frac{1}{N}$ is computed[13] in Van Mieghem (2013) as

$$-\zeta = \frac{\delta}{\sum_{j=1}^{N} \sum_{r=0}^{j-1} \frac{(N-j+r)!}{j(N-j)!} \tau^r} + O\left(\frac{N^2 \log N}{(N\tau)^{2N-1}} \right) \tag{17.30}$$

from which we deduce that $E\left[T\right] \simeq \frac{1}{|\zeta|} = O\left(e^{N \ln \frac{\tau}{\tau_c}}\right)$. Hence, we find for K_N that $b = \ln \frac{\tau}{\tau_c}$ and $a = 1$ in the general estimate $E\left[T\right] = O\left(e^{bN^a}\right)$ of Draief and Massoullié and, moreover, that $a \leq 1$ for any graph, because the complete graph features the slowest decay rate among all graphs. Mountford *et al.* (2013) proved that, above the epidemic threshold in trees with bounded degree, $E\left[T\right] = O(e^{cN})$ for a real number $c > 0$. Moreover, improving a result of Chatterjee and Durrett (2009), they show that for any $\tau > 0$ and large N, the time to absorption or extinction on a power-law graph grows exponentially in N. Combining these lower bound results and the upper bound in (17.30), we may conclude that for almost all graphs, the average time to absorption for $\tau > \tau_c$ is $E\left[T\right] = O\left(e^{cN}\right)$. In summary,

---

[13] In Section 17.6.1, the number of infected nodes in K_N is described by a birth and death process, that possesses an infinitesimal generator of the form (11.1), whose spectral decomposition is analyzed in Appendix A.6.3.

the mean "lifetime" of the epidemic below and above the epidemic phase transition are hugely different, which is a general characteristic of a phase transition.

Finally, we mention:

Theorem 17.3.6 *For any pair (i, j) of nodes in a network, the joint probability of infection satisfies*

$$\Pr[X_i = 1, X_j = 1] \geq \Pr[X_i = 1] \Pr[X_j = 1] \tag{17.31}$$

Inequality (17.31) has been naturally conjectured in Van Mieghem *et al.* (2009), but only recently proved by Cator and Van Mieghem (2013a) using the FKG inequality. Physically, Theorem 17.3.6 means that an infection somewhere in a network node j cannot diminish the infection probability of another node i; in other words, $\Pr[X_i = 1|X_j = 1] \geq \Pr[X_i = 1]$.

17.4 N-Intertwined Mean-Field Approximation (NIMFA)

Since the exact ε-SIS Markov process requires the solution of 2^N linear equations, which is infeasible for real-world networks, we discuss the N-Intertwined Mean-Field Approximation (NIMFA) of Van Mieghem *et al.* (2009), which is currently the most accurate approximation of the $\varepsilon = 0$ SIS model in any network (see e.g. Li *et al.* (2012)).

The number of infected neighbors of node i, $\sum_{j=1}^{N} a_{ij} X_j(t)$ in (17.10), couples or "intertwines" each of the N nodal infection states in the network and causes higher-order joint probabilities. By assuming independence[14], where $E[X_i(t) X_j(t)] = E[X_i(t)] E[X_j(t)]$ and denoting $v_i(t) = \Pr[X_i(t) = 1]$ in this assumption (as opposed to $w_i(t) = \Pr[X_i(t) = 1]$ in (17.8) in the exact SIS epidemics), we observe from (17.11) for $\varepsilon = 0$ that equations for higher-order joint probabilities are no longer needed and we arrive at the single governing equation for a node i in NIMFA,

$$\frac{dv_i(t)}{dt} = -\delta v_i(t) + \beta \left(1 - v_i(t)\right) \sum_{j=1}^{N} a_{ij} v_j(t) \tag{17.32}$$

Comparison of the exact equation (17.10) with NIMFA (17.32) shows that the random variable X_k is replaced by its mean $v_k = E[X_k]$, irrespective of any correlation introduced by terms as $E[X_i X_j]$. Also, ignoring correlations leads to a replacement of the actual number of infected neighbors $\sum_{j=1}^{N} a_{ij} X_j$ by its mean $\sum_{j=1}^{N} a_{ij} v_j$. The mean $\sum_{j=1}^{N} a_{ij} v_j$ is an increasingly accurate estimator of the random variable $\sum_{j=1}^{N} a_{ij} X_j$ when N increases (for weak correlations among $\{X_j\}_{1 \leq j \leq N}$) due to the Central Limit Theorem 6.3.1.

---

[14] This first-order mean-field approximation is the simplest way to "close" the equations – a terminology used in epidemiology. The act of closing the equations, called the *closure*, consists of writing higher-order expectations in terms of lower-order ones as in NIMFA. In most cases, the closure implies an approximation and sometimes, as illustrated in Cator and Van Mieghem (2012), it may lead to inconsistencies.

Theorem 17.4.1 *For each node i, NIMFA upper bounds $\frac{dE[X_i(t)]}{dt} \leq \frac{dv_i(t)}{dt}$ as well as $E[X_i] = w_{i\infty} \leq v_{i\infty}$.*

Proof: We use conditional probabilities (2.48),

$$E[X_i X_j] = \Pr[X_i = 1, X_j = 1] = \Pr[X_i = 1 | X_j = 1] \Pr[X_j = 1]$$

and Theorem 17.3.6. Hence, we have $E[X_i(t) X_j(t)] \geq E[X_i(t)] E[X_j(t)]$. Introduced in (17.10) shows that $\frac{dE[X_i(t)]}{dt} \leq \frac{dv_i(t)}{dt}$ and applied to the steady-state equation (17.16), we obtain $E[X_i] \leq v_{i\infty}$. □

Theorem 17.4.1 and Lemma 17.4.6, in fact, imply Theorem 17.3.1. As shown on p. 658, the infection probability of a node j in a corresponding SIR process provides a lower bound for $\Pr[X_j(t) = 1]$.

Each node in the graph G obeys a differential equation as (17.32),

$$\begin{cases} \frac{dv_1(t)}{dt} = \beta \sum_{j=1}^{N} a_{1j} v_j(t) - v_1(t) \left(\beta \sum_{j=1}^{N} a_{1j} v_j(t) + \delta \right) \\ \frac{dv_2(t)}{dt} = \beta \sum_{j=1}^{N} a_{2j} v_j(t) - v_2(t) \left(\beta \sum_{j=1}^{N} a_{2j} v_j(t) + \delta \right) \\ \quad\quad\quad \vdots \\ \frac{dv_N(t)}{dt} = \beta \sum_{j=1}^{N} a_{Nj} v_j(t) - v_N(t) \left(\beta \sum_{j=1}^{N} a_{Nj} v_j(t) + \delta \right) \end{cases}$$

With $V(t) = \begin{bmatrix} v_1(t) & v_2(t) & \cdots & v_N(t) \end{bmatrix}^T$, the matrix evolution equation of NIMFA is

$$\frac{dV(t)}{dt} = \beta A V(t) - \text{diag}(v_i(t))(\beta A V(t) + \delta u) \tag{17.33}$$

where u is the all-one vector and $\text{diag}(v_i(t))$ is the diagonal matrix with elements $v_1(t), v_2(t), \dots, v_N(t)$. We rewrite (17.33) with $V(t) = \text{diag}(v_i(t)) u$ as

$$\frac{dV(t)}{dt} = (\beta \text{diag}(1 - v_i(t)) A - \delta I) V(t) \tag{17.34}$$

17.4.1 The steady-state of NIMFA

Two different viewpoints are analyzed: the local or nodal steady-state equation (17.32) and the global or matrix equation (17.34).

17.4.1.1 Nodal steady-state from equation (17.32)

Assuming that the steady-state exists, we can calculate the steady-state probabilities of infection for each node. The steady-state, denoted by $v_{j\infty} = \lim_{t \to \infty} v_j(t)$, implies that $\left. \frac{dv_j(t)}{dt} \right|_{t \to \infty} = 0$, and thus we obtain from (17.32) for each node j,

$$\beta \sum_{j=1}^{N} a_{ij} v_{j\infty} - v_{i\infty} \left(\beta \sum_{j=1}^{N} a_{ij} v_{j\infty} + \delta \right) = 0$$

Since all the diagonal elements of the adjacency matrix A are zero, $a_{jj} = 0$, the viral steady-state probability $v_{i\infty}$ is explicitly written as

$$v_{i\infty} = \frac{\beta \sum_{j=1}^{N} a_{ij} v_{j\infty}}{\beta \sum_{j=1}^{N} a_{ij} v_{j\infty} + \delta} = 1 - \frac{1}{1 + \tau \sum_{j=1}^{N} a_{ij} v_{j\infty}} \qquad (17.35)$$

which is the ratio of the (mean) infection rate induced by the node's direct infected neighbors $\sum_{j=1}^{N} a_{ij} v_{j\infty}$ over the total (mean) rate of both the competing infection and the curing process. Equation (17.35) is equal to the steady-state probability in a two-state, continuous Markov chain (see Section 10.6), which exemplifies the local (or nodal) character of NIMFA. The trivial solution of (17.35) is $v_{i\infty} = 0$ for all i, which means that, eventually, all nodes will be healthy.

Lemma 17.4.2 *In a connected graph, either $v_{i\infty} = 0$ for all i nodes, or none of the components $v_{i\infty}$ is zero.*

Proof: If $v_{i\infty} = 0$ for one node i in a connected graph, then (17.35) shows that $\sum_{j=1}^{N} a_{ij} v_{j\infty} = 0$, which is only possible provided $v_{j\infty} = 0$ for all neighbors j of node i. Applying this argument repeatedly to the neighbors of neighbors in a connected graph proves the lemma. □

Apart from the exact steady-state $v_{i\infty} = 0$ for all i, the non-linearity gives rise to a second solution, which corresponds to the metastable state. That second, non-zero solution can be interpreted as the fraction of time that a node is infected while the system is in the metastable state, where there is a long-lived epidemic.

Theorem 17.4.3 *For any effective spreading rate $\tau = \frac{\beta}{\delta} \geq 0$, the non-zero steady-state infection probability of any node i in NIMFA can be expressed as a continued fraction*

$$v_{i\infty} = 1 - \cfrac{1}{1 + \tau d_i - \tau \sum_{j=1}^{N} \cfrac{a_{ij}}{1 + \tau d_j - \tau \sum_{k=1}^{N} \cfrac{a_{jk}}{1 + \tau d_k - \tau \sum_{q=1}^{N} \cfrac{a_{kq}}{1 + \tau d_q - \ddots}}}} \qquad (17.36)$$

where $d_i = \sum_{j=1}^{N} a_{ij}$ is the degree of node i. Consequently, the exact steady-state infection probability of any node i is bounded by

$$0 \leq v_{i\infty} \leq 1 - \frac{1}{1 + \tau d_i} \qquad (17.37)$$

Proof: We rewrite (17.35) as

$$1 - v_{i\infty} = \frac{1}{1 + \tau d_i - \tau \sum_{j=1}^{N} a_{ij} (1 - v_{j\infty})}$$

Omitting the non-negative sum proves (17.37). Next, we define the k-th convergent as

$$w_i(k) = \frac{1}{1 + \tau d_i - \tau \sum_{j=1}^{N} a_{ij} w_j(k-1)} \qquad (17.38)$$

with starting value $w_i(0) = 0$ and $\lim_{k \to \infty} w_i(k) = 1 - v_{i\infty}$. The difference $h_i(k) = w_i(k) - w_i(k-1)$, satisfying

$$h_i(k) = w_i(k) w_i(k-1) \tau \sum_{j=1}^{N} a_{ij} h_i(k-1)$$

with starting value $h_i(1) = (1 + \tau d_i)^{-1}$, is bounded by

$$h_i(k) \geq \frac{\tau \sum_{j=1}^{N} a_{ij} h_i(k-1)}{(1 + \tau d_i)^2} \geq 0$$

which shows that the sequence of convergents

$$w_i(0) \leq w_i(1) \leq \cdots \leq w_i(k-1) \leq w_i(k) \leq \cdots$$

is non-decreasing and leads, in the limit for $k \to \infty$, to the partial fraction (17.36). $\square$

The continued fraction stopped at iteration k includes the effect of virus spread up to the $(k-1)$-hop neighbors of node i. As illustrated in Van Mieghem *et al.* (2009), a few iterations in (17.36) already give an accurate approximation. The accuracy seems worst around $\tau = \tau_c^{(1)}$.

For regular graphs, where each node has degree r, symmetry in the steady-state implies that $v_{i\infty} = v_\infty$ for all nodes i and it follows from (17.35) with the definition of the degree $d_i = \sum_{j=1}^{N} a_{ij}$ that, for $\tau \geq \frac{1}{r}$,

$$v_{\infty;\text{regular}} = y_{\infty;\,\text{regular}}^{(1)}(\tau) = 1 - \frac{1}{\tau r} \tag{17.39}$$

where $y_\infty^{(1)} = \frac{1}{N} \sum_{i=1}^{N} v_{i\infty}$ is the average steady-state fraction of infected nodes in NIMFA.

Lemma 17.4.4 *In a connected graph G with minimum degree $d_{\min}$ and for $\tau \geq \frac{1}{d_{\min}}$, a lower bound of $v_{i\infty}$ for any node i equals*

$$1 - \frac{1}{1 + \frac{d_i}{d_{\min}}(\tau d_{\min} - 1)} \leqslant v_{i\infty} \tag{17.40}$$

Proof: Lemma 17.4.2 and Lemma 17.4.6 below show that, for $\tau > \tau_c$, there exists a non-zero minimum $v_{\min} = \min_{1 \leq i \leq N} \{v_{i\infty}\} > 0$ of steady-state infection probabilities, which obeys (17.35). Assuming that this minimum $v_{\min}$ occurs at node i,

$$v_{\min} = 1 - \frac{1}{1 + \tau \sum_{j=1}^{N} a_{ij} v_{j\infty}} \geq 1 - \frac{1}{1 + \tau \sum_{j=1}^{N} a_{ij} v_{\min}} = 1 - \frac{1}{1 + \tau d_i v_{\min}} \geq 1 - \frac{1}{1 + \tau d_{\min} v_{\min}}$$

The last inequality is equivalent to

$$v_{\min} \geq 1 - \frac{1}{\tau d_{\min}} \tag{17.41}$$

which is only larger than zero provided $\tau > \frac{1}{d_{\min}} \geq \tau_c^{(1)}$. Equality in (17.41) is reached for regular graphs as demonstrated from (17.39). Introducing the bound (17.41), we also have for each node

$$v_i \geq v_{\min} \geq 1 - \frac{1}{1 + \tau d_i v_{\min}} \geqslant 1 - \frac{1}{1 + \frac{d_i}{d_{\min}}(\tau d_{\min} - 1)}$$

which is (17.40). $\square$

By combining the upper bound (17.37) and the lower bound (17.40) for $\tau \geq \frac{1}{d_{\min}}$, we find that $v_{i\infty}$ belongs to the interval

$$1 - \frac{1}{1 + \frac{d_i}{d_{\min}}(\tau d_{\min} - 1)} \leqslant v_{i\infty} \leqslant 1 - \frac{1}{1 + \tau d_i}$$

This shows that for $\tau \to \infty$ variations between all values of v_i for all i will tend to 0.

The following Lemma 17.4.5 is proved in Van Mieghem (2012b), by introducing (17.42) into (17.35) and equating corresponding powers in τ^{-1}.

Lemma 17.4.5 *The Laurent series of the steady-state infection probability*

$$v_{i\infty}(\tau) = 1 + \sum_{m=1}^{\infty} \eta_m(i) \tau^{-m} \tag{17.42}$$

possesses the coefficients $\eta_1(i) = -\frac{1}{d_i}$ *and*

$$\eta_2(i) = \frac{1}{d_i^2}\left(1 - \sum_{j=1}^{N} \frac{a_{ij}}{d_j}\right) \tag{17.43}$$

and for $m \geq 2$, the coefficients obey the recursion

$$\eta_{m+1}(i) = -\frac{1}{d_i}\eta_m(i)\left(1 - \sum_{j=1}^{N}\frac{a_{ij}}{d_j}\right) - \frac{1}{d_i}\sum_{k=2}^{m}\eta_{m+1-k}(i)\sum_{j=1}^{N}a_{ij}\eta_k(j)$$

For regular graphs with degree r, we observe that $\eta_1(i) = \eta_1 = -\frac{1}{r}$ and $\eta_m(i) = 0$ for all $m > 1$, which is in agreement with (17.39).

17.4.1.2 Global steady-state from the matrix equation (17.33)

Additional insight can be gained from the governing equation (17.34) in the steady-state

$$V_{\infty} = \tau \text{diag}(1 - v_{i\infty}) A V_{\infty} \tag{17.44}$$

for finite τ such that $v_{i\infty} < 1$. After left-multiplying both sides by $\frac{1}{\tau}\text{diag}\left(\frac{1}{1-v_{i\infty}}\right)$, we obtain

$$\frac{1}{\tau}\text{diag}\left(\frac{1}{1 - v_{i\infty}}\right) V_{\infty} = A V_{\infty} \tag{17.45}$$

which is a basic equation that is generalized in Section 17.5 and written in terms of a generalized Laplacian. We present reformulations of the steady-state equations, that will play a role in deductions and proofs of NIMFA properties.

After left-multiplication of (17.44) by the vector $V_{\infty}^T \text{diag}\left(v_{i\infty}^{k-1}\right) = \begin{bmatrix} v_{1\infty}^k & v_{2\infty}^k & \cdots & v_{N\infty}^k \end{bmatrix}$, which we denote by $\left(V_{\infty}^k\right)^T$, we obtain the scalar equation

$$\left(V_{\infty}^k\right)^T V_{\infty} = \sum_{j=1}^{N} v_{j\infty}^{k+1} = \tau\left(\left(V_{\infty}^k\right)^T A V_{\infty} - \left(V_{\infty}^{k+1}\right)^T A V_{\infty}\right) \tag{17.46}$$

For $k = 0$ in (17.46), and introducing the all-one vector $u = \lim_{k\to 0} V_{\infty}^k$, we obtain (17.52) below. For $k = 1$ in (17.46), the norm $\|V_{\infty}\|_2^2 = V_{\infty}^T V_{\infty} = \sum_{j=1}^{N} v_{j\infty}^2$ obeys

$$V_{\infty}^T V_{\infty} = \tau\left(V_{\infty}^T A V_{\infty} - V_{\infty}^T \text{diag}(v_{i\infty}) A V_{\infty}\right) \tag{17.47}$$

When summing (17.46) over all k from $m \geq 0$ and taking $|v_{j\infty}| < 1$ into account the telescoping nature of the right-hand side leads to

$$\sum_{k=m}^{\infty} \left(V_{\infty}^k\right)^T V_{\infty} = \sum_{j=1}^{N} \frac{v_{j\infty}^{m+1}}{1 - v_{j\infty}} = \tau \left(V_{\infty}^m\right)^T A V_{\infty} \qquad (17.48)$$

If $m = 0$, then $V_{\infty}^m = u$ and we obtain, with the degree vector $u^T A = D^T$,

$$\frac{1}{\tau} \sum_{j=1}^{N} \frac{v_{j\infty}}{1 - v_{j\infty}} = D^T V_{\infty} = \sum_{j=1}^{N} d_j v_{j\infty} \qquad (17.49)$$

The characteristic structure (17.48) of NIMFA follows more elegantly from (17.45), rewritten as

$$\frac{1}{\tau} \frac{V_{\infty}}{1 - V_{\infty}} = A V_{\infty} \qquad (17.50)$$

where $\left(\frac{V_{\infty}}{1 - V_{\infty}}\right)^T = \left[\begin{array}{cccc} \frac{v_{1\infty}}{1 - v_{1\infty}} & \frac{v_{2\infty}}{1 - v_{2\infty}} & \cdots & \frac{v_{N\infty}}{1 - v_{N\infty}} \end{array}\right]$. By left-multiplication of (17.50) by $(V_{\infty}^m)^T$, we obtain (17.48) again. After expanding each element in (17.50) as $\frac{v_{i\infty}}{1 - v_{i\infty}} = \sum_{k=1}^{\infty} v_{i\infty}^k$, where the geometric series always converges since $v_{i\infty} < 1$, we arrive at yet another form for the steady-state equation

$$\frac{1}{\tau} V_{\infty} + \frac{1}{\tau} \sum_{k=2}^{\infty} V_{\infty}^k = A V_{\infty} \qquad (17.51)$$

17.4.2 Steady-state average fraction $y_{\infty}^{(1)}(\tau)$ of infected nodes

The exact steady-state average fraction of infected nodes in (17.17) with $\varepsilon = 0$,

$$y_{\infty}(\tau) = \frac{\tau}{N} \left(D^T W_{\infty} - E\left[X_{\infty}^T A X_{\infty}\right]\right)$$

has the NIMFA companion

$$y_{\infty}^{(1)}(\tau) = \frac{\tau}{N} \left(D^T V_{\infty} - V_{\infty}^T A V_{\infty}\right) \qquad (17.52)$$

where the vector V_{∞} approximates the vector $W_{\infty} = (E[X_1], E[X_2], \ldots, E[X_N])$. However, the NIMFA companion of (17.18) is

$$y_{\infty}^{(1)}(\tau) = \frac{\tau}{N} \left((u - V_{\infty})^T \Delta V_{\infty} + V_{\infty}^T Q V_{\infty}\right) \qquad (17.53)$$

and not $y_{\infty}^{(1)}(\tau) = \frac{\tau}{N} V_{\infty}^T Q V_{\infty}$ as can be verified from (17.52) or from the governing NIMFA equations (17.33) or (17.46). This is a consequence of replacing a zero-one vector X_{∞} by an approximation V_{∞} of its mean $W_{\infty} = E[X_{\infty}]$ so that (17.53) is formally more involved than the exact equation (17.18).

From the Laurent series (17.42) in Lemma 17.4.5 after the transform $s = \frac{1}{\tau}$, the derivative at $\lim_{s \to 0} \frac{dy_{\infty}^{(1)}(s)}{ds}$ can be computed exactly as

$$\lim_{s \to 0} \frac{dy_{\infty}^{(1)}(s)}{ds} = -\frac{1}{N} \sum_{j=1}^{N} \frac{1}{d_j} = -E\left[\frac{1}{D}\right] \qquad (17.54)$$

which is minus the harmonic mean of the degrees in the graph. The harmonic, geometric and arithmetic mean inequality (5.3) shows[15] that

$$E\left[\frac{1}{D}\right] = \frac{1}{N}\sum_{j=1}^{N}\frac{1}{d_j} \geq \frac{1}{E[D]} \geq \frac{1}{\lambda_1}$$

with equality for the regular graph. Hence, the slope $\lim_{s \to 0} \frac{dy_\infty^{(1)}(s)}{ds}$ is least steep for the regular graph as illustrated in Fig. 17.4. Theoretically – although this is debatable –, one might argue that the effective curing rate $s = \frac{\delta}{\beta} = \frac{1}{\tau}$ is more natural than the effective infection rate $\tau = \frac{\beta}{\delta}$, because a Taylor expansion of $y_\infty^{(1)}(s)$ around $s = 0$ can be deduced, while the corresponding one for $y_\infty^{(1)}(\tau)$ is a Laurent series in $\frac{1}{\tau}$ around $\tau \to \infty$ (see Lemma 17.4.5).

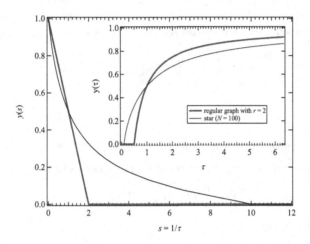

Fig. 17.4. The average steady-state fraction of infected nodes in NIMFA versus $s = \frac{1}{\tau}$ and versus τ in the insert for a cycle (regular graph with degree $r = 2$; in bold) and the star. Both graphs have $N = 100$ nodes and about the same number of links L.

17.4.3 Asymptotics for large τ

We present the *exact* steady-state behavior of the SIS epidemic for large τ. If τ is sufficiently large, the infection probability $w_{j\infty} = \lim_{t \to \infty} \Pr[X_j(t) = 1]$ of a node j with d_j neighbors tends to be independent of the viral state of its d_j neighbors and their neighbors, because the neighbors are infected with overwhelming probability. Hence, the viral state of node j is no longer intertwined with that of its neighbors, but is independent and is exceedingly well described by a two-state continuous Markov process with infection rate βd_j and curing rate δ, where $w_{j\infty} = \frac{\beta d_j}{\delta + \beta d_j} =$

---

[15] Since $f(x) = \frac{1}{x}$ is convex for $x > 0$, Jensen's inequality (5.7) applies as well.

$\frac{1}{1+\frac{1}{\tau d_j}} = \frac{1}{1+\frac{s}{d_j}}$, where $s = \frac{1}{\tau}$. The derivative for large τ or, equivalently $s \to 0$, is

$$\left. \frac{dw_{j\infty}(s)}{ds} \right|_{s=0} = -\frac{1}{d_j}$$

The average steady-state fraction $y_\infty(s) = \frac{1}{N}\sum_{j=1}^{N} w_{j\infty}(s)$ of infected nodes has a derivative at $s = 0$ equal to

$$\left. \frac{dy_\infty}{ds} \right|_{s=0} = -\frac{1}{N}\sum_{j=1}^{N}\frac{1}{d_j} = -E\left[\frac{1}{D}\right]$$

Comparison with (17.54) demonstrates that, for large τ, NIMFA returns both the exact infection probability ($v_{j\infty} \to 1$) and the exact derivative $\left. \frac{dy_\infty}{ds} \right|_{s=0}$.

17.4.4 Epidemic threshold in NIMFA

Lemma 17.4.6 *Let x_1 denote the principal eigenvector belonging to the largest eigenvalue λ_1 of the adjacency matrix A. Then, there exists a value $\tau_c^{(1)} = \frac{1}{\lambda_1} > 0$ and for $\tau < \tau_c^{(1)}$, there is only the trivial steady-state solution $V_\infty = 0$. Beside the $V_\infty = 0$ solution, there is a second, non-zero solution for all $\tau > \tau_c^{(1)}$. For $\tau = \tau_c^{(1)} + \epsilon$, where $\epsilon > 0$ is an arbitrary small constant, the steady-state infection vector equals $V_\infty = \epsilon x_1$.*

Proof: Theorem 17.4.3 shows that the only solution at $\tau = 0$ is the trivial solution $V_\infty = 0$. Let $V_\infty = \epsilon x_1$, where $\epsilon > 0$ is an arbitrary small constant and each component $(x_1)_i \geq 0$. Introduced in (17.51) gives, after division by ϵ,

$$Ax_1 = \frac{1}{\tau}x_1 + \frac{\epsilon}{\tau}x_1^2 + O\left(\epsilon^2\right)$$

For sufficiently small $\epsilon > 0$, the steady-state equations reduce to the eigenvalue equation

$$Ax_1 = \frac{1}{\tau}x_1 \qquad (17.55)$$

which shows that x_1 is an eigenvector of A belonging to the eigenvalue $\frac{1}{\tau}$. Since A is a non-negative matrix, the Perron-Frobenius Theorem (Appendix A.5.1) states that A has a positive largest eigenvalue λ_1 with a corresponding eigenvector whose elements are all positive and there is only one eigenvector of A with non-negative components. Hence, if $\frac{1}{\tau} = \lambda_1 > 0$, then x_1 (and any scaled vector $V_\infty = \epsilon x_1$) is the eigenvector of A belonging to λ_1. If $\tau < \frac{1}{\lambda_1} = \tau_c^{(1)}$, then $\frac{1}{\tau}$ cannot be an eigenvalue of A and the only possible solution is $x_1 = 0$, leading to the trivial solution $V_\infty = 0$. For $\tau > \tau_c^{(1)}$, Theorem 17.4.3 provides the non-zero solution of (17.35). $\square$

For large τ or $s \to 0$, the derivative $\left. \frac{dy_\infty^{(1)}(s)}{ds} \right|_{s=0}$ is given in (17.54). Now, we focus on the other end of the effective infection region for small τ around $\tau_c^{(1)} = \frac{1}{\lambda_1}$, or equivalently $s = \frac{1}{\tau}$ around λ_1. While the right-derivative[16] $\lim_{s\downarrow\lambda_1}\frac{dy_\infty^{(1)}(s)}{ds} = 0$,

---

[16] The notation $\lim_{x\downarrow x_0} f(x)$ is the limit towards x_0 when x approaches x_0 from above, i.e. $x > x_0$.

the left-derivative at $\lim_{s\uparrow\lambda_1}\frac{dy_\infty^{(1)}(s)}{ds}$ can be computed explicitly from the following theorem, proved in Van Mieghem (2012a):

Theorem 17.4.7 *For any graph with spectral radius λ_1 and corresponding eigenvector x_1 normalized such that $x_1^T x_1 = \sum_{j=1}^{N}(x_1)_j^2 = 1$, the steady-state fraction of infected nodes $y_\infty^{(1)}$ obeys*

$$y_\infty^{(1)}(\tau) = \frac{1}{\lambda_1 N}\frac{\sum_{j=1}^{N}(x_1)_j}{\sum_{j=1}^{N}(x_1)_j^3}\left(\frac{1}{\tau_c^{(1)}} - \tau^{-1}\right) + O\left(\frac{1}{\tau_c^{(1)}} - \tau^{-1}\right)^2 \qquad (17.56)$$

when τ approaches the epidemic threshold $\tau_c^{(1)}$ from above.

From (17.56), we have

$$\left.\frac{dy_\infty^{(1)}(s)}{ds}\right|_{s=\lambda_1} = -\frac{1}{\lambda_1 N}\frac{\sum_{j=1}^{N}(x_1)_j}{\sum_{j=1}^{N}(x_1)_j^3} \qquad (17.57)$$

which is shown in Van Mieghem (2012a) to be bounded as

$$-\frac{1}{\lambda_1} \leq \left.\frac{dy_\infty^{(1)}}{ds}\right|_{s\uparrow\lambda_1} \leq \frac{-1}{\lambda_1 N \max_{1\leq j\leq N}(x_1^2)_j} < 0 \qquad (17.58)$$

The method applied to prove Theorem 17.4.7 can be generalized to deduce the complete series of $y_\infty^{(1)}(\tau)$ (and of $v_{j\infty}$) around $\frac{1}{\tau_c^{(1)}} - \tau^{-1}$ as

$$y_\infty^{(1)}(\tau) = \frac{1}{N}\sum_{j=1}^{\infty}\sum_{k=1}^{N} c_j(k)\, u^T x_k \left(\frac{1}{\tau_c^{(1)}} - \tau^{-1}\right)^j \qquad (17.59)$$

which is valid for $\tau \geq \frac{1}{\lambda_1}$. All coefficients $c_j(k)$ are specified in a recursive way in Van Mieghem (2012b). The radius of convergence of the Laurent series (17.42) and of the series (17.59) is, in general, unknown and still an open problem.

Expression (17.56) illustrates that, among all graphs with N nodes and L links, the regular graph with degree $r = \frac{2L}{N}$ has the largest epidemic threshold $\tau_c = \frac{1}{r}$, but also the largest (in absolute value) derivative $\left.\frac{dy_\infty^{(1)}}{ds}\right|_{s\uparrow\lambda_1} = -\frac{1}{\lambda_1} = -\frac{1}{r}$ because equality in (17.58) is reached. This means that a higher effective infection rate τ is needed to cause a non-zero steady-state fraction of nodes in the regular graph to be permanently infected, but that, slightly above that critical rate $\tau_c^{(1)}$, a higher relative fraction of nodes is infected than in other graphs (see Fig. 17.4). In other words, the change in virus conductivity at $\tau = \tau_c^{(1)} + \epsilon$ is highest in regular graphs.

Via extensive simulations, Youssef *et al.* (2011) observed that $y_\infty^{(1)} \simeq \frac{1}{2}$ around $s = \frac{E[D]}{2}$. General bounds that explain this observation are derived in Van Mieghem (2012b).

17.5 Heterogeneous N-intertwined mean-field approximation

The homogenous NIMFA, where the infection and curing rate is the same for each link and node in the network, has been extended to a heterogeneous setting in Van Mieghem and Omic (2008), where an infected node i can infect its neighbors with an infection rate β_i, but it is cured with curing rate δ_i.

Heterogeneity rather than *homogeneity* abounds in real-world networks. For example, in data communications networks, the transmission capacity, age, performance, installed software, security level and other properties of networked computers are generally different. Social and biological networks are very diverse: a population often consists of a mix of weak and strong, or old and young species or of completely different types of species. The network topologies for transport by airplane, car, train and ship are different. More examples (see Section 15.8) can be added to illustrate that homogeneous networks are the exception rather than the rule. This diversity in the "nodes" and "links" of real networks will thus likely affect the spreading pattern of viruses. In previous sections, only a homogeneous virus spread was investigated, where all infection rates $\beta_i = \beta$ and all curing rates $\delta_i = \delta$ were the same for each node. Here, we extend NIMFA to a full heterogeneous setting.

The governing differential equation (17.32) is straightforwardly generalized[17] to

$$\frac{dv_i(t)}{dt} = \sum_{j=1}^{N} \beta_j a_{ij} v_j(t) - v_i(t) \left(\sum_{j=1}^{N} \beta_j a_{ij} v_j(t) + \delta_i \right) \tag{17.60}$$

while the corresponding matrix equation is

$$\frac{dV(t)}{dt} = A \operatorname{diag}(\beta_j) V(t) - \operatorname{diag}(v_i(t)) (A \operatorname{diag}(\beta_j) V(t) + C) \tag{17.61}$$

where $\operatorname{diag}(v_i(t))$ is the diagonal matrix with elements $v_1(t), v_2(t), \ldots, v_N(t)$ and the curing rate vector is $C = (\delta_1, \delta_2, \ldots, \delta_N)$. We note that $A \operatorname{diag}(\beta_i)$ is, in general and opposed to the homogeneous setting, no longer symmetric, unless A and $\operatorname{diag}(\beta_i)$ commute, in which case the eigenvalue $\lambda_i(A \operatorname{diag}(\beta_i)) = \lambda_i(A)\beta_i$ and both β_i and $\lambda_i(A)$ have the same eigenvector x_i.

17.5.1 The steady-state

The metastable steady-state follows from (17.61) as

$$A \operatorname{diag}(\beta_i) V_\infty - \operatorname{diag}(v_{i\infty}) (A \operatorname{diag}(\beta_i) V_\infty + C) = 0$$

---

[17] The most general NIMFA generalization with two states or compartments is

$$\frac{dv_i(t)}{dt} = \sum_{j=1}^{N} \tilde{a}_{ij} v_j(t) - v_i(t) \left(\sum_{j=1}^{N} \tilde{a}_{ij} v_j(t) + \delta_i \right)$$

where $\tilde{A}$ is a weighted adjacency matrix in which the real, non-negative element $\tilde{a}_{ij} = a_{ij}\beta_{ij}$ represents a link specific infection rate.

where $V_\infty = \lim_{t\to\infty} V(t)$. After defining the vector

$$\varpi = A\mathrm{diag}\,(\beta_i)\,V_\infty + C \qquad (17.62)$$

the steady-state equation becomes $\varpi - C = \mathrm{diag}(v_{i\infty})\,\varpi$ or $(I - \mathrm{diag}\,(v_{i\infty}))\,\varpi = C$. Ignoring extreme virus spread conditions (the absence of curing ($\delta_i = 0$) and an infinitely strong infection rate $\beta_i \to \infty$), then the infection probabilities $v_{i\infty}$ cannot be one such that the matrix $(I - \mathrm{diag}\,(v_{i\infty})) = \mathrm{diag}(1 - v_{i\infty})$ is invertible. Hence,

$$\varpi = \mathrm{diag}\left(\frac{1}{1 - v_{i\infty}}\right) C$$

Invoking the definition (17.62) of ϖ, we obtain

$$A\mathrm{diag}\,(\beta_i)\,V_\infty = \mathrm{diag}\left(\frac{\delta_i}{1 - v_{i\infty}}\right) V_\infty \qquad (17.63)$$

that generalizes (17.50). The i-th row of (17.63) yields the nodal steady state equation,

$$\sum_{j=1}^{N} a_{ij}\beta_j v_{j\infty} = \frac{v_{i\infty}\delta_i}{1 - v_{i\infty}} \qquad (17.64)$$

Let $\tilde{V}_\infty = \mathrm{diag}(\beta_i)\,V_\infty$ and the effective spreading rate for node i, $\tau_i = \frac{\beta_i}{\delta_i}$, then we arrive at

$$\mathcal{Q}\left(\frac{1}{\tau_i\,(1 - v_{i\infty})}\right) \tilde{V}_\infty = 0 \qquad (17.65)$$

where the symmetric matrix

$$\begin{aligned}\mathcal{Q}\,(q_i) &= \mathrm{diag}\,(q_i) - A \qquad (17.66)\\ &= \mathrm{diag}\,(q_i - d_i) + Q\end{aligned}$$

can be interpreted as a generalized Laplacian[18], because $\mathcal{Q}\,(d_i) = Q = \Delta - A$, where $\Delta = \mathrm{diag}(d_i)$. The observation that the non-linear set of steady-state equations can be written in terms of the generalized Laplacian $\mathcal{Q}\,(q_i)$ is fortunate, because the powerful theory of the "normal" Laplacian Q applies. Many properties of the generalized Laplacian $\mathcal{Q}\,(q_i)$ are given in Van Mieghem and Omic (2008) that enable us to prove three important theorems. The first theorem is:

Theorem 17.5.1 *The critical threshold is determined by vector*

$$\tau_c^{(1)} = \left(\tau_{1c}^{(1)}, \tau_{2c}^{(1)}, \dots, \tau_{Nc}^{(1)}\right)$$

that obeys $\lambda_{\max}(R) = 1$, *where* $\lambda_{\max}(R)$ *is the largest eigenvalue of the symmetric matrix*

$$R = \mathrm{diag}\left(\sqrt{\tau_i}\right) A\,\mathrm{diag}\left(\sqrt{\tau_i}\right) \qquad (17.67)$$

---

[18] All eigenvalues of the Laplacian $Q = \Delta - A$ in a connected graph are positive, except for the smallest one, which is zero. Hence, Q is positive semi-definite. Many more properties of the Laplacian Q are found in Biggs (1996), Cvetković *et al.* (1995) and Van Mieghem (2011).

whose corresponding eigenvector has positive components if the graph G is connected.

Several bounds for $\lambda_{\max}(R)$ are derived in Van Mieghem and Omic (2008) and $\lambda_{\max}(R)$ for the complete graph K_N is solved exactly. The generalization of Theorem 17.4.3 is:

Theorem 17.5.2 *The non-zero steady-state infection probability of any node i in NIMFA can be expressed as a continued fraction*

$$v_{i\infty} = 1 - \cfrac{1}{1 + \frac{\gamma_i}{\delta_i} - \delta_i^{-1} \sum_{j=1}^N \cfrac{\beta_j a_{ij}}{1 + \frac{\gamma_j}{\delta_j} - \delta_j^{-1} \sum_{k=1}^N \cfrac{\beta_k a_{jk}}{1 + \frac{\gamma_k}{\delta_k} - \delta_k^{-1} \sum_{q=1}^N \frac{a_{kq}\beta_q}{\ddots}}}} \tag{17.68}$$

where the total infection rate of node i, incurred by all neighbors towards node i, is

$$\gamma_i = \sum_{j=1}^N a_{ij}\beta_j = \sum_{j \in \; neighbor(i)} \beta_j \tag{17.69}$$

Consequently, the exact steady-state infection probability of any node i is bounded by

$$0 \le v_{i\infty} \le 1 - \frac{1}{1 + \frac{\gamma_i}{\delta_i}}$$

An important theorem, proved in Van Mieghem and Omic (2008), is:

Theorem 17.5.3 *Given that all curing rates δ_j for $1 \le j \ne i \le N$ are constant and not a function of the infection rates β_j, the non-zero steady-state infection probability $v_{i\infty}(\delta_1, \ldots, \delta_i, \ldots, \delta_N) > 0$ is strictly convex in δ_i, while all other non-zero steady-state infection probabilities $v_{j\infty}(\delta_1, \ldots, \delta_i, \ldots, \delta_N) > 0$ are concave in δ_i.*

A direct consequence of Theorem 17.5.3 to the homogeneous setting is that $y_\infty^{(1)}(s)$ is convex for $s = \frac{1}{\tau} \in [0, \lambda_1)$ (or $y_\infty^{(1)}(\tau)$ is concave for $\tau > \tau_c$). The convexity Theorem 17.5.3 has interesting applications in network protection strategies (Omic *et al.*, 2009; Gourdin *et al.*, 2011).

17.6 Epidemics on the complete graph K_N

We confine ourselves to the complete graph K_N because an exact analysis of the steady-state fraction $y_\infty(\tau)$ of infected nodes is possible. Moreover, the infection behavior in the complete graph is extremal, resulting in upper or lower bounds for other graphs.

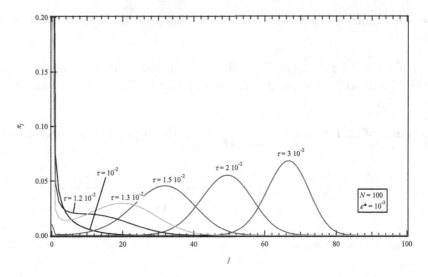

Fig. 17.5. The steady-state probability π_j computed from (17.72) versus j in K_{100} with $\varepsilon^* = 10^{-3}$, for various values of τ around the defined epidemic threshold $\tau_c \simeq 1.2 \times 10^{-2}$ in (17.101).

17.6.1 The number of infected nodes in K_N

The number of infected nodes $M(t)$ at time t in the complete graph K_N is described by a continuous-time Markov process on $\{0, 1, \ldots, N\}$ with the following rates:

$$M \mapsto M + 1 \text{ at rate } (\beta M + \varepsilon)(N - M)$$
$$M \mapsto M - 1 \text{ at rate } \delta M$$

Every infected node heals with rate δ, whereas every healthy node (of which there are $N - M$) has exactly M infected neighbors each actively transferring the virus with rate β in addition to the self-infection rate ε. This Markov process $M(t)$ is, in fact, a birth and death process with birth rate $\lambda_j = (\beta j + \varepsilon)(N - j)$ and death rate $\mu_j = j\delta$, whose steady-state probabilities $\pi_0, \ldots, \pi_N$, where $\pi_j = \lim_{t \to \infty} \Pr[M(t) = j]$, can be computed exactly from (11.16) and (11.17) as

$$\pi_0 = \frac{1}{1 + \sum_{k=1}^{N} \prod_{m=0}^{k-1} \frac{(\beta m + \varepsilon)(N-m)}{(m+1)\delta}} \tag{17.70}$$

$$\pi_j = \pi_0 \prod_{m=0}^{j-1} \frac{(\beta m + \varepsilon)(N - m)}{(m+1)\delta} \qquad 1 \le j \le N \tag{17.71}$$

Using the Gamma function's basic property $\Gamma(z+1) = z\Gamma(z)$, we have

$$\pi_j = \pi_0 \prod_{m=0}^{j-1} \frac{\tau m + \varepsilon^*}{m+1} \prod_{m=0}^{j-1} (N-m) = \frac{\pi_0}{j!} \tau^j \prod_{m=1}^{j} \left(m - 1 + \frac{\varepsilon^*}{\tau}\right) \prod_{m=1}^{j} (N + 1 - m)$$

$$= \frac{\pi_0}{j!} \tau^j \frac{\Gamma\left(\frac{\varepsilon^*}{\tau} + j\right)}{\Gamma\left(\frac{\varepsilon^*}{\tau}\right)} \frac{N!}{(N-j)!} = \pi_0 \binom{N}{j} \tau^j \frac{\Gamma\left(\frac{\varepsilon^*}{\tau} + j\right)}{\Gamma\left(\frac{\varepsilon^*}{\tau}\right)}$$

or

$$\pi_j = \pi_0 \binom{N}{j} \varepsilon^* \tau^{j-1} \frac{\Gamma\left(\frac{\varepsilon^*}{\tau} + j\right)}{\Gamma\left(\frac{\varepsilon^*}{\tau} + 1\right)} \tag{17.72}$$

which shows that, if $\varepsilon^* = 0$, then $\pi_j = 0$ and consequently, $\pi_0 = 1$, because $\sum_{i=0}^{N} \pi_i = 1$ (by the conservation of probability). Further, (17.70) becomes

$$\pi_0 = \frac{1}{1 + \sum_{k=1}^{N} \binom{N}{k} \tau^k \frac{\Gamma\left(\frac{\varepsilon^*}{\tau} + k\right)}{\Gamma\left(\frac{\varepsilon^*}{\tau}\right)}} \tag{17.73}$$

Fig. 17.5 illustrates the Gaussian-like shape of π_j for $\tau > \tau_c$. For $\tau = 0$ and approximately for large $\frac{\varepsilon^*}{\tau}$ because then $\frac{\Gamma\left(\frac{\varepsilon^*}{\tau} + j\right)}{\Gamma\left(\frac{\varepsilon^*}{\tau} + 1\right)} \sim \left(\frac{\varepsilon^*}{\tau}\right)^{j-1}$, we find that $\pi_j = \pi_0 \binom{N}{j} (\varepsilon^*)^j$ and $\pi_0 = \frac{1}{(1+\varepsilon^*)^N}$, from which $y_{\infty;N}(0) = \frac{\varepsilon^*}{1+\varepsilon^*}$, in agreement with the general theory above.

For large τ, we find from (17.72) that $\pi_j \sim \pi_0 \frac{N!}{(N-j)!j} \varepsilon^* \tau^{j-1}$ and from (17.73) that

$$\pi_0^{-1} = 1 + \varepsilon^* \sum_{k=1}^{N} \binom{N}{k} \tau^{k-1} \frac{\Gamma\left(\frac{\varepsilon^*}{\tau} + k\right)}{\Gamma\left(\frac{\varepsilon^*}{\tau} + 1\right)} \sim 1 + \varepsilon^* \sum_{k=1}^{N} \frac{N!}{(N-k)!k} \tau^{k-1} \sim (N-1)! \tau^{N-1} \varepsilon^*$$

Hence, for large τ and fixed ε^*, it holds that

$$\pi_j \sim \frac{N}{(N-j)!j} \tau^{j-N}$$

illustrating that the steady-state probability that j nodes are infected increases with j and ultimately that $\lim_{\tau \to \infty} \pi_j = \delta_{0N} = 1_{\{j=N\}}$ and $\lim_{\tau \to \infty} y_{\infty;N}(\tau) = 1$.

17.6.2 The average steady-state fraction $y_{\infty;N}(\tau)$ of infected nodes

The average steady-state fraction of infected nodes is

$$y_{\infty;N}(\tau) = \frac{1}{N} \sum_{j=0}^{N} j \pi_{j;N} = \frac{\pi_{0;N}}{N} \sum_{j=1}^{N} j \binom{N}{j} \tau^j \frac{\Gamma\left(\frac{\varepsilon^*}{\tau} + j\right)}{\Gamma\left(\frac{\varepsilon^*}{\tau}\right)} \tag{17.74}$$

and $y_{\infty;N}(\tau)$ satisfies the recursion (derived on p. 659)

$$y_{\infty;N}(\tau) = \frac{1}{1 + \frac{1}{\varepsilon^* + (N-1)\tau y_{\infty;N-1}(\tau)}} \tag{17.75}$$

from which we finally deduce

$$y_{\infty;N}(\tau) = 1 - \frac{1}{\frac{\varepsilon^*}{y_{\infty;N}(\tau)} + (N-1)\tau\frac{y_{\infty;N-1}(\tau)}{y_{\infty;N}(\tau)}} \tag{17.76}$$

For sufficiently large N, it holds that $\frac{y_{\infty;N-1}(\tau)}{y_{\infty;N}(\tau)} \lesssim 1$ and (17.76) indicates that, when $\frac{\varepsilon^*}{y_{\infty;N}(\tau)} \leq 1 + \varepsilon^*$ is small enough to neglect, we find that $y_{\infty;N}(\tau) \approx 1 - \frac{1}{(N-1)\tau}$, which supports the simulations in Fig. 17.8. Recall from (17.39) that the average steady-state fraction of infected nodes in NIMFA (where $\varepsilon^* = 0$) for the complete graph K_N equals, for $\tau \geq \tau_c^{(1)} = \frac{1}{N-1}$,

$$y_{\infty;N}^{(1)}(\tau) = 1 - \frac{1}{(N-1)\tau}$$

illustrating how good the first-order mean-field approximation for the complete graph is. When iterating (17.75), a continued fraction for $y_{\infty;N}(\tau)$ is found, which bears resemblance with the continued fraction (17.36) of $v_{j\infty}$ in NIMFA. Since $y_{\infty;N-1}(\tau) \geq y_{\infty;N-1}(0) = \frac{\varepsilon^*}{1+\varepsilon^*}$, each convergent can be used as a lower bound. For example, the first convergent is

$$_1y_{\infty;N}(\tau) = \frac{1}{1 + \frac{1}{\varepsilon^* + (N-1)\tau\frac{\varepsilon^*}{1+\varepsilon^*}}} < y_{\infty;N}(\tau)$$

and the second convergent is

$$_2y_{\infty;N}(\tau) = \frac{1}{1 + \frac{1}{\varepsilon^* + \frac{(N-1)\tau}{1 + \frac{1}{\varepsilon^* + (N-2)\tau\frac{\varepsilon^*}{1+\varepsilon^*}}}}} < y_{\infty;N}(\tau)$$

The variance $\sigma_y^2 = \frac{1}{N^2}\mathrm{Var}\left[\sum_{j=1}^N X_j\right]$ of the steady-state fraction of infected nodes equals

$$\sigma_y^2(\tau) = \frac{1}{N^2}\sum_{j=0}^N j^2\pi_{j;N} - y_{\infty;N}^2(\tau) \tag{17.77}$$

and we can deduce that $\sigma_y^2(0) = \frac{1}{N}\frac{\varepsilon^*}{(1+\varepsilon^*)^2}$.

Fig. 17.3 and 17.6 show the mean (17.74) and variance (17.77) of the steady-state fraction $y_\infty(\tau) = \lim_{t\to\infty} y(t;\tau)$ of infected nodes for various complete graphs K_N as a function of the effective infection rate τ. Fig. 17.6 illustrates that the largest variance occurs in a region around the "epidemic threshold $\tau_c \sim \frac{1}{N}$" and that the variance decreases with N.

Fig. 17.7 for $N = 500$ illustrates the rate at which $y_\infty(\tau)$ tends to the absorbing state with ε^*. When $\varepsilon^* \to 0$, the fraction of infected nodes $y_\infty \to 0$ and no obvious interpretation of an epidemic threshold can be deduced. Fig. 17.7 also illustrates that $\varepsilon^* = \frac{\varepsilon}{\delta}$ is not a perturbation parameter, because arbitrary small, positive values of ε^* have a significant effect on y_∞. However, for large N and $\varepsilon^* = 10^{-a}$ with $3 \leq a \leq 10$, Fig. 17.7 on the linear scale leads us to "estimate" the in

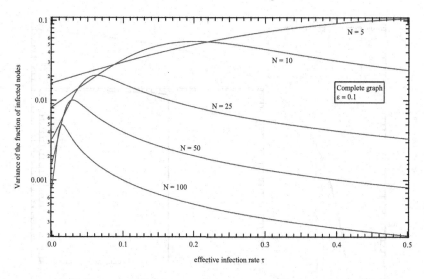

Fig. 17.6. The variance of the steady-state fraction of infected nodes in the complete graph versus the effective infection rate τ for $\varepsilon^* = 10^{-1}$ and for various sizes N.

reality observed threshold fairly well. The left-top insert shows the corresponding highly peaked variance. Fig. 17.7 indicates that the epidemic threshold deduced for $\varepsilon^* = 10^{-6}$ and $\varepsilon^* = 10^{-10}$ only differs moderately. Moreover, much lower values of $\varepsilon^* = 10^{-10}$ are difficult to simulate and seem to be unrealistically small to occur in nature. All these considerations suggest us to propose a *definition* of the epidemic threshold for a small $\varepsilon^* = 10^{-a}$ with $3 \leq a \leq 10$ as the τ-value where $y_\infty(\tau)$ jumps from practically zero to some positive value.

For large τ and large N, NIMFA, yielding $y_{\infty;N}(\tau) \lesssim 1 - \frac{1}{(N-1)\tau}$, is accurate as verified from Fig. 17.8 and 17.7. NIMFA and the exact SIS model are further compared for different graph types in Li *et al.* (2012).

17.7 Non-Markovian SIS epidemics

The continuous-time SIS model on any network, in which the infection time T and the curing or recovery time R have a general distribution, is called the generalized SIS or GSIS model. The classical continuous-time SIS Markov model is a special case where both T and R are exponentially distributed.

If node i in the GSIS model gets infected at time t, we draw independently of everything else a recovery time $R_i(t)$. Given $R_i(t)$, we then draw independently, for each neighbor j of i, a random number $M_{ij}(t)$ of infecting times $T_{ij}^{(1)}(t) \leq \cdots \leq T_{ij}^{(M_{ij}(t))}(t) \leq R_i(t)$ such that at times $t + T_{ij}^{(k)}$ with $1 \leq k \leq M_{ij}(t)$, node i tries to infect node j. If node j is already infected at such a time, nothing happens. Finally, node i recovers at time $t + R_i(t)$ and becomes healthy, but again susceptible to infection. We assume that $R_i \overset{d}{=} R$ and $T_{ij} \overset{d}{=} T$ have the same distribution for each

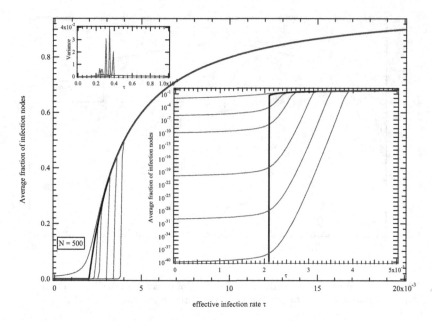

Fig. 17.7. The average fraction $y_\infty(\tau)$ of infected nodes in K_{500} as a function of τ for various small values of ε^*, which can be read off in the inserted log-scale plot because $y_\infty(0) = \frac{\varepsilon^*}{1+\varepsilon^*}$. The thick lines represent NIMFA.

node and each link and that all infection and recovery processes are independent. An exact analysis of the GSIS model on any network is very likely intractable, so that only an approximate treatment seems possible.

17.7.1 Mean-field approximation

We assume that the metastable state exists and that the mean recovery time and number of infection events of a node i is finite, thus $E[R_i] < \infty$ and $E[M_{ij}] < \infty$. We denote by $v_{i\infty}$ the probability that node i is infected in the metastable state. The mean-field approximation in this setting entails that, when we determine the effect of the neighbors on node i, we assume that node i does not influence its neighboring nodes. We will now explain what this assumption implies.

Let node j be a neighbor of node i. In a large time interval $[0, S]$, the number of times node j was infected is asymptotically linear in S by the Elementary Renewal Theorem 8.2.1. Since the length of an infected period equals $E[R]$ (we omit the subscript for the node j, since the expectation is the same for each node), the number of infected periods is equal to $v_{j\infty}S/E[R]$. During each infected period, node j will try to infect node i, on average, $E[M]$ times. This means that the total number of infection attempts from node j to node i asymptotically equals $v_{j\infty}SE[M]/E[R]$. Now we apply the mean-field approximation: the fraction of

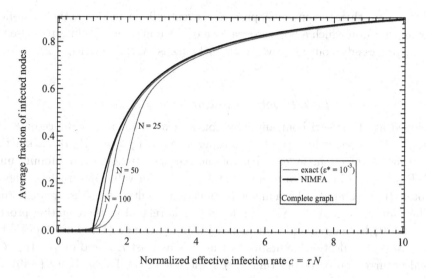

Fig. 17.8. The steady-state average fraction y_∞ (τ) of infected nodes versus the normalized effective infection rate $c = \tau N$ for $N = 25, 50$ and 100. Both the exact $\varepsilon^* = 10^{-3}$ SIS model and NIMFA are shown.

infection events from j to i that is successful (i.e. when node i was healthy at the time of infection) equals $1 - v_{i\infty}$. Hence, the total number of successful infections that node i will receive from all its infected neighbors in the time interval $[0, S]$ will asymptotically be equal to $S \sum_{j=1}^{N} a_{ij} \frac{E[M]}{E[R]} v_{j\infty}(1 - v_{i\infty})$. In the metastable state (where equilibrium holds), this number must equal the number of infected periods of node i in $[0, S]$, so that

$$S \frac{E[M]}{E[R]} \sum_{j=1}^{N} a_{ij} v_{j\infty}(1 - v_{i\infty}) = v_{i\infty} \frac{S}{E[R]}$$

Finally, we arrive, for any node i in the graph G, at

$$E[M](1 - v_{i\infty}) \sum_{j=1}^{N} a_{ij} v_{j\infty} = v_{i\infty} \qquad (17.78)$$

which is exactly the same equation as (17.35) in the NIMFA for the exponential case, if we replace $\tau = \beta/\delta$ by $\theta = E[M]$. The expected number θ of infection events in a Poisson process with intensity β within an exponential recovery time with expectation $1/\delta$ indeed equals the effective infection rate $\tau = \beta/\delta$.

The analogy with the NIMFA equations allows us to transfer the NIMFA analytic framework to the GSIS model. It follows from (17.78) and the epidemic threshold in Lemma 17.4.6 that a *lower bound* for the epidemic threshold in GSIS epidemics satisfies

$$\theta_c = E[M_c] = \frac{1}{\lambda_1} \qquad (17.79)$$

Thus, if $\theta > \theta_c$, then the epidemic process is eventually endemic (in the mean-field approximation), in which a non-zero fraction of the nodes remains infected, else the epidemic process dies out after which the network is overall healthy.

17.7.2 Determination of $\theta = E[M]$

We arrived at (17.78) without any assumption, except for the existence of $E[M]$ and $E[R]$. However, it is natural to consider a renewal process $T_1, T_2, \ldots$ starting at the time of infection, independent of the recovery time R, and running until $T_n \leq R < T_{n+1}$. In that case, we have $M = n$. We will apply renewal theory (Chapter 8) to compute the number of infection events M in the random time R. The waiting time $W_n = \sum_{k=1}^{n} T_k$ for $n \geq 1$ is related to the counting process $\{M(t), t \geq 0\}$ by the renewal equivalence $\{M(t) \geq n\} \Longleftrightarrow \{W_n \leq t\}$. Indeed, the number $M(t)$ of infection events up to time t is at least n if and only if the n-th renewal occurred on or before time t. By the law of total probability (2.49), we condition on the random recovery time R,

$$\Pr[W_n \leq R] = \int_0^\infty \Pr[W_n \leq u | R = u] \frac{d \Pr[R \leq u]}{du} du$$

Since the time W_n of the n-th infection attempt from node j to node i and the recovery time R of node i are independent, we have

$$\Pr[W_n \leq R] = \int_0^\infty \Pr[W_n \leq u] f_R(u) du$$

where $f_R(u)$ denotes the probability density function of the recovery time R. Using the renewal equivalence $\{M(t) \geq n\} \Longleftrightarrow \{W_n \leq t\}$ and the expression (2.37) for the mean in terms of the tail probabilities yields

$$E[M(R)] = \sum_{k=1}^{\infty} \Pr[M(R) \geq k] = \sum_{k=1}^{\infty} \Pr[W_k \leq R]$$

$$= \int_0^\infty \sum_{k=1}^{\infty} \Pr[W_k \leq u] f_R(u) du \qquad (17.80)$$

If $f_T(u)$ is the probability density function of the infection time T and $\varphi_T(z) = E[e^{-zT}]$ the corresponding probability generating function (2.38), then invoking (8.3) yields

$$\Pr[W_k \leq u] = \frac{1}{2\pi i} \int_{c-i\infty}^{c+i\infty} \frac{\varphi_T^k(z)}{z} e^{zu} dz$$

where $c > 0$ and

$$\sum_{k=1}^{\infty} \Pr[W_k \leq u] = \frac{1}{2\pi i} \int_{c-i\infty}^{c+i\infty} \frac{\varphi_T(z)}{1 - \varphi_T(z)} \frac{e^{zu}}{z} dz$$

Writing $E[M] = E[M(R)]$ and substituting the above into (17.80) yields

$$E[M] = \frac{1}{2\pi i} \int_{c-i\infty}^{c+i\infty} \frac{\varphi_T(z)}{1 - \varphi_T(z)} \frac{dz}{z} \int_0^\infty e^{zu} f_R(u) \, du$$

Finally, we arrive at the general expression of the mean number of infection attempts during a healthy, but susceptible period,

$$\theta = E[M] = \frac{1}{2\pi i} \int_{c-i\infty}^{c+i\infty} \frac{\varphi_T(z)\varphi_R(-z)}{1 - \varphi_T(z)} \frac{dz}{z} \tag{17.81}$$

where $\varphi_R(z) = \int_0^\infty e^{-zu} f_R(u) \, du$ is the probability generating function (pgf) of the recovery time R.

Since any pgf obeys $|\varphi_T(z)| < 1$ for $\mathrm{Re}(z) > 0$, the integrand $g(z) = \frac{\varphi_T(z)\varphi_R(-z)}{z(1-\varphi_T(z))}$ is analytic in the positive $\mathrm{Re}(z)$-plane, with the possible exception of the singularities of $\varphi_R(-z)$. Excluding the case of essential singularities or branch cuts of $\varphi_R(-z)$, we can deform the line of integration into a contour over the positive $\mathrm{Re}(z)$-plane (because the integrand $g(z) = g(re^{i\theta})$ vanishes for $r \to \infty$ at all angles $-\frac{\pi}{2} \leq \theta \leq \frac{\pi}{2}$). Thus, (17.81) equals

$$\theta = \frac{1}{2\pi i} \int_C \frac{\varphi_T(z)\varphi_R(-z)}{1 - \varphi_T(z)} \frac{dz}{z} \tag{17.82}$$

where the contour C encloses the whole $\mathrm{Re}(z) > 0$ plane.

We now concentrate on some special cases in which the integral (17.81) or (17.82) for $\theta = E[M]$ can be evaluated.

17.7.2.1 The infection time T is exponential

If the infection time T is exponentially distributed with mean $\frac{1}{\beta}$, then the pgf equals $\varphi_T(z) = \frac{\beta}{z+\beta}$ and (17.81) simplifies to

$$\theta = \frac{\beta}{2\pi i} \int_{c-i\infty}^{c+i\infty} \frac{\varphi_R(-z)}{z^2} dz \qquad (c > 0) \tag{17.83}$$

When we close the contour over the negative $\mathrm{Re}(z)$-plane, in which $\varphi_R(-z)$ is analytic and bounded $|\varphi_R(-z)| \leq 1$, we find, by Cauchy's integral theorem, that

$$\theta = \beta \left. \frac{d\varphi_R(-z)}{dz} \right|_{z=0} = \beta E[R] = \tau$$

because $\varphi_R(-z) = E[e^{zR}]$ and $E[R] = \frac{1}{\delta}$. The result $\theta = \tau$ is intuitively clear. Since the infection time T is exponentially distributed, the infection process is a Poisson process and the mean number of Poisson events in an interval equals its rate β multiplied by the mean length of that interval, which is $E[R]$. Hence, if the infection time T is exponential or the infection follows a Poisson process, then, for *any* distribution of the recovery time R, we find that NIMFA applies.

17.7.2.2 The recovery time R is exponential

If the recovery time R is exponentially distributed with mean $\frac{1}{\delta}$, then the pgf equals $\varphi_R(z) = \frac{\delta}{z+\delta}$ and (17.82) simplifies to

$$\theta = \frac{\delta}{2\pi i} \int_C \frac{\varphi_T(z)}{z(1 - \varphi_T(z))} \frac{dz}{\delta - z}$$

By Cauchy's residue theorem, we find that

$$\theta = \frac{\varphi_T(\delta)}{1 - \varphi_T(\delta)} \qquad (17.84)$$

from which a *lower bound* for the epidemic threshold follows from (17.79) by solving the equation

$$\varphi_T(\delta) = \frac{1}{1 + \lambda_1} \qquad (17.85)$$

The relation (17.84) in case of an exponential recovery time R with rate δ can be derived probabilistically, without resorting to contour integration. At the start of the infection, two cases can happen: either $T_1 < R$ or $T_1 > R$. In the latter case, $M = 0$. In the first case, $M \geq 1$ and at time T_1 everything starts anew: the distribution of $(R - T_1) \mid R > T_1$ is again exponential because of the memoryless property of the exponential distribution! This means that M has a geometric distribution,

$$\Pr[M = k] = p^k(1 - p), \quad \text{with} \quad p = \Pr[R > T_1] \text{ and } k \geq 0,$$

Since all infection times $T_1, T_2, \ldots$ are i.i.d. and have a same distribution as T, we find the mean of a geometric random variable as

$$\theta = \frac{\Pr[R > T]}{\Pr[R \leq T]} = \frac{\Pr[R > T]}{1 - \Pr[R > T]}.$$

This immediately shows that if T has an exponential distribution with parameter β, we find $\theta = \tau$. Furthermore, for a general T, we have a nice connection between these probabilities and the Laplace transform:

$$\Pr[R > T] = \int_0^\infty \Pr[R > s] f_T(s)\, ds = \int_0^\infty e^{-\delta s} f_T(s)\, ds = \varphi_T(\delta)$$

from which (17.84) follows again.

We consider the example, where the curing process is still Poissonean with rate δ, but the infection process at each node infects direct neighbors in a time T, with Weibullean probability density function

$$f_T(x) = \frac{\alpha}{b}\left(\frac{x}{b}\right)^{\alpha-1} e^{-\left(\frac{x}{b}\right)^\alpha} \qquad (17.86)$$

and mean $E[T] = b\Gamma\left(1 + \frac{1}{\alpha}\right)$. For $\alpha = 1$, the Weibull distribution reduces to the exponential distribution. In order to compare the Weibull with the exponential distribution, we fix the mean infection time $E[T]$ to $\frac{1}{\beta}$, so that

$$b = \left(\Gamma\left(1 + \frac{1}{\alpha}\right)\beta\right)^{-1}$$

Thus, the parameter α in (17.86) tunes both the power-law start and the tail of the Weibull distributions that all have the same mean infection time $E\left[T\right] = \frac{1}{\beta}$. Clearly, small α correspond to heavy tails (and large variance Var[T]), while large α correspond to almost deterministic infection times (with Var[T] $\rightarrow$ 0). By inverting (17.85), Cator *et al.* (2013) show that, for large N and any graph, the first-order *mean-field* epidemic threshold (in units of $\tau = \frac{E[R]}{E[T]}$) in a SIS process with Weibullean infection times T with mean $\frac{1}{\beta}$ and exponential recovery times R, is

$$\tau_c^{(1)} = \frac{1}{\Gamma\left(1 + \frac{1}{\alpha}\right)\left(\Gamma\left(\alpha + 1\right)\right)^{\frac{1}{\alpha}} \lambda_1^{\frac{1}{\alpha}}} \tag{17.87}$$

where the superscript (1) refers to the first-order NIMFA mean-field approximation $\tau_c^{(1)}$. When $\alpha = 1$, we find again the classical mean-field epidemic threshold in a Markovian SIS process, $\tau_c^{(1)} = \frac{1}{\lambda_1} \leq \tau_c$, where τ_c is the exact SIS epidemic threshold.

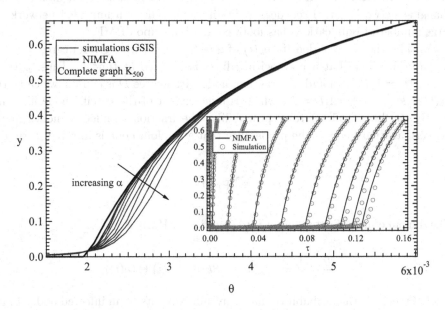

Fig. 17.9. The metastable-state average fraction y of infected nodes in K_{500} versus θ (and versus the effective infection rate τ in the inset) for various values of the parameter α in the Weibull distribution (17.86).

Fig. 17.9 emphasizes that $\theta = E\left[M\right]$ is the more natural parameter, instead of the ratio $\tau = \frac{E[R]}{E[T]}$, because all curves tends to each other, indicating that θ constitutes a proper scaling. The inset illustrates the dramatic influence of the curing time T and the recovery time R: the curves of the average fraction of infected nodes shift with increasing α to the right and heavy tailed distributions (α small) cause a small epidemic threshold in agreement with (17.87). In particular, the heavy tailed regime ($\alpha < 1$) that occurs in reality (Doerr *et al.*, 2013) leads to a high network vulnerability for infection.

Just by a replacement of τ by θ, the NIMFA equations remain valid for GSIS. The shape of the curve of the average metastable state fraction of infected nodes is mainly determined by the underlying topology and can be computed via NIMFA, whereas the mean number θ of infection attempts (17.81) during a recovery time reflects the epidemic details, that allow us to compare viral agents with different epidemic properties (measured via T and R) on the same contact network.

17.8 Heterogeneous mean-field (HMF) approximation

Pastor-Satorras and Vespignani (2001) have proposed the heterogeneous mean-field (HMF) approximation for SIS epidemics, in which only the degree distribution of the underlying topology is taken into account (instead of the complete adjacency matrix as in NIMFA). HMF models exactly the SIS process in an infinitely fast changing network with fixed degree distribution, called an annealed network. A large number of publications has followed and built upon HMF.

Consider the relative density $\rho_k(t)$ of infected nodes with given degree k, which is the probability that a node with k links is infected at time t. Thus, $\rho_k(t) = \Pr[X_n(t) = 1|d_n = k]$ and, if a random node with degree D is chosen, we can write $\rho_k(t) = \Pr[X_n(t) = 1|D = k]$, which is independent of the specific node identifier n. By the law (2.49) of total probability, the fraction of infected nodes in the network at time t equals the probability that a random node is infected at time t,

$$\rho(t) = \sum_{k=1}^{N-1} \Pr[D = k]\rho_k(t)$$

The dynamic mean-field reaction rate equation in HMF is

$$\frac{\partial \rho_k(t)}{\partial t} = -\delta \rho_k(t) + \beta k[1 - \rho_k(t)]\Theta\left(\rho(t)\right)$$

where $\Theta\left(\rho(t)\right)$ is the probability that a given link points to an infected node. In the steady-state, $y_\infty(\tau) = \lim_{t\to\infty} \rho(t)$ is only a function of $\tau = \frac{\beta}{\delta}$, and consequently, so is $\lim_{t\to\infty} \Theta\left(\rho(t)\right) = \Theta(\tau)$. By imposing stationarity, $\lim_{t\to\infty} \frac{\partial \rho_k(t)}{\partial t} = 0$, the relative density (as a function of τ) reduces to

$$\rho_k(\tau) = \frac{\tau k \Theta(\tau)}{1 + k\tau\Theta(\tau)} \tag{17.88}$$

where (see p. 642)

$$\Theta(\tau) = \frac{1}{E[D]} \sum_{k=1}^{N-1} k\Pr[D = k]\rho_k(\tau) \tag{17.89}$$

Clearly, if $\tau = 0$, then $\Theta(0) = 0$. Substituting (17.88) into (17.89) leads to a self-consistent relation, from which $\Theta(\tau)$ can be determined as

$$\Theta(\tau) = \frac{\tau \Theta(\tau)}{E[D]} \sum_{k=1}^{N-1} \frac{k^2 \Pr[D = k]}{1 + k\tau\Theta(\tau)} \tag{17.90}$$

Relation (17.90) has a trivial solution $\Theta(\tau) = 0$. A non-trivial solution $\Theta(\tau) > 0$ must satisfy

$$\frac{E[D]}{\tau} = \sum_{k=1}^{N-1} \frac{k^2 \Pr[D = k]}{1 + k\tau\Theta(\tau)} \tag{17.91}$$

which is similar to the second, non-zero solution in NIMFA as explained in Section 17.4.1.1.

Lemma 17.8.1 *For any graph, the epidemic threshold of the HMF approximation is*

$$\tau_c^{HMF} = \frac{E[D]}{E[D^2]} \tag{17.92}$$

Proof: We introduce the following expansion,

$$\frac{1}{1 + k\tau\Theta(\tau)} = \sum_{j=0}^{\infty} (-1)^j (k\tau\Theta(\tau))^j$$

valid when $k\tau\Theta(\tau) < 1$ for all k, into (17.91)

$$\frac{E[D]}{\tau} = \sum_{j=0}^{\infty} (-1)^j \left\{ \sum_{k=1}^{N-1} \Pr[D = k]k^{j+2} \right\} \tau^j \Theta^j(\tau) = \sum_{j=0}^{\infty} (-1)^j E\left[D^{j+2}\right] \tau^j \Theta^j(\tau)$$

where the latter series converges for $\Theta(\tau) < 1/(d_{max}\tau)$. Since $\tau = 0$ leads to $\Theta(0) = 0$, the non-trivial solution $\Theta(\tau) > 0$ occurs when $\tau > \tau_c^{HMF} \geq 0$ by the definition of the epidemic threshold. When $\Theta(\tau)$ is sufficiently small ($\Theta(\tau) < 1/(d_{max}\tau)$) and $\Theta(\tau) > 0$, we can write the above expansion up to first order as

$$\frac{E[D]}{\tau} = E[D^2] - \tau\Theta(\tau)E[D^3] + O(\Theta(\tau)^2) \tag{17.93}$$

in which $\tau\Theta(\tau)E[D^3] > 0$. Hence, when $\tau > \tau_c^{HMF}$, but $\Theta(\tau)$ is small enough to ignore the second-order terms $O(\Theta(\tau)^2)$, we have from (17.93) that $\frac{E[D]}{\tau} < E\left[D^2\right]$, implying that for all $\tau > \tau_c^{HMF}$, it holds that $\tau > \frac{E[D]}{E[D^2]}$ from which (17.92) follows. $\square$

The HMF epidemic threshold τ_c^{HMF} is compared to $\tau_c^{(1)} = \frac{1}{\lambda_1}$ in Li *et al.* (2012). For a regular graph with degree r, $E[D^2] = E[D]^2 = r^2$, the epidemic threshold is $\tau_c^{HMF} = \frac{1}{r} = \frac{1}{\lambda_1} = \tau_c^{(1)}$. NIMFA is shown in Li *et al.* (2012) to be generally superior to the widely used heterogeneous mean-field model (HMF) of Pastor-Satorras and Vespignani (2001). The mean-field approximations NIMFA and HMF both return $\lim_{s \to 0} y_{\infty}(s)$ and the derivative $\left.\frac{dy_{\infty}(s)}{ds}\right|_{s=0}$ correctly in the large τ-regime (see Section 17.4.3).

17.9 Problems

(i) Using the spectral decomposition of the adjacency matrix (Van Mieghem, 2011, p. 226)

$$A = \sum_{j=1}^{N} \lambda_j z_j z_j^T$$

where z_j is the eigenvector of A belonging to the eigenvalue λ_j, which are ordered as $\lambda_1 \geq \lambda_2 \geq \cdots \geq \lambda_N$, the companion of (17.17) is

$$y_\infty(\tau) = \frac{\varepsilon^*}{1+\varepsilon^*} + \frac{\tau}{(1+\varepsilon^*)N} E\left[D^T X_\infty - \sum_{j=1}^{N} \lambda_j \left(X_\infty^T z_j\right)^2\right] \qquad (17.94)$$

Generalize the analysis of the complete graph on p. 456 to any regular graph.

(ii) The complete bipartite graph $K_{m,n}$ consists of two partitions $\mathcal{N}_m$ with m nodes and $\mathcal{N}_n$ with n nodes (Van Mieghem, 2011), where each node of the one partition is connected to all nodes in the other partition, but nodes of the same partition are not connected. Hence, the total number of nodes and links is $N = m + n$ and $L = mn$, respectively. The adjacency matrix of the complete bipartite graph $K_{m,n}$ is

$$A_{K_{m,n}} = \begin{bmatrix} O_{m \times m} & J_{m \times n} \\ J_{n \times m} & O_{n \times n} \end{bmatrix} \qquad (17.95)$$

The bipartite graph $K_{m,n}$ may represent a set of m servers and n clients. Show that the steady-state infection probability vector in NIMFA equals $V_\infty = \begin{bmatrix} w_{m\infty} u_{m \times 1} \\ w_{n\infty} u_{n \times 1} \end{bmatrix}$, where

$$w_{n\infty} = \frac{mn - \frac{1}{\tau^2}}{\left(\frac{1}{\tau} + m\right)n} \qquad \text{for each node } i \in \mathcal{N}_n \qquad (17.96)$$

$$w_{m\infty} = \frac{mn - \frac{1}{\tau^2}}{\left(\frac{1}{\tau} + n\right)m} \qquad \text{for each node } j \in \mathcal{N}_m \qquad (17.97)$$

Since $y_\infty^{(1)}(s) = \frac{n w_{n\infty} + m w_{m\infty}}{n+m}$, we obtain

$$y_\infty^{(1)}(s) = \frac{\left(mn - \tau^{-2}\right)}{N} \left(\frac{1}{\tau^{-1} + m} + \frac{1}{\tau^{-1} + n}\right) \qquad (17.98)$$

(iii) Prove the general upper bound for the exact average steady-state number of infected nodes in the ε–SIS epidemic process,

$$y_\infty(\tau) \leq \min\left(\frac{\varepsilon^*}{1+\varepsilon^* - \tau\mu_1}, 1\right) \qquad (17.99)$$

Hint: Use the Rayleigh inequalities (Van Mieghem, 2011, p. 223) for a symmetric Laplacian matrix Q with largest and smallest eigenvalue μ_1 and $\mu_N = 0$, respectively, and any real vector w,

$$0 \leq w^T Q w \leq \mu_1 w^T w \qquad (17.100)$$

(iv) *SIR epidemics.* In the SIR model, a node can be in one of the three states. When a node j is healthy, but susceptible to the virus, at time t, its state $Y_j = S$. A node j can be infected, $Y_j = I$, by its direct neighbors that are infected. The infection is modelled by a Poisson process with rate β. Finally, an infected node j can be cured, after which it is removed from the infection process, $Y_j = R$. The curing is modelled by a Poisson process with rate δ. All Poisson processes are independent. This formulation describes a continuous-time SIR process on a graph. Derive the SIR governing equations.

(v) Deduce the recursion (17.75) for the average, exact steady-state fraction $y_{\infty;N}(\tau)$ of infected nodes in the complete graph K_N.

(vi) *Scaling of the epidemic threshold in* K_N. Show that

$$\tau_c^* = \frac{1}{(N+2) - 2\sqrt{(N+1)}} = \frac{1}{N}\left(1 + \frac{2}{\sqrt{N}} + O\left(\frac{1}{N}\right)\right) \qquad (17.101)$$

(vii) Derive the asymptotics for the steady-state probabilities π_j in K_N in Section 17.6.1 when $N \to \infty$.

Notes

Simon *et al.* (2011) studied the same SIS Markov chain with a different labelling that resulted into an infinitesimal generator with a block tri-diagonal matrix structure. Simon *et al.* (2011) also focused on lumping, i.e. reducing the number of linear Markov equations in some special topologies (such as the complete graph and the star) while still maintaining exactness. After removing the absorbing state, the resulting modified SIS steady-state is determined in Cator and Van Mieghem (2013b) for the complete graph and the star graph.

Whereas NIMFA is a continuous-time mean-field approximation for any network, a discrete-time SIS approximation for any network was first sketched in Wang *et al.* (2003), whose paper was later improved in Chakrabarti *et al.* (2008). In those papers and in the work of Ganesh *et al.* (2005), the relation between the epidemic threshold and $\frac{1}{\lambda_1}$ appears, however, Lemma 17.4.6 and the lower bound $\tau_c > \tau_c^{(1)} = \frac{1}{\lambda_1}$ (Theorem 17.3.1) was first rigorously proved in Van Mieghem *et al.* (2009). The mean-field method and analysis of NIMFA has been transferred to the SIR model in Youssef and Scoglio (2011), while in Darabi Sahneh and Scoglio (2011), NIMFA is extended to an SAIS infection, where the Alert (A) state is introduced next to the S ($X_i = 0$) and I ($X_i = 1$) state. A further generalization of NIMFA to m compartments or infection states is proposed in Darabi Sahneh *et al.* (2013), while second-order mean-field improvements are studied in Cator and Van Mieghem (2012). NIMFA is extended to interdependent networks in Wang *et al.* (2013).

Besides the epidemic threshold, Kooij *et al.* (2009) have proposed the *viral conductance*, defined as

$$\psi = \int_0^{\lambda_1} y_\infty(s)\, ds \qquad \text{with } s = \frac{1}{\tau}$$

as a graph metric. The viral conductance has been studied further in Youssef *et al.* (2011) and Van Mieghem (2012b). Theorem 17.3.1 and Lemma 17.4.6 have important practical consequences. Given a network with adjacency matrix A and an imminent infection rate β, a nodal curing rate $\delta > \beta\lambda_1$ can be applied in nodes (in the form of anti-virus software or any other protection scheme) to maintain the network virus-free. A key-point is that the security of each host depends not only on the protection strategies it chooses to adopt but also on those chosen by other hosts in the network. In a heterogeneous setting, explained in Section 17.5, the resulting game-theoretic optimum has been studied in Omic *et al.* (2009), while a different optimization technique in Gourdin *et al.* (2011) minimizes the overall infection in the network by determining the individual curing rates of nodes. When the network can be modified, we possess a much larger number of ways to ban epidemics such as immunization strategies (Chen *et al.*, 2008) and several ways to decrease the spectral radius λ_1 of the network: quarantining by partitioning the network (Omic *et al.*, 2010), degree-preserving rewiring (Van Mieghem *et al.*, 2010c,b; Wang *et al.*, 2011) that changes the assortativity and modularity, and hence, the spectral radius. The optimal strategy to remove m nodes or m links from the network in order to minimize λ_1 is proved in Van Mieghem *et al.* (2011b) to be NP-hard. Consequently, several heuristics are proposed and evaluated in Van Mieghem *et al.* (2011b).

The state of the art in epidemics on networks is reviewed in Pastor-Satorras *et al.* (2014).

18
The efficiency of multicast

The efficiency or gain of multicast in terms of network resources is compared to unicast. Specifically, we concentrate on a one-to-many communication, where a source sends a same message to m different, uniformly distributed destinations along the shortest path. In unicast, this message is sent m times from the source to each destination. Hence, unicast uses on average $f_N(m) = mE[H_N]$ link-traversals or hops, where $E[H_N]$ is the mean number of hops to a uniform location in the graph with N nodes. One of the main properties of multicast is that it economizes on the number of link-traversals: the message is only copied at each branch point of the multicast tree to the m destinations. Let us denote by $H_N(m)$ the number of links in the shortest path tree (SPT) to m uniformly chosen nodes. If we define the multicast gain $g_N(m) = E[H_N(m)]$ as the mean number of hops in the SPT rooted at a source to m randomly chosen distinct destinations, then $g_N(m) \leq f_N(m)$. The purpose here is to quantify the multicast gain $g_N(m)$. We present general results valid for *all* graphs and more explicit results valid for the random graph $G_p(N)$ and for the k-ary tree. The analysis presented here may be valuable to derive a business model for multicast: "How many customers m are needed to make the use of multicast for a service provider profitable?"

Two modeling assumptions are made. First, the multicast process is assumed[1] to deliver packets along the shortest path from a source to each of the m destinations. As most of the current Internet protocols forward packets based on the (reverse) shortest path, the assumption of SPT delivery is quite realistic. The second assumption is that the m multicast group member nodes are uniformly chosen out of the total number of nodes N. This assumption has been discussed by Phillips *et al.* (1999). They concluded that, if m and N are large, deviations from the uniformity assumption are negligibly small. Also the Internet measurements of Chalmers and Almeroth (2001) seem to confirm the validity of the uniformity assumption.

---

[1] The assumption ignores shared tree multicast forwarding such as core-based tree (CBT, see RFC2201).

18.1 General results for $g_N(m)$

Theorem 18.1.1 *For any connected graph with N nodes,*

$$m \le g_N(m) \le \frac{Nm}{m+1} \qquad (18.1)$$

Proof: We need at least one edge for each different user; therefore $g_N(m) \ge m$ and the lower bound is attained in a star topology with the source at the center.

We will next show that an upper bound is obtained in a line topology. It is sufficient to consider trees, because multicast only uses shortest paths without cycles. If the tree has not a line topology, then at least one node has degree 3 or the root has degree 2. Take the node closest to the root with this property and cut one of the branches at this node; we paste that branch to a node at the deepest level. Through this procedure the multicast function $g_N(m)$ stays unaltered or increases. Continuing in this fashion until we reach a line topology demonstrates the claim.

For the line topology we place the source at the origin and the other nodes at the integers $1, 2, \ldots, N-1$. The links of the graph are given by $(i, i+1)$, $i = 0, 1, \ldots, N-2$. The multicast gain $g_N(m)$ equals $E[M]$, where M is the maximum of a sample of size m, without replacement, from the integers $1, 2, \ldots, N-1$. Thus,

$$\Pr[M \le k] = \frac{\binom{k}{m}}{\binom{N-1}{m}}, \quad m \le k \le N-1$$

from which $g_N(m) = E[M]$ is

$$g_N(m) = \sum_{k=m}^{N-1} k \frac{\binom{k}{m} - \binom{k-1}{m}}{\binom{N-1}{m}} = \sum_{k=m}^{N-1} k \frac{\binom{k-1}{m-1}}{\binom{N-1}{m}} = m \sum_{k=m}^{N-1} \frac{\binom{k}{m}}{\binom{N-1}{m}} = \frac{mN}{m+1} \sum_{k=m}^{N-1} \frac{\binom{k}{m}}{\binom{N}{m+1}} = \frac{mN}{m+1}$$

where we have used that $\sum_{k=m}^{N-1} \binom{k}{m} / \binom{N}{m+1} = 1$, because it is a sum of probabilities over *all* possible disjoint outcomes. $\qquad \square$

Figure 18.1 shows the allowable space for $g_N(m)$.

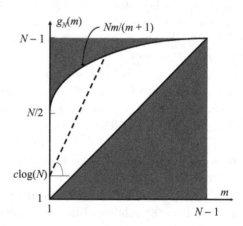

Fig. 18.1. The allowable region (in white) of $g_N(m)$. For exponentially growing graphs, $E[H_N] = c \log N$, implying that the allowable region for these graphs is smaller and bounded at the left (in dotted line) by the straight line $m(c \log N)$.

Theorem 18.1.2 *For any connected graph with N nodes, the map $m \mapsto g_N(m)$ is concave and the map $m \mapsto \frac{g_N(m)}{f_N(m)}$ is decreasing.*

Proof: Define Y_m to be the random variable giving the additional number of hops necessary to reach the m-th user when the first $m-1$ users are already connected. Then we have that

$$E[Y_m] = g_N(m) - g_N(m-1)$$

Moreover, let Y'_m be the random number of additional hops necessary to reach the m-th multicast group member, when we discard all extra hops of the $(m-1)$-th group member. An example is illustrated in Fig. 18.2. The random variable Y'_m has the same distribution as Y_{m-1}, because both the $(m-1)$-th and the m-th group member are chosen uniformly from the remaining $N-m-1$ nodes. In general, $Y'_m \neq Y_{m-1}$, but, for each k, $\Pr[Y'_m = k] = \Pr[Y_{m-1} = k]$ and, hence,

$$E[Y'_m] = E[Y_{m-1}] \tag{18.2}$$

Furthermore, we have by construction that $Y_m \leq Y'_m$ with probability 1, implying that

$$E[Y_m] \leq E[Y'_m] \tag{18.3}$$

Indeed, attaching the m-th group member to the reduced tree takes at least as many hops as attaching that same group member to the non-reduced tree because the former is contained in the latter and the extra hops added by the $(m-1)$-th group member can only help us. Combining (18.2) and (18.3) immediately gives

$$g_N(m) - g_N(m-1) = E[Y_m] \leq E[Y'_m] = g_N(m-1) - g_N(m-2) \tag{18.4}$$

This is equivalent to the concavity of the map $m \mapsto g_N(m)$.

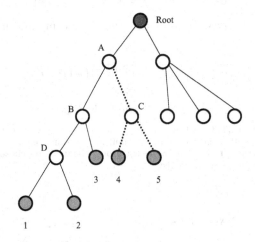

Fig. 18.2. A multicast session with $m = 5$ group members where $Y_5 = 1$ (namely link C-5). To construct Y'_5 the three dotted lines must be removed and we observe that $Y'_5 = 2$ (A-C-5), which is referred to as the reduced tree. In this example, $Y'_5 = Y_4 = 2$ because A-C-4 and A-C-5 both consist of two hops. In general, they are equal in distribution because the role of group members 4 and 5 are identical in the reduced tree.

In order to show that $\frac{g_N(m)}{f_N(m)}$ is decreasing it suffices to show that $m \mapsto \frac{g_N(m)}{m}$ is decreasing, since $f_N(m)$ is proportional to m. Defining $g_N(0) = 0$, we can write $g_N(m)$ as a telescoping sum

$$g_N(m) = \sum_{k=1}^{m} \{g_N(k) - g_N(k-1)\} = \sum_{k=1}^{m} x_k$$

where $x_k = g_N(k) - g_N(k-1)$, $k = 1, \ldots, m$. Then,

$$\frac{g_N(m)}{m} = \frac{1}{m} \sum_{k=1}^{m} x_k$$

is the mean of a sequence of m positive numbers x_k. By (18.4) the sequence $x_k \leq x_{k-1}$ is decreasing and, hence,

$$\frac{g_N(m)}{m} = \frac{1}{m} \sum_{k=1}^{m} x_k \leq \frac{1}{m-1} \sum_{k=1}^{m-1} x_k = \frac{g_N(m-1)}{m-1}$$

This proves that $m \mapsto g_N(m)/m$ is decreasing.   □

Next, we will give a representation for $g_N(m)$ that is valid for all graphs. Let X_i be the number of joint hops that *all* i uniformly chosen and different group members have in common, then the following general theorem holds,

Theorem 18.1.3 *For any connected graph with N nodes,*

$$g_N(m) = \sum_{i=1}^{m} \binom{m}{i} (-1)^{i-1} E[X_i] \qquad (18.5)$$

Note that

$$g_N(1) = f_N(1) = E[X_1] = E[H_N]$$

so that the decrease in mean hops or the "gain" by using multicast over unicast is precisely

$$g_N(m) - f_N(m) = \sum_{i=2}^{m} \binom{m}{i} (-1)^{i-1} E[X_i]$$

However, computing $E[X_i]$ for general graphs is difficult.

Proof of Theorem 18.1.3: Let $A_1, A_2, \ldots, A_m$ be sets where A_i consists of all links that constitute the shortest path from the source to multicast group member i. Denote by $|A_i|$ the number of elements in the set A_i. The multicast group members are chosen uniformly from the set of all nodes except for the root. Hence,

$$E[X_1] = E[|A_i|], \qquad \text{for } 1 \leq i \leq N$$

and

$$E[X_2] = E[|A_i \cap A_j|], \qquad \text{for } 1 \leq i < j \leq N$$

etc.. Now, $g_N(m) = E[|A_1 \cup A_2 \cup \cdots \cup A_m|]$. Since $Q(A) = E[|A|]/\binom{N}{2}$ is a probability measure on the set of all links, we obtain from the inclusion-exclusion formula (2.5) applied to Q and multiplied with $\binom{N}{2}$ afterwards,

$$E[|A_1 \cup A_2 \cup \cdots \cup A_m|] = \sum_{i=1}^{m} E[|A_i|] - \sum_{i<j} E[|A_i \cap A_j|] + \cdots$$

$$+ (-1)^{m-1} E[|A_1 \cap A_2 \cap \cdots \cap A_m|]$$

$$= m E[X_1] - \binom{m}{2} E[X_2] + \cdots + (-1)^{m-1} E[X_m]$$

This proves Theorem 18.1.3.   □

Corollary 18.1.4 *For any connected graph with N nodes,*

$$E\left[X_m\right] = \sum_{i=1}^{m} \binom{m}{i}(-1)^{i-1}g_N(i) \tag{18.6}$$

The corollary is a direct consequence of the inversion formula for the binomial (Riordan, 1968, Chapter 2). Alternatively, in view of the Gregory-Newton interpolation formula (Lanczos, 1988, Chapter 4, Section 2) for $g_N(m) = \sum_{i=1}^{\infty} \binom{m}{i}\Delta^i g_N(0)$, we can write $E\left[X_i\right] = (-1)^{i-1}\Delta^i g_N(0)$, where Δ is the difference operator, $\Delta f(0) = f(1) - f(0)$.

Corollary 18.1.5 *For any connected graph, the multicast efficiency $g_N(m)$ is bounded by*

$$\frac{f_N(m)}{g_N(m)} \le E\left[H_N\right] \tag{18.7}$$

where $E\left[H_N\right]$ is the mean number of hops in unicast.

Proof: We give two demonstrations. (a) From $g_N(N-1) = N-1$ (all nodes, source plus $N-1$ destinations, of the graph are spanned by a tree consisting of $N-1$ links) and the monotonicity of $m \mapsto \frac{g_N(m)}{f_N(m)}$ (see Theorem 18.1.2) we obtain:

$$\frac{g_N(m)}{f_N(m)} \ge \frac{g_N(N-1)}{f_N(N-1)} = \frac{N-1}{(N-1)E\left[H_N\right]} = \frac{1}{E\left[H_N\right]}$$

(b) Alternatively, Theorem 18.1.1 indicates that $g_N(m) \ge m$, which, with the identity $f_N(m) = mE\left[H_N\right]$, immediately leads to (18.7). $\square$

Corollary 18.1.5 means that for any *connected* graph, including the graph describing the Internet, the ratio of the unicast over multicast efficiency is bounded by the expected hopcount in unicast. In order words, the maximum savings in resources an operator can gain by using multicast (over unicast) never exceeds $E\left[H_N\right]$, which is roughly about 15 in the current Internet.

18.2 The random graph $G_p(N)$

In this section, we confine to the class RGU, the random graphs of the class $G_p(N)$ with independent identically and exponentially distributed link weights w with mean $E\left[w\right] = 1$ and where $\Pr[w \le x] = 1 - e^{-x}$, $x > 0$. In Section 16.2, we have shown that the corresponding SPT is, asymptotically, a URT. The analysis below is exact for the complete graph K_N while asymptotically correct for connected random graphs $G_p(N)$.

18.2.1 The hopcount of the shortest path tree

Based on properties of the URT, the complete probability density function of the number of links $H_N(m)$ in the SPT to m uniformly chosen nodes can be determined.

We first derive a recursion for the probability generating function $\varphi_{H_N(m)}(z) = E\left[z^{H_N(m)}\right]$ of the number of links $H_N(m)$ in the SPT to m uniformly chosen nodes in the complete graph K_N.

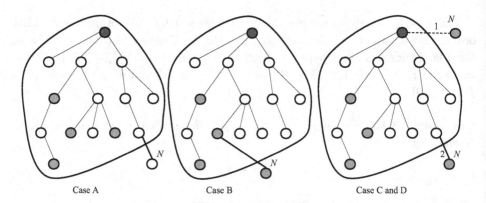

Case A   Case B   Case C and D

Fig. 18.3. The several possible cases in which the N-th node can be attached uniformly to the URT of size $N-1$. The root is darkly shaded while the m multicast member nodes are lightly shaded.

Lemma 18.2.1 *For $N > 1$ and all $1 \leq m \leq N-1$,*

$$\varphi_{H_N(m)}(z) = \frac{(N-m-1)(N-1+mz)}{(N-1)^2}\varphi_{H_{N-1}(m)}(z) + \frac{m^2 z}{(N-1)^2}\varphi_{H_{N-1}(m-1)}(z) \quad (18.8)$$

Proof: To prove (18.8), we use the recursive growth of URTs: a URT of size N is a URT of size $N-1$, where we add an additional link to a uniformly chosen node.

In order to obtain a recursion for $H_N(m)$ we distinguish between the m uniformly chosen nodes all being in the URT of size $N-1$ or not. The probability that they all belong to the tree of size $N-1$ is equal to $1 - \frac{m}{N-1}$ (case A in Fig. 18.3). If they all belong to the URT of size $N-1$, then we have that $H_N(m) = H_{N-1}'(m)$. Thus, we obtain

$$\varphi_{H_N(m)}(z) = \left(1 - \frac{m}{N-1}\right)\varphi_{H_{N-1}(m)}(z) + \frac{m}{N-1}E\left[z^{1+L_{N-1}(m)}\right] \quad (18.9)$$

where $L_{N-1}(m)$ is the number of links in the subtree of the URT of size $N-1$ spanned by $m-1$ uniform nodes and the "one" refers to the link from the added N-th node to its ancestor in the URT of size $N-1$. We complete the proof by investigating the generating function of $L_{N-1}(m)$. Again, there are two cases. In the first case (B in Fig. 18.3), the ancestor of the added N-th node is one of the $m-1$ previous nodes (which can only happen if it is unequal to the root), else we get one of the cases C and D in Fig. 18.3. The probability of the first event equals $\frac{m-1}{N-1}$, the probability of the latter equals $1 - \frac{m-1}{N-1}$. If the ancestor of the added N-th node is one of the $m-1$ previous nodes, then the number of links $L_{N-1}(m)$ equals $H_{N-1}(m-1)$, otherwise the generating function of the number of additional links equals

$$\left(1 - \frac{1}{N-m}\right)\varphi_{H_{N-1}(m)}(z) + \frac{1}{N-m}\varphi_{H_{N-1}(m-1)}(z)$$

The first contribution comes from the case where the ancestor of the added N-th node is not the root, and the second from where it is equal to the root, which has probability $\frac{1}{N-1-(m-1)} = \frac{1}{N-m}$.

Therefore,

$$E\left[z^{L_{N-1}(m)}\right] = \frac{m-1}{N-1}\varphi_{H_{N-1}(m-1)}(z) + \frac{N-m}{N-1}\left(\frac{N-m-1}{N-m}\varphi_{H_{N-1}(m)}(z) + \frac{\varphi_{H_{N-1}(m-1)}(z)}{N-m}\right)$$

$$= \frac{m}{N-1}\varphi_{H_{N-1}(m-1)}(z) + \frac{N-m-1}{N-1}\varphi_{H_{N-1}(m)}(z) \qquad (18.10)$$

Substitution of (18.10) into (18.9) leads to (18.8). $\qquad\qquad\qquad\qquad\qquad\square$

Since $g_N(m) = E[H_N(m)] = \varphi'_{H_N(m)}(1)$, we obtain the recursion for $g_N(m)$,

$$g_N(m) = \left(1 - \frac{m^2}{(N-1)^2}\right)g_{N-1}(m) + \frac{m^2}{(N-1)^2}g_{N-1}(m-1) + \frac{m}{N-1} \qquad (18.11)$$

Theorem 18.2.2 *For all $N \geq 1$ and $1 \leq m \leq N-1$,*

$$\varphi_{H_N(m)}(z) = E\left[z^{H_N(m)}\right] = \frac{m!(N-1-m)!}{((N-1)!)^2}\sum_{k=0}^{m}\binom{m}{k}(-1)^{m-k}\frac{\Gamma(N+kz)}{\Gamma(1+kz)} \qquad (18.12)$$

Consequently,

$$\Pr\left[H_N(m) = j\right] = \frac{m!(-1)^{N-(j+1)}S_N^{(j+1)}S_j^{(m)}}{(N-1)!\binom{N-1}{m}} \qquad (18.13)$$

where $S_N^{(j+1)}$ and $S_j^{(m)}$ denote the Stirling numbers of first and second kind (Abramowitz and Stegun, 1968, Section 24.1).

Proof: By iterating the recursion (18.8) for small values of m, the computations given in van der Hofstad *et al.* (2006, Appendix) suggest the solution (18.12) for (18.8). One can verify that (18.12) satisfies (18.8). This proves (18.12) of Theorem 18.2.2. Using Abramowitz and Stegun (1968, 24.1.3.B), the Taylor expansion around $z = 0$ equals

$$\varphi_{H_N(m)}(z) = \frac{m!N(N-1-m)!}{(N-1)!}\sum_{k=0}^{m}\binom{m}{k}(-1)^{m-k}\left(\frac{\Gamma(N+kz)}{N!\Gamma(1+kz)} - \frac{1}{N}\right)$$

$$= \frac{m!N(N-1-m)!}{(N-1)!}\sum_{k=0}^{m}\binom{m}{k}(-1)^{m-k}\sum_{j=1}^{N-1}\frac{(-1)^{N-(j+1)}S_N^{(j+1)}}{N!}k^j z^j$$

$$= \frac{m!N(N-1-m)!}{(N-1)!}\sum_{j=1}^{N-1}\frac{(-1)^{N-(j+1)}S_N^{(j+1)}}{N!}\left(\sum_{k=0}^{m}\binom{m}{k}(-1)^{m-k}k^j\right)z^j$$

Using the definition of Stirling numbers of the second kind (Abramowitz and Stegun, 1968, 24.1.4.C),

$$m!S_j^{(m)} = \sum_{k=0}^{m}\binom{m}{k}(-1)^{m-k}k^j$$

for which $S_j^{(m)} = 0$ if $j < m$, gives

$$\varphi_{H_N(m)}(z) = \frac{(m!)^2(N-1-m)!}{((N-1)!)^2}\sum_{j=1}^{N-1}(-1)^{N-(j+1)}S_N^{(j+1)}S_j^{(m)}z^j$$

This proves (18.13) and completes the proof of Theorem 18.2.2. $\qquad\qquad\qquad\square$

Figure 18.4 plots the probability density function of $H_{50}(m)$ for different values of m.

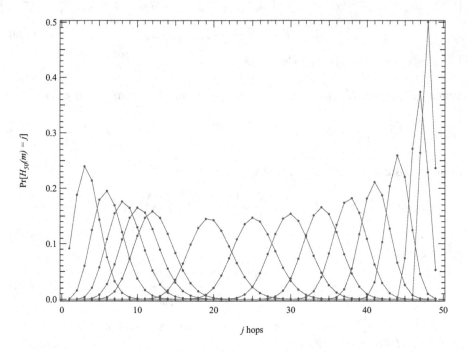

Fig. 18.4. The pdf of $H_{50}(m)$ for $m = 1, 2, 3, 4, 5, 10, 15, 20, 25, 30, 35, 40, 45, 47$.

Corollary 18.2.3 *For all $N \geq 1$ and $1 \leq m \leq N - 1$,*

$$g_N(m) = E\left[H_N(m)\right] = \frac{mN}{N-m}\sum_{k=m+1}^{N}\frac{1}{k} \tag{18.14}$$

and

$$\mathrm{Var}\left[H_N(m)\right] = \frac{N-1+m}{N+1-m}g_N(m) - \frac{g_N^2(m)}{(N+1-m)} - \frac{m^2N^2\sum_{k=m+1}^{N}\frac{1}{k^2}}{(N-m)(N+1-m)} \tag{18.15}$$

The formula (18.14) is proved in two different ways. The earlier proof presented in Section 18.6 below does not rely on the recursion in Lemma 18.2.1 nor on Theorem 18.2.2. The shorter proof is presented here. Formula (18.14) can be expressed in terms of the digamma function $\psi(x)$ as

$$g_N(m) = mN\left(\frac{\psi(N) - \psi(m)}{N-m}\right) - 1 \tag{18.16}$$

Proof of Corollary 18.2.3: The expectation and variance of $H_N(m)$ will not be obtained using the explicit probabilities (18.13), but by rewriting (18.12) as

$$\varphi_{H_N(m)}(z) = \frac{\Gamma(m+1)\Gamma(N-m)}{\Gamma^2(N)} \sum_{k=0}^{m} \binom{m}{k}(-1)^{m-k}\partial_t^{N-1}\left[t^{N-1+kz}\right]_{t=1}$$

$$= \frac{\Gamma(m+1)\Gamma(N-m)}{\Gamma^2(N)}(-1)^m\partial_t^{N-1}\left[t^{N-1}(1-t^z)^m\right]_{t=1} \qquad (18.17)$$

Indeed,

$$E[H_N(m)] = \frac{\Gamma(m+1)\Gamma(N-m)}{\Gamma^2(N)}(-1)^m\partial_z\partial_t^{N-1}\left[t^{N-1}(1-t^z)^m\right]_{t=z=1}$$

$$= \frac{\Gamma(m+1)\Gamma(N-m)}{\Gamma^2(N)}m(-1)^{m-1}\partial_t^{N-1}\left[t^N \log t(1-t)^{m-1}\right]_{t=1},$$

$$E[H_N(m)(H_N(m)-1)] = \frac{\Gamma(m+1)\Gamma(N-m)}{\Gamma^2(N)}(-1)^m\partial_z^2\partial_t^{N-1}\left[t^{N-1}(1-t^z)^m\right]_{t=z=1}$$

$$= \frac{\Gamma(m+1)\Gamma(N-m)}{\Gamma^2(N)}m(-1)^{m-1}$$

$$\times \partial_t^{N-1}\left[t^N \log^2 t(1-t)^{m-2}\left[-(m-1)t+(1-t)\right]\right]_{t=1}$$

We will start with the former. Using $\partial_t^i(1-t)^j|_{t=1} = j!(-1)^j\delta_{i,j}$ and Leibniz' rule, we find

$$E[H_N(m)] = \frac{\Gamma(m+1)\Gamma(N-m)}{\Gamma^2(N)}m!\binom{N-1}{m-1}\partial_t^{N-m}\left[t^N \log t\right]_{t=1}$$

Since

$$\partial_t^k[t^n \log t]_{t=1} = \frac{n!}{(n-k)!}\sum_{j=n-k+1}^{n}\frac{1}{j}$$

we obtain expression (18.14) for $E[H_N(m)]$.

We now extend the above computation to $E[H_N(m)(H_N(m)-1)]$ that we write as

$$E[H_N(m)(H_N(m)-1)] = \frac{\Gamma(m+1)\Gamma(N-m)}{\Gamma^2(N)}(R_1+R_2) \qquad (18.18)$$

where

$$R_1 = m(m-1)(-1)^{m-2}\partial_t^{N-1}\left[t^{N+1}\log^2 t(1-t)^{m-2}\right]_{t=1}$$

$$R_2 = m(-1)^{m-1}\partial_t^{N-1}\left[t^N \log^2 t(1-t)^{m-1}\right]_{t=1}$$

Using

$$\partial_t^k[t^n \log^2 t]_{t=1} = 2\frac{n!}{(n-k)!}\sum_{i=n-k+1}^{n}\sum_{j=i+1}^{n}\frac{1}{ij} = \frac{n!}{(n-k)!}\left[\left(\sum_{i=n-k+1}^{n}\frac{1}{i}\right)^2 - \sum_{i=n-k+1}^{n}\frac{1}{i^2}\right]$$

we obtain,

$$R_1 = \binom{N-1}{m-2}m(m-1)(m-2)!\partial_t^{N-m+1}\left[t^{N+1}\log^2 t\right]_{t=1}$$

$$= (N+1)!\binom{N-1}{m-2}\left[\left(\sum_{k=m+1}^{N+1}\frac{1}{k}\right)^2 - \sum_{k=m+1}^{N+1}\frac{1}{k^2}\right]$$

Similarly,

$$R_2 = \binom{N-1}{m-1}m(m-1)!\partial_t^{N-m}\left[t^N\log^2 t\right]_{t=1} = N!\binom{N-1}{m-1}\left[\left(\sum_{k=m+1}^N\frac{1}{k}\right)^2 - \sum_{k=m+1}^N\frac{1}{k^2}\right]$$

Substitution into (18.18) leads to

$$E[H_N(m)(H_N(m)-1)] = \frac{m^2N^2}{(N+1-m)(N-m)}\left[\left(\sum_{k=m+1}^N\frac{1}{k}\right)^2 - \sum_{k=m+1}^N\frac{1}{k^2}\right]$$
$$+ \frac{2m(m-1)N}{(N+1-m)(N-m)}\sum_{k=m+1}^N\frac{1}{k}$$

From $g_N(m) = E[H_N(m)]$ and $\text{Var}\,[H_N(m)] = E[H_N(m)(H_N(m)-1)] + g_N(m) - g_N^2(m)$, we obtain (18.15). This completes the proof of Corollary 18.2.3. $\qquad\square$

For $N = 1000$, Fig. 18.5 illustrates the typical behavior for large N of the expectation $g_N(m)$ and the standard deviation $\sigma_N(m) = \sqrt{\text{Var}\,[H_N(m)]}$ for all values of m. For any spanning tree, the number of links $H_N(N-1)$ is precisely $N-1$, so that $\text{Var}[H_N(N-1)] = 0$.

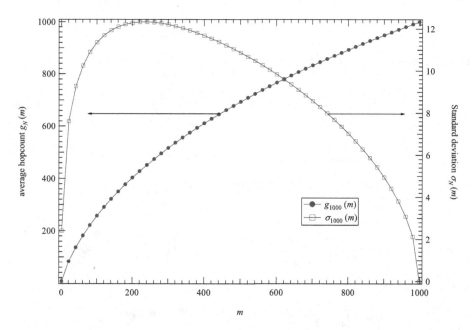

Fig. 18.5. The average number of hops $g_N(m)$ (left axis) in the SPT and the corresponding standard deviation $\sigma_N(m)$ (right axis) as a function of the number m of multicast group members in the complete graph with $N = 1000$.

Figure 18.5 also indicates that the standard deviation $\sigma_N(m)$ of $H_N(m)$ is much smaller than the mean, even for $N = 1000$. In fact, we obtain from (18.15) that

$$\text{Var}[H_N(m)] \leq \frac{N-1+m}{N+1-m} g_N(m) \leq \frac{2N g_N(m)}{N-m} = \frac{2 g_N^2(m)}{m \sum_{k=m+1}^{N} \frac{1}{k}} = o(g_N^2(m))$$

This bound implies that with probability converging to 1 for every $m = 1, \ldots, N-1$,

$$\left| \frac{H_N(m)}{g_N(m)} - 1 \right| \leq \varepsilon$$

In van der Hofstad *et al.* (2006), the scaled random variable $\frac{H_N(m)-g_N(m)}{\sqrt{g_N(m)}}$ is proved to tend to a Gaussian random variable, i.e. $\frac{H_N(m)-g_N(m)}{\sqrt{g_N(m)}} \xrightarrow{d} N(0,1)$, for all $m = o(\sqrt{N})$. For large graphs of the size of the Internet and larger, this observation implies that the mean $g_N(m) = E[H_N(m)]$ is a good approximation for the random variable $H_N(m)$ itself because the variations of $H_N(m)$ around the mean are small. Consequently, it underlines the importance of $g_N(m)$ as a significant measure for multicast.

18.2.2 The weight of the shortest path tree

In this section, we summarize results on the weight $W_N(m)$ of the SPT and omit derivations, but refer to van der Hofstad *et al.* (2006). For all $1 \leq m \leq N-1$, the mean weight of the SPT is

$$E[W_N(m)] = \sum_{j=1}^{m} \frac{1}{N-j} \sum_{k=j}^{N-1} \frac{1}{k} \tag{18.19}$$

In particular, if the shortest path tree spans the whole graph, then for all $N \geq 2$,

$$E[W_N(N-1)] = \sum_{n=1}^{N-1} \frac{1}{n^2} \tag{18.20}$$

from which $E[W_N(N-1)] < \zeta(2) = \frac{\pi^2}{6}$ for any finite N. The variance is

$$\text{Var}[W_N(N-1)] = \frac{4}{N} \sum_{k=1}^{N-1} \frac{1}{k^3} + 4 \sum_{j=1}^{N-1} \frac{1}{j^3} \sum_{k=1}^{j} \frac{1}{k} - 5 \sum_{j=1}^{N-1} \frac{1}{j^4} \tag{18.21}$$

or asymptotically, for large N,

$$\text{Var}[W_N(N-1)] = \frac{4\zeta(3)}{N} + O\left(\frac{\log N}{N^2}\right) \tag{18.22}$$

Asymptotically for large N, the mean weight of a shortest path tree is $\zeta(2) = 1.645\ldots$, while the mean weight of the minimum spanning tree, given by (16.65), is $\zeta(3) < \zeta(2)$. This result has an interesting implication. The Steiner tree is the minimum weight tree that connects a set of m members out of N nodes in the graph.

The Steiner tree problem is NP-complete, which means that it is not feasible to compute for large N. If $m = 2$, the weight of the Steiner tree equals that of the shortest path, $W_{\text{Steiner},N}(2) = W_N$, while for $m = N$, we have $W_{\text{Steiner},N}(N) = W_{\text{MST}}$. Hence, for any $m < N$ and N, $E[W_{\text{Steiner},N}(m)] \leq \zeta(3)$ because the weight of the Steiner tree does not decrease if the number of members m increases. The ratio $\frac{\zeta(2)}{\zeta(3)} = 1.368$ indicates that the use of the SPT (computationally easy) never performs on average more than 37% worse than the optimal Steiner tree (computationally unfeasible). In a broader context and referring to the concept of the "Prize of Anarchy", which is broadly explained in Robinson (2004), the SPT used in a communications network is related to the Nash equilibrium, while the Steiner tree gives the hardly achievable global optimum.

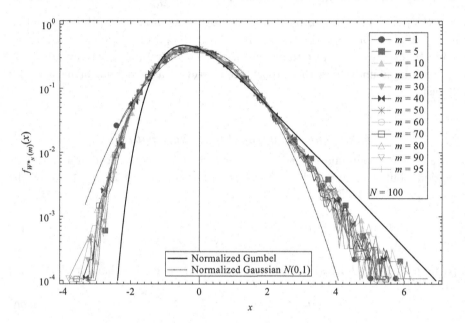

Fig. 18.6. The pdf of the normalized random variable $W_N^*(m)$ for $N = 100$. The exact asymptotics (16.23), properly rescaled, coincides with the $m = 1$ simulated filled circles (and is not shown to avoid overloading).

Simulations – even for small N, which allow us to cover the entire m-range as illustrated in Fig. 18.6 – indicate that the normalized random variable $W_N^*(m) = \frac{W_N(m) - E[W_N(m)]}{\sqrt{\text{Var}[W_N(m)]}}$ lies between a normalized Gaussian $N(0, 1)$ and a normalized Gumbel (see Theorem 6.6.1). For the particular case of $m = 1$, the correct limit law is given in (16.23) and agrees, after proper normalization, very well with the simulation for $N = 100$. For the other extreme for $m = N - 1$, van der Hofstad *et al.* (2007) show that the weight of the shortest path tree tends to a Gaussian,

$$\sqrt{N}\left(W_N(N-1) - \zeta(2)\right) \xrightarrow{d} N\left(0, \sigma_{\text{SPT}}^2\right)$$

with $\sigma_{\text{SPT}}^2 = 4\zeta(3) \approx 4.80823$ as follows from (18.22). This case for $m = N - 1$ shows that simulations for small N do not agree well with the exact asymptotics. Finally, Janson (1995) gave the related result for the minimum spanning tree. He extended Frieze's result (16.65) by proving that the scaled weight of the minimum spanning tree also tends to a Gaussian for large N,

$$\sqrt{N}\left(W_{\text{MST}} - \zeta(3)\right) \xrightarrow{d} N\left(0, \sigma_{\text{MST}}^2\right)$$

where $\sigma_{\text{MST}}^2 \approx 1.6857$.

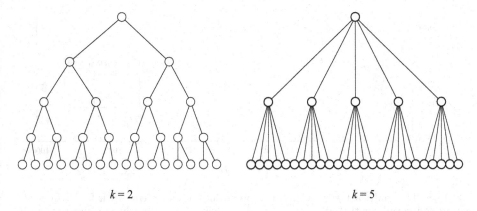

$$k = 2 \qquad\qquad\qquad k = 5$$

Fig. 18.7. The left hand side tree ($k = 2$) has $N = 31$ and $D = 4$, while the right-hand side ($k = 5$) has $N = 31$ and $D = 2$.

18.3 The k-ary tree

In this section, we consider the k-ary tree of depth[2] D with the source at the root of the tree and m receivers at randomly chosen nodes (see Fig. 18.7). In a k-ary tree the total number of nodes satisfies

$$N = 1 + k + k^2 + \cdots + k^D = \frac{k^{D+1} - 1}{k - 1} \tag{18.23}$$

Theorem 18.3.1 *For the k-ary tree,*

$$g_{N,k}(m) = N - 1 - \sum_{j=0}^{D-1} k^{D-j} \frac{\binom{N-1-\frac{k^{j+1}-1}{k-1}}{m}}{\binom{N-1}{m}} \tag{18.24}$$

Proof: See Section 18.7. □

Unfortunately, the j-summation seems difficult to express in closed form. Observe that $g_N(N-1) = N - 1$, because all binomials vanish. The sum extends over all levels $j \leq D - 1$, for which the remaining number of nodes in the lower levels l

---

[2] The depth D is equal to the number of hops from the root to a node at the leaves.

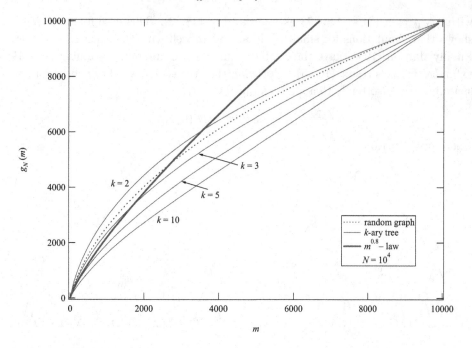

Fig. 18.8. The multicast gain $g_N(m)$ computed for the k-ary tree with four values of k, the random graph (with "effective" $k_{\mathrm{rg}} = e = 2.718...$), and the Chuang-Sirbu power law for $N = 10^4$ on a linear scale where the prefactor $E[H_N]$ is given by (16.11).

(i.e. $D \geq l > j$) is larger than m nodes. In some sense, we may regard (18.24) as an (exact) expansion around $m = N - 1$. Explicitly,

$$g_{N,k}(m) = N - 1 - k^D \left(1 - \frac{m}{N-1}\right) - k^{D-1} \prod_{q=0}^{k} \left(1 - \frac{m}{N-1-q}\right)$$

$$- \sum_{j=2}^{D-1} k^{D-j} \prod_{q=0}^{\frac{k^{j+1}-1}{k-1}-1} \left(1 - \frac{m}{N-1-q}\right) \qquad (18.25)$$

which shows that $g_{N,k}(m)$ is a polynomial in m of degree $\leq \frac{N-1}{k}$. Moreover, the terms in the j-sum rapidly decrease; their ratio equals

$$\frac{\prod_{q=\frac{k^j-1}{k-1}}^{\frac{k^{j+1}-1}{k-1}-1} \left(1 - \frac{m}{N-1-q}\right)}{k} < \frac{1}{k} \left(1 - \frac{m}{N-1-\frac{k^j-1}{k-1}}\right)^{k^j} << 1$$

Figure 18.8 indicates that formula (18.24), although derived subject to (18.23), also seems valid when $D = \left\lfloor \frac{\log[1+N(k-1)]}{\log k} - 1 \right\rfloor$, where $\lfloor x \rfloor$ is the largest integer smaller than or equal to x. This suggests that the deepest level D need not be filled completely to count k^D nodes and that (18.24) may extend to "incomplete" k-ary trees. As further observed from Fig. 18.8, $g_{N,k}(m)$ is monotonously decreasing in k. Hence, it is quite likely that he map $k \mapsto g_{N,k}(m)$ is decreasing in $k \in [1, N-1]$. Intuitively, this conjecture can be understood from Fig. 18.7. Both the $k = 2$ and $k = 5$ trees have an equal number of nodes. We observe that the deeper D (or the smaller k), the more overlap is possible, hence, the larger $g_{N,k}(m)$.

Theorem 18.1.1 can also be deduced from (18.24). The lower bound is attained in a star topology where $k = N - 1$, $D = 1$ and $E[H_N] = 1$. The upper bound is attained in a line topology where $k = 1$, $D = N - 1$ and $E[H_N] = \frac{N}{2}$. Furthermore, for real values of $k \in [1, N-1]$, the set of curves specified by (18.24) covers the total allowable space of $g_{N,k}(m)$, as shown in Fig. 18.1. This suggests to consider (18.24) for estimating k in real topologies.

Since $g_N(1) = E[H_N]$, the mean hopcount in a k-ary tree follows from (18.24) as

$$E[H_N] = N - 1 - \sum_{j=0}^{D-1} k^{D-j} \frac{N - 1 - \frac{k^{j+1}-1}{k-1}}{N-1} = \frac{1}{N-1} \sum_{j=0}^{D-1} k^{D-j} \frac{k^{j+1}-1}{k-1}$$

$$= \frac{ND}{N-1} + \frac{D}{(N-1)(k-1)} - \frac{1}{k-1} \tag{18.26}$$

For large N, we find with

$$D = \left\lfloor \frac{\log[1 + N(k-1)]}{\log k} - 1 \right\rfloor \sim \log_k N + \log_k(1 - 1/k) + O(1/N)$$

that

$$E[H_N] = \log_k N + \log_k(1 - 1/k) - \frac{1}{k-1} + O\left(\frac{\log_k N}{N}\right) \tag{18.27}$$

Comparing (18.27) with the mean hopcount in the random graph (16.11) shows equality to first order if $k_{\mathrm{rg}} = e$. Moreover, both the second-order terms $\gamma - 1 = -0.42$ and $\log(1 - 1/e) - \frac{1}{e-1} = -1.04$ are $O(1)$ and independent of N. This shows that the multicast gain in the random graph is well approximated by $g_{N,e}(m)$.

18.4 The Chuang–Sirbu Law

We discuss the empirical Chuang–Sirbu scaling Law, which states that $g_N(m) \approx E[H_N] m^{0.8}$ for the Internet. Based on Internet measurements, Chuang and Sirbu (1998) observed that $g_N(m) \approx E[H_N] m^{0.8}$. Subsequently, Phillips et $al.$ (1999) dubbed this observation the Chuang–Sirbu Law.

Corollary 18.1.5 implies that the empirical law of Chuang–Sirbu cannot hold true for all $m \le N$. Indeed, if $g_N(m) = E[H_N] m^{0.8}$, we obtain from the inequality (18.7) and the identity $f_N(m) = mE[H_N]$, that $m^{0.2} \le E[H_N]$. Write $m = xN$ for a fixed $0 < x < 1$ and x independent of N. Hence, we have shown that

Corollary 18.4.1 *For all graphs satisfying the condition that $\frac{E[H_N]}{N^{0.2}} \to 0$, for large N, the empirical Chuang–Sirbu Law does not hold in the region $m = xN$ with $0 < x \le 1$ and sufficiently large N.*

The most realistic graph models for the Internet assume that $E[H_N] \approx c \log N$, since this implies that the number of routers that can be reached from any starting destination grows exponentially with the number of hops. For these realistic graphs, Corollary 18.4.1 states that empirical Chuang–Sirbu Law does not hold for all m. On the other hand, there are more regular graphs (such as a d-lattice, where $E[H_N] \simeq \frac{d}{3}N^{1/d}$) with $E[H_N] \sim N^{0.2+\epsilon}$ (and $\epsilon > 0$) for which the mathematical condition $m^{0.2} \leq E[H_N]$ is satisfied for all m and N. As shown in Van Mieghem et al. (2000), however, these classes of graphs, in contrast to random graphs, are not leading to good models for SPTs in the Internet.

18.4.1 Validity range of the Chuang–Sirbu Law

For the random graph $G_p(N)$, the SPT is very close to a URT for large N and with (16.11), we obtain

$$f_N(m) \sim m(\log N + \gamma - 1)$$

From the exact $g_N(m)$ formula (18.16) for the random graph $G_p(N)$, the asymptotic for large N and m follows as

$$g_N(m) \sim \frac{mN}{N-m} \log\left(\frac{N}{m}\right) - \frac{1}{2} \tag{18.28}$$

The above scaling explains the empirical Chuang–Sirbu Law for $G_p(N)$: for m small with respect to N, the graphs of $(\log N + \gamma - 1)m^{0.8}$ and $\frac{mN}{N-m} \log\left(\frac{N}{m}\right) - \frac{1}{2}$ look very alike in a log-log plot, as illustrated in Fig. 18.9.

Using the asymptotic properties of the digamma function ψ, we obtain (18.28) as an excellent approximation for large N (and all m) or, in normalized form with $m = xN$ and $0 < x < 1$,

$$\frac{g_N(xN) + 0.5}{N} \sim \frac{x \log x}{x - 1} \tag{18.29}$$

The normalized Chuang–Sirbu Law is $\frac{g_N(xN)}{N} = \frac{E[H_N]}{N^{0.2}}x^{0.8}$. It is interesting to note that the Chuang–Sirbu Law is "best" if $\frac{E[H_N]}{N^{0.2}} = 1$, since then both endpoints $x = 0$ and $x = 1$ coincide with (18.29). This optimum is achieved when $N \approx 250\,000$, which is of the order of magnitude of the estimated number of routers in the current Internet. This observation may explain the fairly good correspondence on a less sensitive log-log scale with Internet measurements. At the same time, it shows that for a growing Internet, the fit of the Chuang–Sirbu Law will deteriorate. For $N \geq 10^6$, the Chuang–Sirbu Law underestimates $g_N(m)$ for all m.

18.4.2 The effective power exponent $\beta(N)$

For small to moderate values of m, $g_N(m)$ is very close to a straight line in a log-log plot. This power-law behavior implies that $\log g_N(m) \approx \log E(H_N) + \beta(N) \log m$,

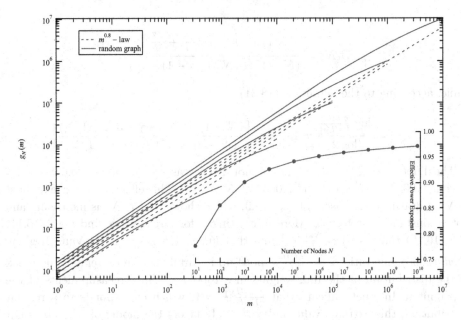

Fig. 18.9. The multicast efficiency for $N = 10^j$ with $j = 3, 4, ..., 7$. The endpoint of each curve $g_N(N-1) = N-1$ determines N. The insert shows the effective power exponent versus N.

which is a first-order Taylor expansion of $\log g_N(m)$ in $\log m$. This observation suggests the computation[3] of the effective power exponent $\beta(N)$ as

$$\beta(N) = \left.\frac{d \log g_N(m)}{d \log m}\right|_{m=1} \qquad (18.30)$$

Only for a straight line, the differential operator can be replaced by the difference operator such that $\beta(N) \equiv \beta^*(N)$, where

$$\beta^*(N) = \frac{\log \frac{g_N(2)}{E[H_N]}}{\log 2} \qquad (18.31)$$

In general, for small m, the effective power exponent (18.30) is not a constant 0.8 as in the Chuang–Sirbu Law, but dependent on N. Since $g_N(m)$ is concave by Theorem 18.1.2, $\beta(N)$ is the maximum possible value for $\frac{d \log g_N(m)}{d \log m}$ at any $m \geq 1$. A direct consequence of Theorem 18.1.1 is that the effective power exponent $\beta(N) \in [\frac{1}{2}, 1]$. From recent Internet measurements, Chalmers and Almeroth (2001) found that $0.66 \leq \beta(N) \leq 0.7$.

The effective power exponent $\beta(N)$ as defined in (18.30) for the random graph

---

[3] Although (18.5) only has meaning for integer m, analytic continuation to a complex variable is possible and, hence, differentiation can be defined.

is

$$\beta(N) = \frac{N\left(\psi(N) + \gamma - \frac{\pi^2}{6} + \frac{\pi^2}{6N}\right)}{(N-1)\left(\psi(N) + (\gamma - 1) + \frac{1}{N}\right)}$$

while, according to the definition (18.31),

$$\beta^*(N) = \frac{\log \frac{g_N(2)}{E[H_N]}}{\log 2} = 1 + \log_2\left[\frac{(N-1)\left(\psi(N) + \gamma - 3/2 + 1/N\right)}{(N-2)\left(\psi(N) + \gamma - 1 + 1/N\right)}\right]$$

The difference $\beta(N) - \beta^*(N)$ monotonously decreases and is largest, 0.048 at $N = 3$ while 0.0083 at $N = 10^5$ and 0.0037 at $N = 10^{10}$. This effective power exponent $\beta(N)$ is drawn in the insert of Fig. 18.9, which shows that $\beta(N)$ is increasing and not a constant close to 0.8. More interestingly, for large N, we find with (16.11) and (16.12) that $\beta(N) \sim \frac{\mathrm{Var}[H_N]}{E[H_N]}$ and that $\lim_{N \to \infty} \beta(N) = 1$. In Van Mieghem *et al.* (2000), the ratio $\alpha = \frac{\mathrm{Var}[H_N]}{E[H_N]}$ pops up naturally as the extreme value index of the distribution of the link weights in a topology. Since measurements of the hopcount in Internet indicate that $\frac{\mathrm{Var}[H_N]}{E[H_N]} \approx 1$, which corresponds to a regular distribution, this extreme value index strongly favors the model of the hopcount based on shortest paths in $G_p(N)$, although random graphs do *not* model the Internet topology well.

Thus, if the number of nodes in the Internet is still growing, we suggest, *only for small to moderate values of* m, the consideration of a power-law approximation for the multicast gain

$$g_N(m) \approx E[H_N] \, m^{\frac{\mathrm{Var}[H_N]}{E[H_N]}}$$

instead of the Chuang–Sirbu Law.

In summary, many properties in nature seem linear on an insensitive log-log scale. However, deriving from these plots simple and attractive power laws for complicated matter, seems a little oversimplified[4].

18.5 Stability of a multicast shortest path tree

We now turn to the problem of quantifying the stability in a multicast tree. Inspired by Poisson arrival processes, at a single instant of time, we assume that either one or zero group members can leave. In the sequel, we do not make any further assumption about the time-dependent process of leaving/joining a multicast group and refrain from dependencies on time. The number of links in the tree that change after one multicast group member leaves the group has been chosen as measure for

---

[4] Many recent articles devote attention to power law behavior but most of them seem prudent: just recall the immense interest (hype?) a few years ago in the long-range and self-similar nature of Internet traffic and the relation to the "simple" power law with only the Hurst parameter (comparable to $\beta(N)$ here) in the exponent.

the stability of the multicast tree. If we denote this quantity by $\Delta_N(m)$, then, by definition of $g_N(m)$, the mean number of changes equals

$$E\left[\Delta_N(m)\right] = g_N(m) - g_N(m-1) \tag{18.32}$$

Since $g_N(m)$ is concave (Theorem 18.1.2), $E\left[\Delta_N(m)\right]$ is always positive and decreasing in m. If the scope of m is extended to real numbers, $E\left[\Delta_N(m)\right] \approx g'_N(m)$ which simplifies further estimates.

The situation where on average less than one link changes if one multicast group member leaves may be regarded as a stable regime. Since $E\left[\Delta_N(m)\right]$ is always positive and decreasing in m, this stable regime is reached when the group size m exceeds m_1, which satisfies $E\left[\Delta_N(m_1)\right] = 1$. For example, for the URT that is asymptotically the SPT for the class RGU defined in Section 16.2.2, this condition approximately follows from (18.28) as

$$E\left[\Delta_N(m)\right] \sim \frac{mN}{N-m}\log\left(\frac{N}{m}\right) - \frac{(m-1)N}{N-m+1}\log\left(\frac{N}{m-1}\right) \tag{18.33}$$

Let $x = \frac{m}{N}$, then $0 < x < 1$ and

$$\frac{E\left[\Delta_N(m)\right]}{N} \sim \frac{-x}{1-x}\log x + \frac{(x-1/N)}{1-(x-1/N)}\log\left(x - \frac{1}{N}\right)$$

After expanding the second term in a Taylor series around x to first order in $\frac{1}{N}$,

$$E\left[\Delta_N(xN)\right] \sim \frac{x-1-\log x}{(1-x)^2} + O\left(\frac{1}{N}\right)$$

For large N, $E\left[\Delta_N(x_1 N)\right] \sim 1$ occurs when $x_1 = 0.3161$, which is the solution in x of $\frac{x-1-\log x}{(1-x)^2} = 1$. For the class RGU, a stable tree as defined above is obtained when the multicast group size m is larger than $m_1 = 0.3161N \approx \frac{N}{3}$. In the sequel, since m_1 is high and of less practical interest, we will focus on multicast group sizes smaller than m_1. The computation of m_1 for other graph types turns out to be difficult. Since, as mentioned above, the comparison with Internet measurement (Van Mieghem *et al.*, 2001a) shows that formula (18.28) provides a fairly good estimate, we expect that $m_1 \approx \frac{N}{3}$ also approximates well the stable regime in the Internet.

The following theorem quantifies the stability in the class RGU.

Theorem 18.5.1 *For sufficiently large N and fixed m, the number of changed edges $\Delta_N(m)$ in a random graph $G_p(N)$ with uniformly distributed link weights tends to a Poisson distribution,*

$$\Pr\left[\Delta_N(m) = k\right] \sim e^{-E[\Delta_N(m)]}\frac{\left(E\left[\Delta_N(m)\right]\right)^k}{k!} \tag{18.34}$$

where $E\left[\Delta_N(m)\right] = g_N(m) - g_N(m-1)$ and $g_N(m)$ is given by (18.16) or approximately by (18.28).

Proof: In Section 16.2.1 we have mentioned that the SPT in the class RGU is a URT for large N. In addition, the random variable for the number of hops H_N from the root to an arbitrary node tends, for large N, to a Poisson random variable with mean $E[H_N] \sim \log N + \gamma - 1$ as shown in Section 16.3.1. Now, $\Delta_N(m) = H_N(m) - H_N(m-1)$ is the positive discrete random variable that counts the absolute value of the difference between the hopcount $h_{R \to m}$ from the root (source) to user m and the hopcount $h_{R \to m-1}$ from the root to the user closest in the tree to m, which we here relabel by $m-1$. Neither users m nor $m-1$ is independent nor are the two random variables $h_{R \to m}$ and $h_{R \to m-1}$ independent in general due to possible overlap in their paths. If the shortest paths from the root to each of the two users m and $m-1$ overlap, there always exists a node in the SPT, say node B as illustrated in Fig. 18.10, that sees the partial shortest paths from itself to m and $m-1$ as non-overlapping and independent. Since the SPT is a URT, the subtree rooted at that node B (enclosed in dotted line in Fig. 18.10) is again a URT as follows from Theorem 16.2.1. With respect to B, the nodes m and $m-1$ are uniformly chosen and the number of links $\Delta_N(m)$ that change if the m-th node leaves is just its hopcount with respect to B (instead of the original root). We denote the unknown number of nodes in that

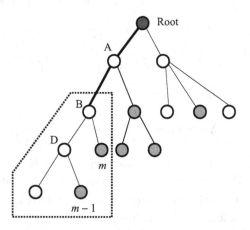

Fig. 18.10. A sketch of a uniform recursive tree, where $h_{R \to m} = 3$ and $h_{R \to m-1} = 4$ and the number of links in common is two (shown in bold Root-A-B).

subtree rooted at B by $\nu(m) \leq N$. We have that $\nu(m) \leq \nu(m-1)$ because by adding a group member, the size of the subtree can only decrease. For large N and small m, $\nu(m)$ is large such that the above-mentioned asymptotic law of the hopcount applies. If both m and N are large, $\nu(m)$ will become too small for the asymptotic law to apply. Thus, for fixed m and large N, this implies that $\Delta_N(m)$ tends to a Poisson random variable with mean $E[\Delta_N(m)]$. □

Simulations in Van Mieghem and Janic (2002) indicate that the Poisson law seems more widely valid than just in the asymptotic regime ($N \to \infty$). The proof can be extended to a general topology. Assume for a certain class of graphs that the pdf of the hopcount $\Pr[H_N = k]$ and the multicast efficiency $g_N(m)$ can be computed for all sizes N. The subtree rooted at B is again a SPT in a subcluster of size $\nu(m)$, which is an unknown random variable. The argument similar as the one in the proof above shows that $\Pr\left[\Delta_N(m) = k\right] = \Pr\left[H_{\nu(m)} = k\right]$. This argument implicitly assumes that all multicast users are uniformly distributed over the graph. By the law of total probability,

$$\Pr\left[H_{\nu(m)} = k\right] = \sum_{n=1}^{N} \Pr\left[H_{\nu(m)} = k|\nu(m) = n\right] \Pr\left[\nu(m) = n\right]$$

$$= \sum_{n=1}^{N} \Pr\left[H_n = k\right] \Pr\left[\nu(m) = n\right]$$

which, unfortunately, shows that the pdf of $\nu(m)$ is needed to specify $\Pr\left[\Delta_N(m) = k\right]$. However, we can proceed further in an approximate way by replacing the unknown random variable $\nu(m)$ by its best estimate, $E\left[\nu(m)\right]$. In that approximation, the mean size $E\left[\nu(m)\right]$ of the shortest path subtree rooted at B can be specified, at least in principle, with the use of (18.32). Indeed, since $E\left[H_{E[\nu(m)]}\right] = \sum_{k=1}^{E[\nu(m)]-1} k \Pr\left[H_{E[\nu(m)]} = k\right]$, by equating

$$E\left[H_{E[\nu(m)]}\right] = g_N(m) - g_N(m-1)$$

a relation in one unknown $E\left[\nu(m)\right]$ is found and can be solved for $E\left[\nu(m)\right]$. In conclusion, we end up with the approximation

$$\Pr\left[\Delta_N(m) = k\right] \approx \Pr\left[H_{E[\nu(m)]} = k\right]$$

which roughly demonstrates that, in general, $\Pr\left[\Delta_N(m) = k\right]$ is likely related to the hopcount distribution in that certain class of graphs.

Unfortunately, for very few types of graphs, both the pdf $\Pr\left[H_N = k\right]$ and the multicast gain $g_N(m)$ can be computed. This fact augments the value of Theorem 18.5.1, although the class RGU is *not* a good model for the *graph* of the Internet. Fortunately, the *shortest path tree* deduced from that class seems a reasonable approximation as shown in Fig. 16.4 and sufficient to provide first-order estimates.

18.6 Proof of (18.16): $g_N(m)$ for random graphs

Before embarking with the proof of formula (18.16), we first prove the following lemma.

Lemma 18.6.1 *For $a > b$,*

$$S(a,b) = \sum_{k=1}^{b} \frac{(a-k)!}{(b-k)!} \frac{1}{k} = \frac{a!}{b!} \left[\psi(a+1) - \psi(a-b+1)\right]$$

and

$$S(b,b) = \sum_{k=1}^{b} \frac{1}{k} = \psi(b+1) + \gamma$$

Proof: We start by writing

$$S(a,b) = \sum_{k=1}^{b} \frac{(a-k)\cdots(b-k+1)}{k} = a \sum_{k=1}^{b} \frac{(a-1-k)\cdots(b-k+1)}{k} - \sum_{k=1}^{b}(a-1-k)\cdots(b-k+1)$$

Since $(a-1-k)\cdots(b-k+1) = (a-b-1)!\binom{a-1-k}{b-k}$ and by the recurrence for the binomial $\sum_{k=1}^{b}\binom{a-1-k}{b-k} = \binom{a-1}{b-1}$, we have that

$$S(a,b) = aS(a-1,b) - \frac{1}{a-b}\frac{(a-1)!}{(b-1)!}$$

After p iterations, we have

$$S(a,b) = a(a-1)\cdots(a-p+1)S(a-p,b) - \frac{a!}{(b-1)!}\sum_{j=0}^{p-1}\frac{1}{(a-j)(a-j-b)}$$

and, if $p = a - b$, the recursions stops with result,

$$S(a,b) = \frac{a!}{b!}\sum_{k=1}^{b}\frac{1}{k} - \frac{a!}{(b-1)!}\sum_{j=0}^{a-b-1}\frac{1}{(a-j)(a-j-b)}$$

$$= \frac{a!}{b!}\sum_{k=1}^{b}\frac{1}{k} - \frac{a!}{b!}\left(\sum_{k=1}^{a-b}\frac{1}{k} - \sum_{k=b+1}^{a}\frac{1}{k}\right) = \frac{a!}{b!}\left(\sum_{k=1}^{a}\frac{1}{k} - \sum_{k=1}^{a-b}\frac{1}{k}\right)$$

from which the lemma follows. □

Proof of equation (18.16): We will investigate $E[X_i] = E\left[X_i^{(N)}\right]$ in the URT with N nodes. Here $E[X_i]$ is the number of joint hops in a multicast SPT from the root to i uniformly chosen nodes in the URT and where all the group member nodes are different from the root. Let $E\left[\tilde{X}_i\right]$ be the same quantity where we allow the group member nodes to be the root. Then,

$$E\left[\tilde{X}_i\right] = \frac{N-i}{N}E[X_i]$$

since there are i possibilities each with probability $\frac{1}{N}$ that one of the nodes equals the root, in which case $X_i = 0$.

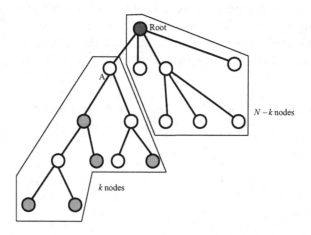

Fig. 18.11. The two contributing clusters leading to the $E\left[\tilde{X}_i^{(N)}\right]$-recursion.

The mean number of joint hops $E\left[\tilde{X}_i\right]$ is deduced from Fig. 18.11, where two clusters are shown each with respectively k and $N-k$ nodes. The first cluster with k nodes does not possess the root (dark shaded), but it contains the i multicast group members (light shaded). There is already at least 1 joint hop because the link between the root and node A, that can be viewed as the root of the first cluster, is used by all i group members lying in the first cluster. Given the size k of the first cluster, the probability that all i uniformly chosen group members belong to the first cluster equals $\frac{k(k-1)\cdots(k-i+1)}{N(N-1)\cdots(N-i+1)}$ because the probability that the first group member belongs to that cluster, which is $\frac{k}{N}$, the probability that the second group member also belongs to the first cluster, which is $\frac{k-1}{N-1}$ and so on. Since the size of the first cluster connected to the root is uniform in between 1 and $N-1$, the probability that the size is k equals $\frac{1}{N-1}$. When all i nodes are in that first cluster of size k, X_i is at least 1, and the problem restarts, but with N replaced by k and A being the root. Hence, if all i group members belong to the first cluster, the mean number of joint hops is $\frac{1}{N-1}\sum_{k=1}^{N-1}\frac{k(k-1)\cdots(k-i+1)}{N(N-1)\cdots(N-i+1)}\left(1+E\left[\tilde{X}_i^{(k)}\right]\right)$ because we must sum over all possible sizes for the first cluster. If *not* all i group member nodes are in the first cluster, the group member nodes are divided over the two clusters. But, in that case, we have no joint overlaps or $X_i = 0$. Thus, if not all i group members nodes are in the first cluster, the only way that there are possible joint overlaps ($X_i > 0$), is that all i group member nodes are in the second cluster. However, by removing the first cluster, we are left again with a uniform recursive tree of size $N-k$. The mean number of joint hops in this case is $\frac{1}{N-1}\sum_{k=1}^{N-1}\frac{(N-k)(N-k-1)\cdots(N-k-i+1)}{N(N-1)\cdots(N-i+1)}E\left[\tilde{X}_i^{(N-k)}\right]$. Adding both contributions results in the recursion formula

$$E\left[\tilde{X}_i^{(N)}\right] = \frac{1}{N-1}\sum_{k=1}^{N-1}\frac{k(k-1)\cdots(k-i+1)}{N(N-1)\cdots(N-i+1)}\left(1+2E\left[\tilde{X}_i^{(k)}\right]\right) \qquad (18.35)$$

We next write

$$\alpha_i^{(N)} = N(N-1)\cdots(N-i+1)E\left[\tilde{X}_i^{(N)}\right] = \frac{N!}{(N-i)!}E\left[\tilde{X}_i^{(N)}\right]$$

then the above recurrence equation (18.35) turns into

$$\alpha_i^{(N)} = \frac{1}{N-1}\sum_{k=1}^{N-1}[k(k-1)\cdots(k-i+1)+2\alpha_i^{(k)}]$$

$$= \frac{1}{N-1}\sum_{k=i}^{N-1}k(k-1)\cdots(k-i+1)+\frac{2}{N-1}\sum_{k=1}^{N-1}\alpha_i^{(k)}$$

Subtracting

$$(N-1)\alpha_i^{(N)} - (N-2)\alpha_i^{(N-1)} = \frac{(N-1)!}{(N-i-1)!}+2\alpha_i^{(N-1)}$$

from which we obtain

$$\frac{\alpha_i^{(N)}}{N} = \frac{(N-2)!}{N(N-i-1)!}+\frac{\alpha_i^{(N-1)}}{N-1} \qquad (18.36)$$

Iterating (18.36) gives

$$\frac{\alpha_i^{(N)}}{N} = \sum_{j=0}^{k-1}\frac{(N-2-j)!}{(N-j)(N-i-1-j)!}+\frac{\alpha_i^{(N-k)}}{N-k}$$

Since $\alpha_i^{(i)} = E\left[\tilde{X}_i^{(i)}\right] = 0$, because the root is then always one of the group member nodes, we finally obtain,

$$\alpha_i^{(N)} = N\sum_{j=0}^{N-i-1}\frac{(N-2-j)!}{(N-j)(N-i-1-j)!} = N\sum_{k=i+1}^{N}\frac{(k-2)!}{k(k-i-1)!} \qquad (18.37)$$

It can be shown that, for large N, $\alpha_i^{(N)} \sim \frac{N}{i-1}\frac{(N-2)!}{(N-i-1)!}$.

Because

$$E\left[X_i^{(N)}\right] = \frac{N}{N-i} E\left[\tilde{X}_i^{(N)}\right] = \frac{(N-i-1)!}{(N-1)!}\alpha_i^{(N)}$$

we have that

$$E\left[X_i^{(N)}\right] = \frac{(N-i-1)!N}{(N-1)!} \sum_{k=i+1}^{N} \frac{(k-2)!}{k(k-i-1)!} \tag{18.38}$$

and, for large N, $E\left[X_i^{(N)}\right] \sim \frac{1}{i-1}\frac{N}{(N-1)} \sim \frac{1}{i-1}$.

Invoking Theorem 18.1.3, the mean number of multicast hops for m uniformly chosen, distinct group members is

$$g_N(m) = \sum_{i=1}^{m} \binom{m}{i}(-1)^{i-1}\frac{(N-i-1)!N}{(N-1)!}\sum_{k=2}^{N}\frac{(k-2)!}{k(k-i-1)!}$$

$$= \frac{-N}{(N-1)!}\sum_{s=0}^{N-2}\frac{(N-2-s)!}{N-s}\sum_{i=1}^{m}\binom{m}{i}(-1)^i\frac{(N-i-1)!}{(N-i-1-s)!}$$

The i-summation can be executed as follows. Consider $x^{N-1}(1-1/x)^m = \sum_{i=0}^{m}\binom{m}{i}(-1)^i x^{N-i-1}$. Differentiating s times yields

$$\sum_{i=0}^{m}\binom{m}{i}(-1)^i\frac{(N-i-1)!}{(N-i-1-s)!}x^{N-i-s-1} = \frac{d^s}{dx^s}\left[x^{N-1-m}(x-1)^m\right]$$

Expanding the right-hand side around $x=1$ gives

$$\frac{d^s}{dx^s}\left[x^{N-1-m}(x-1)^m\right] = \sum_{k=0}^{\infty}\binom{N-1-m}{k}\frac{d^s}{dx^s}(x-1)^{k+m}$$

$$= \sum_{k=0}^{\infty}\binom{N-1-m}{k}\frac{(k+m)!}{(k+m-s)!}(x-1)^{k+m-s}$$

Evaluation at $x=1$ only leads to a non-zero contribution if $k+m-s=0$. Hence,

$$\sum_{i=1}^{m}\binom{m}{i}(-1)^i\frac{(N-i-1)!}{(N-i-1-s)!} = \binom{N-1-m}{s-m}s! - \frac{(N-1)!}{(N-1-s)!}$$

and

$$g_N(m) = \frac{-N(N-1-m)!}{(N-1)!}\sum_{s=0}^{N-2}\frac{s!}{(N-s)(s-m)!(N-1-s)} + N\sum_{s=0}^{N-2}\frac{1}{(N-s)(N-1-s)}$$

$$= \frac{-N(N-1-m)!}{(N-1)!}\left[\sum_{s=m}^{N-2}\frac{s!}{(s-m)!(N-1-s)} - \sum_{s=m}^{N-2}\frac{s!}{(s-m)!(N-s)}\right]$$

$$+ N\left[\sum_{k=1}^{N-1}\frac{1}{k} - \sum_{k=2}^{N}\frac{1}{k}\right]$$

$$= \frac{-N(N-1-m)!}{(N-1)!}\left[\sum_{k=1}^{N-m-1}\frac{(N-k-1)!}{(N-k-1-m)!k} - \sum_{k=2}^{N-m}\frac{(N-k)!}{(N-k-m)!k}\right] + N - 1$$

Rewrite the first summation as

$$\sum_{k=1}^{N-m-1} \frac{(N-k-1)!}{(N-k-1-m)!k} = \frac{(N-2)!}{(N-2-m)!} + \sum_{k=2}^{N-m} \frac{(N-k-1)!(N-k-m)}{(N-k-m)!k}$$

$$= \frac{(N-2)!}{(N-2-m)!} + \sum_{k=2}^{N-m} \frac{(N-k)!}{(N-k-m)!k} - m \sum_{k=2}^{N-m} \frac{(N-k-1)!}{(N-k-m)!k}$$

Then,

$$g_N(m) = \frac{-N(N-1-m)!}{(N-1)!} \left[\frac{(N-2)!}{(N-2-m)!} - m \sum_{k=2}^{N-m} \frac{(N-k-1)!}{k(N-k-m)!} \right] + N - 1$$

$$= \frac{N(m-1)+1}{(N-1)} + \frac{mN(N-1-m)!}{(N-1)!} \sum_{k=2}^{N-m} \frac{(N-k-1)!}{k(N-k-m)!}$$

$$= -1 + \frac{mN(N-1-m)!}{(N-1)!} \sum_{k=1}^{N-m} \frac{(N-k-1)!}{k(N-k-m)!}$$

Using Lemma 18.6.1

$$\sum_{k=1}^{N-m} \frac{(N-k-1)!}{(N-k-m)!} \frac{1}{k} = \frac{(N-1)!}{(N-m)!} \left[\psi(N) - \psi(m) \right] \tag{18.39}$$

finally leads to (18.16). $\qquad\qquad\qquad\qquad\qquad\qquad\qquad\qquad\qquad\qquad\qquad\qquad\square$

18.7 Proof of Theorem 18.3.1: $g_N(m)$ for k-ary trees

Let $\tilde{X}_i$ be the number of joint hops for i *different* multicast group members (we allow the root to be a user in which case $\tilde{X}_i = 0$). Then,

$$\Pr\left[\tilde{X}_i \geq 1\right] = \Pr\left[\text{All group members belong to the same cluster connected to the root}\right]$$

$$= k \cdot \Pr\left[\text{All group members belong to the first cluster connected to the root}\right]$$

$$= k \frac{\binom{(N-1)/k}{i}}{\binom{N}{i}} = k \frac{\binom{1+k+\cdots+k^{D-1}}{i}}{\binom{1+k+\cdots+k^{D}}{i}} \tag{18.40}$$

By self-similarity of k-ary trees we obtain

$$\Pr\left[\tilde{X}_i \geq 2 | \tilde{X}_i \geq 1\right] = p_i^{(D-1)} = k \frac{\binom{1+k+\cdots+k^{D-2}}{i}}{\binom{1+k+\cdots+k^{D-1}}{i}}$$

because each cluster extending from the root is itself a k-ary tree of depth $D-1$. In general, we have $\Pr\left[\tilde{X}_i \geq j\right] = \Pr\left[\tilde{X}_i \geq j | \tilde{X}_i \geq j-1\right] \Pr\left[\tilde{X}_i \geq j-1\right]$. Hence, by iteration,

$$\Pr\left[\tilde{X}_i \geq j\right] = \prod_{n=D-j+1}^{D} p_i^{(n)}, \quad j = 1, 2, \ldots, D-1 \tag{18.41}$$

Note that for $i \geq 2$ the probability $\Pr\left[\tilde{X}_i \geq D\right] = 0$, because if $\tilde{X}_i = D$ some destinations must be identical. From (2.37) we obtain for $i \geq 2$,

$$E\left[\tilde{X}_i\right] = \sum_{j=1}^{D-1} \prod_{n=D-j+1}^{D} p_i^{(n)} = \sum_{j=1}^{D-1} \frac{k^j \binom{1+\cdots+k^{D-j}}{i}}{\binom{1+\cdots+k^{D}}{i}} = \sum_{j=1}^{D-1} \frac{k^{D-j} \binom{1+\cdots+k^j}{i}}{\binom{1+\cdots+k^{D}}{i}} \tag{18.42}$$

Since $E\left[X_i\right] = \frac{N}{N-i}E\left[\tilde{X}_i\right]$, we find

$$E\left[X_i\right] = \frac{N}{N-i}\sum_{j=1}^{D-1}\frac{k^{D-j}\binom{1+\cdots+k^j}{i}}{\binom{1+\cdots+k^D}{i}}, \quad i \geq 2 \tag{18.43}$$

For the value of $E\left[\tilde{X}_1\right]$ and $E\left[X_1\right]$ we find

$$E\left[\tilde{X}_1\right] = \frac{1}{N}\sum_{j=1}^{D}k^j(1+\cdots+k^{D-j}) = \frac{1}{N(k-1)}\left\{Dk^{D+1}-(N-1)\right\}$$

and

$$E\left[X_1\right] = \frac{1}{N-1}\sum_{j=1}^{D}k^j(1+\cdots+k^{D-j}) = \frac{N}{N-1}\sum_{j=1}^{D-1}\frac{k^{D-j}\binom{1+\cdots+k^j}{1}}{\binom{1+\cdots+k^D}{1}} + \frac{k^D}{N-1}$$

Invoking Theorem 18.1.3 yields

$$g_{N,k}(m) = \frac{mk^D}{N-1} - \sum_{i=1}^{m}\binom{m}{i}(-1)^i\frac{N}{N-i}\sum_{j=1}^{D-1}\frac{k^{D-j}\binom{1+\cdots+k^j}{i}}{\binom{1+\cdots+k^D}{i}}$$

Writing $A_j = \frac{k^{j+1}-1}{k-1}$ and reversing the i- and j-summation yields, using (18.23),

$$g_{N,k}(m) = \frac{mk^D}{N-1} - N\sum_{j=1}^{D-1}k^{D-j}\frac{A_j!}{N!}\sum_{i=1}^{m}\binom{m}{i}(-1)^i\frac{(N-i-1)!}{(A_j-i)!}$$

Concentrating on the inner sum with lower sum bound $i = 0$, denoted as S_j, and substituting $k = m - i$, we have

$$S_j = \sum_{k=0}^{m}\binom{m}{k}(-1)^{m-k}\frac{\Gamma(N-m+k)}{\Gamma(A_j-m+k+1)}$$

Invoking the Taylor series of the hypergeometric function (Abramowitz and Stegun, 1968, Section 15.1.1),

$$F(a,b;c;z) = \frac{\Gamma(c)}{\Gamma(a)\Gamma(b)}\sum_{n=0}^{\infty}\frac{\Gamma(a+n)\Gamma(b+n)}{\Gamma(c+n)n!}z^n$$

$m!S_j$ is the coefficient in z^m of the Cauchy product of

$$(1-z)^m = \sum_{k=0}^{\infty}\binom{m}{k}(-1)^k z^k$$

and

$$\frac{\Gamma(N-m)}{\Gamma(A_j-m+1)}F(1,N-m;A_j-m+1;z) = \sum_{k=0}^{\infty}\frac{\Gamma(N-m+k)}{\Gamma(A_j-m+1+k)}z^k$$

Hence,

$$S_j = \frac{1}{m!}\frac{\Gamma(N-m)}{\Gamma(A_j-m+1)}\frac{d^m}{dz^m}\left[(1-z)^m F(1,N-m;A_j-m+1;z)\right]\big|_{z=0}$$

Invoking the differentiation formula (Abramowitz and Stegun, 1968, Section 15.2.7),

$$\frac{d^m}{dz^m}\left[(1-z)^{a+m-1}F(a,b;c;z)\right] = \frac{(-1)^m\Gamma(a+m)\Gamma(c-b+m)\Gamma(c)}{\Gamma(a)\Gamma(c-b)\Gamma(c+m)}(1-z)^{a-1}F(a+m,b;c+m;z)$$

we have, since $a = 1$ and $F(a, b; c; 0) = 1$,

$$S_j = \frac{(-1)^m \Gamma(N-m)\Gamma(A_j + 1 - N + m)}{\Gamma(A_j + 1 - N)\Gamma(A_j + 1)}$$

Thus,

$$g_{N,k}(m) = \frac{mk^D}{N-1} - N \sum_{j=1}^{D-1} k^{D-j} \frac{A_j!}{N!} \left(\frac{(-1)^m (N-m-1)!(A_j - N + m)!}{(A_j - N)! A_j!} - \frac{(N-1)!}{A_j!} \right)$$

$$= \frac{mk^D}{N-1} + \sum_{j=1}^{D-1} k^{D-j} + \frac{(-1)^{m-1}(N-m-1)!}{(N-1)!} \sum_{j=1}^{D-1} k^{D-j} \frac{(A_j - N + m)!}{(A_j - N)!}$$

from which (18.24) is immediate. $\square$

18.8 Problems

(i) Compute the effective power exponent $\beta^*(N)$ for the k-ary tree.

(ii) An ISP wants to announce QoS guarantees for multicast in his network. Therefore, the ISP needs to estimate the probability that the delay from its multimedia server to any of the m multicast members exceeds d^* seconds. The behavior of the largest individual delay $V_N(m)$ among m uniformly distributed multicast members is to a first approximation modeled by considering the shortest path tree in the complete graph K_N with exponentially distributed link delays with mean 1. Derive the pgf of $V_N(m)$. (*Hint:* if $m = 1$, then $\varphi_{V_N(1)}(z)$ is given by (16.20).)

19

The hopcount and weight to an anycast group

In this chapter, the probability density function of the hopcount and of the weight of the shortest path to the most nearby member of an anycast group consisting of m members (e.g. servers or peers) in a graph of N nodes is analyzed.

The results are applied to compute a performance measure η of the efficiency of anycast over unicast and to the server placement problem. The server placement problem asks for the number of (replicated) servers m needed such that any user in the network is not more than j hops away from a server of the anycast group with a certain prescribed probability. As in Chapter 18 on multicast, two types of shortest path trees are investigated: the regular k-ary tree and the irregular uniform recursive tree treated in Chapter 16. Since these two extreme cases of trees indicate that the performance measure $\eta \approx 1 - a \log m$, where the real number a depends on the details of the tree, it is believed that for trees in real networks (as the Internet) a same logarithmic law applies. An order calculus on exponentially growing trees further supplies evidence for the conjecture that $\eta \approx 1 - a \log m$ for small m.

In peer-to-peer (P2P) networks (see e.g. Van Mieghem (2010a, Chapter 13)) such as Napster, Bit Torrent and Tribler, content is often either fully replicated at a peer or is split into chunks and stored over m peers. A major task of a member of the peer group lies in selecting the best peer among those m peers that possess the desired content. Beside searching for that peer that is most nearby in the number of hops, we may select the one that has the smallest delay or latency (or any additive weight of a path). The analysis based on the URT shows how the delay varies as the number m of peers changes and it may set bounds on the minimum number of peers needed to still offer an acceptable content distribution service.

19.1 Introduction

IPv6 possesses a new address type, anycast, that is not supported in IPv4. The anycast address is syntactically identical to a unicast address. However, when a set of interfaces is specified by the same unicast address, that unicast address is called an anycast address. The advantage of anycast is that a group of interfaces at different locations is treated as one single address. For example, the information on servers is often duplicated over several secondary servers at different locations for reasons of robustness and accessibility. Changes are only performed on the primary servers, which are then copied onto all secondary servers to maintain consistency. If both the primary and all secondary servers have a same anycast address, a query from some source towards that anycast address is routed towards the closest server of the group. Hence, instead of routing the packet to the root server (primary server) anycast is more efficient.

Suppose there are m (primary plus all secondary) servers and that these m servers are uniformly distributed over the Internet. The number of hops from the querying device A to the closest server is the minimum number of hops, denoted by $h_N(m)$, of the set of shortest paths from A to these m servers in a network with N nodes. In order to solve the problem, the shortest path tree rooted at node A, the querying device, needs to be investigated. We assume in the sequel that one of the m uniformly distributed servers can possibly coincide with the same router to which the querying machine A is attached. In that case, $h_N(m) = 0$. This assumption is also reflected in the notation, small h, according to the convention made in Section 16.3.2 that capital H for the hopcount excludes the event that the hopcount can be zero.

Clearly, if $m = 1$, the problem reduces to the hopcount of the shortest path from A to one uniformly chosen node in the network and we have that

$$h_N(1) = h_N,$$

where h_N is the hopcount of the shortest path in a graph with N nodes. The other extreme for $m = N$ leads to

$$h_N(N) = 0$$

because all nodes in the network are servers. In between these extremes, it holds that

$$h_N(m) \leq h_N(m-1)$$

since one additional anycast group member (server) can never increase the minimum number of hops from an arbitrary node to that larger group.

The hopcount to an anycast group is a stochastic problem. Even if the network graph is exactly known, an arbitrary node A views the network along a tree. Most often it is a shortest path tree. Although the sequel emphasizes "shortest path trees", the presented theory is equally valid for any type of tree. The node A's perception of the network is very likely different from the view of another node A'. Nevertheless, shortest path trees in the same graph possess to some extent related structural properties that allow us to treat the problem by considering certain types or classes of shortest path trees. Hence, instead of varying the arbitrary node A over all possible nodes in the graph and computing the shortest path tree at each different node, we vary the structure of the shortest path tree rooted at A over all possible shortest path trees of a certain type. Of course, the confinement of the analysis then lies in the type of tree that is investigated. We will only consider the regular k-ary tree and the irregular URT . It seems reasonable to assume that "real" shortest path trees in the Internet possess a structure somewhere in between these extremes and that scaling laws observed in both the two extreme cases may also apply to the Internet.

The presented analysis allows us to address at least three different challenges. First, for a same class of trees, the efficiency of anycast over unicast defined in terms of a performance measure η,

$$\eta = \frac{E\left[h_N(m)\right]}{E\left[h_N(1)\right]} \leq 1$$

is quantified. The performance measure η indicates how much hops (or link traversals or bandwidth consumption) can be saved, on average, by anycast. Alternatively, η also reflects the gain in end-to-end delay or how much faster than unicast, anycast finds the desired information. Second, the so-called *server placement problem* can be treated. More precisely, the question "How many servers m are needed to guarantee that any user request can access the information within j hops with probability $\Pr\left[h_N(m) > j\right] \leq \epsilon$, where ϵ is certain level of stringency?" can be answered. The server placement problem is expected to gain increased interest especially for real-time services where end-to-end QoS (e.g. delay) requirements are desirable. In the most general setting of this server placement problem, all nodes are assumed to be equally important in the sense that users' requests are generated equally likely at any router in the network with N nodes. As mentioned in Chapter 18, the validity of this assumption has been justified by Phillips *et al.* (1999). In the case of uniform user requests, the best strategy is to place servers also uniformly over the network. Computations of $\Pr\left[h_N(m) > j\right] < \epsilon$ for given stringency ϵ and hop j, allow the determination of the minimum number m of servers. Third, fast access to information in content distribution or peer-to-peer networks requires a minimum amount of peers that possess the desired information. If $W_{N;m}$ represents the delay along the shortest path from an arbitrary peer or user to the nearest (in delay) peer in a peer group of size m, then the relation $\Pr\left[W_{N;m} > x\right] < \epsilon$ determines the maximal delay x given m and the strigency ϵ or it lower bounds the peer group size m in case a maximum tolerably delay x is requested.

The solution of the server placement problem as well as the peer group size dimensioning may be regarded as an instance of the general quality of service (QoS) portfolio of an network operator. When the number of servers for a major application offered by the service provider are properly computed, the service provider may announce levels ϵ of QoS (e.g. via $\Pr\left[h_N(m) > j\right] < \epsilon$ or $\Pr\left[W_{N;m} > x\right] < \epsilon$) and may price the use of the application accordingly.

19.2 General analysis of the hopcount

Let us consider a particular shortest path tree T rooted at node A with the level set $L_N = \left\{ X_N^{(k)} \right\}_{1 \le k \le N-1}$ as defined in Section 16.2.2. Suppose that the result of uniformly distributing m anycast group members over the graph leads to a number $m^{(k)}$ of those anycast group member nodes that are k hops away from the root. These $m^{(k)}$ distinct nodes all belong to the k-th level set $\left\{ X_N^{(k)} \right\}$. Similarly as for $X_N^{(k)}$, some relations are immediate. First, $m^{(0)} = 0$ means that none of the m anycast group members coincides with the root node A or $m^{(0)} = 1$ means that one of them (and at most one) is attached to the same router A as the querying device. Also, for all $k > 0$, it holds that $0 \le m^{(k)} \le X_N^{(k)}$ and that

$$\sum_{k=0}^{N-1} m^{(k)} = m \tag{19.1}$$

Given the tree T specified by the level set L_N and the anycast group members specified by the set $\left\{ m^{(0)}, m^{(1)}, \ldots, m^{(N-1)} \right\}$, we will derive the lowest non-empty level $m^{(j)}$, which is equivalent to $h_N(m)$.

Let us denote by e_j the event that all first $j + 1$ levels are not occupied by an anycast group member,

$$e_j = \left\{ m^{(0)} = 0 \right\} \cap \left\{ m^{(1)} = 0 \right\} \cap \cdots \cap \left\{ m^{(j)} = 0 \right\}$$

The probability distribution of the minimum hopcount, $\Pr\left[h_N(m) = j | L_N \right]$, is then equal to the probability of the event $e_{j-1} \cap \left\{ m^{(j)} > 0 \right\}$. Since the event $\left\{ m^{(j)} > 0 \right\} = \left\{ m^{(j)} = 0 \right\}^c$, using the conditional probability yields

$$\Pr\left[h_N(m) = j | L_N \right] = \Pr\left[\left\{ m^{(j)} > 0 \right\} \middle| e_{j-1} \right] \Pr\left[e_{j-1} \right]$$
$$= \left(1 - \Pr\left[\left\{ m^{(j)} = 0 \right\} \middle| e_{j-1} \right] \right) \Pr\left[e_{j-1} \right] \tag{19.2}$$

Since $e_j = e_{j-1} \cap \left\{ m^{(j)} = 0 \right\}$, the probability of the event e_j can be decomposed as

$$\Pr\left[e_j \right] = \Pr\left[\left\{ m^{(j)} = 0 \right\} \middle| e_{j-1} \right] \Pr\left[e_{j-1} \right] \tag{19.3}$$

The assumption that all m anycast group members are uniformly distributed enables to compute $\Pr\left[\left\{ m^{(j)} = 0 \right\} | e_{j-1} \right]$ exactly. Indeed, by the uniform assumption, the probability equals the ratio of the favorable possibilities over the total possible. The total number of ways to distribute m items over $N - \sum_{k=0}^{j-1} X_N^{(k)}$ positions – the latter constraints follows from the condition e_{j-1} – equals $\binom{N - \sum_{k=0}^{j-1} X_N^{(k)}}{m}$. Likewise, the favorable number of ways to distribute m items over the remaining levels higher than j, leads to

$$\Pr\left[\left\{ m^{(j)} = 0 \right\} \middle| e_{j-1} \right] = \frac{\binom{N - \sum_{k=0}^{j} X_N^{(k)}}{m}}{\binom{N - \sum_{k=0}^{j-1} X_N^{(k)}}{m}} \tag{19.4}$$

The recursion (19.3) needs an initialization, given by

$$\Pr\left[e_0\right] = \Pr\left[m^{(0)} = 0\right] = 1 - \frac{m}{N}$$

which follows from $\Pr\left[m^{(0)} = 0\right] = \frac{\binom{N-1}{m}}{\binom{N}{m}}$ and equals $\Pr\left[\{m^{(0)} = 0\}\,|e_{-1}\right]$ (although the event e_{-1} is meaningless). Observe that $\Pr\left[m^{(0)} = 1\right] = \frac{m}{N}$ holds for any tree such that

$$\Pr\left[h_N(m) = 0\right] = \frac{m}{N}$$

By iteration of (19.3), we obtain

$$\Pr\left[e_j\right] = \prod_{s=0}^{j} \frac{\binom{N-\sum_{k=0}^{s} X_N^{(k)}}{m}}{\binom{N-\sum_{k=0}^{s-1} X_N^{(k)}}{m}} = \frac{\binom{N-\sum_{k=0}^{j} X_N^{(k)}}{m}}{\binom{N}{m}} \tag{19.5}$$

where the convention in summation is that $\sum_{k=a}^{b} f_k = 0$ if $a > b$. Finally, combining (19.2) with (19.4) and (19.5), we arrive at the general conditional expression for the minimum hopcount to the anycast group,

$$\Pr\left[h_N(m) = j|L_N\right] = \frac{\binom{N-\sum_{k=0}^{j-1} X_N^{(k)}}{m} - \binom{N-\sum_{k=0}^{j} X_N^{(k)}}{m}}{\binom{N}{m}} \tag{19.6}$$

Clearly, while $\Pr\left[h_N(0) = j|L_N\right] = 0$ since there is no path, we have for $m = 1$,

$$\Pr\left[h_N(1) = j|L_N\right] = \frac{X_N^{(j)}}{N}$$

It directly follows from (19.6) that

$$\Pr\left[h_N(m) \leq n|L_N\right] = 1 - \frac{\binom{N-\sum_{k=0}^{n} X_N^{(k)}}{m}}{\binom{N}{m}} \tag{19.7}$$

If $N - \sum_{k=0}^{n} X_N^{(k)} < m$ or, equivalently, $\sum_{k=n+1}^{N-1} X_N^{(k)} < m$, then equation (19.7) shows that $\Pr\left[h_N(m) > n|L_N\right] = 0$. The maximum possible hopcount of a shortest path to an anycast group strongly depends on the specifics of the shortest path tree or the level set L_N. A general result is worth mentioning:

Theorem 19.2.1 *For any graph, it holds that*

$$\Pr[h_N(m) > N - m] = 0$$

In words, the longest shortest path to an anycast group with m members can never possess more than $N - m$ hops.

Proof: This general theorem follows from the fact that the line topology is the tree with the longest hopcount $N - 1$ and only in the case that all m last positions (with respect to the source or root) are occupied by the m anycast group members, is the maximum hopcount $N - m$. $\square$

For the URT, $\Pr[h_N(m) = N - m]$ is computed exactly in (19.12).

Corollary 19.2.2 *For any graph, it holds that*

$$\Pr[h_N(N-1) = 1] = \frac{1}{N}$$

Proof: This corollary follows from Theorem 19.2.1 and the law of total probability. Alternatively, if there are $N-1$ anycast members in a network with N nodes, the shortest path can only consist of one hop if none of the anycast members coincides with the root node. This probability is precisely $\frac{1}{N}$. □

Using the tail probability formula (2.37) for the mean, it follows from (19.7) that

$$E\left[h_N(m)|L_N\right] = \frac{1}{\binom{N}{m}} \sum_{n=0}^{N-2} \binom{N - \sum_{k=0}^{n} X_N^{(k)}}{m} \tag{19.8}$$

from which we find,

$$E\left[h_N(1)|L_N\right] = \frac{1}{N} \sum_{k=1}^{N-1} k X_N^{(k)}$$

Thus, given L_N, a performance measure η for anycast over unicast can be quantified as

$$\eta = \frac{E\left[h_N(m)|L_N\right]}{E\left[h_N(1)|L_N\right]} \le 1$$

Using the law of total probability, the distribution of the minimum hopcount to the anycast group is

$$\Pr\left[h_N(m) = j\right] = \sum_{\text{all } L_N} \Pr\left[h_N(m) = j|L_N\right] \Pr\left[L_N\right] \tag{19.9}$$

or explicitly,

$$\Pr[h_N(m) = j] = \sum_{\sum_{k=1}^{N-1} x_k = N-1} \frac{\binom{\sum_{k=j}^{N-1} x_k}{m} - \binom{\sum_{k=j+1}^{N-1} x_k}{m}}{\binom{N}{m}} \Pr\left[X_N^{(1)} = x_1,\ldots,X_N^{(N-1)} = x_{N-1}\right]$$

where the integers $x_k \ge 0$ for all k. This expression explicitly shows the importance of the level structure L_N of the shortest path tree T. The level set L_N entirely determines the *shape* of the tree T. Unfortunately, a general form for $\Pr\left[L_N\right]$ or $\Pr\left[h_N(m) = j\right]$ is difficult to obtain. In principle, via extensive trace-route measurements from several roots, the shortest path tree and $\Pr\left[L_N\right]$ can be constructed such that a (rough) estimate of the level set L_N in the Internet can be obtained.

19.3 The k-ary tree

For regular trees, explicit expressions are possible because the summation in (19.9) simplifies considerably. For example, for the k-ary tree defined in Section 18.3,

$$X_N^{(j)} = k^j$$

Provided the set L_N only contains these values of $X_N^{(j)}$ for each j, we have that $\Pr[L_N] = 1$, else it is zero (because then L_N is not consistent with a k-ary tree). Summarizing, for the k-ary tree with $N = \frac{k^{D+1}-1}{k-1}$ and D levels, the distribution of the minimum hopcount to the anycast group is

$$\Pr[h_N(m) = j] = \frac{\binom{N-\frac{k^j-1}{k-1}}{m} - \binom{N-\frac{k^{j+1}-1}{k-1}}{m}}{\binom{N}{m}} \tag{19.10}$$

Extension of the integer k to real numbers in the formula (19.10) is expected to be of value as suggested in Section 18.3. When a k-ary tree was used to fit corresponding Internet multicast measurements (Van Mieghem et al., 2001a), a remarkably accurate agreement was found for the value $k \approx 3.2$, which is about the mean degree of the Internet graph. Hence, if we were to use the k-ary tree as model for the hopcount to an anycast group, we expect that $k \approx 3.2$ is the best value for Internet shortest path trees. However, we feel we ought to mention that the hopcount distribution of the shortest path between two arbitrary nodes is definitely not a k-ary tree, because $\Pr[h_N(1) = j]$ increases with the hopcount j, which is in conflict with Internet trace-route measurements (see, for example, the bell-shaped curve in Fig. 16.4).

Figure 19.1 displays $\Pr[h(m) \le j]$ for a k-ary with outdegree $k = 3$ and $N = 500$. This type of plot allows us to solve the "server placement problem". For example, assuming that the k-ary tree is a good model and the network consists of $N = 500$ nodes, Fig. 19.1 shows that at least $m = 10$ servers are needed to assure that any user is not more than four hops separated from a server with a probability of 93%. More precisely, the equation $\Pr[h_{500}(m) > 4] < 0.07$ is obeyed if $m \ge 10$.

Figure 19.2 gives an idea how the performance measure η decreases with the size of the anycast group in k-ary trees (all with outdegree $k = 3$), but with different size N. For values of m up to around 20% of N, we observe that η decreases logarithmically in m.

19.4 The uniform recursive tree (URT)

Chapter 16 motivates the interest in the URT. The URT is believed to provide a reasonable, first order estimate for the hopcount problem to an anycast group in the Internet.

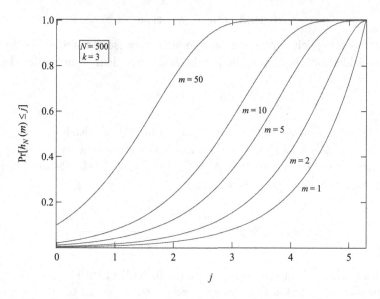

Fig. 19.1. The distribution function of $h_{500}(m)$ versus the hops j for various sizes of the anycast group in a k-ary tree with $k = 3$ and $N = 500$.

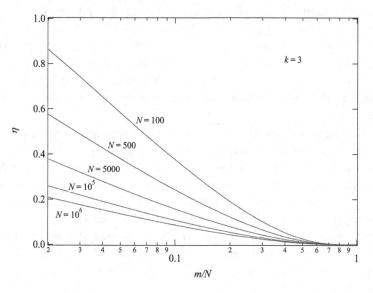

Fig. 19.2. The performance measure η for several sizes of k-ary trees (with $k = 3$) as a function of the ratio of anycast nodes over the total number of nodes.

19.4.1 Recursion for $\Pr[h(m) = j]$

Usually, a combinatorial approach such as (19.9) is seldom successful for URTs while structural properties often lead to results. The basic Theorem 16.2.1 of the URT, applied to the anycast minimum hop problem, is illustrated in Fig. 19.3.

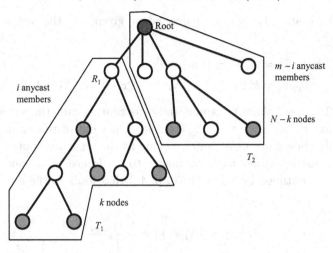

Fig. 19.3. A uniform recursive tree consisting of two subtrees T_1 and T_2 with k and $N-k$ nodes respectively. The first cluster contains i anycast members while the cluster with $N-k$ nodes contains $m-i$ anycast members.

Figure 19.3 shows that any URT can be separated into two subtrees T_1 and T_2 with k and $N-k$ nodes respectively. Moreover, Theorem 16.2.1 states that each subtree is independent of the other and again a URT. Consider now a specific separation of a URT T into $T_1 = t_1$ and $T_2 = t_2$, where the tree t_1 contains k nodes and i of the m anycast members and t_2 possesses $N-k$ nodes and the remaining $m-i$ anycast members. The event $\{h_T(m) = j\}$ equals the union of all possible sizes $N_1 = k$ and subgroups $m_1 = i$ of the event $\{h_{t_1}(i) = j-1\} \cap \{h_{t_2}(m-i) \geq j\}$ and the event $\{h_{t_1}(i) > j-1\} \cap \{h_{t_2}(m-i) = j\}$,

$$\{h_T(m) = j\} = \cup_k \cup_i \{\{h_{t_1}(i) = j-1\} \cap \{h_{t_2}(m-i) \geq j\}\}$$
$$\cup \{\{h_{t_1}(i) > j-1\} \cap \{h_{t_2}(m-i) = j\}\}$$

Because $h_N(0)$ is meaningless, the relation must be modified for the case $i = 0$ to

$$\{h_T(m) = j\} = \{h_{t_2}(m) = j\}$$

and for the case $i = m$ to

$$\{h_T(m) = j\} = \{h_{t_1}(m) = j-1\}$$

This decomposition holds for any URT T_1 and T_2, not only for the specific ones t_1 and t_2. The transition towards probabilities becomes

$$\Pr\left[h_T(m) = j\right] = \sum_{\text{all } t_1, t_2, k, i} \left(\Pr\left[h_{t_1}(i) = j-1\right] \Pr\left[h_{t_2}(m-i) \geq j\right] \right.$$

$$+ \Pr\left[h_{t_1}(i) \geq j-1\right] \Pr\left[h_{t_2}(m-i) = j\right] \left.\right)$$
$$\times \Pr\left[T_1 = t_1, T_2 = t_2, N_1 = k, m_1 = i\right]$$

Since T_1 and T_2 and also m_1 are independent given N_1, the last probability l simplifies to

$$l = \Pr[T_1 = t_1, T_2 = t_2, N_1 = k, m_1 = i]$$
$$= \Pr[T_1 = t_1|N_1 = k]\Pr[T_2 = t_2|N_1 = k]\Pr[m_1 = i|N_1 = k]\Pr[N_1 = k]$$

Theorem 16.2.1 states that N_1 is uniformly distributed over the set with $N - 1$ nodes such that $\Pr[N_1 = k] = \frac{1}{N-1}$. The fact that i out of the m anycast members, uniformly chosen out of N nodes, belong to the recursive subtree T_1 implies that $m - i$ remaining anycast members belong to T_2. Hence, analogous to a combinatorial problem outlined by Feller (1970, p. 43) that leads to the hypergeometric distribution, we have

$$\Pr[m_1 = i|N_1 = k] = \frac{\binom{k}{i}\binom{N-k}{m-i}}{\binom{N}{m}}$$

because all favorable combinations are those $\binom{k}{i}$ to distribute i anycast members in T_1 with k nodes multiplied by all favorable $\binom{N-k}{m-i}$ to distribute the remaining $m - i$ in T_2 containing $N - k$ nodes. The total way to distribute m anycast members over N nodes is $\binom{N}{m}$. Finally, we remark that the hopcount of the shortest path to m anycast members in a URT only depends on its size. This means that the sum over all t_1 of $\Pr[T_1 = t_1|N_1 = k]$, which equals 1, disappears and likewise also the sum over all t_2. Combining the above leads to

$$\Pr[h_N(m) = j] = \sum_{k=1}^{N-1}\sum_{i=1}^{m-1}(\Pr[h_k(i) = j - 1]\Pr[h_{N-k}(m-i) \geq j]$$

$$+ \Pr[h_k(i) > j - 1]\Pr[h_{N-k}(m-i) = j])\frac{\binom{k}{i}\binom{N-k}{m-i}}{(N-1)\binom{N}{m}}$$

$$+ \sum_{k=1}^{N-1}\frac{\binom{N-k}{m}\Pr[h_{N-k}(m) = j] + \binom{k}{m}\Pr[h_k(m) = j - 1]}{(N-1)\binom{N}{m}}$$

By substitution of $k' = N - k$ and $m' = m - i$, we obtain the recursion,

$$\Pr[h_N(m) = j] = \sum_{k=1}^{N-1}\sum_{i=1}^{m-1}\frac{\binom{k}{i}\binom{N-k}{m-i}(\Pr[h_k(i) = j - 1] + \Pr[h_k(i) = j])}{(N-1)\binom{N}{m}}$$

$$\times \sum_{q=j}^{N-k-1}\Pr[h_{N-k}(m-i) = q]$$

$$+ \sum_{k=1}^{N-1}\frac{\binom{k}{m}(\Pr[h_k(m) = j] + \Pr[h_k(m) = j - 1])}{(N-1)\binom{N}{m}} \quad (19.11)$$

This recursion (19.11) is solved numerically for $N = 20$. The result is shown in Fig. 19.4, which demonstrates that $\Pr[h(m) > N - m] = 0$ or that the path with the longest hopcount to an anycast group of m members consists of $N - m$ links.

Since there are $(N-1)!$ possible recursive trees (Theorem 16.2.2) and there is only one line tree with $N-1$ hops where each node has precisely one child node, the probability to have precisely $N-1$ hops from the root is $\frac{1}{(N-1)!}$ (which also is $\Pr[h_N = N-1]$ given in (16.9)). The longest possible hopcount from a root to m anycast members occurs in the line tree where all m anycast members occupy the last m positions. Hence, the probability for the longest possible hopcount equals

$$\Pr[h_N(m) = N - m] = \frac{m!}{(N-1)!\binom{N}{m}} \qquad (19.12)$$

because there are $m!$ possible ways to distribute the m anycast members at the m last positions in the line tree while there are $\binom{N}{m}$ possibilities to distribute m anycast members at arbitrary places in the line tree.

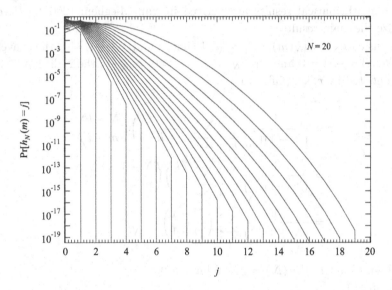

Fig. 19.4. The pdf of $h_N(m)$ in a URT with $N = 20$ nodes for all possible m. Observe that $\Pr[h_N(m) > N - m] = 0$. This relation connects the various curves to the value for m.

Figure 19.4 allows us to solve the "server placement problem". For example, consider the scenario in which a network operator announces that any user request will reach a server of the anycast group in no more than $j = 4$ hops in 99.9% of the cases. Assuming his network has $N = 20$ routers and the shortest path tree is a URT, the network operator has to compute the number of anycast servers m he has to place uniformly spread over the $N = 20$ routers by solving $\Pr[h_{20}(m) > 4] < 10^{-3}$. Figure 19.4 shows that the intersection of the line $j = 4$ and the line $\Pr[h_{20}(m) = 4] = 10^{-3}$ is the curve for $m = 7$. Since the curves for

$m \geq 7$ are exponentially decreasing, $\Pr[h_{20}(m) > 4]$ is safely[1] approximated by $\Pr[h_{20}(m) = 4]$, which leads to the placing of $m = 7$ servers. When following the line $j = 4$, we also observe that the curves for $m = 5, 6, 7, 8$ lie near to that of $m = 7$. This means that placing a server more does not considerably change the situation. It is a manifestation of the law $\eta \approx 1 - a \log m$, which tells us that by placing m servers the gain measured in hops with respect to the single server case is slowly, more precisely logarithmically, increasing. The performance measure η for the URT is drawn for several sizes N in Fig. 19.5.

19.4.2 Analysis of the recursion relation

The product of two probabilities in the double sum in (19.11) seriously complicates a possible analytic treatment. A relation for a generating function of $\Pr[h_N(m) = j]$ and other mathematical results are derived in Van Mieghem (2004). Here, we summarize the main results.

(a) Let us check $\Pr[h_N(m) = 0] = \frac{m}{N}$. Using $\Pr[h_k(i) \geq -1] = 1$, the convention that $\Pr[h_k(i) = -1] = 0$ and $\Pr[h_{N-k}(m-i) = 0] = \frac{m-i}{N-k}$, the right-hand side of (19.11), denoted by r, simplifies to

$$r = \frac{1}{(N-1)\binom{N}{m}} \sum_{k=1}^{N-1} \sum_{i=0}^{m} \frac{m-i}{N-k} \binom{k}{i} \binom{N-k}{m-i}$$

$$= \frac{1}{(N-1)\binom{N}{m}} \sum_{k=1}^{N-1} \sum_{i=0}^{m-1} \binom{k}{i} \binom{N-1-k}{m-1-i}$$

$$= \frac{1}{(N-1)\binom{N}{m}} \sum_{k=1}^{N-1} \binom{N-1}{m-1} = \frac{m}{N}$$

(b) Observe that $\Pr[h_N(N) = j] = 0$ for $j > 0$.

(c) For $m = 1$,

$$\Pr[h_N = j] = \frac{1}{N-1} \sum_{k=1}^{N-1} (\Pr[h_k = j] + \Pr[h_k = j-1]) \frac{k}{N}$$

Multiplying both sides by z^j, summing over all j leads to the recursion for the generating function (16.6)

$$(N+1)\varphi_{N+1}(z) = (z+N)\varphi_N(z)$$

---

[1] More precisely, since $\Pr[h_{20}(4) > 4] = 0.001\,06$ and $\Pr[h_{20}(5) > 4] = 0.000\,32$, only $m = 5$ servers are sufficient.

(d) The case $m = 2$ is solved in Van Mieghem (2004, Appendix) as

$$\Pr\left[h_N(2) = j\right] = \frac{2(-1)^{N-1-j}}{N!} S_N^{(j+1)} + \frac{2(-1)^{N-j}}{N!(N-1)} \sum_{k=1}^{j} (-1)^k \binom{2k-1}{k} S_N^{(k+j+1)}$$

$$+ \frac{2(-1)^{N-j}}{N!(N-1)} \sum_{k=0}^{j-1} \left[\binom{j+k+1}{j} - \binom{2k+1}{k} \right] (-1)^k S_N^{k+j+1}$$

$$(19.13)$$

In van der Hofstad *et al.* (2002b) we have demonstrated that the covariance between the number of nodes at level r and j for $r \le j$ in the URT is

$$E\left[X_N^{(r)} X_N^{(j)}\right] = \frac{(-1)^{N-1}}{(N-1)!} \sum_{k=0}^{r} (-1)^{k+j} \binom{2k+j-r}{k} S_N^{(k+j+1)}$$

For $j - r = 1$, the last term in (19.13) is recognized as $\dfrac{E\left[X_N^{(j-1)} X_N^{(j)}\right]}{\binom{N}{2}}$. Since $\binom{2k-1}{k} = \frac{1}{2}\binom{2k}{k}$, the first sum in (19.13) is

$$\frac{2(-1)^{N-j}}{N!(N-1)} \sum_{k=1}^{j} (-1)^k \binom{2k-1}{k} S_N^{(k+j+1)} = \frac{2(-1)^{N-j-1}}{(N-1)} \frac{S_N^{(j+1)}}{N!} - \frac{E\left[\left(X_N^{(j)}\right)^2\right]}{2\binom{N}{2}}$$

With $\frac{2(-1)^{N-1-j}}{N!} S_N^{(j+1)} = 2 \Pr\left[h_N = j\right]$, we obtain

$$\Pr\left[h_N(2) = j\right] = \frac{2N}{N-1} \Pr\left[h_N = j\right] + \frac{E\left[X_N^{(j-1)} X_N^{(j)}\right]}{\binom{N}{2}} - \frac{E\left[\left(X_N^{(j)}\right)^2\right]}{2\binom{N}{2}}$$

$$+ \frac{2(-1)^{N-1}}{N!(N-1)} \sum_{k=1}^{j} \binom{j+k}{k} (-1)^{k+j} S_N^{k+j}$$

It would be of interest to find an interpretation for the last sum. Without proof[2], we mention the following exact results:

$$\sum_{m=1}^{N} \binom{N}{m} \Pr\left[h_N(m) = N - 2\right] = \sum_{m=1}^{N-1} \binom{N-1}{m} \frac{\Pr\left[h_{N-1}(m) = N - 3\right]}{N-1} + \frac{1}{(N-2)!}$$

For $m \le N - 3$, it holds that

$$\Pr\left[h_N(m) = N - m - 1\right] = \frac{m!}{(N-1)!\binom{N}{m}} \left[\binom{N}{2} + (m-1)(m/2+1) + \sum_{k=2}^{m} \frac{m+1}{k} \right]$$

---

[2] By substitution into the recursion (19.11), one may verify these relations.

19.4.3 Approximate analysis for the URT

Since the general solution (19.9) is in many cases difficult to compute as shown for the URT in Section 19.4, we consider a simplified version of the above problem where each node in the tree has equal probability $p = \frac{m}{N}$ to be a server. Instead of having precisely m servers, the simplified version considers on average m servers and the probability that there are precisely m servers is $\binom{N}{m} p^m (1-p)^{N-m}$. In the simplified version, the associated equations to (19.4) and (19.3) are

$$\Pr\left[\left\{m^{(j)} = 0\right\}\Big| e_{j-1}\right] = \Pr\left[\left\{m^{(j)} = 0\right\}\right] = (1-p)^{X_N^{(j)}}$$

$$\Pr\left[e_j\right] = \prod_{l=0}^{j} \Pr\left[\left\{m^{(j)} = 0\right\}\right] = (1-p)^{\sum_{l=0}^{j-1} X_N^{(l)}}$$

which implies that the probability that there are no servers in the tree is $(1-p)^N$. Since in that case, the hopcount is meaningless, we consider the conditional probability (19.2) of the hopcount given that the level set contains at least one server (which is denoted by $\tilde{h}_N(m)$) is

$$\Pr\left[\tilde{h}_N(m) = j | L_N\right] = \frac{\left(1 - (1-p)^{X_N^{(j)}}\right)(1-p)^{\sum_{l=0}^{j-1} X_N^{(l)}}}{1 - (1-p)^N}$$

Thus,

$$\Pr\left[\tilde{h}_N(m) \le n | L_N\right] = \frac{1 - (1-p)^{\sum_{l=0}^{n} X_N^{(l)}}}{1 - (1-p)^N}$$

Finally, to avoid the knowledge of the entire level set L_N, we use $E\left[X_N^{(l)}\right] = N\Pr\left[h_N(1) = l\right]$ from (16.7) as the best estimate for each $X_N^{(l)}$ and obtain the approximate formula

$$\Pr\left[\tilde{h}_N(m) = j\right] = \frac{\left(1 - (1-p)^{E\left[X_N^{(j)}\right]}\right)(1-p)^{\sum_{l=0}^{j-1} E\left[X_N^{(l)}\right]}}{1 - (1-p)^N} \tag{19.14}$$

In the dotted lines in Fig. 19.5, we have added the approximate result for the URT where $E\left[h_N(m)\right]$ is computed based on (19.14), but where $E[h_N(1)]$ is computed exactly. For $m = 1$, the approximate analysis (19.14) is not well suited: Fig. 19.5 illustrates this deviation in the fact that $\eta_{\mathrm{appr}}(1) = E\left[\tilde{h}_N(1)\right]/E\left[h_N(1)\right] < 1$. For higher values of m we observe a fairly good correspondence. We found that the probability (19.14) reasonably approximates the exact result plotted on a linear scale. Only the tail behavior (on log-scale) and the case for $m = 1$ deviate significantly. In summary for the URT, the approximation (19.14) for $\Pr\left[h_N(m) = j\right]$ is much faster to compute than the exact recursion and it seems appropriate for the computation of η for $m > 1$. However, it is less adequate to solve the server placement problem that requires the tail values $\Pr\left[h_N(m) > j\right]$.

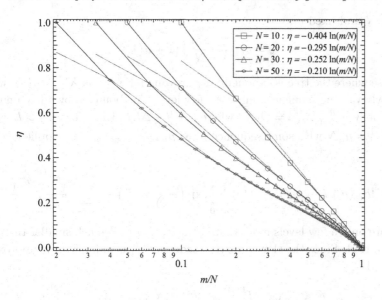

Fig. 19.5. The performance measure η for several sizes N of URTs as a function of the ratio m/N

19.5 The performance measure η in exponentially growing trees

In this section, we investigate the observed law $\eta \approx 1 - a \log m$ for a much larger class of trees, namely the class of exponentially growing trees to which both the k-ary tree and the URT belong. Also most trees in the Internet are exponentially growing trees. A tree is said to grow exponentially in the number of nodes N with degree κ if $\lim_{j\to\infty} \left(X_N^{(j)}\right)^{1/j} = \kappa$ or, equivalently, $X_N^{(j)} \sim \kappa^j$, for large j. The fundamental problem with this definition is that it only holds for infinite graphs $N = \infty$. For real (finite) graphs, there must exist some level $j = l$ for which the sequence $X_N^{(l+1)}, X_N^{(l+2)}, \ldots, X_N^{(N-1)}$ ceases to grow because $\sum_{j=0}^{N-1} X_N^{(j)} = N < \infty$. This boundary effect complicates the definition of exponential growth in finite graphs. The second complication is that even in the finite set $X_N^{(0)}, X_N^{(1)}, \ldots, X_N^{(l)}$ not necessary all $X_N^{(j)}$ with $0 \le j \le l$ need to obey $X_N^{(j)} \sim \kappa^j$, but "enough" should. Without the limit concept, we cannot specify the precise conditions of exponential growth in a finite shortest path tree. If we assume in finite graphs that $X_N^{(j)} \sim \kappa^j$ for $j \le l$, then $\sum_{j=0}^l X_N^{(j)} = \alpha N$ with $0 < \alpha < 1$. Indeed, for $\kappa > 1$, the highest hopcount level l possesses by far the most nodes since $\frac{\kappa^{l+1}-1}{\kappa-1} \approx \kappa^l$, which cannot be larger than a fraction αN of the total number of nodes.

We now present an order calculus to estimate η for exponentially growing trees based on relation (19.8). Let us denote

$$y = \frac{\binom{N-x}{m}}{\binom{N}{m}} = \prod_{j=0}^{m-1} \left(1 - \frac{x}{N-j}\right)$$

For large N and fixed m,

$$y = \exp\left(-\frac{xm}{N}\right)(1+o(1))$$

In the case where the tree is exponentially growing for $j \le l$ as $X_N^{(j)} = \beta_j \kappa^j$ with β_j some slowly varying sequence, only very few levels Δl (bounded by a fixed number) around l obey $\sum_{k=0}^{n} X_N^{(k)} = O(N)$ where $n \in [l - \Delta l, l]$, while for all $j > l$, we have $\sum_{k=0}^{n} X_N^{(k)} = \mu_n N$ with some sequence $\mu_n < \mu_{n+1} < \mu_{\max n} = 1$. Applied to (19.8) where $x = \sum_{k=0}^{n} X_N^{(k)} < N$,

$$E[h_N(m)|L_N] \approx (1+o(1)) \sum_{n=0}^{l} \exp\left(-\frac{m}{N}\beta_n\kappa^n\right) + \sum_{n=l+1}^{N-2} \frac{\binom{(1-\mu_n)N}{m}}{\binom{N}{m}}$$

If there are only a few levels more than l, the last series is much smaller than 1 and can be omitted. Since the slowly varying sequence β_n is unknown, we approximate $\beta_n = \beta$ and

$$\sum_{n=0}^{l} \exp\left(-\frac{m}{N}\beta_n\kappa^n\right) \approx \int_0^l \exp\left(-\frac{m}{N}\beta\kappa^n\right)dn = \frac{1}{\log\kappa}\int_{\frac{m}{N}\beta}^{\frac{m}{N}\beta\kappa^l} \frac{e^{-u}}{u}du$$

$$\approx \frac{1}{\log\kappa}\int_{\frac{m}{N}\beta}^{\infty} \frac{e^{-u}}{u}du - \frac{e^{-m}}{m\log\kappa}$$

$$= \frac{1}{\log\kappa}\left(-\gamma - \frac{e^{-m}}{m} - \log\frac{m}{N}\beta + O\left(\frac{m}{N}\right)\right)$$

where in the last step a series (Abramowitz and Stegun, 1968, Section 5.1.11) for the exponential integral is used. Thus,

$$\eta \approx (1+o(1)) \frac{\left(1 + \frac{-\gamma - \frac{e^{-m}}{m} - \log m - \log\beta}{\log N} + O\left(\frac{m}{N}\right)\right)}{\left(1 - \frac{\gamma + e^{-1} + \log\beta}{\log N} + O\left(\frac{1}{N}\right)\right)}$$

$$= (1+o(1))\left(1 - \frac{\log m}{\log N} - \frac{e^{-1} - \frac{e^{-m}}{m}}{\log N} + O\left(\frac{1}{\log^2 N}\right)\right)$$

Since by definition $\eta = 1$ for $m = 1$, we finally arrive at

$$\eta \approx 1 - \frac{\log m}{\log N} - \frac{e^{-1} - \frac{e^{-m}}{m}}{\log N} + O\left(\frac{1}{\log^2 N}\right)$$

which supplies evidence for the conjecture $\eta \approx 1 - a\log m$ that exponentially growing graphs possess a performance measure η that logarithmically decreases in m, which is rather slow.

Measurement data in the Internet seem to support this $\log m$-scaling law. Apart from the correspondence with figures in the work of Jamin *et al.* (2001), Fig. 6 in Krishnan *et al.* (2000) shows that the relative measured traffic flow reduction decreases logarithmically in the number of caches m.

19.6 The weight to an anycast group

This section focuses on the following problem: given a network of N nodes over which m peers are randomly distributed, what is the delay from a certain peer to the nearest (in delay) peer? In a P2P network, peers are joining and leaving regularly which motivates us to consider such a network to first order as an Erdös-Rényi random graph. The shortest path tree rooted at an arbitrary node to m uniformly chosen peers is the union of the shortest paths from that node to all m different peers in the graph. Thus, the computation of a shortest path requires the knowledge of link weights, such as a monetary cost, the number of hops, a delay, a distance, etc. In most complex networks, link weights are not precisely known. As motivated in Section 16.1, we assign i.i.d. exponentially distributed weights with unit mean to links such that the resulting shortest path tree is approximately a URT.

The weight $W_{N;m}$ of the shortest path from an arbitrary node to the nearest (in weight) peer in a peer group of size m is

$$W_{N;m} = \sum_{k=1}^{N-1} 1_{\{T_N(m)\geq k\}} \tau_k \tag{19.15}$$

where $T_N(m)$ is the number of steps in the continuous Markov discovery process until one peer of the group of m peers is reached, or, $T_N(m)$ is the hitting time to the peer group of size m. Since the random variables $1_{\{T_N(m)\geq k\}}$ and τ_k (for a fixed k) are independent, the mean weight directly follows as

$$E[W_{N;m}] = \sum_{k=1}^{N-1} \Pr[T_N(m) \geq k] E[\tau_k]$$

The event $\{T_N(m) \geq k\}$ implies that the k-th discovered (attached) node in the URT is the first peer out of the m peers that is reached. Hence, the $k-1$ previously discovered nodes do not belong to the peer group. Since the m peers as well as the $k-1$ nodes are uniformly chosen out of the $N-1$ nodes, we have that

$$\Pr[T_N(m) \geq k] = \frac{\binom{N-1-m}{k-1}}{\binom{N-1}{k-1}}$$

such that, with $E[\tau_k] = \frac{1}{k(N-k)}$,

$$E[W_{N;m}] = \frac{(N-1-m)!}{(N-1)!} \sum_{k=1}^{N-1} \frac{(N-1-k)!}{k(N-m-k)!}$$

Invoking the identity, which is similarly proved as Lemma 18.6.1,

$$\sum_{j=n}^{b} \frac{1}{j} \frac{(a-j)!}{(b-j)!} = \frac{a!}{b!} \sum_{j=n}^{b} \frac{1}{j} - \frac{a!}{(b-n)!} \sum_{q=0}^{a-b-1} \frac{1}{a-b-q} \frac{(a-q-n)!}{(a-q)!} \tag{19.16}$$

yields

$$\sum_{k=1}^{N-m} \frac{1}{k} \frac{(N-1-k)!}{(N-m-k)!} = \frac{(N-1)!}{(N-m)!} \sum_{j=1}^{N-m} \frac{1}{j} - \frac{(N-1)!}{(N-m-1)!} \sum_{q=0}^{m-2} \frac{1}{(m-1-q)(N-1-q)}$$

After partial fraction expansion, the q-sum is

$$\sum_{q=0}^{m-2} \frac{1}{(m-1-q)(N-1-q)} = \frac{1}{N-m} \sum_{j=1}^{m-1} \frac{1}{j} - \frac{1}{N-m} \sum_{j=N-m+1}^{N-1} \frac{1}{j}$$

and we arrive at

$$E[W_{N;m}] = \frac{1}{N-m} \left\{ \sum_{j=1}^{N-1} \frac{1}{j} - \sum_{j=1}^{m-1} \frac{1}{j} \right\} = \frac{\psi(N) - \psi(m)}{N-m} \qquad (19.17)$$

where $\psi(x)$ is the digamma function (Abramowitz and Stegun, 1968, Section 6.3). For large N, we have (Abramowitz and Stegun, 1968, Section 6.3.18)

$$E[W_{N;m}] = \frac{\ln \frac{N}{m}}{N-m} + \frac{1}{2Nm} + O\left(N^{-1}m^{-2}\right)$$

Since the sequence of random variables $1_{\{T_N(m)\geq 1\}}, 1_{\{T_N(m)\geq 2\}}, \ldots, 1_{\{T_N(m)\geq N-1\}}$ is obviously not independent, a computation of the probability generating function $\varphi_{W_{N;m}}(z) = E\left[e^{-zW_{N;m}}\right]$ different from straightforwardly using (19.15) is needed. We define $v_k = \sum_{n=1}^{k} \tau_n$ as the weight of the shortest path to the k-th attached (discovered) node. Since the laws of the Markov discovery process and of the URT are independent – this argument is also used in Section 16.5 – we obtain

$$\varphi_{W_{N;m}}(z) = \sum_{k=1}^{N-m} E[e^{-zv_k}] \Pr[Y_m(k)]$$

where $Y_m(k)$ denotes the event that the k-th attached node is the first encountered peer among the m peers in the URT. Now, there are $\binom{N-1}{m}$ ways to distribute the m peers (different from the source node) over the $N-1$ remaining nodes. If the k-th node is the first of the m discovered peers, it implies that the remaining $m-1$ peers need to be discovered later. There are $\binom{N-1-k}{m-1}$ possible ways, whence

$$\Pr[Y_m(k)] = \frac{\binom{N-1-k}{m-1}}{\binom{N-1}{m}}$$

Thus, we obtain

$$\varphi_{W_{N;m}}(z) = \frac{m(N-1-m)!}{(N-1)!} \sum_{k=1}^{N-m} \frac{(N-1-k)!}{(N-m-k)!} \prod_{n=1}^{k} \frac{n(N-n)}{z+n(N-n)} \qquad (19.18)$$

After partial integration of a single-sided Laplace transform $\varphi_X(z) = \int_0^\infty f_X(t)e^{-zt}dt$ and assuming that the derivative $f_X'(t)$ exists, the relation $f_X(0) = \lim_{z\to\infty} z\varphi_X(z)$ is found. Applied to (19.18) yields

$$f_{W_{N;m}}(0) = m \qquad (19.19)$$

Of course, the case $m = 1$ reduces to the weight W_N of an arbitrary shortest path in the URT as studied in Section 16.4.

19.6.1 The asymptotic pgf and pdf of the weight $W_{N,m}$

Although the inverse Laplace transform of (19.18) can be computed, the resulting series for the pdf is hardly appealing. As shown below, an asymptotic analysis leads to an elegant result that, in addition, seems applicable to a relatively small network size N. We write

$$z + n(N - n) = \left(\sqrt{\left(\frac{N}{2}\right)^2 + z} + \frac{N}{2} - n \right) \left(\sqrt{\left(\frac{N}{2}\right)^2 + z} - \left(\frac{N}{2} - n\right) \right)$$

and define $y = \sqrt{(\frac{N}{2})^2 + z}$. Then,

$$\prod_{n=1}^{k} \frac{n(N - n)}{z + n(N - n)} = \frac{k!(N - 1)!}{(N - k - 1)!} \prod_{n=1}^{k} \frac{1}{(y + \frac{N}{2} - n)} \prod_{n=1}^{k} \frac{1}{(y - \frac{N}{2} + n)}$$

$$= \frac{k!(N - 1)!}{(N - k - 1)!} \frac{\Gamma\left(y + \frac{N}{2} - k\right)}{\Gamma\left(y + \frac{N}{2}\right)} \frac{\Gamma\left(y - \frac{N}{2} + 1\right)}{\Gamma\left(y - \frac{N}{2} + k + 1\right)}$$

and, substituted in (19.18), yields

$$\varphi_{W_{N;m}}(z) = \frac{m(N - 1 - m)! \Gamma\left(y - \frac{N}{2} + 1\right)}{\Gamma\left(y + \frac{N}{2}\right)} \sum_{k=1}^{N-m} \frac{k!}{(N - m - k)!} \frac{\Gamma\left(y + \frac{N}{2} - k\right)}{\Gamma\left(y - \frac{N}{2} + k + 1\right)}$$

For large N and $|z| < N$, we have[3] that $y = \sqrt{(\frac{N}{2})^2 + z} \sim \frac{N}{2} + \frac{z}{N}$ such that

$$\varphi_{W_{N;m}}(z) \sim \frac{m(N - 1 - m)! \Gamma\left(\frac{z}{N} + 1\right)}{\Gamma\left(N + \frac{z}{N}\right)} \sum_{k=1}^{N-m} \frac{k!}{(N - m - k)!} \frac{\Gamma\left(N + \frac{z}{N} - k\right)}{\Gamma\left(\frac{z}{N} + k + 1\right)}$$

We now introduce the scaling $z = Nx$, where $|x| < 1$ since $|z| < N$,

$$\varphi_{W_{N;m}}(Nx) \sim \frac{m(N - 1 - m)! \Gamma(x + 1)}{\Gamma(N + x)} \sum_{k=1}^{N-m} \frac{k!}{(N - m - k)!} \frac{\Gamma(N + x - k)}{\Gamma(x + 1 + k)} \tag{19.20}$$

For large N and fixed m and x, the asymptotic order of the sum

$$S = \sum_{k=1}^{N-m} \frac{\Gamma(k + 1)}{\Gamma(k + x + 1)} \frac{\Gamma(N - k + x)}{\Gamma(N - k - m + 1)} \tag{19.21}$$

scales as

$$S = \frac{\Gamma(N + x + 1) N^{-x-1}}{\Gamma(x) \Gamma(N - m)} \frac{\pi}{\sin \pi x} \frac{\Gamma(x + m)}{m!} \left(1 + O\left(N^{-1}\right)\right) \tag{19.22}$$

---

[3] The notation $f(x) \sim g(x)$ for large x means that $\lim_{x \to \infty} \frac{f(x)}{g(x)} = 1$.

This result (19.22) is derived in Van Mieghem and Tang (2008). Substitution of (19.22) into (19.20), leads, for large N, fixed m and $|x| < 1$, to

$$\varphi_{W_{N;m}}(Nx) \sim \frac{\pi x}{\sin \pi x} \frac{\Gamma(x+m)}{\Gamma(m)} N^{-x} \left(1 + O\left(N^{-1}\right)\right)$$

or

$$\lim_{N \to \infty} N^x \varphi_{W_{N;m}}(Nx) = \lim_{N \to \infty} E\left[e^{-(NW_{N;m} - \ln N)x}\right] = \frac{\pi x}{\sin \pi x} \frac{\Gamma(x+m)}{\Gamma(m)} \quad (19.23)$$

Finally, the inverse Laplace transform, computed in Van Mieghem and Tang (2008), is

$$\lim_{N \to \infty} \Pr[NW_{N;m} - \ln N \le t] = me^{-mt}e^{e^{-t}} \int_{e^{-t}}^{\infty} \frac{e^{-u}}{u^{m+1}} du = me^{-mt}e^{e^{-t}} \Gamma\left(-m, e^{-t}\right) \quad (19.24)$$

where $\Gamma\left(-m, e^{-t}\right)$ is the incomplete Gamma function (Abramowitz and Stegun, 1968, Section 6.5). The probability density function follows after derivation of (19.24) with respect to t as

$$\lim_{N \to \infty} f_{NW_{N;m} - \ln N}(t) = m - me^{-mt}\left(e^{-t} + m\right)e^{e^{-t}} \int_{e^{-t}}^{\infty} \frac{e^{-u}}{u^{m+1}} du \quad (19.25)$$

19.6.2 The Fermi-Dirac distribution

For large m and to first order, we have $\frac{\Gamma(x+m)}{\Gamma(m)} = m^x \left(1 + O\left(\frac{1}{m}\right)\right)$, and we can rescale the random variable as

$$\lim_{N \to \infty} E\left[e^{-(NW_{N;m} - \ln N/m)x}\right] = \frac{\pi x}{\sin \pi x} \left(1 + O\left(\frac{1}{m}\right)\right)$$

In view of the mean (19.17) for large N (and fixed m), we can write $NW_{N;m} - \ln N/m = N\left(W_{N;m} - E\left[W_{N;m}\right]\right)$ and we define the scaled asymptotic random

$$\tilde{W} = \lim_{N \to \infty} \left(NW_{N;m} - \ln N/m\right) = N\left(W_{N;m} - E\left[W_{N;m}\right]\right)$$

obtained by first letting $N \to \infty$, followed by $m \to \infty$. The analysis shows that the $NW_{N;m} - \ln N/m$ tends to the asymptotic random variable $\tilde{W}$ as $O\left(\frac{1}{m}\right)$. Using the reflection formula for the Gamma function (Abramowitz and Stegun, 1968, 6.1.17), $\Gamma(1+x)\Gamma(1-x) = \frac{\pi x}{\sin \pi x}$ and the pgf (3.45), we observe that $\tilde{W} = G_1 - G_2$ is the difference of two Gumbel random variables G_1 and G_2 with same parameters. After inverse Laplace transform (2.39) with $0 < c < 1$, we have

$$\Pr\left[\tilde{W} \le t\right] = \frac{1}{2\pi i} \int_{c-i\infty}^{c+i\infty} \frac{\pi e^{tx}}{\sin \pi x} dx$$

For $t > 0$ ($t < 0$), the contour can be closed over the negative (positive) $\operatorname{Re}(x)$-plane, resulting, for any t, in the Fermi-Dirac distribution function

$$\Pr\left[\tilde{W} \le t\right] = \frac{1}{1 + e^{-t}} \quad (19.26)$$

whose symmetric probability density function is

$$f_{\tilde{W}_m}(t) = \frac{1}{4}\operatorname{sech}^2\left(\frac{t}{2}\right)$$

The Fermi-Dirac distribution function frequently appears in statistical physics (see e.g. Ashcroft and Mermin (1981)). Fig. 19.6 plots the scaled, asymptotic pdf (19.25) for various values of m. It is of interest to observe that the deep tails of

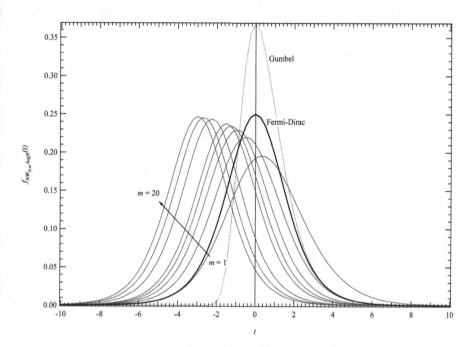

Fig. 19.6. The asymptotic and scaled probability density function (19.25) for various values of $m = 1, 2, 3, 4, 5, 10, 15, 20$. The pdf of the Gumbel and the Fermi-Dirac distribution are also shown.

$f_{NW_{N;m}-\ln N}(t)$ for large $|t|$ decrease as $O\left(e^{-|t|}\right)$. This is different than the decrease of either a Gumbel or Gaussian, which implies that larger variations around the mean can occur. From a practical point of view, if $m > 5$, simulations in Van Mieghem and Tang (2008) indicate that the much simpler Fermi-Dirac distribution is sufficiently accurate. The observation that a peer group of $m \approx 5$ is already close to the asymptotic regime, leads us to conclude that relatively small peer group sizes suffice to offer a good service quality (e.g. small latency) and that increasing the peer group size only logarithmically (i.e. marginally) improves the quality of service of weight related metrics. This logarithmic scaling in m agrees with that of performance measure η.

Appendix A
A summary of matrix theory

This appendix is a summary of our book (Van Mieghem, 2011) and reviews the matrix theory for Markov chains. In-depth analyses are found in classical books by Gantmacher (1959a,b), Wilkinson (1965) and Meyer (2000). The numbers in bold at the start of each paragraph refer to articles, abbreviated by **art.** in this book.

A.1 Eigenvalues and eigenvectors

1. The algebraic eigenproblem consists in the determination of the eigenvalues λ and the corresponding eigenvectors x of an $n \times n$ matrix A for which the set of n homogeneous linear equations in n unknowns,

$$Ax = \lambda x \qquad (A.1)$$

has a non-zero solution. Clearly, the zero vector $x = 0$ is always a solution of (A.1). A non-zero solution of (A.1) is only possible if and only if the matrix $A - \lambda I$ is singular, that is,

$$\det (A - \lambda I) = 0 \qquad (A.2)$$

This determinant can be expanded in a polynomial in λ of degree n,

$$c_A(\lambda) = \sum_{k=0}^{n} c_k \lambda^k = c_n \lambda^n + c_{n-1} \lambda^{n-1} + \cdots + c_1 \lambda + c_0 = 0 \qquad (A.3)$$

which is called the characteristic or eigenvalue polynomial of the matrix A. Apart from $c_n = (-1)^n$, the coefficients for $0 \le k < n$ are

$$c_k = (-1)^k \sum_{all} M_{n-k} \qquad (A.4)$$

and M_k is a principal minor[1]. Since a polynomial of degree n has n complex zeros, the matrix A possesses n eigenvalues $\lambda_1, \lambda_2, \ldots, \lambda_n$, not all necessarily distinct. In

---

[1] A principal minor M_k is the determinant of a principal $k \times k$ submatrix $M_{k \times k}$ obtained by deleting the same $n - k$ rows and columns in A. Hence, the main diagonal elements $(M_{k \times k})_{ii}$ are k elements of main diagonal elements $\{a_{ii}\}_{1 \le i \le n}$ of A.

general, the characteristic polynomials can be written as

$$c_A(\lambda) = \prod_{k=1}^{n} (\lambda_k - \lambda) \tag{A.5}$$

Since $c_A(\lambda) = \det(A - \lambda I)$, it follows from (A.3) and (A.5) that, for $\lambda = 0$,

$$\det A = c_0 = \prod_{k=1}^{n} \lambda_k \tag{A.6}$$

Hence, if $\det A = 0$, there is at least one zero eigenvalue. Also,

$$(-1)^{n-1} c_{n-1} = \sum_{k=1}^{n} \lambda_k = \text{trace}(A) \tag{A.7}$$

and

$$c_1 = -\sum_{k=1}^{n} \prod_{j=1; j \neq k}^{n} \lambda_j = -\det A \sum_{k=1}^{n} \frac{1}{\lambda_k} \tag{A.8}$$

For any eigenvalue λ, the set (A.1) has at least one non-zero eigenvector x. Furthermore, if x is a non-zero eigenvector, also kx is a non-zero eigenvalue. Therefore, eigenvectors are often normalized, for instance, a probabilistic eigenvector has the sum of its components equal to 1 or a norm $\|x\|_1 = 1$ as defined in (A.26).

If the rank of $A - \lambda I$ is less than $n - 1$, there will be more than one independent eigenvector belonging to the eigenvalue λ. Just these cases seriously complicate the eigenvalue problem. In the sequel, we omit the discussion on multiple eigenvalues and refer to Wilkinson (1965).

2. The eigenproblem of the transpose A^T,

$$A^T y = \lambda y \tag{A.9}$$

is of singular importance. Since the determinant of a matrix is equal to the determinant of its transpose, $\det(A^T - \lambda I) = \det(A - \lambda I)$, which shows that the eigenvalues of A and A^T are the same. However, the eigenvectors are, in general, different. Alternatively, transposing (A.9) yields

$$y^T A = \lambda y^T \tag{A.10}$$

The vector y_j^T is therefore called the left-eigenvector of A belonging to the eigenvalue λ_j, whereas x_j is called the right-eigenvector belonging to the same eigenvalue λ_j. An important relation between the left- and right-eigenvectors of a matrix A is, for $\lambda_j \neq \lambda_k$,

$$y_j^T x_k = 0 \tag{A.11}$$

Indeed, left-multiplying (A.1) with $\lambda = \lambda_k$ by y_j^T,

$$y_j^T A x_k = \lambda_k y_j^T x_k$$

and similarly right-multiplying (A.10) with $\lambda = \lambda_j$ by x_k

$$y_j^T A x_k = \lambda_j y_j^T x_k$$

leads, after subtraction to $0 = (\lambda_k - \lambda_j) y_j^T x_k$ and (A.11) follows. Since eigenvectors may be complex in general and since $y_j^T x_k = x_k^T y_j$, the expression $y_j^T x_k$ is not an inner-product that is always real and for which $y_j^T x_k = \left(x_k^T y_j\right)^*$ holds. However, (A.11) expresses that the sets of left- and right-eigenvectors are orthogonal if $\lambda_j \neq \lambda_k$.

3. If A has n distinct eigenvalues, then the n eigenvectors are linearly independent and span the whole n-dimensional space. The proof is by *reductio ad absurdum*. Assume that s is the smallest number of linearly dependent eigenvectors labelled by the first s smallest indices. Linear dependence then means that

$$\sum_{k=1}^{s} \alpha_k x_k = 0 \tag{A.12}$$

where $\alpha_k \neq 0$ for $1 \leq k \leq s$. Left-multiplying by A and using (A.1) yields

$$\sum_{k=1}^{s} \alpha_k \lambda_k x_k = 0 \tag{A.13}$$

On the other hand, multiplying (A.12) by λ_s and subtracting from (A.13) leads to

$$\sum_{k=1}^{s-1} \alpha_k \left(\lambda_k - \lambda_s\right) x_k = 0,$$

which, because all eigenvalues are distinct, implies that there is a smaller set of $s - 1$ linearly depending eigenvectors. This contradicts the initial hypothesis.

This important property has a number of consequences. First, it applies to left- as well as to right-eigenvectors. Relation (A.11) then shows that the sets of left- and right-eigenvectors form a bi-orthogonal system with $y_k^T x_k \neq 0$. For, if x_k were orthogonal to y_k (or $y_k^T x_k = 0$), (A.11) demonstrates that x_k would be orthogonal to all left-eigenvectors y_j. Since the set of left-eigenvectors span the n dimensional vector space, it would mean that the n-dimensional vector x_k would be orthogonal to the whole n-space, which is impossible because x_k is not the null vector. Second, any n-dimensional vector can be written in terms of either the left- or right-eigenvectors.

4. Let us denote by X the matrix with the right-eigenvector x_j in column j and by Y^T the matrix with the left-eigenvector y_k^T in row k. If the right- and left-eigenvectors are scaled such that, for all $1 \leq k \leq n$, $y_k^T x_k = 1$, then

$$Y^T X = I \tag{A.14}$$

Thus, the matrix Y^T is the inverse of the matrix X. Furthermore, for any right-eigenvector, (A.1) holds, rewritten in matrix form, such that

$$AX = X \operatorname{diag}(\lambda_k) \tag{A.15}$$

Left-multiplying by $X^{-1} = Y^T$ yields the similarity transform of matrix A,

$$X^{-1}AX = Y^T AX = \operatorname{diag}(\lambda_k) \tag{A.16}$$

Thus, when the eigenvalues of A are distinct, there exists a similarity transform $H^{-1}AH$ that reduces A to diagonal form. In many applications, similarity transforms are applied to simplify matrix problems. Observe that a similarity transform preserves the eigenvalues, because, if $Ax = \lambda x$, then $\lambda H^{-1}x = H^{-1}Ax = (H^{-1}AH)H^{-1}x$. The eigenvectors are transformed to $H^{-1}x$.

When A has multiple eigenvalues, it may be impossible to reduce A to a diagonal form by similarity transforms. Instead of a diagonal form, the most compact form when A has r distinct eigenvalues each with multiplicity m_j such that $\sum_{j=1}^{r} m_j = n$ is the Jordan canonical form C,

$$C = \begin{bmatrix} C_{m_1}(\lambda_1) & & & & \\ & C_{m_2}(\lambda_1) & & & \\ & & \vdots & & \\ & & & C_{m_{r-1}}(\lambda_{r-1}) & \\ & & & & C_{m_r}(\lambda_r) \end{bmatrix}$$

where $C_m(\lambda)$ is an $m \times m$ submatrix of the form

$$C_m(\lambda) = \begin{bmatrix} \lambda & 1 & 0 & \cdots & & 0 \\ 0 & \lambda & 1 & 0 & & \cdots \\ \vdots & \vdots & \vdots & \vdots & & \vdots \\ 0 & \cdots & 0 & \lambda & & 1 \\ 0 & \cdots & & 0 & 0 & \lambda \end{bmatrix}$$

The number of independent eigenvectors is equal to the number of submatrices. If an eigenvalue λ has multiplicity m, there can be one large submatrix $C_m(\lambda)$, but also a number k of smaller submatrices $C_{b_j}(\lambda)$ such that $\sum_{j=1}^{k} b_j = m$. This illustrates, as mentioned in **art. 1**, the much higher complexity of the eigenproblem in case of multiple eigenvalues. For more details we refer to Wilkinson (1965).

5. When left-multiplying (A.1), we obtain

$$A^2 x = \lambda Ax = \lambda^2 x$$

and, in general for any integer $k \geq 0$,

$$A^k x = \lambda^k x \tag{A.17}$$

Moreover, if A has no zero eigenvalue, i.e., A^{-1} exists, then left-multiplying (A.1) with A^{-1} yields

$$A^{-1}x = \lambda^{-1}x$$

We apply (A.17) to the matrix A^{-1} and conclude that

$$A^{-k}x = \lambda^{-k}x$$

In other words, if the inverse matrix A^{-1} exists, then equation (A.17) is valid for any integer, positive as well as negative.

Combining (A.17) and (A.7) implies that

$$\text{trace}\left(A^k\right) = \sum_{j=1}^{n} \lambda_j^k\left(A\right) \tag{A.18}$$

6. *The Caley-Hamilton Theorem.* Since any eigenvalue λ satisfies its characteristic polynomial $c_A\left(\lambda\right) = 0$, we directly find from (A.17) that the matrix A satisfies its own characteristic equation,

$$c_A(A) = O \tag{A.19}$$

This result is the Caley-Hamilton Theorem. There exist several other proofs of the Caley–Hamilton Theorem.

A.2 Functions of a matrix

7. Consider an arbitrary matrix polynomial in λ,

$$F(\lambda) = \sum_{k=0}^{m} F_k \lambda^k$$

where all F_k are $n \times n$ matrices and $F_m \neq O$. Any matrix polynomial $F(\lambda)$ can be right and left divided by another (non-zero) matrix polynomial $B(\lambda)$ in a unique way as proved in Gantmacher (1959a, Chapter IV). Hence the left-quotient and left-remainder $F(\lambda) = B(\lambda)Q_L(\lambda) + L(\lambda)$ and the right-quotient and right-remainder $F(\lambda) = Q_R(\lambda)B(\lambda) + R(\lambda)$ are unique. Let us concentrate on the right-remainder in the case where $B(\lambda) = \lambda I - A$ is a linear polynomial in λ. Using Euclid's division scheme for polynomials,

$$F(\lambda) = F_m \lambda^{m-1}\left(\lambda I - A\right) + \left(F_m A + F_{m-1}\right)\lambda^{m-1} + \sum_{k=0}^{m-2} F_k \lambda^k$$

$$= \left[F_m \lambda^{m-1} + \left(F_m A + F_{m-1}\right)\lambda^{m-2}\right]\left(\lambda I - A\right)$$

$$+ \left(F_m A^2 + F_{m-1}A + F_{m-2}\right)\lambda^{m-2} + \sum_{k=0}^{m-3} F_k \lambda^k$$

and continuing, we arrive at

$$F(\lambda) = \left[F_m \lambda^{m-1} + \cdots + \lambda^{k-1} \sum_{j=k}^{m} F_j A^{j-k} + \cdots + \sum_{j=1}^{m} F_j A^{j-1} \right] (\lambda I - A) + \sum_{j=0}^{m} F_j A^j$$

In summary, $F(\lambda) = Q_R(\lambda)(\lambda I - A) + R(\lambda)$ (and similarly for the left-quotient and left-remainder) with

$$Q_R(\lambda) = \sum_{k=1}^{m} \lambda^{k-1} \left(\sum_{j=k}^{m} F_j A^{j-k} \right) \quad Q_L(\lambda) = \sum_{k=1}^{m} \lambda^{k-1} \left(\sum_{j=k}^{m} A^{j-k} F_j \right)$$
$$R(\lambda) = \sum_{j=0}^{m} F_j A^j = F(A) \qquad\qquad L(\lambda) = \sum_{j=0}^{m} A^j F_j$$

$$(\text{A.20})$$

and where the right-remainder is independent of λ. The Generalized Bézout Theorem states that the polynomial $F(\lambda)$ is divisible by $(\lambda I - A)$ on the right (left) if and only if $F(A) = O$ ($L(\lambda) = O$).

8.   *The adjoint matrix.* By the Generalized Bézout Theorem, the polynomial $F(\lambda) = g(\lambda)I - g(A)$ is divisible by $(\lambda I - A)$ because $F(A) = g(A)I - g(A) = O$. If $F(\lambda)$ is an ordinary polynomial, the right- and left-quotient and remainder are equal. The Caley–Hamilton Theorem (A.19) states that $c_A(A) = 0$, which indicates that $c_A(\lambda)I = Q(\lambda)(\lambda I - A)$ and also $c_A(\lambda)I = (\lambda I - A)Q(\lambda)$. The matrix

$$Q(\lambda) = (\lambda I - A)^{-1} c_A(\lambda)$$

is called the adjoint matrix of A. Explicitly, from (A.20),

$$Q(\lambda) = \sum_{k=1}^{n} \lambda^{k-1} \left(\sum_{j=k}^{n} c_j A^{j-k} \right)$$

and, with (A.6),

$$Q(0) = - (A)^{-1} \det A = \sum_{j=1}^{n} c_j A^{j-1}$$

The main theoretical interest of the adjoint matrix stems from its definition,

$$c_A(\lambda)I = Q(\lambda)(\lambda I - A) = (\lambda I - A)Q(\lambda)$$

In case $\lambda = \lambda_k$ is an eigenvalue of A, then $(\lambda_k I - A)Q(\lambda_k) = 0$, which indicates by (A.1) and the commutative property $(\lambda I - A)Q(\lambda) = Q(\lambda)(\lambda I - A)$ that every non-zero column(row) of the adjoint matrix $Q(\lambda_k)$ is a right(left)-eigenvector belonging to the eigenvalue λ_k. In addition, by differentiation with respect to λ, we obtain

$$c_A'(\lambda)I = (\lambda I - A)Q'(\lambda) + Q(\lambda)$$

This demonstrates that, if $Q(\lambda_k) \neq O$, the eigenvalue λ_k is a simple root of $c_A(\lambda)$ and, conversely, if $Q(\lambda_k) = O$, the eigenvalue λ_k has higher multiplicity.

The adjoint matrix $Q(\lambda) = (\lambda I - A)^{-1} c_A(\lambda)$ is computed by observing that, on the Generalized Bézout Theorem, $\frac{c_A(\lambda) - c_A(\mu)}{\lambda - \mu}$ is divisible without remainder. By replacing λ and μ in this polynomial by λI and A respectively, $Q(\lambda)$ readily follows.

9. Consider the arbitrary polynomial of degree l,

$$g(x) = g_0 \prod_{j=1}^{l} (x - \mu_j)$$

Substitute x by A, then

$$g(A) = g_0 \prod_{j=1}^{l} (A - \mu_j I)$$

Since $\det(AB) = \det A \det B$ and $\det(kA) = k^n \det A$, we have

$$\det(g(A)) = g_0^n \prod_{j=1}^{l} \det(A - \mu_j I) = g_0^n \prod_{j=1}^{l} c(\mu_j)$$

With (A.5),

$$\det(g(A)) = g_0^n \prod_{j=1}^{l} \prod_{k=1}^{n} (\lambda_k - \mu_j) = \prod_{k=1}^{n} g_0 \prod_{j=1}^{l} (\lambda_k - \mu_j) = \prod_{k=1}^{n} g(\lambda_k)$$

If $h(x) = g(x) - \lambda$, we arrive at the general result: for any polynomial $g(x)$, the eigenvalues of $g(A)$ are $g(\lambda_1), \ldots, g(\lambda_n)$ and the characteristic polynomial is

$$\det(g(A) - \lambda I) = \prod_{k=1}^{n} (g(\lambda_k) - \lambda) \tag{A.21}$$

which is a polynomial in λ of degree at most n. Since the result holds for an arbitrary polynomial, it should not surprise that, under appropriate conditions of convergence, it can be extended to infinite polynomials, in particular to the Taylor series of a complex function. As proved in Gantmacher (1959a, Chapter V), if the power series of a function $f(z)$ around $z = z_0$,

$$f(z) = \sum_{j=1}^{\infty} f_j(z_0)(z - z_0)^j \tag{A.22}$$

converges for all z in the disc $|z - z_0| < R$, then $f(A) = \sum_{j=1}^{\infty} f_j(z_0)(A - z_0 I)^j$ provided that all eigenvalues of A lie within the region of convergence of (A.22), i.e., $|\lambda - z_0| < R$. For example,

$$e^{Az} = \sum_{k=0}^{\infty} \frac{z^k A^k}{k!} \qquad \text{for all } A$$

$$\log A = \sum_{k=1}^{\infty} \frac{(-1)^{k-1}}{k} (A - I)^k \quad \text{for } |\lambda_k - 1| < 1, \text{ all } 1 \le k \le n$$

and, from (A.21), the eigenvalues of e^{Az} are $e^{z\lambda_1}, \ldots, e^{z\lambda_1}$. Hence, the knowledge of the eigenstructure of a matrix A allows us to compute any function of A (under the same convergence restrictions as complex numbers z).

A.3 Hermitian and real symmetric matrices

10. *A Hermitian matrix.* A Hermitian matrix A is a complex matrix that obeys $A^H = \left(A^T\right)^* = A$, where $a^H = (a_{ij})^*$ is the complex conjugate of a_{ij}. The superscript H, in honor of Charles Hermite, means to take the complex conjugate and then a transpose. Hermitian matrices possess a number of attractive properties. A particularly interesting subclass of Hermitian matrices are real, symmetric matrices that obey $A^T = A$. The inner-product of vector y and x is defined as $y^H x$ and obeys $\left(y^H x\right)^* = \left(y^H x\right)^H = x^H y$. The inner-product $x^H x = \sum_{j=1}^{n} |x_j|^2$ is real and positive for all vectors except for the null vector.

11. The eigenvalues of a Hermitian matrix are all real. Indeed, left-multiplying (A.1) by x^H yields

$$x^H A x = \lambda x^H x$$

and, since $\left(x^H A x\right)^H = x^H A^H x = x^H A x$, it follows that $\lambda x^H x = \lambda^H x^H x$ or $\lambda = \lambda^H$ because $x^H x$ is a positive real number. Furthermore, since $A = A^H$, we have

$$A^H x = \lambda x$$

Taking the complex conjugate, yields

$$A^T x^* = \lambda x^*$$

In general, the eigenvectors of a Hermitian matrix are complex, but real for a real symmetric matrix since $A^H = A^T$. Moreover, the left-eigenvector y^T is the complex conjugate of the right-eigenvector x. Hence, the orthogonality relation (A.11) reduces, after normalization, to the inner-product

$$x_k^H x_j = \delta_{kj} \tag{A.23}$$

where δ_{kj} is the Kronecker delta, which is zero if $k \neq j$ and else $\delta_{kk} = 1$. Consequently, (A.14) reduces to

$$X^H X = I \tag{A.24}$$

which implies that the matrix X formed by the eigenvectors is an unitary matrix $(X^{-1} = X^H)$.

For a real symmetric matrix A, the corresponding relation $X^T X = I$ implies that X is an orthogonal matrix $(X^{-1} = X^T)$ obeying

$$X^T X = X X^T = I$$

where the first equality follows from the commutativity of the inverse of a matrix,

$X^{-1}X = XX^{-1}$. Hence, all eigenvectors of a symmetric matrix are orthogonal. Although the arguments so far (see Appendix A.1) have assumed that the eigenvalues of A are distinct, the theorem applies in general as proved in Wilkinson (1965, Section 47): for any Hermitian matrix A, there exists a unitary matrix U such that, for real λ_j,

$$U^H A U = \text{diag}(\lambda_j)$$

and for any real symmetric matrix A, there exists an orthogonal matrix U such that, for real λ_j,

$$U^T A U = \text{diag}(\lambda_j)$$

12. To a real symmetric matrix A, a bilinear form $x^T A y$ is associated, which is a scalar defined as

$$x^T A y = x A y^T = \sum_{i=1}^{n} \sum_{j=1}^{n} a_{ij} x_i y_j$$

We call a bilinear form a quadratic form if $y = x$. A necessary and sufficient condition for a quadratic form to be positive definite, i.e., $x^T A x > 0$ for all $x \neq 0$, is that all eigenvalues of A should be positive. Indeed, **art.** 11 shows the existence of an orthogonal matrix U that transforms A to a diagonal form. Let $x = Uz$, then

$$x^T A x = z^T U^T A U z = \sum_{k=1}^{n} \lambda_k z_k^2 \tag{A.25}$$

which is only positive for all z_k provided $\lambda_k > 0$ for all k. From (A.6), a positive definite quadratic form $x^T A x$ possesses a positive determinant, $\det A > 0$. This analysis shows that the problem of determining an orthogonal matrix U (or the eigenvectors of A) is equivalent to the geometrical problem of determining the principal axes of the hyper-ellipsoid

$$\sum_{i=1}^{n} \sum_{j=1}^{n} a_{ij} x_i y_j = 1$$

Relation (A.25) illustrates that the eigenvalue λ_k^{-1} is the square of the principal axis along the z_k vector. A multiple eigenvalue refers to an indeterminacy of the principal axes. For example if $n = 3$, an ellipsoid with two equal principal axis means that any section along the third axis is a circle. Any two perpendicular diameters of the largest circle orthogonal to the third axis are principal axes of that ellipsoid.

For additional properties of quadratic forms, such as the inertial theorem, we refer to Courant and Hilbert (1953b) and Gantmacher (1959a).

A.4 Vector and matrix norms

13. Vector and matrix norms, denoted by $\|x\|$ and $\|A\|$ respectively, provide a single number reflecting a "size" of the vector or matrix and may be regarded as an extension of the concept of the modulus of a complex number. A norm is a certain function of the vector components or matrix elements. All norms, vector as well as matrix norms, satisfy the three "distance" relations:

(i) $\|x\| > 0$ unless $x = 0$;

(ii) $\|\alpha x\| = |\alpha|\,\|x\|$ for any complex number α;

(iii) $\|x + y\| \le \|x\| + \|y\|$.

In general, the Hölder q-norm of a vector x is defined as

$$\|x\|_q = \left(\sum_{j=1}^{n} |x_j|^q\right)^{1/q} \tag{A.26}$$

For example, the well-known Euclidean norm or length of the vector x is found for $q = 2$ and $\|x\|_2^2 = x^H x$. In probability theory where x denotes a discrete probability density function, the law of total probability states that $\|x\|_1 = \sum_{j=1}^{n} x_j = 1$ and we will write $\|x\|_1 = \|x\|$. Finally, $\max |x_j| = \lim_{q\to\infty} \|x\|_q = \|x\|_\infty$. The unit-spheres $S_q = \{x \mid \|x\|_q = 1\}$ are, in three dimensions $n = 3$, for $q = 1$ an octahedron; for $q = 2$ a ball; and for $q = \infty$ a cube. Furthermore, S_1 fits into S_2, which in turn fits into S_∞, and this implies that $\|x\|_1 \ge \|x\|_2 \ge \|x\|_\infty$ for any x.

The Hölder inequality proved in Section 5.5 states that, for $\frac{1}{p} + \frac{1}{q} = 1$ and real $p, q > 1$,

$$|x^H y| \le \|x\|_p \|y\|_q \tag{A.27}$$

and is given explicitly in (5.18). A special case of the Hölder inequality where $p = q = 2$ is the Cauchy-Schwarz inequality

$$|x^H y| \le \|x\|_2 \|y\|_2 \tag{A.28}$$

The $q = 2$ norm is invariant under a unitary (hence also orthogonal) transformation U, where $U^H U = I$, because $\|Ux\|_2^2 = x^H U^H U x = x^H x = \|x\|_2^2$.

Another example of a non-homogeneous vector norm is the quadratic form

$$\sqrt{\|x\|_A} = \sqrt{x^T A x}$$

provided A is positive definite. Relation (A.25) shows that, if not all eigenvalues λ_j of A are the same, then not all components of the vector x are weighted similarly and, thus, in general, $\sqrt{\|x\|_A}$ is a non-homogeneous norm. The quadratic form $\|x\|_I$ equals the homogeneous Euclidean norm $\|x\|_2^2$.

A.4.1 Properties of norms

14. All norms are equivalent in the sense that there exist positive real numbers c_1 and c_2 such that, for all x,

$$c_1 \|x\|_p \leq \|x\|_q \leq c_2 \|x\|_p$$

For example,

$$\|x\|_2 \leq \|x\|_1 \leq \sqrt{n} \|x\|_2$$
$$\|x\|_\infty \leq \|x\|_1 \leq n \|x\|_\infty$$
$$\|x\|_\infty \leq \|x\|_2 \leq \sqrt{n} \|x\|_\infty$$

By choosing in the Hölder inequality $p = q = 1$, $x_j \to \alpha_j x_j^s$ for real $s > 0$ and $y_j \to \alpha_j > 0$, we obtain with $0 < \theta < 1$ an inequality for the *weighted* q-norm

$$\left(\frac{\sum_{j=1}^n \alpha_j |x_j|^{s\theta}}{\sum_{j=1}^n \alpha_j} \right)^{\frac{1}{s\theta}} \leq \left(\frac{\sum_{j=1}^n \alpha_j |x_j^s|}{\sum_{j=1}^n \alpha_j} \right)^{\frac{1}{s}} \tag{A.29}$$

For $\alpha_j = 1$, the weights α_j disappear such that the inequality for the Hölder q-norm becomes

$$\|x\|_{s\theta} \leq \|x\|_s \, n^{\frac{1}{s}(\frac{1}{\theta}-1)}$$

where $n^{\frac{1}{s}(\frac{1}{\theta}-1)} \geq 1$. On the other hand, with $0 < \theta < 1$ and for real $s > 0$,

$$\frac{\|x\|_s}{\|x\|_{s\theta}} = \frac{\left(\sum_{j=1}^n |x_j|^s \right)^{\frac{1}{s}}}{\left(\sum_{k=1}^n |x_k|^{s\theta} \right)^{\frac{1}{s\theta}}} = \left(\sum_{j=1}^n \frac{|x_j|^s}{\left(\sum_{k=1}^n |x_k|^{s\theta} \right)^{\frac{1}{\theta}}} \right)^{\frac{1}{s}} = \left(\sum_{j=1}^n \left(\frac{|x_j|^{s\theta}}{\sum_{k=1}^n |x_k|^{s\theta}} \right)^{\frac{1}{\theta}} \right)^{\frac{1}{s}}$$

Since $y = \frac{|x_j|^{s\theta}}{\sum_{k=1}^n |x_k|^{s\theta}} \leq 1$ and $\frac{1}{\theta} > 1$, it holds that $y^{\frac{1}{\theta}} \leq y$ and

$$\left(\sum_{j=1}^n \left(\frac{|x_j|^{s\theta}}{\sum_{k=1}^n |x_k|^{s\theta}} \right)^{\frac{1}{\theta}} \right)^{\frac{1}{s}} \leq \left(\sum_{j=1}^n \frac{|x_j|^{s\theta}}{\sum_{k=1}^n |x_k|^{s\theta}} \right)^{\frac{1}{s}} = \left(\frac{\sum_{j=1}^n |x_j|^{s\theta}}{\sum_{k=1}^n |x_k|^{s\theta}} \right)^{\frac{1}{s}} = 1$$

which leads to an opposite inequality,

$$\|x\|_s \leq \|x\|_{s\theta}$$

In summary, if $p > q > 0$, then the general inequality for Hölder q-norm is

$$\|x\|_p \leq \|x\|_q \leq \|x\|_p \, n^{\frac{1}{q}-\frac{1}{p}} \tag{A.30}$$

15. For $m \times n$ matrices A, the most frequently used norms are the Euclidean or Frobenius norm

$$\|A\|_F = \left(\sum_{i=1}^m \sum_{j=1}^n |a_{ij}|^2 \right)^{1/2} \tag{A.31}$$

and the q-norm

$$\|A\|_q = \sup_{x \neq 0} \frac{\|Ax\|_q}{\|x\|_q} \tag{A.32}$$

The second distance relation in **art. 13**, $\frac{\|Ax\|_q}{\|x\|_q} = \left\| A \frac{x}{\|x\|_q} \right\|_q$, shows that

$$\|A\|_q = \sup_{\|x\|_q = 1} \|Ax\|_q \tag{A.33}$$

Furthermore, the matrix q-norm (A.32) implies that

$$\|Ax\|_q \leq \|A\|_q \|x\|_q \tag{A.34}$$

Since the vector norm is a continuous function of the vector components and since the domain $\|x\|_q = 1$ is closed, there must exist a vector x for which equality $\|Ax\|_q = \|A\|_q \|x\|_q$ holds. Since the k-th vector component of Ax is $(Ax)_i = \sum_{j=1}^{n} a_{ij} x_j$, it follows from (A.26) that

$$\|Ax\|_q = \left(\sum_{i=1}^{m} \left| \sum_{j=1}^{n} a_{ij} x_j \right|^q \right)^{1/q}$$

For example, for all x with $\|x\|_1 = 1$, we have that

$$\|Ax\|_1 = \sum_{i=1}^{m} \left| \sum_{j=1}^{n} a_{ij} x_j \right| \leq \sum_{i=1}^{m} \sum_{j=1}^{n} |a_{ij}| \, |x_j| = \sum_{j=1}^{n} |x_j| \sum_{i=1}^{m} |a_{ij}|$$

$$\leq \sum_{j=1}^{n} |x_j| \left(\max_j \sum_{i=1}^{m} |a_{ij}| \right) = \max_j \sum_{i=1}^{m} |a_{ij}|$$

Clearly, there exists a vector x for which equality holds, namely, if k is the column in A with maximum absolute sum, then $x = e_k$, the k-th basis vector with all components zero, except for the k-th one, which is 1. Similarly, for all x with $\|x\|_\infty = 1$,

$$\|Ax\|_\infty = \max_i \left| \sum_{j=1}^{n} a_{ij} x_j \right| \leq \max_i \sum_{j=1}^{n} |a_{ij}| \, |x_j| \leq \max_i \sum_{j=1}^{n} |a_{ij}|$$

Again, if r is the row with maximum absolute sum and $x_j = 1.\mathrm{sign}(a_{rj})$ such that $\|x\|_\infty = 1$, then $(Ax)_r = \sum_{j=1}^{n} |a_{rj}| = \max_i \sum_{j=1}^{n} |a_{ij}| = \|Ax\|_\infty$. Hence, we have proved that

$$\|A\|_\infty = \max_i \sum_{j=1}^{n} |a_{ij}| \tag{A.35}$$

$$\|A\|_1 = \max_j \sum_{i=1}^{m} |a_{ij}| \tag{A.36}$$

from which

$$\left\|A^H\right\|_\infty = \|A\|_1$$

16. The $q = 2$ matrix norm, $\|Ax\|_2$, is obtained differently. Consider

$$\|Ax\|_2^2 = (Ax)^H Ax = x^H A^H Ax$$

Since $A^H A$ is a Hermitian matrix, **art. 11** shows that all eigenvalues are real and non-negative because a norm $\|Ax\|_2^2 \geq 0$. These ordered eigenvalues are denoted as $\sigma_1^2 \geq \sigma_2^2 \geq \cdots \geq \sigma_n^2 \geq 0$. Applying the theorem in **art. 11**, there exists a unitary matrix U such that $x = Uz$ yields

$$x^H A^H Ax = z^H U^H A^H A U z = z^H \operatorname{diag}\left(\sigma_j^2\right) z \leq \sigma_1^2 z^H z = \sigma_1^2 \|z\|_2^2$$

Since the $q = 2$ norm is invariant under a unitary (orthogonal) transform $\|x\|_2 = \|z\|_2$, by the definition (A.32),

$$\|A\|_2 = \sup_{x \neq 0} \frac{\|Ax\|_2}{\|x\|_2} = \sigma_1 \tag{A.37}$$

where the supremum is achieved if x is the eigenvector of $A^H A$ belonging to σ_1^2. Meyer (2000, p. 279) proves the corresponding result for the minimum eigenvalue provided that A is non-singular,

$$\left\|A^{-1}\right\|_2 = \frac{1}{\min\limits_{\|x\|_2=1} \|Ax\|_2} = \sigma_n^{-1}$$

The non-negative quantity σ_j is called the j-th singular value and σ_1 is the largest singular value of A. The importance of this result lies in an extension of the eigenvalue problem to non-square matrices, which is called the singular value decomposition. A detailed discussion is found in Golub and Van Loan (1996) and Horn and Johnson (1991, Chapter 3).

17. The Frobenius norm $\|A\|_F^2 = \operatorname{trace}\left(A^H A\right)$. With (A.7) and the analysis of $A^H A$ above,

$$\|A\|_F^2 = \sum_{k=1}^n \sigma_k^2 \tag{A.38}$$

In view of (A.37), the bounds $\|A\|_2 \leq \|A\|_F \leq \sqrt{n}\,\|A\|_2$ may be attained.

A.4.2 Applications of norms

18. (a) Since $\left\|A^k\right\| = \left\|AA^{k-1}\right\| \leq \|A\|\left\|A^{k-1}\right\|$, by induction, we have for any integer k, that

$$\left\|A^k\right\| \leq \|A\|^k$$

and

$$\lim_{k\to\infty} A^k = 0 \text{ if } \|A\| < 1$$

(b) By taking the norm of the eigenvalue equation (A.1), $\|Ax\| = |\lambda| \, \|x\|$ and with (A.34),

$$|\lambda| \leq \|A\|_q \tag{A.39}$$

Applied to $A^H A$, for any q-norm,

$$\sigma_1^2 \leq \|A^H A\|_q \leq \|A^H\|_q \|A\|_q$$

Choose $q = 1$ and with (A.37),

$$\|A\|_2^2 \leq \|A^H\|_1 \|A\|_1 = \|A\|_\infty \|A\|_1$$

(c) Any matrix A can be transformed by a similarity transform H to a Jordan canonical form C (**art.** 4) as $A = HCH^{-1}$, from which $A^k = HC^k H^{-1}$. A typical Jordan submatrix $(C_m(\lambda))^k = \lambda^{k-2} B$, where B is independent of k. Hence, for large k, $A^k \to 0$ if and only if $|\lambda| < 1$ for all eigenvalues.

A.5 Stochastic matrices

A probability matrix P is reducible if there is a relabeling of the states that leads to

$$\tilde{P} = \begin{bmatrix} P_1 & B \\ O & P_2 \end{bmatrix}$$

where P_1 and P_2 are square matrices and $u^T \tilde{P} = u^T$, the column sum equal to 1. Relabeling amounts to permuting rows and columns in the same fashion. Thus, there exists a similarity transform H such that $P = H\tilde{P}H^{-1}$. Moreover,

$$\tilde{P}^k = \begin{bmatrix} (P_1)^k & Y_k \\ O & (P_2)^k \end{bmatrix}$$

where, for each integer $k \geq 0$,

$$Y_k = \sum_{j=0}^{k-1} (P_1)^{k-1-j} B (P_2)^j$$

Since P_1 is stochastic, whereas P_2 is substochastic (defined on p. 190), the largest eigenvalue of P_1 is $\lambda_1(P_1) = 1$. Assuming that all eigenvalues of P_1 are distinct, then **art.** 4 and **art.** 9 indicate, as in (10.16), that

$$(P_1)^m = (u\pi)^T + \sum_{k=2}^{n} \lambda_k^m y_k x_k^T$$

so that

$$Y_k = \sum_{j=0}^{k-1} \left((u\pi)^T + \sum_{l=2}^{n} \lambda_l^{k-1-j} y_l x_l^T \right) B (P_2)^j$$

$$= (u\pi)^T B \sum_{j=0}^{k-1} (P_2)^j + \sum_{l=2}^{n} \lambda_l^{k-1} y_l x_l^T B \sum_{j=0}^{k-1} \lambda_l^{-j} (P_2)^j$$

Since $|\lambda_l| < 1$, for $l > 1$, while the largest eigenvalue of P_2 is smaller than 1, implying that the j-sum exists for all k, the limit for $k \to \infty$ simplifies to the first term

$$Y_\infty = (u\pi)^T B (I - P_2)^{-1}$$

Hence,

$$\tilde{A} = \lim_{k \to \infty} \tilde{P}^k = \begin{bmatrix} (u\pi)^T & (u\pi)^T B (I - P_2)^{-1} \\ O & O \end{bmatrix}$$

Finally, $\tilde{A}$ is stochastic ($u^T \tilde{A} = u^T$) and $\pi u = 1$, from which

$$u^T = u^T \pi^T u^T B (I - P_2)^{-1} = u^T B (I - P_2)^{-1}$$

which is only possible if $B (I - P_2)^{-1} = I$. This relation is also supported by $u^T \tilde{P} = u^T$, because $u^T B = u^T (I - P_2)$. Thus, the steady-state matrix $\tilde{A}$ consists, just as for an irreducible matrix, of all equal columns, with the first non-zero rows for each column specified by the steady-state vector π of the irreducible matrix P_1.

A.5.1 The eigenstructure

In this section, the basic theorem on the eigenstructure of a stochastic, irreducible matrix will be proved.

Lemma A.5.1 *If P is an irreducible non-negative matrix and if v is a vector with positive components, then the vector $z = (P+I)v$ always has fewer zero components than v.*

Proof: Denote

$$v = \begin{bmatrix} v_1 \\ 0 \end{bmatrix} \quad \text{and} \quad z = \begin{bmatrix} z_1 \\ 0 \end{bmatrix} \quad \text{where } v_1 > 0, z_1 > 0$$

which is always possible by suitable renumbering of the states and

$$P = \begin{bmatrix} P_{11} & P_{12} \\ P_{21} & P_{22} \end{bmatrix}$$

The relation $z = (P + I)v$ is written as

$$\begin{bmatrix} z_1 \\ 0 \end{bmatrix} = \begin{bmatrix} P_{11} v_1 \\ P_{21} v_1 \end{bmatrix} + \begin{bmatrix} v_1 \\ 0 \end{bmatrix}$$

Since P is irreducible, $P_{21} \neq O$, such that $v_1 > 0$ implies that $P_{21} v_1 \neq 0$, which proves the lemma.
□

Observe, in addition, that all components of z are never smaller than those of v. Also, transposing does not alter the result.

Theorem A.5.2 (Frobenius) *The modulus of all eigenvalues λ of an irreducible stochastic matrix P are less than or equal to 1. There is only one real eigenvalue $\lambda = 1$ and the corresponding eigenvector has positive components.*

Proof: The $q = \infty$ norm (A.35) of a probability matrix P with n states defined by (9.7) subject to (9.8) precisely equals $\|P\|_\infty = 1$. From (A.39), it follows that all eigenvalues are, in absolute value, smaller than or equal to 1. Since all elements $P_{ij} \in [0,1]$ and because an irreducible matrix has no zero element rows, $v^T P$ has positive components if v^T has positive components. Thus, there always exists a scalar $0 < \mu_v = \min_{1 \leq k \leq n} \frac{(v^T P)_k}{(v^T)_k}$, such that $\mu_v v^T \leq v^T P$. By Lemma A.5.1, we can always transform the vector v to a vector z by right-multiplying both sides with $(I + P)$ such that

$$\mu_v v^T (I + P) \leq v^T P (I + P)$$
$$\mu_v z^T \leq z^T P$$

and, by definition of μ_v, $\mu_v \leq \mu_z$ since the components of z are never smaller than those of v. Hence, for any arbitrary vector v with positive components, the transform in Lemma A.5.1 leads to an increasing set $\mu_v \leq \mu_z \leq \cdots$, which is bounded by 1 because no eigenvalue can exceed 1. This shows that $\lambda = 1$ is the largest eigenvalue and the corresponding eigenvector y^T has positive components.

This eigenvector y^T is unique. For, if there were another linearly independent eigenvector w^T corresponding to the eigenvalue $\lambda = 1$, any linear combination $z^T = \alpha y^T + \beta w^T$ is also an eigenvector belonging to $\lambda = 1$. But α and β can always be chosen to produce a zero component which the transform method shows to be impossible. The fact that the eigenvector y^T is the only eigenvector belonging $\lambda = 1$, implies that the eigenvalue $\lambda = 1$ is a single zero of the characteristic polynomial of P. $\qquad\square$

The theorem proved for stochastic matrices is a special case of the famous Perron-Frobenius theorem for non-negative matrices (for a proof, see e.g. Gantmacher (1959b, Chapter XIII)). We note that, in the theory of Markov chains, the interest lies in the determination of the left-eigenvector $y^T = \pi$ belonging to $\lambda = 1$, because the right-eigenvector x of P belonging to $\lambda = 1$ equals $u^T = \beta[1 \ 1 \ \cdots \ 1]$, where β is a scalar, because of the constraints (9.8). Recall (A.11) and (A.14), the proper normalization, $y^T u = 1$, precisely corresponds to the axiom 1 on p. 8: $\Pr[\Omega] = 1$. Using the interpretation of Markov chains, an alternative argument is possible. If all eigenvalues were $|\lambda| < 1$, application (c) in Appendix A.4.2 indicates that the steady-state would be non-existent because $P^k \to 0$ for $k \to \infty$. Since this is impossible, there must be at least one eigenvalue with $|\lambda| = 1$. Furthermore, (9.28) shows that at least one eigenvalue corresponding to the steady-state is real and precisely 1.

Corollary A.5.3 *An irreducible probability matrix P cannot have two linearly independent eigenvectors with positive components.*

Proof: Consider, apart from $y^T = \pi$ belonging to $\lambda = 1$, another eigenvector w^T belonging to the eigenvalue $\omega \neq 1$. Since eigenvectors are linearly independent, it holds that $w^T y = 0$, which is only possible if not all components of w^T are positive. $\qquad\square$

The corollary is important because no other eigenvector of P than $y^T = \pi$ can represent a (discrete) probability density. Since the null vector is never an eigenvector, the corollary implies that at least one component in the other eigenvectors must be negative.

Since the characteristic polynomial of P has real coefficients (because P_{ij} is real), the eigenvalues occur in complex conjugate pairs. Since $\lambda = 1$ is an eigenvalue, for an even number of state n, there must be at least another real eigenvalue obeying $-1 \le \lambda < 1$. It has been proved that the boundary of the locations of the eigenvalues inside the unit disc consists of a finite number of points on the unit circle joined by certain curvilinear arcs.

There exist an interesting property of a rank-one update $\bar{P}$ of a stochastic matrix P. The lemma is of a general nature and also applies to reducible Markov chains with several eigenvalues $\lambda_j = 1$ for $1 < j \le k$.

Lemma A.5.4 *If* $\{1, \lambda_2, \lambda_3, \ldots, \lambda_n\}$ *are the eigenvalues of the stochastic matrix* P, *then the eigenvalues of* $\bar{P} = \alpha P + (1 - \alpha) uv^T$, *where* v^T *is any probability vector, are* $\{1, \alpha\lambda_2, \alpha\lambda_3, \ldots, \alpha\lambda_n\}$.

Proof: We start from the eigenvalues equation (A.2)

$$\det\left(\bar{P} - \lambda I\right) = \det\left(\alpha P - \lambda I + (1 - \alpha)uv^T\right)$$
$$= \det\left((\alpha P - \lambda I)\left(I + (\alpha P - \lambda I)^{-1}(1 - \alpha)uv^T\right)\right)$$
$$= \det(\alpha P - \lambda I)\det\left(I + (1 - \alpha)(\alpha P - \lambda I)^{-1}uv^T\right)$$

Applying the formula

$$\det\left(I + cd^T\right) = 1 + d^T c \tag{A.40}$$

which follows, after taking the determinant, from the matrix identity

$$\begin{pmatrix} I & 0 \\ d^T & 1 \end{pmatrix}\begin{pmatrix} I + cd^T & c \\ 0 & 1 \end{pmatrix}\begin{pmatrix} I & 0 \\ -d^T & 1 \end{pmatrix} = \begin{pmatrix} I & c \\ 0 & 1 + d^T c \end{pmatrix}$$

gives

$$\det\left(\bar{P} - \lambda I\right) = \det(\alpha P - \lambda I)\left(1 + v^T(1 - \alpha)(\alpha P - \lambda I)^{-1}u\right)$$

Since the row sum of a stochastic matrix P is 1, we have that $Pu = u$ and, thus, $(\alpha P - \lambda I)u = (\alpha - \lambda)u$ from which $(\alpha P - \lambda I)^{-1}u = (\alpha - \lambda)^{-1}u$. Using this result leads to

$$1 + v^T(1 - \alpha)(\alpha P - \lambda I)^{-1}u = 1 + \frac{1 - \alpha}{\alpha - \lambda}v^T u = 1 + \frac{1 - \alpha}{\alpha - \lambda} = \frac{1 - \lambda}{\alpha - \lambda}$$

because a probability vector is normalized to 1, i.e. $v^T u = 1$. Hence, we end up with

$$\det\left(\bar{P} - \lambda I\right) = \det(\alpha P - \lambda I)\frac{1 - \lambda}{\alpha - \lambda}$$

Invoking (A.21) yields

$$\det\left(\bar{P} - \lambda I\right) = \prod_{k=1}^{n}(\alpha\lambda_k - \lambda)\frac{1 - \lambda}{\alpha - \lambda} = (1 - \lambda)\prod_{k=2}^{n}(\alpha\lambda_k - \lambda)$$

which shows that the eigenvalues of $\bar{P}$ are $\{1, \alpha\lambda_2, \alpha\lambda_3, \ldots, \alpha\lambda_N\}$. $\qquad\square$

A similar property may occur in a special case where a Markov chain is supplemented by an additional state $n+1$ which connects to every other state and to which every other state is connected (such that $\bar{P}$ is irreducible). Then,

$$\bar{P} = \begin{pmatrix} \alpha P & (1-\alpha)u \\ v^T & 0 \end{pmatrix}$$

with corresponding eigenvalues $\{1, \alpha\lambda_2, \alpha\lambda_3, \ldots, \alpha\lambda_n, 0\}$. This result is similarly proved as Lemma A.5.4 using (Meyer, 2000, p. 475)

$$\det \begin{pmatrix} A & B \\ C & D \end{pmatrix} = \det A \det \left(D - CA^{-1}B \right) \tag{A.41}$$

provided A^{-1} exists unless $C = 0$.

A.5.2 Example: the two-state Markov chain

The two-state Markov chain is defined by

$$P = \begin{bmatrix} 1-p & p \\ q & 1-q \end{bmatrix}$$

Observe that $\det P = 1 - p - q$. The eigenvalues of P satisfy the characteristic polynomial $c(\lambda) = \lambda^2 - (2 - p - q)\lambda + \det P = 0$, from which $\lambda_1 = 1$ and $\lambda_2 = 1 - p - q = \det P$. The adjoint matrix $Q(\lambda)$ is computed (**art.** 8) via the polynomial $\frac{c(\lambda) - c(\mu)}{\lambda - \mu}$,

$$\frac{c(\lambda) - c(\mu)}{\lambda - \mu} = \lambda + \mu - (2 - p - q)$$

and after $\lambda \to \lambda I$ and $\mu \to P$

$$Q(\lambda) = \lambda I + P - (2 - p - q)I$$
$$= \begin{bmatrix} \lambda - 1 + q & p \\ q & \lambda - 1 + p \end{bmatrix}$$

The (unscaled) right- (left-)eigenvectors of P follow as the non-zero columns (rows) of $Q(\lambda)$. For $\lambda_1 = 1$, we find $x_1 = (1,1)$ and $y_1^T = (q, p)$. For $\lambda_2 = 1 - p - q$, the eigenvector $x_2 = (-p, q)$ and $y_2^T = (1, -1)$. Normalization (**art.** 4) requires that $y_k^T x_k = 1$ or $x_1 = \frac{1}{p+q}(1,1)$ and $x_2 = \frac{1}{p+q}(p, -q)$. If the eigenvalues are distinct $(p + q \neq 0)$, the matrix P can be written (**art.** 4) as $P = X\mathrm{diag}(\lambda_k)Y^T$,

$$P = \frac{1}{p+q} \begin{bmatrix} 1 & p \\ 1 & -q \end{bmatrix} \begin{bmatrix} 1 & 0 \\ 0 & 1-p-q \end{bmatrix} \begin{bmatrix} q & p \\ 1 & -1 \end{bmatrix}$$

from which any power P^k is immediate as

$$P^k = \frac{1}{p+q} \begin{bmatrix} 1 & p \\ 1 & -q \end{bmatrix} \begin{bmatrix} 1 & 0 \\ 0 & (1-p-q)^k \end{bmatrix} \begin{bmatrix} q & p \\ 1 & -1 \end{bmatrix}$$

$$= \frac{1}{p+q} \begin{bmatrix} q & p \\ q & p \end{bmatrix} + \frac{(1-p-q)^k}{p+q} \begin{bmatrix} p & -p \\ -q & q \end{bmatrix} \qquad (A.42)$$

The steady-state matrix $P^\infty = \lim_{k \to \infty} P^k$ follows as

$$P^\infty = \frac{1}{p+q} \begin{bmatrix} q & p \\ q & p \end{bmatrix} \equiv \begin{bmatrix} \pi \\ \pi \end{bmatrix} \qquad (A.43)$$

because $|1 - p - q| < 1$.

Alternatively, the steady-state vector is a solution of (9.31),

$$\begin{bmatrix} -p & q \\ 1 & 1 \end{bmatrix} \begin{bmatrix} \pi_1 \\ \pi_2 \end{bmatrix} = \begin{bmatrix} 0 \\ 1 \end{bmatrix}$$

Applying Cramer's rule with $D = \det \begin{bmatrix} -p & q \\ 1 & 1 \end{bmatrix} = -(p+q)$, we obtain $\pi_1 = \frac{1}{D} \det \begin{bmatrix} 0 & q \\ 1 & 1 \end{bmatrix}$ and $\pi_2 = \frac{1}{D} \det \begin{bmatrix} -p & 0 \\ 1 & 1 \end{bmatrix}$ or

$$\pi = \begin{bmatrix} \frac{q}{p+q} & \frac{p}{p+q} \end{bmatrix}$$

which indeed agrees with (A.43) and (9.43).

A.5.3 The tendency towards the steady-state

A stochastic matrix P and the corresponding Markov chain is regular if the only eigenvalue with $|\lambda| = 1$ is $\lambda = 1$. It is fully regular if, in addition, $\lambda = 1$ is a simple zero of the characteristic polynomial of P. The Frobenius Theorem A.5.2 indicates that a regular matrix is necessarily reducible. Application (c) in Appendix A.4.2 demonstrates that the steady-state only exists for regular Markov chains. Alternatively, a regular matrix P has the property that $P^k > O$ (for some k), i.e. all elements are strictly positive.

In the sequel, we concentrate on fully regular stochastic matrices P, where all eigenvalues lie within the unit circle, except for the largest one, $\lambda = 1$. If the N eigenvalues of the regular stochastic matrix P are ordered as $\lambda_1 = 1 > |\lambda_2| \geq \cdots \geq |\lambda_N| \geq 0$, the second largest eigenvalue λ_2 will determine the speed of convergence of the Markov chain towards the steady-state.

A.5.3.1 Example: the three-state Markov chain

The three-state Markov chain P is defined by (9.7) with $N = 3$. Assuming that P is irreducible, we determine the eigenvalues. Since the Frobenius Theorem A.5.2

already determines one eigenvalue $\lambda_1 = 1$, the remaining two λ_2 and λ_3 are found from (A.6) and (A.7). They obey the equations

$$\lambda_2 \lambda_3 = \det P$$
$$\lambda_2 + \lambda_3 = P_{11} + P_{22} + P_{33} - 1 = \text{trace}(P) - 1$$

or the quadratic equation $x^2 - (\lambda_2 + \lambda_3)\,x + \lambda_2 \lambda_3 = 0$. The explicit solution is

$$\lambda_2 = \frac{1}{2}\left(\text{trace}(P) - 1\right) + \frac{1}{2}\sqrt{\left(\text{trace}(P) - 1\right)^2 - 4\det P}$$
$$\lambda_3 = \frac{1}{2}\left(\text{trace}(P) - 1\right) - \frac{1}{2}\sqrt{\left(\text{trace}(P) - 1\right)^2 - 4\det P}$$

All eigenvalues are real if the discriminant $\left(\text{trace}(P) - 1\right)^2 - 4\det P$ is non-negative which leads to three cases:

(a) In case $\left(\text{trace}(P) - 1\right)^2 > 4\det P$, the eigenvalues obey $1 > \lambda_2 > \lambda_3$, but not necessarily $1 > |\lambda_2| > |\lambda_3|$. The latter inequality is true if $\text{trace}(P) > 1$, in which case the speed of convergence towards the steady-state is determined by the decay of $(\lambda_2)^k$ as $k \to \infty$. If $\text{trace}(P) = 1$, then $\lambda_2 = -\lambda_3 = \sqrt{-\det P}$ and if $\text{trace}(P) < 1$, $|\lambda_3|$ determines the speed of convergence. Notice that $\lambda_2 > \frac{1}{2}\left(\text{trace}(P) - 1\right) \geq -\frac{1}{2}$.

(b) In case $\left(\text{trace}(P) - 1\right)^2 < 4\det P$, there are two complex conjugate roots $\lambda_2 = \alpha + i\beta$ and $\lambda_3 = \alpha - i\beta$, both with the same modulus $|\lambda_2| = |\lambda_3|$ equal to $\alpha^2 + \beta^2 = \lambda_2 \lambda_3 = \det P$ and with real part $\alpha = \frac{1}{2}\left(\text{trace}(P) - 1\right)$. In this case, we have that $0 \leq \det P < 1$. Hence, the Markov chain converges towards the steady-state as $\left(\sqrt{\det P}\right)^k$ in the discrete-time k.

(c) In case $\left(\text{trace}(P) - 1\right)^2 = 4\det P$, there is a double eigenvalue

$$\lambda = \lambda_2 = \lambda_3 = \frac{1}{2}\left(\text{trace}(P) - 1\right) = \pm\sqrt{\det P}$$

and P cannot be reduced by a similarity transform H to a diagonal matrix (**art.** 4) that but to the Jordan canonical form C such that $P^k = H^{-1}C^k H$. Since (Meyer, 2000, pp. 599–600)

$$C^k = \left(\begin{bmatrix} 1 & 0 & 0 \\ 0 & \lambda & 1 \\ 0 & 0 & \lambda \end{bmatrix}\right)^k = \begin{bmatrix} 1 & 0 & 0 \\ 0 & \lambda^k & k\lambda^{k-1} \\ 0 & 0 & \lambda^k \end{bmatrix}$$

the Markov chain converges towards the steady-state as $k\left(\sqrt{\det P}\right)^{k-1}$ in the discrete-time k. We observe that $-\frac{1}{2} \leq \lambda_2 = \lambda_3 < 1$ because $0 \leq \text{trace}(P) < 3$. If $\text{trace}(P) = 3$, then $P = I$, and P is not irreducible.

The fastest possible convergence occurs when $\lambda_2 = \lambda_3 = 0$ or when $\det P = 0$ and $\text{trace}(P) = 1$ in which case P has rank 1. In any matrix of rank 1, all row vectors are linearly dependent. Since the column sum of a stochastic matrix P is 1 by (9.8), every row in P is precisely the same and (9.6) shows that after one discrete-time step, the steady-state $\pi = \begin{bmatrix} \frac{1}{N} & \frac{1}{N} & \cdots & \frac{1}{N} \end{bmatrix}$ is reached. As shown in Section 9.3.1, a transition probability matrix with constant rows can be regarded as a limit transition probability matrix $A = \lim_{k \to \infty} \check{P}^k$ of a Markov process with transition probability matrix $\check{P}$.

A.6 Special types of stochastic matrices

A.6.1 Doubly stochastic matrices

A doubly stochastic matrix P has both row and column sums equal to 1,

$$\sum_{k=1}^{N} P_{ik} = \sum_{k=1}^{N} P_{kj} = 1 \qquad \text{for all } i, j$$

If P is symmetric, $P = P^T$, then P is doubly stochastic, but the reverse implication is not true. As observed in Appendix A.5.1, the left-eigenvector $y^T = \pi$ and the right-eigenvector $x = u$ belonging to eigenvalue $\lambda = 1$ satisfy $y^T u = 1$. For doubly stochastic matrices, it holds that the role of left- and right-eigenvector can be reversed, which leads to $y = x$ or

$$\pi = \begin{bmatrix} \frac{1}{N} & \frac{1}{N} & \cdots & \frac{1}{N} \end{bmatrix}$$

The example in Appendix A.5.3 illustrates that a steady-state vector equal to $\pi = \begin{bmatrix} \frac{1}{N} & \frac{1}{N} & \cdots & \frac{1}{N} \end{bmatrix}$ does not necessarily imply that P is doubly stochastic.

A.6.2 Tri-diagonal bandmatrices

Since tri-diagonal matrices of the form (11.1) frequently occur in Markov theory, we devote this section to illustrate how far the eigen-analysis can be pushed. If $p_j = p$ and $q_j = q$, the matrix P reduces to a Toeplitz form for which the eigenvalues and eigenvectors can be explicitly written, as shown in Appendix A.6.2.1.

A.6.2.1 Tri-diagonal Toeplitz bandmatrix

A Toeplitz matrix has constant entries on each diagonal parallel to the main diagonal. Of particular interest is the $N \times N$ tri-diagonal Toeplitz matrix,

$$A = \begin{bmatrix} b & a & & & \\ c & b & a & & \\ & \ddots & \ddots & \ddots & \\ & & c & b & a \\ & & & c & b \end{bmatrix}$$

that arises in the Markov chain of the random walk and the birth and death process. Moreover, the eigenstructure of the tri-diagonal Toeplitz matrix A can be expressed in analytic form.

An eigenvector x corresponding to eigenvalue λ satisfies $(A - \lambda I)\,x = 0$ or, written per component,

$$(b - \lambda)x_1 + ax_2 = 0$$
$$cx_{k-1} + (b - \lambda)x_k + ax_{k+1} = 0 \qquad\qquad 2 \le k \le N - 1$$
$$cx_{N-1} + (b - \lambda)x_N = 0$$

We assume that $a \ne 0$ and $c \ne 0$ and rewrite the set with $x_0 = x_{N+1} = 0$ as

$$x_{k+2} + \left(\frac{b-\lambda}{a}\right)x_{k+1} + \left(\frac{c}{a}\right)x_k = 0 \qquad\qquad 0 \le k \le N - 1$$

which are second-order difference equations with constant coefficients. The general solution of these equations is $x_k = ar_1^k + \beta r_2^k$, where r_1 and r_2 are the roots of the corresponding polynomial $x^2 + \left(\frac{b-\lambda}{a}\right)x + \left(\frac{c}{a}\right) = 0$. If $r_1 = r_2$, the general solution is $x_k = ar_1^k + \beta k r_1^k$, which is impossible since it implies that all $x_k = 0$ due to the fact that $x_0 = x_{N+1} = 0$, which forces r_1 to be zero. An eigenvector is never the zero vector. Thus, we have distinct roots $r_1 \ne r_2$ that satisfy

$$r_1 + r_2 = -\left(\frac{b-\lambda}{a}\right)$$

$$r_1 r_2 = \frac{c}{a}$$

The constants α and β follow from the boundary requirement $x_0 = x_{N+1} = 0$ as

$$\alpha + \beta = 0$$
$$\alpha r_1^{N+1} + \beta r_2^{N+1} = 0$$

Rewriting the last equation with $\alpha = -\beta$, yields $\left(\frac{r_1}{r_2}\right)^{N+1} = 1$ or $\frac{r_1}{r_2} = e^{\frac{2\pi i m}{N+1}}$ for some $1 \le m \le N$ (the root $m = 0$ must be rejected since $r_1 \ne r_2$). Substitution of $r_1 = r_2 e^{\frac{2\pi i m}{N+1}}$ into the last root equation yields

$$r_1 = \sqrt{\tfrac{c}{a}}\,e^{\frac{\pi i m}{N+1}} \quad \text{and} \quad r_2 = \sqrt{\tfrac{c}{a}}\,e^{-\frac{\pi i m}{N+1}}$$

The first root equation is only possible for special values of $\lambda = \lambda_m$ with $1 \le m \le N$, which are the eigenvalues,

$$\lambda_m = b + a\sqrt{\frac{c}{a}}\left(e^{-\frac{\pi i m}{N+1}} + e^{\frac{\pi i m}{N+1}}\right) = b + 2\sqrt{ac}\,\cos\left(\frac{\pi m}{N+1}\right)$$

Since there are precisely N different values of m, there are N distinct eigenvalues λ_m. The components x_k of the eigenvector belonging to λ_m are

$$x_k = \alpha\left(\frac{c}{a}\right)^{\frac{k}{2}}\left(e^{\frac{\pi i m k}{N+1}} - e^{-\frac{\pi i m k}{N+1}}\right) = 2i\alpha\left(\frac{c}{a}\right)^{\frac{k}{2}}\sin\left(\frac{\pi m k}{N+1}\right)$$

The scaling constant α follows from the normalization $\|x\|_1 = 1$ or

$$2i\alpha \sum_{k=1}^{N}\left(\frac{c}{a}\right)^{\frac{k}{2}}\sin\left(\frac{\pi m k}{N+1}\right) = 1$$

Since $\sin\left(\frac{\pi mk}{N+1}\right) = \operatorname{Im}\left[e^{\frac{\pi imk}{N+1}}\right]$ we have

$$\sum_{k=1}^{N}\left(\frac{c}{a}\right)^{\frac{k}{2}}\sin\left(\frac{\pi mk}{N+1}\right) = \operatorname{Im}\left[\sum_{k=1}^{N}\left(\sqrt{\frac{c}{a}}e^{\frac{\pi im}{N+1}}\right)^{k}\right] = \operatorname{Im}\left[\frac{1-\left(\sqrt{\frac{c}{a}}e^{\frac{\pi im}{N+1}}\right)^{N+1}}{1-\sqrt{\frac{c}{a}}e^{\frac{\pi im}{N+1}}} - 1\right]$$

$$= \frac{\left(1+(-1)^{m}\left(\sqrt{\frac{c}{a}}\right)^{N+1}\right)\sin\left(\frac{\pi m}{N+1}\right)}{1-2\sqrt{\frac{c}{a}}\cos\left(\frac{\pi m}{N+1}\right)+\frac{c}{a}} - 1$$

from which the scaling constant α is

$$2i\alpha = \left[\frac{\left(1+(-1)^{m}\left(\sqrt{\frac{c}{a}}\right)^{N+1}\right)\sin\left(\frac{\pi m}{N+1}\right)}{1-2\sqrt{\frac{c}{a}}\cos\left(\frac{\pi m}{N+1}\right)+\frac{c}{a}} - 1\right]^{-1}$$

Finally, the components x_k of the eigenvector x belonging to λ_m become, for $1 \le k \le N$,

$$x_k = \frac{\left(\frac{c}{a}\right)^{\frac{k}{2}}\sin\left(\frac{\pi mk}{N+1}\right)}{\dfrac{\left(1+(-1)^{m}\left(\sqrt{\frac{c}{a}}\right)^{N+1}\right)\sin\left(\frac{\pi m}{N+1}\right)}{1-2\sqrt{\frac{c}{a}}\cos\left(\frac{\pi m}{N+1}\right)+\frac{c}{a}} - 1}$$

Observe that for stochastic matrices $a + b + c = 1$ (see the general random walk in Section 11.2) and for the infinitesimal rate matrix $a + b + c = 0$ (see the birth and death process in Section 11.3), which only changes the eigenvalue through b.

A.6.2.2 Tri-diagonal AMS matrix

This section computes the exact spectrum of the tri-diagonal AMS matrix specified in (14.57). The analysis bears some resemblance to that of the birth and death process with constant birth and death rates (see Section 11.3.3).

The eigenvalue equation $D^{-1}Qx = \zeta x$ is rewritten for the j-th component of the right-eigenvector belonging to the eigenvalue ζ as

$$\lambda(N-j+1)\,x_{j-1} - [(\zeta+1-\lambda)j + N\lambda - \zeta c]\,x_j + (j+1)\,x_{j+1} = 0$$

for $0 \le j \le N$. This difference equation has linear coefficients whereas those in Section A.6.2.1 are constant. It is most conveniently solved using generating functions. Let $G(z) = \sum_{j=0}^{N} x_j z^j$, then the difference equation is transformed with $x_j = 0$ if $j \notin [0, N]$ to

$$\lambda N z\left(G(z) - x_N z^N\right) - \lambda z^2\left(G'(z) - N x_N z^{N-1}\right) - (\zeta+1-\lambda)zG'(z) - [N\lambda - \zeta c]G(z) + G'(z) = 0$$

from which the logarithmic derivative is

$$\frac{G'(z)}{G(z)} = \frac{\lambda N z + \zeta c - N\lambda}{\lambda z^2 + (\zeta+1-\lambda)z - 1}$$

The integration of the right-hand side requires a partial faction decomposition,

$$\frac{\lambda N z + \zeta c - N\lambda}{\lambda z^2 + (\zeta+1-\lambda)z - 1} = \frac{c_1}{z - r_1} + \frac{c_2}{z - r_2}$$

where r_1 and r_2 are the roots of the quadratic polynomial $\lambda z^2 + (\zeta + 1 - \lambda)\, z - 1$ and c_1 and c_2 are the residues computed for $k = 1, 2$ as

$$c_k = \lim_{z \to r_k} \frac{(z - r_k)\,(\lambda N z + \zeta c - N\lambda)}{\lambda\,(z - r_1)\,(z - r_2)}$$

and they obey $c_1 + c_2 = N$ and $c_1 r_2 + c_2 r_2 = \frac{\zeta c - N\lambda}{\lambda}$. Explicitly,

$$r_1 = \frac{-(\zeta + 1 - \lambda) + \sqrt{(\zeta + 1 - \lambda)^2 + 4\lambda}}{2\lambda} > 0$$

$$r_2 = \frac{-(\zeta + 1 - \lambda) - \sqrt{(\zeta + 1 - \lambda)^2 + 4\lambda}}{2\lambda} < 0 \tag{A.44}$$

with $r_1 r_2 = -\frac{1}{\lambda}$ and $r_1 + r_2 = -\frac{\zeta + 1 - \lambda}{\lambda}$. Moreover, unless $\zeta = \lambda - 1 \pm 2i\sqrt{\lambda}$ in which case $r_1 = r_2 = \pm\frac{i}{\sqrt{\lambda}}$, the roots are distinct. The residues are

$$c_1 = \frac{\lambda N r_1 + \zeta c - N\lambda}{\lambda\,(r_1 - r_2)} \tag{A.45}$$

$$c_2 = \frac{\lambda N r_2 + \zeta c - N\lambda}{\lambda\,(r_2 - r_1)} = N - c_1$$

Integration now yields

$$\log G(z) = c_1 \log(z - r_1) + c_2 \log(z - r_2) + b$$

or

$$G(z) = e^b\,(z - r_1)^{c_1}\,(z - r_2)^{N - c_1}$$

The integration constant b is obtained from $\lim_{z \to \infty} \frac{G(z)}{z^N} = x_N$. Thus,

$$\lim_{z \to \infty} \frac{G(z)}{z^N} = e^b \lim_{z \to \infty} \left(\frac{z - r_1}{z - r_2}\right)^{c_1} \left(1 - \frac{r_2}{z}\right)^N = e^b$$

such that $e^b = x_N$. The obvious scaling for the eigenvector is to choose $x_N = 1$ and we arrive at

$$G(z) = \sum_{j=0}^{N} x_j z^j = (z - r_1)^{c_1}\,(z - r_2)^{N - c_1} \tag{A.46}$$

which shows that c_1 must be an integer $k \in [0, N]$ for $G(z)$ to be a polynomial of degree N. Expanding the binomials with $c_1 = k$ gives

$$G(z) = \sum_{j=0}^{k} \binom{k}{j} z^j (-r_1)^{k-j} \sum_{n=0}^{N-k} \binom{N-k}{n} z^n (-r_2)^{N-k-n}$$

$$= \sum_{j=0}^{\infty} \sum_{n=0}^{j} \binom{k}{n}\binom{N-k}{j-n}(-r_1)^{k-j}(-r_2)^{N-k-j+n} z^j$$

from which the eigenvector components belonging to $\zeta \to \zeta(k)$ are, for $0 \leq j \leq N$,

$$x_j(k) = (-1)^{N-j} \sum_{n=0}^{j} \binom{k}{n}\binom{N-k}{j-n} r_1^{k-j} r_2^{N-k-j+n} \tag{A.47}$$

The requirement on c_1 also leads to equations for the eigenvalues ζ. Indeed, equating $c_1 = k$ in (A.45) and substituting the explicit expressions for the roots r_1 and r_2, we obtain after squaring the quadratic equations for the eigenvalue $\zeta(k)$ for $0 \leq k \leq N$

$$A(k)\,\zeta^2(k) + B(k)\,\zeta(k) + C(k) = 0 \tag{A.48}$$

where

$$A(k) = (N/2 - k)^2 - (N/2 - c)^2$$
$$B(k) = 2(1 - \lambda)(N/2 - k)^2 - N(1 + \lambda)(N/2 - c)$$
$$C(k) = -(1 + \lambda)^2 [(N/2)^2 - (N/2 - k)^2]$$

Each of the $N+1$ quadratic equations (A.48) has two roots $\zeta_1(k)$ and $\zeta_2(k)$, thus in total $2(N+1)$, while there are only $N+1$ eigenvalues. The coefficients $A(k)$, $B(k)$ and $C(k)$ only depend on k via $(N/2 - k)^2$, which means that the quadratics (A.48) for which $k' = N - k$ are identical. This observation reduces the set $\{\zeta_1(k), \zeta_2(k)\}_{0 \leq k \leq N}$ of roots to precisely $N+1$ and confines the analysis to $0 \leq k \leq N/2$. We will show that all roots are real and distinct (except for $k = N/2$). The discriminant $\Delta(k) = B^2(k) - 4A(k)C(k)$ is with $y = (N/2 - k)^2 \in [0, (N/2)^2]$,

$$\Delta(k) = -16\lambda y^2 + 4(1 + \lambda)(c^2(1 + \lambda) - 2\lambda cN + \lambda N^2) y$$

which shows that $\Delta(k)$ is concave in y because $\frac{d^2\Delta(k)}{dy^2} = -32\lambda < 0$, for $y = 0$, $\Delta(N/2) = 0$ and, for $y = (N/2)^2$, $\Delta(0) = N^2(c(1 + \lambda) - \lambda N)^2 > 0$ and, hence, $\Delta(k) \geq 0$ for $k \in [0, N/2]$. This means that, for $0 \leq k < N/2$, the roots $\zeta_1(k)$ and $\zeta_2(k)$ are real and distinct and, for $k = N/2$ (only if N is even) where $\Delta(N/2) = 0$,

$$\zeta_1(N/2) = \zeta_2(N/2) = \frac{-B(N/2)}{2A(N/2)} = -\frac{1 + \lambda}{1 - 2\frac{c}{N}}$$

For $\kappa < k \leq N/2$, the roots $\{\zeta_1(\kappa), \zeta_2(\kappa)\}$ are different from the roots $\{\zeta_1(k), \zeta_2(k)\}$ because $A(k)z^2 + B(k)z + C(k) < A(\kappa)z^2 + B(\kappa)z + C(\kappa)$ for all z. Indeed, $A(\kappa) - A(k) = (N/2 - \kappa)^2 - (N/2 - k)^2 > 0$ and the discriminant $(B(\kappa) - B(k))^2 - 4(A(\kappa) - A(k))(C(\kappa) - C(k)) < 0$ shows that there are no real solutions. Thus, an extreme eigenvalue occurs for $k = 0$ for which $C(0) = 0$ such that $\zeta_1(0) = 0$ and

$$\zeta_2(0) = -\frac{B(0)}{A(0)} = -\frac{1 + \lambda - \frac{\lambda N}{c}}{1 - \frac{c}{N}} \tag{A.49}$$

The stability requirement $\rho = \frac{N\lambda}{c(1+\lambda)} < 1$ and $c < N$ shows that $\zeta_2(0) < 0$, and thus $\zeta_2(0)$ is the largest negative eigenvalue. The eigenvalues for other $0 < k \leq N/2$ are either larger than 0 or smaller than $\zeta_2(0)$. We need to consider two different cases (a) $c < N/2$ and (b) $c > N/2$ while $C(k) < 0$ for all $k \in [0, N)$.

(a) If $c < N/2$ and if $0 \leq k < c$ and , then $A(k) > 0$. Hence, the product $\zeta_1(k)\zeta_2(k) = \frac{C(k)}{A(k)} < 0$ which means that $\zeta_1(k) > 0 > \zeta_2(k)$ and that there are precisely $[c]$ positive eigenvalues. Similarly, $A(k) < 0$ for $c < k < N/2$, such that $\zeta_1(k)\zeta_2(k) > 0$ while $\zeta_1(k) + \zeta_2(k) = -\frac{B(k)}{A(k)} < 0$ shows that both eigenvalues are negative because $B(k) < 0$. Indeed, if $\lambda \geq 1$ and $c < N/2$, the above expression immediately leads to $B(k) < 0$ while if $\lambda < 1$ and $c < N/2$, the expression

$$B(k) = 2(1 - \lambda)\left[\left(\frac{N}{2} - k\right)^2 - \left(\frac{N}{2} - c\right)^2\right] - 2c\left(\frac{N}{2} - c\right)\left[\frac{\lambda N}{c} + 1 - \lambda\right]$$

shows that both terms are negative.

(b) If $c > N/2$, we see that $A(k) > 0$ for $0 < k < N - c$ leading to $\zeta_1(k) > 0 > \zeta_2(k)$. For $N - c < k < N/2$, we have $A(k) < 0$ and thus $\zeta_1(k)\zeta_2(k) = \frac{C(k)}{A(k)} > 0$ while their same sign follows from $\zeta_1(k) + \zeta_2(k) = -\frac{B(k)}{A(k)}$ requires us to consider the sign of $B(k)$. If $\lambda \leq 1$, then $B(k) > 0$. If $\lambda > 1$, then

$$B(k) = N(1 + \lambda)\left(c - \frac{N}{2}\right) + 2(1 - \lambda)(N/2 - k)^2$$

$$< N(1 + \lambda)\left(c - \frac{N}{2}\right) + 2(1 - \lambda)\left(c - \frac{N}{2}\right)^2 = 2c\left(c - \frac{N}{2}\right)\left(\frac{\lambda(N - c)}{c} + 1\right) > 0$$

which shows that $0 < \zeta_2(k) < \zeta_1(k)$. Hence, there are $N - [c] + 2(N/2 - N + [c]) = [c]$ positive eigenvalues.

In summary, there are $[c]$ positive eigenvalues, one $\zeta_1(0) = 0$ and $N - [c]$ negative eigenvalues. Relabel the eigenvalues as $(\zeta_k, \zeta_{N-k}) = (\zeta_1(k), \zeta_2(k))$ in increasing order $\zeta_{N-[c]-1} < \cdots < \zeta_1 < \zeta_0 < \zeta_N = 0 < \zeta_{N-1} < \cdots < \zeta_{N-[c]}$. This way of writing distinguishes between underload and overload eigenvalues. In terms of the discriminant by $\Delta(k) = B^2(k) - 4A(k)C(k)$, the non-positive eigenvalues are

(a) If $c < N/2$,

$$\zeta_1(k) = \frac{-B(k) - \sqrt{\Delta(k)}}{2A(k)} \qquad 0 \le k \le [c]$$
$$\zeta_{1,2}(k) = \frac{-B(k) \mp \sqrt{\Delta(k)}}{2A(k)} \qquad [c] + 1 \le k \le \frac{N}{2}$$

(b) If $c > N/2$,

$$\zeta_1(k) = \frac{-B(k) - \sqrt{\Delta(k)}}{2A(k)} \qquad 0 \le k \le N - [c] - 1$$

The eigenvector belonging to ζ_j follows from (A.47) where r_1 and r_2 are given in (A.44) and k is determined from (A.45) since $k = c_1$. The eigenvectors for $\zeta_1(k)$ and $\zeta_2(k)$ belonging to a same quadratic k must be different. Especially in this case, the corresponding $k = c_1$ values can be determined from (A.45). For example, for $\zeta_N = 0$, we find $r_1 = 1$, $r_2 = -\frac{1}{\lambda}$ and $k = 0$ and the eigenvector belonging to ζ_N is with (A.47),

$$x_j(0) = (-1)^{N-j} \binom{N}{j} r_1^{-j} r_2^{N-j} = \binom{N}{j} \frac{\lambda^j}{\lambda^N} \tag{A.50}$$

After renormalization such that $\|x(0)\|_1 = 1$, i.e. by dividing each component by $\sum_{j=0}^{N} x_j(0) = \frac{1}{\lambda^N} \sum_{j=0}^{N} \binom{N}{j} \lambda^j = \frac{(1+\lambda)^N}{\lambda^N}$, the steady-state vector (14.58) is obtained. Similarly, for the largest negative eigenvalue ζ_0 in (A.49), we find with $r_1 = 1 - \frac{N}{c}$, $r_2 = \frac{1}{\lambda(\frac{N}{c}-1)}$ and $k = c_1 = N$ such that

$$x_j(N) = (-1)^{N-j} \binom{N}{j} r_1^{N-j} r_2^0 = \binom{N}{j} \left(\frac{N}{c} - 1\right)^{N-j} \tag{A.51}$$

The left-eigenvectors y satisfy (A.10): $y^T D^{-1} Q = \zeta y^T$. The above approach is applicable. However, there is a more elegant method based on the observation that there exists a diagonal matrix $W = diag(W_0, \ldots, W_N)$ for which $W^{-1}QW = (W^{-1}QW)^T$, namely $W_j = \sqrt{\binom{N}{j}} \lambda^j$. Since $W^{-1}QW$ is symmetric, the left- and right-eigenvectors corresponding to the same eigenvalue are the same (**art.** 11). Now $y^T D^{-1} Q = \zeta y^T$ is equivalent to

$$\zeta y^T W = y^T W W^{-1} D^{-1} W W^{-1} Q W = y^T W \left(W^{-1} D W\right)^{-1} \left(W^{-1} Q W\right)$$

With $y_W^T = y^T W$, $D_W = W^{-1} D W$ and $Q_W = W^{-1} Q W = Q_W^T$, we obtain $y_W^T D_W^{-1} Q_W = \zeta y_W^T$. The transpose $\zeta y_W = Q_W D_W^{-1} y_W$ is

$$\zeta W^2 y = Q D^{-1} W^2 y$$

which shows, compared to $D^{-1} Q x = \zeta x$, that $x = W^2 y$ whose vector components are, for $0 \le j \le N$,

$$x_j = \binom{N}{j} \lambda^j y_j \tag{A.52}$$

A.6.3 General tri-diagonal matrices

Here, we consider the general tri-diagonal matrix (11.1) and show how orthogonal polynomials enter the scene. While our approach is more algebraic, Karlin and McGregor (1959) present a different, more probabilistic and function-theoretic method, which is reviewed and complemented by van Doorn and Schrijner (1995).

A.6.3.1 A similarity transform

A similarity transform analogous to the Jacobi matrix for orthogonal polynomials (Van Mieghem, 2011, Section 10.6) is deduced. If there exists a similarity transform that makes the matrix P symmetric, then all eigenvalues of P are real, because a similarity transform (**art.** 4) preserves the eigenvalues. The simplest similarity transform is $H = \text{diag}(h_1, h_2, \ldots, h_{N+1})$ such that $\widetilde{P} = HPH^{-1}$ becomes

$$\widetilde{P} = \begin{bmatrix} r_0 & \frac{h_1}{h_2}p_0 & 0 & 0 & \cdots & 0 & 0 & 0 \\ \frac{h_2}{h_1}q_1 & r_1 & \frac{h_2}{h_3}p_1 & 0 & \cdots & 0 & 0 & 0 \\ 0 & \frac{h_3}{h_2}q_2 & r_2 & \frac{h_3}{h_4}p_2 & \cdots & 0 & 0 & 0 \\ \vdots & \vdots & \vdots & \vdots & \vdots & \vdots & \vdots & \vdots \\ 0 & 0 & 0 & 0 & \cdots & \frac{h_N}{h_{N-1}}q_{N-1} & r_{N-1} & \frac{h_N}{h_{N+1}}p_{N-1} \\ 0 & 0 & 0 & 0 & \cdots & 0 & \frac{h_{N+1}}{h_N}q_N & r_N \end{bmatrix}$$

Thus, in order to produce a symmetric matrix $\widetilde{P} = \widetilde{P}^T$, we need to require that $\left(\widetilde{P}\right)_{i,i-1} = \left(\widetilde{P}\right)_{i-1,i}$ for all $1 \le i \le N$, implying that $\frac{h_{i+1}}{h_i}q_i = \frac{h_i}{h_{i+1}}p_{i-1}$, whence

$$\left(\frac{h_{i+1}}{h_i}\right)^2 = \frac{p_{i-1}}{q_i}$$

Assuming that all p_i and q_i are positive[2], we find that $h_{i+1} = \sqrt{\frac{p_{i-1}}{q_i}}h_i$ for $1 \le i \le N$ and we can choose $h_1 = 1$ such that

$$h_i = \sqrt{\prod_{k=1}^{i-1}\frac{p_{k-1}}{q_k}} \tag{A.53}$$

and

$$\frac{h_{i+1}}{h_i}q_i = \frac{h_i}{h_{i+1}}p_{i-1} = \sqrt{p_{i-1}q_i} = \eta_i$$

---

[2] If $p_i = 0$ (or $q_i = 0$), then an expansion of the determinant of P along the i-th row may be useful. We further omit these more complicated cases.

After the similarity transform H, the matrix $\widetilde{P}$ becomes

$$\widetilde{P} = \begin{bmatrix} r_0 & \sqrt{p_0 q_1} & 0 & 0 & \cdots & 0 & 0 & 0 \\ \sqrt{p_0 q_1} & r_1 & \sqrt{p_1 q_2} & 0 & \cdots & 0 & 0 & 0 \\ 0 & \sqrt{p_1 q_2} & r_2 & \sqrt{p_2 q_3} & \cdots & 0 & 0 & 0 \\ \vdots & \vdots & \vdots & \vdots & \vdots & \vdots & \vdots & \vdots \\ 0 & 0 & 0 & 0 & \cdots & \sqrt{p_{N-2} q_{N-1}} & r_{N-1} & \sqrt{p_{N-1} q_N} \\ 0 & 0 & 0 & 0 & \cdots & 0 & \sqrt{p_{N-1} q_N} & r_N \end{bmatrix}$$

$$\text{(A.54)}$$

In conclusion, if all p_i and q_i are positive, then all eigenvalues of P are real. Rather than solving the eigenvector $\widetilde{x}$ from the eigenvalue equation $\widetilde{P}\widetilde{x} = \lambda \widetilde{x}$, we determine the eigenvector x as a function of λ from the original matrix P for reasons explained below and use the similarity transform $\widetilde{x} = Hx$, where H is independent of λ, later for the left-eigenvectors of P.

A.6.3.2 Eigenvectors of P

The right-eigenvector x of P belonging to eigenvalue λ satisfies $(P - \lambda I)x = 0$ so that

$$\begin{cases} (r_0 - \lambda) x_0 + p_0 x_1 = 0 \\ q_j x_{j-1} + (r_j - \lambda) x_j + p_j x_{j+1} = 0 & 1 \le j < N \\ q_N x_{N-1} + (r_N - \lambda) x_N = 0 \end{cases}$$

We replace the last equation, that breaks the structure, by

$$q_N x_{N-1} + (r_N - \lambda) x_N + p_N x_{N+1} = 0$$

and the condition that $p_N x_{N+1} = 0$. Using $r_j = 1 - q_j - p_j$ for $0 \le j \le N$ with $q_0 = 0$ and $p_N = 0$ and making the dependence on $\xi = \lambda - 1$ explicit, the above set simplifies, subject to the condition $p_N x_{N+1} (\xi) = 0$, to

$$\begin{cases} x_1 (\xi) = \frac{p_0 + \xi}{p_0} x_0 (\xi) \\ x_{j+1} (\xi) = \frac{p_j + q_j + \xi}{p_j} x_j (\xi) - \frac{q_j}{p_j} x_{j-1} (\xi) & 1 \le j < N \end{cases} \qquad \text{(A.55)}$$

For the stochastic matrix P, which obeys $Pu = u$, it holds that $r_N = 1 - q_N$ so that $p_N = 0$ and that the condition $p_N x_{N+1} = 0$ seems to be obeyed. In the theory of orthogonal polynomials (Van Mieghem, 2011, Chapter 10), a similar trick is used where the orthogonal polynomial $x_{N+1} (\xi)$ needs to vanish, because p_N is not necessarily zero in absence of the stochasticity requirement $Pu = u$. The zeros of the orthogonal polynomial $x_{N+1} (\xi)$ are then equal to the eigenvalues of the corresponding Jabobi matrix. Moreover, the powerful interlacing property for the zeros of the set $\{x_j (\xi)\}_{0 \le j \le N+1}$ applies. We will return to the condition $p_N x_{N+1} = 0$ below.

Solving (A.55) iteratively for $j < N$,

$$x_2(\xi) = \frac{x_0(\xi)}{p_0 p_1} \left(\xi^2 + (q_1 + p_1 + p_0)\xi + p_1 p_0\right)$$

$$x_3(\xi) = \frac{x_0(\xi)}{p_2 p_1 p_0} \left(\xi^3 + (q_1 + q_2 + p_2 + p_1 + p_0)\xi^2\right)$$
$$+ (q_2 q_1 + q_2 p_0 + p_2 q_1 + p_2 p_1 + p_2 p_0 + p_1 p_0)\xi + p_2 p_1 p_0$$

reveals that $\frac{x_j(\xi)}{x_0(\xi)}$ is a polynomial of degree j in ξ with positive coefficients, whose zeros are all non-positive[3]. This simple form is the main reason to consider the eigenvector components of P instead of $\tilde{P}$. By inspection, the general form of $x_j(\xi)$ for $1 \le j \le N$ is

$$x_j(\xi) = \frac{x_0(\xi)}{\prod_{m=0}^{j-1} p_m} \sum_{k=0}^{j} c_k(j)\xi^k \qquad (A.56)$$

with[4]

$$c_j(j) = 1; \quad c_{j-1}(j) = \sum_{m=0}^{j-1}(p_m + q_m); \quad c_0(j) = \prod_{m=0}^{j-1} p_m; \qquad (A.57)$$

where $q_0 = p_N = 0$. By substituting (A.56) into (A.55),

$$\sum_{k=1}^{j-1} c_k(j+1)\xi^k = \sum_{k=1}^{j-1} \left[(q_j + p_j)c_k(j) - q_j p_{j-1} c_k(j-1) + c_{k-1}(j)\right]\xi^k$$

and equating the corresponding powers in ξ, a recursion relation for the coefficients $c_k(j)$ for $0 \le k < j$ is obtained with $c_j(j) = 1$,

$$c_k(j+1) = (q_j + p_j)c_k(j) - q_j p_{j-1} c_k(j-1) + c_{k-1}(j) \qquad (A.58)$$

from which all coefficients can be determined as shown in Appendix A.6.3.3. The stochasticity requirement $Pu = u$ implies that the right-eigenvector belonging to the largest eigenvalue $\lambda = 1$, equivalently to $\xi = \lambda - 1 = 0$, equals $x(0) = u$.

We now express the left-eigenvectors of P in terms of the right-eigenvector by using the similarity transform H. Indeed, since $\tilde{P}$ is symmetric, the left- and right-eigenvectors are the same (**art. 11**). The left-eigenvector x of P equals $x = H^{-1}\tilde{x}$, while the right-eigenvector y of P equals $y = H\tilde{x}$ (**art. 4**). Hence, we find that $y = H^2 x$ and explicitly with (A.56) and (A.53),

$$y_j(\xi) = \frac{y_0(\xi)}{\prod_{m=0}^{j-1} q_{m+1}} \sum_{k=0}^{j} c_k(j)\xi^k \qquad (A.59)$$

For any matrix, the left- and right-eigenvectors obey the orthogonality equation

$$x^T(\xi) y(\xi') = x^T(\xi) y(\xi)\delta_{\xi\xi'} \qquad (A.60)$$

---

[3] If P is a stochastic, irreducible matrix, then the Perron-Frobenius Theorem A.5.2 states that the largest (in absolute value) eigenvalue is one, hence $-1 < \lambda = \xi + 1 \le 1$.

[4] We use the convention that $\sum_{k=a}^{b} f(k) = 0$ and $\prod_{k=a}^{b} f(k) = 1$ if $a > b$.

that holds for any pair of eigenvalues $\lambda = \xi + 1$ and $\lambda' = \xi' + 1$ of that matrix (**art. 2**). For symmetric matrices, usually, the normalization

$$\widetilde{x}^T(\xi)\,\widetilde{x}(\xi') = \delta_{\xi\xi'} \tag{A.61}$$

is chosen, which implies, using the transform $H = \operatorname{diag}(h_i)$, that $x^T(\xi)\,y(\xi') = \widetilde{x}^T(\xi)\,\widetilde{x}(\xi') = \delta_{\xi\xi'}$, $x^T(\xi)\,H^2 x(\xi') = \delta_{\xi\xi'}$ and $y^T(\xi)\,H^{-2} y(\xi') = \delta_{\xi\xi'}$. These normalizations of the eigenvector components imply, using (A.56) and (A.59) and with the definition of the polynomial for $0 \le j \le N+1$

$$\rho_j(\xi) = \sum_{k=0}^{j} c_k(j)\xi^k \tag{A.62}$$

that

$$x_0(\xi)\,y_0(\xi) = x_0^2(\xi) = y_0^2(\xi) = \left(1 + \sum_{j=1}^{N} \frac{\rho_j^2(\xi)}{\prod_{m=0}^{j-1} q_{m+1} p_m}\right)^{-1} \tag{A.63}$$

The orthogonality equation (A.60) leads, together with our choice of normalization $\widetilde{x}^T(\xi)\,\widetilde{x}(\xi') = \delta_{\xi\xi'}$, to a couple of important consequences.

Using the general form (A.56), the initially made condition $p_N x_{N+1} = 0$ translates to

$$p_N x_{N+1}(\xi) = \frac{x_0(\xi)\,\rho_{N+1}(\xi)}{\prod_{m=0}^{N-1} p_m} = 0$$

Since $\rho_j(\xi)$ is a polynomial, (A.63) indicates that neither $x_0(\xi)$ nor $y_0(\xi)$ can vanish for finite ξ so that the initial condition is met provided

$$\rho_{N+1}(\xi) = 0 \tag{A.64}$$

which closely corresponds to results in the theory of orthogonal polynomials. Thus, $\rho_j(\xi)$ should be considered as orthogonal polynomial, rather than $x_j(\xi)$ due to the scaling of $x_0(\xi)$, defined in (A.63). For the set of orthogonal polynomials $\{\rho_j(\xi)\}_{0 \le j \le N+1}$ *interlacing* applies, which means that the zeros of $\rho_j(\xi)$ interlace with those of $\rho_l(\xi)$ for all $1 \le l \ne j \le N+1$. Moreover, the eigenvalues of P are equal to the zeros of $\rho_{N+1}(\xi)$ in (A.64).

For stochastic matrices, the left-eigenvector $y(0)$ belonging to $\xi = 0$ equals the steady-state vector π (see Appendix A.5.1). For $\xi = 0$, the orthogonality relation (A.60) becomes $u^T y(\xi') = 0$ and $u^T y(0) = u^T \pi = 1$, from which the j-th component in (A.59) of the left-eigenvector $y(0) = \pi$ follows, for $1 \le j \le N$, as

$$\pi_j = y_j(0) = \frac{y_0(0)\,c_0(j)}{\prod_{m=0}^{j-1} q_{m+1}} = \frac{\prod_{m=0}^{j-1} \frac{p_m}{q_{m+1}}}{1 + \sum_{k=1}^{N} \prod_{m=0}^{k-1} \frac{p_m}{q_{m+1}}}$$

which precisely equal the well-known steady-state probabilities (11.6) of the generalized random walk. For $\xi \ne 0$, the orthogonality relation (A.60) and the fact that $y_0(\xi)$ is non-zero for finite ξ imply that

$$0 = 1 + \sum_{j=1}^{N} \frac{1}{\prod_{m=0}^{j-1} q_{m+1}} \sum_{k=0}^{j} c_k(j)\xi^k = 1 + \sum_{j=1}^{N} \prod_{m=0}^{j-1} \frac{p_m}{q_{m+1}} + \sum_{k=1}^{N} \left(\sum_{j=k}^{N} \frac{c_k(j)}{\prod_{m=0}^{j-1} q_{m+1}} \right) \xi^k$$

We write the right-hand side polynomial as

$$\sum_{k=0}^{N} f_k \xi^k = f_N \prod_{k=1}^{N} (\xi - z_k) \tag{A.65}$$

where $f_0 = \frac{1}{\pi_0}$ by (11.6) and where, for $k > 0$,

$$f_k = \sum_{j=k}^{N} \frac{c_k(j)}{\prod_{m=0}^{j-1} q_{m+1}} \tag{A.66}$$

and, explicitly,

$$f_N = \frac{1}{\prod_{m=0}^{N-1} q_{m+1}}$$

$$f_{N-1} = \frac{1}{\prod_{m=0}^{N-2} q_{m+1}} + \frac{\sum_{m=0}^{N-1}(p_m + q_m)}{\prod_{m=0}^{N-1} q_{m+1}}$$

Relation (A.66) illustrates that all coefficients f_k are non-negative. Moreover, the orthogonality relation (A.60) implies that the polynomial $\sum_{k=0}^{N} f_k \xi^k$ possesses the same zeros as $\frac{c_P(\xi)}{\xi}$, where $c_P(\xi) = \det(P - (\xi+1)I) = \xi \prod_{k=1}^{N}(z_k - \xi)$ is the $N+1$ degree characteristic polynomial of the matrix P, in particular,

$$\frac{c_P(\xi)}{(-1)^N \xi} = \frac{1}{f_N} \sum_{k=0}^{N} f_k \xi^k \tag{A.67}$$

Finally, since the eigenvalues of P also obey (A.64) so that

$$\frac{c_P(\xi)}{(-1)^N \xi} = \frac{1}{f_N} \sum_{k=0}^{N} f_k \xi^k = \sum_{k=1}^{N+1} c_k(N+1)\xi^{k-1}$$

after equating corresponding powers in ξ, we find that

$$c_{k+1}(N+1) = \frac{f_k}{f_N} = \sum_{j=k}^{N} c_k(j) \prod_{m=j}^{N-1} q_{m+1} \tag{A.68}$$

In summary, the stochasticity property of P provides us with an additional relation (A.68) on the coefficients of $c_P(\xi)$, that is not necessarily obeyed for general orthogonal polynomials.

A.6.3.3 Solving the recursion (A.58)

We now propose two different types of solutions of the recursion (A.58) for the coefficients $c_k(j)$ of $x_j(\xi)$ in (A.56).

Theorem A.6.1 *A recursion relation for $c_{j-m}(j)$, valid for $2 \leq m \leq j$, is*

$$c_{j-m}(j) = \sum_{l=0}^{j-1} \left((q_l + p_l)\, c_{l-m+1}(l) - q_l p_{l-1} c_{l-m+1}(l-1) \right) \qquad (A.69)$$

Proof: Letting $k = j - m$ in (A.58) yields

$$c_{j+1-(m+1)}(j+1) = (q_j + p_j)\, c_{j-m}(j) - q_j p_{j-1} c_{j-1-(m-1)}(j-1) + c_{j-(m+1)}(j)$$

With $t_m(j) = c_{j-m}(j)$, the above equation transforms into the difference equation

$$t_{m+1}(j+1) = t_{m+1}(j) + (q_j + p_j)\, t_m(j) - q_j p_{j-1} t_{m-1}(j-1)$$

whose solution is

$$t_{m+1}(j) = \sum_{l=0}^{j-1} \left((q_l + p_l)\, t_m(l) - q_l p_{l-1} t_{m-1}(l-1) \right)$$

With the initial values $t_0(j) = 1$ and $t_1(j) = \sum_{m=0}^{j-1}(p_m + q_m)$ from (A.57), all $t_m(j)$ can be iteratively found from (A.69). $\qquad\square$

Thus, letting $m = 2$ in (A.69) yields

$$c_{j-2}(j) = \sum_{l=0}^{j-1} \left((q_l + p_l) \sum_{m=0}^{l-1}(p_m + q_m) - q_l p_{l-1} \right) \qquad (A.70)$$

Next, for $m = 3$ in (A.69), we have

$$c_{j-3}(j) = \sum_{l=0}^{j-1} (q_l + p_l) \sum_{l_1=0}^{l-1} \left((q_{l_1} + p_{l_1}) \sum_{m=0}^{l_1-1}(p_m + q_m) - q_{l_1} p_{l_1-1} \right)$$
$$- \sum_{l=0}^{j-1} q_l p_{l-1} \sum_{m=0}^{l-2}(p_m + q_m)$$

and so on.

For the polynomial in (A.65), the next general expression will prove more useful.

Theorem A.6.2 *The explicit general expression for the coefficients $c_k(j)$ in terms of $c_{k-1}(l)$ for all $l \geq k - 1$ is*

$$c_k(j) = \prod_{m=k}^{j-1} p_m + \sum_{l=0}^{j-k-1} \prod_{m=k}^{j-l-1} q_m \prod_{m=j-l}^{j-1} p_m + \sum_{m=0}^{k-1}(p_m+q_m) \sum_{l=0}^{j-k-1} \prod_{m=k+1}^{j-l-1} q_m \prod_{m=j-l}^{j-1} p_m$$
$$+ \sum_{l=0}^{j-k-1} \sum_{s=1}^{j-l-k-1} c_{k-1}(j-l-s) \prod_{m=j-l+1-s}^{j-l-1} q_m \prod_{m=j-l}^{j-1} p_m \qquad (A.71)$$

Proof: Rewriting (A.58) as

$$c_k(j+1) - p_j c_k(j) = q_j \{c_k(j) - p_{j-1} c_k(j-1)\} + c_{k-1}(j)$$

and defining $b_k(j) = c_k(j) - p_{j-1} c_k(j-1)$ shows that the second-order recursion (A.58) in j can be decomposed into two first-order recursions in j

$$\begin{cases} c_k(j) = p_{j-1} c_k(j-1) + b_k(j) \\ b_k(j) = q_{j-1} b_k(j-1) + c_{k-1}(j-1) \end{cases}$$

Since $k < j$, the choice for $j = k+1$ yields

$$b_k(k+1) = c_k(k+1) - p_k c_k(k)$$

$$= \sum_{m=0}^{k} (p_m + q_m) - p_k = q_k + \sum_{m=0}^{k-1} (p_m + q_m)$$

Iterating the first recursion downwards yields

$$c_k(j) = p_{j-1} p_{j-2} c_k(j-2) + p_{j-1} b_k(j-1) + b_k(j)$$
$$= p_{j-1} p_{j-2} p_{j-3} c_k(j-3) + p_{j-1} p_{j-2} b_k(j-2) + p_{j-1} b_k(j-1) + b_k(j)$$

from which we deduce that

$$c_k(j) = c_k(j-p) \prod_{m=j-p}^{j-1} p_m + \sum_{l=0}^{p-1} b_k(j-l) \prod_{m=j-l}^{j-1} p_m$$

When $j - p = k$, then $c_k(k) = 1$ and thus

$$c_k(j) = \prod_{m=k}^{j-1} p_m + \sum_{l=0}^{j-k-1} b_k(j-l) \prod_{m=j-l}^{j-1} p_m \tag{A.72}$$

Similarly, we iterate the second recursion downwards,

$$b_k(j) = q_{j-1} q_{j-2} b_k(j-2) + q_{j-1} c_{k-1}(j-2) + c_{k-1}(j-1)$$
$$= q_{j-1} q_{j-2} q_{j-3} b_k(j-3) + q_{j-1} q_{j-2} c_{k-1}(j-3) + q_{j-1} c_{k-1}(j-2) + c_{k-1}(j-1)$$

which suggests that

$$b_k(j) = b_k(j-p) \prod_{m=j-p}^{j-1} q_m + \sum_{l=1}^{p} c_{k-1}(j-l) \prod_{m=j+1-l}^{j-1} q_m$$

For $j - p = k+1$ or $p = j - k - 1$, we have

$$b_k(j) = b_k(k+1) \prod_{m=k+1}^{j-1} q_m + \sum_{l=1}^{j-k-1} c_{k-1}(j-l) \prod_{m=j+1-l}^{j-1} q_m$$

$$= \prod_{m=k}^{j-1} q_m + \prod_{m=k+1}^{j-1} q_m \sum_{m=0}^{k-1} (p_m + q_m) + \sum_{l=1}^{j-k-1} c_{k-1}(j-l) \prod_{m=j+1-l}^{j-1} q_m \tag{A.73}$$

Combining (A.72) and (A.73) yields (A.71). □

For $k = 1$ and using $c_0(j) = \prod_{m=0}^{j-1} p_m$, we find from (A.71) that

$$c_1(j) = \sum_{l=0}^{j-1} \sum_{s=0}^{j-1-l} \prod_{m=0}^{j-2-l-s} p_m \prod_{m=j-l-s}^{j-1-l} q_m \prod_{m=j-l}^{j-1} p_m \tag{A.74}$$

Introducing the expression (A.74) for $c_1(j)$ into (A.71) produces the explicit form for $c_2(j)$,

$$c_2(j) = \prod_{m=2}^{j-1} p_m + (p_0 + p_1 + q_1 + q_2) \sum_{l=0}^{j-3} \prod_{m=3}^{j-l-1} q_m \prod_{m=j-l}^{j-1} p_m$$

$$+ \sum_{l=0}^{j-3} \sum_{s=1}^{j-l-3} \sum_{l_1=0}^{j-l-s-1} \sum_{l_2=0}^{j-l-s-l_1-1} \prod_{m=0}^{j-l-s-2-l_1-l_2} p_m \prod_{m=j-l-s-l_1-l_2}^{j-l-s-1-l_1} q_m$$

$$\times \prod_{m=j-l-s-l_1}^{j-l-s-1} p_m \prod_{m=j-l+1-s}^{j-l-1} q_m \prod_{m=j-l}^{j-1} p_m \tag{A.75}$$

and so on. In this way, all coefficients $c_k(j)$ in the polynomial (A.56) can be explicitly determined[5]. Since all p_j and q_j are probabilities and thus non-negative, the recursion (A.71) together with $c_0(j) = \prod_{m=0}^{j-1} p_m$ illustrates that all coefficients $c_k(j)$ are non-negative.

A.6.3.4 The Christoffel-Darboux formula for eigenvectors of P

We derive the Christoffel-Darboux formula (Van Mieghem, 2011, p. 357) for the matrix P. Indeed, multiply the equation for $x_{j+1}(\xi)$ in (A.55) by $x_j(\omega)$

$$p_j x_{j+1}(\xi) x_j(\omega) = \xi x_j(\xi) x_j(\omega) + (p_j + q_j) x_j(\xi) x_j(\omega) - q_j x_{j-1}(\xi) x_j(\omega)$$

Letting $\xi \to \omega$ in (A.55) and multiply both sides by $x_j(\xi)$,

$$p_j x_{j+1}(\omega) x_j(\xi) = \omega x_j(\xi) x_j(\omega) + (p_j + q_j) x_j(\xi) x_j(\omega) - q_j x_j(\xi) x_{j-1}(\omega)$$

Subtracting both equation yields,

$$(\xi - \omega) x_j(\xi) x_j(\omega) = p_j \{x_{j+1}(\xi) x_j(\omega) - x_{j+1}(\omega) x_j(\xi)\}$$
$$+ q_j \{x_{j-1}(\xi) x_j(\omega) - x_j(\xi) x_{j-1}(\omega)\}$$

Now, we transform to $x_j(\xi) = \frac{\widetilde{x}_j(\xi)}{h_{j+1}}$,

$$\frac{(\xi - \omega)}{h_{j+1}^2} \widetilde{x}_j(\xi) \widetilde{x}_j(\omega) = \frac{p_j}{h_{j+2} h_{j+1}} \{\widetilde{x}_{j+1}(\xi) \widetilde{x}_j(\omega) - \widetilde{x}_{j+1}(\omega) \widetilde{x}_j(\xi)\}$$

$$+ \frac{q_j}{h_j h_{j+1}} \{\widetilde{x}_{j-1}(\xi) \widetilde{x}_j(\omega) - \widetilde{x}_{j-1}(\omega) \widetilde{x}_j(\xi)\}$$

Using (A.53) shows that $\frac{p_j}{h_{j+2} h_{j+1}} = \frac{\sqrt{p_j q_{j+1}}}{h_{j+1}^2}$ and $\frac{q_j}{h_j h_{j+1}} = \frac{\sqrt{p_{j-1} q_j}}{h_{j+1}^2}$ so that

$$g_{j+1} - g_j = (\xi - \omega) \widetilde{x}_j(\xi) \widetilde{x}_j(\omega)$$

where $g_j = \sqrt{p_{j-1} q_j} \{\widetilde{x}_{j-1}(\omega) \widetilde{x}_j(\xi) - \widetilde{x}_{j-1}(\xi) \widetilde{x}_j(\omega)\}$. Summing over $j \in [0, m]$,

$$(\xi - \omega) \sum_{j=0}^{m} \widetilde{x}_j(\xi) \widetilde{x}_j(\omega) = \sum_{j=0}^{m} g_{j+1} - \sum_{j=0}^{m} g_j = g_{m+1} - g_0$$

---

[5] For $j = k$ in (A.71), we find indeed that $c_k(k) = 1$ (based on our convention).

where $g_0 = 0$ because $\widetilde{x}_{-1} = 0$. Hence, we arrive at the Christoffel-Darboux sum for the eigenvectors of $\widetilde{P}$,

$$(\xi - \omega) \sum_{j=0}^{m} \widetilde{x}_j (\xi) \, \widetilde{x}_j (\omega) = \sqrt{p_m q_{m+1}} \left\{ \widetilde{x}_m (\omega) \, \widetilde{x}_{m+1} (\xi) - \widetilde{x}_m (\xi) \, \widetilde{x}_{m+1} (\omega) \right\}$$

which extends the orthogonality relation (A.61). Transformed back to $x_j (\xi)$ using (A.53) yields

$$(\xi - \omega) \sum_{j=0}^{m} h_{j+1}^2 x_j (\xi) \, x_j (\omega) = p_m h_{m+1}^2 \left\{ x_m (\omega) \, x_{m+1} (\xi) - x_m (\xi) \, x_{m+1} (\omega) \right\}$$

$$(\text{A.76})$$

Since $\omega = 0$ is an eigenvalue with corresponding eigenvector $x(0) = \frac{1}{N+1} u$, each other real eigenvalue $\xi \neq 0$ must obey

$$\xi \sum_{j=0}^{m} h_{j+1}^2 x_j (\xi) = p_m h_{m+1}^2 \left\{ x_{m+1} (\xi) - x_m (\xi) \right\}$$

Taking $p_N = 0$ into account, the Christoffel-Darboux formula (A.76) extends (A.61) to all $0 \leq m \leq N$.

A.6.4 A triangular matrix complemented with one subdiagonal

The transition probability matrix P has the structure of a triangular matrix complemented with one subdiagonal,

$$P = \begin{bmatrix} P_{00} & P_{01} & P_{02} & \cdots & \cdots & P_{0N} \\ P_{10} & P_{11} & P_{12} & \cdots & \cdots & P_{1N} \\ 0 & P_{21} & P_{22} & \cdots & \cdots & P_{2N} \\ \vdots & \vdots & \vdots & \cdots & \vdots & \vdots \\ 0 & 0 & \cdots & P_{N-1,N-2} & P_{N-1;N-1} & P_{N-1;N} \\ 0 & 0 & \cdots & 0 & P_{N;N-1} & P_{NN} \end{bmatrix}$$

Besides the normalization $\|\pi\|_1 = 1$, the steady-state vector π obeys the relation $\pi = \pi.P$, or per vector component (9.29),

$$\pi_j = \sum_{k=0}^{j+1} P_{kj} \pi_k$$

because $P_{kj} = 0$ if $k > j + 1$. Immediately we obtain an iterative equation that expresses π_{j+1} (for $j < N$) in terms of the π_k for $0 \leq k \leq j$ as

$$\pi_{j+1} = \left(\frac{1 - P_{jj}}{P_{j+1;j}} \right) \pi_j - \sum_{k=0}^{j-1} \frac{P_{kj}}{P_{j+1;j}} \pi_k$$

Let us consider the eigenvalue equation (A.1) that is written for stochastic matrices as $(P - \lambda I)^T x^T = 0$. The matrix $(P - \lambda I)^T$ is a $(N + 1) \times (N + 1)$ matrix of rank N because $\det(P - \lambda I)^T = 0$ (else all eigenvectors x are zero). Hence, one equation in the set $(P - \lambda I)^T x^T = 0$ can be omitted. When writing this set of equations in terms of x_0, we produce the following set of N equations,

$$
\begin{bmatrix}
P_{10} & 0 & 0 & \cdots & 0 & 0 \\
P_{11} - \lambda & P_{21} & 0 & \cdots & 0 & 0 \\
P_{12} & P_{22} - \lambda & P_{32} & \cdots & & \vdots \\
\cdots & \cdots & \cdots & \ddots & \cdots & \cdots \\
\vdots & \vdots & \vdots & \cdots & P_{N-1;N-2} & 0 \\
P_{1;N-1} & P_{2;N-1} & P_{3;N-1} & \cdots & P_{N-1;N-1} - \lambda & P_{N;N-1}
\end{bmatrix}
\cdot
\begin{bmatrix}
x_1 \\ x_2 \\ x_3 \\ \vdots \\ x_{N-1} \\ x_N
\end{bmatrix}
=
\begin{bmatrix}
\lambda - P_{00} \\ -P_{01} \\ -P_{02} \\ \vdots \\ -P_{0;N-2} \\ -P_{0;N-1}
\end{bmatrix}
x_0
$$

where the $(N+1)$-th equation is omitted. Since the right-hand side matrix is a triangular matrix, the determinant equals the product of the diagonal elements or $\prod_{k=0}^{N-1} P_{k+1;k}$. By Cramer's rule, we find that

$$
\frac{x_j}{x_0} = \frac{\det
\begin{bmatrix}
P_{10} & 0 & \cdots & 0 & \lambda - P_{00} & 0 & \cdots & 0 \\
P_{11} - \lambda & P_{21} & \cdots & 0 & -P_{01} & 0 & \cdots & 0 \\
P_{12} & P_{22} - \lambda & \ddots & \vdots & \vdots & \vdots & \cdots & \vdots \\
\vdots & \vdots & \cdots & P_{j-1;j-2} & -P_{0,j-2} & \vdots & \cdots & \vdots \\
\vdots & \vdots & \cdots & P_{j-1;j-1} - \lambda & -P_{0;j-1} & 0 & \cdots & \vdots \\
P_{1j} & P_{2j} & \cdots & P_{j-1;j} & -P_{0j} & P_{j+1;j} & \cdots & \vdots \\
\vdots & \vdots & \vdots & \vdots & \vdots & \vdots & \ddots & 0 \\
P_{1;N-1} & P_{2;N-1} & \cdots & P_{j-1;N-1} & -P_{0;N-1} & P_{j+1;N-1} & \cdots & P_{N;N-1}
\end{bmatrix}}{\prod_{k=0}^{N-1} P_{k+1;k}}
$$

The above determinant is of the form (Meyer, 2000, p. 467)

$$
\det
\begin{bmatrix}
A_{j \times j} & O_{j \times N-j} \\
B_{N-j \times j} & C_{N-j \times N-j}
\end{bmatrix}
= \det C \det A
$$

where $\det C = \prod_{k=j}^{N-1} P_{k+1;k}$. In the determinant $\det A$, we can change the j-th column with the $(j-1)$-th, and subsequently, the $(j-1)$-th with the $(j-2)$-th and so on until the last column is permuted to the first column, in total $j - 1$ permutations. After changing the sign of that first column, the result is that $\det A = (-1)^j \det (P_{j \times j} - \lambda I_{j \times j})$, where $P_{j \times j}$ is the original transition probability matrix limited to j states (instead of $N + 1$). Hence, for $1 \leq j \leq N$,

$$
x_j = \frac{x_0 (-1)^j \det (P_{j \times j} - \lambda I_{j \times j})}{\prod_{k=0}^{j-1} P_{k+1;k}}
$$

and the normalization of eigenvectors $\|x\|_1 = 1$ determines x_0 as

$$
x_0 = \frac{1}{1 + \sum_{j=1}^{N} \frac{(-1)^j \det(P_{j \times j} - \lambda I_{j \times j})}{\prod_{k=0}^{j-1} P_{k+1;k}}}
$$

If the $N + 1$ eigenvalues λ are known, we observe that all eigenvectors can be expressed in terms of the original matrix P in the same way.

Appendix B

Solutions of problems

B.1 Probability theory (Chapter 2)

(i) *The Quizmaster problem.* There is one car behind one of the n doors. Since this is your only information, you choose one door, say with label n_1, at random. The probability that there is a car behind door n_1 equals $\frac{1}{n}$ as given by (2.1). Equivalently, the probability that there is a car behind one of the $n-1$ other doors is $1 - \frac{1}{n}$, which is the complement event as explained in Section 2.1. The quizmaster now opens door $n_2 \neq n_1$, behind which there is no car. Hence, additional information is given: behind one of the $n-1$ other doors, there is no car. The initial probability p_0 that there is a car behind one, say n_3, of the $n-1$ other doors equals the probability of two mutually exclusive events: (a) the probability that the car is placed behind a door different than yours n_1, which is $1 - \frac{1}{n}$; and (b) the probability that the car is behind door n_3 of the remaining $n-1$ doors, which is $\frac{1}{n-1}$. Combined yields $p_0 = \left(1 - \frac{1}{n}\right)\frac{1}{n-1} = \frac{1}{n}$. The events are exclusive because the car can only be placed behind one of the two groups of doors, either behind your door n_1 or behind the $n-1$ other doors. After the additional information provided by the quizmaster, the probability that there is a car behind door $n_3 \neq n_2$ is still equal to the probability of two mutually exclusive events. However, the probability in case (b) is now equal to $\frac{1}{n-2}$, because we surely know that there is no car behind door n_2. Nothing else is changed: the car is not replaced and the number of doors is still the same. Hence, the modified probability p that there is a car behind one of the other doors equals

$$p = \frac{1}{n-2}\left(1 - \frac{1}{n}\right) = \frac{1}{n}\left(1 + \frac{1}{n-2}\right) > \frac{1}{n} = p_0$$

Since $p > p_0$, you should always change your initial choice. For example, if $n = 3$, $p = \frac{2}{3} = 2p_0$ being the double of the initial probability p_0, while for $n = 4$, $p = \frac{3}{8}$ and p is decreasing in n, but always larger than p_0 for a finite number of doors.

(ii) *The birthday paradox.* Suppose that there are T days in a year and that birthdays of people are independent from each other. We order the n birthdays and talk about the first, second, etc. birthday. The probability p that there are at least two persons out of n with the same birthday equals one minus the complementary probability y, namely, that none of the n birthdays is the same. The computation of y follows formula (2.1) and an iteration of the conditional probability (2.48). Indeed, let B_k denote the event that the k-th birthday is different from all the previous birthdays. Then, $y = \Pr\left[\cap_{k=1}^n B_k\right]$, the probability that all n birthdays are different. We use (2.48),

$$\Pr\left[\cap_{k=1}^n B_k\right] = \Pr\left[B_n\mid \cap_{k=1}^{n-1} B_k\right]\Pr\left[\cap_{k=1}^{n-1} B_k\right]$$

Again invoking (2.48) yields

$$\Pr\left[\cap_{k=1}^n B_k\right] = \Pr\left[B_n\mid \cap_{k=1}^{n-1} B_k\right]\Pr\left[B_{n-1}\mid \cap_{k=1}^{n-2} B_k\right]\Pr\left[\cap_{k=1}^{n-2} B_k\right]$$

575

and after iterating, we arrive at

$$y = \Pr\left[B_n | \cap_{k=1}^{n-1} B_k\right] \Pr\left[B_{n-1} | \cap_{k=1}^{n-2} B_k\right] \dots \Pr\left[B_2 | B_1\right] \Pr\left[B_1\right]$$

where $\Pr\left[B_1\right] = 1$ by definition or because the first birthday can be freely chosen without conflict. Given the first birthday, the probability that the birthday of the second person is different equals $\Pr\left[B_2 | B_1\right] = \frac{T-1}{T} = 1 - \frac{1}{T}$. Next, given the first and second birthday, the probability that the third birthday is different from the first and second equals $\Pr\left[B_3 | B_1 \cap B_2\right] = \frac{T-2}{T} = 1 - \frac{2}{T}$, and so on. Thus,

$$y = 1\frac{T-1}{T}\frac{T-2}{T}\dots\frac{T-n+1}{T} = \frac{\Gamma(T)}{T^{n-1}\Gamma(T-(n-1))}$$

$$= \left(1 - \frac{1}{T}\right)\left(1 - \frac{2}{T}\right)\dots\left(1 - \frac{n-1}{T}\right)$$

so that

$$p = 1 - \prod_{j=1}^{n-1}\left(1 - \frac{j}{T}\right)$$

In order to explain the paradox, we will upper bound the probability y by using the inequality (5.11), $e^{-x} > 1 - x$ for real $x \neq 0$. Then,

$$y = \prod_{j=1}^{n-1}\left(1 - \frac{j}{T}\right) < \prod_{j=1}^{n-1} e^{-j/T} = \exp\left(-\frac{1}{T}\sum_{j=1}^{n-1} j\right)$$

Invoking $\sum_{j=1}^{n-1} j = \frac{(n-1)n}{2} \simeq \frac{n^2}{2}$ yields $y \lesssim e^{-\frac{n^2}{2T}}$ and $p \gtrsim 1 - e^{-\frac{n^2}{2T}}$. In order to have a chance higher than p that in a group of n people there is at least one pair with the same birthday requires that the group size is larger than

$$n \geq \sqrt{2T |\log(1-p)|}$$

For $T = 365$ days, we have with $p = 99\%$ that $n \geq 58$. For $p = 90\%$, the group should be larger than $n > 41$, whereas $p = 50\%$ requires that $n \geq 23$. Alternatively, if the group size is $n = 50$ and $n = 71$, then there is a pair of people with the same birthday with 97% and 99.9% chance, respectively. Hence, the size n of the population is surprisingly small compared to the number of days, which is, at first glance and intuitively, unexpected.

(iii) *Proof of the inclusion-exclusion formula*[1]: Let $A = \cup_{k=1}^{n-1} A_k$ and $B = A_n$ such that $A \cup B = \cup_{k=1}^{n-1} A_k$ and $A \cap B = A_n \cap \left(\cup_{k=1}^{n-1} A_k\right) = \cup_{k=1}^{n-1} A_k \cap A_n$ by the distributive law in set theory, then application of (2.4) yields the recursion in n

$$\Pr\left[\cup_{k=1}^{n} A_k\right] = \Pr\left[\cup_{k=1}^{n-1} A_k\right] + \Pr\left[A_n\right] - \Pr\left[\cup_{k=1}^{n-1} A_k \cap A_n\right] \tag{B.1}$$

---

[1] Another proof (Grimmett and Stirzacker, 2001, p. 56) uses the indicator function defined in Section 2.2.1. Useful indicator function relations are

$$1_{A \cap B} = 1_A 1_B$$
$$1_{A^c} = 1 - 1_A$$
$$1_{A \cup B} = 1 - 1_{(A \cup B)^c} = 1 - 1_{A^c \cap B^c} = 1 - 1_{A^c} 1_{B^c}$$
$$= 1 - (1 - 1_A)(1 - 1_B) = 1_A + 1_B - 1_A 1_B = 1_A + 1_B - 1_{A \cap B}$$

Generalizing the last relation yields

$$1_{\cup_{k=1}^{n} A_k} = 1 - \prod_{k=1}^{n}(1 - 1_{A_k})$$

Multiplying out and taking the expectations using (2.13) leads to (2.5).

By direct substitution of $n \to n - 1$, we have

$$\Pr\left[\cup_{k=1}^{n-1} A_k\right] = \Pr\left[\cup_{k=1}^{n-2} A_k\right] + \Pr\left[A_{n-1}\right] - \Pr\left[\cup_{k=1}^{n-2} A_k \cap A_{n-1}\right]$$

while substitution in this formula of $A_k \to A_k \cap A_n$ gives

$$\Pr\left[\cup_{k=1}^{n-1} A_k \cap A_n\right] = \Pr\left[\cup_{k=1}^{n-2} A_k \cap A_n\right] + \Pr\left[A_{n-1} \cap A_n\right] - \Pr\left[\cup_{k=1}^{n-2} A_k \cap A_n \cap A_{n-1}\right]$$

Substitution of the last two terms into (B.1) yields

$$\Pr\left[\cup_{k=1}^{n} A_k\right] = \Pr\left[A_{n-1}\right] + \Pr\left[A_n\right] - \Pr\left[A_{n-1} \cap A_n\right] + \Pr\left[\cup_{k=1}^{n-2} A_k\right]$$
$$- \Pr\left[\cup_{k=1}^{n-2} A_k \cap A_{n-1}\right] - \Pr\left[\cup_{k=1}^{n-2} A_k \cap A_n\right] + \Pr\left[\cup_{k=1}^{n-2} A_k \cap A_n \cap A_{n-1}\right]$$
$$\text{(B.2)}$$

Similarly, in a next iteration we use (B.1) after suitable modification in the right-hand side of (B.2) to lower the upper index in the union,

$$\Pr\left[\cup_{k=1}^{n-2} A_k\right] = \Pr\left[\cup_{k=1}^{n-3} A_k\right] + \Pr\left[A_{n-2}\right] - \Pr\left[\cup_{k=1}^{n-3} A_k \cap A_{n-2}\right]$$
$$\Pr\left[\cup_{k=1}^{n-2} A_k \cap A_{n-1}\right] = \Pr\left[\cup_{k=1}^{n-3} A_k \cap A_{n-1}\right] + \Pr\left[A_{n-2} \cap A_{n-1}\right]$$
$$- \Pr\left[\cup_{k=1}^{n-3} A_k \cap A_{n-1} \cap A_{n-2}\right]$$
$$\Pr\left[\cup_{k=1}^{n-2} A_k \cap A_n\right] = \Pr\left[\cup_{k=1}^{n-3} A_k \cap A_n\right] + \Pr\left[A_{n-2} \cap A_n\right] - \Pr\left[\cup_{k=1}^{n-3} A_k \cap A_n \cap A_{n-2}\right]$$
$$\Pr\left[\cup_{k=1}^{n-2} A_k \cap A_n \cap A_{n-1}\right] = \Pr\left[\cup_{k=1}^{n-3} A_k \cap A_n \cap A_{n-1}\right] + \Pr\left[A_{n-2} \cap A_n \cap A_{n-1}\right]$$
$$- \Pr\left[\cup_{k=1}^{n-3} A_k \cap A_n \cap A_{n-1} \cap A_{n-2}\right]$$

The result is

$$\Pr\left[\cup_{k=1}^{n} A_k\right] = \Pr\left[A_{n-2}\right] + \Pr\left[A_{n-1}\right] + \Pr\left[A_n\right] + - \Pr\left[A_{n-2} \cap A_{n-1}\right] - \Pr\left[A_{n-2} \cap A_n\right]$$
$$- \Pr\left[A_{n-1} \cap A_n\right] + \Pr\left[A_{n-2} \cap A_{n-1} \cap A_n\right] + \Pr\left[\cup_{k=1}^{n-3} A_k\right]$$
$$- \Pr\left[\cup_{k=1}^{n-3} A_k \cap A_{n-2}\right] - \Pr\left[\cup_{k=1}^{n-3} A_k \cap A_{n-1}\right] - \Pr\left[\cup_{k=1}^{n-3} A_k \cap A_n\right]$$
$$+ \Pr\left[\cup_{k=1}^{n-3} A_k \cap A_{n-1} \cap A_{n-2}\right] + \Pr\left[\cup_{k=1}^{n-3} A_k \cap A_n \cap A_{n-2}\right]$$
$$+ \Pr\left[\cup_{k=1}^{n-3} A_k \cap A_n \cap A_{n-1}\right] - \Pr\left[\cup_{k=1}^{n-3} A_k \cap A_n \cap A_{n-1} \cap A_{n-2}\right]$$

which starts revealing the structure of (2.5). Rather than continuing the iterations, we prove the validity of the inclusion-exclusion formula (2.5) via induction. In case $n = 2$, the basic expression (2.4) is found. Assume that (2.5) holds for n, then the case for $n + 1$ must obey (B.1) where $n \to n + 1$,

$$\Pr\left[\cup_{k=1}^{n+1} A_k\right] = \Pr\left[\cup_{k=1}^{n} A_k\right] + \Pr\left[A_{n+1}\right] - \Pr\left[\cup_{k=1}^{n} A_k \cap A_{n+1}\right]$$

Substitution of (2.5) into the above expression yields, after suitable grouping of the terms,

$$\Pr\left[\cup_{k=1}^{n+1} A_k\right] = \Pr[A_{n+1}] + \sum_{k_1=1}^{n} \Pr[A_{k_1}] - \sum_{k_1=1}^{n}\sum_{k_2=k_1+1}^{n} \Pr[A_{k_1} \cap A_{k_2}] - \sum_{k_1=1}^{n}\Pr[A_{k_1} \cap A_{n+1}]$$

$$+ \sum_{k_1=1}^{n}\sum_{k_2=k_1+1}^{n}\sum_{k_3=k_2+1}^{n} \Pr[A_{k_1} \cap A_{k_2} \cap A_{k_3}] + \sum_{k_1=1}^{n}\sum_{k_2=k_1+1}^{n} \Pr[A_{k_1} \cap A_{k_2} \cap A_{n+1}]$$

$$+ \cdots + (-1)^{n-1} \sum_{k_1=1}^{n}\sum_{k_2=k_1+1}^{n} \cdots \sum_{k_n=k_{n-1}+1}^{n} \Pr\left[\cap_{j=1}^{n} A_{k_j}\right]$$

$$+ \cdots + (-1)^{n} \sum_{k_1=1}^{n}\sum_{k_2=k_1+1}^{n} \cdots \sum_{k_n=k_{n-1}+1}^{n} \Pr\left[\cap_{j=1}^{n} A_{k_j} \cap A_{n+1}\right]$$

and

$$\Pr\left[\cup_{k=1}^{n+1} A_k\right] = \sum_{k_1=1}^{n+1} \Pr\left[A_{k_1}\right] - \sum_{k_1=1}^{n+1}\sum_{k_2=k_1+1}^{n+1} \Pr\left[A_{k_1} \cap A_{k_2}\right]$$

$$+ \sum_{k_1=1}^{n+1}\sum_{k_2=k_1+1}^{n+1}\sum_{k_3=k_2+1}^{n+1} \Pr\left[A_{k_1} \cap A_{k_2} \cap A_{k_3}\right]$$

$$+ \cdots + (-1)^{n} \sum_{k_1=1}^{n+1}\sum_{k_2=k_1+1}^{n+1} \cdots \sum_{k_{n+1}=k_n+1}^{n+1} \Pr\left[\cap_{j=1}^{n} A_{k_j} \cap A_{n+1}\right]$$

which proves (2.5).
(iv) We compute the variance of S_N from (2.79) and (2.27). First, we have that

$$\frac{d\varphi_{S_N}(z)}{dz} = \varphi'_N(\varphi_X(z)) \frac{d\varphi_X(z)}{dz}$$

and

$$\frac{d^2\varphi_{S_N}(z)}{dz^2} = \varphi''_N(\varphi_X(z)) \left(\frac{d\varphi_X(z)}{dz}\right)^2 + \varphi'_N(\varphi_X(z)) \frac{d^2\varphi_X(z)}{dz^2}$$

Then, introduced into (2.27) yields

$$\mathrm{Var}[S_N] = \varphi''_N(1)\left(\varphi'_X(1)\right)^2 + \varphi'_N(1)\varphi''_X(1) + \varphi'_N(1)\varphi'_X(1) - \left(\varphi'_N(1)\varphi'_X(1)\right)^2$$
$$= \left\{\varphi''_N(1) - \left(\varphi'_N(1)\right)^2\right\}\left(\varphi'_X(1)\right)^2 + \varphi'_N(1)\left\{\varphi''_X(1) + \varphi'_X(1)\right\}$$

Again using (2.27), we observe that

$$\mathrm{Var}[S_N] = (E[X])^2 \left(\mathrm{Var}[N] - E[N]\right) + E[N]\left(\mathrm{Var}[X] + (E[X])^2\right)$$

which leads, after simplification, to (2.83).

B.2 Probability theory (Chapter 3)

(i) Using the general formula (2.12) for a non-zero random variable X, we have

$$E[\log X] = \sum_{k=1}^{\infty} \log k \Pr[X = k]$$

while (2.18), $\varphi_X(z) = \sum_{k=0}^{\infty} \Pr[X = k] z^k$, shows that we need to express $\log k$ in terms of z^k. A possible solution starts from the double integral with $0 < a \le b$,

$$\int_a^b dx \int_0^\infty dt e^{-tx} = \int_0^\infty dt \int_a^b dx e^{-tx}$$

where the reversal of integration is justified by absolute convergence (Titchmarsh, 1964, Section 1.8). Since $\int_0^\infty dt e^{-tx} = \frac{1}{x}$, the left-hand side integral equals

$$\int_a^b dx \int_0^\infty dt e^{-tx} = \log \frac{b}{a}$$

while the integral at the right-hand side is

$$\int_0^\infty dt \int_a^b dx e^{-tx} = \int_0^\infty \frac{e^{-ta} - e^{-tb}}{t} dt$$

hence,

$$\log k = \int_0^\infty \frac{e^{-t} - e^{-tk}}{t} dt$$

Multiplying both sides by $\Pr[X = k]$ and summing over k, we obtain (reversal in operators is justified on absolute convergence)

$$\sum_{k=1}^{\infty} \log k \Pr[X = k] = \int_0^\infty \frac{dt}{t} \sum_{k=1}^{\infty} \left(e^{-t} - e^{-tk}\right) \Pr[X = k]$$

$$= \int_0^\infty \frac{dt}{t} \left(e^{-t} \sum_{k=1}^{\infty} \Pr[X = k] - \sum_{k=1}^{\infty} e^{-tk} \Pr[X = k]\right)$$

which finally gives with (2.18)

$$E[\log X] = \int_0^\infty \frac{e^{-t}(1 - \varphi_X(0)) - \varphi_X(e^{-t}) + \varphi_X(0)}{t} dt$$

(ii) (a) The pdf of the k-th smallest order statistic follows from (3.43) for an exponential distribution as

$$f_{X_{(k)}}(x) = m\alpha \binom{m-1}{k-1} \left(1 - e^{-\alpha x}\right)^{k-1} e^{-\alpha(m-k+1)x}$$

The probability generating function (2.38) is

$$\varphi_{X_{(k)}}(z) = E\left[e^{-zX_{(k)}}\right] = m\alpha \binom{m-1}{k-1} \int_0^\infty \left(1 - e^{-\alpha t}\right)^{k-1} e^{-(z+\alpha(m-k+1))t} dt$$

Let $u = e^{-\alpha t}$ and $\beta = z + \alpha(m - k + 1)$, then the integral reduces to the well-known Beta function (Abramowitz and Stegun, 1968, Section 6.2)

$$\int_0^\infty \left(1 - e^{-\alpha t}\right)^{k-1} e^{-\beta t} dt = \frac{1}{\alpha} \int_0^1 (1 - u)^{k-1} u^{\beta/\alpha - 1} dt = \frac{1}{\alpha} B(k, \beta/\alpha) = \frac{1}{\alpha} \frac{\Gamma(k) \Gamma(\beta/\alpha)}{\Gamma(k + \beta/\alpha)}$$

Hence,

$$\varphi_{X_{(k)}}(z) = \frac{m!}{(m-k)!} \frac{\Gamma\left(\frac{z}{\alpha} + m + 1 - k\right)}{\Gamma\left(\frac{z}{\alpha} + m + 1\right)} = \frac{m!}{(m-k)!} \prod_{j=0}^{k-1} \frac{1}{\frac{z}{\alpha} + m - j}$$

The mean follows from $E\left[X_{(k)}\right] = -L'_{w(k)}(0)$, where L_X is the logarithm of the generating function (2.44), as

$$E\left[X_{(k)}\right] = \frac{1}{\alpha} \sum_{j=0}^{k-1} \frac{1}{m-j} \qquad (B.3)$$

(b) For a polynomial probability density function $f_X(x) = \alpha x^{\alpha-1}1_{x\in[0,1]}$ with $\alpha > 0$, we have with (3.43) for $x \in [0,1]$ that

$$f_{X_{(k)}}(x) = \alpha m \binom{m-1}{k-1} x^{\alpha k-1}(1-x^\alpha)^{m-k}$$

with mean

$$E\left[X_{(k)}\right] = \alpha m \binom{m-1}{k-1}\int_0^1 x^{\alpha k}(1-x^\alpha)^{m-k}\,dx$$

$$= m\binom{m-1}{k-1}\int_0^1 t^{k+\frac{1}{\alpha}-1}(1-t)^{m-k}\,dt = \frac{m!}{(k-1)!}\frac{\Gamma\left(k+\frac{1}{\alpha}\right)}{\Gamma\left(m+1+\frac{1}{\alpha}\right)}$$

If $\alpha \to \infty$, then $E\left[X_{(k)}\right] = 1$, while for $\alpha \to 0$, $E\left[X_{(k)}\right] = 0$. For a uniform distribution where $\alpha = 1$, the result is $E\left[X_{(k)}\right] = \frac{k}{m+1}$. Indeed, the m independently chosen uniform random variables divide, after ordering, the line segment $[0,1]$ into $m+1$ subintervals. The length L of each subinterval has a same distribution, which more easily follows by symmetry if the line segment is replaced by a circle of unit perimeter. Since the length L of each subinterval is equal in distribution, one can consider the first subinterval $[0, X_{(1)}]$ whose length L exceeds a value $x \in (0,1)$ if and only if all m uniform random variables belong to $[x,1]$. The latter event has probability equal to $(1-x)^m$ such that $\Pr[L > x] = (1-x)^m$ and, with (2.36), $E[L] = \frac{1}{m+1}$.

(iii) *Histogram.* If X were a discrete random variable, then $\Pr[X=k] \approx \frac{n_k}{n}$, where n_k is the number of values in the set $\{x_1, x_2, \ldots, x_n\}$ that is equal to k. For a continuous random variable X, the values are generally real numbers ranging from $x_{\min} = \min_{1\le j\le n} x_j$ until $x_{\max} = \max_{1\le j\le n} x_j$. We first construct a histogram H from the set $\{x_1, x_2, \ldots, x_n\}$ by choosing a bin size $\Delta x = \frac{x_{\max}-x_{\min}}{m}$, where m is the number of bins (abscissa points). The choice of $1 < m < n$ is in general difficult to determine. However, most computer packages allow us to experiment with m and the human eye proves sensitive enough to make a good choice of m: if m is too small, we loose details, while a high m may lead to high irregularities due to the stochastic nature of X. Once m is chosen, the histogram consists of the set $\{h_0, h_1, \ldots, h_{m-1}\}$, where h_j equals the number of X values in the set $\{x_1, x_2, \ldots, x_n\}$ that lies in the interval $[x_{\min}+j\Delta x, x_{\min}+(j+1)\Delta x]$ for $0 \le j \le m-1$. By construction, $\sum_{j=0}^{m-1} h_j = n$.

The histogram H approximates the probability density function $f_X(x)$ after dividing each value h_j by $n\Delta x$ because

$$1 = \int_{x_{\min}}^{x_{\max}} f_X(x)dx = \lim_{\Delta x\to 0}\sum_{j=0}^{\frac{x_{\max}-x_{\min}}{\Delta x}-1} f_X(\xi_j)\Delta x \approx \sum_{j=0}^{m-1}\frac{h_j}{n\Delta x}\Delta x = 1$$

where in the Riemann sum ξ_j denotes a real number $\xi_j \in [x_{\min}+j\Delta x, x_{\min}+(j+1)\Delta x]$. Alternatively from (2.31) we obtain

$$f_X(\xi_j) = \lim_{\Delta x\to 0}\frac{\Pr[\xi_j < X \le \xi_j+\Delta x]}{\Delta x} \approx \frac{\Pr[x_{\min}+j\Delta x < X \le x_{\min}+(j+1)\Delta x]}{\Delta x}$$

such that

$$f_X(\xi_j) \approx \frac{h_j}{n\Delta x}$$

which reduces to the discrete case where $\Delta x = 1$.

Distribution function. Based on the set of realizations $x_1, x_2, \ldots, x_n$ of the random variable X and again using (2.1), we have that

$$\Pr[X \le u] \approx \frac{1}{n}\sum_{k=1}^{n} 1_{\{x_k\le u\}}$$

which is equally valid for discrete as well as continuous random variables. We sort the set of realizations (simulations or measurements) in increasing order so that $x_{(1)} \le x_{(2)} \le \cdots \le$

$x_{(n)}$. Since the sum over all realizations is the same as the sum over all ordered realizations, we obtain

$$\Pr\left[X \le u\right] \approx \frac{1}{n}\sum_{k=1}^{n} 1_{\left\{x_{(k)} \le u\right\}}$$

from which, for $u = x_{(j)}$, we find

$$\Pr\left[X \le x_{(j)}\right] \approx \frac{j}{n}$$

Hence, if we plot on the abscissa the sorted vector $\left(x_{(1)}, x_{(2)}, \ldots, x_{(n)}\right)$ against the vector $\left(\frac{1}{n}, \frac{2}{n}, \ldots, 1\right)$ on the ordinate axis, the *empirical* distribution function $\widetilde{F}_X(u)$ of X is found (avoiding the problem of choosing the adequate bin-size, see p. 138). Fitting the distribution function is less sensitive (due to the monotonous increasing, sigmoid shape of $F_X(u)$) than fitting the pdf $f_X(u) = \frac{dF_X(u)}{du}$, but numerical derivation of $\widetilde{F}_X(u)$ to determine the empirical pdf $\widetilde{f}_X(u)$ magnifies potential errors embedded in the realization x_j. These considerations result in a trade-off between the bin-size problem and the accurate fitting of $\widetilde{F}_X(u)$. Clauset *et al.* (2009) have discussed, in more detail, how to estimate the pdf from empirical data.

(iv) The density of mobile nodes in the circle with radius r equals $\nu = \frac{N}{\pi r^2}$. Let R denote the (random) position of a mobile node. The probability that there is a mobile node between distance x and $x + dx$ (and $x \le r$) is $\Pr\left[x \le R \le x + dx\right] = \frac{2\pi x \, dx}{\pi r^2}$. From (2.31), the pdf of R equals $f_R(x) = \frac{2x}{r^2} 1_{x \le r}$ and the distribution function follows by integration as $F_R(x) = \frac{x^2}{r^2} 1_{x \le r} + 1_{x > r}$. The (random) position $R_{(m)}$ of the m-th nearest mobile node to the center is given by (3.43)

$$f_{R_{(m)}}(x) = N f_R(x) \binom{N-1}{m-1} \left(F_R(x)\right)^{m-1} \left(1 - F_R(x)\right)^{N-m}$$

Written in terms of the density ν for $x \le r$,

$$f_{R_{(m)}}(x) = 2x\pi\nu \binom{N-1}{m-1} \left(\frac{\pi\nu x^2}{N}\right)^{m-1} \left(1 - \frac{\pi\nu x^2}{N}\right)^{N-1-(m-1)}$$

we recognize, apart from the prefactor $2x\pi\nu$, a binomial distribution (3.3) with $p = \frac{\pi\nu x^2}{N}$. Similar to the derivation of the law of rare events in Section 3.1.4, this binomial distribution tends, for large N but constant density ν, to a Poisson distribution with $\lambda = \pi\nu x^2$. Hence, asymptotically, the pdf of the position $R_{(m)}$ of the m-th nearest mobile node to the center is, for $x \le r$,

$$f_{R_{(m)}}(x) = 2x\pi\nu \frac{\left(\pi\nu x^2\right)^{m-1}}{(m-1)!} e^{-\pi\nu x^2}$$

(v) We use the law of total probability (2.49) first assuming that W is discrete,

$$\Pr[V - W \le x] = \sum_k \Pr[V - W \le x | W = k] \Pr[W = k]$$

and, by independence, $\Pr[V - W \le x | W = k] = \Pr[V \le k + x]$. Hence,

$$\Pr[V - W \le x] = \sum_k \Pr[V \le k + x] \Pr[W = k]$$

If W is continuous, the general formula is

$$\Pr[V - W \le x] = \int_{-\infty}^{\infty} \Pr[V \le x + y] \frac{d\Pr[W \le y]}{dy} dy \qquad (B.4)$$

from which the pdf follows by differentiation

$$f_{V-W}(x) = \int_{-\infty}^{\infty} f_V(x+y)f_W(y)dy$$

This resembles the convolution integral (2.70). If both V and W have the same distribution, direct integration of (B.4) yields

$$\Pr[V \leq W] = \Pr[W \leq V] = \frac{1}{2}$$

This equation confirms the intuitive result that two independent random variables with same density function have equal probability to be larger or smaller than the other.

(vi) The probability density function of the product Z of two independent uniform random variables U_1 and U_2 on $[a, b]$ and $[c, d]$, respectively, is computed from (2.67) as

$$f_{Z;\text{uniform}}(w) = \frac{1}{(b-a)(d-c)} \int_{-\infty}^{\infty} 1_{\{u \in [a,b]\}} 1_{\{\frac{w}{u} \in [c,d]\}} \frac{du}{|u|} = \frac{1}{(b-a)(d-c)} \int_{\max(a, \frac{w}{d})}^{\min(b, \frac{w}{c})} \frac{du}{|u|}$$

The latter integral is only finite provided the integration range excludes the point $u = 0$. Thus, for $a > 0$ and $c > 0$ and $w > 0$, we find

$$f_{Z;\text{uniform}}(w) = \frac{1}{(b-a)(d-c)} \ln\left(\frac{\min\left(b, \frac{w}{c}\right)}{\max\left(a, \frac{w}{d}\right)} \right)$$

(vii) We use the law of total probability (2.49) together with $\Pr[Y = j] = \frac{1}{l-k+1} 1_{j \in [k,l]}$,

$$\Pr[Z \leq x] = \sum_{j=k}^{l} \Pr[X + Y \leq x | Y = j] \Pr[Y = j] = \frac{1}{l-k+1} \sum_{j=k}^{l} \Pr[X \leq x - j | Y = j]$$

Since X and Y are independent, we have that $\Pr[X \leq x - j | Y = j] = \Pr[X \leq x - j]$. The continuous uniform random variable X has the distribution (see Section 3.2.1)

$$\Pr[X \leq x] = \Pr[\{X < a\} \cup \{a \leq X \leq x\}] = \Pr[a \leq X \leq x] = \frac{x-a}{b-a} 1_{x \in [a,b]} + 1_{x>b}$$

such that

$$\Pr[Z \leq x] = \frac{1}{(l-k+1)(b-a)} \sum_{j=k}^{l} (x-j-a) 1_{x-j \in [a,b]} + \frac{1}{l-k+1} \sum_{j=k}^{l} 1_{x-j>b}$$

The condition $x - j \in [a, b]$ implies that $x - b \leq j \leq x - a$, which simplifies the sum as

$$\Pr[Z \leq x] = \frac{1}{(l-k+1)(b-a)} \sum_{j=\max(k, \lceil x-b \rceil)}^{\min(l, \lfloor x-a \rfloor)} (x-j-a) + \frac{1}{l-k+1} \sum_{j=k}^{\min(l, \lfloor x-b \rfloor)} 1$$

where the floor $\lfloor x \rfloor$ is the largest integer smaller than x and the ceiling $\lceil x \rceil$ is the smallest integer larger than x. Denoting $A = \min(l, \lfloor x - a \rfloor)$, $B = \max(k, \lceil x - b \rceil)$ and $C = \min(l, \lfloor x - b \rfloor)$ and assuming that $A \geq B$ else the sum vanishes and $\Pr[Z \leq x] = 0$, we obtain

$$\Pr[Z \leq x] = \frac{1}{(l-k+1)(b-a)} \sum_{j=B}^{A} (x-j-a) + \frac{C-k+1}{l-k+1} 1_{C \geq k}$$

$$= \frac{A-B+1}{(l-k+1)(b-a)} \left(x - a - \frac{A+B}{2} \right) 1_{A \geq B} + \frac{C-k+1}{l-k+1} 1_{C \geq k}$$

where we have used $\sum_{j=m}^{n} j = \frac{(n+m)(n-m+1)}{2}$.

(viii) The total number of bits $B = \sum_{j=1}^{N} K_j$ is the sum of independent, geometric random variables K_i in which the number of terms N in the sum B is also a geometric random variable. Combining (3.6) and (2.79) yields

$$\varphi_B(z) = \varphi_N(\varphi_K(z)) = \frac{q\left(\frac{pz}{1-(1-p)z}\right)}{1-(1-q)\left(\frac{pz}{1-(1-p)z}\right)} = \frac{pqz}{1-(1-pq)z}$$

which is again a pgf (3.6) of a geometric random variable with mean $\frac{1}{pq}$.

(ix) *Random variables whose moments are also random variables.*
(a) Applying the law of total probability (2.49) to the probability density function of a continuous random variable X gives

$$f_X(x) = \int_0^\infty f_{X|\sigma^2}(t)\, d\Pr\left[\sigma^2 \le t\right]$$

where the variance σ^2 is a random variable with pdf $f_{\sigma^2}(t) = \lambda e^{-\lambda t}$. With the lognormal pdf (3.54) with mean $\mu = 0$, we have

$$f_X(x) = \int_0^\infty \frac{e^{-\frac{(\log x)^2}{2t}}}{x\sqrt{2\pi t}} \lambda e^{-\lambda t}\, dt$$

The integral is a special case, for $s = \frac{1}{2}$, $a = \lambda$ and $b = \frac{(\log x)^2}{2}$, of (3.31), which reduces for $s = \frac{1}{2}$ to $K_{1/2}(z) = \sqrt{\frac{\pi}{2z}} e^{-z}$. Thus,

$$f_X(x) = \frac{\lambda}{x\sqrt{2\pi}} \int_0^\infty t^{-\frac{1}{2}} e^{-\lambda t} e^{-\frac{(\log x)^2}{2t}}\, dt = \sqrt{\frac{\lambda}{2}} \frac{e^{-|\log x|\sqrt{2\lambda}}}{x}$$

which simplifies to

$$f_X(x) = \begin{cases} \sqrt{\frac{\lambda}{2}} x^{-1+\sqrt{2\lambda}} & \text{for } x \le 1 \\ \sqrt{\frac{\lambda}{2}} x^{-1-\sqrt{2\lambda}} & \text{for } x \ge 1 \end{cases}$$

This demonstrates that X behaves as a "double" power-law random variable that is different for $x > 1$ and $x < 1$.
(b) We can follow the arguments that led to (14.23) so that

$$\Pr[X = n] = \frac{(-1)^n}{n!} \left. \frac{d^n \varphi_\Lambda(z)}{dz^n} \right|_{z=1}$$

Thus, if the mean Λ is exponentially distributed with mean μ, then $f_\Lambda(t) = \mu e^{-\mu t}$ and $\varphi_\Lambda(z) = \frac{\mu}{\mu+z}$ such that $\Pr[X = n] = \mu(\mu+1)^{-n-1}$. Hence, X is a geometric random variable with mean $\frac{1}{\mu}$.

(x) The conditional probability $\Pr[Y = k|X = l]$ has a binomial distribution (3.3),

$$\Pr[Y = k|X = l] = \binom{l}{k} p^k q^{l-k}$$

The probability that the original flow has length l given that the sampled flow has length k is with (2.47)

$$\Pr[X = l|Y = k] = \frac{\Pr[X = l, Y = k]}{\Pr[Y = k]} = \frac{\Pr[Y = k|X = l]\Pr[X = l]}{\Pr[Y = k]}$$

where by the law of total probability (2.49),

$$\Pr[Y = k] = \sum_{l=0}^\infty \Pr[Y = k|X = l]\Pr[X = l] = \sum_{l=0}^\infty \binom{l}{k} p^k q^{l-k}\Pr[X = l]$$

Since the length of an original flow is geometric (see Section 3.1.3), which means that $\Pr[X = l] = (1-g)g^l$ for any $l \geq 0$ and $0 < g < 1$, then

$$\Pr[Y = k] = (1-g) \sum_{l=k}^{\infty} \binom{l}{k} p^k q^{l-k} g^l = \frac{(1-g)\,(pg)^k}{(1-qg)^{k+1}}$$

where the series $(1-x)^{-k-1} = \sum_{l=k}^{\infty} \binom{l}{k} x^{l-k}$ for $|x| < 1$ has been used (Abramowitz and Stegun, 1968, 24.1.1). Finally, the asked conditional probability is

$$\Pr[X = l | Y = k] = \binom{l}{k} (gq)^{l-k} (1-qg)^{k+1}$$

which is, apart from a factor $(1-qg)$, a binomial distribution with mean lpq.

(xi) *Uniform sampling.* We compute the probability p_k that the k-th item uniformly drawn from a set of N items of which m belong to a special set $\mathcal{M}$, belongs to $\mathcal{M}$. For $k = 1$, we have that the first item n_1 has probability $p_1 = p = \frac{m}{N}$ to belong to $\mathcal{M}$, because the fraction of type A items in the initial set is just $\frac{m}{N}$, i.e. the number of favorable possibilities to draw a type A over the total possible number in which an item can be drawn. For $k = 2$, the total number of items is now $N - 1$ and

$$p_2 = \frac{m}{N-1} \Pr[n_1 \notin \mathcal{M}] + \frac{m-1}{N-1} \Pr[n_1 \in \mathcal{M}]$$

$$= \frac{m}{N-1}(1-p_1) + \frac{m-1}{N-1}p_1 = \frac{m-p_1}{N-1} = \frac{m-\frac{m}{N}}{N-1} = p$$

For $k = 3$, a similar reasoning yields

$$p_3 = \frac{m}{N-2} \Pr[\{n_1, n_2\} \notin \mathcal{M}] + \frac{m-1}{N-2} \Pr[n_1 \in \mathcal{M}, n_2 \notin \mathcal{M}]$$

$$+ \frac{m-1}{N-2} \Pr[n_2 \in \mathcal{M}, n_1 \notin \mathcal{M}] + \frac{m-2}{N-2} \Pr[\{n_1, n_2\} \in \mathcal{M}]$$

Since the way in which the previous nodes are selected are independent from each other, $\Pr[\{n_1, n_2\} \notin \mathcal{M}] = \Pr[n_1 \notin \mathcal{M}] \Pr[n_2 \notin \mathcal{M}]$ and similar for the others such that

$$p_3 = \frac{m}{N-2}(1-p)^2 + 2\frac{m-1}{N-2}p(1-p) + \frac{m-2}{N-2}p^2$$

$$= \frac{m - 2mp + mp^2 + 2mp - 2mp^2 - 2p + 2p^2 + mp^2 - 2p^2}{N-2}$$

$$= \frac{m-2p}{N-2} = \frac{m-2\frac{m}{N}}{N-2} = p$$

We now infer that $p_k = p$. Let $p_k = p$ be the induction argument. It holds for $k \leq 3$. Assume that it holds for $k = n$, then $k = n+1$, enumerating all possibilities and taking into independence into account gives

$$p_{n+1} = \sum_{j=0}^{n} \frac{m-j}{N-n}\binom{n}{j}p^j(1-p)^{n-j} = \frac{m\sum_{j=0}^{n}\binom{n}{j}p^j(1-p)^{n-j}}{N-n} - \frac{\sum_{j=0}^{n}j\binom{n}{j}p^j(1-p)^{n-j}}{N-n}$$

$$= \frac{m-np}{N-n} = \frac{m-n\frac{m}{N}}{N-n} = p$$

which demonstrates the induction. Hence, $p_k = p$. In summary, the uniform selection of the items from a set with two types maintains the ratio among the types. Each new, independent and uniformly chosen item has probability p to be of type A.

(xii) (a) Denote by $S_n = \sum_{j=n}^{h} w_{(j)}$ with $1 \leq n \leq h$, then $S_1 = W_h$. Applying the law of total probability (2.49) yields

$$\Pr\left[S_1 \le x\right] = \int_0^x \Pr\left[w_{(1)} + S_2 \le x \,\middle|\, w_{(1)} = t_1\right] f_{w_{(1)}}(t_1)\,dt_1$$

$$= \int_0^x \Pr\left[S_2 \le x - t_1 \,\middle|\, w_{(1)} = t_1\right] f_{w_{(1)}}(t_1)\,dt_1$$

After writing $S_2 = S_3 + w_{(2)}$, we apply again the law of total probability and obtain,

$$\Pr\left[S_1 \le x\right] = \int_0^x dt_1 \int_{t_1}^{x-t_1} dt_2 \Pr\left[S_3 \le x - t_1 - t_2 \,\middle|\, w_{(1)} = t_1, w_{(2)} = t_2, w_{(1)} < w_{(2)}\right]$$

$$\times f_{w_{(1)},w_{(2)}}(t_1, t_2)$$

Repeating this method j times results in

$$\Pr\left[S_1 \le x\right] = \int_0^x dt_1 \int_{t_1}^{x-t_1} dt_2 \cdots \int_{t_{j-1}}^{x-\sum_{k=1}^{j-1} t_k} \Pr\left[S_{j+1} \le x - \sum_{k=1}^{j} t_k \,\middle|\, w_{(1)} = t_1, \ldots, w_{(j)} = t_j\right]$$

$$\times f_{\{w_{(k)}\}_{1\le k \le j}}(t_1, t_2, \ldots, t_j)\,dt_j$$

where the order statistics condition $w_{(1)} < w_{(2)} < \cdots < w_{(j)}$ is obeyed by the lower limits in the integral. For $j = h - 1$, it holds that $S_h = w_{(h)}$ such that

$$\Pr\left[S_1 \le x\right] = \int_0^x dt_1 \int_{t_1}^{x-t_1} dt_2 \cdots \int_{t_{h-2}}^{x-\sum_{k=1}^{h-2} t_k} dt_{h-1} f_{\{w_{(k)}\}_{1\le k \le h-1}}(t_1, t_2, \ldots, t_{h-1})$$

$$\times \Pr\left[w_{(h)} \le x - \sum_{k=1}^{h-1} t_k \,\middle|\, w_{(1)} = t_1, \ldots, w_{(h-1)} = t_{h-1}\right]$$

One additional conditioning, where $w_{(h)} = t_h$ for $t_{h-1} < t_h < x - \sum_{k=1}^{h-1} t_k$, makes $w_{(h)} \le x - \sum_{k=1}^{h-1} t_k$ an always true event. Hence, we arrive at

$$\Pr\left[S_1 \le x\right] = \int_0^x dt_1 \int_{t_1}^{x-t_1} dt_2 \cdots \int_{t_{h-1}}^{x-\sum_{k=1}^{h-1} t_k} dt_h\, f_{\{w_{(k)}\}_{1\le k \le h}}(t_1, t_2, \ldots, t_h)$$

The joint density $f_{\{w_{(k)}\}_{1\le k \le h}}(t_1, t_2, \ldots, t_h)$, where $t_1 < t_2 < \cdots < t_h$, involves h of the m i.i.d. random variables that need to be placed at positions t_k for $1 \le k \le h$, while all other $m - h$ random variables must be larger than t_h. Using the multinomial coefficient and independence, we obtain

$$f_{\{w_{(k)}\}_{1\le k \le h}}(t_1, t_2, \ldots, t_h) = \frac{m!}{(m-h)!} \prod_{k=1}^{h} f_w(t_k)\left(1 - F_w(t_h)\right)^{m-h}$$

which reduces to (3.41) if $h = m$. The integration region is determined by the inequalities

$$\begin{cases} 0 \le t_1 \le x \\ t_1 \le t_2 \le x - t_1 \\ \cdots \\ t_{h-1} \le t_h \le x - t_1 - \cdots - t_{h-1} \end{cases} \quad \text{equivalent to} \quad \begin{cases} 0 \le t_1 \le x \\ 0 \le t_2 - t_1 \le x - 2t_1 \\ \cdots \\ 0 \le t_h - t_{h-1} \le x - t_1 - \cdots - 2t_{h-1} \end{cases}$$

Since all $t_j \ge 0$, the last inequality is the most confining equation and equivalent to $t_{h-1} \le \frac{1}{2}\left(x - \sum_{k=1}^{h-2} t_k\right)$. But, the smallest possible value of $t_{h-1} = t_{h-2}$ such that $t_{h-2} \le \frac{1}{2}\left(x - \sum_{k=1}^{h-2} t_k\right)$ or $t_{h-2} \le \frac{1}{3}\left(x - \sum_{k=1}^{h-3} t_k\right)$. By the same argument, the smallest possible value of $t_{h-2} = t_{h-3}$, which leads to $t_{h-3} \le \frac{1}{4}\left(x - \sum_{k=1}^{h-4} t_k\right)$. Iterating this process yields $t_{h-j+1} \le \frac{1}{j}\left(x - \sum_{k=1}^{h-j} t_k\right)$ for all $1 \le j \le h$, which shows that, for $j = h$, $t_1 \le \frac{x}{h}$.

Replacing these more stringent bounds finally proves (3.64). In fact, the integration over t_h can be done explicitly,

$$I_h = \int_{t_{h-1}}^{x - \sum_{k=1}^{h-1} t_k} dt_h f_w\left(t_h\right)\left(1 - F_w\left(t_h\right)\right)^{m-h}$$

$$= \frac{\left(1 - F_w\left(t_{h-1}\right)\right)^{m-h+1} - \left(1 - F_w\left(x - \sum_{k=1}^{h-1} t_k\right)\right)^{m-h+1}}{m - h + 1}$$

The first term can again be integrated. After h integrations, its contribution to $\Pr\left[S_1 \le x\right]$ as two of the 2^h terms is $1 - \left(1 - F_w\left(\frac{x}{h}\right)\right)^m$.

(b) For an exponential distribution $f_w\left(x\right) = a e^{-ax}$ and $F_w\left(x\right) = 1 - e^{-ax}$, we have that

$$\Pr\left[W_h \le x\right] = \frac{m! a^h}{(m - h)!} \int_0^{\frac{x}{h}} dt_1 \int_{t_1}^{\frac{x - t_1}{h-1}} dt_2 \cdots \int_{t_{h-1}}^{x - \sum_{k=1}^{h-1} t_k} dt_h \prod_{k=1}^{h-1} e^{-a t_k} e^{-a(m-h+1)t_h}$$

For $h = 1$, we find

$$\Pr\left[W_1 \le x\right] = \frac{m! a}{(m - 1)!} \int_0^x dt_1\, e^{-amt_1} = 1 - e^{-amx}$$

For $h = 2$, we have

$$\Pr\left[W_2 \le x\right] = \frac{m! a^2}{(m - 2)!} \int_0^{\frac{x}{2}} dt_1 \int_{t_1}^{x - t_1} dt_2\, e^{-at_1} e^{-a(m-1)t_2}$$

where the integration area is a triangle in the (t_1, t_2)-plane with corner points $(0, 0), \left(\frac{x}{2}, \frac{x}{2}\right)$ and $(0, x)$. After integrating over t_2,

$$\Pr\left[W_2 \le x\right] = \frac{m! a}{(m - 2)!} \int_0^{\frac{x}{2}} dt_1 e^{-at_1} \frac{e^{-a(m-1)t_1} - e^{-a(m-1)(x-t_1)}}{m - 1}$$

$$= \frac{m!}{(m - 1)!} \left(\frac{1 - e^{-\frac{amx}{2}}}{m} - e^{-a(m-1)x} \frac{e^{a(m-2)\frac{x}{2}} - 1}{m - 2} \right)$$

Hence,

$$\Pr\left[W_2 \le x\right] = 1 - \frac{2(m - 1)}{(m - 2)} e^{-\frac{amx}{2}} + \frac{m}{(m - 2)} e^{-a(m-1)x}$$

For $h = 3$, the integral is

$$\Pr\left[W_3 \le x\right] = \frac{m! a^3}{(m - 3)!} \int_0^{\frac{x}{3}} dt_1 \int_{t_1}^{\frac{x - t_1}{2}} dt_2 \int_{t_2}^{x - t_1 - t_2} dt_3\, e^{-a(t_1 + t_2)} e^{-a(m-2)t_3}$$

After integration over t_3, followed by an integration over t_2 and t_1, we find (after simplification)

$$\Pr\left[W_3 \le x\right] = 1 - \frac{m(m - 1)}{2(m - 3)^2} e^{-a(m-2)x} + \frac{4m(m - 2)}{(m - 3)^2} e^{-a\left(\frac{m-1}{2}\right)x} - \frac{9(m - 1)(m - 2) e^{-\frac{max}{3}}}{2(m - 3)^2}$$

Similarly, for $h = 4$, we find

$$\Pr\left[W_4 > x\right] = \frac{-m!}{(m - 4)!\,(m - 4)^3} \left[\frac{e^{-(m-3)ax}}{6(m - 3)} - 4\frac{e^{-\frac{(m-2)ax}{2}}}{(m - 2)} + 27\frac{e^{-\frac{(m-1)ax}{3}}}{2(m - 1)} - 32\frac{e^{-\frac{max}{4}}}{3m} \right]$$

(c) By partial summation, attributed to Niels Abel, we can rewrite the sum $W_h = \sum_{j=1}^{h} w_{(j)}$ in terms of the spacings y_k. Consider the more general sum $\sum_{k=1}^{n} a_k b_k$, where $n \ge 1$, and denote the partial sums, $s_j = \sum_{k=1}^{j} a_k$, where $j \ge 1$. The basic observation is that

$a_k = s_k - s_{k-1}$ for $k > 1$ and the difference does not dependent on a fixed constant K added to s_j. Thus, with $s_j = K + \sum_{k=1}^{j} a_k$,

$$\sum_{k=1}^{n} a_k b_k = a_1 b_1 + \sum_{k=2}^{n} (s_k - s_{k-1}) b_k = a_1 b_1 + \sum_{k=2}^{n} s_k b_k - \sum_{k=1}^{n-1} s_k b_{k+1}$$

$$= a_1 b_1 + \sum_{k=2}^{n-1} s_k (b_k - b_{k+1}) + s_n b_n - s_1 b_2$$

Since $s_1 = a_1 + K$, Abel's generalized partial summation is found,

$$\sum_{k=1}^{n} a_k b_k = -b_1 K + \sum_{k=1}^{n-1} s_k (b_k - b_{k+1}) + s_n b_n \qquad (B.5)$$

which reduces for $K = 0$ to the usual formula

$$\sum_{k=1}^{n} a_k b_k = \sum_{k=1}^{n-1} \left(\sum_{j=1}^{k} a_j \right) (b_k - b_{k+1}) + \left(\sum_{j=1}^{n} a_j \right) b_n$$

By choosing $K = -\sum_{k=1}^{n} a_k$ such that $s_n = 0$ and $s_k = -\sum_{j=k+1}^{n} a_j$, we arrive at

$$\sum_{k=1}^{n} a_k b_k = b_1 \sum_{k=1}^{n} a_k + \sum_{k=1}^{n-1} \left(\sum_{j=k+1}^{n} a_j \right) (b_{k+1} - b_k)$$

We apply the latter formula to $W_h = \sum_{j=1}^{h} w_{(j)}$ for $a_k = 1$ and $b_k = w_{(k)}$ and obtain

$$W_h = \sum_{j=1}^{h} w_{(j)} = h w_{(1)} + \sum_{k=1}^{h-1} (h-k) (w_{(k+1)} - w_{(k)}) = \sum_{k=0}^{h-1} (h-k) y_{k+1}$$

such that

$$E[W_h] = \sum_{k=0}^{h-1} (h-k) E[y_{k+1}] = \frac{1}{a} \sum_{k=0}^{h-1} \frac{h-k}{m-k} = \frac{1}{a} \left(h - (m-h) \sum_{j=m-h+1}^{m} \frac{1}{j} \right)$$

where $\frac{h-k}{m-k} = \frac{h-m}{m-k} + 1$ is used. For large m, we obtain

$$E[W_h] \approx \frac{h}{a} - \frac{m-h}{a} \log \frac{m}{m-h+1}$$

$$= \frac{h}{a} - \frac{m-h}{a} \log \frac{1}{1 - \frac{h-1}{m}} \approx \frac{h}{a} - \left(1 - \frac{h}{m}\right) \frac{h-1}{a} \approx \frac{h^2}{m} \frac{1}{a}$$

The generating function of W_h, defined as $\varphi_{W_h}(z) = E\left[e^{-zW_h}\right]$, is

$$\varphi_{W_h}(z) = E\left[e^{-z\sum_{k=0}^{h-1}(h-k)y_{k+1}}\right] = E\left[\prod_{k=0}^{h-1} e^{-z(h-k)y_{k+1}}\right] = \prod_{k=0}^{h-1} E\left[e^{-z(h-k)y_{k+1}}\right]$$

where, in the last step, independence of the y_k has been invoked. With $E\left[e^{-zy_{k+1}}\right] = \frac{(m-k)a}{z+(m-k)a}$, we find

$$\varphi_{W_h}(z) = \prod_{k=0}^{h-1} \frac{(m-k)a}{z(h-k)+(m-k)a} = \frac{a^h}{(z+a)^h} \prod_{k=0}^{h-1} \frac{(m-k)}{\frac{zh+ma}{z+a} - k}$$

$$= \frac{a^h}{(z+a)^h} \frac{\Gamma(m+1)}{\Gamma(m-h+1)} \frac{\Gamma(\frac{zh+ma}{z+a} - h + 1)}{\Gamma(\frac{zh+ma}{z+a} + 1)} = \prod_{k=0}^{h-1} \frac{(m-k)a}{h-k} \frac{1}{\prod_{k=0}^{h-1} z + \frac{(m-k)a}{h-k}}$$

By inverse Laplace transform (2.39),

$$f_{W_h}(x) = \frac{\prod_{k=0}^{h-1} \frac{(m-k)a}{h-k}}{2\pi i} \int_{c-i\infty}^{c-i\infty} \frac{e^{zx}}{\prod_{k=0}^{h-1} z + \frac{(m-k)a}{h-k}} dz = \sum_{j=0}^{h-1} \frac{\prod_{k=0}^{h-1} \frac{(m-k)a}{h-k} \exp\left(-\frac{(m-j)a}{h-j}x\right)}{\prod_{k=0;k\neq j}^{h-1} \left(\frac{(m-k)a}{h-k} - \frac{(m-j)a}{h-j}\right)}$$

Slightly rewritten,

$$f_{W_h}(x) = a\prod_{j=1}^{h} \frac{(m-h+j)}{j} \sum_{j=1}^{h} \frac{\exp\left(-\frac{(m-h+j)a}{j}x\right)}{\prod_{k=1;k\neq j}^{h} \left(\frac{(m-h+k)}{k} - \frac{(m-h+j)}{j}\right)}$$

$$= a\frac{m!e^{-ax}}{h!(m-h)!} \sum_{j=1}^{h} \frac{\exp\left(-\frac{(m-h)ax}{j}\right)}{(m-h)^{h-1} \prod_{k=1;k\neq j}^{h} \left(\frac{j-k}{kj}\right)}$$

Since

$$\prod_{k=1;k\neq j}^{h} \left(\frac{j-k}{kj}\right) = \prod_{k=1}^{j-1} \left(\frac{j-k}{kj}\right) \prod_{k=j+1}^{h} \left(\frac{j-k}{kj}\right) = \frac{(-1)^{h-j}j!(j-1)!(h-j)!}{j^{h-1}(j-1)!h!}$$

we finally arrive at (3.65). The distribution function is

$$\Pr[W_h \leq x] = \frac{a\binom{m}{h}}{(m-h)^{h-1}} \sum_{j=1}^{h} (-1)^{h-j} j^{h-1} \binom{h}{j} \int_0^x e^{-\frac{(m-h+j)au}{j}} du$$

$$= \frac{\binom{m}{h}}{(m-h)^{h-1}} \sum_{j=1}^{h} (-1)^{h-j} j^h \binom{h}{j} \frac{1 - e^{-\frac{(m-h+j)ax}{j}}}{m-h+j}$$

Since $\lim_{x\to\infty} \Pr[W_h \leq x] = 1$, we find the formula

$$\sum_{j=1}^{h} (-1)^{h-j} \binom{h}{j} \frac{j^h}{m-h+j} = \frac{(m-h)^{h-1}}{\binom{m}{h}}$$

that is related to the Stirling number of the second kind,

$$\mathcal{S}_m^{(k)} = \frac{1}{k!} \sum_{j=0}^{k} (-1)^{k-j} \binom{k}{j} j^m \tag{B.6}$$

Finally, we obtain

$$\Pr[W_h > x] = \frac{\binom{m}{h}e^{-ax}}{(m-h)^{h-1}} \sum_{j=1}^{h} (-1)^{h-j} j^h \binom{h}{j} \frac{e^{-\frac{(m-h)ax}{j}}}{m-h+j}$$

from which the few explicit forms computed in (b) are readily verified.

(xiii) (a) As usual, the moments of X can be computed from $\varphi_X(z)$. An alternative evaluation is

$$E[X^m] = K^{-1}(\lambda) \int_0^\infty \frac{x^m e^{x\ln\lambda}}{\Gamma(x+1)} dx = K^{-1}(\lambda) \int_0^\infty \frac{dx}{\Gamma(x+1)} \frac{d^m e^{xy}}{dy^m}\Big|_{y=\ln\lambda}$$

$$= K^{-1}(\lambda) \frac{d^m}{dy^m} \int_0^\infty \frac{e^{xy} dx}{\Gamma(x+1)}\Big|_{y=\ln\lambda}$$

which leads, with (3.66), to (3.68).

(b) The relation (3.70) illustrates the power and beauty of contour integration. The Laplace transform of $K(\lambda)$ is

$$L(s) = \int_0^\infty K(\lambda) e^{-\lambda s} d\lambda$$

Introducing (3.66) gives the double integral

$$L(s) = \int_0^\infty d\lambda \int_0^\infty \frac{\lambda^x e^{-\lambda s}}{\Gamma(x+1)} dx = \int_0^\infty \frac{dx}{\Gamma(x+1)} \int_0^\infty \lambda^x e^{-\lambda s} d\lambda$$

where the reversal of integration is justified by absolute convergence. With $\int_0^\infty \lambda^x e^{-\lambda s} d\lambda = \frac{\Gamma(x+1)}{s^{x+1}}$, we arrive, for $\mathrm{Re}(s) > 1$, at

$$L(s) = \int_0^\infty \frac{dx}{s^{x+1}} = \frac{1}{s \ln s}$$

The inverse Laplace transform (2.39) returns a second, complex integral for $K(\lambda)$,

$$K(\lambda) = \frac{1}{2\pi i} \int_{c-i\infty}^{c+i\infty} \frac{e^{\lambda s}}{s \ln s} ds \qquad (B.7)$$

where $c > 1$ because $L(s)$ is only analytic for $\mathrm{Re}(s) > 1$ due to the pole at $s = 1$. Indeed, the Taylor expansion of $\ln s$ around $s = 1$, valid for $0 < s < 2$,

$$\ln s = \ln(1 + (s-1)) = \sum_{k=0}^\infty \frac{(-1)^k (s-1)^{k+1}}{k+1} = (s-1)(1 + O(s-1))$$

demonstrates that $L(s)$ has a pole at $s = 1$ with residue 1. We now deform the contour in (B.7) over the negative $\mathrm{Re}(s)$-plane around the branch cut of $\ln(s)$, which is the negative real axis. Thus, we consider the contour C that consists of the line at $c > 1$, the quarter of a circle with infinite radius from $\frac{\pi}{2}$ to $\pi - \varepsilon$, the line segment above the real negative axis from minus infinity to $s = 0$, the circle around the origin $s = 0$ from $\pi - \varepsilon$ back to $-\pi - \varepsilon$ with radius δ, the line segment below the real negative axis from $s = 0$ towards minus infinity, the quarter circle with infinite radius back to close the contour C. This contour encloses the pole at $s = 1$, whose residue is $\lim_{s \to 1} \frac{e^{\lambda s}(s-1)}{s \ln s} = e^\lambda$. Cauchy's Residue Theorem (Titchmarsh, 1964) results in

$$\frac{1}{2\pi i} \int_C \frac{e^{\lambda s}}{s \ln s} ds = e^\lambda$$

while the evaluation of the contour C yields

$$\frac{1}{2\pi i} \int_C \frac{e^{\lambda s}}{s \ln s} ds = K(\lambda) + \frac{1}{2\pi i} \int_\infty^0 \frac{e^{-\lambda x} d(-x)}{-x \ln(x e^{i(\pi-\varepsilon)})} + \frac{1}{2\pi i} \int_0^\infty \frac{e^{-\lambda x} d(-x)}{-x \ln(x e^{i(-\pi-\varepsilon)})}$$

since the parts of C along the circles vanish. Hence, we obtain

$$K(\lambda) = e^\lambda + \frac{1}{2\pi i} \int_0^\infty \frac{e^{-\lambda x}}{x\{i\pi + \ln x\}} dx - \frac{1}{2\pi i} \int_0^\infty \frac{e^{-\lambda x}}{x\{-i\pi + \ln x\}} dx$$

Finally, with $\frac{1}{i\pi + \ln x} - \frac{1}{-i\pi + \ln x} = \frac{-2i\pi}{(\pi^2 + (\ln x)^2)}$, we arrive at (3.70).

(xiv) (a) The event $\{X \le x\} = \{Y^{-2} \le x\}$ only exists for $x > 0$ because a square can never be negative. As shown in Section 3.3.5, it is equivalent to $\{Y \le -\frac{1}{\sqrt{x}}\} \cup \{Y \ge \frac{1}{\sqrt{x}}\}$ and the complement of this event is $\{-\frac{1}{\sqrt{x}} < Y < \frac{1}{\sqrt{x}}\}$, which is equivalent to $\{X > x\}$. Since

$Y = N\left(0, \sigma^2\right)$, defined in (3.19), we have, for $x > 0$,

$$\Pr\left[X > x\right] = \Pr\left[-\frac{1}{\sqrt{x}} < Y < \frac{1}{\sqrt{x}}\right] = \frac{1}{\sigma\sqrt{2\pi}}\int_{-\frac{1}{\sqrt{x}}}^{\frac{1}{\sqrt{x}}} \exp\left[-\frac{t^2}{2\sigma^2}\right] dt$$

$$= \frac{1}{\sigma}\sqrt{\frac{2}{\pi}}\int_0^{\frac{1}{\sqrt{x}}} \exp\left[-\frac{t^2}{2\sigma^2}\right] dt$$

By differentiation of $F_X\left(x\right) = 1 - \Pr\left[X > x\right]$ with respect to x, we find (3.72).
(b) The k-th moment of X is

$$E\left[X^k\right] = \int_0^\infty t^k f_X\left(t\right) dt = \frac{1}{\sigma\sqrt{2\pi}}\int_0^\infty t^{k-3/2} e^{-\frac{1}{2\sigma^2 t}} dt = \frac{1}{\sigma\sqrt{2\pi}}\int_0^\infty u^{-k-\frac{1}{2}} e^{-\frac{u}{2\sigma^2}} du$$

This integral Gamma integral only converges $k < \frac{1}{2}$. Hence, the mean and higher moments do not exist, the case $k = 0$ corresponds with the normalization of the probability, $E\left[X^0\right] = 1$. The probability generating function (2.38) is

$$\varphi_X(z) = \frac{1}{\sigma\sqrt{2\pi}}\int_0^\infty e^{-zt}\frac{1}{t^{3/2}} e^{-\frac{1}{2\sigma^2 t}} dt$$

By using (3.31) and $K_s\left(z\right) = K_{-s}\left(z\right)$, we obtain

$$\varphi_X(z) = \frac{1}{\sigma}\sqrt{\frac{2}{\pi}}\left(2z\sigma^2\right)^{1/4} K_{\frac{1}{2}}\left(\sqrt{\frac{2z}{\sigma^2}}\right)$$

After invoking (3.34) for $n = 0$, which leads to $K_{\frac{1}{2}}\left(z\right) = \sqrt{\frac{\pi}{2z}}e^{-z}$, we arrive at (3.73). Since $\varphi_X(z) = E\left[e^{-zX}\right] = e^{-\frac{1}{\sigma}\sqrt{2z}}$ is not analytic at $z = 0$, there does not exist a Taylor series around $z = 0$ and we conclude again from (2.41) that no moments exist.
(c) We apply the formula (3.40) for maximum of a set of i.i.d random variables,

$$\Pr\left[X_{(n)} \le \alpha x\right] = \left(\Pr[X \le \alpha x]\right)^n = \left(1 - \frac{1}{\sigma}\sqrt{\frac{2}{\pi}}\int_0^{\frac{1}{\sqrt{\alpha x}}} \exp\left[-\frac{t^2}{2\sigma^2}\right] dt\right)^n$$

We determine α as a function of n in order to have a finite limit for $\lim_{n\to\infty} \Pr\left[X_{(n)} \le \alpha x\right]$. For large α, the integral tends to zero and a Taylor series expansion of $\int_0^y \exp\left[-\frac{t^2}{2\sigma^2}\right] dt$ around $y = 0$ yields

$$\int_0^y \exp\left[-\frac{t^2}{2\sigma^2}\right] dt = y - \frac{1}{6\sigma^2}y^3 + O\left(y^5\right)$$

such that

$$1 - \frac{1}{\sigma}\sqrt{\frac{2}{\pi}}\int_0^{\frac{1}{\sqrt{\alpha x}}} \exp\left[-\frac{t^2}{2\sigma^2}\right] dt = 1 - \frac{1}{\sigma}\sqrt{\frac{2}{\pi x}}\frac{1}{\sqrt{\alpha}} + O\left(\frac{1}{\alpha^{3/2}}\right)$$

Let $y = \left(\Pr[X \le \alpha x]\right)^n$, then $\log y = n \log \Pr[X \le \alpha x]$ and

$$\log y = n\log\left(1 - \frac{1}{\sigma}\sqrt{\frac{2}{\pi x}}\frac{1}{\sqrt{\alpha}} + O\left(\frac{1}{\alpha^{3/2}}\right)\right) = -\frac{1}{\sigma}\sqrt{\frac{2}{\pi x}}\frac{n}{\sqrt{\alpha}} + O\left(\frac{n}{\alpha^{3/2}}\right)$$

In order for $\lim_{n\to\infty}\left(\Pr[X \le \alpha x]\right)^n$ to be finite, $\lim_{n\to\infty} \log y = c$, where the constant $c < 0$. This condition is satisfied if we choose $\alpha = n^2$ and we obtain

$$\lim_{n\to\infty}\left(\Pr[X \le n^2 x]\right)^n = \exp\left(-\frac{1}{\sigma}\sqrt{\frac{2}{\pi x}}\right)$$

which proves (3.74).

(d) Applying the probability generating function (2.75) of the sum of i.i.d. random variables to the pgf (3.73) yields

$$\varphi_{S_n}(z) = e^{-\frac{n}{\sigma}\sqrt{2z}}$$

Similar to (c) and to the proof of the Central Limit Theorem 6.3.1, we introduce the scaling α as a function of n to deduce a limit law. We consider

$$\log \varphi_{S_n}(\alpha z) = -\frac{n\sqrt{\alpha}}{\sigma}\sqrt{2z}$$

and observe that $\lim_{n\to\infty} \log \varphi_{S_n}(\alpha z) = -\frac{1}{\sigma}\sqrt{2z}$ provided $\alpha = n^{-2}$. Since $\varphi_{S_n}(\alpha z) = E\left[e^{-z\alpha S}\right] = \varphi_{\alpha S_n}(z)$, we find that

$$\lim_{n\to\infty} \varphi_{\frac{S_n}{n^2}}(z) = e^{-\frac{1}{\sigma}\sqrt{2z}} = \varphi_X(z)$$

The Continuity Theorem 6.1.3 then proves the limit law (3.75).

B.3 Correlation (Chapter 4)

(i) We explain two different methods. Each of these methods can be extended to higher dimensions when $f_{X_1 X_2 \ldots X_N}(x_1, \ldots, x_N)$ is given.
(a) Applying the conditional probability (2.48),

$$\Pr[X \leq x, Y \leq y] = \Pr[Y \leq y | X \leq x] \Pr[X \leq x]$$

shows that we can first generate the random variable X and then the random variable Y given X. The marginal distribution of X is obtained as

$$\Pr[X \leq x] = F_X(x) = \int_{-\infty}^{x} du \int_{-\infty}^{\infty} f_{XY}(u, v) dv$$

The conditional distribution of Y given X follows from (2.54) as

$$\Pr[Y \leq y | X = x] = F_{Y|X}(y|x) = \int_{-\infty}^{y} \frac{f_{XY}(x, v)}{f_X(x)} dv$$

The random variable X is now generated from a uniform random variable U_1 after the transform $X = F_X^{-1}(U_1)$ as explained in Section 3.2.1. To each realization of $U_1 = u_{1i} \in [0, 1]$, there corresponds a realization of $X = x_i$. Given $X = x_i$, a realization of $Y = y_i$ is computed after a similar transformation of a second uniform random variable U_2 on $[0, 1]$, that is independent of U_1, with the result $Y = F_{Y|X}^{-1}(U_2|x_i) = y_i$. In this way, a realization of (x_i, y_i) is obtained. By repeating the process of generating two independent realizations (u_{1i}, u_{2i}) of the uniform random variables U_1 and U_2, each time a realization (x_i, y_i) of X and Y with joined pdf $f_{XY}(x, y)$ is found.
(b) The second method assumes a finite support of X and Y, i.e. $a \leq X \leq b$ and $c \leq Y \leq d$. In addition, also the joint density needs to be bounded such that $\max f_{XY}(x, y) = z > 0$ is finite. Consider three independent uniform random variables W_1 on $[a, b]$, W_2 on $[c, d]$ and W_3 on $[0, z]$. We accept a realization $(x_i, y_i) = (w_{1i}, w_{2i})$ if and only if $w_{3i} \leq f_{XY}(x_i, y_i)$. If $w_{3i} > f_{XY}(x_i, y_i)$, we reject the realization (x_i, y_i). By repeating this recipe, n realizations can be found. However, the test may fail an arbitrarily number of times, which may reduce the computational efficiency. We end by demonstrating that the above recipe indeed generates a realization (x_i, y_i) of X and Y with joined pdf $f_{XY}(x, y)$. The recipe leads to a success with

$$\Pr[\text{success}] = \frac{1}{(b-a)(d-c)z} \int_a^b du \int_c^d dv\, f_{XY}(u, v) = \frac{1}{(b-a)(d-c)z}$$

because $\int_a^b du \int_c^d dv\, f_{XY}(u, v) = 1$ and the three independent coordinates w_{1i}, w_{2i} and w_{3i} are confined to the volume $V = (b-a)(d-c)z$, whereas success is only awarded in case a realization of (w_{1i}, w_{2i}, w_{3i}) lies in a volume below the surface $f_{XY}(x, y)$ enclosed by the volume V. Further, the chance of a particular, successful realization is related to the conditional probability

$$R = \Pr\left[x \le X \le x + dx, y \le Y \le y + dy | \text{success}\right]$$

$$= \frac{\Pr\left[\{x \le X \le x + dx, y \le Y \le y + dy\} \cap \{\text{success}\}\right]}{\Pr\left[\text{success}\right]} = \frac{f_{XY}(x,y)\frac{dx\,dy}{V}}{\frac{1}{V}} = f_{XY}(x,y)\,dx\,dy$$

which shows that, if the realization (x_i, y_i) is accepted, it possesses the desired joined pdf $f_{XY}(x,y)$. The analysis also shows that we may choose an arbitrary volume V provided that it encloses the total volume under the surface $f_{XY}(x,y)$. This observation removes the restriction of the finite extent of z and allows that density functions are unbounded. Since $\int_{-\infty}^{\infty}\int_{-\infty}^{\infty} f_{XY}(u,v)\,du\,dv = 1$ and $f_{XY}(u,v) \ge 0$, it is always possible to find an enclosing surface $g(u,v) \ge f_{XY}(u,v)$ with finite volume V. The efficiency of the recipe clearly depends on $\Pr\left[\text{success}\right] = \frac{1}{V}$, which shows that the smaller the volume, the more efficient the algorithm.

(ii) The linear correlation coefficient (2.61) is

$$\rho(X_{\min}, X) = \frac{E\left[X_{\min}X\right] - E\left[X_{\min}\right]E\left[X\right]}{\sqrt{\text{Var}\left[X_{\min}\right]}\sqrt{\text{Var}\left[X\right]}}$$

The distribution of $X_{\min}$ is given by (3.39). The mean can be expressed using the tail probability formula (2.36) as

$$E\left[X_{\min}\right] = \mu_{\min} = \int_0^{\infty} (\Pr\left[X > t\right])^m\,dt$$

The variance $\text{Var}[X_{\min}] = E\left[X_{\min}^2\right] - \mu_{\min}^2$ is deduced from the second moment

$$E\left[X_{\min}^2\right] = \int_0^{\infty} t^2 f_{X_{\min}}(t)\,dt$$

which is, after partial integration,

$$E\left[X_{\min}^2\right] = 2\int_0^{\infty} x\Pr\left[X_{\min} > x\right]\,dx = 2\int_0^{\infty} x\,(\Pr\left[X > t\right])^m\,dx$$

Using (2.64), we now concentrate on

$$E\left[X_{\min}X\right] = \int_0^{\infty} x\,dx \int_0^{\infty} y f_{XX_{\min}}(x,y)\,dy$$

Partial integration with respect to y yields

$$\int_0^{\infty} y f_{XX_{\min}}(x,y)\,dy = -y \int_y^{\infty} f_{XX_{\min}}(x,u)\,du \bigg|_0^{\infty} + \int_0^{\infty} dy \int_y^{\infty} f_{XX_{\min}}(x,u)\,du$$

$$= \int_0^{\infty} dy \int_y^{\infty} f_{XX_{\min}}(x,u)\,du$$

Written in terms of the distribution function (2.62) yields

$$\int_0^{\infty} y f_{XX_{\min}}(x,y)\,dy = \int_0^{\infty} dy \int_y^{\infty} \frac{\partial^2}{\partial x \partial u}\Pr\left[X \le x, X_{\min} \le u\right]\,du$$

$$= \int_0^{\infty} dy \left(\frac{\partial}{\partial x}\Pr\left[X \le x\right] - \frac{\partial}{\partial x}\Pr\left[X \le x, X_{\min} \le y\right]\right)$$

We may write $\frac{\partial}{\partial x}\Pr\left[X \le x, X_{\min} \le y\right] = \Pr\left[X_{\min} \le y, X = x\right]$ and using the definition of conditional probability (2.47), we have

$$\Pr\left[X_{\min} \le y, X = x\right] = \Pr\left[X_{\min} \le y | X = x\right]\Pr\left[X = x\right]$$

Now, if $y > x$, then $\Pr\left[X_{\min} \le y | X = x\right] = 1$ because the minimum is always smaller than or equal to any of the m random variables $\{X_k\}_{1 \le k \le m}$. If $y \le x$, the event that $\{X_{\min} \le y\}$

is equivalent to the fact that at least one of the remaining $m-1$ random variables is smaller or equal to y. Hence, we have that

$$\Pr\left[X_{\min} \leq y | X = x\right] = \begin{cases} 1 & \text{if } y > x \\ 1 - \left(\Pr\left[X > y\right]\right)^{m-1} & \text{if } y \leq x \end{cases}$$

such that

$$\int_0^\infty y f_{XX_{\min}}(x, y) dy = f_X(x) \int_0^\infty dy \left(1 - \Pr\left[X_{\min} \leq y | X = x\right]\right)$$

$$= f_X(x) \int_0^x dy \left(\Pr\left[X > y\right]\right)^{m-1}$$

Thus,

$$E\left[X_{\min} X\right] = \int_0^\infty dx \, x f_X(x) \int_0^x dy \left(\Pr\left[X > y\right]\right)^{m-1}$$

$$\leq \int_0^\infty dx \, x f_X(x) \int_0^\infty dy \left(\Pr\left[X > y\right]\right)^{m-1} = E\left[X\right] E\left[X_{\min;m-1}\right]$$

In general for finite m, it is not easy to determine whether $\rho\left(X_{\min}, X\right)$ is positive or negative. One may verify for an exponential distribution $f_X(x) = e^{-x}$, that $\rho\left(X_{\min}, X\right) = \frac{1}{m-1}\left(1 - \frac{1}{m}\right) \geq 0$ for all m and that the correlation between X and $X_{\min}$ is weak, tending to $\rho\left(X_{\min}, X\right) \to 0$ for large m.

B.4 Inequalities (Chapter 5)

(i) If all $a_k > 0$, then the event $\{a_1 X_1 + a_2 X_2 + \cdots + a_n X_n > t\}$ is decomposed as the union of $\cup_{k=1}^n \left\{a_k X_k > \frac{t}{n}\right\}$. Applying Boole's inequality (2.9) then leads to (5.45).

(ii) We start from the equality

$$1 = \frac{1}{n} \sum_{k=1}^n \frac{x_k}{x_k + y_k} + \frac{1}{n} \sum_{k=1}^n \frac{y_k}{x_k + y_k}$$

and apply the harmonic, geometric and arithmetic mean inequality (5.3) to each sum so that

$$1 \geq \sqrt[n]{\prod_{k=1}^n \frac{x_k}{x_k + y_k}} + \sqrt[n]{\prod_{k=1}^n \frac{y_k}{x_k + y_k}} = \frac{\sqrt[n]{\prod_{k=1}^n x_k} + \sqrt[n]{\prod_{k=1}^n y_k}}{\sqrt[n]{\prod_{k=1}^n x_k + y_k}}$$

from which the inequality (5.46) follows.

(iii) Applying the functional relation $f(\theta x) = \theta^p f(x)$ with $\theta = \frac{1}{h(x)}$ shows that, with the definition of $h(x) = (f(x))^{1/p}$ that

$$f\left(\frac{x}{h(x)}\right) = \frac{f(x)}{h^p(x)} = 1 \tag{B.8}$$

To prove convexity, we consider the inequality (5.6) for $0 \leq \theta \leq 1$

$$h\left(\theta \frac{x}{h(x)} + (1-\theta) \frac{y}{h(y)}\right) = \left(f\left(\theta \frac{x}{h(x)} + (1-\theta) \frac{y}{h(y)}\right)\right)^{1/p}$$

Since f is convex on the interval I,

$$f\left(\theta \frac{x}{h(x)} + (1-\theta) \frac{y}{h(y)}\right) \leq \theta f\left(\frac{x}{h(x)}\right) + (1-\theta) f\left(\frac{y}{h(y)}\right) = 1$$

where the last inequality follows from (B.8). Hence,

$$h\left(\theta \frac{x}{h(x)} + (1-\theta) \frac{y}{h(y)}\right) \leq 1$$

and, by choosing $\theta = \frac{h(x)}{h(x)+h(y)}$ for all nonzero $x \in I$, then $0 \le \theta \le 1$ and

$$h\left(\frac{x+y}{h(x)+h(y)}\right) \le 1$$

Finally, since $h(\theta x) = \theta h(x)$, we find $h(x+y) \le h(x) + h(y)$, which demonstrates that h obeys the inequality (5.6), after letting $x = \theta u$ and $y = (1-\theta)v$.

B.5 Limit Laws (Chapter 6)

(i) The basic law (15.4) for the degree implies that the nodal degrees are not independent. However, if the graph is sufficiently large, this dependence is weak and the nodal degrees may be considered as independent random variables which allows us to apply the theory of Section 6.6. If D_k is the degree of node k and all nodal degrees are i.i.d. random variables with probability density function (3.52), then the maximum degree $D_{\max} = \max_{1 \le k \le N} D_k$ has the distribution, for $x > \tau$,

$$\Pr[D_{\max} \le x] = \left(\left(\frac{x}{\tau}\right)^{-\alpha}\right)^N$$

According to the theory of Section 6.6, we need to find a sequence of real numbers x_N such that $N\left(\frac{x_N}{\tau}\right)^{-\alpha} \to \xi$ if $N \to \infty$. Equivalently, $x_N = \tau\left(\frac{\xi}{N}\right)^{-\frac{1}{\alpha}} = \tau N^{\frac{1}{\alpha}}\xi^{-\frac{1}{\alpha}}$ such that,

$$\lim_{N\to\infty} \Pr\left[D_{\max} \le \tau N^{\frac{1}{\alpha}}\xi^{-\frac{1}{\alpha}}\right] = e^{-\xi}$$

With $y = \xi^{-\frac{1}{\alpha}} \ge 0$, the scaled random variable $X = \frac{D_{\max}}{\tau N^{\frac{1}{\alpha}}}$ has a Fréchet distribution

$$\lim_{N\to\infty} \Pr\left[\frac{D_{\max}}{\tau N^{\frac{1}{\alpha}}} \le y\right] = e^{-y^{-\alpha}}$$

from which we find that the maximum degree in a power-law graph with degree distribution $\Pr[D \ge x] = \left(\frac{x}{\tau}\right)^{-\alpha}$ scales as $D_{\max} = O\left(N^{\frac{1}{\alpha}}\right)$.

B.6 Poisson process (Chapter 7)

(i) Let Y be a binomial random variable with parameters N and p, where N is a Poisson random variable with parameter λ. The probability density function of Y is obtained by applying the law of total probability (2.49),

$$\Pr[Y = k] = \sum_{n=0}^{\infty} \Pr[Y = k|N = n]\Pr[N = n]$$

With (3.3) and (3.9), we have

$$\Pr[Y = k] = \sum_{n=0}^{\infty} \binom{n}{k} p^k q^{(n-k)} \frac{\lambda^n e^{-\lambda}}{n!} = \frac{p^k}{k!} e^{-\lambda} \sum_{n=k}^{\infty} \frac{q^{(n-k)}\lambda^n}{(n-k)!}$$

$$= \frac{p^k \lambda^k}{k!} e^{-\lambda} \sum_{n=0}^{\infty} \frac{q^n \lambda^n}{n!} = \frac{(p\lambda)^k}{k!} e^{-\lambda+q\lambda}$$

Since $q = 1 - p$, we arrive at $\Pr[Y = k] = \frac{(p\lambda)^k}{k!} e^{-\lambda p}$, which means that Y is a Poisson random variable with mean $p\lambda$. If a sufficient sample of test strings defined above is sent and received, the mean number of "one bits" at receiver divided by the mean number of bits at the sender gives the probability p (if errors occur indeed independently).

(ii) Since the counting process of a sum of a Poisson process is again a Poisson counting process with rate equal to $\sum_{j=1}^{4} \lambda_j$, the mean number of packets of the four classes in the router's buffers during interval T is $\lambda = T \sum_{j=1}^{4} \lambda_j$. Hence, the probability density function for the total number N of arrivals is $\Pr[N = n] = \frac{\lambda^n}{n!} e^{-\lambda}$.

(iii) Theorem 7.3.3 states the $N(t)$ is a Poisson counting process with rate $\lambda_1 + \lambda_2$. Then,

$$\Pr[X_1(t) = 1 | X(t) = 1] = \frac{\Pr[\{X_1(t) = 1\} \cap \{X(t) = 1\}]}{\Pr[X(t) = 1]} = \frac{\Pr[\{X_1(t) = 1\} \cap \{X_2(t) = 0\}]}{\Pr[X(t) = 1]}$$

$$= \frac{\Pr[X_1(t) = 1] \Pr[X_2(t) = 0]}{\Pr[X(t) = 1]} = \frac{\lambda_1}{\lambda_1 + \lambda_2}$$

since the Poisson random variables X_1 and X_2 are independent. As an application we can consider a Poissonean arrival flow of packets at a router with rate λ. If the packets are marked randomly with probability $p = \frac{\lambda_1}{\lambda}$, the resulting flow consists of two types, those marked and those not. Each of these flows is again a Poisson flow, the marked flow with rate $\lambda_1 = p\lambda$ and the non-marked flow with $\lambda_2 = (1 - p)\lambda$. Actually, this procedure leads to a decomposition of the Poisson process into two independent Poisson processes and leads to the reverse of Theorem 7.3.3.

(iv) (a) Applying the solution of previous exercise immediately gives $\frac{\lambda_1}{\lambda_1 + \lambda_2 + \lambda_3}$.

(b) Since the three Poisson processes are independent, the total number of cars on the three lanes, denoted by X, is also a Poisson process (Theorem 7.3.3) with rate $\lambda = \lambda_1 + \lambda_2 + \lambda_3$. Hence, $\Pr[X = n] = \frac{\lambda^n}{n!} e^{-\lambda}$.

(c) Let us denote the Poisson process in lane j by X_j. Then, using the independence between the X_j,

$$\Pr[X_1 = n, X_2 = 0, X_3 = 0] = \Pr[X_1 = n] \Pr[X_2 = 0] \Pr[X_3 = 0] = \frac{\lambda_1^n}{n!} e^{-\lambda}$$

(v) (a) The player relies on the fact that during the time (s, T) there is exactly one arrival. Since the game rules mention that he should identify the last signal in $(0, T)$, signals arriving during $(0, s)$ do not influence his chance to win because of the memoryless property of the Poisson process. The number of arrivals in the interval (s, T) obeys a Poisson distribution with parameter $\lambda(T - s)$. The probability that precisely one signal arrives in the interval (s, T) is $\Pr[N(T) - N(s) = 1] = \lambda(T - s) e^{-\lambda(T-s)}$.

(b) Maximizing this winning probability with respect to s (by equating the first derivative to zero) yields

$$\frac{d}{ds} \Pr[N(T) - N(s) = 1] = -\lambda e^{-\lambda(T-s)} + \lambda^2 (T - s) e^{-\lambda(T-s)} = 0$$

with solution $\lambda(T - s) = 1$ or $s = T - 1/\lambda$. This maximum (which is readily verified by checking that $\frac{d^2}{ds^2} \Pr[N(T) - N(s) = 1] < 0$) lies inside the allowed interval $(0, T)$. The maximum probability of winning is $\Pr[N(T) - N(T - 1/\lambda) = 1] = 1/e$.

(vi) (a) We apply the general formula (7.1) for the pdf of a Poisson process with mean $E[X(t)] = \lambda t = 1$. Then, $\Pr[X(t + s) - X(s) = 0] = e^{-\lambda t} = \frac{1}{e}$.

(b) $\Pr[X(t + s) - X(s) > 10] = 1 - \Pr[X(t + s) - X(s) \le 10] = 1 - \frac{1}{e} \sum_{k=0}^{10} \frac{1}{k!}$.

(c) Each minute is equally probable as follows from Theorem 7.4.1.

(vii) This exercise is an application of randomly marking in a Poisson flow as explained in solution (iii) above. The total flow of packets can be split up into an ACK stream, a Poisson process N_1 with rate $p\lambda = 3 s^{-1}$ and a data flow, an independent Poisson process N_2 with rate $(1 - p)\lambda = 7 s^{-1}$. Then,

(a) $\Pr[N_1 \ge 1] = 1 - \Pr[N_1 = 0] = 1 - e^{-3}$,

(b) The mean number is $E[N_1 + N_2 | N_1 = 5] = E[N_1 | N_1 = 5] + E[N_2 | N_1 = 5] = 5 + E[N_2] = 5 + 7 = 12$ packets.

(c) $\Pr[N_1 = 2 | N_1 + N_2 = 8] = \frac{\Pr[N_1 = 2, N_1 + N_2 = 8]}{\Pr[N_1 + N_2 = 8]} = \frac{\frac{3^2}{2!} e^{-3} \frac{7^6}{6!} e^{-7}}{\frac{10^8}{8!} e^{-10}} \approx 29.65\%$.

(viii) (a) Since the three Poisson arrival processes are independent, the total number of requests will also be a Poisson process with the parameter $\lambda = \lambda_1 + \lambda_2 + \lambda_3 = 20$ requests/hour

(Theorem 7.3.3). The expected number of requests during an 8-hour working day is $E\left[N\right] = \lambda t = 20 \times 8 = 160$ requests.

(b) If we denote arrival processes of requests with different ADSL problems each with a random variable X_i for $i = 1, 2$ and 3, then due to their mutual independence

$$\Pr\left[X_1 = 0, X_2 = k, X_3 = 0\right] = \Pr\left[X_1 = 0\right]\Pr\left[X_2 = k\right]\Pr\left[X_3 = 0\right] = e^{-\lambda_1 t}\frac{\left(\lambda_2 t\right)^k e^{-\lambda_2 t}}{k!}e^{-\lambda_3 t}$$

from which $\Pr\left[X_1 = 0, X_2 = 3, X_3 = 0\right] = e^{-\frac{8}{3}}\dfrac{\left(\frac{6}{3}\right)^3 e^{-\frac{6}{3}}}{3!}e^{-\frac{6}{3}} = 1.7 \times 10^{-3}$.

(c) If we denote the total number of requests by X then $\Pr\left[X = 0\right] = e^{-\lambda t} = e^{-\frac{20}{4}} = 6.7 \times 10^{-3}$.

(d) The precise time is irrelevant for Poisson processes, only the duration of the interval matters. Here intervals are overlapping and we need to compute the probability

$$p = \Pr\left[\{X\left(0.2\right) = 1\} \cap \{X\left(0.5\right) - X\left(0.1\right) = 2\}\right]$$

$$= \sum_{k=0}^{1} \Pr\left[\{X\left(0.1\right) = k\} \cap \{X\left(0.2\right) - X\left(0.1\right) = 1 - k\} \cap \{X\left(0.5\right) - X\left(0.2\right)\} = 1 + k\right]$$

$$= \sum_{k=0}^{1} \Pr\left[X\left(0.1\right) = k\right]\Pr\left[X\left(0.2\right) - X\left(0.1\right) = 1 - k\right]\Pr\left[X\left(0.5\right) - X\left(0.2\right) = 1 + k\right]$$

$$= \sum_{k=0}^{1} e^{-2}\frac{\left(2\right)^k}{k!}e^{-2}\frac{\left(2\right)^{1-k}}{\left(1 - k\right)!}e^{-6}\frac{\left(6\right)^{1+k}}{\left(1 + k\right)!} = 48e^{-10} = 2.18 \times 10^{-3}$$

(e) Given that at the moment $t + s$ there are $k + m$ requests, the probability that there were k requests at the moment t is

$$\Pr\left[X\left(t\right) = k \mid X\left(t + s\right) = k + m\right] = \frac{\Pr\left[\{X\left(t\right) = k\} \cap \{X\left(t + s\right) = k + m\}\right]}{\Pr\left[X\left(t + s\right) = k + m\right]}$$

$$= \frac{\Pr\left[X\left(t\right) = k\right]\Pr\left[X\left(t + s\right) - X\left(t\right) = m\right]}{\Pr\left[X\left(t + s\right) = k + m\right]}$$

$$= \frac{\dfrac{\left(\lambda t\right)^k e^{-\lambda t}}{k!}\dfrac{\left(\lambda s\right)^m e^{-\lambda s}}{m!}}{\dfrac{\left(\lambda\left(t + s\right)\right)^{k+m} e^{-\lambda\left(t+s\right)}}{\left(k + m\right)!}} = \binom{k + m}{k}\left(\frac{t}{t + s}\right)^k\left(\frac{s}{t + s}\right)^m$$

(ix) (a) The number of attacks that are arriving to the PC is a Poisson random variable $X\left(t\right)$ with rate $\lambda = 6$. The probability that exactly one $\left(k = 1\right)$ attack during one $\left(t = 1\right)$ hour follows from (7.1) as $\Pr\left[X(1) = 1\right] = 6e^{-6}$.

(b) Applying (7.2), the expected amount of time that the PC has been on is $t = \frac{E[X(t)]}{\lambda} = \frac{60}{6} = 10$ hours.

(c) The arrival time of the fifth attack is denoted by T. Given that there are six attacks in one hour $\left(t = 1\right)$, we compute the probability $\Pr[T < t \mid X(1) = 6]$ that either five attacks arrive in the interval $(0, t)$ and one arrives in $(t, 1)$ or all six attacks arrive in $(0, t)$ and none arrives in the interval $(t, 1)$. Hence, for $0 \leq t < 1$,

$$F_T(t) = \Pr[T < t \mid X(1) = 6]$$

$$= \frac{\Pr[\{X(t) = 5\} \cap \{X(1) = 6\}] + \Pr[\{X(t) = 6\} \cap \{X(1) = 6\}]}{\Pr[X(1) = 6]}$$

$$= \frac{\Pr[X(t) = 5]\Pr[X(1) - X(t) = 1] + \Pr[X(t) = 6]\Pr[X(1) - X(t) = 0]}{\Pr[X(1) = 6]}$$

$$= \frac{((\lambda t)^5/5!)e^{-\lambda t}\lambda(1 - t)e^{-\lambda(1-t)} + ((\lambda t)^6/6!)e^{-\lambda t}e^{-\lambda(1-t)}}{(\lambda^6/6!)e^{-\lambda}} = 6t^5 - 5t^6.$$

The probability that the fifth attack will arrive between 1:30 p.m. and 2 p.m. is $F_T(1) - F_T\left(\frac{1}{2}\right) = 1 - \frac{7}{64} = \frac{57}{64}$.

(d) The expectation of T given $X(1) = 6$ follows from (2.34) as $E[T|X(1)] = \int_0^1 x f_T(x)dx$ where $f_T(t) = \frac{dF_T(t)}{dt}$ derived in (c). Alternatively, the expectation can be computed from (2.36), $E[T|X(1)] = \int_0^1 \left(1 - (6x^5 - 5x^6)\right)dx = \frac{5}{7}$. Hence the expected arrival time of the fifth attack between 1 p.m. and 2 p.m. is about 1:43 p.m.

(x) (a) For a Poisson process $\{X(t), t \geq 0\}$ with rate λ, the distribution of arrival time W_n of the n-th event or the waiting time until the n-th event is given for $t \geq 0$ by the Erlang distribution (7.6). The corresponding probability density function $f_{W_n}(t)$ for $t \geq 0$ is found by derivation of (7.6) with respect to t as

$$f_{W_n}(t) = -\sum_{j=n}^{\infty} \lambda e^{-\lambda t} \frac{(\lambda t)^j}{j!} + \sum_{j=n}^{\infty} \lambda e^{-\lambda t} \frac{(\lambda t)^{j-1}}{(j-1)!} = \lambda e^{-\lambda t} \frac{(\lambda t)^{n-1}}{(n-1)!}$$

We denote the m-th event or the waiting time until the m-th event in the Poisson process Y by V_m. Using the continuous version of the law of total probability (2.49) we compute

$$\Pr[W_k \leq V_m] = \int_0^{\infty} \Pr[W_k \leq t | V_m = t] f_{V_m}(t)dt$$

Since the process X and Y are independent, the conditional probability is

$$\Pr[W_k \leq t | V_m = t] = \Pr[W_k \leq t]$$

and we obtain

$$\Pr[W_k \leq V_m] = \int_0^{\infty} \sum_{j=k}^{\infty} \frac{(\lambda t)^j e^{-\lambda t}}{j!} \mu e^{-\mu t} \frac{(\mu t)^{m-1}}{(m-1)!} dt = \sum_{j=k}^{\infty} \frac{\lambda^j \mu^m}{j!(m-1)!} \int_0^{\infty} t^{j+m-1} e^{-(\lambda+\mu)t} dt$$

Using the Gamma function, $\Gamma(s) = \int_0^{\infty} t^{s-1} e^{-t}dt$, we arrive at

$$\Pr[W_k \leq V_m] = \frac{\mu^m}{(m-1)!(\lambda+\mu)^m} \sum_{j=k}^{\infty} \frac{\lambda^j (j+m-1)!}{j!(\lambda+\mu)^j}$$

Alternatively, we can reformulate the law of total probability as

$$\Pr[W_k \leq V_m] = \int_0^{\infty} \Pr[t \leq V_m | W_k = t] f_{W_k}(t)dt$$

Again using independence and

$$\Pr[V_m \geq t] = 1 - \Pr[V_m \leq t] = 1 - e^{-\mu t} \sum_{j=m}^{\infty} \frac{(\mu t)^j}{j!} = e^{-\mu t} \sum_{j=0}^{m-1} \frac{(\mu t)^j}{j!}$$

we arrive at

$$\Pr[W_k \leq V_m] = \int_0^{\infty} e^{-\mu t} \sum_{j=0}^{m-1} \frac{(\mu t)^j}{j!} \lambda e^{-\lambda t} \frac{(\lambda t)^{k-1}}{(k-1)!} dt$$

$$= \sum_{j=0}^{m-1} \frac{\mu^j \lambda^k}{j!(k-1)!} \int_0^{\infty} t^{j+k-1} e^{-(\lambda+\mu)t} dt = \sum_{j=0}^{m-1} \frac{\mu^j \lambda^k (j+k-1)!}{j!(k-1)!(\lambda+\mu)^{j+k}}$$

Finally, for any integer $k, m \geq 1$, we find two expressions representing the point of view of the second and first process,

$$\Pr[W_k \leq V_m] = \frac{\mu^m}{(m-1)!(\lambda+\mu)^m} \sum_{j=k}^{\infty} \frac{\lambda^j (j+m-1)!}{j!(\lambda+\mu)^j}$$

$$= \frac{\lambda^k}{(k-1)!(\lambda+\mu)^k} \sum_{j=0}^{m-1} \frac{\mu^j (j+k-1)!}{j!(\lambda+\mu)^j}$$

In particular, the probability that an event in X occurs before an event in Y equals

$$\Pr\left[W_1 \le V_1\right] = \Pr\left[\tau_X \le \tau_Y\right] = \frac{\lambda}{\lambda + \mu}$$

which also equals the steady-state probability in a two-state Markov chain (see Section 10.6), when X is Poisson process to escape from state 1 to state 2 (and vice versa for Y).
(b) The probability that the number of X-events in $[0, t]$ is larger than the number of Y-events in $[0, u]$ is

$$\Pr\left[X\left(t\right) > Y\left(u\right)\right] = \sum_{j=0}^{\infty} \Pr\left[X\left(t\right) > j | Y\left(u\right) = j\right] \Pr\left[Y\left(u\right) = j\right]$$

By independence of the X and Y process,

$$\Pr\left[X\left(t\right) > Y\left(u\right)\right] = \sum_{j=0}^{\infty} \Pr\left[X\left(t\right) > j\right] \Pr\left[Y\left(u\right) = j\right] = e^{-a-b} \sum_{j=0}^{\infty} \left(\sum_{k=j+1}^{\infty} \frac{a^k}{k!} \right) \frac{b^j}{j!}$$

where $a = E\left[X\left(t\right)\right] = \lambda t$ and $b = E\left[Y\left(u\right)\right] = \mu u$. Analogously,

$$\Pr\left[X\left(t\right) = Y\left(u\right)\right] = \sum_{j=0}^{\infty} \Pr\left[X\left(t\right) = j\right] \Pr\left[Y\left(u\right) = j\right] = e^{-a-b} \sum_{j=0}^{\infty} \frac{(ab)^j}{(j!)^2}$$

Since $f\left(n\right) = \int_0^x u^n e^{-u} du$ obeys (via partial integration) the recursion $f\left(n\right) = nf\left(n-1\right) - x^n e^{-x}$ with $f\left(0\right) = 1 - e^{-x}$, we find, after p iterations that,

$$f\left(n\right) = \frac{n!}{(n-p)!} f\left(n-p\right) - e^{-x} \sum_{j=0}^{p-1} \frac{n!}{(n-j)!} x^{n-j}$$

Choosing $p = n$ yields

$$\frac{1}{n!} \int_0^x u^n e^{-u} du = e^{-x} \left(\sum_{k=0}^{\infty} \frac{x^k}{k!} - \sum_{k=0}^{n} \frac{x^k}{k!} \right) = e^{-x} \sum_{k=n+1}^{\infty} \frac{x^k}{k!}$$

Hence,

$$\Pr\left[X\left(t\right) > Y\left(u\right)\right] = e^{-b} \int_0^a du \, e^{-u} \sum_{j=0}^{\infty} \frac{(ub)^j}{(j!)^2}$$

Using the modified Bessel function (Abramowitz and Stegun, 1968, 9.6.10)

$$I_0\left(z\right) = \sum_{j=0}^{\infty} \frac{\left(\left(\frac{z}{2}\right)^2\right)^j}{(j!)^2}$$

leads to

$$\Pr\left[X\left(t\right) > Y\left(u\right)\right] = e^{-b} \int_0^a e^{-u} I_0\left(2\sqrt{bu}\right) du$$

The curious point is that, if $a = b$, then $\lim_{a \to \infty} \Pr\left[X\left(t\right) > Y\left(u\right)\right] = \frac{1}{2}$, although the other extreme $\lim_{a \to 0} \Pr\left[X\left(t\right) > Y\left(u\right)\right] = 0$ and $\Pr\left[X\left(t\right) > Y\left(u\right)\right]$ increases from 0 to 0.5 with a. When $b = 0$, which means that $\Pr\left[Y\left(u\right) = 0\right] = 1$, then

$$\Pr\left[Y\left(u\right) > X\left(t\right)\right] = e^{-a} = \Pr\left[X\left(t\right) = 0\right]$$

Analogously, the probability that two independent Poisson processes generate precisely the same number of events during given intervals is

$$\Pr\left[X\left(t\right) = Y\left(u\right)\right] = \sum_{j=0}^{\infty} \Pr\left[X\left(t\right) = j\right] \Pr\left[Y\left(u\right) = j\right] = e^{-a-b} \sum_{j=0}^{\infty} \frac{(ab)^j}{(j!)^2} = e^{-a-b} I_0\left(2\sqrt{ab}\right)$$

Thus, if $a = b$, then $\Pr[X(t) = Y(u)] = e^{-2a} I_0(2a)$, which is $\Pr[X(t) = Y(u)] = 1 - 2a + O(a^2)$ for small a. For large a, using the asymptotic expansion (Abramowitz and Stegun, 1968, 9.7.1),

$$\Pr[X(t) = Y(u)] = \frac{1}{2\sqrt{\pi a}}\left(1 + \frac{1}{16a} + O(a^{-2})\right)$$

(xi) Let X and X_j denote the lifetime of system and subsystem j respectively. For a series of subsystems with independent lifetimes X_j is the event $\{X > t\} = \cap_{j=1}^{n}\{X_j > t\}$ and $\Pr[X > t] = \prod_{j=1}^{n} \Pr[X_j > t]$. Recall with (3.39) that $\Pr[X > t] = \Pr\left[\min_{1 \leq j \leq n} X_j > t\right]$. Using the definition of the reliability function (7.8) then yields

$$R_{\text{series}}(t) = \prod_{j=1}^{n} R_j(t)$$

(xii) The probability that the system S shown in Fig. 7.7 fails is determined by the subsystem with longest lifetime or $X = \max_{1 \leq j \leq n} X_j$. Invoking relation (3.40) combined with the definition of the reliability function (7.8) leads to

$$R_{\text{parallel}}(t) = 1 - \prod_{j=1}^{n}(1 - R_j(t))$$

(xiii) (a) The probability generating function of the interference is expressed in terms of the conditional expectation (2.81) as

$$\varphi_Y(z) = E\left[e^{-zY}\right] = E_X\left[E\left[e^{-zY}\Big| X\right]\right]$$

Since the location of the terminals are independent, the pair (R_j, W_j) is independent of (R_k, W_k) for $k \neq j$, so that

$$E\left[e^{-zY}\Big| X\right] = E\left[e^{-z\sum_{j=1}^{X} g(R_j, W_j)}\right] = \prod_{j=1}^{X} E\left[e^{-zg(R_j, W_j)}\right]$$

Moreover, both R_j and W_j are identically distributed as R and W respectively such that

$$E_Y\left[e^{-zY}\Big| X\right] = \left(E\left[e^{-zg(R,W)}\right]\right)^{X}$$

Hence,

$$\varphi_Y(z) = E_X\left[\left(E\left[e^{-zg(R,W)}\right]\right)^{X}\right]$$

The probability generating function of the number of terminals X is, with (3.10), $\varphi_X(z) = E\left[z^X\right] = e^{\lambda A(z-1)}$. Invoking the pgf of X, we end up with

$$\varphi_Y(z) = \exp\left[\lambda A\left(f(z) - 1\right)\right] \tag{B.9}$$

where

$$f(z) = E\left[e^{-zg(R,W)}\right] = \int dr \int dw \, e^{-zg(r,w)} f_{RW}(r,w)$$

and where $f_{RW}(r,w)$ is the joint probability density function of R and W and the integral is over all possible values of R and W. The mean interference follows from $E[Y] = -\varphi_Y'(0)$ with $f(0) = 1$ as

$$E[Y] = \lambda V E[g(R,W)] \tag{B.10}$$

We assume further that the position R of a terminals is independent from the power fluctuations W of the emitted signal at position R. In that case, the expression for $f(z)$ further simplifies with $f_{RW}(r,w) = f_R(r) f_W(w)$ to

$$f(z) = \int_0^{\infty} f_R(r)\, dr \int dw \, f_W(w)\, e^{-zg(r,w)}$$

It is common to assume that the region A is a circle with radius a. Since the position R is uniformly distributed over that circle, the pdf $f_R(r) = \frac{2r}{a^2}$ if $r \le a$, else $f_R(r) = 0$. Thus,

$$f(z) = \frac{2}{a^2} \int_0^a r\,dr \int dw\, f_W(w)\, e^{-zg(r,w)}$$

In order to proceed further, we must specify the function g and the distribution of W.

(b) The assumption $g(r,w) = g(r) + h(w)$ simplifies $f(z)$ as

$$f(z) = \frac{2}{a^2} \int_0^a r e^{-zg(r)} dr \int_{-\infty}^\infty dw\, f_W(w)\, e^{-zh(w)}$$

Moreover, if the stochastic variations in the radio signal are negligibly small, it means that $h \to 0$ and $\int_{-\infty}^\infty dw\, f_W(w)\, e^{-zh(w)} \to 1$. We thus obtain

$$f(z) = \frac{2}{a^2} \int_0^a r e^{-zg(r)} dr$$

Integrating by parts yields

$$f(z) = e^{-zg(a)} + \frac{z}{a^2} \int_0^a r^2 g'(r)\, e^{-zg(r)} dr$$

such that (B.9) becomes

$$\varphi_Y(z) = \exp\left[\lambda \pi a^2 \left(e^{-zg(a)} - 1\right)\right] \exp\left[z\lambda\pi \int_0^a r^2 g'(r)\, e^{-zg(r)} dr\right]$$

It is convenient to make the substitution $u = g(r)$ because $du = g'(r)\,dr$ such that

$$\int_0^a r^2 g'(r)\, e^{-zg(r)} dr = \int_{g(0)}^{g(a)} \left(g^{-1}(u)\right)^2 e^{-zu}\, du$$

In general, $\lim_{a\to\infty} g(a) = 0$ as power decreases with distance. When assuming that the area is large, we can expand $e^{-zg(a)} - 1 = zg(a) + O(g^2(a))$, but to have the first factor $\exp\left[\lambda\pi a^2\left(e^{-zg(a)} - 1\right)\right]$ finite, we need to require that $a^2 g(a) \to c$ if $a \to \infty$. The choice $g(r) = r^{-\eta}$ and $\eta > 2$ – in free space $\eta = 2$, in practice $\eta > 2$ – simplifies the pgf of Y further when assuming $a \to \infty$,

$$\varphi_Y(z) = \exp\left[-z\lambda\pi \int_0^\infty u^{-\frac{2}{\eta}} e^{-zu}\, du\right] = \exp\left[-z^{\frac{2}{\eta}} \lambda\pi\Gamma\left(1 - \frac{2}{\eta}\right)\right]$$

valid for $\mathrm{Re}\,(z) > 0$ and $\mathrm{Re}\,(\eta) > 2$. The pdf follows from the inverse Laplace transform (2.39) with $c > 0$ as

$$f_Y(t) = \frac{1}{2\pi i} \int_{c-i\infty}^{c+i\infty} \exp\left[-z^{\frac{2}{\eta}} \lambda\pi\Gamma\left(1 - \frac{2}{\eta}\right) + zt\right] dz$$

This integral can be approximated well by the steepest descent method. Only for $\eta = 4$, the integral can be computed exactly (Abramowitz and Stegun, 1968, Section 29.3.82) as

$$f_Y(t) = \frac{\pi\lambda}{2} t^{-\frac{3}{2}} \exp\left(\frac{-\pi^3\lambda^2}{4t}\right)$$

(xiv) (a) The mean number of TV viewers from 19:00 to 20:00 on a certain day is $\lambda = 1000/\text{hour}$. Each viewer has a probability of $p = 0.1$ to choose the "Discovery" TV channel. Let the random variable X represent the number of "Discovery" TV channel viewers, while Y represents the number of TV viewers, which follows a Poisson distribution

$$\Pr[Y = i] = \frac{\lambda^i}{i!} e^{-\lambda}, \qquad i = 0, 1, 2, \ldots$$

The conditional probability, that among i TV viewers, there are k that watch the "Discovery" TV channel is binomially distributed

$$\Pr[X = k | Y = i] = \binom{i}{k} p^k (1 - p)^{i-k}$$

because each "Discovery" viewer has, independently from the others, a probability p to successfully grasp that channel. By the law of total probability (2.49), we have

$$\Pr[X = k] = \sum_{i=0}^{\infty} \Pr[X = k | Y = i] \Pr[Y = i] = \sum_{i=0}^{\infty} \binom{i}{k} p^k (1 - p)^{i-k} \frac{\lambda^i}{i!} e^{-\lambda}$$

$$= \frac{(\lambda p)^k}{k!} e^{-\lambda} \sum_{i=k}^{\infty} \frac{[\lambda(1 - p)]^{i-k}}{(i - k)!} = \frac{(\lambda p)^k}{k!} e^{-\lambda} e^{\lambda(1-p)} = \frac{(\lambda p)^k}{k!} e^{-\lambda p}$$

Thus, the probability $\Pr[X = k]$ that k viewers watch the "Discovery" TV channel follows a Poisson distribution with parameter λp.

(b) Clearly, with $\lambda p = 100$, the probability to have fewer than four "Discovery" viewers from 19:00 to 20:00 on a certain day is

$$\Pr[X < 4]) = \sum_{k=0}^{3} \Pr[X = k] = \sum_{k=0}^{3} \frac{(\lambda p)^k}{k!} e^{-\lambda p} = 6.39 \times 10^{-39}$$

which means that this event almost never happens.

B.7 Renewal theory (Chapter 8)

(i) The equivalence $\{N(t) \geq n\} \Longleftrightarrow \{W_n \leq t\}$ indicates

$$\Pr[W_{N(t)} \leq x] = \sum_{n=0}^{\infty} \Pr[\{W_n \leq x\} \cap \{W_{n+1} > t\}]$$

$$= \Pr[W_0 \leq x, W_1 > t] + \sum_{n=1}^{\infty} \Pr[W_n \leq x, W_{n+1} > t]$$

The convention $W_0 = 0$ reduces $\Pr[W_0 \leq x, W_1 > t] = \Pr[W_1 > t] = \Pr[\tau_1 > t] = 1 - F_\tau(t)$.
Furthermore, by the law of total probability,

$$\Pr[W_n \leq x, W_{n+1} > t] = \int_0^{\infty} \Pr[W_n \leq x, W_{n+1} > t | W_n = u] \frac{d\Pr[W_n \leq u]}{du} du$$

$$= \int_0^x \Pr[W_{n+1} > t | W_n = u] d\Pr[W_n \leq u]$$

A renewal process restarts after each renewal from scratch (due to the stationarity and the independent increments of the renewal process). This implies that $\Pr[W_{n+1} > t | W_n = u] = \Pr[\tau_{n+1} > t - u] = 1 - F_\tau(t - u)$ because the interarrival times are i.i.d. random variables. Combined,

$$\Pr[W_{N(t)} \leq x] = \Pr[\tau > t] + \sum_{n=1}^{\infty} \int_0^x \Pr[\tau > t - u] d\Pr[W_n \leq u]$$

$$= \Pr[\tau > t] + \int_0^x \Pr[\tau > t - u] d\left(\sum_{n=1}^{\infty} \Pr[W_n \leq u]\right)$$

With the basic equivalence (8.6) and the definition (8.7) of the renewal function $m(t)$, we arrive at

$$\Pr[W_{N(t)} \leq x] = \Pr[\tau > t] + \int_0^x \Pr[\tau > t - u] dm(u)$$

This equation holds for all x. If $x = t$, we can use the renewal equation (8.9),

$$\int_0^t \Pr[\tau > t - u]\, dm(u) = m(t) - \int_0^t \Pr[\tau \le t - u]\, dm(u)$$

$$= m(t) - m(t) + F_\tau(t) = F_\tau(t)$$

which indeed confirms $\Pr[W_{N(t)} \le t] = 1$.

(ii) The generating function of the number of renewals in the interval $[0, t]$ is with (8.10)

$$\varphi_{N(t)}(z) = E\left[z^{N(t)}\right] = \sum_{k=0}^{\infty} \Pr[N(t) = k]\, z^k$$

$$= \Pr[N(t) = 0] + \int_0^t \left(\sum_{k=1}^{\infty} \Pr[N(t - s) = k - 1]\, z^k\right) f_\tau(s)\, ds$$

$$= \Pr[N(t) = 0] + z\int_0^t \left(\sum_{k=0}^{\infty} \Pr[N(t - s) = k]\, z^k\right) f_\tau(s)\, ds$$

$$= \Pr[N(t) = 0] + z\int_0^t \varphi_{N(t-s)}(z)\, dF_\tau(s)$$

From (8.6), we have that $\Pr[N(t) = 0] = 1 - F_\tau(t)$ and

$$\varphi_{N(t)}(z) = 1 - F_\tau(t) + z\int_0^t \varphi_{N(t-s)}(z)\, dF_\tau(s)$$

By derivation with respect to z, we arrive at the differential-integral equation for the derivative of the generating function,

$$\varphi'_{N(t)}(z) = \int_0^t \varphi_{N(t-s)}(z)\, dF_\tau(s) + z\int_0^t \varphi'_{N(t-s)}(z)\, dF_\tau(s)$$

$$= \frac{\varphi_{N(t)}(z) - 1 + F_\tau(t)}{z} + z\int_0^t \varphi'_{N(t-s)}(z)\, dF_\tau(s)$$

which reduces to the renewal equation (8.9) for $z = 1$ since $\varphi'_{N(t)}(1) = m(t)$. The second derivative

$$\varphi''_{N(t)}(z) = 2\int_0^t \varphi'_{N(t-s)}(z)\, dF_\tau(s) + z\int_0^t \varphi''_{N(t-s)}(z)\, dF_\tau(s)$$

$$= \frac{2}{z}\varphi'_{N(t)}(z) - \frac{2}{z}\int_0^t \varphi_{N(t-s)}(z)\, dF_\tau(s) + z\int_0^t \varphi''_{N(t-s)}(z)\, dF_\tau(s)$$

evaluated at $z = 1$, is

$$\varphi''_{N(t)}(1) = 2m(t) - 2F_\tau(t) + \int_0^t \varphi''_{N(t-s)}(1)\, dF_\tau(s)$$

The variance $\mathrm{Var}[N(t)]$ follows from (2.27) as

$$\mathrm{Var}[N(t)] = \varphi''_{N(t)}(1) + \varphi'_{N(t)}(1) - \left(\varphi'_{N(t)}(1)\right)^2$$

$$= 3m(t) - m^2(t) - 2F_\tau(t) + \int_0^t \varphi''_{N(t-s)}(1)\, dF_\tau(s)$$

(iii) Every time an IP packet is launched by TCP, a renewal occurs and the reward is that 2000 km are travelled in each renewal, thus $R_n = 2000$ km. The speed in a trip that suffers from congestion is, on average, 40 000 km/s, while the speed without congestion is 120 000 km/s. Since congestion only occurs in 1/5 cases, the mean length (in s) of a renewal period is

$$E[\tau] = \frac{2000}{40\,000} \times \frac{1}{5} + \frac{2000}{120\,000} \times \frac{4}{5} = \frac{7}{300}$$

The average speed of an IP packet (in km/s) then follows from (8.22) as

$$\lim_{t\to\infty} \frac{R(t)}{t} = \frac{E\,[R]}{E\,[\tau]} = \frac{2000}{\frac{7}{300}} = 85\,714.3$$

(iv) Every transmission of an ATM cell is a renewal with mean length of the renewal interval equal to $E\,[\tau] = N/r$, where $1/r$ is the mean interarrival time for a voice sample. If τ_n is the time between the n-th and $n+1$-th arrival of sample, then the mean total cost per ATM cell transmission equals

$$E\,[R] = E\left[\sum_{n=1}^{N} nc \times \tau_n\right] + K = c\sum_{n=1}^{N} nE\,[\tau_n] + K = \frac{Nc}{r}\frac{N(N+1)}{2} + K$$

Hence, the mean cost per unit time incurred in UMTS is $\frac{E[R]}{E[\tau]} = c\frac{N(N+1)}{2} + \frac{Kr}{N}$.

(v) (a) The replacement of a router is a renewal process where the time at which router R_j is replaced is $W_j = \min(X_j, T)$, and

$$R_j = \begin{cases} A, & \text{if } X_j \le T \\ B, & \text{if } X_j > T \end{cases}$$

The mean cost per renewal period is $E\,[R] = A\Pr\,[X_j \le T] + B\Pr\,[X_j > T]$ and the mean length of a renewal interval equals

$$E\,[X] = \int_0^\infty \Pr\,[W_j > t]\,dt = \int_0^\infty \Pr\,[\min\,(X_j, T) > t]\,dt = \int_0^T \Pr\,[X_j > t]\,dt$$

The time average cost rate of the policy CHANGEROUTER is $C = \dfrac{E\,[R]}{E\,[X]}$.

(b) For $A = 10000$, $B = 7000$, $\Pr\,[X_j \le T] = 1 - e^{-\alpha T}$ with mean lifetime $\dfrac{1}{\alpha} = 10$ years and $T = 5$, we have

$$E\,[R] = A\Pr\,[X_j \le T] + B\Pr\,[X_j > T] = 10000 \times \left(1 - e^{-1/2}\right) + 7000 \times e^{-1/2} \simeq 8200$$

and

$$E\,[X] = \int_0^T \Pr\,[X_j > t]\,dt = \int_0^5 e^{-0.1t}dt = 10 \times \left(1 - e^{-1/2}\right) \simeq 4$$

such that time average cost rate of the policy CHANGEROUTER is $C \simeq \dfrac{8200}{4} = 2050$.

(vi) (a) The renewal process consists of a repeating sequence of telephone calls and each telephone call has a duration with pdf $f_D\,(x) = \frac{1}{\mu}e^{-x/\mu}$. In other words, the interarrival time of renewals are exponentially distributed with mean μ and the renewals reflect the opportunities that a busy calling person has just ended the previous call. Given that the telephone is occupied, the distribution of the waiting to reach the person equals the residual lifetime $R\,(t)$ of that telephone call as illustrated in Fig. 8.2. Assuming a steady-state situation, the steady-state distribution of the residual waiting time follows from (8.19) with (2.30) as $f_R\,(x) = \frac{1}{\mu}\,(1 - F_\tau\,(x)) = \frac{1}{\mu}e^{-x/\mu}$, which is precisely equal to $f_D\,(x)$. Clearly, this is an instance of the inspection paradox. Fig. 8.2 shows that, in steady-state, $R + A = D$. In addition, in the steady-state, the age A and the residual time R have the same distribution, $A \overset{d}{=} R$. This information is not sufficient to derive the pdf of R. For, if $R = A$, then $D \overset{d}{=} 2R$ such that $\varphi_D\,(z) = E\left[e^{-2zR}\right] = \varphi_R\,(2z)$ and

$$\varphi_R\,(z) = \varphi_D\left(\frac{z}{2}\right) = \frac{1}{\frac{\mu}{2}z + 1}$$

from which $f_R\,(x) = f_A\,(x) = \frac{2}{\mu}e^{-2x/\mu}$ and $E\,[R] = \frac{E[D]}{2} = \frac{\mu}{2}$. In that case, the inspection time t always lies in the middle of a renewal period. If A and R were independent – which is certainly not the case because $R = D - A$ implying that A and R are perfectly negatively correlated – then we would find that

$$\varphi_D(z) = E\left[e^{-zA}e^{-zR}\right] = E\left[e^{-zA}\right]E\left[e^{-zR}\right] = \left(E\left[e^{-zR}\right]\right)^2 = \varphi_R^2(z)$$

and, thus,

$$\varphi_R(z) = \sqrt{\varphi_D(z)} = \frac{1}{\sqrt{\mu z + 1}}$$

whose inverse Laplace transform is not known in closed form.

(b) From (a), we know that $R \overset{d}{=} D$. We are asked to compute the conditional distribution $\Pr[R > t + T | R > T]$. Since R is an exponential distribution which possesses the memoryless property (see Section 3.2.2), we have that

$$\Pr[R > t + T | R > T] = \Pr[R > t]$$

Hence, the waiting time is again exponential with mean μ. If we redial after the mean residual waiting time μ, the second mean waiting time is again μ, and the same holds for subsequent redials.

(c) If R is a uniform random variable on $[0, 2\mu]$, then

$$\Pr[R > t + T | R > T] = \frac{\Pr[R > t + T]}{\Pr[R > T]} = \frac{2\mu - t - T}{2\mu - T} 1_{t+T \le 2\mu}$$

and the mean second waiting time is computed via (2.36) as

$$E[R_2 | R_1 > T] = \int_0^\infty \frac{2\mu - t - T}{2\mu - T} 1_{t+T \le 2\mu} \, dt = \frac{2\mu - T}{2}$$

If we redial after the mean waiting time μ, the second mean waiting time is $\frac{\mu}{2}$, and, the next $\frac{\mu}{4}$, etc.

(vii) The server is considered to be always busy (because of the high load). This implies that the video download process can be treated as a renewal process: the start of a new download is the renewal, while the lengths of the download periods are the interarrival times τ between renewals. With probability $1/20 = 0.05$, the download time is equal to $\tau_1 = \frac{10^6 \times 8 \text{ bits}}{2 \, 10^5 \text{ bits/s}} = 40s$ and with probability $19/20 = 0.95$, it is $\tau_2 = \frac{10^6 \times 8 \text{ bits}}{10^6 \text{ bits/s}} = 8s$. The mean interarrival time is $E[\tau] = \frac{\tau_1}{20} + \frac{19\tau_2}{20} = 9.6s$. The waiting time of the brother equals the mean remaining waiting time $E[R]$ given in (8.20). It remains to compute $\text{Var}[\tau]$ as

$$\text{Var}[\tau] = E[\tau^2] - (E[\tau])^2 = \frac{\tau_1^2}{20} + \frac{19\tau_2^2}{20} - (E[\tau])^2 = 48.64s^2$$

such that $E[R] = \frac{9.6s}{2} + \frac{48.64s^2}{2 \times 9.6s} = 7.33s$.

B.8 Discrete-time Markov chains (Chapter 9)

(i) (a) The Markov chain is drawn in Fig. B.1.

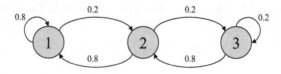

Fig. B.1. Three-state Markov chain.

(b) The steady-state vector π is computed via (9.30). The sequence P^{2^k} rapidly converges to yield three correct digits after four multiplications

$$P^2 = \begin{bmatrix} 0.800 & 0.160 & 0.040 \\ 0.640 & 0.320 & 0.040 \\ 0.640 & 0.160 & 0.200 \end{bmatrix} \qquad P^4 = \begin{bmatrix} 0.768 & 0.168 & 0.046 \\ 0.742 & 0.211 & 0.046 \\ 0.742 & 0.186 & 0.072 \end{bmatrix}$$

$$P^8 = \begin{bmatrix} 0.762 & 0.190 & 0.048 \\ 0.761 & 0.191 & 0.048 \\ 0.761 & 0.190 & 0.048 \end{bmatrix} \qquad P^{16} = \begin{bmatrix} 0.762 & 0.190 & 0.048 \\ 0.762 & 0.190 & 0.048 \\ 0.762 & 0.190 & 0.048 \end{bmatrix}$$

from which we find that the row vector in P^{16} equals $\pi = \begin{bmatrix} 0.762 & 0.190 & 0.048 \end{bmatrix}$. The second method consists in solving the set (9.31) by Cramer's method. Hence,

$$\det M = \begin{vmatrix} -0.2 & 0.8 & 0.0 \\ 0.2 & -1.0 & 0.8 \\ 1 & 1 & 1 \end{vmatrix} = 0.84 \qquad \pi_1 = \frac{\begin{vmatrix} 0 & 0.8 & 0.0 \\ 0 & -1 & 0.8 \\ 1 & 1 & 1 \end{vmatrix}}{\det M} = 0.762$$

$$\pi_2 = \frac{\begin{vmatrix} -0.2 & 0 & 0.0 \\ 0.2 & 0 & 0.8 \\ 1 & 1 & 1 \end{vmatrix}}{\det M} = 0.19 \qquad \pi_3 = \frac{\begin{vmatrix} -0.2 & 0.8 & 0 \\ 0.2 & -1 & 0 \\ 1 & 1 & 1 \end{vmatrix}}{\det M} = 0.048$$

The third method relies on the specific structure of the Markov chain, a discrete birth and death process or general random walk with constant $p_k = p$ and $q_k = q$. Applying formula (11.7), taking into account that $N = 2$, $\rho = \frac{0.2}{0.8} = \frac{1}{4}$ yields $\pi_1 = \frac{1-\frac{1}{4}}{1-\frac{1}{4^3}} = \frac{16}{21} = 0.762$, $\pi_2 = \pi_1 \rho = \frac{4}{21} = 0.190$ and $\pi_3 = \pi_2 \rho = \frac{1}{21} = 0.048$.

(ii) The Markov chain is shown in Fig. B.2. The state 1 is an absorbing state. From (9.29), the

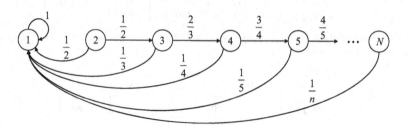

Fig. B.2. A recurrent Markov chain with positive drift.

steady-state vector components are found as $\pi_1 = \sum_{k=1}^{N} \frac{\pi_k}{k}$, $\pi_2 = 0$ and

$$\pi_j = \frac{j-2}{j-1}\pi_{j-1} \qquad 2 \le j \le N$$

Thus, $\pi_1 = 1$ and $\pi_j = 0$ for $j > 1$. Hence, the steady-state vector exists, and is different from $\pi = 0$, which demonstrates that the Markov chain is positive recurrent only in state 1. All other states are transient. However, the drift for states $j > 1$ (because $j = 1$ is absorbing) is

$$E[X_{k+1} - X_k | X_k = j] = 1 - \frac{1}{j} - \frac{1}{j} = 1 - \frac{2}{j}$$

which is always positive for $j > 2$. Hence, given an initial state $j > 2$, the Markov chain will, on average, move to the right (higher states).

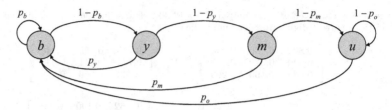

Fig. B.3. Markov chain of the growth process of trees in a forest during a period of 15 years.

(iii) (a) The Markov chain is shown in Fig. B.3.

(b) The evolution of the Markov process is defined by

$$
\begin{bmatrix} b[k+1] \\ y[k+1] \\ m[k+1] \\ u[k+1] \end{bmatrix} = \begin{bmatrix} p_b & p_y & p_m & p_o \\ 1-p_b & 0 & 0 & 0 \\ 0 & 1-p_y & 0 & 0 \\ 0 & 0 & 1-p_m & 1-p_o \end{bmatrix} \cdot \begin{bmatrix} b[k] \\ y[k] \\ m[k] \\ u[k] \end{bmatrix}
$$

(c) The number of trees in each category after 15 years (one period) is

$$
\begin{bmatrix} b[1] \\ y[1] \\ m[1] \\ u[1] \end{bmatrix} = \begin{bmatrix} 0.1 & 0.2 & 0.3 & 0.4 \\ 0.9 & 0 & 0 & 0 \\ 0 & 0.8 & 0 & 0 \\ 0 & 0 & 0.7 & 0.6 \end{bmatrix} \cdot \begin{bmatrix} 5000 \\ 0 \\ 0 \\ 0 \end{bmatrix} = \begin{bmatrix} 500 \\ 4500 \\ 0 \\ 0 \end{bmatrix}
$$

and after 30 years (two periods)

$$
\begin{bmatrix} b[2] \\ y[2] \\ m[2] \\ u[2] \end{bmatrix} = \begin{bmatrix} 0.1 & 0.2 & 0.3 & 0.4 \\ 0.9 & 0 & 0 & 0 \\ 0 & 0.8 & 0 & 0 \\ 0 & 0 & 0.7 & 0.6 \end{bmatrix} \cdot \begin{bmatrix} 500 \\ 4500 \\ 0 \\ 0 \end{bmatrix} = \begin{bmatrix} 950 \\ 450 \\ 3600 \\ 0 \end{bmatrix}
$$

(d) The steady-state vector π obeys equation (9.28) or, equivalently, (9.31). Applying a variant of (9.31), we have

$$
\begin{bmatrix} 1 \\ 0 \\ 0 \\ 0 \end{bmatrix} = \begin{bmatrix} 1 & 1 & 1 & 1 \\ 1-p_b & -1 & 0 & 0 \\ 0 & 1-p_y & -1 & 0 \\ 0 & 0 & 1-p_m & -p_o \end{bmatrix} \cdot \begin{bmatrix} \pi_b \\ \pi_y \\ \pi_m \\ \pi_0 \end{bmatrix}
$$

The determinant is $\det P = -p_0 - (1-p_b)(1+2p_0 - p_y - p_m + p_y p_m - p_0 p_y)$ and via Cramer's method we have

$$
\pi_b = \frac{1}{\det P} \det \begin{bmatrix} 1 & 1 & 1 & 1 \\ 0 & -1 & 0 & 0 \\ 0 & 1-p_y & -1 & 0 \\ 0 & 0 & 1-p_m & -p_o \end{bmatrix} = \frac{\det \begin{bmatrix} -1 & 0 & 0 \\ 1-p_y & -1 & 0 \\ 0 & 1-p_m & -p_o \end{bmatrix}}{\det P} = \frac{-p_o}{\det P}
$$

With the numerical values given in (c), $\pi_b = 0.25773$. After a similar calculation for the other categories, the total number of trees in steady growth is

$$
\begin{bmatrix} 5000\pi_b \\ 5000\pi_y \\ 5000\pi_m \\ 5000\pi_0 \end{bmatrix} \simeq \begin{bmatrix} 1289 \\ 1160 \\ 928 \\ 1624 \end{bmatrix}
$$

(iv) (a) The clustered error pattern is modeled as a two-state discrete Markov chain. When a bit is received incorrectly, the system is in state 0 else it is in state 1. The Markov chain is shown in Fig. 9.2, where $p = 1 - 0.95 = 0.05$ and $q = 1 - 0.999 = 0.001$. The transition probability matrix is $P = \begin{bmatrix} 0.95 & 0.05 \\ 0.001 & 0.999 \end{bmatrix}$.

(b) There is only one communicating class because both states 0 and 1 are reachable from each other. The Markov chain is therefore irreducible.

(c) The steady-state vector follows from (9.43) as

$$\pi = \begin{bmatrix} \frac{1}{51} & \frac{50}{51} \end{bmatrix} = \begin{bmatrix} 0.0196 & 0.9804 \end{bmatrix}$$

The fraction of correctly received bits in the long-run is 98.04% and the fraction of incorrectly received bits is 1.96%.

(d) After repair, the system operates correctly in 99.9% of the cases, which implies that $\pi_1 = 0.999$ and $\pi_0 = 0.001$. Formula (9.43) indicates that $\frac{p}{p+q} = 0.999$ and $\frac{q}{p+q} = 0.001$ or $p = 999q$. The test sequence shows that

$$\Pr[X_0 = 1, \ldots, X_{11} = 1] = (1 - q)^{10} \Pr[X_0 = 1] = (1 - q)^{10} = 0.9999$$

which leads to $q \simeq 10^{-5}$ and thus, $p = 0.009\,99$. A correctly (incorrectly) received bit is followed by a next correctly (incorrectly) received bit with probability $1 - q = 0.999\,99$, respectively $1 - p = 0.990\,01$.

(v) Consider the j-th component of the equation $xP \geq \lambda x$ which is $\sum_{k=1}^{N} x_k P_{kj} \geq \lambda x_j$. After summing both sides over all components, we obtain

$$\lambda \sum_{j=1}^{N} x_j \leq \sum_{j=1}^{N} \sum_{k=1}^{N} x_k P_{kj} = \sum_{k=1}^{N} x_k \left(\sum_{j=1}^{N} P_{kj} \right) = \sum_{k=1}^{N} x_k$$

where we have invoked (9.8). Since $x > 0$ which implies that $\sum_{k=1}^{N} x_k > 0$, we find that $\lambda \leq 1$. Hence, the inequality $xP \geq \lambda x$ is only possible for $\lambda \leq 1$ and the eigenvalue equation $xP \geq \lambda x$ is a special case of this inequality. Incidentally, the argument supports (but does not prove – see Corollary A.5.3) the fact that the only vector with positive components possesses eigenvalue $\lambda = 1$. For, if not all components $x_k \geq 0$, we cannot assure that $\sum_{k=1}^{N} x_k \neq 0$, and, hence, that $\lambda \leq 1$.

(vi) We refer to Karlin and Taylor (1981, p. 4) for a proof.

(vii) For $m = 1$ and using (9.18), the recursion (9.17) gives

$$\Pr[T_j = 1 | X_0 = i] = \sum_{k \neq j} P_{ik} \Pr[T_j = 0 | X_0 = k] = \sum_{k \neq j} P_{ik} \delta_{jk} = P_{ij} 1_{i \neq j}$$

The case $m = 2$ where $i \neq j$ follows as

$$\Pr[T_j = 2 | X_0 = i] = \sum_{k \neq j} P_{ik} \Pr[T_j = 1 | X_0 = k] = \sum_{k \neq j} P_{ik} P_{kj} = \left(P^2 \right)_{ij} - P_{ij} P_{jj}$$

By iterating a few times, we observe that the general expression is of the form

$$\Pr[T_j = m | X_0 = i] = \sum_{l=1}^{m} a_{ml} \left(P^l \right)_{ij}$$

where the coefficients a_{ml} are found by introducing this expression into the recursion (9.17),

$$\sum_{l=1}^{m} a_{ml} \left(P^l \right)_{ij} = \sum_{k \neq j} P_{ik} \sum_{l=1}^{m-1} a_{m-1;l} \left(P^l \right)_{kj} = \sum_{l=1}^{m-1} a_{m-1;l} \sum_{k \neq j} P_{ik} \left(P^l \right)_{kj}$$

The k-sum is $\sum_{k \neq j} P_{ik} \left(P^l \right)_{kj} = \left(P^{l+1} \right)_{ij} - P_{ij} \left(P^l \right)_{jj}$ and

$$\sum_{l=1}^{m} a_{ml} \left(P^l \right)_{ij} = \sum_{l=1}^{m-1} a_{m-1;l} \left(\left(P^{l+1} \right)_{ij} - P_{ij} \left(P^l \right)_{jj} \right) = \sum_{l=2}^{m} a_{m-1;l-1} \left(P^l \right)_{ij} - P_{ij} \sum_{l=1}^{m-1} a_{m-1;l} \left(P^l \right)_{jj}$$

Equating corresponding powers in $\left(P^l\right)_{ij}$ yields

$$a_{ml} = a_{m-1;l-1}$$

$$a_{m1} = -\sum_{l=1}^{m-1} a_{m-1;l} \left(P^l\right)_{jj}$$

Since $\Pr[T_j = 1 | X_0 = i] = P_{ij}$, the initial value is $a_{11} = 1$ which suffices to iteratively determine all a_{ml}. The first equation implies that $a_{ml} = a_{m-l+1;1}$ such that only the coefficients a_{m1} are needed. Hence,

$$\Pr[T_j = m | X_0 = i] = \sum_{l=1}^{m} a_{m+1-l;1} \left(P^l\right)_{ij} \qquad \text{(B.11)}$$

Iterating the second equation leads to the explicit form of a_{m1} for $m \geq 1$,

$$a_{m1} = \sum_{q=1}^{m-1} (-1)^q \sum_{\sum_{i=1}^{q} n_i = m-1; n_i > 0} \prod_{i=1}^{q} \left(P^{n_i}\right)_{jj}$$

and it can be shown that the form $a_{m+1;1}$ is equal to the Taylor coefficient around $z = 0$ of $\frac{1}{\varphi_P(j,j;z)}$, where $\varphi_P(i,j;z) = \sum_{n=0}^{\infty} (P^n)_{ij} z^n$ is defined in (9.45). Indeed, define the generating function of the coefficients a_{m1} by $\check{A}(z) = \sum_{m=2}^{\infty} a_{m1} z^m$. After multiplying the recursion for a_{m1} by z^m and summing over all $m \geq 1$, we obtain with $a_{11} = 1$,

$$\check{A}(z) = -\sum_{m=2}^{\infty} \left(\sum_{l=1}^{m-1} a_{m-l;1} \left(P^l\right)_{jj}\right) z^m = -\sum_{m=1}^{\infty} \sum_{l=1}^{m} a_{m+1-l;1} \left(P^l\right)_{jj} z^{m+1}$$

$$= -\sum_{l=1}^{\infty} \left(\sum_{m=l}^{\infty} a_{m+1-l;1} z^{m+1}\right) \left(P^l\right)_{jj} = -\sum_{l=1}^{\infty} \left(\sum_{m=1}^{\infty} a_{m;1} z^{m+l}\right) \left(P^l\right)_{jj}$$

$$= -\sum_{m=1}^{\infty} a_{m;1} z^m \sum_{l=1}^{\infty} z^l \left(P^l\right)_{jj} = -\left(z + \check{A}(z)\right) \sum_{l=1}^{\infty} z^l \left(P^l\right)_{jj}$$

Thus,

$$\check{A}(z) = \frac{-z \sum_{l=1}^{\infty} z^l \left(P^l\right)_{jj}}{1 + \sum_{l=1}^{\infty} z^l \left(P^l\right)_{jj}} = \frac{-z \left(\sum_{l=1}^{\infty} z^l \left(P^l\right)_{jj} + 1 - 1\right)}{1 + \sum_{l=1}^{\infty} z^l \left(P^l\right)_{jj}} = -z + \frac{z}{1 + \sum_{l=1}^{\infty} z^l \left(P^l\right)_{jj}}$$

or,

$$\frac{\check{A}(z)}{z} = \sum_{m=1}^{\infty} a_{m+1;1} z^m = -1 + \frac{1}{1 + \sum_{l=1}^{\infty} z^l \left(P^l\right)_{jj}}$$

Including $a_{11} = 1$ and taking into account that $P^0 = I$ such that $\left(P^0\right)_{jj} = 1$, finally gives, with a new definition of $A(z)$,

$$A(z) = \sum_{m=0}^{\infty} a_{m+1;1} z^m = \frac{1}{\sum_{k=0}^{\infty} z^k \left(P^k\right)_{jj}} \qquad \text{(B.12)}$$

which implies by Taylor's theorem that, for $m \geq 0$,

$$a_{m+1;1} = \frac{1}{m!} \frac{d^m}{dz^m} \frac{1}{\sum_{k=0}^{\infty} z^k \left(P^k\right)_{jj}} \bigg|_{z=0}$$

After substituting this expression for the coefficients $a_{m+1;1}$ into (B.11) we obtain (9.21).

We list a few m values in case the Markov chain has no self-transitions, i.e. $P_{jj} = 0$,

$$\Pr[T_j = 2|X_0 = i] = \left(P^2\right)_{ij}$$

$$\Pr[T_j = 3|X_0 = i] = \left(P^3\right)_{ij} - P_{ij}\left(P^2\right)_{jj}$$

$$\Pr[T_j = 4|X_0 = i] = \left(P^4\right)_{ij} - \left(P^2\right)_{ij}\left(P^2\right)_{jj} - P_{ij}\left(P^3\right)_{jj}$$

$$\Pr[T_j = 5|X_0 = i] = \left(P^5\right)_{ij} - \left(P^3\right)_{ij}\left(P^2\right)_{jj} - \left(P^2\right)_{ij}\left(P^3\right)_{jj} - P_{ij}\left(\left(P^4\right)_{jj} - \left(P^2\right)_{jj}^2\right)$$

$$\Pr[T_j = 6|X_0 = i] = \left(P^6\right)_{ij} - \left(P^4\right)_{ij}\left(P^2\right)_{jj} - \left(P^3\right)_{ij}\left(P^3\right)_{jj} - \left(P^2\right)_{ij}\left(\left(P^4\right)_{jj} - \left(P^2\right)_{jj}^2\right)$$
$$- P_{ij}\left(\left(P^5\right)_{jj} - 2\left(P^2\right)_{jj}\left(P^3\right)_{jj}\right)$$

Finally, the recursion (9.17) is bounded by

$$\Pr[T_j = m|X_0 = i] = \sum_{k \neq j} P_{ik}\Pr[T_j = m-1|X_0 = k] \leq \sum_{k=1}^{N} P_{ik}\Pr[T_j = m-1|X_0 = k]$$

Since the general solution of $X[m,i,j] = \sum_{k=1}^{N} P_{ik}X[m-1,k,j]$ is $X[m,i,j] = \left(P^m\right)_{ij}$, we conclude that

$$\Pr[T_j = m|X_0 = i] \leq \left(P^m\right)_{ij}$$

This inequality is also understood from the fact that the event $\{T_j = m|X_0 = 1\} \subset \{X_m = j|X_0 = i\}$ because the latter event does not restrict that the process visits the state j at discrete time(s) $k < m$, while the definition of the hitting time excludes these transitions into state j at earlier time(s) $k < m$.

(viii) The appearing powers of the stochastic matrix P suggests us to consider the spectrum. For simplicity, we assume further that all eigenvalues of P are different. The derivation below is similar, but slightly more general, than for the random walk on a graph where $P = \Delta^{-1}A$ and whose spectral decomposition is derived in Van Mieghem (2011) and Lovász (1993). With **art.** 4 and (A.17) we know that

$$P^k = X\text{diag}\left(1, \lambda_2^k, \ldots, \lambda_N^k\right)Y^T$$

where $X = \begin{bmatrix} x_1 & x_2 & \cdots & x_N \end{bmatrix}$ and $Y = \begin{bmatrix} y_1 & y_2 & \cdots & y_N \end{bmatrix}$ contain as columns the right- and left-eigenvectors of P, respectively. Hence,

$$\left(P^k\right)_{ij} = \sum_{n=1}^{N} \lambda_n^k x_{ni}y_{nj} = \sum_{n=1}^{N} \lambda_n^k \left(x_n y_n^T\right)_{ij}$$

where[2] $x_{ki} = (x_k)_i$ is the i-th component of vector x_k. Substituted in (B.11) and using the fact that $\lambda_1 = 1$ with corresponding eigenvectors $x_1 = u$ and $y_1 = \pi$ yields

$$\Pr[T_j = m|X_0 = i] = \sum_{n=1}^{N}\left(\sum_{l=1}^{m} a_{m+1-l;1}\lambda_n^l\right)\left(x_n y_n^T\right)_{ij}$$

$$= \pi_j\sum_{l=1}^{m} a_{m+1-l;1} + \sum_{n=2}^{N}\left(\sum_{l=1}^{m} a_{m+1-l;1}\lambda_n^l\right)\left(x_n y_n^T\right)_{ij}$$

which relates the hitting T_j to the steady-state probability π_j that the process is in state j. From (B.12), we have, provided that $|xz| < 1$,

$$\frac{A(z)}{1-xz} = \sum_{m=0}^{\infty} x^m z^m \sum_{m=0}^{\infty} a_{m+1;1}z^m = \sum_{m=0}^{\infty}\sum_{l=0}^{m} x^l a_{m-l+1;1}z^m$$

---

[2] This notation is different from the usual matrix notation for X because the emphasis lies on the eigenvector rather than on individual matrix elements.

such that

$$\sum_{m=0}^{\infty}\sum_{l=0}^{m} x^l a_{m-l+1;1} z^m = \frac{1}{(1-xz)\sum_{k=0}^{\infty} z^k \left(P^k\right)_{jj}}$$

It follows, for $m \geq 0$, by Taylor's theorem that

$$\sum_{l=0}^{m} x^l a_{m+1-l;1} = \frac{1}{m!}\frac{d^m}{dz^m}\left(\frac{1}{(1-xz)\sum_{k=0}^{\infty} z^k \left(P^k\right)_{jj}}\right)\Bigg|_{z=0}$$

With $\sum_{l=1}^{m} a_{m+1-l;1}\lambda_n^l = \lambda_n \sum_{l=0}^{m-1} a_{m-l;1}\lambda_n^l$, we find that

$$\Pr\left[T_j = m | X_0 = i\right] = \frac{1}{(m-1)!}\frac{d^{m-1}}{dz^{m-1}}\left(\frac{1}{\sum_{k=0}^{\infty} z^k \left(P^k\right)_{jj}}\sum_{n=1}^{N}\frac{\lambda_n\left(x_n y_n^T\right)_{ij}}{1-\lambda_n z}\right)\Bigg|_{z=0}$$

$$= \frac{1}{2\pi i}\int_{C(0)}\frac{dz}{z^m \sum_{k=0}^{\infty} z^k \left(P^k\right)_{jj}}\sum_{n=1}^{N}\frac{\lambda_n\left(x_n y_n^T\right)_{ij}}{1-\lambda_n z}$$

where the last relation follows from (2.20), which also shows that the probability generating function $\varphi_{T_j | X_0 = i}(z)$ of T_j equals

$$\varphi_{T_j | X_0 = i}(z) = \sum_{m=0}^{\infty}\Pr\left[T_j = m+1 | X_0 = i\right] z^m = \frac{1}{\sum_{k=0}^{\infty} z^k \left(P^k\right)_{jj}}\sum_{n=1}^{N}\frac{\lambda_n\left(x_n y_n^T\right)_{ij}}{1-\lambda_n z}$$

The function $\varphi_P(j,j;z) = \sum_{k=0}^{\infty}\left(P^k\right)_{jj} z^k$ is analytic inside the unit circle as shown in Section 9.5.2, but the series diverges for $z \to 1$ for recurrent states and only for a transient state $\sum_{k=1}^{\infty}\left(P^k\right)_{jj}$ is finite (see Section 9.2.4). Using (9.52), the probability generating function $\varphi_{T_j | X_0 = i}(z)$ becomes

$$\varphi_{T_j | X_0 = i}(z) = \frac{\sum_{n=1}^{N}\frac{\lambda_n\left(x_n y_n^T\right)_{ij}}{1-\lambda_n z}}{\sum_{n=1}^{N}\frac{\left(x_n y_n^T\right)_{jj}}{1-z\lambda_n}} = \frac{\frac{\pi_j}{1-z}+\sum_{n=2}^{N}\frac{\lambda_n\left(x_n y_n^T\right)_{ij}}{1-\lambda_n z}}{\frac{\pi_j}{1-z}+\sum_{n=2}^{N}\frac{\left(x_n y_n^T\right)_{jj}}{1-z\lambda_n}}$$

which shows that $\varphi_{T_j | X_0 = i}(1) = 1$, $\varphi_{T_j | X_0 = i}(0) = P_{ij} = \Pr\left[T_j = 1 | X_0 = i\right]$. After some tedious calculations using (2.26), we obtain the mean hitting time into state j given $X_0 = i$,

$$E\left[T_j | X_0 = i\right] = \frac{1}{\pi_j}\sum_{n=2}^{N}\frac{\left(x_n y_n^T\right)_{jj} - \lambda_n\left(x_n y_n^T\right)_{ij}}{1-\lambda_n}$$

Since $\varphi_{T_j | X_0 = i}\left(\lambda_n^{-1}\right) = \frac{\lambda_n\left(x_n y_n^T\right)_{ij}}{\left(x_n y_n^T\right)_{jj}}$, the $N-1$ zeros of the polynomial in the denominator are, generally, different from the eigenvalues of P and different from the zeros of the polynomial in the numerator. Since

$$\varphi_{T_j | X_0 = i}(z) = \frac{q(z)}{v(z)} = \frac{\pi_j \prod_{k=2}^{N}(1-\lambda_k z) + (1-z)\sum_{n=2}^{N}\lambda_n\left(x_n y_n^T\right)_{ij}\prod_{k=2;k\neq n}^{N}(1-\lambda_k z)}{\pi_j \prod_{k=2}^{N}(1-\lambda_k z) + (1-z)\sum_{n=2}^{N}\left(x_n y_n^T\right)_{jj}\prod_{k=2;k\neq n}^{N}(1-\lambda_k z)}$$

is a ratio of two $N - 1$ degree polynomials $q(z)$ and $v(z)$,

$$
\lim_{z \to \infty} \varphi_{T_j | X_0 = i}(z) = \frac{\pi_j (-1)^{N-1} \det P + \sum_{n=2}^{N} \lambda_n \left(x_n y_n^T \right)_{ij} (-1)^{N-1} \frac{\det P}{\lambda_n}}{\pi_j (-1)^{N-1} \det P + \sum_{n=2}^{N} \left(x_n y_n^T \right)_{jj} (-1)^{N-1} \frac{\det P}{\lambda_n}}
$$

$$
= \frac{\pi_j + \sum_{n=2}^{N} \left(x_n y_n^T \right)_{ij}}{\pi_j + \sum_{n=2}^{N} \frac{\left(x_n y_n^T \right)_{jj}}{\lambda_n}} \qquad \frac{(P^0)_{ij}}{\pi_j + \sum_{n=2}^{N} \frac{\left(x_n y_n^T \right)_{jj}}{\lambda_n}} = 0
$$

since the analysis has assumed that $i \neq j$. Hence, the contour in

$$
\Pr[T_j = m | X_0 = i] = \frac{1}{2\pi i} \int_{C(0)} \frac{\varphi_{T_j | X_0 = i}(z)}{z^m} dz
$$

can be closed over the whole complex plane excluding the origin because the integrand vanishes for $m > 0$ if $z \to \infty$. By Cauchy residue theorem, the integral is equal to the sum of the residues at the zeros of $v(z)$ different from those of $q(z)$. For large m, the asymptotic behavior is determined, as explained in Section 5.7, by the dominant pole of $\varphi_{T_j | X_0 = i}(z)$ which is the zero $z_{\min}$ of $v(z)$ with smallest absolute value,

$$
\Pr[T_j = m | X_0 = i] \sim \frac{q(z_{\min})}{v'(z_{\min})} z_{\min}^{-m}
$$

Since any pgf is analytic inside and on the unit circle, we have that $|z_{\min}| > 1$. Hence, we expect that $\Pr[T_j = m | X_0 = i]$ decreases, in general, exponentially fast, which means that the hitting time T_j is close to a geometric random variable with parameter $z_{\min}$.

(ix) If $X_0 = i \in A$, then $T_A = 0$ and $\Pr[T_A < \infty | X_0 = i] = 1$. If $X_0 = i \notin A$, then $T_A \geq 1$ and, by the law of total probability (2.49),

$$
\Pr[T_A < \infty | X_0 = i] = \sum_{j \in S} \Pr[\{T_A < \infty | X_0 = i\} \cap \{X_1 = j\}]
$$

$$
= \sum_{j \in S} \Pr[T_A < \infty | X_1 = j, X_0 = i] \Pr[X_1 = j | X_0 = i]
$$

The Markov property (9.1) tells us that

$$
\Pr[T_A < \infty | X_1 = j, X_0 = i] = \Pr[T_A < \infty | X_1 = j]
$$

and since $T_A \geq 1$, we have that $\Pr[T_A < \infty | X_0 = j] = \Pr[T_A < \infty | X_1 = j]$ due to stationarity, which proves (9.53) using the stationary transition probabilities (9.4). Similarly, if $X_0 = i \in A$, then $T_A = 0$ and $E[T_A | X_0 = i] = 0$. If $X_0 = i \notin A$, then using (2.82) yields

$$
E[T_A | X_0 = i] = E_{X_1} \left[E\left[\{T_A | X_0 = i\} | X_1 \right] \right] = E_{X_1} \left[E[T_A | X_0 = i, X_1] \right]
$$

$$
= \sum_{j \in S} E[T_A | X_0 = i, X_1 = j] \Pr[X_1 = j | X_0 = i]
$$

where we have invoked (2.50). It now follows from the Markov property (9.1) and stationarity that, if $i \notin A$ and thus $T_A \geq 1$,

$$
E[T_A | X_0 = i, X_1 = j] = 1 + E[T_A | X_1 = j] = 1 + E[T_A | X_0 = j]
$$

while, if $i \in A$, then $E[T_A | X_0 = i, X_1 = j] = 0$. Together with (9.4), we arrive at (9.54).

(x) The download time T of a file of size B can be defined as

$$
T = \min_{t > 0} \left\{ \int_0^t C(u) \, du \geq B \right\} \tag{B.13}
$$

and T can be regarded as the first hitting for the process $C(t)$ to reach the level B. If $C(t)$ is continuous in time t as well, then the equality sign is possible such that

$$
B = \int_0^T C(u) \, du
$$

(a) If the process $C(t)$ is independent and identically distributed over time and, in addition, if $C(t)$ is continuous, then the mean download time is

$$B = E\left[\int_0^T C(u)\,du\right] = E_T\left[E\left[\int_0^T C(u)\,du\,\Big|\,T=\tau\right]\right]$$

where the conditional expectation is used. The linearity of the expectation gives

$$E\left[\int_0^T C(u)\,du\,\Big|\,T=\tau\right] = E\left[\int_0^\tau C(u)\,du\,\Big|\,T=\tau\right] = \int_0^\tau E\left[C(u)|T=\tau\right]du$$

while independence implies that $E\left[C(u)|T=\tau\right] = E[C]$ such that

$$E\left[\int_0^T C(u)\,du\,\Big|\,T=\tau\right] = E[C]\,T$$

and $B = E[C]\,E[T]$, which is Wald's identity for continuous i.i.d. random variables.
(b) If $C(t) = C$, where C is uniformly distributed over $[c_1, c_2]$, then $B = C.T$ and from (2.35), we obtain

$$E[T] = E\left[\frac{B}{C}\right] = B\int_{c_1}^{c_2}\frac{dx}{x} = B\log\frac{c_2}{c_1} \tag{B.14}$$

This case corresponds to perfect correlation over time because, at each point in time $C(t) = C(t+dt)$. In case of independence and $E[C] = \frac{c_1+c_2}{2}$, we would have $E[T] = \frac{B}{E[C]} = \frac{2B}{c_1+c_2}$. In case $c_1 = 0$, we observe from (B.14) that the mean download time $E[T] \to \infty$, which heavily contrasts with an i.i.d. available capacity process. Thus, if the available capacity can be zero but also be very high and equal to $c_2 = \max(C)$ with *equal probability*, the mean file download time can grow unboundedly large.
(c) If the available capacity $C(t) = C[k\Delta]$, where Δ is a unit time during which $C(t)$ is constant, then the definition (B.13) can be written as an equivalence between the events

$$\{T < j\} \iff \{C[\Delta] + C[2\Delta] + \cdots + C[j\Delta] \geq B\}$$

which reminds us of the basic equivalence in the renewal theory in Chapter 8. However, in renewal theory, the random variables $C[k\Delta]$ were all independent, while here, independence does not hold in general. Further,

$$\sum_{k=1}^{[T]} C[k\Delta] \leq B < \sum_{k=1}^{[T]+1} C[k\Delta]$$

and, assuming equality, we have that

$$B = \sum_{k=1}^{T} C[k\Delta] = \sum_{k=1}^{\infty} C[k\Delta]\,1_{\{T\geq k\}}$$

Taking the mean yields

$$B = \sum_{k=1}^{\infty} E\left[C[k\Delta]\,1_{\{T\geq k\}}\right]$$

Only if we assume that correlation is known (e.g. a positive correlation meaning that, if $C[j\Delta]$ is large, also $C[(j+1)\Delta]$ is large with high probability), then the above equivalence results in

$$E\left[C[k\Delta]\,1_{\{T\geq k\}}\right] \leq E[C[k\Delta]]\,E\left[1_{\{T\geq k\}}\right] = E[C]\Pr[T \geq k]$$

where $E[C[k\Delta]] = E[C]$ because the process is stationary, a further analysis is possible leading to

$$B = \sum_{k=1}^{\infty} E\left[C[k\Delta]\,1_{\{T\geq k\}}\right] \leq E[C]\sum_{k=1}^{\infty}\Pr[T \geq k]$$

Finally, using the tail probability expression (2.37),

$$E[T] \geq \frac{B}{E[C]}$$

Thus, the mean download time of a positively correlated process $C[k\Delta]$ over discrete time k is larger than in an i.i.d. process.

(xi) (a) If state k is the only absorbing state, then $P_{kk} = 1$ and, from (9.8), we find that $P_{kj} = 0$. Since the Markov chain is recurrent, i.e. all states are reachable, not all P_{jk} can be zero, otherwise the Markov chain would be reducible and consisting of two non-communicating equivalence classes. From the k-th component equation (9.29) of the steady-state vector, we obtain

$$\sum_{j=1;j\neq k}^{N} P_{jk}\pi_j = 0$$

Since both P_{jk} and π_j, for all j, are non-negative and not all P_{jk} are zero, the only possible solution is that $\pi_j = 0$ for all $j \in J\backslash\{k\}$ for which $P_{jk} \neq 0$. All the component equations (9.29) of elements of the set of states $J\backslash\{k\}$ with a direct transition to the absorbing state k return a same kind of equation as above. By a same argument, all direct neighbors of the set of states in $J\backslash\{k\}$ also have a zero steady-state vector component. Since the Markov chain is recurrent, eventually, all states will be effected such that $\pi_j = 0$ for all states $j \neq k$. The normalization condition $\|\pi\|_1 = 1$ of the steady-state vector π shows that $\pi_k = 1$. This proves the claim.

(b) The fact that $\pi_j = 0$ for all other states, implies that they are either transient or null recurrent. In a recurrent Markov chain, there is a non-zero probability that, in a finite time, there is a transition from state j to k. Once in the absorbing state k, there is no return to any other state j. Since there is an absorbing state k, the number of visits to a state $j \neq k$ cannot be infinite. Only transient states have a finite number of visits, which differentiates transient from null recurrent states. Hence, all the other states are transient states.

Considerably more can be deduced about a Markov chain with transient and absorbing states for which we refer to Taylor and Karlin (1984, p. 116-119).

(xii) (a) Since the choice of the peer A is random, the probability to select a peer that is already off-line is determined by the number of inactive peers remaining in the list. The number of active peers in the list are unchanged each time, while X_k only depends on the number of inactive peers in the list at the $(k-1)$ time, which satisfies the Markov property (9.2). Let n be the number of inactive peers in the list at the $(k-1)$-th timeslot. At the random selection at time k, the list either remains at n inactives or is reduced by one, if A selects an inactive peer. The transition probabilities are

$$\Pr[X_k = n-1 | X_{k-1} = n] = \frac{n}{14+n}$$

and

$$\Pr[X_k = n | X_{k-1} = n] = 1 - \Pr[X_k = n-1 | X_{k-1} = n]$$

Thus, for $0 \leq n \leq 6$ inactive peers, the transition probability matrix is

$$P = \begin{pmatrix} 1 & 0 & 0 & 0 & 0 & 0 & 0 \\ \frac{1}{15} & \frac{14}{15} & 0 & 0 & 0 & 0 & 0 \\ 0 & \frac{2}{16} & \frac{14}{16} & 0 & 0 & 0 & 0 \\ 0 & 0 & \frac{3}{17} & \frac{14}{17} & 0 & 0 & 0 \\ 0 & 0 & 0 & \frac{4}{18} & \frac{14}{18} & 0 & 0 \\ 0 & 0 & 0 & 0 & \frac{5}{19} & \frac{14}{19} & 0 \\ 0 & 0 & 0 & 0 & 0 & \frac{6}{20} & \frac{14}{20} \end{pmatrix}$$

(b) X_k enters the steady-state when all inactive peers are deleted. Equation (9.30) shows that, to compute $\lim_{\to\infty} P^k$, we need to calculate the state vector π, that we solve from

(9.31). The left-hand side matrix in (9.31) is

$$
M = \begin{pmatrix}
0 & \frac{1}{15} & 0 & 0 & 0 & 0 & 0 \\
0 & -\frac{1}{15} & \frac{2}{16} & 0 & 0 & 0 & 0 \\
0 & 0 & -\frac{2}{16} & \frac{3}{17} & 0 & 0 & 0 \\
0 & 0 & 0 & -\frac{3}{17} & \frac{4}{18} & 0 & 0 \\
0 & 0 & 0 & 0 & -\frac{4}{18} & \frac{5}{19} & 0 \\
0 & 0 & 0 & 0 & 0 & -\frac{5}{19} & \frac{6}{20} \\
1 & 1 & 1 & 1 & 1 & 1 & 1
\end{pmatrix}
$$

By Cramer's method, we have

$$
\pi_1 = \frac{\det \begin{pmatrix}
0 & \frac{1}{15} & 0 & 0 & 0 & 0 & 0 \\
0 & -\frac{1}{15} & \frac{2}{16} & 0 & 0 & 0 & 0 \\
0 & 0 & -\frac{2}{16} & \frac{3}{17} & 0 & 0 & 0 \\
0 & 0 & 0 & -\frac{3}{17} & \frac{4}{18} & 0 & 0 \\
0 & 0 & 0 & 0 & -\frac{4}{18} & \frac{5}{19} & 0 \\
0 & 0 & 0 & 0 & 0 & -\frac{5}{19} & \frac{6}{20} \\
1 & 1 & 1 & 1 & 1 & 1 & 1
\end{pmatrix}}{\det M} = 1
$$

Similarly, we have,

$$
\pi_2 = \frac{\det \begin{pmatrix}
0 & 0 & 0 & 0 & 0 & 0 & 0 \\
0 & 0 & \frac{2}{16} & 0 & 0 & 0 & 0 \\
0 & 0 & -\frac{2}{16} & \frac{3}{17} & 0 & 0 & 0 \\
0 & 0 & 0 & -\frac{3}{17} & \frac{4}{18} & 0 & 0 \\
0 & 0 & 0 & 0 & -\frac{4}{18} & \frac{5}{19} & 0 \\
0 & 0 & 0 & 0 & 0 & -\frac{5}{19} & \frac{6}{20} \\
1 & 1 & 1 & 1 & 1 & 1 & 1
\end{pmatrix}}{\det M} = 0
$$

because two columns are the same. The same situation occurs for all other components of π. Thus, the steady-state vector is $\pi = [1, 0, 0, 0, 0, 0, 0]$. Therefore,

$$
\lim_{k \to \infty} P^k = \begin{pmatrix}
1 & 0 & 0 & 0 & 0 & 0 & 0 \\
1 & 0 & 0 & 0 & 0 & 0 & 0 \\
1 & 0 & 0 & 0 & 0 & 0 & 0 \\
1 & 0 & 0 & 0 & 0 & 0 & 0 \\
1 & 0 & 0 & 0 & 0 & 0 & 0 \\
1 & 0 & 0 & 0 & 0 & 0 & 0 \\
1 & 0 & 0 & 0 & 0 & 0 & 0
\end{pmatrix}
$$

In fact, the observation that $X_k = 0$ is the only absorbing state, immediately leads to $\pi = [1, 0, 0, 0, 0, 0, 0]$.

B.9 Continuous-time Markov processes (Chapter 10)

(i) (a) The failure rate for each processor is $\lambda = 0.001$ per hour. The repair rate is $\mu = 0.01$ per hour. The Markov chain is shown in Fig. B.4.

(b) The infinitesimal generator is

$$
Q = \begin{pmatrix}
-2\lambda & 2\lambda & 0 \\
\mu & -(\lambda + \mu) & \lambda \\
0 & 2\mu & -2\mu
\end{pmatrix}
$$

If the state probability vector is denoted by $s(t)$, we can also write $s(t)Q = \frac{d}{dt}(s(t))$, or

$$
[s_1(t) \quad s_2(t) \quad s_3(t)] \cdot \begin{pmatrix}
-2\lambda & 2\lambda & 0 \\
\mu & -(\lambda + \mu) & \lambda \\
0 & 2\mu & -2\mu
\end{pmatrix} = [s_1'(t) \quad s_2'(t) \quad s_3'(t)]
$$

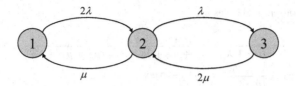

Fig. B.4. The Markov chain for the three states: (1) both processors work, (2) one processor is damaged and (3) both processors are damaged.

(c) The steady-state $\pi = \lim_{t\to\infty} s(t)$ obeys the equation (10.21)

$$\begin{bmatrix} \pi_1 & \pi_2 & \pi_3 \end{bmatrix} \cdot \begin{pmatrix} -2\lambda & 2\lambda & 0 \\ \mu & -(\lambda+\mu) & \lambda \\ 0 & 2\mu & -2\mu \end{pmatrix} = \begin{bmatrix} 0 & 0 & 0 \end{bmatrix}$$

Since $\pi_1+\pi_2+\pi_3 = 1$, we find that $\pi_1 = \left(\frac{\mu}{\lambda+\mu}\right)^2$, $\pi_2 = \frac{2\lambda\mu}{(\lambda+\mu)^2}$ and $\pi_3 = \left(\frac{\lambda}{\lambda+\mu}\right)^2$. From the balance equation, we know that the probability flux from state 1 to state 2 should precisely equal that in the opposite direction such that $2\lambda\pi_1 = \mu\pi_2$ and similar for the transitions $2 \to 3$, $\lambda\pi_2 = 2\mu\pi_3$. Using $\pi_1 + \pi_2 + \pi_3 = 1$ leads faster to the solution. With $\lambda = 0.001$ and $\mu = 0.01$, the values are $\pi_1 = 0.8264$, $\pi_2 = 0.1653$ and $\pi_3 = 0.0083$.

(d) The availability in case (i) is $\pi_1 = 0.8264$. The availability in case (ii) is $\pi_1 + \pi_2 = 0.9917$.

(ii) (a) In state 0, both servers are damaged, state 1 refers to one server down and one operating while in state 2, both servers are operating. The corresponding Markov chain is shown in Fig. B.5.

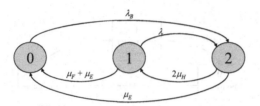

Fig. B.5. The Markov chain is specified by $\lambda = \frac{1}{15}\mathrm{h}^{-1} = 6.66 \times 10^{-2}\ \mathrm{h}^{-1}$, $\lambda_B = \frac{1}{20}\mathrm{h}^{-1} = 5 \times 10^{-2}\ \mathrm{h}^{-1}$, $\mu_H = 3 \times 10^{-4}\ \mathrm{h}^{-1}$, $\mu_F = 7 \times 10^{-4}\ \mathrm{h}^{-1}$ and $\mu_E = 6 \times 10^{-5}\ \mathrm{h}^{-1}$.

(b) The infinitesimal generator $Q = \begin{bmatrix} -\lambda_B & 0 & \lambda_B \\ \mu_F + \mu_E & -\mu_F - \mu_E - \lambda & \lambda \\ \mu_E & 2\mu_H & -\mu_E - 2\mu_H \end{bmatrix}$

(c) The steady-state vector π obeys (10.21). The solution of $\pi Q = 0$ is

$$\begin{bmatrix} \pi_0 & \pi_1 & \pi_2 \end{bmatrix} \begin{bmatrix} -\lambda_B & 0 & \lambda_B \\ \mu_F + \mu_E & -\mu_F - \mu_E - \lambda & \lambda \\ \mu_E & 2\mu_H & -\mu_E - 2\mu_H \end{bmatrix} = \begin{bmatrix} 0 & 0 & 0 \end{bmatrix}$$

Since this linear set of equation is undetermined, we remove an arbitrary equation and add the normalization condition $\sum_{j=0}^{2} \pi_j = 1$,

$$\begin{bmatrix} \pi_0 & \pi_1 & \pi_2 \end{bmatrix} \begin{bmatrix} -\lambda_B & 0 & 1 \\ \mu_F + \mu_E & -\mu_F - \mu_E - \lambda & 1 \\ \mu_E & 2\mu_H & 1 \end{bmatrix} = \begin{bmatrix} 0 & 0 & 1 \end{bmatrix}$$

The steady-state probabilities are

$$\pi_2 = \frac{\lambda_B \left(\mu_F + \mu_E + \lambda\right)}{\left(\mu_E + \lambda_B\right)\left(\mu_F + \mu_E + \lambda\right) + 2\mu_H \left(\mu_F + \mu_E + \lambda_B\right)} = 0.9898$$

$$\pi_1 = \frac{2\mu_H}{\left(\mu_F + \mu_E + \lambda\right)}\pi_2 = 0.0088$$

$$\pi_0 = 1 - \pi_2 - \pi_1 = 0.0013$$

(d) Theorem 10.2.3 states that the mean lifetime of state j is $E\left[\tau_j\right] = \dfrac{1}{q_j}$. This yields

$$E\left[\tau_0\right] = \frac{1}{q_0} = \frac{1}{\lambda_B} = 20 \text{ h}$$

$$E\left[\tau_1\right] = \frac{1}{q_1} = \frac{1}{\mu_F + \mu_E + \lambda} = 14.9 \text{ h}$$

$$E\left[\tau_2\right] = \frac{1}{q_2} = \frac{1}{\mu_E + 2\mu_H} = 1515 \text{ h}$$

(e) A repair takes place when the system transfers from state 1 to 2. When the system jumps from state 0 to state 2, two repairs take place. The fraction of time during which both servers are damaged is π_0 and the fraction of time in which one server is operating is π_1. The rate of repairs will be the rate of changing from state 1 to 2, plus two times the rate of changing from state 0 to state 2: $f_r = \pi_1 q_{12} + 2\pi_0 q_{02} = \pi_1 \lambda + 2\pi_0 \lambda_B = 7.17 \ 10^{-4}$. If we denote with X the random variable of the number of total failures over the period of 1 year, then the mean value of X will be $E\left[X\right] = f_r \times 24 \times 365 = 6.28$.

(iii) (a) The Markov chain is drawn in Figure B.6.

The failure rate (when both routers are working) for the first router is $\mu_1 = \frac{1}{100h} = 0.01 \ h^{-1}$.

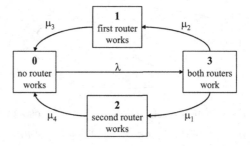

Fig. B.6. The continuous-time Markov chain of failures and repairs in the two access router problem.

The failure rate (when both routers are working) for the second router is $\mu_2 = \frac{1}{200h} = 0.005$ h^{-1}. The failure rate (when only one routers is working) for the first router is $\mu_3 = \frac{1}{50h} = 0.02 \ h^{-1}$. The failure rate (when only one routers is working) for the second router is $\mu_4 = \frac{1}{100h} = 0.01 \ h^{-1}$. The repair rate for both damaged routers is $\lambda = \frac{1}{50h} = 0.02 \ h^{-1}$. The infinitesimal generator matrix Q is

$$Q = \begin{bmatrix} -\lambda & 0 & 0 & \lambda \\ \mu_3 & -\mu_3 & 0 & 0 \\ \mu_4 & 0 & -\mu_4 & 0 \\ 0 & \mu_2 & \mu_1 & -\mu_2 - \mu_1 \end{bmatrix}$$

(b) The steady-state vector is a solution of (10.21). In order to solve this set of equations, the normalization condition $\sum_{i=0}^{3} \pi_i = 1$ is added and we obtain

$$
\begin{bmatrix} \pi_0 & \pi_1 & \pi_2 & \pi_3 \end{bmatrix}
\begin{bmatrix}
-\lambda & 0 & 0 & 1 \\
\mu_3 & -\mu_3 & 0 & 1 \\
\mu_4 & 0 & -\mu_4 & 1 \\
0 & \mu_2 & \mu_1 & 1
\end{bmatrix}
= \begin{bmatrix} 0 & 0 & 0 & 1 \end{bmatrix}
$$

from which

$$
\begin{bmatrix} \pi_0 & \pi_1 & \pi_2 & \pi_3 \end{bmatrix} = \frac{1}{a} \begin{bmatrix} \mu_3\mu_4(\mu_1+\mu_2) & \lambda\mu_2\mu_4 & \lambda\mu_1\mu_3 & \lambda\mu_3\mu_4 \end{bmatrix}
$$

where $a = \lambda\mu_3\mu_4 + \lambda\mu_1\mu_3 + \lambda\mu_2\mu_4 + \mu_3\mu_4(\mu_1+\mu_2) = 12.10^{-6}$. Thus,

$$
\begin{bmatrix} \pi_0 & \pi_1 & \pi_2 & \pi_3 \end{bmatrix} = \begin{bmatrix} \frac{1}{4} & \frac{1}{12} & \frac{1}{3} & \frac{1}{3} \end{bmatrix}
$$

(c) The steady-state probability that the first router is working equals $\pi_1 + \pi_3 = \frac{1}{12} + \frac{1}{3} = \frac{5}{12}$.

(d) Theorem 10.2.3 states that the mean lifetime of a state j is $E[\tau_j] = \frac{1}{q_j}$. Thus, on average the second router is down for $\frac{1}{q_0} + \frac{1}{q_1} = \frac{1}{\lambda} + \frac{1}{\mu_3} = 50h + 50h = 100h$.

(iv) (a) Each state is has a 3-tuple (x, y, z) label, where x and y represent the status of the primary routes of c_1 and c_2, respectively, and z represents which connection is using the backup route b. The possibilities for x and y are either "up" (U) and "down" (D), while z is 0 (when none of the connections is backed-up), or $z = c_1$ or c_2. Figure B.7 represents the Markov chain with states $S_1 = (U, U, 0)$, $S_2 = (D, U, c_1)$, $S_3 = (U, D, c_2)$, $S_4 = (D, D, c_1)$ and $S_5 = (D, D, c_2)$.

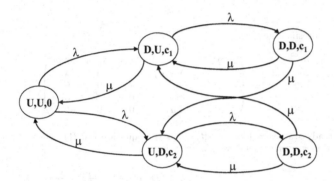

Fig. B.7. The Markov chain of the continuous-time back-up path problem.

(b) The infinitesimal generator is

$$
Q = \begin{pmatrix}
-2\lambda & \lambda & \lambda & 0 & 0 \\
\mu & -(\lambda+\mu) & 0 & \lambda & 0 \\
\mu & 0 & -(\lambda+\mu) & 0 & \lambda \\
0 & \mu & \mu & -2\mu & 0 \\
0 & \mu & \mu & 0 & -2\mu
\end{pmatrix}
$$

(c) The steady-state $\pi = \lim_{t\to\infty} s[t]$ obeys (10.21). Since $\det Q = 0$, we can replace one the rows of Q by the normalization condition $\pi_1 + \pi_2 + \pi_3 + \pi_4 + \pi_5 = 1$. The solution of that linear set is

$$
\begin{bmatrix} \pi_1 & \pi_2 & \pi_3 & \pi_4 & \pi_5 \end{bmatrix} = \frac{\begin{bmatrix} \mu^2 & \mu\lambda & \mu\lambda & \frac{\lambda^2}{2} & \frac{\lambda^2}{2} \end{bmatrix}}{\lambda^2 + 2\lambda\mu + \mu^2}
$$

(c) The conditional probability that c_1 is using the backup route, given that both c_1 and c_2 have failed is $\frac{\pi_4}{\pi_4+\pi_5} = 0.5$, which also follows from the equal role of c_1 and c_2 in this problem.

(v) (a) The continuous Markov chain is drawn in Fig. B.8. Each state is represented by the tuple (x, y). The first entry $x \in \{0, 1, 2\}$ represents the number of paths available. The second entry $y \in \{0, 1, 2\}$ represents the type of failure along the path on which the connection is setup, where 0 denotes no failure, 1 is a node failure and 2 is a link failure. The following represents the Markov chain: $S_1 = (2, 0)$ represents the state in which the connection is up and there is no failure in the other path, $S_2 = (1, 0)$ represents the state in which the connection is up and there is a failure in the other path, in $S_3 = (1, 1)$, the connection is down due to node failure along the path on which the connection is setup, but it has not been sensed yet and in state $S_4 = (1, 2)$, the connection is down due to node failure along the path on which the connection is setup, but it has not been sensed yet. Finally, $S_5 = (0, 0)$ represents the state in which both paths have failed.

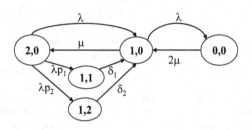

Fig. B.8. The graph of the continuous Markov chain of the two link-disjoint path failure problem

(b) The infinitesimal generator is

$$
Q = \begin{pmatrix}
-2\lambda & \lambda & \lambda p_1 & \lambda p_2 & 0 \\
\mu & -(\lambda + \mu) & 0 & 0 & \lambda \\
0 & \delta_1 & -\delta_1 & 0 & 0 \\
0 & \delta_2 & 0 & -\delta_2 & 0 \\
0 & 2\mu & 0 & 0 & -2\mu
\end{pmatrix}
$$

(c) The steady-state $\pi = \lim_{t \to \infty} s[t]$ obeys the equation (10.21), whose solution, using the normalization $\pi_1 + \pi_2 + \pi_3 + \pi_4 + \pi_5 = 1$, is

$$
\begin{bmatrix} \pi_1 & \pi_2 & \pi_3 & \pi_4 & \pi_5 \end{bmatrix} = \frac{\left[1 \quad \frac{2\lambda}{\mu} \quad \frac{\lambda p_1}{\delta_1} \quad \frac{\lambda p_2}{\delta_2} \quad \frac{\lambda^2}{\mu^2} \right]}{1 + \frac{2\lambda}{\mu} + \frac{\lambda p_1}{\delta_1} + \frac{\lambda p_2}{\delta_2} + \frac{\lambda^2}{\mu^2}}
$$

(d) The connection is available only in states S_1 and S_2. Thus, the steady-state connection availability is

$$
\pi_1 + \pi_2 = \frac{1 + \frac{2\lambda}{\mu}}{\left(1 + \frac{2\lambda}{\mu} + \frac{\lambda p_1}{\delta_1} + \frac{\lambda p_2}{\delta_2} + \frac{\lambda^2}{\mu^2} \right)}
$$

B.10 Applications of Markov chains (Chapter 11)

(i) In both cases we apply the general formulae (11.16) and (11.17) for the steady-state of a general birth and death process.
(a) Using the notation $\rho = \frac{\lambda}{\mu}$, we first compute

$$
\prod_{m=0}^{j-1} \frac{\lambda_m}{\mu_{m+1}} = \prod_{m=0}^{j-1} \frac{\lambda}{(m+1)\mu} = \frac{\lambda^j}{\mu^j} \prod_{m=1}^{j} \frac{1}{m} = \frac{\rho^j}{j!}
$$

Then, with (11.17),

$$\pi_j = \frac{\frac{\rho^j}{j!}}{1 + \sum_{j=1}^{\infty} \frac{\rho^j}{j!}} = \frac{\rho^j}{j!} e^{-\rho} \qquad j \geq 0$$

which demonstrates that the steady-state probability that the birth and death process is in state j is Poisson distributed with mean ρ.

(b) Similarly, we first compute with $\lambda_m = \frac{\lambda}{(m+1)}$ and $\mu_m = \mu$,

$$\prod_{m=0}^{j-1} \frac{\lambda_m}{\mu_{m+1}} = \prod_{m=0}^{j-1} \frac{\lambda}{(m+1)\mu} = \frac{\rho^j}{j!}$$

which leads to precisely the same steady-state as in (a). Indeed, the steady-state is only a function of the ratios $\frac{\lambda_m}{\mu_{m+1}}$, which are the same in both (a) and (b).

(ii) After multiplying (11.31) by x^k and summing over all k, we have

$$\sum_{k=0}^{\infty} s_k'(t) x^k = \lambda(t) \sum_{k=0}^{\infty} (k-1) s_{k-1}(t) x^k - (\lambda(t) + \mu(t)) \sum_{k=0}^{\infty} k s_k(t) x^k + \mu(t) \sum_{k=0}^{\infty} (k+1) s_{k+1}(t) x^k$$

$$= \lambda(t) x^2 \sum_{k=-1}^{\infty} k s_k(t) x^{k-1} - (\lambda(t) + \mu(t)) x \sum_{k=0}^{\infty} k s_k(t) x^{k-1} + \mu(t) \sum_{k=1}^{\infty} k s_k(t) x^{k-1}$$

Taking into account that $s_k(t) = 0$ for $k < 0$ and using the partial derivative of the generating function (11.32),

$$\frac{\partial \varphi}{\partial x} = \sum_{k=0}^{\infty} k s_k(t) x^{k-1}$$

yields

$$\frac{\partial \varphi}{\partial t} = \left\{ \lambda(t) x^2 - (\lambda(t) + \mu(t)) x + \mu(t) \right\} \frac{\partial \varphi}{\partial x}$$

Since $\lambda(t) x^2 - (\lambda(t) + \mu(t)) x + \mu(t) = \{\lambda(t) x - \mu(t)\}(x-1)$, we have established (11.33).

(iii) All stations in slotted Aloha operate independently and each has probability $p_t = 0.12$ to transmit in a timeslot. A station is successful in one slot with probability $p_s = p_t(1-p_t)^{N-1}$ where the number of stations $N = 8$. Thus, $p_s = 0.049$. The waiting time W to transmit one packet is a geometric random variable with parameter p_s from which (Section 3.1.3) the mean $E[W] = \frac{1}{p_s}$. Alternatively, $E[W]$ obeys the equation

$$E[W] = p_s + (1 - p_s)(1 + E[W])$$

because the mean waiting time equals 1 timeslot with probability p_s plus 1 timeslot increased with the mean waiting time with probability $1 - p_s$. Solving that equation again yields $E[W] = \frac{1}{p_s} = 20.39$ timeslots. The mean transmission time for 7 packets is $7E[W] = 142.7$ timeslots.

B.11 Branching processes (Chapter 12)

(i) Consider $\psi(x) = \frac{\varphi_Y(x-a)}{\varphi_Y(x)}$ with real $a \in (0,1)$ and real x. If

$$\frac{d\psi(x)}{dx} = \frac{\varphi_Y(x) \varphi_Y'(x-a) - \varphi_Y(x-a) \varphi_Y'(x)}{\varphi_Y^2(x)} \geq 0$$

then $\varphi_Y(x) \varphi_Y'(x-a) \geq \varphi_Y(x-a) \varphi_Y'(x)$. Let $x = 1$ and $x - a = \pi_0 = \varphi_Y(\pi_0)$, then $\varphi_Y'(\pi_0) \geq \pi_0 \mu$. By using Rolle's Theorem as in the proof of Lemma 12.5.1, it follows that

$1 \geq \varphi'_Y (\pi_0)$ if $\mu > 1$. The fact that $\frac{d\psi(x)}{dx} \geq 0$ implies that

$$\frac{d}{dz} \log \varphi_Y (z) \bigg|_{z=x-a} \geq \frac{d}{dz} \log \varphi_Y (z) \bigg|_{z=x}$$

If $\log \varphi_Y (z)$ is concave for real $z \in [0,1]$, then $\frac{d}{dz} \log \varphi_Y (z)$ is non-increasing for real $z \in [0,1]$, and the inequality (12.43) is a consequence of concavity of $\log \varphi_Y (x)$ for real $x \in [0,1]$. The Poisson generating function $\varphi_Y (x) = e^{\mu(z-1)}$ is an example where $\frac{d\psi(x)}{dx} = 0$.

(ii) The probability of extinction in a time-dependent branching process

$$\pi_0 = \lim_{t \to \infty} \Pr[Z(t) = 0]$$

is written in terms of the pgf as $\pi_0 = \lim_{t \to \infty} \varphi_{Z(t)} (0)$. Evaluating the Bellman-Harris equation (12.32) at $s = 0$ yields

$$\varphi_{Z(t)} (0) = \int_0^t \varphi_Y \left(\varphi_{Z(t-y)} (0) \right) dG(y)$$

Assuming the existence of the limit $\lim_{t \to \infty} \varphi_{Z(t)} (0) = \pi_0$ (and interchanging the limit operator with the integral and pgf as justified in Section 12.3), we obtain

$$\pi_0 = \int_0^\infty \varphi_Y (\pi_0) dG(y) = \varphi_Y (\pi_0)$$

which is, indeed, (12.18).

(iii) The mean and variance of the Gamma distribution (where $\alpha = \frac{1}{\theta}$) are obtained from (2.40) and (3.26) as $E[L] = \beta\theta$ and $\text{Var}[L] = \beta\theta^2$, so that $\beta = \frac{(E[L])^2}{\text{Var}[L]}$, illustrating that β is small if the variance of the lifetime L is large, given a fixed mean lifetime. Using the Laplace transform (3.26) of the Gamma distribution

$$\int_0^\infty e^{-\alpha u} dG(u) = \frac{1}{(\alpha\theta + 1)^\beta}$$

into (12.36), the Malthusian parameter is $\alpha = \frac{\mu^{\frac{1}{\beta}} - 1}{\theta}$, which reduces to the Malthusian parameter for an exponential lifetime when $\beta = 1$.

B.12 Queueing (Chapter 14)

(i) Let us denote the number of packets in the server by N_x. Since a router either serves 0 or 1 packet, the problem states that $\Pr[N_x = 1] = 0.8$, and also that $E[N_x] = 0.8$. For any queue, it holds that $N_S = N_Q + N_x$ and $T = w + x$, the number in the system equals the number in the buffer and the number that is being served. From Little's Theorem (13.24), it follows with $E[x] = \frac{1}{\mu}$ that $E[N_x] = \frac{\lambda}{\mu}$ or $\lambda = \mu E[N_x]$. Substituted into Little's Law for the waiting time in the buffer, $E[N_Q] = \lambda E[w]$, and using $E[N_Q] = 3.2$ gives

$$E[w] = \frac{E[N_Q]}{E[N_x]} \frac{1}{\mu} = \frac{4}{\mu}$$

(ii) In a M/M/m/m queue, the number of busy servers equals the number (of packets) in the system N_S. From (14.17) and the definition (2.11), the mean number of busy servers equals

$$E[N_S] = \sum_{j=0}^m j \Pr[N_S = j] = \frac{1}{\sum_{j=0}^m \frac{\lambda^j}{j!\mu^j}} \sum_{j=1}^m \frac{\lambda^j}{(j-1)!\mu^j}$$

The sum can be rewritten as

$$\sum_{j=1}^m \frac{\lambda^j}{(j-1)!\mu^j} = \frac{\lambda}{\mu} \sum_{j=0}^{m-1} \frac{\lambda^j}{j!\mu^j} = \frac{\lambda}{\mu} \left[\sum_{j=0}^m \frac{\lambda^j}{j!\mu^j} - \frac{\lambda^m}{m!\mu^m} \right]$$

such that

$$E\left[N_S\right] = \frac{\lambda}{\mu}\left[1 - \frac{\frac{\lambda^m}{m!\mu^m}}{\sum_{j=0}^m \frac{\lambda^j}{j!\mu^j}}\right] = \frac{\lambda}{\mu}\left(1 - \Pr\left[N_S = m\right]\right)$$

where the last probability is recognized as the Erlang B formula (14.18).

(iii) (a) Since the mean service rate $\mu = 2$ s^{-1}, the mean response time (mean system time) follows from (14.3) as $E\left[T_{M/M/1}\right] = \frac{1}{2-\lambda}$.

(b) If $E\left[T_{M/M/1}\right] = 2.5$ s, then it follows from (a) that $\lambda = 1.6$ s^{-1}. Hence, the number of jobs/s that can be processed for a given mean response time of 2.5s equals 1.6 jobs/s (c) A 10% increase in arrival rate corresponds to $\lambda = 1.76$ s^{-1} and from (a) we obtain that $E\left[T_{M/M/1}\right] = \frac{1}{0.24} = 4.17$ s, which is with respect to 2.5s an increase in the mean response time of 67%.

(iv) We know that when the mean call holding time is $1/\mu = 10$ min, the time blocking probability $P_{Bt} = \frac{1}{10}$. Additionally, for Poisson call arrivals, the time blocking probability P_{Bt} equals the call blocking probability P_B on the PASTA property. The number of channels is $m = 2$. The arrival intensity λ can be calculated from the Erlang B formula (14.18)

$$P_B = \frac{r^2/2}{1 + r + r^2/2}$$

where $r = \lambda/\mu$. Solving this equation for $r = 2\rho$ and taking into account that the traffic intensity $\rho \in [0, 1]$ yields $\rho = \frac{P_B + \sqrt{2P_B - P_B^2}}{1 - P_B}$. For $P_B = \frac{1}{10}$, we have that $r = \frac{1 + \sqrt{19}}{9}$ from which $\lambda = \frac{1 + \sqrt{19}}{90}$. The blocking probability (14.18) corresponding to a mean call holding time $1/\mu = 15$ min for which $r = \lambda/\mu = \frac{1 + \sqrt{19}}{90} \times 15$ is $P_{Bt} \approx 0.174$.

(v) The queue 1 is a M/M/1 queue. By Burke's Theorem 14.1.1, the departure process from queue 1 is a Poisson with rate λ. By assumption, this departure process, which is the arrival process to the second queue, is independent of the service process at queue 2. Therefore, queue 2, viewed in isolation, is also a M/M/1 queue. We know that the queueing processes in both queues are stable because the load $\rho_1 = \frac{\lambda}{\mu_1} < 1$ and $\rho_2 = \frac{\lambda}{\mu_2} < 1$. The steady-state distribution of the number of customers in queue 1 and queue 2 follow from (14.1) as $\Pr[n$ at queue 1$] = \rho_1^n(1 - \rho_1)$ and $\Pr[m$ at queue 2$] = \rho_2^m(1 - \rho_2)$. The number of customers in queue 1 is independent of the sequence of earlier arrivals at queue 2 and, therefore, also of the number of customers in queue 2. This implies that

$$\Pr[n \text{ at queue } 1, m \text{ at queue } 2] = \Pr[n \text{ at queue } 1]\Pr[m \text{ at queue } 2] = \rho_1^n(1 - \rho_1)\rho_2^m(1 - \rho_2)$$

(vi) The mean system times $E[T]$ for the three different queueing systems are immediate. From (14.3), for system A, we have $E[T_A] = \frac{1}{k\mu(1-\rho)}$. For each of the k subqueues of system B, (14.3) gives $E[T_B] = \frac{1}{\mu(1-\rho)}$, while for the M/M/k queue, (14.15) yields

$$E[T_C] = \frac{k(1 - \rho) + \Pr[N_S \geq k]}{k\mu(1 - \rho)}$$

Clearly, $E[T_B] = kE[T_A]$ shows that by replacing k small systems by one larger system with same processing capability, the mean system time decreases by a factor k.

From the relation $E[T_C] = f(k, \rho)E[T_A]$ with $f(k, \rho) = k(1 - \rho) + \Pr[N_S \geq k]$, it is more complicated to decide where $f(k, \rho)$ is larger or smaller than 1. The extreme values of $f(k, \rho)$ are known: $f(k, 0) = k$ and $f(k, 1) = 1$. Since $\frac{\partial f(k,\rho)}{\partial \rho} = -k + \frac{\partial \Pr[N_S \geq k]}{\partial \rho}$ and $\frac{\partial \Pr[N_S \geq k]}{\partial \rho} > 0$, it cannot be concluded that $\frac{\partial f(k,\rho)}{\partial \rho}$ is monotonously decreasing from k to 1 in which case we would have $f(k, \rho) > 1$. Assuming k real, we observe that $\frac{\partial f(k,\rho)}{\partial k} = (1 - \rho) + \frac{\partial \Pr[N_S \geq k]}{\partial k} > 0$ for all $\rho < 1$, which implies that $f(2, \rho) < f(3, \rho) < \cdots$ and allows us to concentrate only on $f(2, \rho)$. Numerical results show that $f(2, \rho) < 1$ if $\rho \geq 0.85$, but $f(3, \rho) \geq 1$. This leads us to the conclusion that for $k > 2$, system A always outperforms system C; only if $k = 2$ and in the heavy traffic regime $\rho \geq 0.85$, system C leads to a slightly shorter mean system time of maximum 1.7%. Hence, by replacing $k > 2$ processing units (servers) by one with same processing capability, always lowers the total time spent in the system. Of course, all conclusions only apply to systems that can be well modeled as M/M/m queueing systems. To first order a computing device (processor) may be regarded as a M/M/1 queue. Then, the analysis shows that replacing an old processor by a k times faster one is faster (on average) than installing k old processors in parallel.

(vii)   The waiting process for aeroplanes is modeled as a M/D/1 queue because the arrival process is Poissonean with rate $\lambda = \frac{1}{10}$ arrivals/minute, it consists of a single queue as 1 aeroplane can land at a time and the service process (the landing process) takes precisely $x = 5$ minutes (constant service time). Thus, $E[x] = 5$ minutes, $\text{Var}[x] = 0$ and $\rho = \frac{5}{10}$. Since the M/D/1 process is a special case of the M/G/1, we can apply the general formula (14.29) for the mean waiting time in the queue of an M/G/1 system

$$E[w] = \frac{\lambda E[x^2]}{2(1-\rho)} = \frac{\frac{1}{10} \cdot 5^2}{2 \cdot \frac{1}{2}} = 2.5 \text{ minutes}$$

(viii) *Mobile Communications.* (a) We know that the arrival intensity of new calls to the cell is $\lambda_f = 20$ calls/min. Let λ_h denote the arrival rate of the handover calls. The mean time spent by a call in the cell is $E[T] = 1.64$ minutes and the mean number of ongoing calls is $E[N] = 52$. Furthermore, the blocking rate is $P_B = 0.02$. The total arrival rate of calls that are carried by the base station is

$$\lambda_{\text{carried}} = (1 - P_B)\,\lambda_{\text{offered}} = (1 - P_B)\,(\lambda_f + \lambda_h)$$

Little's formula (13.24) states that $E[N] = \lambda_{\text{carried}} E[T]$. Note that only the carried calls have an influence on the state of the system. We can solve the asked λ_h from these two equations as

$$\lambda_h = \frac{E[N]}{E[T](1 - P_B)} - \lambda_f = 12.35 \text{ calls/minute}$$

(b) The arrival intensity of lost calls is $\lambda_{\text{lost}} = P_B\,(\lambda_f + \lambda_h) = 0.647$ calls/minute. If only the new calls are blocked, the asked blocking rate is

$$P_{Bf} = \frac{\lambda_{\text{lost}}}{\lambda_f} = 3.24\%$$

(ix)   (a) The derivation is given in the study of the M/G/1 queue in Section 14.3.1 where $A(z)$ given by (14.26) should be replace by $E\left[z^N\right]$.
        (b) Use in (14.26) the Laplace transform of an exponential random variable with mean $\frac{1}{\mu}$ given in (3.16) so that

$$E\left[z^N\right] = \varphi_x\left(\lambda(1-z)\right) = \frac{\mu}{(1-z)\lambda + \mu} = \frac{\mu/(\lambda + \mu)}{1 - (\lambda/(\lambda + \mu))z}$$

$$= \left(1 - \frac{\lambda}{\lambda + \mu}\right)\sum_{k=0}^{\infty}\left(\frac{\lambda}{\lambda + \mu}\right)^k z^k$$

from which the probability density function

$$\Pr[N = k] = \left(1 - \frac{\lambda}{\lambda + \mu}\right)\left(\frac{\lambda}{\lambda + \mu}\right)^k$$

follows. Thus, $\Pr[N = k]$ is recognized as a geometric random variable with mean $E[N] = \frac{\lambda}{\mu}$.

(x)   The queueing system is modeled by a birth and death process. The death rate is obvious and equal to μ. The arrival rate into state j equals the arrival rate of customers λ multiplied by the probability of really going to that state j which is $\frac{1}{j+1}$. Hence, $\lambda_j = \frac{\lambda}{j+1}$. The steady-state equation of this birth and death process is a Poisson process with rate $\rho = \frac{\lambda}{\mu}$ as derived above in Section B.10 solution (i).

(xi) *The M/M/m/m/s queue (The Engset formula).* The arrival rate in Engset model is proportional to the size of the "still demanding subgroup" and the number of arrivals is exponential. The holding time of a line is exponentially distributed with mean $\frac{1}{\mu}$.

The Engset model is described as a birth–death process where each state k refers to the size of the "served subgroup". Since the total of customers is s, the "still demanding subgroup" consists of $s - k$ members. The birth rate is $\lambda_j = \alpha(s - k)$ and the death rate $\mu_k = k\mu$. The proportionality factor α can be interpreted as the arrival rate per "still demanding customer". The Markov graph is depicted in Fig. B.9. Application of the general birth–death formulae

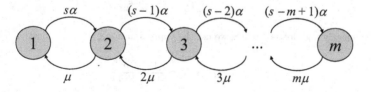

Fig. B.9. The Markov chain of the Engset loss model.

for the steady-state vector (11.17) or $\Pr[N_S = j]$ yields, with $r = \frac{\alpha}{\mu}$,

$$\pi_j = \frac{\prod_{k=1}^{j} \frac{\alpha(s-k+1)}{k\mu}}{1 + \sum_{n=1}^{m} \prod_{k=1}^{n} \frac{\alpha(s-k+1)}{k\mu}} = \frac{\binom{s}{j} r^j}{\sum_{n=0}^{m} \binom{s}{n} r^n}$$

The computation of the blocking probability is more complex than for the Erlang B formula, because the arrival process is not a Poisson process. Indeed, due to the finite number of customers s, the largest number of possible arrivals is finite and the arrival rate depends on the state. Hence, the PASTA property cannot be applied. For a small time interval Δt, the blocking probability P_B equals the ratio of the $p_b(\Delta t)$, the probability of blocking in Δt, over $p_a(\Delta t)$, the probability of an arrival in Δt. Since the arrival rates depend on the state, the probability of an arrival in Δt is not equal to π_m as for the Erlang B model. Instead, we have

$$p_a(\Delta t) = \Delta t \sum_{n=0}^{m} (s-n)\alpha \Pr[N_S = n] = \alpha\Delta t \frac{\sum_{n=0}^{m} (s-n)\binom{s}{n} r^n}{\sum_{n=0}^{m} \binom{s}{n} r^n} = \alpha s \Delta t \frac{\sum_{n=0}^{m} \binom{s-1}{n} r^n}{\sum_{n=0}^{m} \binom{s}{n} r^n}$$

Furthermore, blocking is only caused if $N_S = m$ and if at least one of the $s - m$ customers of the "still demanding group" generates an arrival. However, since the interval Δt can be made arbitrarily small[3], a generation of more than 1 arrival has probability $o(\Delta t)$ such that it suffices to consider only one call attempt. Hence,

$$p_b(\Delta t) = \Delta t(s - m) \Pr[N_S = m] = \alpha s \Delta t \frac{\binom{s-1}{m} r^m}{\sum_{n=0}^{m} \binom{s}{n} r^n}$$

The Engset call blocking probability $P_B = \frac{p_b(\Delta t)}{p_a(\Delta t)}$ becomes

$$P_B = \frac{\binom{s-1}{m} r^m}{\sum_{n=0}^{m} \binom{s-1}{n} r^n} \tag{B.15}$$

Observe that $P_B = \Pr[N_S = m]$ in a system with $s - 1$ instead of s customers: an entering customer observes a system with $s - 1$ customers ignoring himself. At last, if we denote $\lambda = \alpha s$, the Engset call blocking formula (B.15) can be rewritten as

$$P_B = \frac{\frac{1}{m!} \left(\frac{\lambda}{\mu}\right)^m}{\sum_{n=0}^{m} \frac{(s-1-m)! s^{m-n}}{n!(s-1-n)!} \left(\frac{\lambda}{\mu}\right)^n}$$

---

[3] Similar arguments are used in Chapter 7 when studying the Poisson process.

The ratio $\frac{(s-1-m)!}{(s-1-n)!}$ is a polynomial in s of degree $n-m$ such that $\lim_{s\to\infty} \frac{(s-1-m)!s^{m-n}}{(s-1-n)!} = 1$. In conclusion, if $\lambda = \alpha s$ and $s \to \infty$, the Engset call blocking probability reduces to the Erlang B formula (14.18).

(xii) Although for a M/D/1 the exact expression of the overflow probability (14.45) exists, this series converges slowly for high traffic intensities $\rho = \lambda$ so that fast executable expressions are desirable. Substituting (14.46) with $\rho = \lambda$ into (14.73) gives

$$clr(\lambda) \approx \frac{1-\rho}{\rho} \frac{\frac{1-\rho}{\rho\zeta-1}\zeta^{-K}}{1 - \frac{1-\rho}{\rho\zeta-1}\zeta^{-K}}$$

For sufficiently high loads $\rho > 0.8$, we use the approximation $\zeta \approx \lambda^{-2}$ of Section 5.7 to obtain

$$clr_{\mathrm{M/D/1/K}} \simeq \frac{(1-\rho)\rho^{2K}}{1 - \rho^{2K+1}} \tag{B.16}$$

Comparing with (14.21) in the M/M/1/K queue,

$$clr_{\mathrm{M/M/1/K}} \simeq \frac{(1-\rho)\rho^{K}}{1 - \rho^{K+1}}$$

the M-server (in continuous-time) needs *approximately twice* as much buffer places to guarantee the same cell loss ratio as in the corresponding D-server (in discrete-time). Further combining (14.4) and (14.39) shows that $\mu_{\mathrm{M/M/1}}E\left[w_{\mathrm{M/M/1}}\right] = 2\mu_{\mathrm{M/D/1}}E\left[w_{\mathrm{M/D/1}}\right]$ or, the mean waiting time in the queue (normalized to the mean service time) for the M/M/1 queue is *exactly twice* as long as for the M/D/1 queue. The variability of the service in the M-server causes these rather large differences in performance. Furthermore, the simple formula (B.16) is particularly useful to engineer ATM buffers or to dimension simple queueing networks. If the number of individual flows that constitute the aggregate flow are large enough and none of the individual flows is dominant, the aggregate arrival process is quite well approximated by a Poisson process. Given as a QoS requirement a stringent cell loss ratio clr^*, the input flow ρ can be limited such that $clr_{\mathrm{M/D/1/K}} < clr^*$. Alternatively, the buffer size K can be derived from (B.16) subject to $clr_{\mathrm{M/D/1/K}} = clr^*$ for an aggregate Poisson input flow $\lambda = 0.9$. As long as the input flow is limited to $\lambda < 0.9$, the thus found buffer size K always guarantees a cell loss ratio below clr^* provided the input flow can be approximated as a Poisson arrival process.

(xiii) (a) The arrival rate at the first processor is given $\lambda_1 = \lambda$ and $\lambda = 1200\ \mathrm{s}^{-1}$. Burke's Theorem 14.1.1 states that, for the M/M/1 queue, the arrival and departure processes are both Poisson processes with the same mean. Hence, the arrival rate at processor 2 obeys

$$\lambda_2 = \frac{1}{2}\lambda_1 + \frac{1}{3}\lambda_3 \tag{B.17}$$

and similar for processor 3,

$$\lambda_3 = \frac{1}{2}\lambda_1 + \lambda_2 \tag{B.18}$$

Substituting (B.17) into (B.18) eliminates λ_2 resulting in $\lambda_3 = \frac{3}{2}\lambda_1 = 1800\ \mathrm{s}^{-1}$. Using the values of λ_1 and λ_3 into (B.17) yields $\lambda_2 = \lambda_1 = 1200\ \mathrm{s}^{-1}$. Since all processors are identical, the mean service rate is the same for all and equal to

$$\mu = \frac{1}{E\left[x\right]} = \frac{1}{500\ 10^{-6}\ \mathrm{s}} = 2000\ \mathrm{s}^{-1}$$

The mean number of packets (which is dimensionless, it is just a non-negative real number) in a processor i is $E[N_i] = \frac{\rho_i}{1-\rho_i}$. Hence,

$$\rho_1 = \frac{\lambda}{\mu} = \frac{1200}{2000} = 0.6 \rightarrow E[N_1] = \frac{0.6}{1-0.6} = 1.5$$

$$\rho_2 = \frac{\lambda_2}{\mu} = \frac{1200}{2000} = 0.6 \rightarrow E[N_2] = \frac{0.6}{1-0.6} = 1.5$$

$$\rho_3 = \frac{\lambda_3}{\mu} = \frac{1800}{2000} = 0.9 \rightarrow E[N_3] = \frac{0.9}{1-0.9} = 9$$

(b) By Little's Law (13.24), the mean time spent in the system is $E[T] = \frac{E[N_S]}{\lambda}$, where λ is the mean rate into the system, which equals λ_1. The number of packets in the system is the sum of the number of packets in each processor

$$E[N_S] = E\left[\sum_{i=1}^{3} N_i\right] = \sum_{i=1}^{3} E[N_i] = 1.5 + 1.5 + 9 = 12$$

from which

$$E[T] = \frac{E[N_S]}{\lambda} = \frac{12}{1200 \text{ s}^{-1}} = 0.01 \text{ s}$$

(c) $\Pr[N_S = j] = (1 - \rho)\rho^j$ is the probability that the M/M/1 system contains j packets. Hence, the joint probability distribution of the number of packets in different and independent processors is

$$\Pr[N_1 = k, N_2 = l, N_3 = m] = \Pr[N_1 = k]\Pr[N_2 = l]\Pr[N_3 = m]$$
$$= (1 - \rho_1)\rho_1^k (1 - \rho_2)\rho_2^l (1 - \rho_3)\rho_3^m$$
$$= 0.4 \times 0.6^k \times 0.4 \times 0.6^l \times 0.1 \times 0.9^m = 0.016 \times 0.6^{k+l}0.9^m$$

(xiv) (a) The mean waiting time in a M/G/1 queue is given by (14.29), where $\rho = \lambda E[x]$,

$$E[w] = \frac{\lambda E[x^2]}{2(1-\rho)}$$

The probability distribution function of the service time is

$$f_x(t) = \alpha\delta(t - d) + (1 - \alpha)\frac{1}{h}e^{-\frac{1}{h}t}$$

where $\delta(t)$ is the Dirac delta function. Since formula (14.29) requires both $E[x]$ and $E[x^2]$, it is convenient to compute the Laplace transform (2.38) of the service time,

$$\varphi_x(z) = \alpha e^{-zd} + (1 - \alpha)\frac{\frac{1}{h}}{\frac{1}{h} + z}$$

The mean service time of the packets is

$$E[x] = -\varphi_x'(0) = \alpha d + (1 - \alpha)h$$

while the second moment is

$$E[x^2] = \varphi_x''(0) = \alpha d^2 + (1 - \alpha)2h^2$$

The mean waiting time then follows from (14.29) as

$$E[w] = \frac{\lambda(\alpha d^2 + (1 - \alpha)2h^2)}{2(1 - \lambda(\alpha d + (1 - \alpha)h))}$$

(b) The mean system time is

$$E[T] = E[w] + E[x] = \frac{\lambda(\alpha d^2 + (1 - \alpha)2h^2)}{2(1 - \lambda(\alpha d + (1 - \alpha)h))} + \alpha d + (1 - \alpha)h$$

(c) The mean queue size is found from Little's Law (13.24) as

$$E\left[N_Q\right] = \lambda E\left[w\right] = \frac{\lambda^2\left(\alpha d^2 + (1-\alpha)\,2h^2\right)}{2\left(1 - \lambda\left(\alpha d + (1-\alpha)\,h\right)\right)}$$

(xv) (a) We first compute the generating function (2.18) of the system content using (14.20),

$$\varphi_{N_S}(z) = \sum_{j=0}^{K} \Pr\left[N_S = j\right] z^j = \frac{(1-\rho)}{1-\rho^{K+1}} \sum_{j=0}^{K} (\rho z)^j = \frac{(1-\rho)}{1-\rho^{K+1}} \frac{1 - (\rho z)^{K+1}}{1-\rho z}$$

The mean (14.74) follows then from $E\left[N_{S;M/M/1/K}\right] = \varphi'_{N_S}(1)$.
(b) The limit of (14.74) for $\rho \to 1$ is

$$\lim_{\rho \to 1} E\left[N_{S;M/M/1/K}\right] = \lim_{\rho \to 1} \frac{K\rho^{K+2} - (K+1)\,\rho^{K+1} + \rho}{\rho^{K+2} - \rho^{K+1} - \rho + 1}$$

Using the l'Hospital's rule twice, we obtain

$$\lim_{\rho \to 1} E\left[N_{S;M/M/1/K}\right] = \lim_{\rho \to 1} \frac{K\left(K+2\right)\rho^{K+1} - (K+1)^2\,\rho^K + 1}{(K+2)\,\rho^{K+1} - (K+1)\,\rho^K - 1}$$

$$= \lim_{\rho \to 1} \frac{K\left(K+2\right)(K+1)\,\rho^K - (K+1)^2\,K\rho^{K-1}}{(K+2)\,(K+1)\,\rho^K - (K+1)\,K\rho^{K-1}} = \frac{K}{2}$$

(c) The mean queue size is

$$E\left[N_{Q;M/M/1/K}\right] = \sum_{j=1}^{K} (j-1)\Pr\left[N_{S;M/M/1/K} = j\right]$$

$$= E\left[N_{S;M/M/1/K}\right] - \left(1 - \Pr\left[N_{S;M/M/1/K} = 0\right]\right)$$

Substituting (14.74) and (14.20) yields

$$E\left[N_{Q;M/M/1/K}\right] = \frac{\rho\left(1 - (K+1)\,\rho^K + K\rho^{K+1}\right)}{(1-\rho)\left(1-\rho^{K+1}\right)} - \left(1 - \frac{(1-\rho)}{1-\rho^{K+1}}\right)$$

$$= \frac{\rho\left(\rho - K\rho^K + (K-1)\,\rho^{K+1}\right)}{(1-\rho)\left(1-\rho^{K+1}\right)}$$

(d) Little's Law (13.24) states that $E\left[N_{S;M/M/1/K}\right] = \lambda E\left[T_{S;M/M/1/K}\right]$, where λ is the arrival rate *seen* by the system! The effective rate of packets entering the system is $\lambda\left(1-r\right)$ where the rejection rate r is specified by (14.21). Hence, we obtain (14.75) from

$$E\left[T_{S;M/M/1/K}\right] = \frac{1}{\lambda\left(1-r\right)} E\left[N_{S;M/M/1/K}\right]$$

(xvi) (a) The telephony system can be modeled by a M/M/m system, with $m = 10$. The arrival process is Poissonean with $\lambda = \frac{90\times 7}{8\times 60} = 1.75$ calls/minute. The calls are exponentially distributed with mean of 4 minutes per phone call, $\mu = \frac{1}{4} = 0.25$ calls/minute such that $\lambda/\mu = 7$. The load at the server is then

$$\rho = \frac{\lambda}{m\,\mu} = \frac{1.75}{10 \times 0.25} = 0.70$$

The probability that the system is empty, $\Pr[N_S = 0]$ is given by (14.7),

$$\Pr[N_S = 0] = \left(1 + 7 + \frac{7^2}{2} + \frac{7^3}{6} + \frac{7^4}{24} + \frac{7^5}{150} + \frac{7^6}{720} + \frac{7^7}{5040} + \right.$$

$$\left. \frac{7^8}{40\,320} + \frac{7^9}{362\,880} + \frac{7^{10}}{3\,628\,800}\frac{1}{1-\rho}\right)^{-1} = 0.000\ 854\ 537$$

The probability that all lines are occupied is then given by the Erlang C formula (14.10),

$$\Pr[N_S \geq m] = \frac{\Pr[N_S = 0]\,7^{10}}{10! \times (1 - \rho)} = 0.221\ 731 \simeq 0.22$$

(b) The probability that employee needs to wait more than one minute follows from (14.12) as

$$\Pr[W \geq t] = \Pr[N_S \geq m]\,e^{-(1-\rho)m\,\mu\,t} = 0.22\,e^{-0.3 \times 10 \times 0.25 \times 1} \simeq 0.10$$

(c) The probability, that all positions in the server are occupied, is, by the Erlang B formula (14.18),

$$\Pr[N_S = 10] = \frac{7^{10}}{10!}\left(1 + 7 + \frac{7^2}{2} + \frac{7^3}{6} + \frac{7^4}{24} + \frac{7^5}{150} + \frac{7^6}{720} + \right.$$
$$\left. \frac{7^7}{5\ 040} + \frac{7^8}{40\ 320} + \frac{7^9}{362\ 880} + \frac{7^{10}}{3\ 628\ 800} \right)^{-1} = 0.078\ 741$$

Hence $7,87\%$ of the calls is redirected.

(xvii) (a) The problem can be regarded as a variant of the M/M/1 queue, because we have only one server with exponentially distributed interarrival and service time. The queue is not infinitely long but contains at most $K = 3$ machines to be repaired. Hence, the Kendal notation is M/M/1/3.
(b) The failure rate of one machine is $\lambda = 1/4$ per day and the repair rate per machine is $\mu = 1$ per day. The corresponding continuous-time Markov chain is presented in Fig. B.10 and the infinitesimal generator is

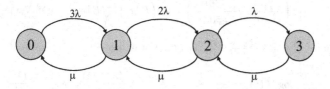

Fig. B.10. Markov chain for the three copy machine failure problem.

$$Q = \begin{bmatrix} -3\lambda & 3\lambda & 0 & 0 \\ \mu & -(2\lambda + \mu) & 2\lambda & 0 \\ 0 & \mu & -(\lambda + \mu) & \lambda \\ 0 & 0 & \mu & -\mu \end{bmatrix}$$

(c) Using the normalization $\sum_{j=0}^{3} \pi_j = 1$, the steady-state equations follow from (10.21) after transpose as

$$\begin{bmatrix} -3\lambda & \mu & 0 & 0 \\ 3\lambda & -(2\lambda + \mu) & \mu & 0 \\ 0 & 2\lambda & -(\lambda + \mu) & \mu \\ 1 & 1 & 1 & 1 \end{bmatrix}\begin{bmatrix} \pi_0 \\ \pi_1 \\ \pi_2 \\ \pi_3 \end{bmatrix} = \begin{bmatrix} 0 \\ 0 \\ 0 \\ 1 \end{bmatrix}$$

whose solution is

$$\begin{bmatrix} \pi_0 \\ \pi_1 \\ \pi_2 \\ \pi_3 \end{bmatrix} = \frac{1}{\mu^3 + 3\lambda\mu^2 + 6\lambda^2\mu + 6\lambda^3}\begin{bmatrix} \mu^3 \\ 3\lambda\mu^2 \\ 6\lambda^2\mu \\ 6\lambda^3 \end{bmatrix}$$

Numerically, with $\lambda = \frac{1}{4}$ and $\mu = 1$, we find that $\pi = \left(\frac{32}{71}, \frac{24}{71}, \frac{12}{71}, \frac{3}{71}\right)$. The steady-state probability that all machines are broken is $\pi_3 = 4.23\%$ of the time.
(d) On average, $E[N_s] = \sum_{j=0}^{3} j\pi_j = \frac{57}{71} = 0.803$ machines are broken.
(e) According to Little's Law (13.24), the mean time that a machine is broken is just the

mean system time of a machine in the queueing system $E[T] = \frac{E[N_s]}{E[\lambda]}$, where $E[\lambda]$ is the mean arrival rate,

$$E[\lambda] = 3\lambda\pi_0 + 2\lambda\pi_1 + \lambda\pi_2 = 39/71 = 0.549 \text{ per day}$$

Hence, $E[T] = \frac{57}{39} = 1.463$ day.

(xviii) The variance $\text{Var}[N_S]$ is computed here based on (2.27). The logarithm of the probability generating function $S(z) = E[z^{N_S}]$ of the *Pollaczek-Khinchin equation* (14.27) is

$$\log S(z) = \log(1-\rho) + \log\varphi_x(\lambda - \lambda z) + \log\left(\frac{z-1}{z - \varphi_x(\lambda - \lambda z)}\right)$$

where $\varphi_x(z) = E[e^{zx}]$ is the probability generating function of the service time. The first derivative is

$$\frac{d\log S(z)}{dz} = \frac{1}{z-1} - \lambda\frac{\varphi_x'(\lambda - \lambda z)}{\varphi_x(\lambda - \lambda z)} - \frac{1 + \lambda\varphi_x'(\lambda - \lambda z)}{z - \varphi_x(\lambda - \lambda z)}$$

from which we can deduce (14.28). The second derivative is

$$\frac{d^2\log S(z)}{dz^2} = \lambda^2\left(\frac{\varphi_x''(\lambda - \lambda z)}{\varphi_x(\lambda - \lambda z)} - \frac{(\varphi_x'(\lambda - \lambda z))^2}{\varphi_x^2(\lambda - \lambda z)}\right) + \frac{d^2}{dz^2}\log\left(\frac{z-1}{z - \varphi_x(\lambda - \lambda z)}\right)$$

The limit of the first term in brackets is, with $\varphi_x^{(k)}(0) = (-1)^k E[x^k]$, equal to

$$\lambda^2\left(\frac{\varphi_x''(0)}{\varphi_x(0)} - \frac{(\varphi_x'(0))^2}{\varphi_x^2(0)}\right) = \lambda^2\left(E[x^2] - (E[x])^2\right) = \lambda^2\,\text{Var}[x]$$

Let us denote $f(z) = \frac{z - \varphi_x(\lambda - \lambda z)}{z-1}$, then the last derivative can be executed as

$$-\frac{d^2}{dz^2}\log(f(z)) = -\frac{1}{(z-1)^2} + \frac{(1 + \lambda\varphi_x'(\lambda - \lambda z))^2}{(z - \varphi_x(\lambda - \lambda z))^2} + \frac{\lambda^2\varphi_x''(\lambda - \lambda z)}{z - \varphi_x(\lambda - \lambda z)}$$

Rather than straightforwardly computing the limit by invoking de l'Hospital's rule several times, we prefer a series approach. For any function $f(z)$, it holds that

$$\frac{d^2}{dz^2}\log(f(z))\bigg|_{z=x} = \frac{f''(x)f(x) - (f'(x))^2}{f^2(x)}$$

Here $x = 1$ and we expand $f(z) = \sum_{k=0}^{\infty} f_k(z-1)^k$ as a Taylor series around $z = 1$. Using the Taylor series of $\varphi_x(\lambda - \lambda z)$ around zero, we have

$$z - \sum_{k=0}^{\infty}\frac{\varphi_x^{(k)}(0)}{k!}\lambda^k(1-z)^k = \sum_{k=0}^{\infty} f_k(z-1)^{k+1}$$

After rearranging with $\varphi_x(0) = 1$, $\varphi_x'(0) = -\frac{1}{\mu}$ and $\rho = \frac{\lambda}{\mu}$ as

$$(1-\rho)(z-1) + \sum_{k=2}^{\infty}\frac{(-1)^{k-1}\varphi_x^{(k)}(0)}{k!}\lambda^k(z-1)^k = f_0(z-1) + \sum_{k=2}^{\infty} f_{k-1}(z-1)^k$$

we find, by equating corresponding powers in $(z-1)$, that $f_0 = 1 - \rho$ and $f_{k-1} = \frac{(-1)^{k-1}\varphi_x^{(k)}(0)}{k!}\lambda$ for $k \geq 2$. Since $f_k = \frac{f^{(k)}(1)}{k!}$, we can now determine

$$\frac{d^2}{dz^2}\log(f(z))\bigg|_{z=1} = \frac{2f_0 f_2 - f_1^2}{f_0^2}$$

With $f_k = \frac{-E\left[x^{k+1}\right]}{(k+1)!}\lambda^{k+1}$ for $k \geq 1$ and $f_0 = 1 - \rho$, we arrive at

$$\left.\frac{d^2 \log S(z)}{dz^2}\right|_{z=1} = \lambda^2 \text{ Var}\left[x\right] - \left.\frac{d^2}{dz^2}\log\left(f\left(z\right)\right)\right|_{z=1}$$

$$= \lambda^2 \text{ Var}\left[x\right] + \frac{\frac{\lambda^4}{4}\left(E\left[x^2\right]\right)^2 + \frac{1}{3}\left(1-\rho\right)\lambda^3 E\left[x^3\right]}{\left(1-\rho\right)^2}$$

Finally, from (2.27) and $\left.\frac{d\log S(z)}{dz}\right|_{z=1} = E\left[N_S\right] = \rho + \frac{\lambda^2 E[x^2]}{2(1-\rho)}$, we find that

$$\text{Var}\left[N_S\right] = \left.\frac{d^2 \log S(z)}{dz^2}\right|_{z=1} + \left.\frac{d\log S(z)}{dz}\right|_{z=1}$$

and

$$\text{Var}\left[N_S\right] = \rho + \lambda^2 \text{ Var}\left[x\right] + \frac{\frac{\lambda^2}{2}\left(1-\rho\right)E\left[x^2\right] + \frac{\lambda^3}{3}\left(1-\rho\right)E\left[x^3\right] + \frac{\lambda^4}{4}\left(E\left[x^2\right]\right)^2}{\left(1-\rho\right)^2} \qquad (B.19)$$

Using (14.28) and $\rho = \lambda E\left[x\right]$, we can rewrite the variance (B.19) as

$$\text{Var}\left[N_S\right] = \left(E\left[N_S\right]\right)^2 + \frac{\frac{\lambda^3}{3}E\left[x^3\right]}{1-\rho} + \left(3 - 4\rho\right)E\left[N_S\right] - 2\rho\left(1-\rho\right) \qquad (B.20)$$

Clearly, as $\rho \to 1$, $\text{Var}[N_S] = O\left((1-\rho)^{-2}\right)$. In contrast to (B.19), where all terms are positive, the third term in (B.20) changes sign and becomes negative when $\rho > \frac{3}{4}$; the last term is never smaller than $-\frac{1}{2}$. Thus, $\sqrt{\text{Var}\left[N_S\right]}$ exceeds $E\left[N_S\right]$, when the last three terms are positive (for $\rho \geq \frac{3}{4}$) or when $E\left[N_S\right] > \frac{2\rho(1-\rho)}{3-4\rho}$ (for $\rho < \frac{3}{4}$).

As a verification, for an exponential service time (see Section 3.2.2), it holds that $E\left[x^k\right] = \frac{k!}{\mu^k}$ and (B.19) reduces, indeed, after a tedious computation to (14.2), illustrating that $\sqrt{\text{Var}\left[N_{S;M/M/1}\right]} > E\left[N_{S;M/M/1}\right]$. For the M/D/1 queue, where $E\left[x^k\right] = \frac{1}{\mu^k}$ and $\rho^k = \lambda^k E\left[x^k\right]$, (B.19) gives

$$\text{Var}\left[N_{S;M/D/1}\right] = \rho + \frac{\rho^2\left(1 - \frac{\rho}{3} - \frac{\rho^2}{6}\right)}{2\left(1-\rho\right)^2} < \rho + \frac{\rho^2}{2\left(1-\rho\right)^2} < \left(E\left[N_{S;M/D/1}\right]\right)^2$$

which shows that, for the M/D/1 queue, the standard deviation $\sqrt{\text{Var}[N_S]}$ is always smaller than the mean $E\left[N_S\right]$.

B.13 General characteristics of graphs (Chapter 15)

(i) In one dimension ($d = 1$), the hopcount h_N of the shortest path between two uniformly chosen points x_A and x_B equals the distance between x_A and x_B. We allow the hopcount to be zero which is reflected by the small h while capital H refers to the case where the source A and the destination B are different. Thus, $\Pr[|x_A - x_B| = k] = \frac{1}{Z}1_{k=0} + \frac{2(Z-k)}{Z^2}1_{1\leq k\leq Z-1}$ with corresponding generating function

$$\varphi_Z(x) = \sum_{k=0}^{Z-1} \Pr[|x_A - x_B| = k]x^k = \frac{Z - Zx^2 + 2x(x^Z - 1)}{Z^2(x - 1)^2}$$

Since the nodes are uniformly chosen, all coordinate dimensions are independent and the generating function of the hopcount of the shortest path in a d-lattice is[4] $\varphi_Z^d(x)$. From (2.26) and (2.27), the mean number of hops is immediate as $E[h_N] = \frac{d}{3Z}(Z^2 - 1)$ and the variance

---

[4] If the sizes of the hypercube are not identical, the pgf is $\prod_{j=1}^d \varphi_{Z_j}(x)$.

as $\text{Var}[h_N] = \frac{d(Z^2-1)(Z^2+2)}{18Z^2}$. The total number of nodes in the d-lattice is $N = Z^d$ such that, for large N, we obtain $E[h_N] \simeq \frac{2}{3}N^{1/d}$ and $\text{Var}[h_N] \simeq \frac{d}{18}N^{2/d}$, both increasing in $d > 1$ (for constant N) as in N (for constant d). For a two-dimensional lattice, the mean hopcount scales as $O\left(\sqrt{N}\right)$.

(ii) Using the definition (15.14) of the clustering coefficient and applying the law of total probability (2.49) yields

$$\Pr\left[c_{G_p(N)} \le x\right] = \sum_{k=0}^{N-1} \Pr\left[\frac{2y}{d_v(d_v - 1)} \le x \middle| d_v = k\right] \Pr[d_v = k]$$

The degree distribution $\Pr[d_v = k]$ in the random graph is given by (15.19) and

$$\Pr\left[\frac{2y}{d_v(d_v-1)} \le x \middle| d_v = k\right] = \Pr\left[y \le \binom{k}{2}x \middle| d_v = k\right] = \sum_{j=0}^{\left[\binom{k}{2}x\right]} \binom{\binom{k}{2}}{j} p^j (1-p)^{\binom{k}{2}-j}$$

because y is the number of links between the $d_v = k$ neighbors of v, which is binomially distributed with parameter p. Combined gives

$$\Pr\left[c_{G_p(N)} \le x\right] = \sum_{k=0}^{N-1} \binom{N-1}{k} p^k (1-p)^{N-1-k} \sum_{j=0}^{\left[\binom{k}{2}x\right]} \binom{\binom{k}{2}}{j} p^j (1-p)^{\binom{k}{2}-j}$$

The mean $E\left[c_{G_p(N)}\right] = \int_0^1 \Pr\left[c_{G_p(N)} > x\right] dx$ is computed via (2.36) as

$$E\left[c_{G_p(N)}\right] = \sum_{k=0}^{N-1} \binom{N-1}{k} p^k (1-p)^{N-1-k} \int_0^1 \sum_{j=\left[\binom{k}{2}x\right]+1}^{\binom{k}{2}} \binom{\binom{k}{2}}{j} p^j (1-p)^{\binom{k}{2}-j} dx$$

Let $t = \binom{k}{2}x$, then

$$\int_0^1 \sum_{j=\left[\binom{k}{2}x\right]}^{\binom{k}{2}} \binom{\binom{k}{2}}{j} p^j (1-p)^{\binom{k}{2}-j} dx = \frac{1}{\binom{k}{2}} \int_0^{\binom{k}{2}} \sum_{j=[t]+1}^{\binom{k}{2}} \binom{\binom{k}{2}}{j} p^j (1-p)^{\binom{k}{2}-j} dt$$

$$= \frac{1}{\binom{k}{2}} \sum_{t=0}^{\binom{k}{2}} \sum_{j=t+1}^{\binom{k}{2}} \binom{\binom{k}{2}}{j} p^j (1-p)^{\binom{k}{2}-j}$$

Reversing the t- and j- sums yields

$$\int_0^1 \sum_{j=\left[\binom{k}{2}x\right]}^{\binom{k}{2}} \binom{\binom{k}{2}}{j} p^j (1-p)^{\binom{k}{2}-j} dx = \frac{1}{\binom{k}{2}} \sum_{j=1}^{\binom{k}{2}} \sum_{t=0}^{j-1} \binom{\binom{k}{2}}{j} p^j (1-p)^{\binom{k}{2}-j}$$

$$= \frac{1}{\binom{k}{2}} \sum_{j=1}^{\binom{k}{2}} j \binom{\binom{k}{2}}{j} p^j (1-p)^{\binom{k}{2}-j} = p$$

Hence, we find that

$$E\left[c_{G_p(N)}\right] = p \tag{B.21}$$

which directly follows from the independence of links: if a node v has d_v neighbors, they are interconnected by $\sum_{l=1}^{\binom{d_v}{2}} 1_{\{l \text{ exists}\}}$ links. Using conditional expectation (2.82) yields

$$
E\left[c_{G_p(N)}\right] = E\left[\frac{1}{\binom{d_v}{2}} \sum_{l=1}^{\binom{d_v}{2}} 1_{\{l \text{ exists}\}}\right] = E_{d_v}\left[E_{ex}\left[\frac{1}{\binom{d_v}{2}} \sum_{l=1}^{\binom{d_v}{2}} 1_{\{l \text{ exists}\}} \middle| d_v\right]\right]
$$

$$
= E_{d_v}\left[\frac{1}{\binom{d_v}{2}}\left[\sum_{l=1}^{\binom{d_v}{2}} E_{ex}\left[1_{\{l \text{ exists}\}}\right]\right]\right] = E_{d_v}\left[\frac{1}{\binom{d_v}{2}} p\binom{d_v}{2}\right] = p
$$

Along the same lines, we find that the generating function $\varphi_c(z)$ of the clustering coefficient $c_{G_p(N)}$ is

$$
\varphi_c(z) = \sum_{k=0}^{N-1} \binom{N-1}{k} p^k (1-p)^{N-1-k} \left(1 - p + pe^{-\frac{z}{\binom{k}{2}}}\right)^{\binom{k}{2}}
$$

The variance is computed from (2.46) as

$$
\text{Var}\left[c_{G_p(N)}\right] = (p - p^2) \sum_{k=2}^{N-1} \binom{N-1}{k} \frac{p^k (1-p)^{N-1-k}}{\binom{k}{2}}
$$

(iii) The mean number of trees T_k of size k in $G_p(N)$ is computed by considering a set of k nodes, that needs to be connected by precisely $k-1$ links to form a tree, and the remaining set of $N-k$ nodes. To avoid that the tree connects more than k nodes, no links between the set of k nodes and the set of the remaining $N-k$ others are allowed, thus, in total $k(N-k)$ links. In addition, the k nodes are only connected by $k-1$ links, which implies that we need to prohibit $\binom{k}{2} - (k-1) = \frac{(k-1)(k-2)}{2}$ links among the k nodes of the tree T_k. In each set of k nodes, Caley's Theorem states that precisely k^{k-2} (labelled) trees can be constructed. There are $\binom{N}{k}$ ways of choosing a set of k nodes out of N nodes. Finally, since all links are independent and appear with probability p, the expected number of trees T_k in $G_p(N)$ equals

$$
E[T_k] = \binom{N}{k} k^{k-2} p^{k-1} (1-p)^{k(N-k) - \frac{(k-1)(k-2)}{2}}
$$

(iv) Let T_N denote the total number of trees that span N nodes. Since any tree T_N consists of precisely $N-1$ links, by removing one link in that tree, we obtain two disjoint other trees T_{N-k} and T_k containing each $N-k$ and k nodes respectively, but combined they both span N nodes. In any tree T_N, we can remove one link in $N-1$ ways. Hence, in total there are $(N-1)T_N$ ways in which two disjoint trees can be constructed. On the other hand, there are $k(N-k)$ ways in which two disjoint trees, T_{N-k} and T_k, can be connected to form a tree that spans N nodes. In addition, the total number of such disjoint tree pairs equals $\binom{N-1}{k-1} T_k T_{N-k}$, all possible combinations to select $k-1$ links for T_k out of a total of $N-1$ possible links. Combining all contribution yields the basis recursion for T_N,

$$
(N-1)T_N = \sum_{k=1}^{N-1} k(N-k) \binom{N-1}{k-1} T_k T_{N-k} \tag{B.22}
$$

which we rewrite as

$$
\frac{T_N}{(N-2)!} = \sum_{k=1}^{N-1} \frac{kT_k}{(k-1)!} \frac{T_{N-k}}{(N-k-1)!}
$$

Let us define the generating function for the total number of spanning trees by

$$
g_T(z) = \sum_{k=1}^{\infty} \frac{T_k}{(k-1)!} z^k \tag{B.23}
$$

Multiplying the recursion with z^{N-2} and summing over all $N \geq 2$ yields

$$\sum_{N=2}^{\infty} \frac{T_N}{(N-2)!} z^{N-2} = \sum_{N=2}^{\infty} \left(\sum_{k=1}^{N-1} \frac{kT_k}{(k-1)!} \frac{T_{N-k}}{(N-k-1)!} \right) z^{N-2}$$

$$= \sum_{k=1}^{\infty} \frac{kT_k}{(k-1)!} z^{k-1} \sum_{j=1}^{\infty} \frac{T_j}{(j-1)!} z^{j-1}$$

where the last equation follows from the Cauchy product. Written in terms of the generating function gives

$$\frac{d}{dz} \left(\frac{g_T(z)}{z} \right) = \frac{d(g_T(z))}{dz} \frac{g_T(z)}{z}$$

or

$$\frac{d(g_T(z))}{dz} = \frac{z}{g_T(z)} \frac{d}{dz} \left(\frac{g_T(z)}{z} \right) = \frac{d \log \frac{g_T(z)}{z}}{dz}$$

After integrating both sides from $z = 0$ to $z = x$

$$\log \frac{g_T(x)}{x} - \log \frac{g_T(z)}{z} \Bigg|_{z=0} = g_T(x)$$

With $\lim_{z \to 0} \frac{G_T(z)}{z} = 1$ since $T_1 = 1$, we arrive at the functional equation for the generating function,

$$g_T(x) = x \exp(g_T(x))$$

Using the contour integral (2.20), the coefficient of the Taylor series in (B.23) can be obtained, for $k > 0$, as

$$\frac{T_k}{(k-1)!} = \frac{1}{2\pi i} \int_{C(0)} \frac{g_T(z)}{z^{k+1}} dz$$

Since we attempt to use the functional equation as $z = g_T(z) e^{-g_T(z)}$, it is more appropriate to apply (2.20) to the derivative of $g_T(x)$. Hence,

$$\frac{kT_k}{(k-1)!} = \frac{1}{2\pi i} \int_{C(0)} \frac{dz}{z^k} \frac{d(g_T(z))}{dz} = \frac{1}{2\pi i} \int_{C(0)} \frac{d(g_T(z))}{\left(g_T(z) e^{-g_T(z)} \right)^k}$$

The transform $w = g_T(z)$ maps the contour $C(0)$ around zero into another contour $C^*(0)$ around zero, because $g_T(0) = 0$ and $g_T(z)$ is analytic around $z = 0$ (since the power series (B.23) exists and converges for arbitrary small, but non-zero values of z). Thus, we obtain

$$\frac{kT_k}{(k-1)!} = \frac{1}{2\pi i} \int_{C^*(0)} \frac{dw}{(we^{-w})^k} = \frac{1}{2\pi i} \int_{C^*(0)} \frac{e^{wk} dw}{w^k} = \frac{\frac{d^{k-1}}{dw^{k-1}} e^{wk} \Big|_{w=0}}{(k-1)!} = \frac{k^{k-1}}{(k-1)!}$$

where, again, Cauchy's integral for the k-th derivative has been invoked. Finally, we arrive at Caley's formula

$$T_k = k^{k-2} \tag{B.24}$$

There is an interesting relation between Caley's formula and Abel's identity (14.53). After taking the derivative of Abel's identity (14.53) with respect to u, the evaluation at $u = 0$ with $n = N - 1$, $z = -1$ and $y = N$ yields

$$(N-1) N^{N-2} = \sum_{k=1}^{N-1} \binom{N-1}{k} k^{k-1} (N-k)^{N-1-k}$$

After changing the variable $j = N - k$ in the recursion (B.22), using $\binom{N-1}{N-j-1} = \binom{N-1}{j}$ and relabeling $j \to k$, we find

$$(N-1)\,T_N = \sum_{k=1}^{N-1} \binom{N-1}{k} k\,T_k\,(N-k)\,T_{N-k}$$

which shows that the recursion (B.22) with (B.24) is an instance of Abel's identity.

(v) The degree $D|d_j = k$ of an arbitrary neighbor of a node j with degree k in the Erdős-Rényi random graph $G_p(N)$ is $1+$ Binom$(N-2, p)$. Indeed, since all links are independent in $G_p(N)$, the k neighbors of a random node j with degree k are just k randomly chosen nodes in the random graph G_p^* that is obtained from $G_p(N)$ by removing the node j and the links incident to node j. Thus, G_p^* is also an Erdős-Rényi random graph with $(N-1)$ nodes and link density p. The degrees of these k neighbors of j are independent, identically distributed binomial random variables with mean $(N-2)p$ and parameter p, denoted as Binom$(N-2, p)$. Since each of the neighbors has surely one link that connects to node j, we must add 1 to Binom$(N-2, p)$. The pdf of degree $D|d_j = k$ is also known as the second-order or two-points degree correlation function.

(vi) We assume that, in the Erdős-Rényi random graph $G_p(N)$, all nodal degrees are i.i.d. random variables with binomial distribution (15.19). Consider the probability generating function of both sides of the basic law of the degree (15.4),

$$E\left[z^{\sum_{j=1}^{N} d_j}\right] = E\left[z^{2L}\right]$$

By the assumed independence of the d_j, we obtain

$$E\left[z^{\sum_{j=1}^{N} d_j}\right] = \prod_{j=1}^{N} E\left[z^{d_j}\right] = \left(E\left[z^{D_{\mathrm{rg}}}\right]\right)^N$$

Since each node j is i.i.d. with D_{rg}, and applying (15.19) and (3.2), we end up with

$$E\left[z^{\sum_{j=1}^{N} d_j}\right] = \left(E\left[z^{D_{\mathrm{rg}}}\right]\right)^N = (1 - p + pz)^{N(N-1)}$$

The pgf of the number of links in $G_p(N)$ in the right-hand side is

$$E\left[z^{2L}\right] = E\left[\left(z^2\right)^L\right] = (1 - p + pz^2)^{\binom{N}{2}}$$

Hence, the assumption of independence in the basic law of the degree (15.4) amounts to

$$(1 - p + p\sqrt{z})^{N(N-1)} \overset{?}{=} (1 - p + pz)^{\binom{N}{2}}$$

At first glance, the conclusion is dramatic: Due to $\sqrt{z}$, the left-hand side is not analytic at $z = 0$ and inside the unit circle. This means that the expression on the left-hand side is not a probability generating function illustrating that the assumption of independence is not correct. However, the mean and variance of the both sides (computed e.g. from (2.26) and (2.27)) is exactly equal! While we cannot deduce meaningful results for $\Pr[L = k]$, we surprisingly can for the variance Var$[L]$. That the mean $E[L]$ is correct, was to be expected since mean of the sum of dependent random variables is the sum of their means, by the linearity of the expectation operator. The basic law of the degree (15.4) is not at all suited for the independence assumption because each link is counted twice!

(vii) We first compute the joint probability $\Pr[D_i = k, D_j = m]$, where node i and node j are random nodes in $G_p(N)$. The dependence lies in the possible direct link between nodes i and j. By the law of total probability (2.49), we have

$$\Pr[D_i(N) = k, D_j(N) = m] = \Pr[D_i(N) = k, D_j(N) = m | a_{ij} = 1] \Pr[a_{ij} = 1]$$
$$+ \Pr[D_i(N) = k, D_j(N) = m | a_{ij} = 0] \Pr[a_{ij} = 0]$$

where a_{ij} is the matrix element of the adjacency matrix A. Since $\Pr[a_{ij} = 1] = p$, and, in absence of the direct link, D_i and D_j are independent, we obtain

$$\Pr[D_i(N) = k, D_j(N) = m] = p\Pr[D_i(N) = k, D_j(N) = m | a_{ij} = 1]$$
$$+ (1-p)\Pr[D_i(N-1) = k]\Pr[D_j(N-1) = m]$$

Further, given the existence of the direct link, the direct link is counted both in D_i and in D_j such that

$$\Pr[D_i(N) = k, D_j(N) = m | a_{ij} = 1] = \Pr[D_i(N-1) = k-1]\Pr[D_j(N-1) = m-1]$$

Combining the contributions and introducing the binomial density (15.19) of $\Pr[D_i(N) = k]$, we obtain

$$\Pr[D_i(N) = k, D_j(N) = m] = p\binom{N-2}{k-1}p^{k-1}(1-p)^{N-1-k}\binom{N-2}{m-1}p^{m-1}(1-p)^{N-1-k}$$
$$+ (1-p)\binom{N-2}{k}p^k(1-p)^{N-2-k}\binom{N-2}{m}p^m(1-p)^{N-2-m}$$

The joint expectation is

$$E[D_i(N)D_j(N)] = \sum_{k=0}^{N-1}\sum_{m=0}^{N-1}mk\Pr[D_i(N) = k, D_j(N) = m]$$

$$= p\sum_{k=0}^{N-1}k\binom{N-2}{k-1}p^{k-1}(1-p)^{N-1-k}\sum_{m=0}^{N-1}m\binom{N-2}{m-1}p^{m-1}(1-p)^{N-1-k}$$

$$+ (1-p)\sum_{k=0}^{N-1}k\binom{N-2}{k}p^k(1-p)^{N-2-k}\sum_{m=0}^{N-1}m\binom{N-2}{m}p^m(1-p)^{N-2-}$$

$$= p(1 + (N-2)p)^2 + (1-p)((N-2)p)^2 = p - 2Np^2 + N^2p^2$$

The covariance is

$$\mathrm{Cov}[D_i(N), D_j(N)] = E[D_i(N)D_j(N)] - E[D_i(N)]E[D_j(N)]$$
$$= p - 2Np^2 + N^2p^2 - (N-1)^2p^2 = p(1-p)$$

Since the variance $\mathrm{Var}[D_i(N)] = \mathrm{Var}[D_j(N)] = (N-1)p(1-p)$, we find that the correlation coefficient, defined in (2.61), is (15.46), which shows that, although arbitrary degrees $D_i(N)$ and $D_j(N)$ are not independent, their degree of correlation decreases with N. Hence, for large N, both $D_i(N)$ and $D_j(N)$ can be regarded as almost independent.

(viii) The complement of $G_p(N)$ is $(G_p(N))^c = G_{1-p}(N)$, because a link in $G_p(N)$ is present with probability p and absent with probability $1 - p$ and $(G_p(N))^c$ is also an ER random graph. The minimum degree $D_{\min}(p)$ in $G_p(N)$ equals $N - 1$ minus the maximum degree in $(G_p(N))^c$. This demonstrates the property (15.47).

(ix) In general, a matrix has a zero eigenvalue if its determinant is zero. A determinant is zero if two rows are identical or if some of the rows are linearly dependent. For example, two rows are identical if two distinct nodes are connected to a same set of nodes. Since the elements a_{ij} of an adjacency matrix A are only 0 or 1, linear dependence of rows here occurs every time the sum of a set of rows equals another row in the adjacency matrix. For example, consider the sum of two rows. If n_1 is connected to the set S_1 of nodes and n_2 is connected to the distinct set S_2, where $S_1 \cap S_2 = \varnothing$ and $n_1 \neq n_2$, then the graph has a zero eigenvalue if another node $n_3 \neq n_2 \neq n_1$ is connected to $S_1 \cup S_2$. These zero eigenvalues occur when a graph possesses a "local bipartiteness". In real networks, this type of interconnection often occurs.

(x) An adjacency matrix A has an eigenvalue $\lambda(A) = -1$ every time nodes n_1 and n_2 in the graph are connected to a same set S of different nodes and n_1 and n_2 are also connected. Indeed, without loss of generality, we can relabel the nodes such that $n_1 = 1$ and $n_2 = 2$. In that case, the first two rows in A are of the form

$$
\begin{array}{cccccc}
0 & 1 & a_{13} & a_{14} & \cdots & a_{1N} \\
1 & 0 & a_{13} & a_{14} & \cdots & a_{1N}
\end{array}
$$

and the corresponding rows in $\det\left(A - \lambda I\right)$ of the characteristic polynomial are

$$
\begin{array}{cccccc}
-\lambda & 1 & a_{13} & a_{14} & \cdots & a_{1N} \\
1 & -\lambda & a_{13} & a_{14} & \cdots & a_{1N}
\end{array}
$$

If two rows are identical, the determinant is zero. In order to make these rows identical, it suffices to take $\lambda = -1$ and $\det\left(A + I\right) = 0$ which shows that $\lambda = -1$ is an eigenvalue of A with this particular form. This observation generalizes to a graph where k nodes are fully meshed and, in addition, all these k nodes are connected to a same set S of different nodes. Again, we may relabel nodes such that the first k rows describe these k nodes in a complete graph configuration, also called a clique. Let x denote a $(N - k) \times 1$ vector, then ux^T is a matrix with all rows identical and equal to x. The structure of $\det\left(A - \lambda I\right)$ is

$$
\det\left(A - \lambda I\right) = \left| \begin{array}{cc} (J - (\lambda + 1)\,I)_{k \times k} & u.x^T \\ B_{(N-k) \times k} & (C - \lambda I)_{(N-k) \times (N-k)} \end{array} \right|
$$

which shows that the first k rows are identical is $\lambda = -1$, implying that the multiplicity of this eigenvalue is $k - 1$. Observe that the spectrum of the complete graph K_N where $k = N$, indeed contains an eigenvalue $\lambda = -1$ with multiplicity $N - 1$.

(xi) The mean number of nodes in a circular area with radius r is $\delta \pi r^2$, where the density $\delta = \frac{N}{\pi R^2}$. Since all these nodes are in the radio range of the node most near to the center, the mean degree of that node is $\delta \pi r^2 = \frac{Nr^2}{R^2}$. Since (a) the nodes are placed independently of each other; (b) two or more nodes are not placed at the same location; and (c) the nodes are placed with constant density, the nodal placement processes is a (point) Poisson process. By placing nodes in such an i.i.d. manner, the resulting graph is, for large N where boundary effects can be ignored, very close to an Erdős-Rényi random graph $G_p\left(N\right)$. The mean degree in $G_p\left(N\right)$ is $p\left(N - 1\right)$. Hence, for $r > r_c$, we find that

$$
r = R\sqrt{\frac{p\left(N - 1\right)}{N}}
$$

As shown in (15.28), the random graph $G_p\left(N\right)$ is connected, almost surely, for link density $p > p_c \sim \frac{\log N + x}{N}$ for any constant x and large N. Hence, for large N, the critical radio range is $r_c \sim R\sqrt{\frac{\log N}{N}} + O\left(\frac{1}{\sqrt{N \log N}}\right)$.

(xii) *Watts-Strogatz Small World Graph.*
(a) In case $p_r = 0$, the hopcount from a node to another arbitrary node is computed as follows. Without loss of generality, we may label the source node by 1. Any other node $u \leq \frac{N}{2}$ is at $H_u = \left\lceil \frac{u-1}{\kappa} \right\rceil$ hops from 1, where $\lceil x \rceil$ is the integer equal or just larger than x. The nodes with label $u > \frac{N}{2}$ are reached in the other direction. We further assume that all u are uniformly distributed between 1 and $\frac{N}{2}$. We thus neglect the small difference of one node in the interval $\left[1, \frac{N}{2}\right]$ and $\left[\frac{N}{2} + 1, N\right]$ when N is even or odd, which is equivalent in treating $\frac{N}{2}$ as an integer. Then,

$$
\Pr\left[H = j\right] = \Pr\left[\left\lceil \frac{u-1}{\kappa} \right\rceil = j\right]
$$

Using the law of total probability (2.49) yields

$$
\Pr\left[H = j\right] = \sum_{l=1}^{\frac{N}{2}} \Pr\left[\left\lceil \frac{u-1}{\kappa} \right\rceil = j \middle| u = l\right] \Pr\left[u = l\right] = \frac{2}{N} \sum_{l=1}^{\frac{N}{2}} \Pr\left[\left\lceil \frac{l-1}{\kappa} \right\rceil = j\right]
$$

The event $\left\{\left\lceil \frac{l-1}{\kappa} \right\rceil = j\right\}$ is only true if $l = \kappa j - \kappa + 2, \kappa j - \kappa + 3, \ldots, \kappa j + 1$, i.e. in κ cases. The largest hopcount is $H_{\max} = \left\lceil \frac{N/2 - 1}{\kappa} \right\rceil$ and it appears in a cases. We thus obtain an almost uniformly distributed hopcount when $p_r = 0$,

$$
\Pr\left[H = j\right] = \left\{ \begin{array}{ll} \frac{2}{N}\kappa & 1 \leq j < H_{\max} \\ \frac{2}{N}a & j = H_{\max} \end{array} \right.
$$

Since $\sum_{j=1}^{H_{\max}} \Pr[H = j] = 1$, the integer number a equals $\frac{N}{2} - \kappa \left\lceil \frac{N/2-1}{\kappa} \right\rceil + \kappa$. The mean hopcount is approximately equal to $E[H] \simeq \frac{H_{\max}-1}{2} = \frac{1}{2} \left\lceil \frac{N/2-1}{\kappa} \right\rceil$. For large N, this reduces to $E[H] \sim \frac{N}{4\kappa}$. This points to a big world: reaching an arbitrary node in case $p_r = 0$ consumes, on average, much hops when N is large.

(b) We compute the degree distribution when $p_r > 0$, assuming N available nodes for each rewiring. Since the κ counter-clockwise pointing links of the 2κ initial links of each node are not modified, the degree D_{p_r} of a random node can be expressed as $D_{p_r} = \kappa + R$, where $0 \leq R \leq \kappa$. The random variable R is influenced by two processes: a node n can maintain a number R_{no} of links that is not rewired and it can gain links $R_{\mathrm{new}} = R - R_{\mathrm{no}}$ from other nodes after a link is rewired towards the node n. By the law of total probability (2.49),

$$\Pr[D_{p_r} = k] = \sum_{j=0}^{\kappa} \Pr[D_{p_r} = k | R_{\mathrm{no}} = j] \Pr[R_{\mathrm{no}} = j]$$

$$= \sum_{j=0}^{\kappa} \Pr[D_{p_r} = \kappa + R_{\mathrm{new}} + R_{\mathrm{no}} = k | R_{\mathrm{no}} = j] \Pr[R_{\mathrm{no}} = j]$$

Since, by independence of two processes,

$$\Pr[D_{p_r} = k | R_{\mathrm{no}} = j] = \Pr[D = \kappa + R_{\mathrm{new}} + R_{\mathrm{no}} = k | R_{\mathrm{no}} = j]$$
$$= \Pr[R_{\mathrm{new}} = k - j - \kappa]$$

we need to compute

$$\Pr[D_{p_r} = k] = \sum_{j=0}^{\kappa} \Pr[R_{\mathrm{new}} = k - j - \kappa] \Pr[R_{\mathrm{no}} = j]$$

First, $R_{\mathrm{no}} \leq \kappa$ is the number of the clockwise pointing links that is not rewired. Since each rewiring is a Bernoulli process with parameter p_r, the sum is binomially distributed as explained in Section 3.1.2,

$$\Pr[R_{\mathrm{no}} = j] = \binom{\kappa}{j} p_r^{\kappa-j} (1 - p_r)^j$$

The computation of the remaining R_{new} is a little more involved. Each of the rewired links has equal probability of $p = \frac{1}{N} \times p_r$ to be rewired towards node n (by our assumption). The total number of links that can be rewired is κN. Thus, R_{new} is a binomial random variable with distribution

$$\Pr[R_{\mathrm{new}} = j] = \binom{\kappa N}{j} p^j (1 - p)^{\kappa N - j}$$

and corresponding probability generating function $\varphi_{R_{\mathrm{new}}}(z) = \left(1 + \frac{p_r}{N}(z - 1)\right)^{\kappa N}$. For large N, we have that $\lim_{N \to \infty} \varphi_{R_{\mathrm{new}}}(z) = e^{\kappa p_r (z-1)}$ and, using the Continuity Theorem 6.1.3, we find for large N, that R_{new} tends to a Poisson random variable,

$$\lim_{N \to \infty} \Pr[R_{\mathrm{new}} = j] = \frac{(\kappa p_r)^j}{j!} e^{-\kappa p_r}$$

Finally, after combining all, we arrive, for large N, at

$$\Pr[D_{p_r} = k] \stackrel{N \text{ large}}{\to} \sum_{j=0}^{\kappa} \frac{(\kappa p_r)^{k-j-\kappa}}{(k - j - \kappa)!} e^{-\kappa p_r} \binom{\kappa}{j} p_r^{\kappa-j} (1 - p_r)^j \qquad (\mathrm{B.25})$$

which shows that $\Pr[D_{p_r} = k] = 0$ if $k < \kappa$ (because, in that case, $\frac{1}{(k-j-\kappa)!} = 0$).

(c) The probability generating function corresponding to (B.25) is

$$\varphi_{D|p_r}(z) = \sum_{k=0}^{\infty} \sum_{j=0}^{\kappa} \frac{(\kappa p_r)^{k-j-\kappa}}{(k-j-\kappa)!} e^{-\kappa p_r} \binom{\kappa}{j} p_r^{\kappa-j} (1-p_r)^j z^k$$

$$= e^{-\kappa p_r} \sum_{j=0}^{\kappa} \binom{\kappa}{j} p_r^{\kappa-j} (1-p_r)^j \sum_{k=j+\kappa}^{\infty} \frac{(\kappa p_r)^{k-j-\kappa}}{(k-j-\kappa)!} z^k$$

With $\sum_{k=j+\kappa}^{\infty} \frac{(\kappa p_r)^{k-j-\kappa} z^k}{(k-j-\kappa)!} = z^{j+\kappa} \sum_{k=0}^{\infty} \frac{(\kappa p_r)^k z^k}{k!} = z^{j+\kappa} e^{\kappa p_r z}$, we obtain

$$\varphi_{D|p_r}(z) = z^{\kappa} e^{\kappa p_r (z-1)} \sum_{j=0}^{\kappa} \binom{\kappa}{j} p_r^{\kappa-j} (1-p_r)^j z^j$$

Finally, with $\sum_{j=0}^{\kappa} \binom{\kappa}{j} p_r^{\kappa-j} (1-p_r)^j z^j = (p_r + (1-p_r) z)^{\kappa}$, we arrive at (15.48).

(d) If $p_r \to 1$, then we obtain from (B.25)

$$\Pr[D_{p_r=1} = k] \xrightarrow{N \text{ large}} \frac{(\kappa)^{k-\kappa}}{(k-\kappa)!} e^{-\kappa}$$

which is a shifted Poisson distribution. The corresponding pgf follows from (15.48) as $\varphi_{D|p_r=1}(z) = z^{\kappa} e^{\kappa(z-1)}$. From (2.26) and (2.27), we find that $E[D|p_r=1] = 2\kappa$ and $\mathrm{Var}[D|p_r=1] = \kappa$. The degree distribution (15.19) of the Erdős-Rényi random graph $G_p(N)$ tends to a Poisson distribution $\Pr[D_{\mathrm{rg}} = k] = \frac{\lambda^k e^{-\lambda}}{k!}$ for large N, provided $p = \frac{\lambda}{N}$ as shown in Section 3.1.4. Hence, if the mean degree $E[D_{\mathrm{rg}}] = \lambda$ equals 2κ, the variance of degree in the Watts-Strogatz graph is a factor of 2 smaller. Moreover, $D|p_r=1 \geq \kappa$, while D_{rg} can be smaller than κ, even disconnectivity and $D_{\mathrm{rg}} = 0$ are possible. The Watts-Strogatz graph is κ-connected: at least κ links need to be removed to disconnect the graph. Complete rewiring ($p_r = 1$) makes the Watts-Strogatz graph, for large N and not too small κ, similar but not identical, to the Erdős-Rényi random graph $G_{\frac{2\kappa}{N}}(N)$.

(xiii) The clustering coefficient (15.14) of the Watts-Strogatz graph, without rewiring $p_r = 0$, requires us to compute the number of links between all the neighbors of a node. If we label that node by zero, then each node $1 \leq j \leq \kappa$ to the right (clockwise sense in Fig. 15.10), has precisely κ links pointing to neighbors of node 0 to its left (to lower labels) and $\kappa - j$ links pointing to neighbors of node 0 to its right. The situation for nodes to the left of node 0 is symmetric. Hence, counting a link only once, the total number y of links between neighbors of node 0 equals $y = \sum_{j=1}^{\kappa} (2\kappa - j) - \kappa$, where the last κ refers to the link $(j, 0)$ for each $1 \leq j \leq \kappa$, that needs to be excluded. With $\sum_{j=1}^{\kappa} (2\kappa - j) = 2\kappa^2 - \frac{\kappa(\kappa+1)}{2} = \frac{3\kappa^2 - \kappa}{2}$, we arrive at $y = \frac{3\kappa(\kappa-1)}{2}$. Hence, the clustering coefficient of node 0 (and $p_r = 0$) is

$$c_{WS}(0) = \frac{3\kappa(\kappa-1)}{2\kappa(2\kappa-1)}$$

Due to the cyclic symmetry, all nodes have a same clustering coefficients, given by (15.49).

(xiv) *Price's citation graph.* (a) A new paper (node $n+1$) refers to paper $j \in [1, n]$ with probability proportional to its number of citations (the in-degree q_j of node j) plus the constant a,

$$\Pr[n+1 \text{ cites } j] = \alpha (q_i + a) \tag{B.26}$$

and since $\sum_{j=1}^{n} \Pr[n+1 \text{ cites } j] = 1$ as a citation must refers to some older paper, we find that

$$\alpha^{-1} = \sum_{j=1}^{n} (q_i + a) = \sum_{j=1}^{n} q_i + na$$

In a directed graph, the in-degree d_j^{in} of a node j equals $d_j^{in} = \sum_{k=1}^{n} a_{kj}$, where the adjacency element a_{kj} refers to a link from node k to node j. The out-degree equals $d_j^{out} = \sum_{k=1}^{n} a_{jk}$. Since the total number of directed links $L = \sum_{j=1}^{n} \sum_{k=1}^{n} a_{kj}$, it holds that $\sum_{j=1}^{n} d_j^{in} = \sum_{j=1}^{n} d_j^{out}$, meaning that the mean in-degree in a directed graph equals the mean out-degree. Applied to Price's directed citation graph, where each paper cites, on average c older ones, we have that $c = \frac{1}{n} \sum_{j=1}^{n} d_j^{out}$ and also, that $\sum_{j=1}^{n} d_j^{in} = \sum_{j=1}^{n} q_i = nc$. Hence, $\alpha^{-1} = n(c+a)$.

The expected number of new citations to papers with precisely q citations (in-degree q) equals the number $np_q(n)$ of those papers multiplied by $c\alpha(q+a)$. The number of papers with in-degree q after the addition of the new paper $n+1$ equals $(n+1)p_q(n+1)$. New citations increase papers with in-degree $q-1$ to in-degree q with probability $np_{q-1}(n)c\alpha(q-1+a)$, but papers already with in-degree q can attain in-degree $q+1$ with probability $np_q(n)c\alpha(q+a)$, leading, with $\alpha^{-1} = n(c+a)$, to the balance equation for $q > 0$,

$$(n+1)p_q(n+1) = np_q(n) + p_{q-1}(n)\frac{c(q-1+a)}{c+a} - p_q(n)\frac{c(q+a)}{c+a} \tag{B.27}$$

The mean number of papers with in-degree q after addition of the new paper $n+1$ equals their mean number just before increased by the mean number of papers that obtain in-degree q minus those that lost in-degree q (and have now in-degree $q+1$). If $q = 0$, the second term in (B.27) does not exist, but each new added paper has in-degree $q = 0$, because a new paper cannot have citations when it appears. Hence,

$$(n+1)p_0(n+1) = np_0(n) + 1 - p_0(n)\frac{ca}{c+a} \tag{B.28}$$

The set of equations (B.27) and (B.28) can be solved for each $n > 1$ and $0 \le q \le n-1$.
(b) For large n, we assume that

$$\lim_{n\to\infty} (n+1)p_q(n+1) - np_q(n) = \lim_{n\to\infty} n\{p_q(n+1) - p_q(n)\} + \lim_{n\to\infty} p_q(n+1)$$
$$= p_q$$

so that (B.27) and (B.28) become asymptotically,

$$\begin{cases} p_q = \frac{c}{c+a}((q-1+a)p_{q-1} - (q+a)p_q) & \text{for } q > 0 \\ p_0 = 1 - p_0\frac{ca}{c+a} & \text{for } q = 0 \end{cases}$$

These equations are readily rewritten as a recursion

$$p_q = \frac{q-1+a}{q+a+1+\frac{a}{c}}p_{q-1}$$

with initial start $p_0 = \frac{c+a}{c+a+ca} = \frac{1+\frac{a}{c}}{1+a+\frac{a}{c}}$. After m iterations, we find

$$p_q = \frac{q-1+a}{q+a+1+\frac{a}{c}}\frac{q-2+a}{q+a+\frac{a}{c}}\frac{q-3+a}{q-1+a+\frac{a}{c}}\cdots\frac{q-m+a}{q-m+2+a+\frac{a}{c}}p_{q-m}$$

Using the functional equation of the Gamma function $\Gamma(x+1) = x\Gamma(x)$ yields

$$p_q = \frac{(q-1+a)\cdots(q-m+a)\,\Gamma(q-m+a)\,\Gamma\left(q-m+2+a+\frac{a}{c}\right)p_{q-m}}{\left(q+a+1+\frac{a}{c}\right)\cdots\left(q-m+2+a+\frac{a}{c}\right)\Gamma\left(q-m+2+a+\frac{a}{c}\right)\Gamma(q-m+a)}$$
$$= \frac{\Gamma(q+a)}{\Gamma\left(q+a+2+\frac{a}{c}\right)}\frac{\Gamma\left(q-m+2+a+\frac{a}{c}\right)}{\Gamma(q-m+a)}p_{q-m}$$

When $m = q$, then

$$p_q = \frac{\Gamma(q+a)}{\Gamma\left(q+a+2+\frac{a}{c}\right)}\frac{\Gamma\left(2+a+\frac{a}{c}\right)}{\Gamma(a)}\frac{1+\frac{a}{c}}{1+a+\frac{a}{c}}$$
$$= \left(1+\frac{a}{c}\right)\frac{\Gamma(q+a)}{\Gamma\left(q+a+2+\frac{a}{c}\right)}\frac{\Gamma\left(1+a+\frac{a}{c}\right)}{\Gamma(a)}$$

Finally, for large q and using $\frac{\Gamma(q+b)}{\Gamma(q+g)} \sim q^{b-g}$ (Abramowitz and Stegun, 1968, 6.1.47), we find that $p_q \sim \left(1+\frac{a}{c}\right)\frac{\Gamma\left(1+a+\frac{a}{c}\right)}{\Gamma(a)}q^{-2-\frac{a}{c}}$.

(c) In the BA-random graph model, each time a node is added with precisely m links to older, existing ones. In Price's model, a node is added with on average c links to older nodes. Hence, c does not need to be an integer. Price's model generates a directed graph, while the BA-random graph model is undirected. Consequently, in the BA-random graph model, there are no nodes with degree less than m.

In fact, the BA-random graph model is a special case of Price's citation graph. First assume during the creation of the BA-random graph directed links, then the out-degree is exactly $c = m$. The total degree is the sum of the in- and out-degrees, thus $d_i = q_i + c$. Preferential attachment is proportional to d_i, which means that we must choose $a = c$ in Price's model (B.26), so that we obtain $p_q \sim q^{-3}$ as shown differently in Section 15.7.2.

(xv) *Chromatic number of an Erdős-Rényi graph.* A set of nodes is independent if each pair of nodes is not adjacent. The probability that there is no link between two nodes in $G_p(N)$ is $q = 1 - p$. The probability that a set $\mathcal{U}_k \subseteq \mathcal{N}$ of precisely k nodes in an Erdős-Rényi graph is independent is $q^{\binom{k}{2}}$. Moreover, there are $\binom{N}{k}$ ways to construct such k-sets in any graph. Let the number of nodes in the largest set of independent nodes of G is denoted as $\alpha(G) = \max_k |\mathcal{U}_k|$. Then, the probability that $\alpha(G)$ is not smaller than k, is not larger than

$$\Pr[\alpha(G) \geq k] \leq \binom{N}{k} q^{\binom{k}{2}}$$

Clearly, if k is fixed and not a function of N, for large N, the right-hand side tends to infinity, which results in a useless bound. In order to deduce a sharp bound, i.e. $\Pr[\alpha(G) \geq k] \to 0$ for large N, we need to determine k as a function of N. We upper bound $\binom{N}{k} = \frac{1}{k!} N(N-1)\ldots(N-k+1) < \frac{1}{k!} N^k < N^k$ such that

$$\Pr[\alpha(G) \geq k] < N^k q^{\binom{k}{2}} = q^{k \frac{\log N}{\log q} + \frac{1}{2} k(k-1)} = q^{\frac{k}{2}\left\{ -\frac{2\log N}{\log q^{-1}} + (k-1) \right\}}$$

If $\frac{2\log N}{\log q^{-1}} < k - 1$, then $\lim_{N \to \infty} \Pr[\alpha(G) \geq k] = 0$. It suffices to take $k = \frac{(2+\epsilon)\log N}{\log q^{-1}}$, where $\epsilon > 0$ is arbitrarily small, to establish $\lim_{N \to \infty} \Pr[\alpha(G) \geq k] = 0$ because k increases with N. In that case, almost every graph $G \in G_p(N)$ has a largest set of independent nodes $\alpha(G)$ that is not larger than k. Hence, for any node coloring of $G_p(N)$, no more than k nodes can have the same color, which implies that more than $\frac{N}{k}$ colors must be used in the color set $\mathcal{S}$. Thus, the chromatic number $\chi(G)$ is larger than $\frac{N}{k}$ for almost every graph $G \in G_p(N)$, which proves the inequality (15.50).

(xvi) *Friendship paradox.* (a) Consider an undirected friendship graph G with N nodes, where each node j represents an individual and a link between nodes represents a friendship relation. The number of friends of each friend of an individual j translates to the sum S_j of the degrees of the neighboring nodes of j, which is

$$S_j = \sum_{k \in \text{ neighbor}(j)} d_k = \sum_{k=1}^{N} a_{jk} d_k$$

where we have used adjacency matrix elements. The total number of friends of a friend equals

$$s = \sum_{j=1}^{N} S_j = \sum_{j=1}^{N} \sum_{k=1}^{N} a_{jk} d_k = \sum_{k=1}^{N} d_k \sum_{j=1}^{N} a_{jk} = \sum_{k=1}^{N} d_k^2$$

while the total number of friendship relations in G equals $\sum_{k=1}^{N} d_k = 2L$, by (15.4). Hence, the mean number of friends of friends $\xi = \frac{s}{2L}$ is

$$\xi = \frac{\sum_{k=1}^{N} d_k^2}{\sum_{k=1}^{N} d_k} = \frac{E[D^2]}{E[D]} = E[D] + \frac{\text{Var}[D]}{E[D]} \geq E[D]$$

Hence, the mean number of friends of friends is at least as large as the mean number of friends. Equality is only possible in regular friendship graphs, where each individual has an equal number of friends. The higher the variability in an individual's amount of friends, the larger the variance of the degree D of an arbitrary node in the graph G and the stronger the intuitive impression that the expected number of friends of your friends is larger than that of your friends. Feld (1991) gave a different derivation (without using graph theory) by noting that each individual j is a friend d_j times and has d_j friends, so that this individual contributes d_j friends' friends d_j times, a total of d_j^2 friends' friends. The total number of friends' friends is simply this quantity summed over all individuals equal to s (above). Actually, the paradox arises because of a biased sampling: a friend is not counted once, but d_j times. More details and a deeper discussion is given by Feld (1991).

(b) In an Erdős-Rényi graph, we may consider that the degree of each node j is (almost) independent from the degree of any other node. The number of neighbors of the neighbors of an arbitrary node with degree D is $S = \sum_{k=1}^{D} D_k$, where all random variables D_k are (almost) independent and identically distributed, specified by (15.19). If all degrees are independent, then Wald's inequality (2.78) is applicable and $E[S] = (E[D])^2$ so that $\xi = \frac{E[S]}{E[D]} = E[D]$ and there is no paradox. However, if all random variables D_k are distributed as D and there is dependence, then $S \approx D^2$, in which case Wald's inequality is not applicable and the friendship paradox arises. This example shows that, when N in $S_N = \sum_{k=1}^{N} X_k$ is depending on $\{X_k\}_{1 \leq k}$, then Wald's inequality (2.78) is not valid, in particular, in the derivation leading to (2.78), $\Pr[S_N = x | N = k]$ is not simply equal to $\Pr[S_k = x]$.

(xvii) *Degree distribution after adding or removing links.* (a) By the law of total probability (2.49), it holds that

$$\Pr[D = k] = \sum_{j=0}^{N-1} \Pr[D = k | D_{G_0} = j] \Pr[D_{G_0} = j] \tag{B.29}$$

and the remainder will consist in determining the conditional probability $\Pr[D = k | D_{G_0} = j]$. A newly added link belongs to the complement G_0^c of G_0. Since each of the m links is added randomly and independently of the others to the graph G_0, the probability that an empty place in the adjacency matrix of G_0 is occupied by a link equals $p_a = \frac{m}{\frac{N(N-1)}{2} - L}$. Randomly adding n links to an arbitrary node is equivalent to increasing its degree by $\sum_{j=1}^{n} X_j$, where X_j is a Bernoulli random variable with mean p and all $X_1, X_2, \ldots, X_n$ are independent. Thus, the probability that a random node, given that its degree is $D_{G_0} = j$, finally has degree k is a binomial distribution (which is a sum of Bernoulli random variables, see Section 3.1.2),

$$\Pr[D = k | D_{G_0} = j] = \binom{N-1-j}{k-j} p_a^{k-j} (1 - p_a)^{N-1-k}$$

In the available $N - 1 - j$ empty positions of the row of the adjacency matrix corresponding to the randomly chosen node, we can add links in $\binom{N-1-j}{k-j}$ and precisely $k-j$ are added with probability p_a, while the remaining (possible) links are not added with probability $1 - p_a$. In summary, we arrive at[5]

$$\Pr[D = k] = (1 - p_a)^{N-1-k} \sum_{j=0}^{N-1} \binom{N-1-j}{k-j} p_a^{k-j} \Pr[D_{G_0} = j] \tag{B.30}$$

The limiting case for $p_a \to 1$, that leads to a complete graph, (B.30) indeed shows that $\Pr[D = k] \to 0$, for all $k < N - 1$, and $\Pr[D = N - 1] \to 1$. The result (B.30) can be rephrased in term of the pgf $\varphi_{D_{G_0}} = E\left[z^{D_{G_0}}\right]$ as

$$\Pr[D = k] = \frac{(1 - p_a)^{N-1-k}}{(N - 1 - k)!} \left. \frac{d^{N-1-k}\left\{z^{N-1}\varphi_{D_{G_0}}\left(z^{-1}\right)\right\}}{dz^{N-1-k}} \right|_{z = p_a}$$

Probabilistically, $z^{N-1}\varphi_{D_{G_0}}\left(z^{-1}\right) = E\left[z^{N-1-D_{G_0}}\right]$ is the pgf of the degree in the complement of G_0.

(b) If we are interested in removing m links from a graph with L links uniformly at random, then $p_r = \frac{m}{L}$ is the probability that a random link is removed, which is different from the p_a

---

[5] Since $\binom{n}{k} = 0$ when the integer $k < 0$ (provided that n is also an integer), we can lower the upper bound from $N - 1$ to k.

above! The analogous process of removing links with probability p_r in an initial graph G_0 leads, in the same way as above, to

$$\Pr\left[D = k | D_{G_0} = j\right] = \binom{j}{j-k} p_r^{j-k} (1-p_r)^k$$

because from the possible j links, we remove precisely $j - k$ to reach a degree $D = k$. The number of ways to remove $k - j$ links out of j links is $\binom{j}{j-k} = \binom{j}{k}$ and each removed link has probability p_r, while the remaining k links each have probability $1 - p_r$ to not be removed. After introducing this result in (B.29), we find

$$\Pr\left[D = k\right] = (1-p_r)^k \sum_{j=0}^{N-1} \binom{j}{k} p_r^{j-k} \Pr\left[D_{G_0} = j\right] \tag{B.31}$$

Similarly as above, when $p_r \to 1$ in (B.31), we find that $\Pr\left[D = 0\right] = 1$, while $\Pr\left[D = k\right] \to 0$ for $k > 0$. Indeed, removing from a graph G_0 links with probability 1 is equivalent to removing all links, which leads to the empty graph. The corresponding generating function $\varphi_D\left(z\right) = \sum_{k=0}^{N-1} \Pr\left[D = k\right] z^k$ is with (B.31)

$$\varphi_D\left(z\right) = \sum_{k=0}^{N-1} \sum_{j=k}^{N-1} \binom{j}{k} (1-p_r)^k p_r^{j-k} \Pr\left[D_{G_0} = j\right] z^k$$

$$= \sum_{j=0}^{N-1} \left(\sum_{k=0}^{j} \binom{j}{k} ((1-p_r) z)^k p_r^{j-k} \right) \Pr\left[D_{G_0} = j\right]$$

$$= \sum_{j=0}^{N-1} ((1-p_r) z + p_r)^j \Pr\left[D_{G_0} = j\right]$$

and

$$\varphi_D\left(z\right) = \varphi_{D_{G_0}}\left((1-p_r) z + p_r\right) = \varphi_{D_{G_0}}\left(1 + q\left(z-1\right)\right) \tag{B.32}$$

where $q = 1 - p_r$. Finally, removing n arbitrary nodes from G_0 results in precisely the same degree equations as for arbitrary link removals, except that p_r is replaced by $p_n = \frac{n}{N}$.

(xviii) *Degree distribution of the end node of an arbitrarily chosen link.* (a) Let l denote an arbitrarily (i.e. uniformly at random) chosen link in the graph G and $l^+ = i$ and $l^- = j$ are the two end nodes of link l. In contrast to the degree of a randomly chosen node, a randomly chosen link has higher probability to connect to a high-degree node. Indeed, consider the adjacency matrix A in which any one element in A represents a link. Suppose that we choose an arbitrary link l, which is the same as picking a one entry in A at random, and suppose that link l lies in the row of a node n with degree k. The probability of choosing a link l belonging to a degree k node is proportional to k, the number of one entries on row of node n, and the number of rows in A in which a node has degree k, hence, $\Pr\left[D_{l^+} = k\right] = \alpha k \Pr\left[D = k\right]$ and the proportionality factor α is obtained from the normalization $\sum_{k=0}^{N-1} \Pr\left[D_{l^+} = k\right] = 1$ as $\alpha^{-1} = \sum_{k=0}^{N-1} k \Pr\left[D = k\right] = E\left[D\right]$. Alternatively, by the law of total probability (2.49), we have

$$\Pr\left[D_{l^+} = k\right] = \sum_{n=1}^{N} \Pr\left[D_{l^+} = k | l^+ \text{ is node } n\right] \Pr\left[l^+ \text{ is node } n\right]$$

Since node n has degree $d_n = k$, an arbitrarily chosen link l with end node l^+ in G equal to node n is

$$\Pr\left[l^+ \text{ is node } n\right] = \frac{d_n}{\sum_{j=1}^{N} d_j} = \frac{k}{2L}$$

With $\Pr\left[D_{l^+} = k | l^+ \text{ is node } n\right] = \Pr\left[D_n = k\right]$ and since node n is a random node with

$\Pr[D_n = k] = \Pr[D = k]$, we arrive at

$$\Pr[D_{l+} = k] = \sum_{n=1}^{N} \Pr[D = k] \frac{k}{2L} = \frac{N}{2L} k \Pr[D = k] = \frac{k \Pr[D = k]}{E[D]}$$

The corresponding generating function, given $\varphi_D(z) = \sum_{k=0}^{N-1} \Pr[D = k] z^k$, is

$$\varphi_{D_{l+}}(z) = \frac{1}{E[D]} \sum_{k=0}^{N-1} k \Pr[D = k] z^k = \frac{z}{E[D]} \sum_{k=0}^{N-1} k \Pr[D = k] z^{k-1}$$

so that

$$\varphi_{D_{l+}}(z) = \frac{z \varphi_D'(z)}{E[D]} = z \frac{\varphi_D'(z)}{\varphi_D'(1)} \tag{B.33}$$

Finally, the pgf of the number of links of node l^+ reached by following an arbitrary link l to that end node l^+ is simply $E\left[z^{D_{l+}-1}\right] = \frac{\varphi_D'(z)}{\varphi_D'(1)}$ and $E[D_{l+} - 1] = \frac{E[D^2] - E[D]}{E[D]}$ is also called the branching factor of a network.

(b) While the sum of the neighbors of a node is its degree, the total number of neighbors of the neighbors of a node n equals $d_n^{(2)} = \sum_{l \in \text{neighbors}(n)} (d_{l+} - 1)$, where $l^- = n$. Since we choose an arbitrary node n whose degree is a random variable $D_n = D$, the pgf of the total number $D^{(2)}$ of neighbors up to the second hop (along the shortest path tree rooted at n) is

$$\varphi_{D^{(2)}}(z) = E\left[z^{D^{(2)}}\right] = E\left[z^{\sum_{l \in \text{neighbors}(n)} (D_{l+} - 1)}\right]$$

Using the law of total probability (2.49) yields

$$\varphi_{D^{(2)}}(z) = \sum_{k=0}^{N-1} E\left[z^{\sum_{l \in \text{neighbors}(n)} (D_{l+} - 1)} \,\middle|\, D_n = k\right] \Pr[D_n = k]$$

$$= \sum_{k=0}^{N-1} E\left[z^{\sum_{j=1}^{k}\left(D_{l_j^+} - 1\right)}\right] \Pr[D = k]$$

where l_j^+ is a neighbor of n, which is reached by following link l_j starting from $l_j^- = n$. While all links $l_1, l_2, \ldots, l_k$ incident to node n with degree k, are independent (non-overlapping), the random variables $D_{l_j^+}$ for $1 \le j \le k$ are, in generally, not independent because the second-nearest neighbors can overlap, i.e. the set of neighbors of l_j^+ and l_m^+ may contain common nodes. By ignoring this dependence or overlap and assuming that all nodes are different, which is correct in trees and approximately true for sparse, large graphs as well, we have

$$E\left[z^{\sum_{j=1}^{k}\left(D_{l_j^+} - 1\right)}\right] = E\left[\prod_{j=1}^{k} z^{\left(D_{l_j^+} - 1\right)}\right] \simeq \prod_{j=1}^{k} E\left[z^{\left(D_{l_j^+} - 1\right)}\right]$$

Since $D_{l_j^+}$ is distributed as D_{l+} and $E\left[z^{(D_{l+}-1)}\right] = \frac{1}{z} E\left[z^{D_{l+}}\right] = \frac{\varphi_{D_{l+}}(z)}{z}$, we obtain by using (B.33)

$$E\left[z^{\sum_{j=1}^{k}\left(D_{l_j^+} - 1\right)}\right] \simeq \left(\frac{\varphi_D'(z)}{\varphi_D'(1)}\right)^k$$

so that

$$\varphi_{D^{(2)}}(z) = \sum_{k=0}^{N-1} \left(\frac{\varphi_D'(z)}{\varphi_D'(1)}\right)^k \Pr[D = k] \simeq \varphi_D\left(\frac{\varphi_D'(z)}{\varphi_D'(1)}\right)$$

from which $E\left[D^{(2)}\right] = \varphi_D''(1) = E\left[D^2\right] - E[D]$. When counting the number $D^{(m)}$ of m-hop neighbors along the shortest path tree rooted at node n, we find that $\varphi_{D^{(m)}}(z) = \varphi_{D^*}\left(\varphi_{D_{l+}^*}\left(\varphi_{D_{l+}^*}(\cdots)\right)\right)$, where D^* is the degree distribution of a random node in the shortest path tree rooted at n. In that case, where possible loops and overlap are excluded, the branching theory of Chapter 12 applies.

B.14 The shortest path problem (Chapter 16)

(i) The relative error r, defined as 1 minus the simulated value over the exact value at hop k given in (16.15), versus the number of hops k is shown in Fig. B.11. The insert in Fig. B.11 illustrates that, on a linear scale, the difference between simulation and theory (full line) is not distinguishable for $n \geq 10^5$ iterations. The mean $E\left[r_n\right]$ and standard deviation $\sigma\left[r_n\right]$ of

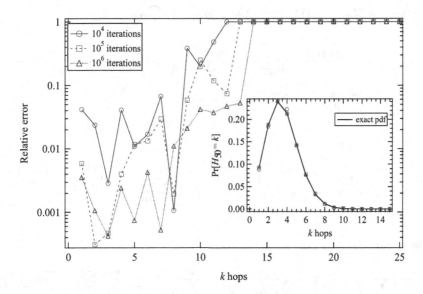

Fig. B.11. The relative error of the simulations of the hopcount in the complete graph with exponential link weight versus the hopcount for $10^4, 10^5$ and 10^6 iterations.

the relative error for n iterations versus the hops k are

$$
\begin{array}{ll}
E\left[r_{10^4}\right] = 0.12 & \sigma\left[r_{10^4}\right] = 0.17 \\
E\left[r_{10^5}\right] = 0.047 & \sigma\left[r_{10^5}\right] = 0.073 \\
E\left[r_{10^6}\right] = 0.017 & \sigma\left[r_{10^6}\right] = 0.02
\end{array}
$$

where the range of k values has been limited for $n = 10^4$ to 10 hops, for $n = 10^5$ to 11 hops, and for $n = 10^6$ to 12 hops. For larger hops, the simulations return zeros because the tail probability $\Pr\left[H_N > k\right]$ decreases as $O\left(1/k!\right)$ and simulating such a rare event requires on average at least as many simulations as $\left(\Pr\left[H_N = k\right]\right)^{-1}$. The table roughly shows that the mean error over the non-zero returned values decreases as $O\left(\frac{1}{\sqrt{n}}\right)$, which is in agreement with the Central Limit Theorem 6.3.1. Each iteration of the simulation can be regarded as an independent trial and the histogram sums in a particular way the number of these trials.

(ii) Using (2.46), we have

$$\text{Var}\,[W_N] = \varphi''_{W_N}(0) - \left(\varphi'_{W_N}(0)\right)^2 = \varphi''_{W_N}(0) - (E\,[W_N])^2$$

$$= \frac{1}{N-1}\sum_{k=1}^{N-1}\frac{d^2}{dz^2}\prod_{n=1}^{k}\frac{n(N-n)}{z+n(N-n)}\Bigg|_{z=0} - \left(\frac{1}{N-1}\sum_{n=1}^{N-1}\frac{1}{n}\right)^2$$

where $E\,[W_N]$ is given in (16.21). The derivatives of the product

$$g\,(z) = \prod_{n=1}^{k}\frac{n(N-n)}{z+n(N-n)}$$

are elegantly computed via the logarithmic derivative $\frac{dg(z)}{dz} = g\,(z)\,\frac{d\log g(z)}{dz}$. The second derivative is $\frac{d^2 g(z)}{dz^2} = g\,(z)\left(\frac{d\log g(z)}{dz}\right)^2 + g\,(z)\,\frac{d^2\log g(z)}{dz^2}$. With

$$\frac{d\log g\,(z)}{dz} = \frac{d}{dz}\sum_{n=1}^{k}\log\frac{n(N-n)}{z+n(N-n)} = -\sum_{n=1}^{k}\frac{1}{z+n(N-n)}$$

$$\frac{d^2\log g\,(z)}{dz^2} = \sum_{n=1}^{k}\frac{1}{(z+n(N-n))^2}$$

we obtain since $g(0) = 1$,

$$\text{Var}\,[W_N] = \frac{1}{N-1}\sum_{k=1}^{N-1}\left(\sum_{n=1}^{k}\frac{1}{n(N-n)}\right)^2 + \frac{1}{N-1}\sum_{k=1}^{N-1}\sum_{n=1}^{k}\frac{1}{n^2(N-n)^2} - \left(\frac{\sum_{n=1}^{N-1}\frac{1}{n}}{N-1}\right)^2$$

$$\text{(B.34)}$$

The first sum is

$$\sum_{k=1}^{N-1}\left(\sum_{n=1}^{k}\frac{1}{n(N-n)}\right)^2 = \sum_{k=1}^{N-1}\sum_{n=1}^{k}\frac{1}{n(N-n)}\sum_{j=1}^{k}\frac{1}{j(N-j)} = \sum_{k=1}^{N-1}\frac{\sum_{k=n}^{N-1}\sum_{j=1}^{k}\frac{1}{j(N-j)}}{n(N-n)}$$

and, with $\sum_{k=n}^{N-1}\sum_{j=1}^{k}\frac{1}{j(N-j)} = \sum_{k=1}^{N-1}\sum_{j=1}^{k}\frac{1}{j(N-j)} - \sum_{k=1}^{n-1}\sum_{j=1}^{k}\frac{1}{j(N-j)}$,

$$\sum_{k=1}^{N-1}\left(\sum_{n=1}^{k}\frac{1}{n(N-n)}\right)^2 = \sum_{n=1}^{N-1}\frac{1}{n(N-n)}\left(\sum_{k=1}^{N-1}\sum_{j=1}^{k}\frac{1}{j(N-j)} - \sum_{k=1}^{n-1}\sum_{j=1}^{k}\frac{1}{j(N-j)}\right)$$

$$= \sum_{n=1}^{N-1}\frac{1}{n(N-n)}\left(\sum_{j=1}^{N-1}\frac{1}{j(N-j)}\sum_{k=j}^{N-1}1 - \sum_{j=1}^{n-1}\frac{1}{j(N-j)}\sum_{k=j}^{n-1}1\right)$$

$$= \sum_{n=1}^{N-1}\frac{1}{n(N-n)}\sum_{j=1}^{N-1}\frac{1}{j} - \sum_{n=1}^{N-1}\frac{1}{(N-n)}\sum_{j=1}^{n-1}\frac{1}{j(N-j)}$$

$$+ \sum_{n=1}^{N-1}\frac{1}{n(N-n)}\sum_{j=1}^{n-1}\frac{1}{(N-j)}$$

Furthermore, since $\sum_{n=1}^{k} \frac{1}{n(N-n)} = \frac{1}{N} \sum_{n=1}^{k} \frac{1}{n} + \frac{1}{N} \sum_{n=N-k}^{N-1} \frac{1}{n}$, we have

$$\sum_{n=1}^{N-1} \frac{1}{(N-n)} \sum_{j=1}^{n-1} \frac{1}{j(N-j)} = \frac{1}{N} \sum_{n=1}^{N-1} \frac{1}{(N-n)} \sum_{k=1}^{n-1} \frac{1}{k} + \frac{1}{N} \sum_{n=1}^{N-1} \frac{1}{(N-n)} \sum_{k=N-n+1}^{N-1} \frac{1}{k}$$

$$= \frac{1}{N} \sum_{j=1}^{N-1} \frac{1}{j} \sum_{k=1}^{N-j-1} \frac{1}{k} + \frac{1}{N} \sum_{j=1}^{N-1} \frac{1}{j} \sum_{k=j+1}^{N-1} \frac{1}{k}$$

and

$$\sum_{n=1}^{N-1} \frac{1}{n(N-n)} \sum_{j=1}^{n-1} \frac{1}{(N-j)} = \frac{1}{N} \sum_{n=1}^{N-1} \left(\frac{1}{n} + \frac{1}{N-n} \right) \sum_{j=1}^{n-1} \frac{1}{(N-j)}$$

$$= \frac{1}{N} \sum_{j=1}^{N-1} \frac{1}{j} \sum_{k=N-j+1}^{N-1} \frac{1}{k} + \frac{1}{N} \sum_{j=1}^{N-1} \frac{1}{j} \sum_{k=j+1}^{N-1} \frac{1}{k}$$

Hence,

$$\sum_{k=1}^{N-1} \left(\sum_{n=1}^{k} \frac{1}{n(N-n)} \right)^2 = \frac{2}{N} \left(\sum_{n=1}^{N-1} \frac{1}{n} \right)^2 - \frac{1}{N} \sum_{j=1}^{N-1} \frac{1}{j} \sum_{k=1}^{N-j-1} \frac{1}{k} + \frac{1}{N} \sum_{j=1}^{N-1} \frac{1}{j} \sum_{k=N-j+1}^{N-1} \frac{1}{k}$$

$$= \frac{2}{N} \left(\sum_{n=1}^{N-1} \frac{1}{n} \right)^2 - \frac{1}{N} \sum_{j=1}^{N-1} \frac{1}{j} \left(\sum_{k=1}^{N-1} \frac{1}{k} - \sum_{k=N-j}^{N-1} \frac{1}{k} \right)$$

$$+ \frac{1}{N} \sum_{j=1}^{N-1} \frac{1}{j} \sum_{k=N-j+1}^{N-1} \frac{1}{k}$$

$$= \frac{1}{N} \left(\sum_{n=1}^{N-1} \frac{1}{n} \right)^2 + \frac{1}{N} \sum_{j=1}^{N-1} \frac{1}{j} \frac{1}{N-j} + \frac{2}{N} \sum_{j=1}^{N-1} \frac{1}{j} \sum_{k=N-j+1}^{N-1} \frac{1}{k}$$

$$= \frac{1}{N} \left(\sum_{n=1}^{N-1} \frac{1}{n} \right)^2 + \frac{2}{N^2} \sum_{j=1}^{N-1} \frac{1}{j} + \frac{2}{N} \sum_{j=1}^{N-1} \frac{1}{j} \sum_{k=N-j+1}^{N-1} \frac{1}{k}$$

Substitution into (B.34) yields

$$\text{Var}\,[W_N] = -\frac{\left(\sum_{n=1}^{N-1} \frac{1}{n} \right)^2}{(N-1)^2\,N} + \frac{2 \sum_{j=1}^{N-1} \frac{1}{j}}{(N-1)\,N^2} + \frac{2 \sum_{j=1}^{N-1} \frac{1}{j} \sum_{k=N-j+1}^{N-1} \frac{1}{k}}{N\,(N-1)}$$

$$+ \frac{\sum_{k=1}^{N-1} \sum_{n=1}^{k} \frac{1}{n^2(N-n)^2}}{N-1}$$

Further,

$$\sum_{k=1}^{N-1} \sum_{n=1}^{k} \frac{1}{n^2(N-n)^2} = \sum_{n=1}^{N-1} \frac{1}{n^2(N-n)^2} \sum_{k=n}^{N-1} 1 = \sum_{n=1}^{N-1} \frac{1}{n^2(N-n)}$$

The partial fraction expansion of $\frac{1}{n^2(N-n)} = \frac{1}{N^2 n} + \frac{1}{N n^2} + \frac{1}{N^2(N-n)}$ such that

$$\sum_{k=1}^{N-1} \sum_{n=1}^{k} \frac{1}{n^2(N-n)^2} = \frac{2}{N^2} \sum_{n=1}^{N-1} \frac{1}{n} + \frac{1}{N} \sum_{n=1}^{N-1} \frac{1}{n^2}$$

Combined,

$$\text{Var}\,[W_N] = -\frac{\left(\sum_{n=1}^{N-1} \frac{1}{n} \right)^2}{(N-1)^2\,N} + \frac{4 \sum_{n=1}^{N-1} \frac{1}{n}}{(N-1)\,N^2} + \frac{2}{N\,(N-1)} \sum_{j=1}^{N-1} \frac{1}{j} \sum_{k=N-j+1}^{N-1} \frac{1}{k} + \frac{\sum_{n=1}^{N-1} \frac{1}{n^2}}{N(N-1)}$$

Invoking the identity

$$\sum_{j=1}^{N-1} \frac{1}{N-j} \sum_{k=j}^{N-1} \frac{1}{k} = \sum_{n=1}^{N-1} \frac{1}{n^2}$$

(which can be verified by induction) yields

$$\sum_{j=1}^{N-1} \frac{1}{j} \sum_{k=N-j+1}^{N-1} \frac{1}{k} = \sum_{j=1}^{N-1} \frac{1}{N-j} \sum_{k=j+1}^{N-1} \frac{1}{k} = \sum_{j=1}^{N-1} \frac{1}{N-j} \left(\sum_{k=j}^{N-1} \frac{1}{k} - \frac{1}{j} \right)$$

$$= \sum_{j=1}^{N-1} \frac{1}{N-j} \sum_{k=j}^{N-1} \frac{1}{k} - \sum_{j=1}^{N-1} \frac{1}{j(N-j)} = \sum_{n=1}^{N-1} \frac{1}{n^2} - \frac{2}{N} \sum_{n=1}^{N-1} \frac{1}{n}$$

Finally, we arrive at (16.22).

(iii) The probability $\Pr[H_N = 2]$ is determined by the intersection of two independent events. First, there is no direct path between nodes A and B. This event has a chance proportional to $1 - p$. Second, there is at least one path with two hops. All $N - 2$ possible two-hop paths between A and B have the structure $(A \to j)(j \to B)$ and they have no links in common, i.e. they are mutually independent and independent from the direct link. The probability of the second event equals $1 - P_2$, where P_2 is the probability that there is no path with two hops. Hence, we have that $\Pr[H_N = 2] = (1 - p)(1 - P_2)$ and it remains to compute P_2. The event of no path with two hops is

$$\left(\cup_{j=1}^{N-2} 1_{(A \to j)(j \to B)} \right)^c = \cap_{j=1}^{N-2} 1_{((A \to j)(j \to B))^c}$$

such that

$$P_2 = \Pr\left[\left(\cup_{j=1}^{N-2} 1_{(A \to j)(j \to B)} \right)^c \right] = \Pr\left[\cap_{j=1}^{N-2} 1_{((A \to j)(j \to B))^c} \right]$$

$$= \prod_{j=1}^{N-2} \Pr\left[1_{((A \to j)(j \to B))^c} \right] = \prod_{j=1}^{N-2} \left(1 - \Pr\left[1_{((A \to j)(j \to B))} \right] \right) = \left(1 - p^2 \right)^{N-2}$$

which demonstrates (16.55).

(iv) When the intermediate nodes of the shortest path between a source A and a destination B are removed from K_N, we obtain again a complete graph with $N - h_N + 1$ nodes. The resulting graph contains link weights that are no longer perfectly exponentially distributed nor are they perfectly independent, because we have removed a special set of nodes rather than a random set. But, since we have removed at each node of the shortest path, apart from the shortest link, $N - 3$ other links, we assume that the dependence between K_N and the reduced graph is ignorably small. Under these assumptions, the shortest node-disjoint path in K_N is a shortest path in K_{N-H_N+1} with exponential link weight with mean 1. The distribution of hopcount h_N^{nd} of that shortest node-disjoint path is

$$\Pr\left[h_N^{nd} = k \right] \simeq \sum_{j=0}^{N-1} \Pr\left[h_{N-j+1} = k | h_N = j \right] \Pr\left[h_N = j \right]$$

The hopcount H_N of the shortest path in the complete graph K_N with independent exponential link weights with mean 1 is given in (16.9). With the assumption that

$$\Pr\left[h_{N-j+1} = k | h_N = j \right] = \Pr\left[h_{N-j+1} = k \right]$$

we obtain

$$\Pr\left[h_N^{nd} = k \right] \simeq \frac{(-1)^{k+1}}{N!} \sum_{j=0}^{N-1} \frac{S_{N-j+1}^{(k+1)} S_N^{(j+1)}}{(N-j+1)!}$$

For large N, we can use the Poisson approximation (16.14)

$$\Pr\left[h_N^{nd} = k \right] \approx \frac{1}{Nk!} \sum_{j=0}^{N-1} \frac{(\log(N-j+1))^k}{N-j+1} \frac{(\log N)^j}{j!}$$

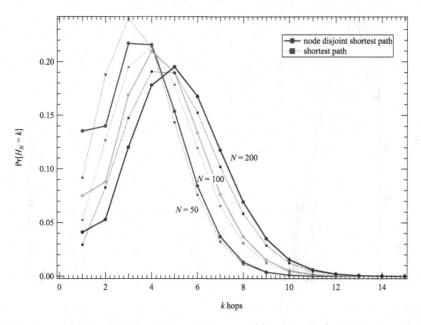

Fig. B.12. The pdf of the hopcount of the shortest path (thin lines) and the shortest node-disjoint path (bold lines).

Since $(\log(N - j + 1))^k = \log^k N - \frac{k(j-1)}{N}\log^{k-1} N + O\left(\frac{\log^{k-1}}{N^2}\right)$ and $\frac{1}{N-j+1} = \frac{1}{N} + O\left(\frac{1}{N^2}\right)$, we have to highest order in N,

$$\Pr\left[h_N^{nd} = k\right] \approx \frac{1}{Nk!}\sum_{j=0}^{N-1}\left(\frac{\log^k N}{N} + O\left(\frac{\log^{k-1} N}{N^2}\right)\right)\frac{(\log N)^j}{j!}$$

$$\approx \frac{1}{k!}\left(\frac{\log^k N}{N} + O\left(\frac{\log^{k-1} N}{N^2}\right)\right) \approx \Pr\left[h_N = k\right]$$

For large N, we expect that the hopcount of the shortest and shortest node-disjoint path have approximately the same distribution. The validity of the assumption is illustrated in Fig. B.12 for relatively small values of $N = 50, 100$, and 200. Each simulation consisted of $n = 10^6$ iterations. The corresponding weight of the shortest and node-disjoint shortest paths are drawn in Fig. B.13. The weight of the node-disjoint shortest path is evidently always larger than that of the shortest path in the same graph. Nevertheless, for large N, the simulations suggest that both pdfs tend to each other.

(v) (a) Applying the law of total probability (2.49), the probability that k red links appear in the shortest path is

$$\Pr[R = k] = \sum_{j=1}^{N-1}\Pr[R = k|H_N = j]\Pr[H_N = j]$$

In case of a random distribution of red links and given the hopcount H_N of the shortest path, the probability distribution of the number of red/blue links that appear in the shortest path has the binomial distribution,

$$\Pr[R = k|H_N = j] = \binom{j}{k}p_R^k(1 - p_R)^{j-k}$$

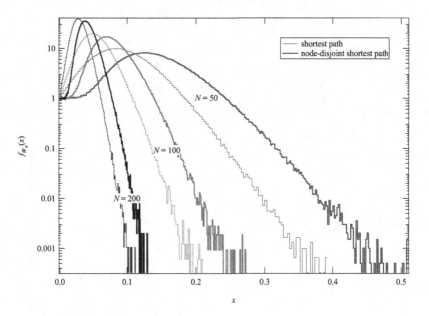

Fig. B.13. The pdf of the weight of the shortest path (thin lines) and the node-disjoint shortest path (bold lines).

Other conditional distributions can be envisaged, but most of them will complicate the analysis. Moreover, in practice, sufficiently accurate information of possible other conditional distributions is difficult to obtain. Hence, assuming a random distribution of red links in the network, we obtain

$$\Pr[R = k] = \sum_{j=1}^{N-1} \binom{j}{k} p_R^k (1 - p_R)^{j-k} \Pr[H_N = j]$$

$$= \frac{p_R^k}{(1 - p_R)^k} \sum_{j=1}^{N-1} \binom{j}{k} \Pr[H_N = j](1 - p_R)^j$$

Based on the definition of a probability generating function $\varphi_H(z) = \sum_{j=1}^{N-1} \Pr[H_N = j]\, z^j$, we find that

$$\frac{d^k}{dz^k} \varphi_H(z) = \sum_{j=1}^{N-1} j(j-1)\cdots(j-k+1)\Pr[H_N = j]\, z^{j-k} = \frac{k!}{z^k} \sum_{j=1}^{N-1} \binom{j}{k} \Pr[H_N = j]\, z^j$$

Using this expression in that of $\Pr[R = k]$ yields

$$\Pr[R = k] = \frac{p_R^k}{k!} \frac{d^k}{dz^k} \varphi_H(z) \Big|_{z=1-p_R} \tag{B.35}$$

In order to proceed, we need to specify the probability generating function of the hopcount in the graph. Since the hopcount distribution in the complete graph with exponential links is known and, analytically, relatively simple, we confine to that case. The generating function of the hopcount $\varphi_H(z)$ is given in (16.6). As mentioned in Section 16.2.2, formula (16.6) also describes asymptotically (i.e. for large N) the hopcount in the Erdős-Rényi random graph $G_p(N)$. Since the exponential distribution is a regular distribution, we claim, in addition, that (16.6) is applicable for any regular distribution in small-world graphs (Watts, 1999). The

Erdős-Rényi random graph $G_p(N)$ is an example of a small-world graph. Internet trace route measurements indicate that this simple model is a surprisingly good first-order approximation for the hopcount of Internet paths.

Applying (B.35) to the probability generating function in (16.6) gives

$$\Pr[R = k] = \begin{cases} \frac{N}{N-1}\left(\frac{\Gamma(N+1-p_R)}{N!\Gamma(2-p_R)} - \frac{1}{N}\right), & k = 0 \\ \frac{N}{(N-1)N!} \frac{p_R^k}{k!} \frac{d^k}{dz^k} \left.\frac{\Gamma(N+z)}{\Gamma(z+1)}\right|_{z=1-p_R}, & 0 < k < N \end{cases} \tag{B.36}$$

These derivatives are most conveniently computed via the logarithmic derivative. In this way, we can subsequently calculate $\frac{d^k}{dz^k}\varphi_H(z)$ and finally get $\Pr[R = k]$. For example, for $k \le 3$, we explicitly obtain

$$\Pr[R = 0] = \frac{N}{N-1}\left(\frac{\Gamma(N+1-p_R)}{N!\Gamma(2-p_R)} - \frac{1}{N}\right)$$

$$\Pr[R = 1] = p_R \frac{N}{N-1}\frac{\Gamma(N+1-p_R)}{N!\Gamma(2-p_R)} \cdot \sum_{j=2}^{N}\frac{1}{j-p_R}$$

$$\Pr[R = 2] = \frac{p_R^2}{2}\frac{N}{N-1}\frac{\Gamma(N+1-p_R)}{N!\Gamma(2-p_R)}\left(\left(\sum_{j=2}^{N}\frac{1}{j-p_R}\right)^2 - \sum_{j=2}^{N}\frac{1}{(j-p_R)^2}\right)$$

For large N and small z, it holds approximately that

$$\frac{\Gamma(N+z)}{N!\Gamma(z+1)} \sim \frac{N^{z-1}}{\Gamma(z+1)} \approx e^{(z-1)\log N} \tag{B.37}$$

from which

$$\Pr[R = k] \approx \frac{(p_R \log N)^k}{k!}e^{-p_R \log N}$$

This means that, in sufficiently large networks, the probability that red links appear in the shortest path is Poisson distributed with mean $p_R \log N$. The relation with the Poisson distribution immediately gives additional information such as $\mathrm{Var}[R] = E[R] \approx p_R \log N$ and several other elegant relations.

Fig. B.14 shows the simulation result of the probability distribution of the number of red/blue links that appear in the shortest path in K_{25}. R/B stands for the number of red/blue links that appear in the shortest path. As shown in the table below, for $p_R = 0.1$ and $N = 25$, the analytic results are confirmed by the simulations and correspond to three points (triangles) in Fig. B.14.

	analytic result	simulation result
$\Pr[R = 0]$	0.7419	0.7411
$\Pr[R = 1]$	0.2256	0.2261
$\Pr[R = 2]$	0.0287	0.0302

The probability that there appear red links in the shortest path is $1 - \Pr[R = 0]$ with

$$1 - \Pr[R = 0] = \frac{N}{N-1}\left(1 - \frac{\Gamma(N+1-p_R)}{N!\Gamma(2-p_R)}\right) \tag{B.38}$$

The minimum of the Gamma function for positive real values is attained at $x_{\min} = 1.461\,63$ with $\Gamma(x_{\min}) = 0.885\,603$. Since $\Gamma(2) = \Gamma(1) = 1$, we have for any $p_R \in [0,1]$ that $0.885\,603 \le \Gamma(2-p_R) \le 1$. For large N, employing this bound and (B.37), we thus find that

$$N^{-p_R} \le \Pr[R = 0] \le 1.129 N^{-p_R}$$

The probability (B.38) is shown for different density p_R in Fig. B.15.

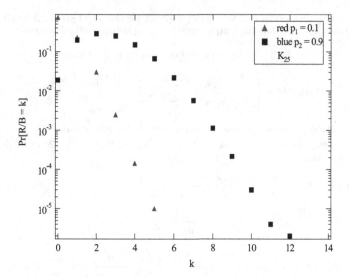

Fig. B.14. Probability distribution of the number of red/blue links that appear in the shortest path.

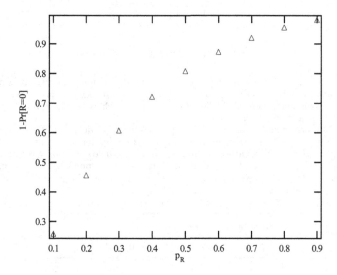

Fig. B.15. The probability that red links with density p_R appear in the shortest path.

(b) This analysis has practical interest. For example, the red links can be considered as links that have failures. Our analysis quantifies via $1 - \Pr[R = 0]$ the effect of link failures on the shortest path. A practical application may arise in ad-hoc networks where nodes may possess a different radio range or where link failures occur due to the lack of battery energy. Another example may apply to cases where links are expensive with probability p_R: the cost of communication may depend on the number of red (expensive) links in the communication path. If the fraction p_R of the routers that support IPv6 (or, in general, any other new protocol that includes e.g. QoS or enhanced security features) can be approximated as roughly uniformly distributed in the Internet, the analysis computes the chance that a communication can benefit from these enhancements with respect to legacy IPv4 protocols.

(vi) (a) Let us consider hot potato routing in a graph $G(N, L)$ from node n to node m. Given that the hot potato routing process is at the node i at discrete time k, the next hop j at discrete time $k+1$ is determined completely by the state X_k at time k. Indeed, the uniform selection among the neighbors of node i results in the transition probability

$$\Pr[X_{k+1} = j | X_k = i] = \frac{a_{ij}}{d_i} = \frac{a_{ij}}{\sum_{n=1}^{N} a_{in}}$$

where a_{ij} is an element of the adjacency matrix A and the d_i is the degree of node i. Moreover, the transition probability matrix P is independent of the discrete time k and equal (Van Mieghem, 2011, p. 63) to $P = \Delta^{-1}A$. If the hot potato routing process starts in node 1, then the initial state vector $s[0] = \begin{bmatrix} 1 & 0 & \dots & 0 \end{bmatrix}$ and the evolution over time is described by equation (9.10). The probability density function of the hopcount of a path in hot potato routing is precisely equal to the hitting time of the Markov process defined by $P = \Delta^{-1}A$,

$$\Pr[H = k] = \Pr[T_m = k | X_0 = n]$$

The general expression of $\Pr[T_m = k | X_0 = n]$ is of the form (B.11) and explicitly computed in (9.21).

(b) We now compute the hopcount of hot potato routing in the complete graph whose adjacency matrix is $A = J - I$. The transition probability matrix is $P = \frac{1}{N-1}(J - I)$ because $d_i = \frac{1}{N-1}$ for any node i. We need to compute P^k. Since $J^2 = NJ$, it follows by induction that $J^m = N^{m-1}J$. Hence,

$$(J - I)^k = \sum_{m=0}^{k} \binom{k}{m} J^m (-1)^{k-m} I^{k-m} = (-1)^k I + J \sum_{m=1}^{k} \binom{k}{m} N^{m-1} (-1)^{k-m}$$

$$= (-1)^k \left(I + \frac{J}{N} \left((1 - N)^k - 1 \right) \right)$$

Since $P^k = \frac{1}{(N-1)^k}(J - I)^k$, we arrive at $P^k = \frac{J}{N} \left(1 - \frac{(-1)^k}{(N-1)^k} \right) + \frac{(-1)^k}{(N-1)^k} I$ from which

$$P_{ij}^k = \frac{1}{N} \left(1 - \frac{(-1)^k}{(N-1)^k} \right) + \frac{(-1)^k}{(N-1)^k} 1_{\{i=j\}}$$

or

$$P_{ij}^k = \frac{1}{N} \left(1 - \frac{(-1)^k}{(N-1)^k} \right) \qquad \text{if } i \neq j$$

$$P_{jj}^k = \frac{1}{N} \left(1 - \frac{(-1)^{k-1}}{(N-1)^{k-1}} \right)$$

Applied to (B.11) gives $\Pr[H = m] = \sum_{l=1}^{m} a_{m+1-l;1} P_{ij}^l$, where $a_{m;1}$ is conveniently computed from the Taylor series of $(1 + f(z))^{-1}$, where $f(z) = \sum_{k=1}^{\infty} P_{jj}^k z^k$. First, observe that

$$f(z) = \frac{1}{N} \sum_{k=1}^{\infty} \left(1 - \frac{(-1)^{k-1}}{(N-1)^{k-1}} \right) z^k = \frac{1}{N} \frac{z}{1-z} - \frac{1}{N} \frac{z}{1 + \frac{z}{N-1}} = \frac{1}{N} \frac{z}{1-z} - \frac{1}{N} \frac{(N-1)z}{N-1+z}$$

so that

$$(1 + f(z))^{-1} = \left(1 + \frac{1}{N} \frac{z}{1-z} - \frac{1}{N} \frac{(N-1)z}{N-1+z} \right)^{-1} = \frac{(1-z)(N-1+z)}{(N-1) - (N-2)z}$$

The Taylor series of $(1 + f(z))^{-1}$ is

$$(1 + f(z))^{-1} = \frac{(1-z)(N-1+z)}{N-1\left(1 - \frac{N-2}{N-1}z\right)} = \frac{-z^2 - (N-2)z + (N-1)}{N-1} \sum_{k=0}^{\infty} \left(\frac{N-2}{N-1}\right)^k z^k$$

$$= -\sum_{k=0}^{\infty} \frac{1}{N-1} \left(\frac{N-2}{N-1}\right)^k z^{k+2} - \sum_{k=0}^{\infty} \left(\frac{N-2}{N-1}\right)^{k+1} z^{k+1} + \sum_{k=0}^{\infty} \left(\frac{N-2}{N-1}\right)^k z^k$$

$$= 1 - \sum_{k=2}^{\infty} \frac{1}{N-1} \left(\frac{N-2}{N-1}\right)^{k-2} z^k$$

Hence, for $k \geq 1$, the Taylor coefficients are $\frac{1}{N-1}\left(\frac{N-2}{N-1}\right)^{k-2} = a_{k+1,1}$ such that

$$a_{m;1} = -\frac{1}{N-1} \left(\frac{N-2}{N-1}\right)^{m-3} \qquad \text{for } m > 1$$

$$a_{11} = 1 \text{ and } a_{21} = 0$$

and

$$\Pr[H = m] = P_{ij}^m + \sum_{l=1}^{m-2} a_{m+1-l;1} P_{ij}^l$$

$$= \frac{1}{N}\left(1 - \frac{(-1)^m}{(N-1)^m}\right) - \frac{1}{N-1}\frac{1}{N} \sum_{l=1}^{m-2} \left(\frac{N-2}{N-1}\right)^{m-l-2} \left(1 - \frac{(-1)^l}{(N-1)^l}\right)$$

$$= \frac{1}{N}\left[1 - \frac{1}{(N-1)^2}\right]\left(\frac{N-2}{N-1}\right)^{m-2}$$

Finally,

$$\Pr[H = m] = \frac{1}{N-1}\left(1 - \frac{1}{N-1}\right)^{m-1} \tag{B.39}$$

which means by (3.5) that the hopcount H of hot potato routing in K_N is precisely a geometric random variable with parameter $\frac{1}{N-1}$. Indeed, any node in K_N has precisely $N-1$ neighbors. At each discrete time $j < k$, there is, by symmetry, precisely a chance $\frac{1}{N-1}$ to uniformly choose the destination. The hopcount is only equal to k if there are precisely $k-1$ failures to choose the destination followed by one success at time k. The mean hopcount is equal to $E[H] = N - 1$. Since the longest possible path equals $N - 1$ hops, hot potato routing may visit intermediate nodes more than once. The probability that there is surely a loop is

$$\Pr[H > N - 1] = \sum_{m=N}^{\infty} \Pr[H = m] = \left(1 - \frac{1}{N-1}\right)^{N-1} \to \frac{1}{e}$$

(c) Let $\mathcal{P}_{a \to b} = n_a \to n_2 \to \cdots \to n_b$ denote a shortest path in a graph between a and b. The probability that a random walk RW that starts in node n_a follows that shortest path $\mathcal{P}_{a \to b}$ is

$$\Pr[RW = \mathcal{P}_{a \to b}] = \frac{1}{d_a} \prod_{n \in \mathcal{P}_{a \to b}} \frac{1}{d_n - 1}$$

where d_n is the degree of node n. The product follows by independence of the random walk strategy. Only at the starting node a, all possible links (i.e. d_a) towards a random next neighbor can be used whereas for each intermediate n (and destination node b), only $d_n - 1$ links can be used since the link, where the RW entered that node, needs to be excluded. If all link weights are one, then all shortest hop paths are shortest paths. Since they generally may overlap, it is difficult to find a general, closed expression for the probability that RW follows one of the shortest paths.

(d) The new variant of hot potato routing with memory restricts the possibility that the

Markov process returns to a previously visited intermediate state (node in the network). The process starts at discrete time 0 in state i with transition probability matrix $P[1]$ defined by

$$P_{ik}[1] = \Pr[X_1 = k | X_0 = i] = \frac{a_{ik}}{\sum_{n=1}^{N} a_{in}}$$

If the process moves in the first step to state k which is not the final node, at discrete time 2, the process moves to another state l with transition probability matrix $P[2]$ that is constructed from the adjacency matrix A by removing the row i and column i and renormalizing all remaining rows,

$$P_{kl}[2] = \Pr[X_2 = l | X_1 = k, X_0 = i] = \frac{a_{kl}}{\sum_{n=1;n\neq i}^{N} a_{kn}}$$

In the next step, the transition probability matrix $P[3]$ is determined from the adjacency matrix A by removing the state i and k followed by a renormalization, and so on. Thus, instead of considering walks, hot potato routing with memory operates with paths. Paths guarantee that after $N-1$ hops the selection process stops, i.e. $\Pr\left[\breve{H} > N-1\right] = 0$. However, hot potato routing with memory cannot guarantee that, after $N-1$ hops, the destination is found because deadlocks may occur, e.g. when the connected graph consists of three clusters that are interconnected by one link at one node per cluster. If the source and destination do not lie in the same cluster and hot potato routing with memory selects the third cluster to which neither the source nor the destination belong, the selection process can no longer leave that "wrong" cluster.

In view of (15.1), the pdf of the hopcount $\breve{H}$ in hot potato routing with memory is

$$\Pr\left[\breve{H} = m\right] = \sum_{r_1 \neq \{i,j\}} \sum_{r_2 \neq \{i,r_1,j\}} \cdots \sum_{r_{m-1} \neq \{i,r_1,\ldots,r_{m-2},j\}} P_{ir_1}[1] P_{r_1 r_2}[2] \cdots P_{r_{m-1}j}[m]$$

where the sum over all possible m-hop paths consists of the likelihood

$$P_{ir_1}[1] P_{r_1 r_2}[2] \cdots P_{r_{m-1}j}[m]$$

that the particular path $i \to r_1 \to r_2 \to \ldots \to j$ will be selected. The structure of each transition probability matrix depends on all previous transitions because of the renormalization,

$$P_{r_{m-1}j}[m] = \Pr[X_m = j | X_{m-1} = r_{m-1}, X_{m-2} = r_{m-2}, \ldots, X_0 = i]$$

$$= \frac{a_{r_{m-1};j}}{\sum_{n=1;n\neq\{i,r_1,\ldots,r_{m-2}\}}^{N} a_{r_{m-1};n}}$$

This implies, in fact, that hot potato routing with memory is not a Markov process. Applied to the complete graph where $P_{ij}[k] = \frac{1}{N-k}1_{i\neq j}$ gives

$$\Pr\left[\breve{H} = m\right] = \sum_{r_1 \neq \{i,j\}} \sum_{r_2 \neq \{i,r_1,j\}} \cdots \sum_{r_{m-1} \neq \{i,r_1,\ldots,r_{k-2},j\}} \frac{1}{(N-1)(N-2)\cdots(N-m)}$$

Using (15.2) yields

$$\Pr\left[\breve{H} = m\right] = \frac{(N-m-1)!}{(N-1)!} \frac{(N-2)!}{(N-m-1)!} = \frac{1}{N-1}$$

which means that the hopcount in hot potato routing with memory is a uniform random variable on $[1, N-1]$. This result can be found directly by observing that, after removing a node in the complete graph K_N, again a complete graph K_{N-1} appears. Hence, the probability that the hopcount in hot potato routing with memory in the complete graph equals m hops is

$$\Pr\left[\breve{H} = m\right] = \left(1 - \frac{1}{N-1}\right) \cdots \left(1 - \frac{1}{N-m+1}\right) \frac{1}{N-m} = \frac{1}{N-1}$$

Indeed, we need $m-1$ unsuccessful uniform node selections out of $[1, N-k]$ in each step $1 \leq k \leq m-1$, followed by a successful selection in step m.

(vii) (a) By taking the expectation of both sides in (16.67), using the linearity of the expectation and the moments $E\left[V_j^k\right] = \frac{1}{j^k}$ of Bernoulli random variables, we have $E[h_N] = \sum_{j=2}^{N} E[V_j] = \sum_{j=2}^{N} \frac{1}{j}$. Since $V_2, ..., V_N$ are independent random variables and $\text{Var}[V_j] = \frac{1}{j}\left(1 - \frac{1}{j}\right)$, the variance of h_N follows from (2.60) as

$$\text{Var}[h_N] = \sum_{j=2}^{N} \text{Var}[V_j] = \sum_{j=2}^{N} \frac{1}{j}\left(1 - \frac{1}{j}\right)$$

(b) The hopcount h_N from the root to an arbitrary node in the URT also equals the hopcount of an arbitrary path starting from the root, which we label by 1. The number of hops in an arbitrary path at the root can be described by the URT growth process and by the selection process of the end node of that path. Only if the end node of the path equals the root itself, the hopcount is zero. By definition of the URT, at each stage j, a new node j is attached uniformly to one of the existing $j-1$ nodes until the total number of nodes is equal to N. After attaching the node j to the URT, the hopcount h_j can only increase in one case when the node j is attached to the end node of the previously (in stage $j-1$) longest hop path. In all other cases, h_j remains the same as h_{j-1}. The probability that h_j increases consists of two independent events: (1) the j-th node is attached to the correct node of the longest hopcount path at stage $j-1$ and (2) the endnode of the arbitrary path is not the root. Hence, the probability that h_j increases equals $\frac{1}{j-1}\left(1 - \frac{1}{j}\right) = \frac{1}{j}$. Indeed, correct attachment in the URT has probability $\frac{1}{j-1}$, while in the resulting URT with j nodes, the root should not be selected as end node which has probability $1 - \frac{1}{j}$. In summary, at stage j, the hopcount h_j of an arbitrary path rooted at 1 only increases by 1 with probability $\frac{1}{j}$ and remains the same as h_{j-1} with probability $1 - \frac{1}{j}$. This observation can be written as $h_j = h_{j-1} + V_j$ where V_j is a Bernoulli random variable defined by $\Pr[V_j = 1] = \frac{1}{j}$ and $\Pr[V_j = 0] = 1 - \frac{1}{j}$. This explains that h_N is, eventually, equal to the sum (16.67) of $N-1$ Bernoulli random variables.

(viii) The complete uniform recursive tree has N nodes. The subtree $T_j^{(N)}$ of the complete uniform recursive tree rooted at the j-th attached node has at least one node, namely the node j itself and at most $N-1$ node, because the j-th attached node is never equal to the root of the complete URT. The subtree $T_j^{(N)}$ contains precisely k nodes, if $k-1$ nodes are attached to the subtree $T_j^{(N)}$ rooted at j, and if the remaining $N-j-(k-1)$ nodes are attached to other subtrees rooted at the other $j-1$ nodes that were attached to the URT prior to the node j. There are precisely $\binom{N-j}{k-1}$ ways to select $k-1$ nodes out of the $N-j$ nodes that still need to be attached after the attachment of the j-th node. Since each new node attachment to the URT is independent from the other attachments and since each new node has equal probability to be attached to any node already in the URT, it is irrelevant that in the mean time nodes are attached to any subtree rooted at the $j-1$ other nodes. The number of ways to attach the $k-1$ nodes to the subtree $T_j^{(N)}$ rooted at j is $1.2. \ldots (k-1) = (k-1)!$. The remaining other $l = N-j-(k-1)$ nodes are attached to any of the subtrees rooted at the $j-1$ other nodes in $(j-1)j(j+1)\ldots(j-1+l-1)$ ways. Similarly, the total number of ways to attach $N-j$ nodes to the URT, just after the j-th node is attached, is $j(j+1)\ldots(j+N-j-1)$. The number of favorable ways to construct a subtree $T_j^{(N)}$ of size k over the total number of possible ways leads, for $k \geq 1$, to

$$\Pr\left[|T_j^{(N)}| = k\right] = \binom{N-j}{k-1}\frac{(k-1)!(j-1)j\cdots(N-k-1)}{j(j+1)\cdots(N-1)}$$

which equals (16.68).

(ix) The main argument why the reasoning is wrong is that uniformity applies to a property of the URT that is independent of its growth or structure. For example, it may be of interest to know the color of a node in the URT if we attach nodes in red or blue. The fact that they are red or blue has nothing in common with the form of the URT. In this derivation, the property of attaching nodes to a subpart of the URT is related to the structure of the URT. The probability of attaching the $(j+1)$-th node to the j-th is $\frac{1}{j}$. But, the probability

$\Pr\left[n_{j+2} \in T_j^{(N)} \middle| n_{j+1} \in T_j^{(N)}\right]$ is not $\frac{2}{j+1}$, but more complicated. The outcome $\frac{2}{j+1}$ holds for any set of two independent nodes, but the $(j+1)$-th and j-th nodes are not independent, they are coupled in the URT by precisely one link.

(x) Using the law of total probability (2.49), we have for the URT that

$$\Pr[B_l = k(N-k)] = \sum_{j=2}^{N} \Pr[B_l = k(N-k) \,|\, l = j] \Pr[l = j], \quad 1 \le k < \left\lfloor \frac{N}{2} \right\rfloor$$

A random link l is the j-th attached link to the URT with probability $\Pr[l = j] = \frac{1}{N-1}$. The conditional probability

$$\Pr[B_l = k(N-k) \,|\, l = j] = \Pr\left[\left|T_j^{(N)}\right| = k\right] + \Pr\left[\left|T_j^{(N)}\right| = N - k\right]$$

because only if the size of the subtree rooted at node j is of size $\left|T_j^{(N)}\right| = k$ or of size $\left|T_j^{(N)}\right| = N - k$, the betweenness of the link $l = j$ equals $k\,(N - k)$. Combining both yields

$$\Pr[B_l = k(N-k)] = \frac{1}{N-1} \sum_{j=2}^{N} \Pr\left[\left|T_j^{(N)}\right| = k\right] + \Pr\left[\left|T_j^{(N)}\right| = N - k\right]$$

Substituting (16.68) gives, denoting $r = \Pr[B_l = k(N - k)]$,

$$r = \frac{(N-k-1)!}{(N-1)\,(N-1)!} \sum_{j=2}^{N} \frac{(j-1)(N-j)!}{(N-j-k+1)!} + \frac{(k-1)!}{(N-1)(N-1)!} \sum_{j=2}^{N} \frac{(j-1)(N-j)!}{(k+1-j)!}$$

We use the identity

$$\sum_{j=n}^{m} j \binom{a-j}{b-j} = n \binom{a+1-n}{b-n} + \binom{a+1-n}{b-1-n} - m \binom{a-m}{b-1-m} - \binom{a+1-m}{b-1-m}$$

and obtain

$$\sum_{j=2}^{N} \frac{(j-1)(N-j)!}{(N-j-k+1)!} = (k-1)! \sum_{j=1}^{N-1} j \binom{N-1-j}{N-k-j} = (k-1)! \left(\binom{N-1}{N-k-1} + \binom{N-1}{N-k-2} \right)$$

$$= (k-1)! \binom{N}{N-k-1} = \frac{(k-1)!N!}{(k+1)!(N-k-1)!}$$

Similarly,

$$\sum_{j=2}^{N} \frac{(j-1)(N-j)!}{(k+1-j)!} = (N-1-k)! \sum_{j=1}^{N-1} j \binom{N-1-j}{k-j} = (N-1-k)! \left(\binom{N-1}{k-1} + \binom{N-1}{k-2} \right)$$

$$= (N-1-k)! \binom{N}{k-1} = \frac{(N-1-k)!N!}{(k-1)!(N-k+1)!}$$

Hence,

$$r = \frac{(N-k-1)!}{(N-1)\,(N-1)!} \frac{(k-1)!N!}{(k+1)!(N-k-1)!} + \frac{(k-1)!}{(N-1)(N-1)!} \frac{(N-1-k)!N!}{(k-1)!(N-k+1)!}$$

$$= \frac{N}{(N-1)} \left(\frac{1}{(k+1)k} + \frac{1}{(N-k+1)(N-k)} \right)$$

$$= \frac{N}{(N-1)\,k(N-k)} \left(\frac{N-k}{k+1} + \frac{k}{N-k+1} \right)$$

(xi) The shortest path tree from a node A to all the other routers is a URT with N nodes. The shortest path tree from A to the m nearest routers is just the sub-URT when only the first m nodes, which are the first m discovered nodes as explained in Section 16.2.1, have been attached. A sub-URT is also a URT. Hence, $E[h]$ is equal to the mean hopcount $E[H_N]$ of a shortest path in a URT with $N = m$ nodes. According to (16.17), we have

$$E[h] = E[H_{N=m}] = \frac{m}{m-1} \sum_{l=2}^{m-1} \frac{1}{l}$$

B.15 Epidemics in networks (Chapter 17)

(i) Applying (17.94) to a regular graph, where $D = r.u$, $\lambda_1 = r$ and $z_1 = \frac{u}{\sqrt{N}}$, so that

$$\lambda_1 E\left[\left(X_\infty^T z_1\right)^2\right] = \frac{r}{N} E\left[\left(X_\infty^T u\right)^2\right] = rN\left(y_\infty^2\left(\tau\right) + \mathrm{Var}\left[S_N\right]\right)$$

we obtain, after some tedious manipulations, the quadratic equation in $y_\infty\left(\tau\right)$,

$$\tau r y_\infty^2\left(\tau\right) + (1 + \varepsilon^* - \tau r)\, y_\infty\left(\tau\right) + \tau\left(r\mathrm{Var}\left[S_N\right] + \frac{1}{N}\sum_{j=2}^N \lambda_j E\left[\left(X_\infty^T z_j\right)^2\right]\right) - \varepsilon^* = 0$$

Confining to the case without self-infection, $\varepsilon^* = 0$, the solution is

$$y_{\infty;\mathrm{regular}}\left(\tau\right) = \frac{\tau r - 1 \pm \sqrt{(\tau r - 1)^2 - 4\tau^2 r\left(r\mathrm{Var}\left[S_N\right] + \frac{1}{N}\sum_{j=2}^N \lambda_j E\left[\left(X_\infty^T z_j\right)^2\right]\right)}}{2\tau r}$$

where the negative sign is meaningful, because we must find that $\lim_{\tau \to \infty} y_\infty\left(\tau\right) = 0$. Since $y_\infty \geq 0$, the negative sign implies that $r\mathrm{Var}[S_N] + \frac{1}{N}\sum_{j=2}^N \lambda_j E\left[\left(X_\infty^T z_j\right)^2\right] \geq 0$ that is not obvious[6] because $\sum_{j=2}^N \lambda_j = -r$. For the complete graph[7], $\lambda_j = -1$ for $j > 1$, it holds that $\sum_{j=2}^N \lambda_j E\left[\left(X_\infty^T z_j\right)^2\right] \leq 0$. Further the discriminant is non-negative,

$$\left(\frac{1}{\tau\sqrt{r}} - \sqrt{r}\right)^2 \geq 4r\mathrm{Var}\left[S_N\right] + \frac{4}{N}\sum_{j=2}^N \lambda_j E\left[\left(X_\infty^T z_j\right)^2\right]$$

which leads to a sharper bound for the threshold point $\tau^* = \frac{1}{r}$ deduced above,

$$\tau \geq \frac{1}{r\left(1 - 2\sqrt{\mathrm{Var}\left[S_N\right] + \frac{1}{Nr}\sum_{j=2}^N \lambda_j E\left[(X_\infty^T z_j)^2\right]}\right)} \geq \frac{1}{r}$$

Assume that $\lim_{\tau \to \infty} X_\infty = u$, which contradicts that the steady-state is the overall-healthy state. Since u is the principal eigenvector in a regular graph, the sum vanishes because eigenvectors are orthogonal. Only in the case when $X_\infty = u$, we find the threshold point (17.22) for the complete graph.

---

[6] Since $\|X_\infty\|_2 \leq N$ and $\|z_j\|_2 = 1$,

$$\frac{1}{N}\sum_{j=2}^N \lambda_j E\left[\left(X_\infty^T z_j\right)^2\right] < \frac{\max_{2 \leq j \leq N} E\left[\left(X_\infty^T z_j\right)^2\right]}{N}\sum_{j=2}^N |\lambda_j| < \sum_{j=1}^N |\lambda_j| - r$$

where $\sum_{j=1}^N |\lambda_j|$ is known as the graph energy (Van Mieghem, 2011, p. 201).

[7] Specifically, the spectral decomposition of the adjacency $A_{K_N} = u.u^T - I$ is $A_{K_N} = \frac{(N-1)}{N}u.u^T - \sum_{j=2}^N z_j z_j^T$ so that $\sum_{j=2}^N z_j z_j^T = I - \frac{1}{N}J$.

(ii) The equation (17.34) for the bipartite graph becomes

$$\frac{dV(t)}{dt} = \beta \operatorname{diag}(1 - v_i(t)) \begin{bmatrix} O_{m \times m} & J_{m \times n} \\ J_{n \times m} & O_{n \times n} \end{bmatrix} \begin{bmatrix} V_{m \times 1} \\ V_{n \times 1} \end{bmatrix}$$

$$- \delta \begin{bmatrix} I_{m \times m} & O_{m \times n} \\ O_{n \times m} & I_{n \times n} \end{bmatrix} \begin{bmatrix} V_{m \times 1} \\ V_{n \times 1} \end{bmatrix}$$

After some manipulations, we find

$$\frac{1}{\beta} \frac{d}{dt} \begin{bmatrix} V_{m \times 1} \\ V_{n \times 1} \end{bmatrix} = \begin{bmatrix} -\frac{1}{\tau} V_{m \times 1} + \operatorname{diag}(1 - v_i)_m J_{m \times n} V_{n \times 1} \\ \operatorname{diag}(1 - v_i)_n J_{n \times m} V_{m \times 1} - \frac{1}{\tau} V_{n \times 1} \end{bmatrix}.$$

With $J_{m \times n} = u_{m \times 1} u_{1 \times n}$, we rewrite

$$\operatorname{diag}(1 - v_i)_m J_{m \times n} V_{n \times 1} = \operatorname{diag}(1 - v_i)_m u_{m \times 1} u_{1 \times n} V_{n \times 1} = (u_{m \times 1} - V_{m \times 1}) u_{1 \times n} V_{n \times 1}$$

With $N_{y_n} = u_{1 \times n} V_{n \times 1}$, the first m rows

$$\frac{1}{\beta} \frac{d}{dt} V_{m \times 1} = -\frac{1}{\tau} V_{m \times 1} + (u_{m \times 1} - V_{m \times 1}) N_{y_n} = -\left(\frac{1}{\tau} + N_{y_n}\right) V_{m \times 1} + N_{y_n} u_{m \times 1}$$

reduce to m identical equations, from which it is tempting to conclude that $v_i = w_m$ for all $1 \le i \le m$ and for all t. However, this assumption is only valid if all initial conditions $v_i(0)$ are the same. Only in that case,

$$\frac{1}{\beta} \frac{dw_m}{dt} = -\left(\frac{1}{\tau} + N_{y_n}\right) w_m + N_{y_n}$$

Similarly for the n last equations, we have with $v_i = w_n$ for all $m + 1 \le i \le N$, that

$$\frac{1}{\beta} \frac{dw_n}{dt} = -\left(\frac{1}{\tau} + N_{y_m}\right) w_n + N_{y_m}$$

With $N_{y_n} = u_{1 \times n} V_{n \times 1} = n w_n$ and $N_{y_m} = u_{1 \times m} V_{m \times 1} = m w_m$, we arrive at

$$\begin{cases} \frac{1}{\beta} \frac{dw_m}{dt} = -\left(\frac{1}{\tau} + n w_n\right) w_m + n w_n \\ \frac{1}{\beta} \frac{dw_n}{dt} = -\left(\frac{1}{\tau} + m w_m\right) w_n + m w_m \end{cases} \tag{B.40}$$

The steady-state obeys

$$\begin{cases} 0 = -\left(\frac{1}{\tau} + n w_{n\infty}\right) w_{m\infty} + n w_{n\infty} \\ 0 = -\left(\frac{1}{\tau} + m w_{m\infty}\right) w_{n\infty} + m w_{m\infty} \end{cases}$$

These equations hold in general for $K_{m,n}$ because the steady-state does not depend on the initial conditions. Substituting $w_{m\infty} = \frac{n w_{n\infty}}{\left(\frac{1}{\tau} + n w_{n\infty}\right)}$ from the first equation into the second, yields (17.96) and, introduced in the first equation, we find (17.97).

(iii) Invoking the Rayleigh inequality (17.100), we find from (17.18) that

$$y_\infty(\tau) \le \frac{\varepsilon^*}{1 + \varepsilon^*} + \frac{\tau \mu_1 E\left[X_\infty^T X_\infty\right]}{(1 + \varepsilon^*) N}$$

Since $E\left[X_\infty^T X_\infty\right] = E\left[\sum_{k=1}^N X_k^2\right] = E\left[\sum_{k=1}^N X_k\right] = N y_\infty(\tau) \le N$, we obtain (17.99). Only when[8] $\tau < \frac{1}{\mu_1} \le \frac{1}{d_{\max} + 1}$, the bound (17.99) can be useful. The upper bound (17.99) is, in general, weak because the largest eigenvector x_1 belonging to μ_1 is orthogonal to u (the eigenvector of Q belonging to $\mu_N = 0$), with all components non-negative, just like X_∞. In other words, X_∞ is closer aligned to u than to x_1 implying that the Rayleigh upper bound in (17.100) is not tight.

---

[8] As shown in Van Mieghem (2011, p. 82), it holds that $\min(N, 2d_{\max}) \ge \mu_1 \ge d_{\max} + 1$.

(iv) *SIR epidemics.* The SIR governing equation for the probability that a node j is infected reads

$$\frac{d\Pr\left[Y_j = I\right]}{dt} = E\left[-\delta 1_{\{Y_j=I\}} + 1_{\{Y_j=S\}}\beta \sum_{k=1}^{N} a_{kj} 1_{\{Y_k=I\}}\right] \qquad (B.41)$$

where the time-dependence of $Y_j(t)$ has been omitted for simplicity. In words, the change in the probability that a node j is infected at time t equals the expectation of (a) the rate β times the number of infected neighbors (specified by the adjacency matrix element a_{kj}), given that node j is susceptible minus (b) the rate δ given that the infected node is cured (and thereafter removed). Next, the dynamic process that removes nodes satisfies

$$\frac{d\Pr\left[Y_j = R\right]}{dt} = E\left[\delta 1_{\{Y_j=I\}}\right] = \delta\Pr\left[Y_j = I\right] \qquad (B.42)$$

which says that the time-derivative of the probability that a node j is removed from the process equals the expectation of the rate δ, given that node j is infected. Finally, a node is either healthy but susceptible, infected, or cured (and removed); in other words, $1_{\{Y_j=S\}} + 1_{\{Y_j=I\}} + 1_{\{Y_j=R\}} = 1$.

The first equation (B.41) is complicating due to the interaction with other infected nodes in the network, but (B.41) is of exactly the same form as the corresponding SIS governing equation (17.10). However, in the SIS process, there are only two nodal states (or compartments) possible so that $1_{\{X_j=S\}} + 1_{\{X_j=I\}} = 1$, which leads to fewer equations than in the SIR process. We proceed by rewriting equation (B.41) using $E\left[1_{\{Y_j=S\}\cap\{Y_k=I\}}\right] = \Pr\left[Y_j = S, Y_k = I\right]$,

$$\frac{d\Pr\left[Y_j = I\right]}{dt} = -\delta\Pr\left[Y_j = I\right] + \beta \sum_{k=1}^{N} a_{kj} \Pr\left[Y_j = S, Y_k = I\right]$$

After invoking the law of total probability (2.49),

$$\Pr\left[Y_k = I\right] = \Pr\left[Y_j = S, Y_k = I\right] + \Pr\left[Y_j = I, Y_k = I\right] + \Pr\left[Y_j = R, Y_k = I\right]$$

the SIR governing equation (B.41) becomes

$$\frac{d\Pr[Y_j = I]}{dt} = \beta \sum_{k=1}^{N} a_{kj}\Pr[Y_k = I] - \delta\Pr[Y_j = I] \qquad (B.43)$$

$$- \beta \sum_{k=1}^{N} a_{kj}\{\Pr[Y_j = I, Y_k = I] + \Pr[Y_j = R, Y_k = I]\}$$

The first two terms on the right-hand side in (B.43) describe the spread of the infection from infected neighbors minus the nodal curing, while the third term excludes infection spread to an infected or removed node j. This last terms grows over time, because (B.42) illustrates that the probability to become removed is non-decreasing over time. Relation (B.43) explains the bell-shape of $\Pr\left[Y_j(t) = I\right]$ as a function of time t: initially the third term is small and near to exponential growth arises from the first and second terms. As the number of removed nodes increases over time, the third term counteracts the initial growth and forces its decline towards extinction (for large t). The SIS differential equation corresponding to (B.43) is (17.11), which shows that the SIR process lower bounds the SIS infection process, i.e. $\Pr[Y_j = I] \leq \Pr[X_j = I]$.

There two ways to proceed from (B.43): either we deduce the governing equations for the two-pair probabilities as in Section 17.3, followed by higher-order joint probabilities until all 2^N SIS and 3^N SIR linear Markov equations are established or we try to "close" the equations (Newman, 2010, p. 653-654), as coined in epidemiology.

(v) Using $j\binom{N}{j} = N\binom{N-1}{j-1}$ into (17.74) yields

$$\frac{y_{\infty;N}(\tau)}{\pi_{0;N}} = \sum_{j=1}^{N}\binom{N-1}{j-1}\tau^{j}\frac{\Gamma\left(\frac{\varepsilon^{*}}{\tau}+j\right)}{\Gamma\left(\frac{\varepsilon^{*}}{\tau}\right)} = \tau\sum_{j=0}^{N-1}\binom{N-1}{j}\tau^{j}\frac{\Gamma\left(\frac{\varepsilon^{*}}{\tau}+j+1\right)}{\Gamma\left(\frac{\varepsilon^{*}}{\tau}\right)}$$

$$= \varepsilon^{*} + \tau\sum_{j=1}^{N-1}\binom{N-1}{j}\tau^{j}\frac{\Gamma\left(\frac{\varepsilon^{*}}{\tau}+j+1\right)}{\Gamma\left(\frac{\varepsilon^{*}}{\tau}\right)}$$

Further, using $\Gamma(z+1) = z\Gamma(z)$ in the last sum gives

$$R = \sum_{j=1}^{N-1}\binom{N-1}{j}\tau^{j}\frac{\left(\frac{\varepsilon^{*}}{\tau}+j\right)\Gamma\left(\frac{\varepsilon^{*}}{\tau}+j\right)}{\Gamma\left(\frac{\varepsilon^{*}}{\tau}\right)}$$

$$= \frac{\varepsilon^{*}}{\tau}\sum_{j=1}^{N-1}\binom{N-1}{j}\tau^{j}\frac{\Gamma\left(\frac{\varepsilon^{*}}{\tau}+j\right)}{\Gamma\left(\frac{\varepsilon^{*}}{\tau}\right)} + \sum_{j=1}^{N-1}j\binom{N-1}{j}\tau^{j}\frac{\Gamma\left(\frac{\varepsilon^{*}}{\tau}+j\right)}{\Gamma\left(\frac{\varepsilon^{*}}{\tau}\right)}$$

We recognize from (17.73) that the first sum equals

$$\sum_{j=1}^{N-1}\binom{N-1}{j}\tau^{j}\frac{\Gamma\left(\frac{\varepsilon^{*}}{\tau}+j\right)}{\Gamma\left(\frac{\varepsilon^{*}}{\tau}\right)} = \frac{1}{\pi_{0;N-1}} - 1$$

while (17.74) indicates that the last sum equals

$$\sum_{j=1}^{N-1}j\binom{N-1}{j}\tau^{j}\frac{\Gamma\left(\frac{\varepsilon^{*}}{\tau}+j\right)}{\Gamma\left(\frac{\varepsilon^{*}}{\tau}\right)} = (N-1)\frac{y_{\infty;N-1}(\tau)}{\pi_{0;N-1}}$$

so that a recursion relation for $\frac{y_{\infty;N}(\tau)}{\pi_{0;N}}$,

$$\frac{y_{\infty;N}(\tau)}{\pi_{0;N}} = \frac{\varepsilon^{*}}{\pi_{0;N-1}} + (N-1)\tau\frac{y_{\infty;N-1}(\tau)}{\pi_{0;N-1}} \tag{B.44}$$

is found. Next, we use the binomial recursion $\binom{N}{k} = \binom{N-1}{k-1} + \binom{N-1}{k}$ in the denominator F_{d} of

$$y_{\infty;N}(\tau) = \frac{\sum_{k=1}^{N}\binom{N-1}{k-1}\tau^{k}\frac{\Gamma\left(\frac{\varepsilon^{*}}{\tau}+k\right)}{\Gamma\left(\frac{\varepsilon^{*}}{\tau}\right)}}{\sum_{k=0}^{N}\binom{N}{k}\tau^{k}\frac{\Gamma\left(\frac{\varepsilon^{*}}{\tau}+k\right)}{\Gamma\left(\frac{\varepsilon^{*}}{\tau}\right)}}$$

and obtain, invoking the recursion (B.44) and (17.73),

$$F_{d} = \sum_{k=0}^{N}\left\{\binom{N-1}{k-1} + \binom{N-1}{k}\right\}\tau^{k}\frac{\Gamma\left(\frac{\varepsilon^{*}}{\tau}+k\right)}{\Gamma\left(\frac{\varepsilon^{*}}{\tau}\right)}$$

$$= \frac{\varepsilon^{*}}{\pi_{0;N-1}} + (N-1)\tau\frac{y_{\infty;N-1}(\tau)}{\pi_{0;N-1}} + \frac{1}{\pi_{0;N-1}}$$

which leads to the recursion (17.75) for $y_{\infty;N}(\tau)$.

(vi) *Scaling of the epidemic threshold in K_{N}.* Let us consider the ratio $\frac{\pi_{j+1}}{\pi_{j}} = \frac{(\tau j+\varepsilon^{*})(N-j)}{j+1}$ of the steady-state probabilities in (11.17). The ratio is maximal for

$$j^{*} = \sqrt{(N+1)\left(1-\frac{\varepsilon^{*}}{\tau}\right)} - 1$$

and

$$\left(\frac{\pi_{j+1}}{\pi_j}\right)_{max} = \left((N+2) - \frac{\varepsilon^*}{\tau} - 2\sqrt{(N+1)\left(1 - \frac{\varepsilon^*}{\tau}\right)}\right)\tau$$

while the requirement $\frac{\pi_{j+1}}{\pi_j} = 1$ yields

$$j_\pm = \frac{1}{2}\left(N - \frac{\varepsilon^* + 1}{\tau}\right) \pm \frac{1}{2}\sqrt{\left(N - \frac{\varepsilon^* + 1}{\tau}\right)^2 + \frac{4(N\varepsilon^* - 1)}{\tau}}$$

If $\varepsilon^* > \frac{1}{N}$, there is only one (non-negative) index for j at which $\frac{\pi_{j+1}}{\pi_j} = 1$. Thus, $\frac{\pi_{j+1}}{\pi_j} < 1$, implying that π_j decreases with j, when $j > j_+ \approx N - \frac{\varepsilon^*+1}{\tau}$. In other words, π_j increases with j when $j < j_+$. The more interesting case appears if $\varepsilon^* < \frac{1}{N}$, then $j_- \approx \frac{1-N\varepsilon^*}{\tau\left(N - \frac{\varepsilon^*+1}{\tau}\right)}$ and π_j decreases with j when $j < j_-$ and $j > j_+$. Now, when $j_- = j_+$ there is only one value for which $\frac{\pi_{j+1}}{\pi_j} = 1$ and this is the maximum value $\left(\frac{\pi_{j+1}}{\pi_j}\right)_{max}$. This means that π_j decreases with j nearly everywhere, except in a small region around $j = j_+ = j_-$. We can consider this region as the onset of the epidemic which may *define* the epidemic threshold of the SIS epidemics (for negligibly small ε^*). Hence, if we choose $\left(\frac{\pi_{i+1}}{\pi_i}\right)_{max} = 1$ and let $\varepsilon^* = 0$, then we arrive at the scaling for this *defined* epidemic threshold (17.101). We observe that $\tau_c^* > \tau_c^{(1)} = \frac{1}{N-1} = \frac{1}{N}\left(1 + \frac{1}{N} + O\left(\frac{1}{N^2}\right)\right)$, in line with the fact that NIMFA upper bounds the viral node probability (Theorem 17.4.1).

(vii) With $\prod_{j=1}^i (N - j) = N^i \exp\left(\sum_{j=1}^i \log(1 - \frac{j}{N})\right)$ and

$$\prod_{j=1}^i (N-j) = N^i \exp\left(-\sum_{j=1}^i \frac{j}{N} + O(i^3 N^{-2})\right) = N^i \exp\left(-\frac{i^2}{2N} + O(iN^{-1}) + O(i^3 N^{-2})\right)$$

we replace the binomial in (17.72),

$$\pi_j = \frac{\pi_0}{j!}\varepsilon^* \tau^{j-1} \frac{\Gamma\left(\frac{\varepsilon^*}{\tau} + j\right)}{\Gamma\left(\frac{\varepsilon^*}{\tau} + 1\right)} N \prod_{m=1}^{j-1}(N - m)$$

$$= \frac{\pi_0}{j!}\frac{\varepsilon^*}{\tau}\frac{\Gamma\left(\frac{\varepsilon^*}{\tau} + j\right)}{\Gamma\left(\frac{\varepsilon^*}{\tau} + 1\right)}(\tau N)^j \exp\left(-\frac{j^2}{2N} + O(jN^{-1}) + O(j^3 N^{-2})\right)$$

$$= \pi_0 \frac{\varepsilon^*}{\tau}\frac{\Gamma\left(\frac{\varepsilon^*}{\tau} + j\right)}{j!\Gamma\left(\frac{\varepsilon^*}{\tau} + 1\right)} \exp\left(-\frac{j^2}{2N} + j\log(\tau N)\right)(1 + O(jN^{-1} + j^3 N^{-2}))$$

and

$$\pi_j = \pi_0 \frac{\varepsilon^*}{\tau}\frac{\Gamma\left(\frac{\varepsilon^*}{\tau} + j\right)}{j!\Gamma\left(\frac{\varepsilon^*}{\tau} + 1\right)} e^{\frac{N}{2}(\log \tau N)^2} \exp\left(-\frac{1}{2}\frac{(j - N\log \tau N)^2}{N}\right)(1 + O(jN^{-1} + j^3 N^{-2}))$$

For large N and small $\frac{\varepsilon^*}{\tau}$ so that $\frac{\Gamma\left(\frac{\varepsilon^*}{\tau} + j\right)}{j!\Gamma\left(\frac{\varepsilon^*}{\tau} + 1\right)} \approx \frac{1}{j}$, the appropriate scaling for τ is found by requiring that $N(\log \tau N)^2 = C^2$, where C is independent of N. Hence,

$$\tau = \frac{1}{N}\exp\left(\frac{C}{\sqrt{N}}\right) = \frac{1}{N}\left\{1 + \frac{C}{\sqrt{N}} + O(N^{-1})\right\}$$

and

$$\pi_j \simeq \pi_0 \frac{\varepsilon^*}{j\tau} e^{\frac{1}{2}C^2} \exp\left(-\frac{1}{2}\frac{(j - C\sqrt{N})^2}{N}\right)(1 + O(jN^{-1} + j^3 N^{-2}))$$

illustrating that, asymptotically, π_j resembles approximately a Gaussian with mean $C\sqrt{N}$ and standard deviation $\sigma = \sqrt{N}$, in agreement with Fig. 17.5.

B.16 The efficiency of multicast (Chapter 18)

(i) Using (18.24), we obtain

$$
\begin{aligned}
g_{N,k}(2) &= N - 1 - \sum_{j=0}^{D-1} k^{D-j} \frac{\left(N - 1 - \frac{k^{j+1}-1}{k-1}\right)\left(N - 2 - \frac{k^{j+1}-1}{k-1}\right)}{(N-1)(N-2)} \\
&= \frac{(2N-3)DN}{(N-1)(N-2)} + \frac{(2N-3)D}{(N-1)(N-2)(k-1)} - \frac{(2N-3)}{(N-2)(k-1)} + \\
&\quad - \frac{N(N-1-2D)}{(N-1)(N-2)(k-1)} - \frac{(N-1-2D)}{(N-1)(N-2)(k-1)^2} - \frac{1}{(N-2)(k-1)^2}
\end{aligned}
$$

or, for large N,

$$
g_{N,k}(2) \sim 2D - \frac{3}{k-1} + O\left(\frac{\log_k N}{N}\right)
$$

the effective power exponent $\beta^*(N)$ as defined in (18.31), equals for the k-ary tree and large N,

$$
\begin{aligned}
\beta^*(N) &\sim \frac{\log\left[\frac{2D - \frac{3}{k-1}}{D - \frac{1}{k-1}}\right]}{\log 2} = 1 + \log_2\left[1 - \frac{1}{2(k-1)\left(\log_k N + \log_k(1-1/k) - \frac{1}{k-1}\right)}\right] \\
&= 1 + \log_2\left[1 - \frac{1}{2(k-1)E[H_N]}\right] \sim 1 - \frac{1}{(\log 4)(k-1)E[H_N]}
\end{aligned}
$$

which shows, for large N, that $\beta^*(N) < 1$, but that $\beta^*(N) \to 1$ if $k \to \infty$.

(ii) The largest individual delay $V_N(m)$ among m uniformly distributed multicast members equals the sum of the interattachement times τ_j until the last multicast member is attached to the URT. Analogous to the derivation of W_N in Section 16.4, the pgf of $V_N(m)$ is

$$
\varphi_{V_N(m)}(z) = \sum_{k=1}^{N-1} E\left[e^{-zv_k}\right] \Pr[\text{the } m\text{-th member is the } k\text{-th attached node in URT}]
$$

where $E\left[e^{-zv_k}\right]$ is specified in (16.19) and the m-th multicast member is the last member reached by the source, or equivalently, the last multicast member attached to the URT. Further,

$$
\Pr[\text{the } m\text{-th member is the } k\text{-th attached node in URT}] = \frac{\binom{1}{1}\binom{k-1}{m-1}}{\binom{N-1}{m}}
$$

because, if the m-th multicast member is attached at time v_k, all $m-1$ other members are already attached to the URT. Since they are uniformly distributed over the nodes in the graph, there are precisely $\binom{k-1}{m-1}$ possible ways in which these earlier members are attached to $k-1$ nodes – the root with label 0 and the k-th node are excluded – of the URT of size $k+1$. Hence, we arrive at

$$
\varphi_{V_N(m)}(z) = \frac{1}{\binom{N-1}{m}} \sum_{k=m}^{N-1} \prod_{n=1}^{k} \frac{n(N-n)}{z + n(N-n)} \binom{k-1}{m-1}
$$

We note that $\varphi_{V_N(m)}(0) = 1$ because $\sum_{k=m}^{N-1} \binom{k-1}{m-1} = \binom{N-1}{m}$. As in Section 16.4, the mean of the largest individual delay is

$$E\left[V_N\left(m\right)\right] = \frac{1}{\binom{N-1}{m}} \sum_{k=1}^{N-1} \sum_{n=1}^{k} \frac{1}{n(N-n)} \binom{k-1}{m-1} = \frac{1}{\binom{N-1}{m}} \sum_{n=1}^{N-1} \frac{1}{n(N-n)} \sum_{k=n}^{N-1} \binom{k-1}{m-1}$$

Since $\binom{k-1}{m-1} = 0$ if $k < m$, we split the sum into two parts

$$E\left[V_N\left(m\right)\right] = \frac{1}{\binom{N-1}{m}} \sum_{n=1}^{m} \frac{1}{n(N-n)} \sum_{k=m}^{N-1} \binom{k-1}{m-1} + \frac{1}{\binom{N-1}{m}} \sum_{n=1+m}^{N-1} \frac{1}{n(N-n)} \sum_{k=n}^{N-1} \binom{k-1}{m-1}$$

Using

$$\sum_{k=n}^{N-1} \binom{k-1}{m-1} = \sum_{k=m}^{N-1} \binom{k-1}{m-1} - \sum_{k=m}^{n-1} \binom{k-1}{m-1} = \binom{N-1}{m} - \binom{n-1}{m}$$

we obtain

$$E\left[V_N\left(m\right)\right] = \sum_{n=1}^{m} \frac{1}{n(N-n)} + \frac{1}{\binom{N-1}{m}} \sum_{n=m+1}^{N-1} \frac{\binom{N-1}{m} - \binom{n-1}{m}}{n(N-n)}$$

$$= \sum_{n=1}^{N-1} \frac{1}{n(N-n)} - \frac{1}{\binom{N-1}{m}} \sum_{n=m+1}^{N-1} \frac{\binom{n-1}{m}}{n\left(N-n\right)}$$

We recognize with (16.34) that the first term is equal to the mean of the flooding time T_N, which also equals $E\left[V_N\left(N-1\right)\right]$.

References

Abramowitz, M. and Stegun, I. A. (1968). *Handbook of Mathematical Functions.* (Dover Publications, Inc., New York).

Addario-Berry, L., Broutin, N., Goldschmidt, C., and Miermont, G. (2013). The scaling limit of the minimum spanning tree of the complete graph. *arXiv:1301.1664v1.*

Addario-Berry, L., Broutin, N., and Reed, B. (2009). Critical random graphs and the structure of a minimum spanning tree. *Random Structures and Algorithms* 35, 3 (October), 323–347.

Aldous, D. (2001). The $\zeta(2)$ limit in the random assignment problem. *Random Structures and Algorithms* 18, 4, 381–418.

Allen, A. O. (1978). *Probability, Statistics, and Queueing Theory.* Computer Science and Applied Mathematics, (Academic Press, Inc., Orlando).

Almkvist, G. and Berndt, B. C. (1988). Gauss, Landen, Ramanuyan, the Arithmic-Geometric Mean, Ellipses, π and the Ladies Diary. *American Mathematical Monthly* 95, 585–608.

Anderson, R. M. and May, R. M. (1991). *Infectious Diseases of Humans: Dynamics and Control.* (Oxford University Press, Oxford, U.K.).

Anick, D., Mitra, D., and Sondhi, M. M. (1982). Stochastic theory of a data-handling system with multiple sources. *The Bell System Technical Journal* 61, 8 (October), 1871–1894.

Anupindi, R., Chopra, S., Deshmukh, S. D., Van Mieghem, J. A., and Zemel, E. (2006). *Managing Business Flows. Principles of Operations Management*, 2nd edn. (Prentice Hall, Upper Saddle River).

Ashcroft, N. W. and Mermin, N. D. (1981). *Solid State Physics.* (Holt-Saunders International Editions, Tokyo).

Athreya, K. B. and Ney, P. E. (1972). *Branching Processes.* (Springer-Verlag, Berlin).

Bailey, N. T. J. (1975). *The Mathematical Theory of Infectious Diseases and its Applications*, 2nd edn. (Charlin Griffin & Company, London).

Bak, P. (1996). *How Nature Works: The Science of Self-organized Criticality.* (Copernicus, Springer-Verlag, New York).

Barabási, A. L. (2002). *Linked, The New Science of Networks.* (Perseus, Cambridge, MA).

Barabási, A. L. and Albert, R. (1999). Emergence of scaling in random networks. *Science* 286, 509–512.

Baran, P. (2002). The beginnings of packet switching - some underlying concepts: The Franklin Institute and Drexel University seminar on the evolution of packet switching and the Internet. *IEEE Communications Magazine*, 2–8.

Barrat, A., Bartelemy, M., and Vespignani, A. (2008). *Dynamical Processes on Complex Networks.* (Cambridge University Press, Cambridge, U.K.).

Berger, M. A. (1993). *An Introduction to Probabiliy and Stochastic Processes.* (Springer-Verlag, New York).

663

Bertsekas, D. and Gallager, R. (1992). *Data Networks*, 2nd edn. (Prentice-Hall International Editions, London).

Bhamidi, S., van der Hofstad, R., and Hooghiemstra, G. (2012). Universality for first passage percolation on sparse random graphs. *arXiv:1210.6839*.

Biggs, N. (1996). *Algebraic Graph Theory*, 2nd edn. (Cambridge University Press, Cambridge, U.K.).

Billingsley, P. (1995). *Probability and Measure*, 3rd edn. (John Wiley & Sons, New York).

Bisdikian, C., Lew, J. S., and Tantawi, A. N. (1992). On the tail approximation of the blocking probability of single server queues with finite buffer capacity. *Queueing Networks with Finite Capacity, Proc. 2nd Int. Conf.*, 267–280.

Bollobas, B. (2001). *Random Graphs*, 2nd edn. (Cambridge University Press, Cambridge).

Bollobas, B. and Riordan, O. (2001). Mathematical results on scale-free random graphs. *Handbook of Graphs and Networks, ed. S. Bornholdt and H. G. Schuster; Wiley-VCH*, 1–34.

Borovkov, A. A. (1976). *Stochastic Processes in Queueing Theory*. (Springer-Verlag, New York).

Boyd, S. and Vandenberghe, L. (2004). *Convex Optimization*. (Cambridge University Press, New York).

Breda, D., Diekmann, O., de Graaf, W. F., Pugliese, A., and Vermiglio, R. (2012). On the formulation of epidemic models (an appraisal of Kermack and McKendrick). *Journal of Biological Dynamics* 6, Supplement 2, 103–117.

Brockmeyer, E., Halstrom, H. L., and Jensen, A. (1948). *The Life and Works of A. K. Erlang*. (Academy of Technical Sciences, Copenhagen).

Buldyrev, S. V., Parshani, R., Paul, G., Stanley, H. E., and Havlin, S. (2010). Catastrophic cascade of failures in interdependent networks. *Nature Letters* 464, 1025–1028.

Carlitz, L. (1963). The inverse of the error function. *Pacific Journal of Mathematics* 13, 459–470.

Cator, E., van de Bovenkamp, R., and Van Mieghem, P. (2013). Susceptible-Infected-Susceptible epidemics on networks with general infection and curing times. *Physical Review E* 87, 6 (June), 062816.

Cator, E. and Van Mieghem, P. (2012). Second order mean-field SIS epidemic threshold. *Physical Review E* 85, 5 (May), 056111.

Cator, E. and Van Mieghem, P. (2013a). Nodal infection in Markovian SIS and SIR epidemics on networks are non-negatively correlated.

Cator, E. and Van Mieghem, P. (2013b). Susceptible-Infected-Susceptible epidemics on the complete graph and the star graph: Exact analysis. *Physical Review E* 87, 1 (January), 012811.

Chakrabarti, D., Wang, Y., Wang, C., Leskovec, J., and Faloutsos, C. (2008). Epidemic thresholds in real networks. *ACM Transactions on Information and System Security (TISSEC)* 10, 4, 1–26.

Chalmers, R. C. and Almeroth, K. C. (2001). Modeling the branching characteristics and efficiency gains in global multicast trees. *IEEE INFOCOM2001*.

Chatterjee, S. and Durrett, R. (2009). Contact process on random graphs with degree power-law distribution have critical value zero. *Annals of Probability* 37, 2332–2356.

Chen, L. Y. (1975). Poisson approximation for dependent trials. *The Annals of Probability* 3, 3, 534–545.

Chen, W.-K. (1971). *Applied Graph Theory*. (North-Holland Publishing Company, Amsterdam).

Chen, Y., Paul, G., Havlin, S., Liljeros, F., and Stanley, H. E. (2008). Finding a better immunization strategy. *Physical Review Letters* 101, 058701.

Chuang, J. and Sirbu, M. A. (1998). Pricing multicast communication: A cost-based approach. *Proceedings of the INET'98*.

Clauset, A., Shalizi, C. R., and Newman, M. E. J. (2009). Power-law distributions in empirical data. *SIAM Review* 51, 4, 661–703.

Cohen, J. W. (1969). *The Single Server Queue*. (North-Holland Publishing Company, Amsterdam).

Cohen, R. and Havlin, S. (2010). *Complex Networks: Structure, Robustness and Function*. (Cambridge University Press, Cambridge, U.K.).

Cohen-Tannoudji, C., Diu, B., and Laloë, F. (1977). *Mécanique Quantique*. Vol. I and II. (Hermann, Paris).

Comtet, L. (1974). *Advanced Combinatorics*, revised and enlarged edn. (D. Riedel Publishing Company, Dordrecht, Holland).

Coppersmith, D. and Sorkin, G. B. (1999). Constructive bounds and exact expectations for the random assignment problem. *Random Structures and Algorithms* 15, 2, 113–144.

Cormen, T. H., Leiserson, C. E., and Rivest, R. L. (1991). *An Introduction to Algorithms*. (MIT Press, Boston).

Courant, R. and Hilbert, D. (1953a). *Methods of Mathematical Physics*, first English edition, translated and revised from the German original of Springer in 1937 edn. Wiley Classic Library, vol. II. (Interscience, New York).

Courant, R. and Hilbert, D. (1953b). *Methods of Mathematical Physics*, first English edition, translated and revised from the German original of Springer in 1937 edn. Wiley Classic Library, vol. I. (Interscience, New York).

Cover, T. M. and Thomas, J. A. (1991). *Elements of Information Theory*. Wiley Series in Telecommunications, (John Wiley & Sons, New York).

Cramér, H. (1999). *Mathematical Methods of Statistics*, 19th edn. (Princeton Landmarks in Mathematics, Princeton, N.J.).

Crow, E. L. and Shimizu, K. (1988). *Lognormal distributions, Theory and Applications*. (Marcel Dekker, Inc., New York).

Cvetković, D. M., Doob, M., and Sachs, H. (1995). *Spectra of Graphs, Theory and Applications*, third edn. (Johann Ambrosius Barth Verlag, Heidelberg).

Da F. Costa, L., Rodrigues, F. A., Travieso, G., and Boas, P. R. V. (2007). Characterization of complex networks: A survey of measurements. *Advances in Physics* 56, 1 (Februari), 167–242.

Daley, D. J. and Gani, J. (1999). *Epidemic modelling: An Introduction*. (Cambridge University Press, Cambridge, U.K.).

Darabi Sahneh, F. and Scoglio, C. (2011). Epidemic spread in human networks. *50th IEEE Conference on Decision and Control, Orlando, FL, USA; also on arXiv:1107.2464v1*.

Darabi Sahneh, F., Scoglio, C., and Van Mieghem, P. (2013). Generalized epidemic mean-field model for spreading processes over multi-layer complex networks. *IEEE/ACM Transaction on Networking* 21, 5 (October), 1609–1620.

Dehmer, M. and Emmert-Streib, F. (2009). *Analysis of Complex Networks*. (Wiley-VCH Verlag GmbH& Co., Weinheim).

Diekmann, O., Heesterbeek, H., and Britton, T. (2012). *Mathematical Tools for Understanding Infectious Disease Dynamics*. (Princeton University Press, Princeton, USA).

Doerr, C., Blenn, N., and Van Mieghem, P. (2013). Lognormal infection times of online information spread. *PLoS ONE* 8, 5 (May), e64349.

Dorogovtsev, S. N. and Mendes, J. F. F. (2003). *Evolution of Networks, From Biological Nets to the Internet and WWW*. (Oxford University Press, Oxford).

Draief, M. and Massoulié, L. (2010). *Epidemics and Rumours in Complex Networks*. London Mathematical Society Lecture Node Series: 369, (Cambridge University Press, Cambridge, UK).

Embrechts, P., Klüppelberg, C., and Mikosch, T. (2001a). *Modelling Extremal Events for Insurance and Finance*, 3rd edn. (Springer-Verlag, Berlin).

Embrechts, P., McNeil, A., and Straumann, D. (2001b). *Correlation and Dependence in Risk Management: Properties and Pitfalls*. Risk Management: Value at Risk and Beyond, ed. M. Dempster and H. K. Moffatt, (Cambridge University Press, Cambridge, UK).

Erdős, P. and Rényi, A. (1959). On random graphs. *Publicationes Mathematicae Debrecen* 6, 290–297.

Erdős, P. and Rényi, A. (1960). On the evolution of random graphs. *Magyar Tud. Akad. Mat. Kutato Int. Kozl.* 5, 17–61.

Estrada, E. (2012). *The Structure of Complex Networks.* (Oxford University Press, Oxford, U.K.).

Faloutsos, M., Faloutsos, P., and Faloutsos, C. (1999). On power-law relationships of the Internet Topology. *Proceedings of ACM SIGCOMM'99, Cambridge, MA*, 251–262.

Feld, S. L. (1991). Why your friends have more friends than you do. *American Journal of Sociology* 96, 6 (May), 1464–1477.

Feller, W. (1970). *An Introduction to Probability Theory and Its Applications*, 3rd edn. Vol. 1. (John Wiley & Sons, New York).

Feller, W. (1971). *An Introduction to Probability Theory and Its Applications*, 2nd edn. Vol. 2. (John Wiley & Sons, New York).

Fetter, A. L. and Walecka, J. D. (1971). *Quantum Theory of Many-particle Systems.* (McGraw-Hill, San Francisco).

Floyd, S. and Paxson, V. (2001). Difficulties in simulating the Internet. *IEEE Transactions on Networking* 9, 4 (August), 392–403.

Fortz, B. and Thorup, M. (2000). Internet traffic engineering by optimizing OSPF weights. *IEEE INFOCOM2000.*

Frieze, A. M. (1985). On the value of a random minimum spanning tree problem. *Discrete Applied Mathematics* 10, 47–56.

Gallager, R. G. (1996). *Discrete Stochastic Processes.* (Kluwer Academic Publishers, Boston).

Ganesh, A., Massoulié, L., and Towsley, D. (2005). The effect of network topology on the spread of epidemics. *IEEE INFOCOM2005.*

Gantmacher, F. R. (1959a). *The Theory of Matrices.* Vol. I. (Chelsea Publishing Company, New York).

Gantmacher, F. R. (1959b). *The Theory of Matrices.* Vol. II. (Chelsea Publishing Company, New York).

Gauss, C. F. (1809). *Theoria motus corporum coelestium in sectionibus conicis solem ambientium.* (Hamburgi sumtibus Frid. Perthes et I. H. Besser).

Gauss, C. F. (1821). Theoria combinationis observationum erroribus minimus obnoxiae. Pars prior. *Gauss Werke* 4, 3–26.

Gibrat, R. (1930). Une loi des répartitions economique: l'effect proportionnel. *Bulletin de la Statistique Générale de la France* 19, 469–514.

Gilbert, E. N. (1956). Enumeration of labelled graphs. *Canadian Journal of Mathematics* 8, 405–411.

Gnedenko, B. V. and Kovalenko, I. N. (1989). *Introduction to Queuing Theory*, 2nd edn. (Birkhauser, Boston).

Golub, G. H. and Van Loan, C. F. (1996). *Matrix Computations*, 3rd edn. (The John Hopkins University Press, Baltimore).

Goulden, I. P. and Jackson, D. M. (1983). *Combinatorial Enumeration.* (John Wiley & Sons, New York).

Gourdin, E., Omic, J., and Mieghem, P. V. (2011). Optimization of network protection against virus spread. *8th International Workshop on Design of Reliable Communication Networks (DRCN 2011), Krakow, Poland.*

Grimmett, G. R. (1989). *Percolation.* (Springer-Verlag, New York).

Grimmett, G. R. and Stirzacker, D. (2001). *Probability and Random Processes*, 3rd edn. (Oxford University Press, Oxford).

Guo, D., Trajanovski, S., van de Bovenkamp, R., Wang, H., and Van Mieghem, P. (2013). Epidemic threshold and topological structure of Susceptible-Infectious-Susceptible epidemics in adaptive networks. *Physical Review E* 88, 4 (October), 042802.

Haccou, P., Jagers, P., and Vatutin, V. A. (2005). *Branching processes: Variation, Growth and Extinction of Populations.* (Cambridge University Press, Cambridge, UK).

Hardy, G. H. (1948). *Divergent Series.* (Oxford University Press, London).

Hardy, G. H. (1978). *Ramanujan*, 3rd edn. (Chelsea Publishing Company, New York).

Hardy, G. H., Littlewood, J. E., and Polya, G. (1999). *Inequalities*, 2nd edn. (Cambridge University Press, Cambridge, UK).

Hardy, G. H. and Wright, E. M. (1968). *An Introduction to the Theory of Numbers*, 4th edn. (Oxford University Press, London).

Harris, T. E. (1963). *The Theory of Branching Processes*. (Springer-Verlag, Berlin).

Harrison, J. M. (1990). *Brownian Motion and Stochastic Flow Systems*. (Krieger Publishing Company, Malabar, Florida).

Hooghiemstra, G. and Koole, G. (2000). On the convergence of the power series algorithm. *Performance Evaluation* 42, 21–39.

Hooghiemstra, G. and Van Mieghem, P. (2001). Delay distributions on fixed Internet paths. Delft University of Technology, Report20011020 (www.nas.ewi.tudelft.nl/people/Piet/TUDelftReports).

Hooghiemstra, G. and Van Mieghem, P. (2005). On the mean distance in scale free graphs. *Methodology and Computing in Applied Probability (MCAP)* 7, 285–306.

Hooghiemstra, G. and Van Mieghem, P. (2008). The weight and hopcount of the shortest path in the complete graph with exponential weights. *Combinatorics, Probability and Computing* 17, 4, 537–548.

Horn, R. A. and Johnson, C. R. (1991). *Topics in Matrix Analysis*. (Cambridge University Press, Cambridge, U.K.).

Iribarren, J. L. and Moro, E. (2011). Branching dynamics of viral information spreading. *Physical Review E* 84, 04116.

Jamin, S., C. Jin, A. R. Kurc, D. R., and Shavitt, Y. (2001). Constrained mirror placement on the Internet. *IEEE INFOCOM2001*.

Janic, M., Kuipers, F. A., Zhou, X., and Van Mieghem, P. (2002). Implications for QoS provisioning based on traceroute measurements. *Proceedings of 3rd International Workshop on Quality of Future Internet Services, QofIS2002, ed. B. Stiller et al., Zurich, Switzerland; Springer Verlag LNCS 2511*, 3–14.

Janson, S. (1995). The minimal spanning tree in a complete graph and a functional limit theorem for trees in a random graph. *Random Structures and Algorithms* 7, 4 (December), 337–356.

Janson, S. (2002). On concentration of probability. *Contemporary Combinatorics; ed. B. Bollobas, Bolyai Soc. Math. Stud. 10, Janos Bolyai Mathematical Society, Budapest*, 289–301.

Janson, S., Knuth, D. E., Luczak, T., and Pittel, B. (1993). The birth of the giant component. *Random Structures and Algorithms* 4, 3, 233–358.

Karlin, S. and McGregor, J. (1959). Random walks. *Illinois Journal of Mathematics* 3, 1, 66–81.

Karlin, S. and Taylor, H. M. (1975). *A First Course in Stochastic Processes*, 2nd edn. (Academic Press, San Diego).

Karlin, S. and Taylor, H. M. (1981). *A Second Course in Stochastic Processes*. (Academic Press, San Diego).

Karrer, B. and Newman, M. E. J. (2010). A message passing approach for general epidemic models,. *Physical Review E* 82, 016101.

Keeling, M. J. and Rohani, P. (2008). *Modeling Infectious diseases in Humans and Animals*. (Princeton University Press, Princeton, USA).

Kelly, F. P. (1991). Special invited paper: Loss networks. *The Annals of Applied Probability* 1, 3, 319–378.

Kendall, D. G. (1948). On the generalized birth-and-death process. *Annals of Mathematical Statistics* 19, 1, 1–15.

Kermack, W. O. and McKendrick, A. G. (1927). A contribution to the mathematical theory of epidemics. *Proceedings of the Royal Society London, A* 115, 700–721.

Kleinrock, L. (1975). *Queueing Systems*. Vol. 1 – Theory. (John Wiley and Sons, New York).

Kleinrock, L. (1976). *Queueing Systems*. Vol. 2 – Computer Applications. (John Wiley and Sons, New York).

Kooij, R., Schumm, P., Scoglio, C., and Youssef, M. (2009). A new metric for robustness with respect to virus spread. *Networking 2009, LNCS 5550*, 562–572.

Krishnan, P., Raz, D., and Shavitt, Y. (2000). The cache location problem. *IEEE/ACM Transactions on Networking* 8, 5 (October), 586–582.

Kuipers, F. A. and Van Mieghem, P. (2003). The impact of correlated link weights on QoS routing. *IEEE INFOCOM2003*.

Lanczos, C. (1988). *Applied Analysis*. (Dover Publications, Inc., New York).

Langville, A. N. and Meyer, C. D. (2005). Deeper inside PageRank. *Internet Mathematics* 1, 3 (Februari), 335–380.

Leadbetter, M. R., Lindgren, G., and Rootzen, H. (1983). *Extremes and Related Properties of Random Sequences and Processes*. (Springer-Verlag, New York).

Lehmann, E. L. (1999). *Elements of Large-Sample Theory*. (Springer-Verlag, New York).

Leon-Garcia, A. (1994). *Probability and Random Processes for Electrical Engineering*, 2nd edn. (Addison-Wesley, Reading, Massachusetts).

Li, C., van de Bovenkamp, R., and Van Mieghem, P. (2012). Susceptible-infected-susceptible model: A comparison N-intertwined and heterogeneous mean-field approximations. *Physical Review E* 86, 2 (August), 026116.

Linusson, S. and Wästlund, J. (2004). A proof of Parisi's conjecture on the random assignment problem. *Probability Theory and Related Fields* 128, 419–440.

Lovász, L. (1993). Random walks on graphs: A survey. *Combinatorics* 2, 1–46.

Mandelbrot, B. (1977). *Fractal Geometry of Nature*. (W. H. Freeman, New York).

Markushevich, A. I. (1985). *Theory of Functions of a Complex Variable*. Vol. I – III. (Chelsea Publishing Company, New York).

Marlow, N. A. (1967). A normal limit theorem for power sums of normal random variables. *The Bell System Technical Journal* 46, 9 (November), 2081–2089.

Meyer, C. D. (2000). *Matrix Analysis and Applied Linear Algebra*. (Society for Industrial and Applied Mathematics (SIAM), Philadelphia).

Mitra, D. (1988). Stochastic theory of a fluid model of producers and consumers coupled by a buffer. *Advances in Applied Probability* 20, 646–676.

Molloy, M. and Reed, B. (1995). A critical point for random graphs with a given degree sequence. *Random Structures and Algorithms* 6, 2 and 3, 161– 179.

Morse, P. M. and Feshbach, H. (1978). *Methods of Theoretical Physics*. (McGraw-Hill Book Company, New York).

Mountford, T., Mourrat, J.-C., Valesin, D., and Yao, Q. (2013). Exponential extinction time of the contact process on finite graphs. *arXiv:1203.2972v1*.

Nair, C., Prabhakar, B., and Sharma, M. (2005). Proofs of the Parisi and Coppersmith-Sorkin random assignment conjectures. *Random Structures and Algorithms* 27, 4, 413–444.

Nelsen, R. B. (2006). *An Introduction to Copulas*, 2nd edn. (Springer, New York).

Neuts, M. F. (1989). *Structured Stochastic Matrices of the M/G/1 Type and Their Applications*. (Marcel Dekker Inc., New York).

Newman, M. E. J. (2002). The spread of epidemic disease on networks. *Physical Review E* 66, 016128.

Newman, M. E. J. (2003). Mixing patterns in networks. *Physical Review E* 67, 026126.

Newman, M. E. J. (2006). Modularity and community structure in networks. *Proceedings of the National Academy of Sciences of the United States of America (PNAS)* 103, 23 (June), 8577–8582.

Newman, M. E. J. (2010). *Networks: An Introduction*. (Oxford University Press, Oxford, U. K.).

Newman, M. E. J. and Girvan, M. (2004). Finding and evaluating community structure in networks. *Physical Review E* 69, 026113.

Newman, M. E. J., Strogatz, S. H., and Watts, D. J. (2001). Random graphs with arbitrary degree distributions and their applications. *Physical Review E* 64, 026118.

Norros, I. (1994). A storage model with self-similar input. *Queueing Systems* 16, 3-4, 387–396.

Omic, J., Martin Hernandez, J., and Van Mieghem, P. (2010). Network protection against worms and cascading failures using modularity partitioning. *22nd International Tele-traffic Congress (ITC 22), Amsterdam, Netherlands.*

Omic, J., Van Mieghem, P., and Orda, A. (2009). Game theory and computer viruses. *IEEE Infocom2009.*

Papoulis, A. and Unnikrishna Pillai, S. (2002). *Probability, Random Variables, and Stochastic Processes,* 4th edn. (McGraw-Hill, Boston).

Parisi, G. (1998). A conjecture on random bipartite matching. *arXiv:cond-mat/9801176.*

Pascal, B. (1954). *Oeuvres Completes.* Bibliothèque de la Pléade, (Gallimard, Paris).

Pastor-Satorras, R., Castellano, C., Van Mieghem, P., and Vespignani, A. (2014). Epidemic processes in complex networks. *Review of Modern Physics.*

Pastor-Satorras, R. and Vespignani, A. (2001). Epidemic dynamics and endemic states in complex networks. *Physical Review E* 63, 066117.

Paxson, V. (1997). End-to-end Routing Behavior in the Internet. *IEEE/ACM Transactions on Networking* 5, 5 (October), 601–615.

Phillips, G., Schenker, S., and Tangmunarunkit, H. (1999). Scaling of multicast trees: Comments on the chuang-sirbu scaling law. *ACM Sigcomm99.*

Pietronero, L. and Schneider, W. (1990). Invasion percolation as a fractal growth problem. *Physica A* 170, 81–104.

Press, W. H., Teukolsky, S. A., Vetterling, W. T., and Flannery, B. P. (1992). *Numerical Recipes in C,* 2nd edn. (Cambridge University Press, New York).

Radicchi, F., Fortunato, S., and Castellano, C. (2008). Universality of citation distributions: Toward an objective measure of scientific impact. *Proceedings of the National Academy of Science of the USA (PNAS)* 105, 45, 17268–17272.

Rainville, E. D. (1960). *Special Functions.* (Chelsea Publishing Company, New York).

Riordan, J. (1968). *Combinatorial Identities.* (John Wiley & Sons, New York).

Roberts, J. W. (1991). *Performance Evaluation and Design of Multiservice Networks.* Information Technologies and Sciences, vol. COST 224. (Commission of the European Communities, Luxembourg).

Robinson, S. (2004). The prize of anarchy. *SIAM News* 37, 5 (June), 1–4.

Ross, S. M. (1996). *Stochastic Processes,* 2nd edn. (John Wiley & Sons, New York).

Royden, H. L. (1988). *Real Analysis,* 3rd edn. (Macmillan Publishing Company, New York).

Sansone, G. and Gerretsen, J. (1960). *Lectures on the Theory of Functions of a Complex Variable.* Vol. 1 and 2. (P. Noordhoff, Groningen).

Schoutens, W. (2000). *Stochastic Processes and Orthogonal Polynomials.* (Springer-Verlag, New York).

Shockley, W. (1957). On the statistics of individual variations of productivity in research laboratories. *Proceedings of the Institute of Radio Engineers (IRE)* 45, 3 (March), 279–290.

Siganos, G., Faloutsos, M., Faloutsos, P., and Faloutsos, C. (2003). Power laws and the AS-level Internet topology. *IEEE/ACM Transactions on Networking* 11, 4 (August), 514–524.

Simon, P. L., Taylor, M., and Kiss, I. Z. (2011). Exact epidemic models on graphs using graph-automorphism driven lumping. *Mathematical Biology* 62, 479–507.

Smythe, R. T. and Mahmoud, H. M. (1995). A survey of recursive trees. *Theory of Probability and Mathematical Statistics* 51, 1–27.

Son, S.-W., Bizhani, G., Christensen, C., Grassberger, P., and Paczuski, M. (2012). Percolation theory on interdependent networks based on epidemic spreading. *Europhysics Letters (EPL)* 97, 16006.

Steyaert, B. and Bruneel, H. (1994). Analytic derivation of the cell loss probability in finite multiserver buffers, from infinite buffer results. *Proceedings of the second workshop on performance modelling and evaluation of ATM networks, Bradford UK,* 18.1–11.

Strecok, A. J. (1968). On the calculation of the inverse of the error function. *Mathematics of Computation* 22, 144–158.

Strogatz, S. H. (2001). Exploring complex networks. *Nature* 410, 8 (March), 268–276.

Syski, R. (1986). *Introduction to Congestion Theory in Telephone Systems*, 2nd edn. Studies in Telecommunication, vol. 4. (North-Holland, Amsterdam).

Tang, S., Blenn, N., Doerr, C., and Van Mieghem, P. (2011). Digging in the Digg online social network. *IEEE Transactions on Multimedia* 13, 5 (October), 1163–1175.

Taylor, H. M. and Karlin, S. (1984). *An Introduction to Stochastic Modeling*. (Academic Press, Boston).

Titchmarsh, E. C. (1948). *Introduction to the Theory of Fourier Integrals*, 2nd edn. (Oxford University Press, Ely House, London W. I).

Titchmarsh, E. C. (1964). *The Theory of Functions*. (Oxford University Press, Amen House, London).

Titchmarsh, E. C. and Heath-Brown, D. R. (1986). *The Theory of the Zeta-function*, 2nd edn. (Oxford Science Publications, Oxford).

Trajanovski, S., Kuipers, F. A., Martin-Hernandez, J., and Van Mieghem, P. (2013). Generating graphs that approach a prescribed modularity. *Computer Communications* 36, 363–372.

Trajanovski, S., Wang, H., and Van Mieghem, P. (2012). Maximum modular graphs. *The European Physical Journal B* 85, 7, 244: 1–14.

van den Broek, J. and Heesterbeek, H. (2007). Nonhomogeneous birth and death models for epidemic outbreak data. *Biostatistics* 8, 2, 453–467.

van der Hofstad, R. (2013). *Random Graphs and Complex Networks*, notes at www.win.tue/ rhofstad/NotesRGCN.pdf edn.).

van der Hofstad, R., Hooghiemstra, G., and Van Mieghem, P. (2001). First passage percolation on the random graph. *Probability in the Engineering and Informational Sciences (PEIS)* 15, 225–237.

van der Hofstad, R., Hooghiemstra, G., and Van Mieghem, P. (2002a). The flooding time in random graphs. *Extremes* 5, 2 (June), 111–129.

van der Hofstad, R., Hooghiemstra, G., and Van Mieghem, P. (2002b). On the covariance of the level sizes in recursive trees. *Random Structures and Algorithms* 20, 519–539.

van der Hofstad, R., Hooghiemstra, G., and Van Mieghem, P. (2005). Distances in random graphs with finite variance degree. *Random Structures and Algorithms* 27, 1 (August), 76–123.

van der Hofstad, R., Hooghiemstra, G., and Van Mieghem, P. (2006). Size and weight of shortest path trees with exponential link weights. *Combinatorics, Probability and Computing* 15, 903–926.

van der Hofstad, R., Hooghiemstra, G., and Van Mieghem, P. (2007). The weight of the shortest path tree. *Random Structures and Algorithms* 30, 3, 359–379.

van der Hofstad, R. and Litvak, N. (2013). Degree-degree dependencies in random graphs with heavy-tailed degrees. *arXiv:1202.3071v5*.

van Doorn, E. A. and Schrijner, P. (1995). Geometric ergodicity and quasi-stationarity in discrete-time birth-death processes. *Journal of the Australian Mathematical Society, Series B* 37, 121–144.

Van Mieghem, P. (1996). The asymptotic behaviour of queueing systems: Large deviations theory and dominant pole approximation. *Queueing Systems* 23, 27–55.

Van Mieghem, P. (2001). Paths in the simple random graph and the Waxman graph. *Probability in the Engineering and Informational Sciences (PEIS)* 15, 535–555.

Van Mieghem, P. (2004). The probability distribution of the hopcount to an anycast group. Delft University of Technology, Report 2003605 (www.nas.ewi.tudelft.nl/people/Piet/TUDelftReports).

Van Mieghem, P. (2005). The limit random variable W of a branching process. Delft University of Technology, Report 20050206 (www.nas.ewi.tudelft.nl/people/Piet/TUDelftReports).

Van Mieghem, P. (2010a). *Data Communications Networking*, 2nd edn. (Piet Van Mieghem, ISBN 978-94-91075-01-8, Delft).

Van Mieghem, P. (2010b). Weight of a link in a shortest path tree and the Dedekind Eta function. *Random Structures and Algorithms* 36, 3 (May), 341–371.

Van Mieghem, P. (2011). *Graph Spectra for Complex Networks*. (Cambridge University Press, Cambridge, U.K.).

Van Mieghem, P. (2012a). Epidemic phase transition of the SIS-type in networks. *Europhysics Letters (EPL)* 97, 48004.

Van Mieghem, P. (2012b). Viral conductance of a network. *Computer Communications* 35, 12 (July), 1494–1509.

Van Mieghem, P. (2013). Decay towards the overall-healthy state in SIS epidemics on networks. *arXiv:1310.3980*.

Van Mieghem, P. (2014). Exact Markovian SIR and SIS epidemics on networks and an upper bound for the epidemic threshold. Delft University of Technology, Report20140210 (www.nas.ewi.tudelft.nl/people/Piet/TUDelftReports); arXiv:1402.1731.

Van Mieghem, P., Blenn, N., and Doerr, C. (2011a). Lognormal distribution in the Digg online social network. *European Physical Journal B* 83, 2, 252–261.

Van Mieghem, P. and Cator, E. (2012). Epidemics in networks with nodal self-infections and the epidemic threshold. *Physical Review E* 86, 1 (July), 016116.

Van Mieghem, P., Doerr, C., Wang, H., Hernandez, J. M., Hutchison, D., Karaliopoulos, M., and Kooij, R. E. (2010a). A framework for computing topological network robustness. Delft University of Technology, Report20101218 (www.nas.ewi.tudelft.nl/people/Piet/TUDelftReports).

Van Mieghem, P., Ge, X., Schumm, P., Trajanovski, S., and Wang, H. (2010b). Spectral graph analysis of modularity and assortativity. *Physical Review E* 82, 5 (November), 056113.

Van Mieghem, P., Hooghiemstra, G., and van der Hofstad, R. (2000). A scaling law for the hopcount in the Internet. Delft University of Technology, Report2000125 (www.nas.ewi.tudelft.nl/people/Piet/TUDelftReports).

Van Mieghem, P., Hooghiemstra, G., and van der Hofstad, R. (2001a). On the efficiency of multicast. *IEEE/ACM Transactions on Networking* 9, 6 (December), 719–732.

Van Mieghem, P., Hooghiemstra, G., and van der Hofstad, R. W. (2001b). Stochastic model for the number of traversed routers in Internet. *Proceedings of Passive and Active Measurement: PAM-2001, April 23-24, Amsterdam*.

Van Mieghem, P. and Janic, M. (2002). Stability of a multicast tree. *IEEE INFOCOM2002* 2, 1099–1108.

Van Mieghem, P. and Magdalena, S. M. (2005). A phase transition in the link weight structure of networks. *Physical Review E* 72, 5 (November), 056138.

Van Mieghem, P. and Omic, J. (2008). In-homogeneous virus spread in networks. Delft University of Technology, Report2008081 (www.nas.ewi.tudelft.nl/people/Piet/TUDelftReports); arXiv:1306.2588.

Van Mieghem, P., Omic, J., and Kooij, R. E. (2009). Virus spread in networks. *IEEE/ACM Transactions on Networking* 17, 1 (February), 1–14.

Van Mieghem, P., Stevanović, D., Kuipers, F. A., Li, C., van de Bovenkamp, R., Liu, D., and Wang, H. (2011b). Decreasing the spectral radius of a graph by link removals. *Physical Review E* 84, 1 (July), 016101.

Van Mieghem, P. and Tang, S. (2008). Weight of the shortest path to the first encountered peer in a peer group of size m. *Probability in the Engineering and Informational Sciences (PEIS)* 22, 37–52.

Van Mieghem, P. and Wang, H. (2009). The observable part of a network. *IEEE/ACM Transactions on Networking* 17, 1, 93–105.

Van Mieghem, P., Wang, H., Ge, X., Tang, S., and Kuipers, F. A. (2010c). Influence of assortativity and degree-preserving rewiring on the spectra of networks. *The European Physical Journal B* 76, 4, 643–652.

Vazquez, A., Rácz, B., Lukács, A., and Barabási, A.-L. (2007). Impact of non-Poissonian activity patterns on spreading processes. *Physical Review Letters* 98, 158702.

Veres, A. and Boda, M. (2000). The chaotic nature of TCP congestion control. *IEEE INFOCOM2000*.

Walrand, J. (1988). *An Introduction to Queueing Networks*. (Prentice-Hall, New York).

Walrand, J. (1998). *Communication Networks, A First Course*, 2nd edn. (McGraw-Hill, Boston).

Wang, H., Li, Q., D'Agostino, G., Havlin, S., Stanley, H. E., and Van Mieghem, P. (2013). Effect of the interconnected network structure on the epidemic threshold. *Physical Review E* 88, 2 (022801).

Wang, H. and Van Mieghem, P. (2010). Sampling networks by the union of m shortest path trees. *Computer Networks* 54, 1042–1053.

Wang, H., Winterbach, W., and Van Mieghem, P. (2011). Assortativity of complementary graphs. *The European Physical Journal B* 83, 2, 203–214.

Wang, Y., Chakrabarti, D., Wang, C., and Faloutsos, C. (2003). Epidemic spreading in real networks: An eigenvalue viewpoint. *22nd International Symposium on Reliable Distributed Systems (SRDS'03); IEEE Computer*, 25–34.

Wästlund, J. (2006). Random assignment and shortest path problems. *Proceedings of the Fourth Colloquium on Mathematics and Computer Science, Algorithms, Trees, Combinatorics and Probabilities; Institut Elie Cartan, Nancy, France*.

Watson, G. N. (1995). *A Treatise on the Theory of Bessel Functions*, Cambridge Mathematical Library edn. (Cambridge University Press, Cambridge, UK).

Watts, D. J. (1999). *Small Worlds, The Dynamics of Networks between Order and Randomness*. (Princeton University Press, Princeton, New Jersey).

Watts, D. J. and Strogatz, S. H. (1998). Collective dynamics of "small-worlds" networks. *Nature* 393, 440–442.

Waxman, B. M. (1988). Routing of multipoint connections. *IEEE Journal on Selected Areas in Communications* 6, 9 (December), 1617–1622.

Weibull, W. (1951). A statistical distribution function of wide applicability. *ASME Journal of Applied Mechanics*, 293–297.

Whittaker, E. T. and Watson, G. N. (1996). *A Course of Modern Analysis*, Cambridge Mathematical Library edn. (Cambridge University Press, Cambridge, UK).

Wilkinson, J. H. (1965). *The Algebraic Eigenvalue Problem*. (Oxford University Press, New York).

Woess, W. (2000). *Random Walks on Infinite Graphs and Groups*. (Cambridge University Press, Cambridge, UK).

Woess, W. (2009). *Denumerable Markov Chains*. (European Mathematical Society, Zurich, Switzerland).

Wolff, R. W. (1982). Poisson arrivals see time averages. *Operations Research* 30, 2 (April), 223–231.

Wolff, R. W. (1989). *Stochastic Modeling and the Theory of Queues*. (Prentice-Hall International Editions, New York).

Youssef, M., Kooij, R. E., and Scoglio, C. (2011). Viral conductance: Quantifying the robustness of networks with respect to spread of epidemics. *Journal of Computational Science* 2, 3 (August), 286–298.

Youssef, M. and Scoglio, C. (2011). An individual-based approach to SIR epidemics in contact networks. *Journal of Theoretical Biology* 283, 136–144.

Index

Printed in the United States
By Bookmasters